Ford Full-size Vans Automotive Repair Manual

by Curt Choate
and John H Haynes
Member of the Guild of Motoring Writers

Models covered:

All full-size models from 1969 thru 1991
with 240 or 300 cu in inline six cylinder engines and
302, 351, 400 or 460 cu in V8 engines

(3D14 - 36090)
(344)

ABCDE
FG

2

Haynes Publishing Group
Sparkford Nr Yeovil
Somerset BA22 7JJ England

Haynes North America, Inc
861 Lawrence Drive
Newbury Park
California 91320 USA

Acknowledgements

We are grateful to the Ford Motor Company for their assistance with technical information, certain illustrations and vehicle photos.

A book in the **Haynes Automotive Repair Manual Series**

Printed in the U.S.A.

ISBN 1 56392 029 8

Library of Congress Catalog Card Number 92-70523

Contents

1981 Econoline Club Wagon

About this manual

Its purpose

The purpose of this manual is to help you get the best value from your vehicle. It can do so in several ways. It can help you decide what work must be done, even if you choose to have it done by a dealer service department or a repair shop; it provides information and procedures for routine maintenance and servicing; and it offers diagnostic and repair procedures to follow when trouble occurs.

It is hoped that you will use the manual to tackle the work yourself. For many simpler jobs, doing it yourself may be quicker than arranging an appointment to get the vehicle into a shop and making the trips to leave it and pick it up. More importantly, a lot of money can be saved by avoiding the expense the shop must pass on to you to cover its labor and overhead costs. An added benefit is the sense of satisfaction and accomplishment that you feel after having done the job yourself.

Using the manual

The manual is divided into Chapters. Each Chapter is divided into numbered Sections, which are headed in bold type between horizontal lines. Each Section consists of consecutively numbered paragraphs.

At the beginning of each numbered section you will be referred to any illustrations which apply to the procedures in that section. The reference numbers used in illustration captions pinpoint the pertinent Section and the Step within that section. That is, illustration 3.2 means the illustration refers to Section 3 and Step (or paragraph) 2 within that Section.

Procedures, once described in the text, are not normally repeated. When it is necessary to refer to another Chapter, the reference will be given as Chapter and Section number i.e. Chapter 1/16). Cross references given without use of the word "Chapter" apply to Sections and/or paragraphs in the same Chapter. For example, "see Section 8" means in the same Chapter.

Reference to the left or right side of the vehicle is based on the assumption that one is sitting in the driver's seat, facing forward.

Even though extreme care has been taken during the preparation of this manual, neither the publisher nor the author can accept responsibility for any errors in, or omissions from, the information given.

NOTE

A Note provides information necessary to properly complete a procedure or information which will make the steps to be followed easier to understand.

CAUTION

A Caution indicates a special procedure or special steps which must be taken in the course of completing the procedure in which the **Caution** is found which are necessary to avoid damage to the assembly being worked on.

WARNING

A Warning indicates a special procedure or special steps which must be taken in the course of completing the procedure in which the **Warning** is found which are necessary to avoid injury to the person performing the procedure.

Introduction to the Ford full-size vans

The Ford Econoline was first introduced in 1961 as the E-100 Series. It was available with either a van, club wagon or pick-up body and a choice of three six-cylinder engines of various displacements.

This manual covers all models equipped with inline six-cylinders and V8 engines from 1969 to the present. The inline six-cylinder engine has a displacement of either 240 or 300 cubic inches and is standard equipment. An optional V8 engine, with a displacement of 302, 351, 400 or 460 cubic inches, is also available, depending on the model year.

The standard transmission on all models is a three-speed manual, with mechanical or hydraulic clutch. An automatic transmission is available as an option.

The front suspension uses twin I beam axles pivoted at the inner ends and attached to the chassis by coil springs and radius rods. The rear suspension is a conventional semi-elliptical leaf spring arrangement.

Short or long wheelbase versions are available, with a gross vehicle weight capacity ranging from 2300 to 11,000 lbs.

1986 Econoline Club Wagon XLT

Vehicle identification numbers

Modifications are a continuing and unpublicized process in vehicle manufacturing. Since spare parts manuals and lists are compiled on a numerical basis, the individual vehicle numbers are essential to correctly identify the component required.

Vehicle identification number (VIN)

This very important identification number is located on a plate attached to the top left corner of the dashboard of the vehicle. The VIN also appears on the Vehicle Certificate of Title and Registration. It contains valuable information such as where and when the vehicle was manufactured, the model year and the body style.

Body identification plate

This label is located on the drivers door lock pillar. Like the VIN, it contains valuable information concerning the production of the vehicle as well as information about the way in which the vehicle is equipped.

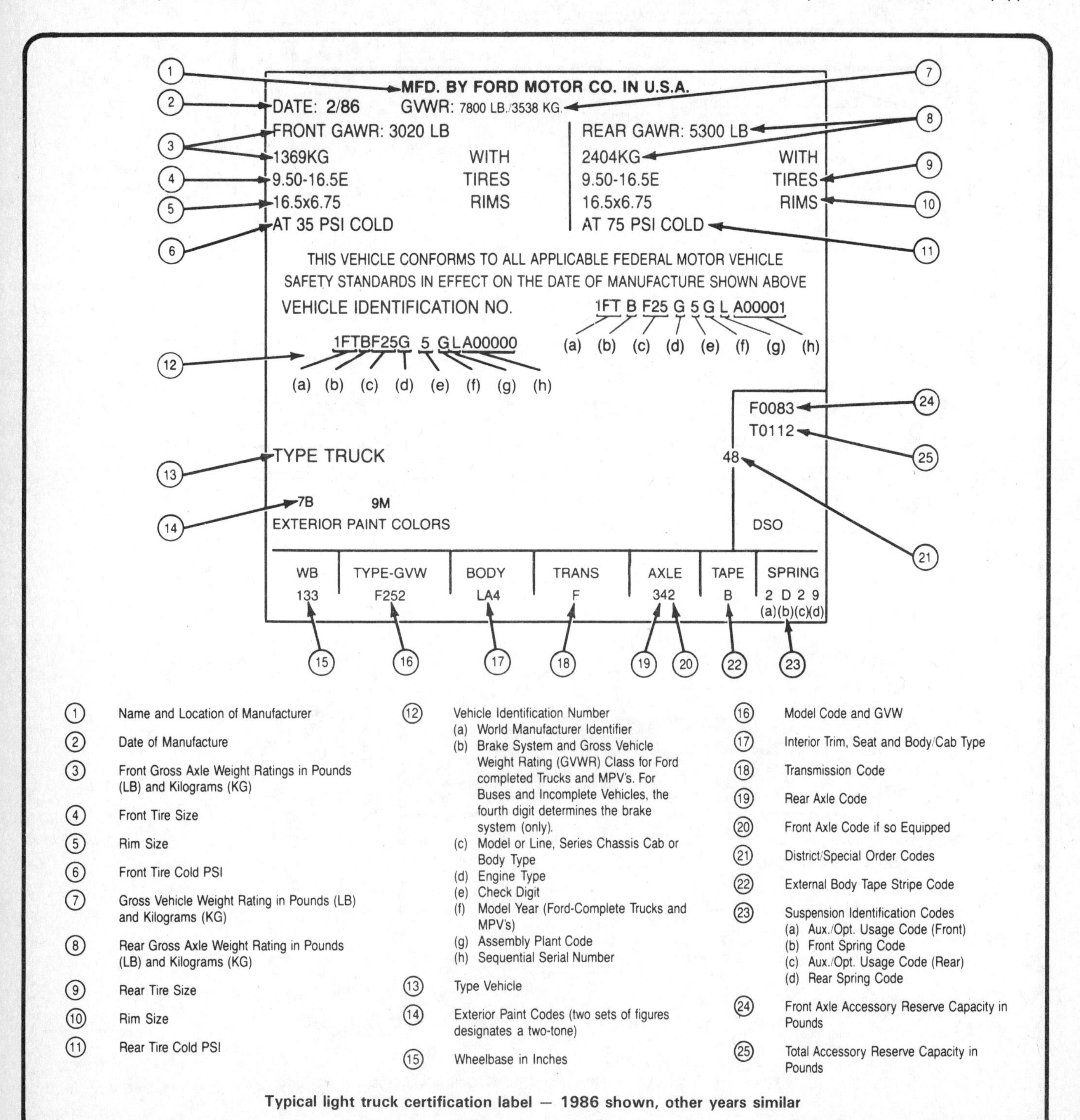

1. Name and Location of Manufacturer
2. Date of Manufacture
3. Front Gross Axle Weight Ratings in Pounds (LB) and Kilograms (KG)
4. Front Tire Size
5. Rim Size
6. Front Tire Cold PSI
7. Gross Vehicle Weight Rating in Pounds (LB) and Kilograms (KG)
8. Rear Gross Axle Weight Rating in Pounds (LB) and Kilograms (KG)
9. Rear Tire Size
10. Rim Size
11. Rear Tire Cold PSI
12. Vehicle Identification Number
    - (a) World Manufacturer Identifier
    - (b) Brake System and Gross Vehicle Weight Rating (GVWR) Class for Ford completed Trucks and MPV's. For Buses and Incomplete Vehicles, the fourth digit determines the brake system (only).
    - (c) Model or Line, Series Chassis Cab or Body Type
    - (d) Engine Type
    - (e) Check Digit
    - (f) Model Year (Ford-Complete Trucks and MPV's)
    - (g) Assembly Plant Code
    - (h) Sequential Serial Number
13. Type Vehicle
14. Exterior Paint Codes (two sets of figures designates a two-tone)
15. Wheelbase in Inches
16. Model Code and GVW
17. Interior Trim, Seat and Body/Cab Type
18. Transmission Code
19. Rear Axle Code
20. Front Axle Code if so Equipped
21. District/Special Order Codes
22. External Body Tape Stripe Code
23. Suspension Identification Codes
    - (a) Aux./Opt. Usage Code (Front)
    - (b) Front Spring Code
    - (c) Aux./Opt. Usage Code (Rear)
    - (d) Rear Spring Code
24. Front Axle Accessory Reserve Capacity in Pounds
25. Total Accessory Reserve Capacity in Pounds

Typical light truck certification label — 1986 shown, other years similar

This plate is especially useful for matching the color and type of paint during repair work.

Engine identification numbers

The engine number is located on a label usually located on the rocker arm cover. Engine identification can also be done by referring to the Emissions Control Information label and the actual VIN number on the door pillar.

Vehicle Emissions Control Information label

The Emissions Control Information label is attached to the engine rocker arm cover on early models and to the radiator support bracket on later models.

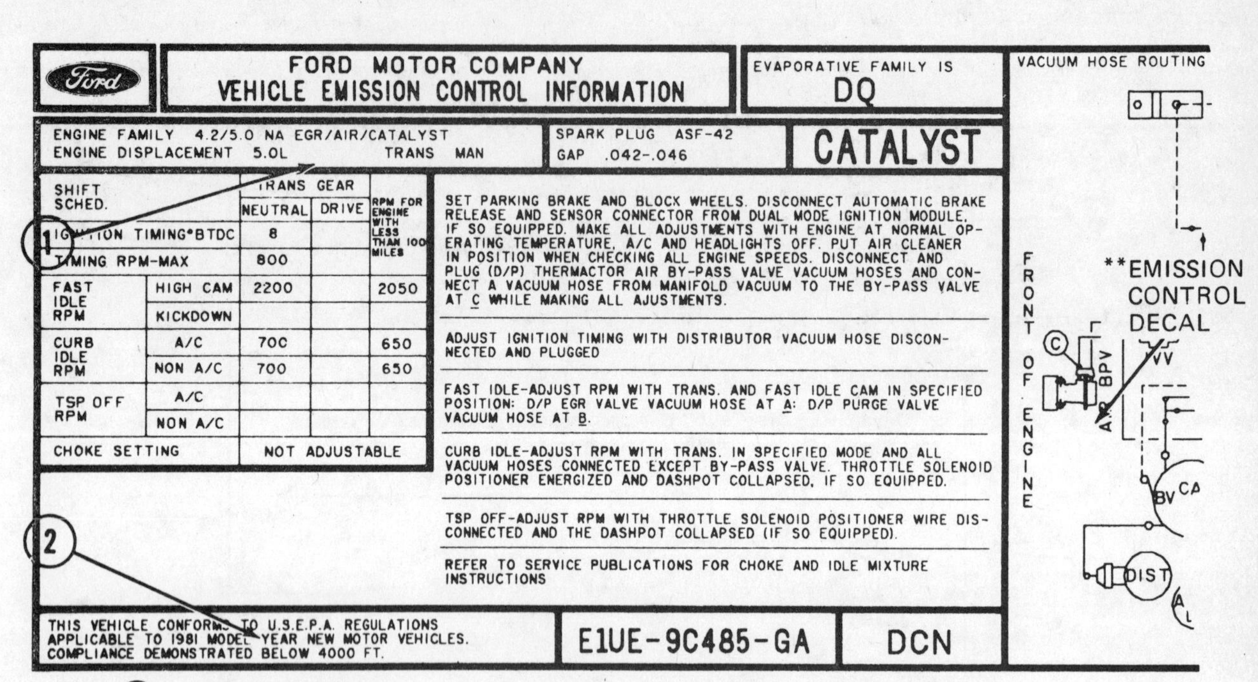

1. DETERMINE ENGINE DISPLACEMENT FROM EMISSION CONTROL DECAL ENTER IN BLANK "A" OF EXAMPLE SHOWN BELOW.
2. DETERMINE VEHICLE MODEL YEAR FROM EMISSION CONTROL DECAL ENTER LAST DIGIT IN BLANK "B". SEE NOTE BELOW.
3. DETERMINE CALIBRATION BASE NUMBER FROM ENGINE CODE LABEL ENTER NUMBERS & LETTERS IN BLANK "C".
4. DETERMINE REVISION LEVEL FROM ENGINE CODE LABEL* ENTER IN BLANK "D"

EXAMPLE: ENGINE: 3.3L (200 CID) (A) CALIBRATION 1 (B) – 93J (C) – R0 (D)

*Located on engine valve cover (Black Letters on White Background)

Typical Emission Control Information and engine calibration label and Engine Code Label — 1981 shown, other years similar

Buying parts

Replacement parts are available from many sources, which generally fall into one of two categories – authorized dealer parts departments and independent retail auto parts stores. Our advice concerning these parts is as follows:

Retail auto parts stores: Good auto parts stores will stock frequently needed components which wear out relatively fast, such as clutch components, exhaust systems, brake parts, tune-up parts, etc. These stores often supply new or reconditioned parts on an exchange basis, which can save a considerable amount of money. Discount auto parts stores are often very good places to buy materials and parts needed for general vehicle maintenance such as oil, grease, filters, spark plugs, belts, touch-up paint, bulbs, etc. They also usually sell tools and general accessories, have convenient hours, charge lower prices and can often be found not far from home.

Authorized dealer parts department: This is the best source for parts which are unique to the vehicle and not generally available elsewhere (such as major engine parts, transmission parts, trim pieces, etc.).

Warranty information: If the vehicle is still covered under warranty, be sure that any replacement parts purchased – regardless of the source – do not invalidate the warranty!

To be sure of obtaining the correct parts, have engine and chassis numbers available and, if possible, take the old parts along for positive identification.

Maintenance techniques, tools and working facilities

Maintenance techniques

There are a number of techniques involved in maintenance and repair that will be referred to throughout this manual. Application of these techniques will enable the home mechanic to be more efficient, better organized and capable of performing the various tasks properly, which will ensure that the repair job is thorough and complete.

Fasteners

Fasteners are nuts, bolts, studs and screws used to hold two or more parts together. There are a few things to keep in mind when working with fasteners. Almost all of them use a locking device of some type, either a lockwasher, locknut, locking tab or thread adhesive. All threaded fasteners should be clean and straight, with undamaged threads and undamaged corners on the hex head where the wrench fits. Develop the habit of replacing all damaged nuts and bolts with new ones. Special locknuts with nylon or fiber inserts can only be used once. If they are removed, they lose their locking ability and must be replaced with new ones.

Rusted nuts and bolts should be treated with a penetrating fluid to ease removal and prevent breakage. Some mechanics use turpentine in a spout-type oil can, which works quite well. After applying the rust penetrant, let it work for a few minutes before trying to loosen the nut or bolt. Badly rusted fasteners may have to be chiseled or sawed off or removed with a special nut breaker, available at tool stores.

If a bolt or stud breaks off in an assembly, it can be drilled and removed with a special tool commonly available for this purpose. Most automotive machine shops can perform this task, as well as other repair procedures, such as the repair of threaded holes that have been stripped out.

Flat washers and lockwashers, when removed from an assembly, should always be replaced exactly as removed. Replace any damaged washers with new ones. Never use a lockwasher on any soft metal surface (such as aluminum), thin sheet metal or plastic.

Fastener sizes

For a number of reasons, automobile manufacturers are making wider and wider use of metric fasteners. Therefore, it is important to be able to tell the difference between standard (sometimes called U.S. or SAE) and metric hardware, since they cannot be interchanged.

All bolts, whether standard or metric, are sized according to diameter, thread pitch and length. For example, a standard 1/2 – 13 x 1 bolt is 1/2 inch in diameter, has 13 threads per inch and is 1 inch long. An M12 – 1.75 x 25 metric bolt is 12 mm in diameter, has a thread pitch of 1.75 mm (the distance between threads) and is 25 mm long. The two bolts are nearly identical, and easily confused, but they are not interchangeable.

In addition to the differences in diameter, thread pitch and length, metric and standard bolts can also be distinguished by examining the bolt heads. To begin with, the distance across the flats on a standard bolt head is measured in inches, while the same dimension on a metric bolt is sized in millimeters (the same is true for nuts). As a result, a standard wrench should not be used on a metric bolt and a metric wrench should not be used on a standard bolt. Also, most standard bolts have slashes radiating out from the center of the head to denote the grade or strength of the bolt, which is an indication of the amount of torque that can be applied to it. The greater the number of slashes, the greater the strength of the bolt. Grades 0 through 5 are commonly used on automobiles. Metric bolts have a property class (grade) number, rather than a slash, molded into their heads to indicate bolt strength. In this case, the higher the number, the stronger the bolt. Property class numbers 8.8, 9.8 and 10.9 are commonly used on automobiles.

Strength markings can also be used to distinguish standard hex nuts from metric hex nuts. Many standard nuts have dots stamped into one side, while metric nuts are marked with a number. The greater the number of dots, or the higher the number, the greater the strength of the nut.

Metric studs are also marked on their ends according to property class (grade). Larger studs are numbered (the same as metric bolts),

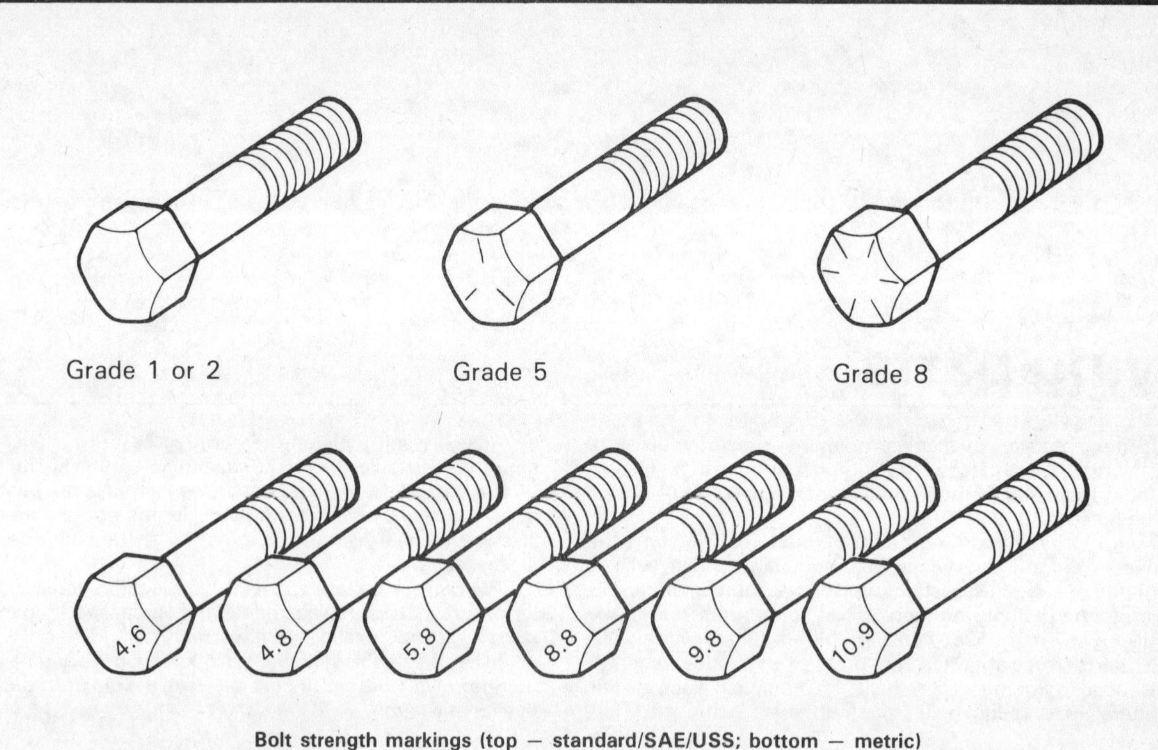

Bolt strength markings (top – standard/SAE/USS; bottom – metric)

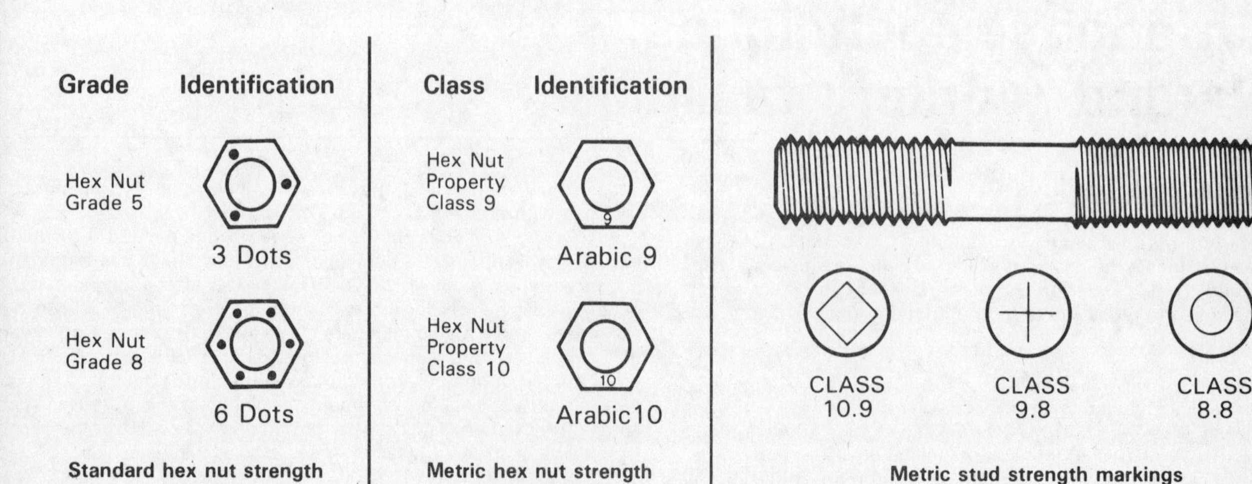

Standard hex nut strength markings

Metric hex nut strength markings

Metric stud strength markings

while smaller studs carry a geometric code to denote grade.

It should be noted that many fasteners, especially Grades 0 through 2, have no distinguishing marks on them. When such is the case, the only way to determine whether it is standard or metric is to measure the thread pitch or compare it to a known fastener of the same size.

Standard fasteners are often referred to as SAE, as opposed to metric. However, it should be noted that SAE technically refers to a non-metric *fine thread* fastener only. Coarse thread non-metric fasteners are referred to as USS sizes.

Since fasteners of the same size (both standard and metric) may have different strength ratings, be sure to reinstall any bolts, studs or nuts removed from your vehicle in their original locations. Also, when replacing a fastener with a new one, make sure that the new one has a strength rating equal to or greater than the original.

Tightening sequences and procedures

Most threaded fasteners should be tightened to a specific torque value (torque is the twisting force applied to a threaded component such as a nut or bolt). Overtightening the fastener can weaken it and cause it to break, while undertightening can cause it to eventually come loose. Bolts, screws and studs, depending on the material they are made of and their thread diameters, have specific torque values, many of which are noted in the Specifications at the beginning of each Chapter. Be sure to follow the torque recommendations closely. For fasteners not assigned a specific torque, a general torque value chart is presented here as a guide. As was previously mentioned, the size and grade of a fastener determine the amount of torque that can safely be applied to it. The figures listed here are approximate for Grade 2 and Grade 3 fasteners. Higher grades can tolerate higher torque values.

Metric thread sizes	**Ft-lb**	**Nm/m**
M-6	6 to 9	9 to 12
M-8	14 to 21	19 to 28
M-10	28 to 40	38 to 54
M-12	50 to 71	68 to 96
M-14	80 to 140	109 to 154
Pipe thread sizes		
1/8	5 to 8	7 to 10
1/4	12 to 18	17 to 24
3/8	22 to 33	30 to 44
1/2	25 to 35	34 to 47
U.S. thread sizes		
1/4 — 20	6 to 9	9 to 12
5/16 — 18	12 to 18	17 to 24
5/16 — 24	14 to 20	19 to 27
3/8 — 16	22 to 32	30 to 43
3/8 — 24	27 to 38	37 to 51
7/16 — 14	40 to 55	55 to 74
7/16 — 20	40 to 60	55 to 81
1/2 — 13	55 to 80	75 to 108

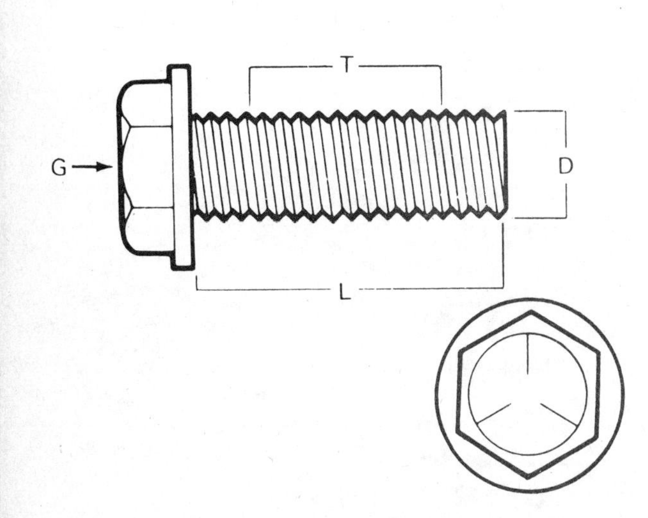

Standard (SAE and USS) bolt dimensions/grade marks

G Grade marks (bolt strength)
L Length (in inches)
T Thread pitch (number of threads per inch)
D Nominal diameter (in inches)

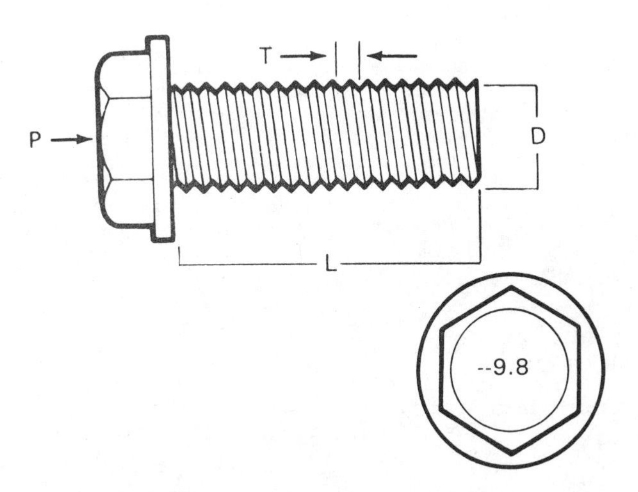

Metric bolt dimensions/grade marks

P Property class (bolt strength)
L Length (in millimeters)
T Thread pitch (distance between threads in millimeters)
D Diameter

Fasteners laid out in a pattern, such as cylinder head bolts, oil pan bolts, differential cover bolts, etc., must be loosened or tightened in sequence to avoid warping the component. This sequence will normally be shown in the appropriate Chapter. If a specific pattern is not given, the following procedures can be used to prevent warping.

Initially, the bolts or nuts should be assembled finger-tight only. Next, they should be tightened one full turn each, in a criss-cross or diagonal pattern. After each one has been tightened one full turn, return to the first one and tighten them all one-half turn, following the same pattern. Finally, tighten each of them one-quarter turn at a time until each fastener has been tightened to the proper torque. To loosen and remove the fasteners, the procedure would be reversed.

Component disassembly

Component disassembly should be done with care and purpose to help ensure that the parts go back together properly. Always keep track of the sequence in which parts are removed. Make note of special characteristics or marks on parts that can be installed more than one way, such as a grooved thrust washer on a shaft. It is a good idea to lay the disassembled parts out on a clean surface in the order that they were removed. It may also be helpful to make sketches or take instant photos of components before removal.

When removing fasteners from a component, keep track of their locations. Sometimes threading a bolt back in a part, or putting the washers and nut back on a stud, can prevent mix-ups later. If nuts and bolts cannot be returned to their original locations, they should be kept in a compartmented box or a series of small boxes. A cupcake or muffin tin is ideal for this purpose, since each cavity can hold the bolts and nuts from a particular area (i.e. oil pan bolts, valve cover bolts, engine mount bolts, etc.). A pan of this type is especially helpful when working on assemblies with very small parts, such as the carburetor, alternator, valve train or interior dash and trim pieces. The cavities can be marked with paint or tape to identify the contents.

Whenever wiring looms, harnesses or connectors are separated, it is a good idea to identify the two halves with numbered pieces of masking tape so they can be easily reconnected.

Gasket sealing surfaces

Throughout any vehicle, gaskets are used to seal the mating surfaces between two parts and keep lubricants, fluids, vacuum or pressure contained in an assembly.

Many times these gaskets are coated with a liquid or paste-type gasket sealing compound before assembly. Age, heat and pressure can sometimes cause the two parts to stick together so tightly that they are very difficult to separate. Often, the assembly can be loosened by striking it with a soft-face hammer near the mating surfaces. A regular hammer can be used if a block of wood is placed between the hammer and the part. Do not hammer on cast parts or parts that could be easily damaged. With any particularly stubborn part, always recheck to make sure that every fastener has been removed.

Avoid using a screwdriver or bar to pry apart an assembly, as they can easily mar the gasket sealing surfaces of the parts, which must remain smooth. If prying is absolutely necessary, use an old broom handle, but keep in mind that extra clean up will be necessary if the wood splinters.

After the parts are separated, the old gasket must be carefully scraped off and the gasket surfaces cleaned. Stubborn gasket material can be soaked with rust penetrant or treated with a special chemical to soften it so it can be easily scraped off. A scraper can be fashioned from a piece of copper tubing by flattening and sharpening one end. Copper is recommended because it is usually softer than the surfaces to be scraped, which reduces the chance of gouging the part. Some gaskets can be removed with a wire brush, but regardless of the method used, the mating surfaces must be left clean and smooth. If for some reason the gasket surface is gouged, then a gasket sealer thick enough to fill scratches will have to be used during reassembly of the components. For most applications, a non-drying (or semi-drying) gasket sealer should be used.

Hose removal tips

Warning: *If the vehicle is equipped with air conditioning, do not disconnect any of the A/C hoses without first having the system depressurized by a dealer service department or an air conditioning specialist.*

Hose removal precautions closely parallel gasket removal precautions. Avoid scratching or gouging the surface that the hose mates against or the connection may leak. This is especially true for radiator hoses. Because of various chemical reactions, the rubber in hoses can bond itself to the metal spigot that the hose fits over. To remove a hose, first loosen the hose clamps that secure it to the spigot. Then, with slip-joint pliers, grab the hose at the clamp and rotate it around the spigot. Work it back and forth until it is completely free, then pull it off. Silicone or other lubricants will ease removal if they can be applied between the hose and the outside of the spigot. Apply the same lubricant to the inside of the hose and the outside of the spigot to simplify installation.

As a last resort (and if the hose is to be replaced with a new one anyway), the rubber can be slit with a knife and the hose peeled from the spigot. If this must be done, be careful that the metal connection is not damaged.

If a hose clamp is broken or damaged, do not reuse it. Wire-type clamps usually weaken with age, so it is a good idea to replace them with screw-type clamps whenever a hose is removed.

Tools

A selection of good tools is a basic requirement for anyone who plans to maintain and repair his or her own vehicle. For the owner who has few tools, the initial investment might seem high, but when compared to the spiraling costs of professional auto maintenance and repair, it is a wise one.

Micrometer set

Dial indicator set

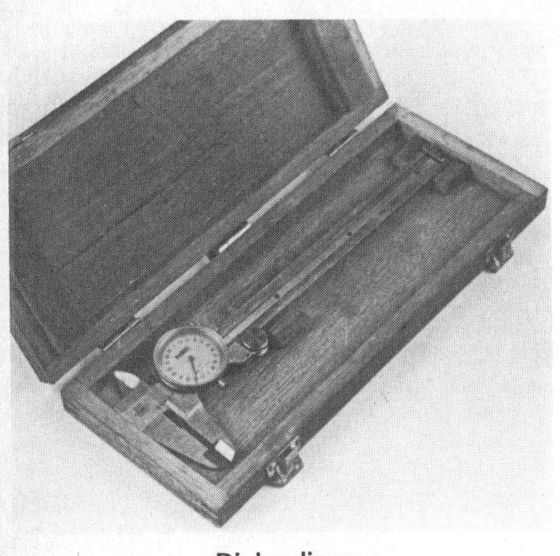

Dial caliper

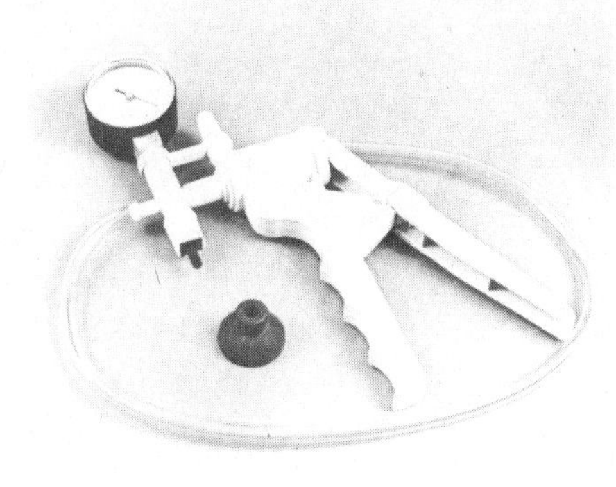

Hand-operated vacuum pump

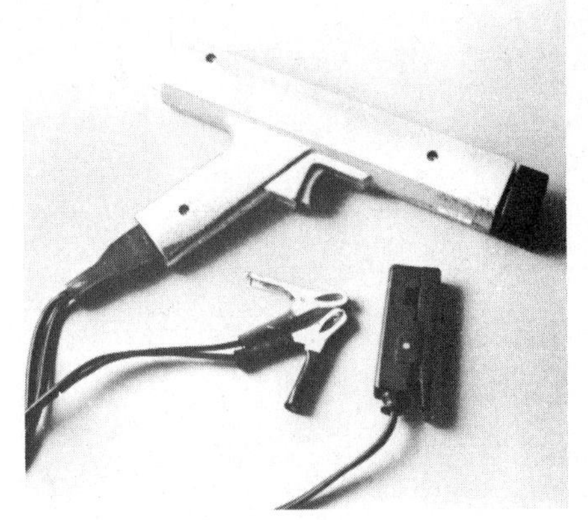

Timing light

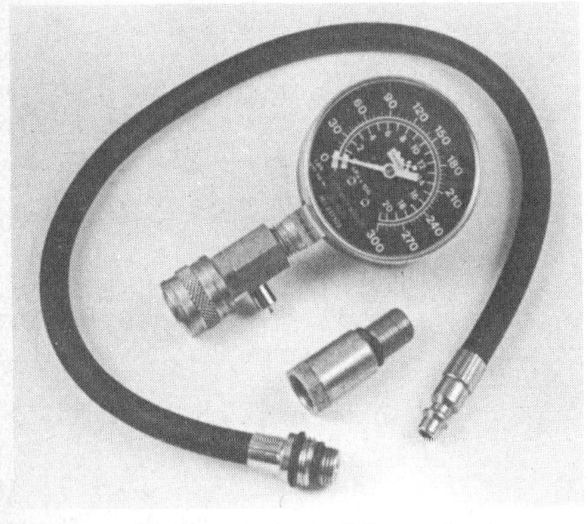

Compression gauge with spark plug hole adapter

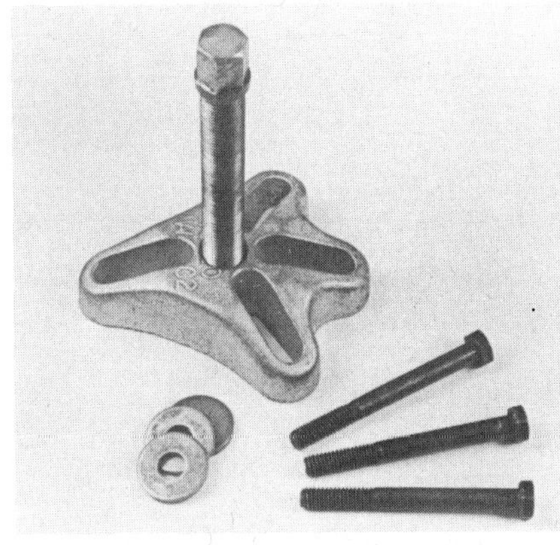

Damper/steering wheel puller

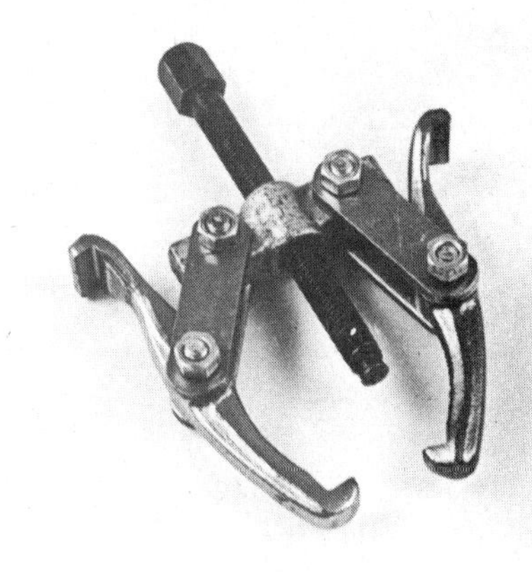

General purpose puller

Hydraulic lifter removal tool

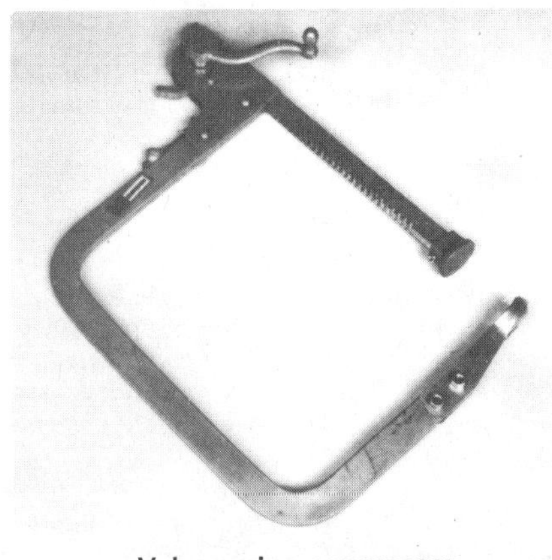

Valve spring compressor

Valve spring compressor

Ridge reamer

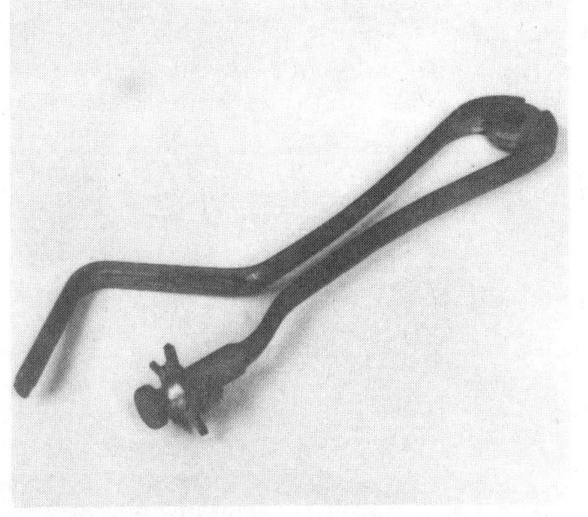

Piston ring groove cleaning tool

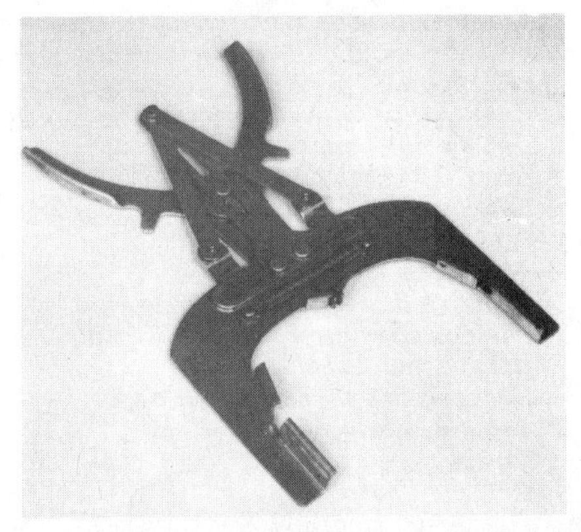

Ring removal/installation tool

Ring compressor

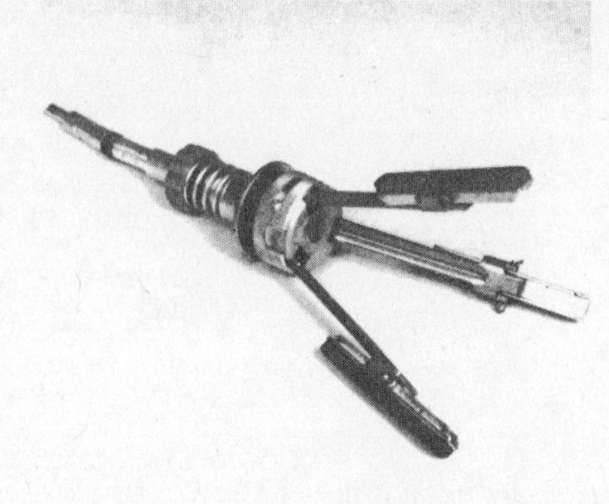
Cylinder hone

Brake hold-down spring tool

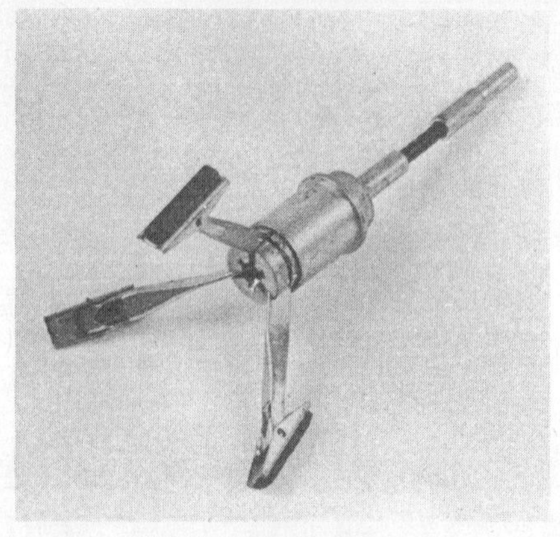
Brake cylinder hone

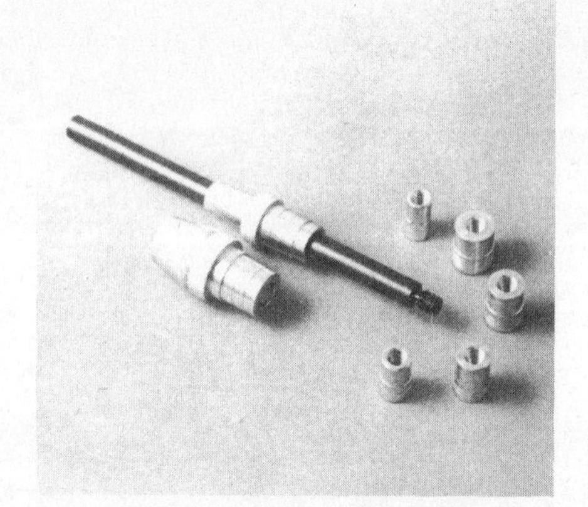
Clutch plate alignment tool

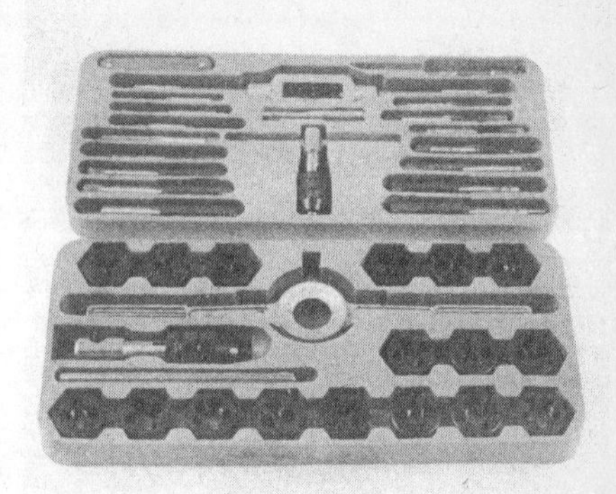
Tap and die set

To help the owner decide which tools are needed to perform the tasks detailed in this manual, the following tool lists are offered: *Maintenance and minor repair, Repair/overhaul* and *Special*.

The newcomer to practical mechanics should start off with the maintenance and minor repair tool kit, which is adequate for the simpler jobs performed on a vehicle. Then, as confidence and experience grow, the owner can tackle more difficult tasks, buying additional tools as they are needed. Eventually the basic kit will be expanded into the repair and overhaul tool set. Over a period of time, the experienced do-it-yourselfer will assemble a tool set complete enough for most repair and overhaul procedures and will add tools from the special category when it is felt that the expense is justified by the frequency of use.

Maintenance and minor repair tool kit

The tools in this list should be considered the minimum required for performance of routine maintenance, servicing and minor repair work. We recommend the purchase of combination wrenches (box-end and open-end combined in one wrench). While more expensive than open end wrenches, they offer the advantages of both types of wrench.

Combination wrench set (1/4-inch to 1 inch or 6 mm to 19 mm)
Adjustable wrench, 8 inch
Spark plug wrench with rubber insert
Spark plug gap adjusting tool
Feeler gauge set
Brake bleeder wrench
Standard screwdriver (5/16-inch x 6 inch)
Phillips screwdriver (No. 2 x 6 inch)
Combination pliers — 6 inch
Hacksaw and assortment of blades
Tire pressure gauge
Grease gun
Oil can
Fine emery cloth
Wire brush
Battery post and cable cleaning tool
Oil filter wrench
Funnel (medium size)
Safety goggles
Jackstands (2)
Drain pan

Note: *If basic tune-ups are going to be part of routine maintenance, it will be necessary to purchase a good quality stroboscopic timing light and combination tachometer/dwell meter. Although they are included in the list of special tools, it is mentioned here because they are absolutely necessary for tuning most vehicles properly.*

Repair and overhaul tool set

These tools are essential for anyone who plans to perform major repairs and are in addition to those in the maintenance and minor repair tool kit. Included is a comprehensive set of sockets which, though expensive, are invaluable because of their versatility, especially when various extensions and drives are available. We recommend the 1/2-inch drive over the 3/8-inch drive. Although the larger drive is bulky and more expensive, it has the capacity of accepting a very wide range of large sockets. Ideally, however, the mechanic should have a 3/8-inch drive set and a 1/2-inch drive set.

Socket set(s)
Reversible ratchet
Extension — 10 inch
Universal joint
Torque wrench (same size drive as sockets)
Ball peen hammer — 8 ounce
Soft-face hammer (plastic/rubber)
Standard screwdriver (1/4-inch x 6 inch)
Standard screwdriver (stubby — 5/16-inch)
Phillips screwdriver (No. 3 x 8 inch)
Phillips screwdriver (stubby — No. 2)

Pliers — vise grip
Pliers — lineman's
Pliers — needle nose
Pliers — snap-ring (internal and external)
Cold chisel — 1/2-inch
Scribe
Scraper (made from flattened copper tubing)
Centerpunch
Pin punches (1/16, 1/8, 3/16-inch)
Steel rule/straightedge — 12 inch
Allen wrench set (1/8 to 3/8-inch or 4 mm to 10 mm)
A selection of files
Wire brush (large)
Jackstands (second set)
Jack (scissor or hydraulic type)

Note: *Another tool which is often useful is an electric drill motor with a chuck capacity of 3/8-inch and a set of good quality drill bits.*

Special tools

The tools in this list include those which are not used regularly, are expensive to buy, or which need to be used in accordance with their manufacturer's instructions. Unless these tools will be used frequently, it is not very economical to purchase many of them. A consideration would be to split the cost and use between yourself and a friend or friends. In addition, most of these tools can be obtained from a tool rental shop on a temporary basis.

This list primarily contains only those tools and instruments widely available to the public, and not those special tools produced by the vehicle manufacturer for distribution to dealer service departments. Occasionally, references to the manufacturer's special tools are inluded in the text of this manual. Generally, an alternative method of doing the job without the special tool is offered. However, sometimes there is no alternative to their use. Where this is the case, and the tool cannot be purchased or borrowed, the work should be turned over to the dealer service department or an automotive repair shop.

Valve spring compressor
Piston ring groove cleaning tool
Piston ring compressor
Piston ring installation tool
Cylinder compression gauge
Cylinder ridge reamer
Cylinder surfacing hone
Cylinder bore gauge
Micrometers and/or dial calipers
Hydraulic lifter removal tool
Balljoint separator
Universal-type puller
Impact screwdriver
Dial indicator set
Stroboscopic timing light (inductive pick-up)
Hand operated vacuum/pressure pump
Tachometer/dwell meter
Universal electrical multimeter
Cable hoist
Brake spring removal and installation tools
Floor jack

Buying tools

For the do-it-yourselfer who is just starting to get involved in vehicle maintenance and repair, there are a number of options available when purchasing tools. If maintenance and minor repair is the extent of the work to be done, the purchase of individual tools is satisfactory. If, on the other hand, extensive work is planned, it would be a good idea to purchase a modest tool set from one of the large retail chain stores. A set can usually be bought at a substantial savings over the individual tool prices, and they often come with a tool box. As additional tools are needed, add-on sets, individual tools and a larger tool box can be purchased to expand the tool selection. Building a tool set gradually allows the cost of the tools to be spread over a longer period of time and gives the mechanic the freedom to choose only those tools that will actually be used.

Tool stores will often be the only source of some of the special tools that are needed, but regardless of where tools are bought, try to avoid cheap ones, especially when buying screwdrivers and sockets, because they won't last very long. The expense involved in replacing cheap tools will eventually be greater than the initial cost of quality tools.

Care and maintenance of tools

Good tools are expensive, so it makes sense to treat them with respect. Keep them clean and in usable condition and store them properly when not in use. Always wipe off any dirt, grease or metal chips before putting them away. Never leave tools lying around in the work area. Upon completion of a job, always check closely under the hood for tools that may have been left there so they won't get lost during a test drive.

Some tools, such as screwdrivers, pliers, wrenches and sockets, can be hung on a panel mounted on the garage or workshop wall, while others should be kept in a tool box or tray. Measuring instruments, gauges, meters, etc. must be carefully stored where they cannot be damaged by weather or impact from other tools.

When tools are used with care and stored properly, they will last a very long time. Even with the best of care, though, tools will wear out if used frequently. When a tool is damaged or worn out, replace it. Subsequent jobs will be safer and more enjoyable if you do.

Working facilities

Not to be overlooked when discussing tools is the workshop. If anything more than routine maintenance is to be carried out, some sort of suitable work area is essential.

It is understood, and appreciated, that many home mechanics do not have a good workshop or garage available, and end up removing an engine or doing major repairs outside. It is recommended, however, that the overhaul or repair be completed under the cover of a roof.

A clean, flat workbench or table of comfortable working height is an absolute necessity. The workbench should be equipped with a vise that has a jaw opening of at least four inches.

As mentioned previously, some clean, dry storage space is also required for tools, as well as the lubricants, fluids, cleaning solvents, etc. which will soon become necessary.

Sometimes waste oil and fluids, drained from the engine or cooling system during normal maintenance or repairs, present a disposal problem. To avoid pouring them on the ground or into a sewage system, pour the used fluids into large containers, seal them with caps and take them to an authorized disposal site or recycling center. Plastic jugs, such as old antifreeze containers, are ideal for this purpose.

Always keep a supply of old newspapers and clean rags available. Old towels are excellent for mopping up spills. Many mechanics use rolls of paper towels for most work because they are readily available and disposable. To help keep the area under the vehicle clean, a large cardboard box can be cut open and flattened to protect the garage or shop floor.

Whenever working over a painted surface, such as when leaning over a fender to service something under the hood, always cover it with an old blanket or bedspread to protect the finish. Vinyl covered pads, made especially for this purpose, are available at auto parts stores.

Booster battery (jump) starting

Certain precautions must be observed when using a booster battery to jump start a vehicle.

a) Before connecting the booster battery, make sure that the ignition switch is in the Off position.
b) Turn off the lights, heater and other electrical loads.
c) The eyes should be shielded. Safety goggles are recommended.
d) Make sure the booster battery is the same voltage as the dead one in the vehicle.
e) The two vehicles must not touch each other.
f) Make sure the transmission is in Neutral (manual transmission) or Park (automatic transmission).
g) If the booster battery is not a maintenance-free type, remove the vent caps and lay a cloth over the vent holes.

Connect the red jumper cable to the *Positive* (+) terminals of each battery.

Connect one end of the black jumper cable to the *Negative* (−) terminal of the booster battery. The other end of this cable should be connected to a good ground on the vehicle to be started, such as a bolt or bracket on the engine block. Use caution to ensure that the cable will not come into contact with the fan, drivebelts or other moving parts of the engine.

Start the engine using the booster battery, then, with the engine running at idle speed, disconnect the jumper cables in the reverse order of connection.

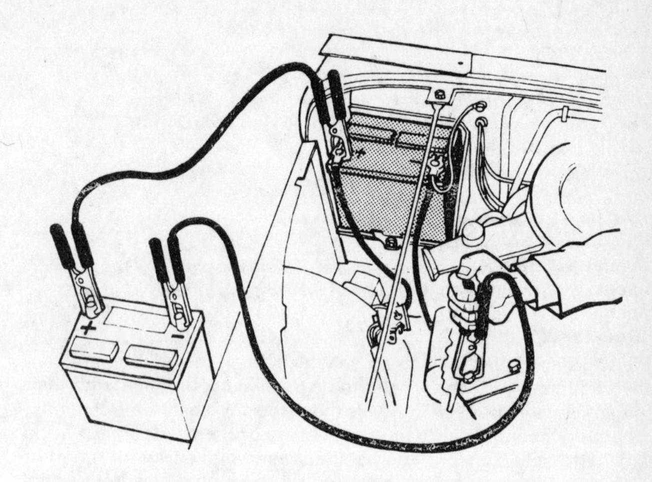

Booster battery cable connections (note that the negative cable is *not* attached to the negative terminal of the dead battery)

Jacking and towing

Jacking

The jack supplied with the vehicle should only be used for raising the vehicle when changing a tire or placing jackstands under the frame. **Warning:** *Never work under the vehicle or start the engine while this jack is being used as the only means of support.*

The vehicle should be on level ground with the wheels blocked and the transmission in Park (automatic) or Reverse (manual). If the tire is to be changed, pry off the hub cap (if equipped) using the tapered end of the lug wrench. If the wheel is being replaced, loosen the wheel nuts one-half turn and leave them in place until the wheel is raised off the ground.

Place the jack under the vehicle using the illustration to determine the jack position, depending on your vehicle type. Operate the jack with a slow, smooth motion until the wheel is raised off the ground.

After the wheel is changed, lower the vehicle, remove the jack and tighten the nuts (if loosened or removed) in a criss-cross sequence by turning the wrench clockwise. Replace the hub cap (if equipped) by placing it in position and using the heel of your hand or a rubber mallet to seat it.

Towing

The vehicle can be towed with all four wheels on the ground, provided that speeds do not exceed 30 mph and the distance is not over 15 miles, otherwise transmission damage can result.

Towing equipment specifically designed for this purpose should be used and should be attached to the main structural members of the vehicle and not the bumper or brackets.

Safety is a major consideration when towing and all applicable state and local laws must be obeyed. A safety chain system must be used for all towing.

If the vehicle must be towed over 30 mph or farther than 15 miles, then the rear wheels must be lifted off the ground or the driveshaft must be disconnected.

While towing the parking brake should be released and the transmission should be in Neutral. The steering must be unlocked (ignition switch in the Off position). Remember that power steering and power brakes will not work with the engine off.

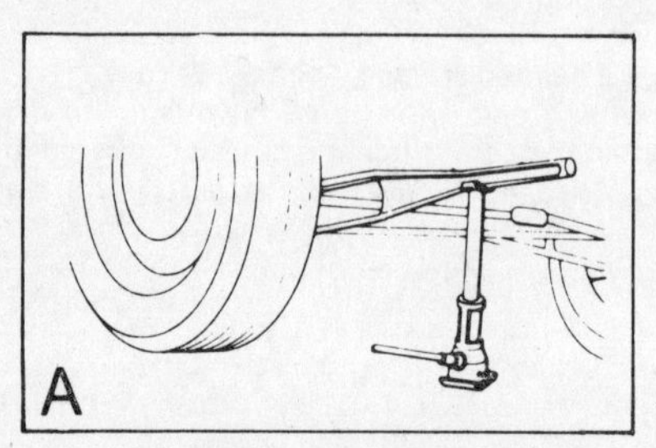

A Front jacking point for Series E100 and 150

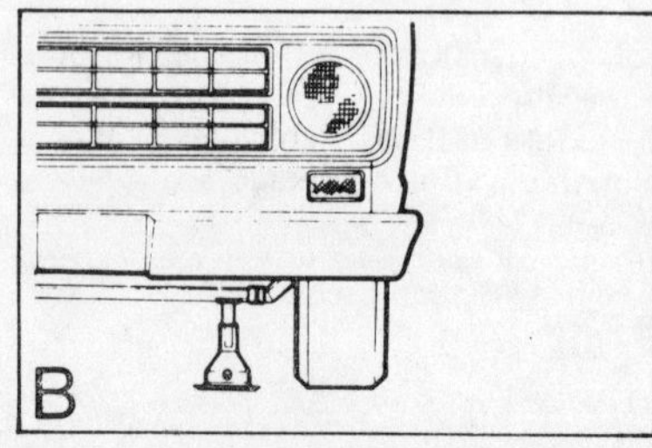

B Front jacking point for Series E250 and 350

C Rear jacking point

Econoline jacking points; all series, front and rear

Conversion factors

Length (distance)					
Inches (in)	X 25.4	= Millimetres (mm)	X	0.0394	= Inches (in)
Feet (ft)	X 0.305	= Metres (m)	X	3.281	= Feet (ft)
Miles	X 1.609	= Kilometres (km)	X	0.621	= Miles
Volume (capacity)					
Cubic inches (cu in; in³)	X 16.387	= Cubic centimetres (cc; cm³)	X	0.061	= Cubic inches (cu in; in³)
Imperial pints (Imp pt)	X 0.568	= Litres (l)	X	1.76	= Imperial pints (Imp pt)
Imperial quarts (Imp qt)	X 1.137	= Litres (l)	X	0.88	= Imperial quarts (Imp qt)
Imperial quarts (Imp qt)	X 1.201	= US quarts (US qt)	X	0.833	= Imperial quarts (Imp qt)
US quarts (US qt)	X 0.946	= Litres (l)	X	1.057	= US quarts (US qt)
Imperial gallons (Imp gal)	X 4.546	= Litres (l)	X	0.22	= Imperial gallons (Imp gal)
Imperial gallons (Imp gal)	X 1.201	= US gallons (US gal)	X	0.833	= Imperial gallons (Imp gal)
US gallons (US gal)	X 3.785	= Litres (l)	X	0.264	= US gallons (US gal)
Mass (weight)					
Ounces (oz)	X 28.35	= Grams (g)	X	0.035	= Ounces (oz)
Pounds (lb)	X 0.454	= Kilograms (kg)	X	2.205	= Pounds (lb)
Force					
Ounces-force (ozf; oz)	X 0.278	= Newtons (N)	X	3.6	= Ounces-force (ozf; oz)
Pounds-force (lbf; lb)	X 4.448	= Newtons (N)	X	0.225	= Pounds-force (lbf; lb)
Newtons (N)	X 0.1	= Kilograms-force (kgf; kg)	X	9.81	= Newtons (N)
Pressure					
Pounds-force per square inch (psi; lbf/in²; lb/in²)	X 0.070	= Kilograms-force per square centimetre (kgf/cm²; kg/cm²)	X	14.223	= Pounds-force per square inch (psi; lbf/in²; lb/in²)
Pounds-force per square inch (psi; lbf/in²; lb/in²)	X 0.068	= Atmospheres (atm)	X	14.696	= Pounds-force per square inch (psi; lbf/in²; lb/in²)
Pounds-force per square inch (psi; lbf/in²; lb/in²)	X 0.069	= Bars	X	14.5	= Pounds-force per square inch (psi; lbf/in²; lb/in²)
Pounds-force per square inch (psi; lbf/in²; lb/in²)	X 6.895	= Kilopascals (kPa)	X	0.145	= Pounds-force per square inch (psi; lbf/in²; lb/in²)
Kilopascals (kPa)	X 0.01	= Kilograms-force per square centimetre (kgf/cm²; kg/cm²)	X	98.1	= Kilopascals (kPa)
Torque (moment of force)					
Pounds-force inches (lbf in; lb in)	X 1.152	= Kilograms-force centimetre (kgf cm; kg cm)	X	0.868	= Pounds-force inches (lbf in; lb in)
Pounds-force inches (lbf in; lb in)	X 0.113	= Newton metres (Nm)	X	8.85	= Pounds-force inches (lbf in; lb in)
Pounds-force inches (lbf in; lb in)	X 0.083	= Pounds-force feet (lbf ft; lb ft)	X	12	= Pounds-force inches (lbf in; lb in)
Pounds-force feet (lbf ft; lb ft)	X 0.138	= Kilograms-force metres (kgf m; kg m)	X	7.233	= Pounds-force feet (lbf ft; lb ft)
Pounds-force feet (lbf ft; lb ft)	X 1.356	= Newton metres (Nm)	X	0.738	= Pounds-force feet (lbf ft; lb ft)
Newton metres (Nm)	X 0.102	= Kilograms-force metres (kgf m; kg m)	X	9.804	= Newton metres (Nm)
Power					
Horsepower (hp)	X 745.7	= Watts (W)	X	0.0013	= Horsepower (hp)
Velocity (speed)					
Miles per hour (miles/hr; mph)	X 1.609	= Kilometres per hour (km/hr; kph)	X	0.621	= Miles per hour (miles/hr; mph)
Fuel consumption*					
Miles per gallon, Imperial (mpg)	X 0.354	= Kilometres per litre (km/l)	X	2.825	= Miles per gallon, Imperial (mpg)
Miles per gallon, US (mpg)	X 0.425	= Kilometres per litre (km/l)	X	2.352	= Miles per gallon, US (mpg)

Temperature

Degrees Fahrenheit = (°C x 1.8) + 32

Degrees Celsius (Degrees Centigrade; °C) = (°F - 32) x 0.56

**It is common practice to convert from miles per gallon (mpg) to litres/100 kilometres (l/100km), where mpg (Imperial) x l/100 km = 282 and mpg (US) x l/100 km = 235*

Safety first!

Regardless of how enthusiastic you may be about getting on with the job at hand, take the time to ensure that your safety is not jeopardized. A moment's lack of attention can result in an accident, as can failure to observe certain simple safety precautions. The possibility of an accident will always exist, and the following points should not be considered a comprehensive list of all dangers. Rather, they are intended to make you aware of the risks and to encourage a safety conscious approach to all work you carry out on your vehicle.

Essential DOs and DON'Ts

DON'T rely on a jack when working under the vehicle. Always use approved jackstands to support the weight of the vehicle and place them under the recommended lift or support points.
DON'T attempt to loosen extremely tight fasteners (i.e. wheel lug nuts) while the vehicle is on a jack — it may fall.
DON'T start the engine without first making sure that the transmission is in Neutral (or Park where applicable) and the parking brake is set.
DON'T remove the radiator cap from a hot cooling system — let it cool or cover it with a cloth and release the pressure gradually.
DON'T attempt to drain the engine oil until you are sure it has cooled to the point that it will not burn you.
DON'T touch any part of the engine or exhaust system until it has cooled sufficiently to avoid burns.
DON'T siphon toxic liquids such as gasoline, antifreeze and brake fluid by mouth, or allow them to remain on your skin.
DON'T inhale brake lining dust — it is potentially hazardous (see *Asbestos* below)
DON'T allow spilled oil or grease to remain on the floor — wipe it up before someone slips on it.
DON'T use loose fitting wrenches or other tools which may slip and cause injury.
DON'T push on wrenches when loosening or tightening nuts or bolts. Always try to pull the wrench toward you. If the situation calls for pushing the wrench away, push with an open hand to avoid scraped knuckles if the wrench should slip.
DON'T attempt to lift a heavy component alone — get someone to help you.
DON'T rush or take unsafe shortcuts to finish a job.
DON'T allow children or animals in or around the vehicle while you are working on it.
DO wear eye protection when using power tools such as a drill, sander, bench grinder, etc. and when working under a vehicle.
DO keep loose clothing and long hair well out of the way of moving parts.
DO make sure that any hoist used has a safe working load rating adequate for the job.
DO get someone to check on you periodically when working alone on a vehicle.
DO carry out work in a logical sequence and make sure that everything is correctly assembled and tightened.
DO keep chemicals and fluids tightly capped and out of the reach of children and pets.
DO remember that your vehicle's safety affects that of yourself and others. If in doubt on any point, get professional advice.

Asbestos

Certain friction, insulating, sealing, and other products — such as brake linings, brake bands, clutch linings, torque converters, gaskets, etc. — contain asbestos. *Extreme care must be taken to avoid inhalation of dust from such products since it is hazardous to health.* If in doubt, assume that they *do* contain asbestos.

Fire

Remember at all times that gasoline is highly flammable. Never smoke or have any kind of open flame around when working on a vehicle. But the risk does not end there. A spark caused by an electrical short circuit, by two metal surfaces contacting each other, or even by static electricity built up in your body under certain conditions, can ignite gasoline vapors, which in a confined space are highly explosive. Do not, under any circumstances, use gasoline for cleaning parts. Use an approved safety solvent.

Always disconnect the battery ground (–) cable *at the battery* before working on any part of the fuel system or electrical system. Never risk spilling fuel on a hot engine or exhaust component.

It is strongly recommended that a fire extinguisher suitable for use on fuel and electrical fires be kept handy in the garage or workshop at all times. Never try to extinguish a fuel or electrical fire with water.

Fumes

Certain fumes are highly toxic and can quickly cause unconsciousness and even death if inhaled to any extent. Gasoline vapor falls into this category, as do the vapors from some cleaning solvents. Any draining or pouring of such volatile fluids should be done in a well ventilated area.

When using cleaning fluids and solvents, read the instructions on the container carefully. Never use materials from unmarked containers.

Never run the engine in an enclosed space, such as a garage. Exhaust fumes contain carbon monoxide, which is extremely poisonous. If you need to run the engine, always do so in the open air, or at least have the rear of the vehicle outside the work area.

If you are fortunate enough to have the use of an inspection pit, never drain or pour gasoline and never run the engine while the vehicle is over the pit. The fumes, being heavier than air, will concentrate in the pit with possibly lethal results.

The battery

Never create a spark or allow a bare light bulb near the battery. The battery normally gives off a certain amount of hydrogen gas, which is highly explosive.

Always disconnect the battery ground (–) cable *at the battery* before working on the fuel or electrical systems.

If possible, loosen the filler caps or cover when charging the battery from an external source. Do not charge at an excessive rate or the battery may burst.

Take care when adding water and when carrying a battery. The electrolyte, even when diluted, is very corrosive and should not be allowed to contact clothing or skin.

Always wear eye protection when cleaning the battery to prevent the caustic deposits from entering your eyes.

Household current

When using an electric power tool, inspection light, etc., which operates on household current, always make sure that the tool is correctly connected to its plug and that, where necessary, it is properly grounded. Do not use such items in damp conditions and, again, do not create a spark or apply excessive heat in the vicinity of fuel or fuel vapor.

Secondary ignition system voltage

A severe electric shock can result from touching certain parts of the ignition system (such as the spark plug wires) when the engine is running or being cranked, particularly if components are damp or the insulation is defective. In the case of an electronic ignition system, the secondary system voltage is much higher and could prove fatal.

Automotive chemicals and lubricants

A number of automotive chemicals and lubricants are available for use during vehicle maintenance and repair. They include a wide variety of products ranging from cleaning solvents and degreasers to lubricants and protective sprays for rubber, plastic and vinyl.

Cleaners

Carburetor cleaner and choke cleaner is a strong solvent for gum, varnish and carbon. Most carburetor cleaners leave a dry-type lubricant film which will not harden or gum up. Because of this film it is not recommended for use on electrical components.

Brake system cleaner is used to remove grease and brake fluid from the brake system where clean surfaces are absolutely necessary. It leaves no residue and often eliminates brake squeal caused by contaminants.

Electrical cleaner removes oxidation, corrosion and carbon deposits from electrical contacts, restoring full current flow. It can also be used to clean spark plugs, carburetor jets, voltage regulators and other parts where an oil-free surface is desired.

Demoisturants remove water and moisture from electrical components such as alternators, voltage regulators, electrical connectors and fuse blocks. It is non-conductive, non-corrosive and non-flammable.

Degreasers are heavy-duty solvents used to remove grease from the outside of the engine and from chassis components. They can be sprayed or brushed on, and, depending on the type, are rinsed off either with water or solvent.

Lubricants

Motor oil is the lubricant formulated for use in engines. It normally contains a wide variety of additives to prevent corrosion and reduce foaming and wear. Motor oil comes in various weights (viscosity ratings) from 5 to 80. The recommended weight of the oil depends on the season, temperature and the demands on the engine. Light oil is used in cold climates and under light load conditions. Heavy oil is used in hot climates and where high loads are encountered. Multi-viscosity oils are designed to have characteristics of both light and heavy oils and are available in a number of weights from 5W-20 to 20W-50.

Gear oil is designed to be used in differentials, manual transaxles and other areas where high-temperature lubrication is required.

Chassis and wheel bearing grease is a heavy grease used where increased loads and friction are encountered, such as for wheel bearings, balljoints, tie rod ends and universal joints.

High temperature wheel bearing grease is designed to withstand the extreme temperatures encountered by wheel bearings in disc brake equipped vehicles. It usually contains molybdenun disulfide (moly), which is a dry-type lubricant.

White grease is a heavy grease for metal to metal applications where water is a problem. White grease stays soft under both low and high temperatures (usually from −100 °F to +190 °F), and will not wash off or dilute in the presence of water.

Assembly lube is a special extreme pressure lubricant, usually containing moly, used to lubricate high-load parts such as main and rod bearings and cam lobes for initial start-up of a new engine. The assembly lube lubricates the parts without being squeezed out or washed away until the engine oiling system begins to function.

Silicone lubricants are used to protect rubber, plastic, vinyl and nylon parts.

Graphite lubricants are used where oils cannot be used due to contamination problems, such as in locks. The dry graphite will lubricate metal parts while remaining uncontaminated by dirt, water, oil or acids. It is electrically conductive and will not foul electrical contacts in locks such as the ignition switch.

Moly penetrants loosen and lubricate frozen, rusted and corroded fasteners and prevent future rusting or freezing.

Heat-sink grease is a special electrically non-conductive grease that is used for mounting HEI ignition modules where it is essential that heat be transferred away from the module.

Sealants

RTV sealant is one of the most widely used gasket compounds. Made from silicone, RTV is air curing, it seals, bonds, waterproofs, fills surface irregularities, remains flexible, doesn't shrink, is relatively easy to remove, and is used as a supplementary sealer with almost all low and medium temperature gaskets.

Anaerobic sealant is much like RTV in that it can be used either to seal gaskets or to form gaskets by itself. It remains flexible, is solvent resistant and fills surface imperfections. The difference between an anaerobic sealant and an RTV-type sealant is in the curing. RTV cures when exposed to air, while an anaerobic sealant cures only in the absence of air. This means that an anaerobic sealant cures only after the assembly of parts, sealing them together.

Thread and pipe sealant is used for sealing hydraulic and pneumatic fittings and vacuum lines. It is usually made from a teflon compound, and comes in a spray, a paint-on liquid and as a wrap-around tape.

Chemicals

Anti-seize compound prevents seizing, galling, cold welding, rust and corrosion in fasteners. High temperature anti-seize, usually made with copper and graphite lubricants, is used for exhaust system and manifold bolts.

Anaerobic locking compounds are used to keep fasteners from vibrating or working loose, and cure only after installation, in the absence of air. Medium strength locking compound is used for small nuts, bolts and screws that you expect to be removing later. High strength locking compound is for large nuts, bolts and studs which you don't intend to be removing on a regular basis.

Oil additives range from viscosity index improvers to chemical treatments that claim to reduce internal engine friction. It should be noted that most oil manufacturers caution against using additives with their oils.

Gas additives perform several functions, depending on their chemical makeup. They usually contain solvents that help dissolve gum and varnish that build up on carburetor and intake parts. They also serve to break down carbon deposits that form on the inside surfaces of the combustion chambers. Some additives contain upper cylinder lubricants for valves and piston rings, and others chemicals to remove condensation from the gas tank.

Other

Brake fluid is specially formulated hydraulic fluid that can withstand the heat and pressure encountered in brake systems. Care must be taken that this fluid does not come in contact with painted surfaces or plastics. An opened container should always be resealed to prevent contamination by water or dirt.

Weatherstrip adhesive is used to bond weatherstripping around doors, windows and trunk lids. It is sometimes used to attach trim pieces.

Undercoating is a petroleum-based tar-like substance that is designed to protect metal surfaces on the underside of the vehicle from corrosion. It also acts as a sound-deadening agent by insulating the bottom of the vehicle.

Waxes and polishes are used to help protect painted and plated surfaces from the weather. Different types of paint may require the use of different types of wax and polish. Some polishes utilize a chemical or abrasive cleaner to help remove the top layer of oxidized (dull) paint on older vehicles. In recent years many non-wax polishes that contain a wide variety of chemicals such as polymers and silicones have been introduced. These non-wax polishes are usually easier to apply and last longer than conventional waxes and polishes.

Troubleshooting

Contents

This section provides an easy reference guide to the more common problems which may occur during the operation of your vehicle. These problems and possible causes are grouped under various components or systems; i.e. Engine, Cooling system, etc., and also refer to the Chapter which deals with the problem.

Remember that successful troubleshooting is not a mysterious black art practiced only by professional mechanics. It is simply the result of a bit of knowledge combined with an intelligent, systematic approach to the problem. Always work by a process of elimination, starting with the simplest solution and working through to the most complex — and never overlook the obvious. Anyone can forget to fill the gas tank or leave the lights on overnight.

Finally, always get clear in your mind why a problem has occurred and take steps to ensure that it doesn't happen again. If the electrical system fails because of a poor connection, check all other connections in the system to make sure that they don't fail as well. If a particular fuse continues to blow, find out why — don't just go on replacing fuses. Remember, failure of a small component can often be indicative of potential failure or incorrect functioning of a more important component or system.

Engine

Note: *When diagnosing engine problems that may be fuel system related on EFI (electronic fuel injection) equipped vehicles, special tools and equipment may be required to pinpoint the causes of hard starting, rough running or poor performance. The following information applies only to carburetor equipped vehicles.*

1 Engine will not rotate when attempting to start

1 Battery terminal connections loose or corroded. Check the cable terminals at the battery. Tighten the cable or remove the corrosion as necessary.
2 Battery discharged or faulty. If the cable connections are clean and tight on the battery posts, turn the key to the On position and switch on the headlights and/or windshield wipers. If they fail to function, the battery is discharged.
3 Automatic transmission not completely engaged in Park, standard transmission clutch not completely depressed or neutral safety switch malfunction (Chapter 7).
4 Broken, loose or disconnected wiring in the starting circuit. Inspect all wiring and connectors at the battery, starter solenoid and ignition switch.
5 Starter motor pinion jammed in flywheel ring gear. If equipped with a manual transmission, place the transmission in gear and rock the vehicle to manually turn the engine. Remove the starter and inspect the pinion and flywheel at the earliest opportunity.
6 Starter solenoid faulty (Chapter 5).
7 Starter motor faulty (Chapter 5).
8 Ignition switch faulty (Chapter 12).
9 On vehicles equipped with a hydraulic clutch release system, the starter interlock switch on the clutch pedal bracket may be malfunctioning, the wire harness may be disconnected or adjustment may be required (Chapter 8).

2 Engine rotates but will not start

1 Fuel tank empty.
2 Battery discharged (engine rotates slowly). Check the operation of the electrical components as described in the previous Section.
3 Battery terminal connections loose or corroded. See the previous Section.
4 Carburetor flooded and/or fuel level in the carburetor incorrect. This will usually be accompanied by a strong fuel odor from under the hood. Wait a few minutes, depress the accelerator pedal all the way to the floor and attempt to start the engine.
5 Choke control inoperative (Chapter 1).
6 Fuel not reaching the carburetor. With the ignition switch in the Off position, open the hood, remove the top plate of air cleaner assembly and observe the top of the carburetor (manually move the choke plate back if necessary). Have an assistant depress the accelerator pedal and check that fuel spurts into the carburetor. If not, check the fuel filter (Chapter 1), fuel lines and fuel pump (Chapter 4).
7 Excessive moisture on, or damage to, ignition components (Chapter 5).
8 Worn, faulty or incorrectly gapped spark plugs (Chapter 1).
9 Broken, loose or disconnected wiring in the starting circuit (see the previous Section).
10 Distributor loose, causing the ignition timing to change. Turn the distributor as necessary to start the engine, then set the ignition timing properly as soon as possible (Chapter 1).
11 Defective condenser or worn or dirty contact breaker points (not all models — see Chapters 1 and 5).
12 Broken, loose or disconnected wires at the ignition coil or faulty coil (Chapter 5).

3 Starter motor operates without rotating the engine

1 Starter pinion sticking. Remove the starter (Chapter 5) and inspect.
2 Starter pinion or flywheel teeth worn or broken. Remove the cover at the rear of the engine and inspect.

4 Engine hard to start when cold

1 Battery discharged or low. Check as described in Section 1.
2 Choke control inoperative or out of adjustment (Chapter 4).
3 Carburetor flooded (see Section 2).
4 Fuel supply not reaching the carburetor (see Section 2).
5 Carburetor system in need of overhaul (Chapter 4).
6 Distributor rotor carbon tracked and/or mechanical advance mechanism rusted (Chapter 5).

5 Engine hard to start when hot

1 Choke sticking in the closed position (Chapter 1).
2 Carburetor flooded (see Section 2).
3 Air filter clogged (Chapter 1).
4 Fuel not reaching the carburetor (see Section 2).
5 Poor engine ground. Clean the ground cable.

6 Starter motor noisy or excessively rough in engagement

1 Pinion or flywheel gear teeth worn or broken. Remove the cover at the rear of the engine (if so equipped) and inspect.
2 Starter motor mounting bolts loose or missing.

7 Engine starts but stops immediately

1 Loose or faulty electrical connections at the distributor, coil or alternator.
2 Insufficient fuel reaching the carburetor. Disconnect the fuel line at the carburetor and remove the filter (Chapter 1). Place a container under the disconnected fuel line. Observe the flow of fuel from the line. If little or none at all, check for blockage in the lines and/or replace the fuel pump (Chapter 4).
3 Vacuum leak at the gasket surfaces of the intake manifold and/or carburetor. Make sure that all mounting bolts and nuts are tightened securely and that all vacuum hoses connected to the carburetor/fuel injection unit and manifold are positioned properly and in good condition.

8 Oil puddle under engine

1 Oil pan gasket and/or oil plug seal leaking. Check and replace if necessary.
2 Oil pressure sending unit leaking. Replace the unit or seal the threads with teflon tape (Chapter 5).
3 Rocker cover gaskets leaking at the front or rear of the engine.
4 Engine oil seals leaking at the front or rear of the engine.

9 Engine lopes while idling or idles erratically

1 Vacuum leakage. Check the mounting bolts and nuts at the carburetor and intake manifold for tightness. Make sure that all vacuum hoses are connected and in good condition. Use a stethoscope or a length of fuel hose held against your ear to listen for vacuum leaks

while the engine is running. A hissing sound will be heard. **Warning:** *Do not get the stethoscope, hose, clothing or parts of your body entangled in moving engine parts such as drivebelts, the fan, etc., or burn yourself on hot engine parts such as the exhaust manifold, EGR components, etc.*
2 Leaking EGR valve or plugged PCV valve (Chapters 1 and 6).
3 Air filter clogged (Chapter 1).
4 Fuel pump not delivering sufficient fuel to the carburetor (see Section 7).
5 Carburetor out of adjustment (Chapter 4).
6 Leaking head gasket. If this is suspected, take the vehicle to a repair shop or dealer where the engine can be pressure checked.
7 Timing chain and/or gears worn (Chapter 2).
8 Camshaft lobes worn (Chapter 2).

10 Engine misses at idle speed

1 Spark plugs worn or not gapped properly (Chapter 1).
2 Faulty cap, rotor or spark plug wires (Chapter 1).
3 Choke not operating properly (Chapter 1).

11 Engine misses throughout the driving speed range

1 Fuel filter clogged and/or impurities in the fuel system (Chapter 1). Also check fuel output at the carburetor (see Section 7).
2 Faulty or incorrectly gapped spark plugs (Chapter 1).
3 Incorrect ignition timing (Chapter 1).
4 Check for a cracked distributor cap, disconnected distributor wires and damaged distributor components (Chapter 1).
5 Leaking spark plug wires (Chapter 1).
6 Faulty emissions system components (Chapter 6).
7 Low or uneven cylinder compression pressures. Remove the spark plugs and test the compression with gauge (Chapter 1).
8 Weak or faulty ignition system (Chapter 5).
9 Vacuum leaks at the carburetor, intake manifold or vacuum hoses (see Section 8).

12 Engine stumbles on acceleration

1 Spark plugs fouled (Chapter 1). Clean or replace.
2 Carburetor needs adjustment or repair (Chapter 4).
3 Fuel filter clogged. Replace the filter.
4 Incorrect ignition timing (Chapter 1).
5 Intake manifold air leak (Chapters 4 and 6).

13 Engine stalls

1 Idle speed incorrect (Chapter 1).
2 Fuel filter clogged and/or water and impurities in the fuel system (Chapter 1).
3 Choke improperly adjusted or sticking (Chapter 1).
4 Distributor components damp or damaged (Chapter 5).
5 Faulty emissions system components (Chapter 6).
6 Faulty or incorrectly gapped spark plugs (Chapter 1). Also check the spark plug wires (Chapter 1).
7 Vacuum leak at the carburetor, intake manifold or vacuum hoses. Check as described in Section 8.

14 Engine lacks power

1 Incorrect ignition timing (Chapter 1).
2 Excessive play in the distributor shaft. At the same time, check for a worn rotor, faulty distributor cap, wires, etc. (Chapters 1 and 5).
3 Faulty or incorrectly gapped spark plugs (Chapter 1).
4 Carburetor not adjusted properly or excessively worn (Chapter 4).
5 Faulty coil or condenser (Chapters 1 and 5).
6 Brakes binding (Chapter 1).
7 Automatic transmission fluid level incorrect (Chapter 1).
8 Clutch slipping (Chapter 8).
9 Fuel filter clogged and/or impurities in the fuel system (Chapter 1).
10 Emissions control system not functioning properly (Chapter 6).
11 Use of substandard fuel. Fill tank with proper octane fuel.
12 Low or uneven cylinder compression pressures. Test with a compression tester, which will detect leaking valves and/or a blown head gasket (Chapter 1).

15 Engine backfires

1 Emissions system not functioning properly (Chapter 6).
2 Ignition timing incorrect (Chapter 1).
3 Faulty secondary ignition system. Cracked spark plug insulator, faulty plug wires, distributor cap and/or rotor (Chapters 1 and 5).
4 Carburetor in need of adjustment or worn excessively (Chapter 4).
5 Vacuum leak at the carburetor/fuel injection unit, intake manifold or vacuum hoses. Check as described in Section 8.
6 Lifter pumping up and/or valves sticking (Chapter 2).

16 Pinging or knocking engine sounds during acceleration or uphill

1 Incorrect grade of fuel. Fill tank with fuel of the proper octane rating.
2 Ignition timing incorrect (Chapter 1).
3 Carburetor in need of adjustment (Chapter 4).
4 Incorrect spark plugs. Check the plug type against the Emissions Control Information label located in engine compartment. Also check the plugs and wires for damage (Chapter 1).
5 Worn or damaged distributor components (Chapter 5).
6 Faulty emissions system (Chapter 6).
7 Vacuum leak. Check as described in Section 8.

17 Engine runs with oil pressure light on

1 Low oil level. Check the oil level and add oil if necessary (Chapter 1).
2 Idle rpm below specification (Chapter 1).
3 Short in the wiring circuit. Repair or replace the damaged wire.
4 Faulty oil pressure sender. Replace the sender.
5 Worn engine bearings and/or oil pump.

18 Engine diesels (continues to run) after switching off

1 Idle speed too high (Chapter 1).
2 Electrical solenoid at side of carburetor not functioning properly (not all models, see Chapter 4).
3 Ignition timing incorrectly adjusted (Chapter 1).
4 Thermo-controlled air cleaner heat valve not operating properly (Chapter 6).
5 Excessive engine operating temperature. Probable causes of this are a malfunctioning thermostat, clogged radiator or a faulty water pump (Chapter 3).
6 EGR system malfunction.
7 Carbon buildup on piston crowns.

Engine electrical system

19 Battery will not hold a charge

1 Alternator drivebelt defective or not adjusted properly (Chapter 1).
2 Electrolyte level low or battery discharged (Chapter 1).
3 Battery terminals loose or corroded (Chapter 1).
4 Alternator not charging properly (Chapter 5).
5 Loose, broken or faulty wiring in the charging circuit (Chapter 5).
6 Short in the vehicle wiring causing a continual drain on the battery.
7 Battery defective internally.

20 Alternator light fails to go out

1 Fault in the alternator or charging circuit (Chapter 5).
2 Alternator drivebelt defective or not properly adjusted (Chapter 1).
3 Alternator voltage regulator inoperative (Chapter 5).

21 Ignition light fails to come on when the key is turned on

1 Warning light bulb defective (Chapter 12).
2 Alternator faulty (Chapter 5).
3 Fault in the printed circuit, dash wiring or bulb holder (Chapter 12).

Fuel system

Note: *The following information applies only to carburetor equipped vehicles. Vehicles equipped with EFI (electronic fuel injection) are covered in Chapter 4.*

22 Excessive fuel consumption

1 Dirty or clogged air filter element (Chapter 1).
2 Incorrectly set ignition timing (Chapter 1).
3 Choke sticking or improperly adjusted (Chapter 1).
4 Emissions system not functioning properly (not all vehicles, see Chapter 6).
5 Carburetor idle speed and/or mixture not adjusted properly (Chapter 1).
6 Carburetor internal parts excessively worn or damaged (Chapter 4).
7 Low tire pressure or incorrect tire size (Chapter 1).

23 Fuel leakage and/or fuel odor

1 Leak in a fuel feed or vent line (Chapter 4).
2 Tank overfilled. Fill only to automatic shut off.
3 Emissions system filter clogged (Chapter 1).
4 Vapor leaks from system lines (Chapter 4).
5 Carburetor internal parts excessively worn or out of adjustment (Chapter 4).

Cooling system

24 Overheating

1 Insufficient coolant in the system (Chapter 1).
2 Water pump drivebelt defective or not adjusted properly (Chapter 1).
3 Radiator core blocked or radiator grille dirty and restricted (Chapter 3).
4 Thermostat faulty (Chapter 3).
5 Fan blades broken or cracked (Chapter 3).
6 Radiator cap not maintaining proper pressure. Have the cap pressure tested by a gas station or repair shop.
7 Ignition timing incorrect (Chapter 1).
8 Defective water pump (Chapter 3).

25 Overcooling

1 Thermostat faulty (Chapter 3).
2 Inaccurate temperature gauge (Chapter 12).

26 External coolant leakage

1 Deteriorated or damaged hoses or loose clamps. Replace the hoses and/or tighten the clamps at the hose connections (Chapter 1).
2 Water pump seals defective. If this is the case, water will drip from the weep hole in the water pump body (Chapter 1).
3 Leakage from the radiator core or header tank. This will require professional radiator repair (see Chapter 3 for removal procedures).
4 Engine drain plugs or water jacket core plugs leaking (Chapter 2).

27 Internal coolant leakage

Note: *Internal coolant leaks can usually be detected by examining the oil. Check the dipstick and the inside of the rocker arm cover for water deposits and an oil consistency like that of a milkshake.*

1 Leaking cylinder head gasket. Have the cooling system pressure tested.
2 Cracked cylinder bore or cylinder head. Dismantle the engine and inspect (Chapter 2).
3 Loose cylinder head bolts (Chapter 2).

28 Coolant loss

1 Too much coolant in the system (Chapter 1).
2 Coolant boiling away due to overheating (see Section 16).
3 Internal or external leakage (see Sections 25 and 26).
4 Faulty radiator cap. Have the cap pressure tested.

29 Poor coolant circulation

1 Inoperative water pump. A quick test is to pinch the top radiator hose closed with your hand while the engine is idling, then let it loose. You should feel the surge of coolant if the pump is working properly (Chapter 1).
2 Restriction in the cooling system. Drain, flush and refill the system (Chapter 1). If necessary, remove the radiator (Chapter 3) and have it reverse flushed.
3 Water pump drivebelt defective or not adjusted properly (Chapter 1).
4 Thermostat sticking (Chapter 3).

Clutch

30 Fails to release (pedal pressed to the floor – shift lever does not move freely in and out of Reverse)

1 Improper linkage free play adjustment (Chapter 1).
2 Clutch fork off the ball stud. Look under the vehicle, on the left side of transmission.
3 Clutch plate warped or damaged (Chapter 8).
4 On vehicles equipped with a hydraulic clutch system, the control or slave cylinders may not be functioning properly (Chapter 8).
5 Leak in the hydraulic system. Check the clutch fluid level (Chapter 1), then refer to Chapter 8.

31 Clutch slips (engine speed increases with no increase in vehicle speed)

1 Linkage out of adjustment (Chapter 1).
2 Clutch plate oil soaked or lining worn. Remove the clutch (Chapter 8) and inspect.
3 Clutch plate not seated. It may take 30 or 40 normal starts for a new one to seat.
4 Warped flywheel (Chapter 2).

32 Grabbing (chattering) as the clutch is engaged

1 Oil on clutch plate lining. Remove (Chapter 8) and inspect. Correct any leakage source.
2 Worn or loose engine or transmission mounts. These units move slightly when the clutch is released. Inspect the mounts and bolts.
3 Worn splines on the clutch plate hub. Remove the clutch com-

ponents (Chapter 8) and inspect.
4 Warped pressure plate or flywheel. Remove the clutch components and inspect.
5 Hardened or warped clutch disc lining.

33 Squeal or rumble with the clutch fully engaged (pedal released)

1 Improper adjustment — no free play (Chapter 1).
2 Release bearing binding on the transmission bearing retainer. Remove the clutch components (Chapter 8) and check the bearing. Remove any burrs or nicks, clean and relubricate before reinstallation.
3 Weak linkage return spring. Replace the spring.
4 Cracked clutch disc.

34 Squeal or rumble with the clutch fully disengaged (pedal depressed)

1 Worn, defective or broken release bearing (Chapter 8).
2 Worn or broken pressure plate springs or diaphragm fingers (Chapter 8).

35 Clutch pedal stays on the floor when disengaged

1 Bind in the linkage or release bearing. Inspect the linkage or remove the clutch components as necessary.
2 Linkage springs being overextended. Adjust the linkage for proper free play. Make sure the proper pedal stop (bumper) is installed.

Manual transmission

36 Noisy in Neutral with the engine running

1 Input shaft bearing worn.
2 Damaged main drive gear bearing.
3 Worn countershaft bearings.
4 Worn or damaged countershaft end play shims.

37 Noisy in all gears

1 Any of the above causes, and/or:
2 Insufficient lubricant (see the checking procedures in Chapter 1).
3 Worn or damaged output shaft or bearings.

38 Noisy in one particular gear

1 Worn, damaged or chipped gear teeth for that particular gear.
2 Worn or damaged synchronizer for that particular gear.

39 Slips out of gear

1 Transmission loose on the clutch housing (Chapter 7).
2 Shift rods interfering with the engine mounts or the clutch lever (Chapter 7).
3 Shift rods not working freely (Chapter 7).
4 Damaged mainshaft pilot bearing.
5 Dirt between the transmission case and engine or misalignment of the transmission (Chapter 7).
6 Worn or improperly adjusted linkage (Chapter 7).
7 Worn synchro units.

40 Difficulty in engaging gears

1 Clutch not releasing completely (see clutch adjustment in Chapter 8).
2 Loose, damaged or out-of-adjustment shift linkage. Make a thorough inspection, replacing parts as necessary (Chapter 7).

41 Oil leakage

1 Excess lubricant in the transmission (see Chapter 1 for the correct checking procedures). Drain lubricant as required.
2 Side cover loose or gasket damaged.
3 Rear oil seal or speedometer oil seal in need of replacement (Chapter 7).

Automatic transmission

Note: *Due to the complexity of the automatic transmission, it is difficult for the home mechanic to properly diagnose and service this component. For problems other than the following, the vehicle should be taken to a dealer or reputable mechanic.*

42 Fluid leakage

1 Automatic transmission fluid is a deep red color. Fluid leaks should not be confused with engine oil, which can easily be blown by air flow to the transmission.
2 To pinpoint a leak, first remove all built-up dirt and grime from around the transmission. Degreasing agents and/or steam cleaning will achieve this. With the underside clean, drive the vehicle at low speeds so air flow will not blow the leak far from its source. Raise the vehicle and determine where the leak is coming from. Common areas of leakage are:

a) Pan: Tighten the mounting bolts and/or replace the pan gasket as necessary (see Chapters 1 and 7).
b) Filler pipe: Replace the rubber seal where the pipe enters transmission case.
c) Transmission oil lines: Tighten the connectors where the lines enter the transmission case and/or replace the lines.
d) Speedometer connector: Replace the O-ring where the speedometer cable enters transmission case (Chapter 7).

43 Transmission fluid brown or has a burned smell

Transmission low on fluid. Replace the fluid. Do not overfill.

44 General shift mechanism problems

1 Chapter 7B deals with checking and adjusting the shift linkage on automatic transmissions. Common problems which may be attributed to poorly adjusted linkage are:

a) Engine starting in gears other than Park or Neutral.
b) Indicator on shifter pointing to a gear other than the one actually being used.
c) Vehicle moves when in Park.

2 Refer to Chapter 7B to adjust the linkage.

45 Transmission will not downshift with the accelerator pedal pressed to the floor

Chapter 7B deals with adjusting the throttle valve (TV) cable to enable the transmission to downshift properly.

46 Engine will start in gears other than Park or Neutral

Chapter 7B deals with adjusting the neutral start switches used on automatic transmissions.

47 Transmission slips, shifts rough, is noisy or has no drive in forward or reverse gears

1 There are many probable causes for the above problems, but the home mechanic should be concerned with only one possibility — fluid level.
2 Before taking the vehicle to a repair shop, check the level and condition of the fluid as described in Chapter 1. Correct the fluid level as necessary or change the fluid and filter if needed. If the problem persists, have a professional diagnose the probable cause.

Driveshaft

48 Leakage of fluid at front of driveshaft

Defective transmission rear oil seal. See Chapter 7 for replacement procedures. While this is done, check the splined yoke for burrs or a rough condition which may be damaging the seal. If found, these can be removed with emery cloth or a fine abrasive stone.

49 Knock or clunk when the transmission is put under initial load (just after transmission is put into gear)

1 Loose or disconnected rear suspension components. Check all mounting bolts and bushings (Chapter 10).
2 Loose driveshaft bolts. Inspect all bolts and nuts and tighten to specification (Chapter 8).
3 Worn or damaged universal joint bearings. Test for wear (Chapter 8).
4 Worn mainshaft splines.

50 Metallic grating sound consistent with road speed

Pronounced wear in the universal joint bearings. Test for wear (Chapter 8).

51 Vibration

Note: *Before it can be assumed that the driveshaft is at fault, make sure the tires are perfectly balanced and perform the following test.*
1 Install a tachometer inside the vehicle to monitor engine speed as the vehicle is driven. Drive the vehicle and note the engine speed at which the vibration is most pronounced. Now shift the transmission to a different gear and bring the engine speed to the same point.
2 If the vibration occurs at the same engine speed (rpm) regardless of which gear the transmission is in, the driveshaft is *Not* at fault since the driveshaft speed varies.
3 If the vibration decreases or is eliminated when the transmission is in a different gear at the same engine speed, refer to the following probable causes:
4 Bent or dented driveshaft. Inspect and replace as necessary (Chapter 8).
5 Undercoating or built-up dirt, etc. on the driveshaft. Clean the shaft thoroughly and test.
6 Worn universal joint bearings. Remove and inspect (Chapter 8).
7 Driveshaft and/or companion flange out of balance. Check for missing weights on the shaft. Remove the driveshaft and reinstall 180° from the original position. Retest. Have the driveshaft professionally balanced if the problem persists.
8 Driveshaft improperly installed (Chapter 8).
9 Worn transmission rear bushing (Chapter 7).

Rear axle

52 Noise

1 Road noise. No corrective procedures available.
2 Tire noise. Inspect the tires and check tire pressures (Chapter 1).
3 Front wheel bearings loose, worn or damaged (Chapter 10).
4 Insufficient differential oil (whining noise consistent with vehicle speed changes).

53 Vibration

See probable causes under *Driveshaft*. Proceed under the guidelines listed for the driveshaft. If the problem persists, check the rear wheel bearings by raising the rear of the car and spinning the wheels by hand. Listen for evidence of rough bearings. Remove and inspect (Chapter 8).

54 Oil leakage

1 Pinion oil seal damaged (Chapter 8).
2 Axleshaft oil seals damaged (Chapter 8).
3 Differential cover leaking. Tighten the mounting bolts or replace the gasket as required (Chapter 8).

Brakes

Note: *Before assuming that a brake problem exists, make sure that the tires are in good condition and inflated properly (see Chapter 1), that the front end alignment is correct and that the vehicle is not loaded with weight in an unequal manner.*

55 Vehicle pulls to one side during braking

1 Defective, damaged or oil contaminated disc brake pads on one side. Inspect as described in Chapter 9.
2 Excessive wear of brake pad material or disc on one side. Inspect and correct as necessary.
3 Loose or disconnected front suspension components. Inspect and tighten all bolts to the specified torque (Chapter 10).
4 Defective caliper assembly. Remove the caliper and inspect for a stuck piston or other damage (Chapter 9).

56 Noise (high-pitched squeal with the brakes applied)

Disc brake pads worn out. The noise comes from the wear sensor rubbing against the disc (does not apply to all vehicles). Replace the pads with new ones immediately (Chapter 9).

57 Excessive brake pedal travel

1 Partial brake system failure. Inspect the entire system (Chapter 9) and correct as required.
2 Insufficient fluid in the master cylinder. Check (Chapter 1), add fluid and bleed the system if necessary (Chapter 9).
3 Rear brakes not adjusting properly. Make a series of starts and stops while the vehicle is in Reverse. If this does not correct the situation, remove the drums and inspect the self-adjusters (Chapter 9).

58 Brake pedal feels spongy when depressed

1 Air in the hydraulic lines. Bleed the brake system (Chapter 9).
2 Faulty flexible hoses. Inspect all system hoses and lines. Replace parts as necessary.
3 Master cylinder mounting nuts or bolts loose.
4 Master cylinder defective (Chapter 9).

59 Excessive effort required to stop vehicle

1 Power brake booster not operating properly (Chapter 9).
2 Excessively worn linings or pads. Inspect and replace if necessary (Chapter 9).

3 One or more caliper pistons or wheel cylinders seized or sticking. Inspect and rebuild as required (Chapter 9).
4 Brake linings or pads contaminated with oil or grease. Inspect and replace as required (Chapter 9).
5 New pads or shoes installed and not yet seated. It will take a while for the new material to seat against the drum or rotor.

60 Pedal travels to the floor with little resistance

Little or no fluid in the master cylinder reservoir caused by leaking wheel cylinder(s), leaking caliper piston(s), loose, damaged or disconnected brake lines. Inspect the entire system and correct as necessary.

61 Brake pedal pulsates during brake application

1 Wheel bearings not adjusted properly or in need of replacement (Chapter 1).
2 Caliper not sliding properly due to improper installation or obstructions. Remove and inspect (Chapter 9).
3 Rotor defective. Remove the rotor (Chapter 9) and check for excessive lateral runout and parallelism. Have the rotor resurfaced or replace it with a new one.

62 Parking brake does not hold

Mechanical parking brake linkage improperly adjusted. Adjust according to the procedure in Chapter 9.

Suspension and steering systems

63 Vehicle pulls to one side

1 Tire pressures uneven (Chapter 1).
2 Defective tire (Chapter 1).
3 Excessive wear in suspension or steering components (Chapter 10).
4 Front end in need of alignment.
5 Front brakes dragging. Inspect the brakes as described in Chapter 9.
6 Wheel bearings improperly adjusted.

64 Shimmy, shake or vibration

1 Tire or wheel out-of-balance or out-of-round. Have professionally balanced.
2 Loose, worn or out-of-adjustment wheel bearings (Chapters 1 and 8).
3 Shock absorbers and/or suspension components worn or damaged (Chapter 10).

65 Excessive pitching and/or rolling around corners or during braking

1 Defective shock absorbers. Replace as a set (Chapter 10).
2 Broken or weak springs and/or suspension components. Inspect as described in Chapter 10.

66 Excessively stiff steering

1 Lack of fluid in power steering fluid reservoir (Chapter 1).
2 Incorrect tire pressures (Chapter 1).
3 Lack of lubrication at steering joints (Chapter 1).
4 Front end out of alignment.
5 Air in power steering system
6 Low tire pressure.

67 Excessive play in steering

1 Loose front wheel bearings (Chapter 1).
2 Excessive wear in suspension or steering components (Chapter 10).
3 Steering gearbox out of adjustment (Chapter 10).

68 Lack of power assistance

1 Steering pump drivebelt faulty or not adjusted properly (Chapter 1).
2 Fluid level low (Chapter 1).
3 Hoses or lines restricted. Inspect and replace parts as necessary.
4 Air in power steering system. Bleed the system (Chapter 10).

69 Excessive tire wear (not specific to one area)

1 Incorrect tire pressures (Chapter 1).
2 Tires out of balance. Have professionally balanced.
3 Wheels damaged. Inspect and replace as necessary.
4 Suspension or steering components excessively worn (Chapter 10).

70 Excessive tire wear on outside edge

1 Inflation pressures incorrect (Chapter 1).
2 Excessive speed in turns.
3 Front end alignment incorrect (excessive toe-in). Have professionally aligned.
4 Suspension arm bent or twisted (Chapter 10).

71 Excessive tire wear on inside edge

1 Inflation pressures incorrect (Chapter 1).
2 Front end alignment incorrect (toe-out). Have professionally aligned.
3 Loose or damaged steering components (Chapter 10).

72 Tire tread worn in one place

1 Tires out of balance.
2 Damaged or buckled wheel. Inspect and replace if necessary.
3 Defective tire (Chapter 1).

Chapter 1 Tune-up and routine maintenance

Refer to Chapter 13 for Specifications and information on 1987 and later models

Contents

Specifications

Note: *Additional specifications and torque recommendations can be found in each individual Chapter.*

Recommended lubricants and fluids

Note: *Listed here are manufacturer recommendations at the time this manual was written. Manufacturers occasionally upgrade their fluid and lubricant specifications, so check with your local auto parts store for current recommendations.*

Engine oil type	SAE grade SE or better
Engine oil viscosity	Consult your owner's manual or local dealer for recommendations on the particular oil viscosity to use for your area, special driving conditions and climate
Automatic transmission fluid type	
C6 and ADD transmissions	Dexron 11 Series D (Ford no. ESP-M2C138-CJ)
all others	Type F (Ford no. ESW-M2C33-F)
Brake and clutch master cylinders	DOT 3 heavy duty brake fluid (Ford no. ESA-M6C25-A)
Power steering reservoir	Type F automatic transmission fluid (Ford no. ESW-M2C33-F)
Manual transmission lubricant	SAE 140W gear oil (Ford no. ESO-M2C83-C)
Engine coolant	Ethylene-glycol based antifreeze (Ford no. ESE-M97B18-C)
Differential lubricant	
Dana and Ford (non-locking)	SAE 90W gear oil (Ford no. ESW-M2C 105-A)
Dana (limited slip)	Add friction modifier EST-M2C 11 8-A to the above oil during refill
Ford (Traction-Lok)	SAE 90W gear oil (Ford no. ESW-M2C 11 9-A)
Chassis lubrication, clutch linkage, parking brake linkage, steering column U-joints, driveshaft U-joints, front and rear wheel bearings and transmission linkage	Lithium based grease, NLGI no. 2 (Ford no. ESA-MIC75-B)
Manual steering gear	Ford no. ESW-MIC87-A steering gear lubricant or equivalent

Engine

Engine compression	The lowest reading must be within 75% of the highest reading
Drivebelt tension (with Burroughs-type gauge)	
1/4-inch belts	
during maintenance	30 lbs
installation used (over 10 minutes engine operation)	60 lbs
installation new	80 lbs
3/8, 15/32 and 1/2-inch belts	
during maintenance	50 lbs
installation used (over 10 minutes engine operation)	110 lbs
installation new	140 lbs

Ignition system *(See page 55 for cylinder numbering and firing order diagrams)*

Spark plug type and gap	See tune-up decal or Emissions Control Information label in engine compartment
Distributor type	
1969 thru 1974	Mechanical breaker-point type
1975 thru 1986	Breakerless, Dura-spark
Distributor direction of rotation	
six-cylinder engines	Clockwise
V8 engines	Counterclockwise
Firing order	
six-cylinder engines	1-5-3-6-2-4
302 and 460 V8 engines	1-5-4-2-6-3-7-8
351 and 400 V8 engines	1-3-7-2-6-5-4-8
Ignition timing	See tune-up decal or Emissions Control Information label in engine compartment
Breaker-point gap	
six-cylinder engines	0.027 in
V8 engines	0.017 in
Dwell angle	
six-cylinder engines	35 to 39 degrees
V8 engines	24 to 30 degrees

Clutch

Pedal height	7-1/2 to 7-3/4 in
Pedal free play	3/4 to 1-1/2 in

Brakes

Front disc brake pad lining minimum thickness	1/8 in
Brake shoe minimum thickness (from surface to rivet head)	1/8 in

Torque specifications	**Ft-lbs**
Spark plugs	10 to 15
Wheel lug nut	
1/2 inch nut	90
9/16 inch nut (single rear wheels)	145
9/16 inch nut (dual rear wheels)	220
Front wheel bearing adjusting nut	22 to 25 (back-off 1/8-turn)
Rear wheel bearing adjusting nut (full-floating axles)	120 to 140 (back-off 1/8-to-1/4-turn on 1975 thru 1980 models; 1/8-to-3/8-turn on 1981 and later models)
Axleshaft retaining bolts (full-floating axles)	40 to 50
Oil pan drain plug	15 to 25
Transmission filler plug (manual)	10 to 20
Transmission pan bolts (automatic)	12 to 16
Filter screen-to-main body (automatic)	40 to 55 in-lbs

1 Introduction to routine maintenance

This Chapter was designed to help the home mechanic maintain his or her vehicle with the goals of maximum performance, economy, safety and durability in mind.

On the following pages you will find a maintenance schedule, along with procedures which deal specifically with each item on the schedule. Included are visual checks, adjustments and item replacements.

Servicing your vehicle following the time/mileage maintenance schedule and the step-by-step procedures will result in a planned maintenance program. Keep in mind that it is an all-inclusive plan. Maintaining only a few items at the specified intervals will not produce the desired results.

You will find as you service your vehicle that many of the procedures can, and should, be grouped together, due to their nature. Examples of this are as follows:

If the vehicle is raised for chassis lubrication, for example, check the exhaust, suspension, steering and fuel systems.

If the tires and wheels are removed, as during a routine tire rotation, check the brakes and wheel bearings at the same time.

If you must borrow or rent a torque wrench, replace the spark plugs and check the intake manifold bolt mounting torque as well to save time and money.

The first step in this, or any, maintenance plan is to prepare yourself before the actual work begins. Read through the appropriate Sections for all work that is to be performed before you begin. Gather together all the necessary parts and tools. If it appears that you could have a problem during a particular job, don't hesitate to seek advice from your local parts man or dealer service department.

2 Routine maintenance intervals

The following recommendations are based on the assumption that the vehicle owner will be doing the maintenance or service work, as opposed to having a dealer service department do the work. The time/mileage intervals are based on factory recommendations. However, subject to the preference of the individual owner interested in keeping his or her vehicle in peak condition at all times and with the vehicle's ultimate resale in mind, many of the operations may be performed more often. We encourage such owner initiative.

When the vehicle is new it should be serviced initially by a factory authorized dealer service department to protect the factory warranty. In many cases the initial maintenance check is done at no cost to the owner.

Every 250 miles (400 km), weekly and before long trips

Check the tire pressures (cold)
Check the steering for smooth and accurate operation
Inspect the tires for wear and damage
Check the power steering reservoir fluid level
Check the brake fluid level. If the amount of fluid has dropped noticeably since the last check, inspect all brake lines and hoses for leaks and damage
Check for satisfactory brake operation
Check the operation of the windshield wipers and washers
Check the windshield wiper condition
Check the radiator coolant level
Check the battery electrolyte level and add distilled water as needed
Check the engine oil level

Every 3000 miles (5000 km) or 3 months, whichever comes first

Change the engine oil and filter
Check and if necessary replace the air filter element
Check the idle speed
Check the engine drivebelts for condition and tension
Check the exhaust heat control valve operation

Every 5000 miles (8000 km) or 5 months, whichever comes first

Check the fuel deceleration valve (Chapter 6)
Check the throttle solenoid operation (Chapter 6)

Every 6000 miles (10,000 km) or 6 months, whichever comes first

Check the cooling system
Check the exhaust system for loose and damaged components
Lubricate the steering linkage
Lubricate the clutch linkage (if applicable)
Check the brake linings and pads
Check the differential oil level and add oil as necessary
Check the manual transmission oil level and add oil as necessary
Adjust the automatic transmission bands (initial adjustment — adjust every 12,000 miles or 12 months after the initial adjustment)
Check the clutch pedal free play and adjust if necessary
Check the idle fuel/air mixture (if applicable)
Check the ignition points
Check/adjust the ignition timing (except for electronic ignition)
Replace the fuel filter

Every 12,000 miles (20,000 km) or 12 months, whichever comes first

Lubricate the parking brake linkage
Check the spark plug wires
Replace the ignition points
Replace the spark plugs (engines using leaded fuel)
Check and adjust the valve clearances
Check and if necessary replace the distributor cap and rotor (except for electronic ignition)
Check the operation of the EGR system and delay valve (if so equipped)
Check and clean the crankcase breather cap
Inspect the brake lines and hoses
Check the PCV valve
Adjust the automatic transmission bands
Check the operation of the thermo-controlled air cleaner
Check the spark control system and delay valve (Chapter 6)
Check and service the battery
Check the suspension and steering systems

Every 15,000 miles (24,000 km) or 15 months, whichever comes first

Check the Thermactor system (Chapter 6)
Check the air cleaner temperature control and delay valves (Chapter 6)

Every 18,000 miles (30,000 km) or 18 months, whichever comes first

Check the ignition timing (electronic ignition only)
Check the distributor cap and rotor (electronic ignition only)
Tighten the intake manifold bolts/nuts to the specified torque
Replace the crankcase ventilation filter
Replace the automatic transmission fluid
Replace the spark plugs (engines using unleaded fuel)
Check the carburetor choke for proper operation
Add oil to the distributor shaft oil cup (if equipped)
Check and repack the front wheel bearings

Every 24,000 miles (40,000 km) or 24 months, whichever comes first

Drain and refill the cooling system
Drain and refill the manual transmission
Drain and refill the differential
Check the fuel filler cap
Check the evaporative emissions system canister and replace as necessary
Check the engine compression
Replace the underhood hoses with new ones

Severe operating conditions

Severe operating conditions are defined as follows:
Extended periods of idling or low speed operation
Towing trailers up to 1000 lbs (450 kg) for long distances
Operating when the outside temperatures remain below 10 degrees F (−12 degrees C) for 60 days or more
When most trips you make are less than ten miles in duration
Operation in severe dust conditions

If your vehicle is operated under the severe conditions described above, the maintenance schedule must be amended as follows:

a) Change the engine oil and filter every 2 months or 2000 miles (3200 km)
b) Check, clean and regap the spark plugs every 4000 miles (6400 km)
c) Service the automatic transmission bands every 5000 miles (8000 km) and drain and refill the transmission with fresh fluid every 2000 miles (32000 km)
d) If the vehicle is operated in severe dust conditions, check the air filter element every 1000 miles
f) Check and replace the fuel filter frequently when operating in dusty conditions

3 Tune-up sequence

The term tune-up is loosely used for any general operation that puts the engine back into proper running condition. A tune-up is not a specific operation, but rather a combination of individual operations, such as replacing the spark plugs, adjusting the idle speed, setting the ignition timing, etc.

If, from the time the vehicle is new, the routine maintenance schedule (Section 2) is followed closely and frequent checks are made of fluid levels and high wear items, as suggested throughout this manual, the engine will be kept in relatively good running condition and the need for a comprehensive tune-up will be minimized.

More likely than not, however, there will be times when the engine is running poorly due to lack of regular maintenance. This is even more likely if a used vehicle, which has not received regular and frequent maintenance checks, is purchased. In such cases an engine tune-up will be needed outside of the regular routine maintenance intervals.

The following operations are those most often needed to bring a generally poor running engine back into a proper state of tune.

Minor tune-up

Clean, inspect and test the battery
Tighten the carburetor/throttle body mounting bolts/nuts
Check all engine related fluids
Check and adjust the drivebelts
Replace the spark plugs
Inspect the distributor cap and rotor
Inspect the spark plug and coil wires
Check and adjust the idle speed
Check and adjust the ignition timing
Check the air and PCV filters
Replace the fuel filter
Check the PCV valve
Clean and lubricate the throttle linkage
Check all underhood hoses

Major tune-up

All operations listed under Minor tune-up, plus:
Check the EGR system (Chapter 6)
Check the charging system (Chapter 5)
Check the fuel system
Test the battery
Check the engine compression
Check the cooling system
Replace the distributor cap and rotor
Replace the spark plug wires
Replace the air and PCV filters

4 Tire and tire pressure checks

Refer to illustrations 4.3 and 4.8

1 Periodically inspecting the tires may not only prevent you from being stranded with a flat tire, but can also give you clues as to possible problems with the steering and suspension systems before major damage occurs.

2 Proper tire inflation adds miles to the lifespan of the tires, allows the vehicle to achieve maximum miles per gallon of gasoline and contributes to the overall quality of the ride.

3 When inspecting the tires, first check for tread wear (see illustration). Irregularities in the tread pattern such as cupping, flat spots, more wear on one side than the other, are indications of front end alignment and/or balance problems. If any of these conditions are noted, take the vehicle to a repair shop to correct the problem.

4 Check the tread area for cuts and punctures. Many times a nail or tack will embed itself in the tire tread and yet the tire will hold air pressure for a short time. In most cases a repair shop or gas station can repair the punctured tire.

5 It is important to check both the inner and outer sidewalls of each tire. Check for deteriorated rubber, cuts and punctures. Inspect the inner side of the tire for signs of brake fluid leakage, indicating that a thorough brake inspection is needed immediately.

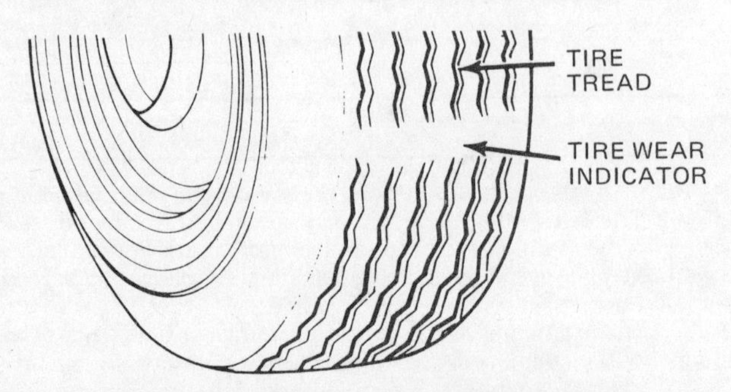

4.3 Tire wear indicators will appear as bars across the tread when the tire has worn to the point where replacement is required

4.8 Press the gauge securely onto the valve stem to obtain an accurate tire pressure reading

6 Incorrect tire pressure cannot be determined merely by looking at the tire. This is especially true for radial tires. A tire pressure gauge must be used. If you do not already have a reliable gauge, purchase one and keep it in the glove compartment. Built-in pressure gauges at gas stations are often inaccurate.

7 Always check tire inflation when the tires are cold. Cold, in this case, means the vehicle has not been driven more than one mile after sitting for three hours or more. It is normal for the pressure to increase four to eight pounds when the tires are hot.

8 Unscrew the valve cap protruding from the wheel or hubcap and press the gauge firmly onto the valve stem (see illustration). Note the reading on the gauge and compare the figure to the recommended tire pressure listed on the tire placard. The tire placard is usually attached to the driver's door post or glove compartment door.

9 Check all tires and add air as necessary to bring them up to the recommended pressure levels. Don't forget the spare tire. Be sure to reinstall the valve caps, which will keep dirt and moisture out of the valve stem mechanism.

5 Fluid level checks

Refer to illustrations 5.1, 5.2, 5.4, 5.6, 5.9, 5.15, 5.17, 5.20a, 5.20b, 5.26, 5.33, 5.37, 5.44, 5.52a and 5.52b

1 There are a number of components on a vehicle which rely on the use of fluids to perform their job (see illustration). During normal operation of the vehicle these fluids are used up and must be replenished before damage occurs. See *Recommended lubricants and fluids* at the front of this Chapter for specific fluids to be used when addition is required. When checking fluid levels it is important to have the vehicle on a level surface.

Engine oil

2 The engine oil level is checked with a dipstick, which is visible just above the top of the radiator when the hood is opened (see illustration). The dipstick travels through a tube and into the oil pan at the bottom of the engine.

3 The oil level should be checked before the vehicle has been driven, or about 15 minutes after the engine has been shut off. If the oil is checked immediately after driving the vehicle, some of the oil will remain in the upper engine components, resulting in an inaccurate reading on the dipstick.

4 Pull the dipstick from the tube and wipe all the oil from the end with a clean rag. Insert the clean dipstick all the way back into the

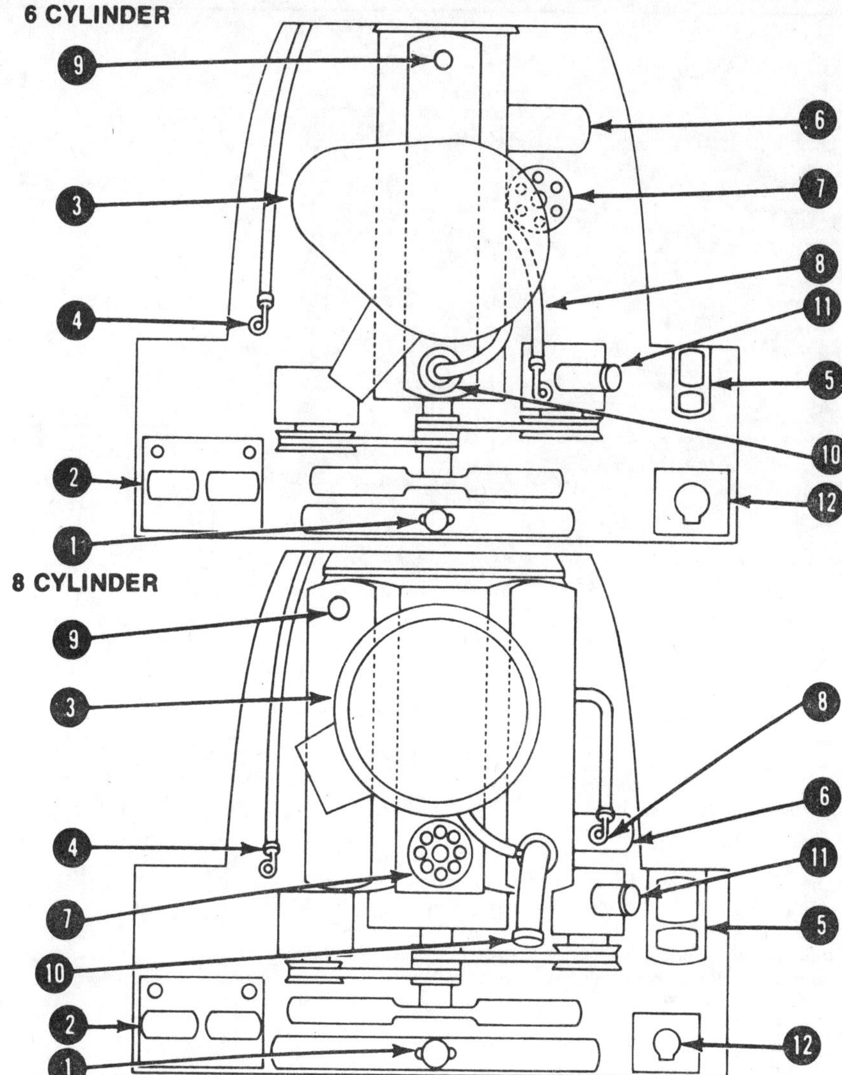

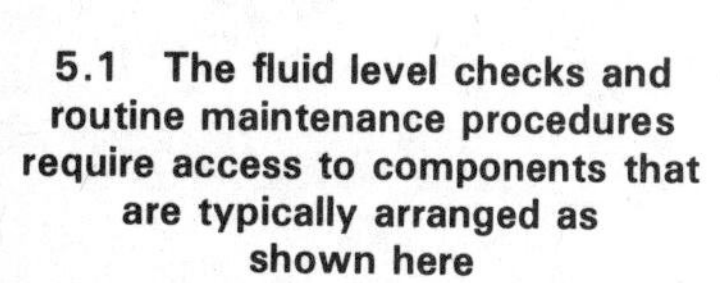
5.1 The fluid level checks and routine maintenance procedures require access to components that are typically arranged as shown here

1 *Radiator filler cap*
2 *Battery*
3 *Air cleaner*
4 *Automatic transmission dipstick*
5 *Brake master cylinder*
6 *Engine oil filter*
7 *Distributor*
8 *Engine oil dipstick*
9 *PCV valve*
10 *Engine oil filler cap*
11 *Power steering reservoir*
12 *Windshield washer reservoir*

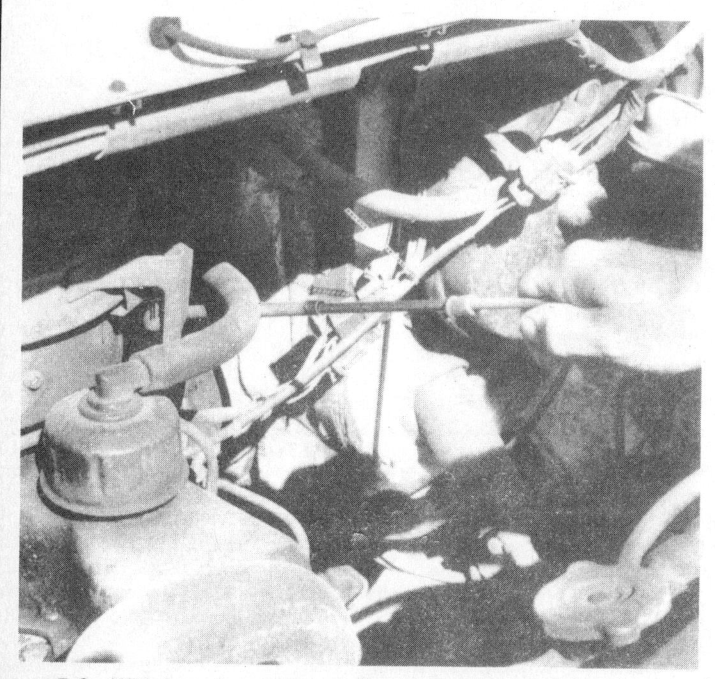

5.2 Withdraw the engine oil dipstick to check the oil level

ADD 2 | ADD 1 →| SAFE |←WARRANTY

ADD →| SAFE |← WARRANTY

ADD |←SAFE→| |←MAX. OVERFILL

5.4 The oil dipstick may be marked in one of several ways, but the safe oil level will be obvious when looking at it

5.6 Oil is added to the engine through the filler hole or tube on the rocker arm cover (V8 engine shown)

tube and pull it out again. Note the oil level at the end of the dipstick. At its highest point, the level should be between the marks that indicate the safe area (see illustration). **Note:** *Dipsticks may be marked in one of several different ways to indicate the safe oil level. Some have Add and Full marks, some have L (low) and F (full) marks and still others may have Add and Safe labels.*

5 Do not allow the level to drop below the safe area or engine damage due to oil starvation may occur. On the other hand, do not overfill the engine by adding oil above the safe area, since it may result in oil fouled spark plugs, oil leaks or oil seal failures.

6 Oil is added to the engine after removing a twist-off cap attached to a tube leading to the rocker arm cover or to the rocker arm cover itself (see illustration). An oil can spout or funnel will help prevent spills as the oil is poured in.

7 Checking the oil level can be an important preventative maintenance step. If you find the oil level dropping abnormally, it is an indication of oil leakage or internal engine wear which should be corrected. If there are water droplets in the oil, or if the oil is milky looking, component failure has occurred and the engine should be checked immediately. The condition of the oil can be checked along with the level. While the dipstick is out of the engine, take your thumb and index finger and wipe the oil up the dipstick, looking for small dirt or metal particles which will cling to the dipstick. Their presence is an indication that the oil should be drained and fresh oil added.

Engine coolant

8 Vehicles covered by this manual are equipped with either a pressurized coolant recovery system or an overflow system. In a recovery system, a clear or translucent coolant reservoir, attached to the fender panel near the radiator, is connected by a hose to the radiator neck. As the engine heats up during operation, coolant is forced from the radiator, through the connecting tube and into the reservoir. As the engine cools, the coolant is drawn back into the radiator to keep the level correct. On an overflow system there isn't any tank to catch excess coolant; it simply runs out the tube onto the ground.

9 The coolant level on a recovery system should be checked when the engine is hot. Note the level of fluid in the reservoir; it should be at or near the Full mark on the side of the reservoir. On an overflow system, the coolant level should be checked with the engine cold to avoid injury from escaping steam and coolant. Remove the radiator cap and note the coolant level in the tank. It should be between 3/4 and 1-1/2 inches from the filler neck (see illustration).

10 **Warning:** *The radiator cap should not, under any circumstances, be removed when the engine is hot, as escaping steam and coolant could cause serious injury.* Wait until the system has cooled completely, then wrap a thick cloth around the cap and turn it to the first stop. If any steam escapes, wait until the system has cooled longer, then remove the cap.

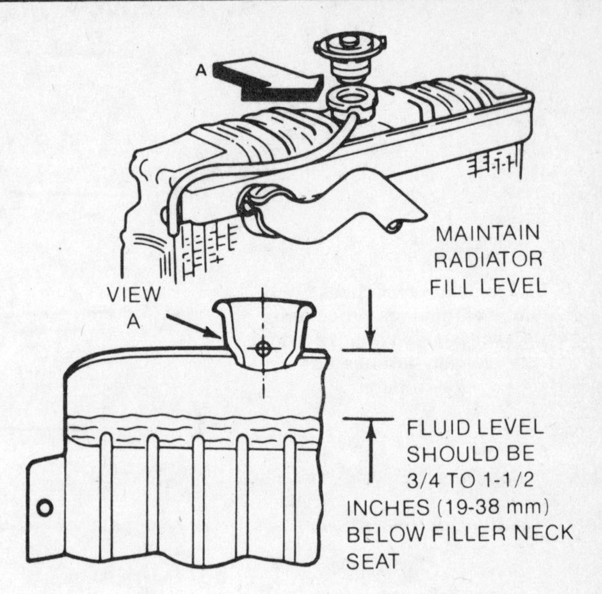

5.9 On vehicles not equipped with a coolant reservoir, the level should be 3/4 to 1-1/2 inches below the filler neck in the radiator

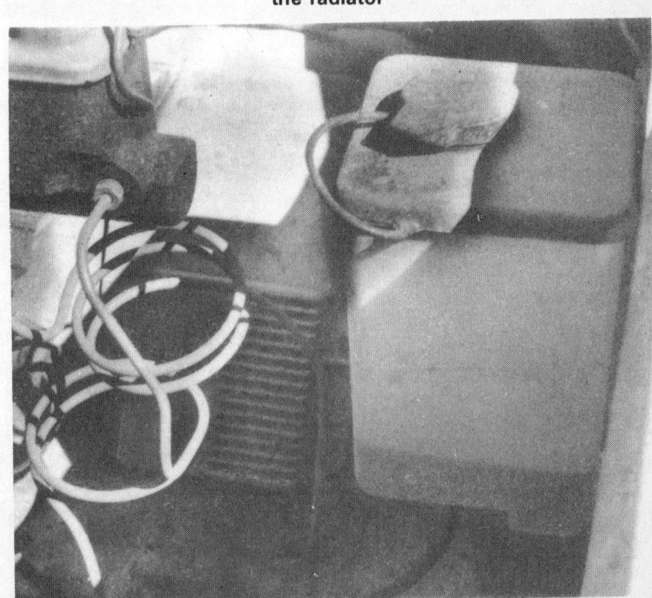

5.15 Do not confuse the coolant reservoir and the windshield washer fluid reservoir (this one is for the windshield washer)

11 If only a small amount of coolant is required to bring the system up to the proper level, plain water can be used. However, to maintain the proper antifreeze/water mixture in the system, antifreeze should be mixed with the water to replenish a low level. High-quality antifreeze offering protection to -20°F should be mixed with water in the proportion specified on the antifreeze container. **Warning:** *Do not allow antifreeze to come in contact with your skin or painted surfaces of the vehicle. Flush affected areas immediately with plenty of water. Antifreeze can be fatal to children and pets (they are attracted to its sweet taste). Just a few licks can cause death. Wipe up garage floor and drip pan coolant puddles immediately. Keep antifreeze containers sealed and repair cooling system leaks as soon as they are noticed.*

12 On systems with a reservoir, coolant should be added to the reservoir after removing the cap. Coolant should be added directly to the radiator on systems without a reservoir.

13 As the coolant level is checked, note the condition of the coolant. It should be relatively clear. If it is brown or a rust color, the system should be drained, flushed and refilled.

14 If the cooling system requires repeated additions to maintain the

5.17 Checking the electrolyte level in the battery (not necessary with maintenance-free battery)

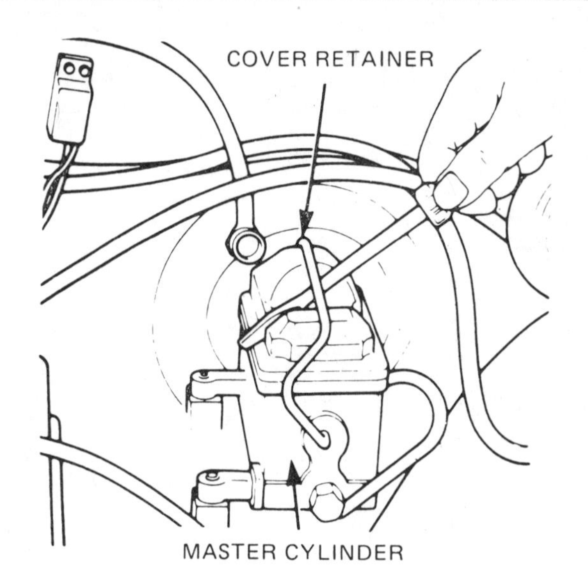

5.20a Use a screwdriver to unsnap the retainer, then carefully remove the master cylinder cover . . .

5.20b . . . and make sure the brake fluid level is within 1/4-inch of the top in both reservoirs

proper level, have the radiator cap sealing ability checked. Also check for leaks in the system such as cracked hoses, loose hose connections, leaking gaskets, etc.

Windshield washer fluid

15 Fluid for the windshield washer system is located in a plastic reservoir mounted on the firewall or radiator support (see illustration).
16 The level in the reservoir should be maintained at the Full mark, except during periods when freezing temperatures are expected. The fluid level should then be maintained no higher than 3/4 full to allow for expansion should the fluid freeze. The use of windshield washer fluid instead of water will prevent freezing and result in better cleaning of the windshield surface. **Caution:** *Do not use antifreeze in the windshield washer system as it will damage the vehicle's paint.*

Battery electrolyte

17 To check the electrolyte level in the battery, remove all the cell caps (see illustration). Note that maintenance-free batteries do not require this check. If the electrolyte level is low, add distilled water until the level is above the plates. There is usually a split-ring indicator in each cell to help you judge when enough water has been added — do not overfill.

Brake fluid

18 The master cylinder is mounted on the left side of the engine compartment firewall and has a cover which must be removed to check the fluid level.
19 Remove all dirt, moisture and oily residue from the master cylinder before removing the cover. Cover all painted surfaces near the master cylinder to avoid damage to the paint from spilled brake fluid.
20 Snap the retainer to the side with a screwdriver and carefully remove the cover (see illustration). The fluid level should be within 1/4-inch of the top edge of each reservoir (see illustration). If a low level is found, fluid must be added.
21 When adding fluid, pour it carefully into the reservoir, taking care not to spill any onto surrounding painted surfaces. Be sure the specified fluid is used, since mixing different types of brake fluid can cause damage to the system. See *Recommended lubricants and fluids* or your owner's manual.
22 At this time the fluid and master cylinder can be inspected for contamination. Normally, the brake system will not need periodic draining and refilling, but if rust deposits, dirt particles or water droplets are seen in the fluid, the system should be dismantled, drained and refilled with fresh fluid.
23 After filling the reservoir to the proper level, make sure the cover is properly seated to prevent fluid leakage, then reposition the retainer to hold the cover in place.
24 The brake fluid level in the master cylinder will drop slightly as the brake shoes or pads at each wheel wear down during normal operation. If the master cylinder requires repeated replenishing to keep the fluid at the proper level, there may be a leak in the brake system, which should be repaired immediately. Check all brake lines and connections, along with the wheel cylinders and master cylinder.
25 If, upon checking the master cylinder fluid level, you discover one or both reservoirs empty or nearly empty, the brake system should be bled (Chapter 9).

Manual transmission oil

26 Manual shift transmissions do not have a dipstick. The oil level is checked by removing a plug from the side of the transmission (see

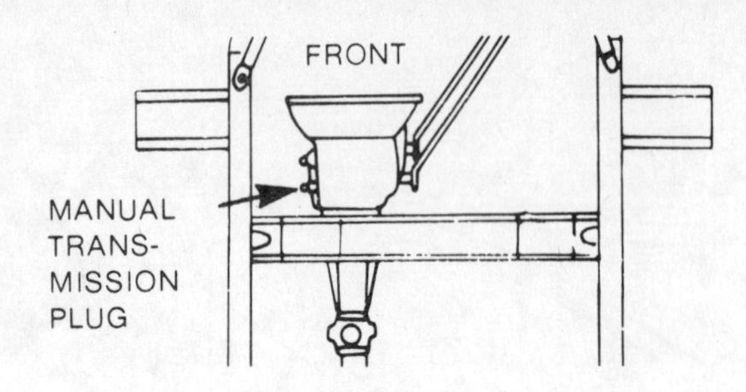

5.26 The manual transmission has a check/filler plug, which must be removed to check the oil level

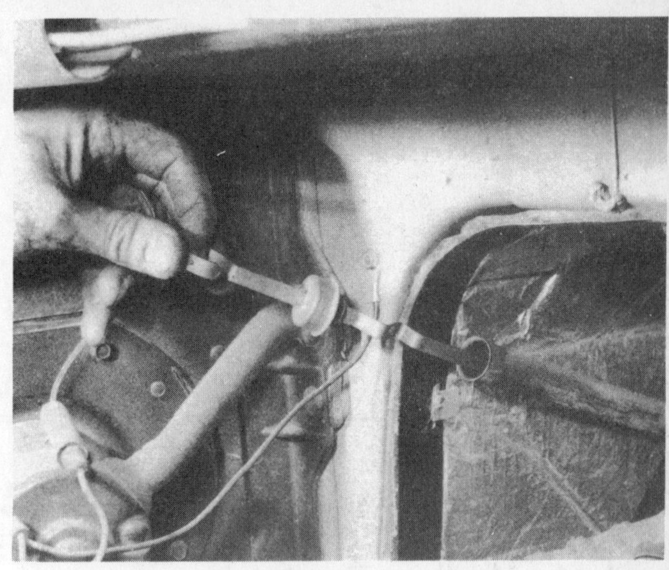

5.33 The automatic transmission dipstick (shown here) looks a great deal like the engine oil dipstick — don't confuse the two!

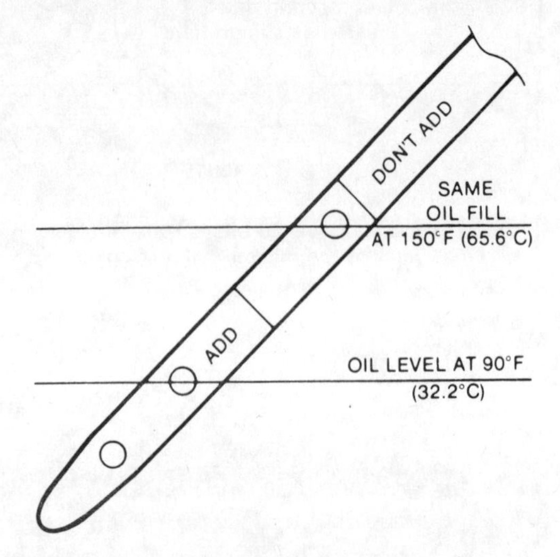

5.37 The automatic transmission fluid dipstick will typically be marked as shown here — don't overfill the transmission!

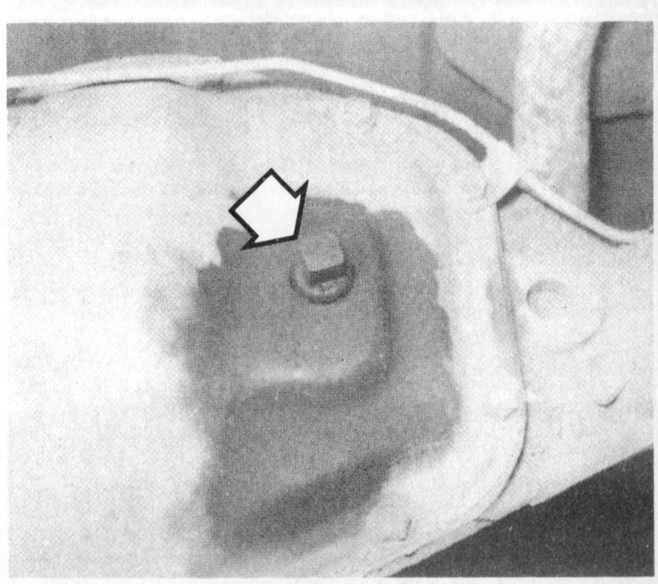

5.44 The differential check/filler plug may be located on the differential cover as shown here or on the front side of the carrier assembly

illustration). The engine must be cold when this check is performed and the vehicle must be level. If the vehicle must be raised to gain access to the plug, be sure to support it securely on jackstands — do not crawl under a vehicle supported only by a jack. Clean the plug and the area around it with a rag, then remove the plug with a wrench.
27 If oil immediately starts leaking out, thread the plug back into the transmission because the oil level is correct. If no oil leaks out, completely remove the plug and reach into the hole with your little finger. The oil level should be even with the bottom of the plug hole.
28 If the transmission needs more oil, use a syringe or squeeze bottle to add the recommended lubricant until the level is correct.
29 Thread the plug back into the transmission and tighten it securely.
30 Drive the vehicle a short distance, then check to make sure the plug is not leaking.

Automatic transmission fluid

31 The automatic transmission fluid level should be carefully maintained. Low fluid level can lead to slipping or loss of drive, while overfilling can cause foaming and loss of fluid.
32 With the parking brake set, start the engine, then move the shift lever through all the gears, ending in Park. The fluid level must be checked with the vehicle level and the engine at normal operating temperature (running at idle). **Note:** *Incorrect fluid level readings will result if the vehicle has just been driven at high speeds for an extended period, in city traffic during hot weather or while pulling a trailer. If any of these conditions apply, wait until the fluid has cooled slightly (about 30 minutes).*
33 Locate the dipstick in the engine compartment and pull it out of the filler tube (see illustration). Don't confuse the transmission dipstick with the engine oil dipstick — the fluid in the transmission is a reddish color.
34 Carefully touch the end of the dipstick to determine if the fluid is cool (about room temperature), warm, or hot (uncomfortable to the touch).
35 Wipe the fluid from the dipstick with a clean rag, then push the dipstick back into the tube until the cap seats.
36 Pull the dipstick out again and note the fluid level.
37 If the fluid felt cool, the level should be 1/8 to 3/8-inch below the Add mark. The two dimples below the Add mark indicate this range (see illustration).
38 If the fluid felt warm, the level should be close to the Add mark (just above or below it).
39 If the fluid felt hot, the level should be near the Don't Add mark. **Note:** *In some cases, specific instructions may be found on the dipstick itself. If they differ from the procedure described here, follow the instructions on the dipstick.*
40 If necessary, add just enough of the recommended fluid to fill the transmission to the proper level. It takes about one pint to raise the level from the Add mark to the Don't Add mark, so add the fluid a little

5.52a The power steering reservoir is attached to the pump and the dipstick is attached to the filler cap

at a time and keep checking the level until it is correct.

41 The condition of the fluid should be checked along with the level. If the fluid at the end of the dipstick is a dark reddish-brown color, or smells burned, the transmission fluid should be drained and replaced. If you are in doubt about the condition of the fluid, purchase some new fluid and compare the two for color and smell.

Hydraulic clutch fluid

42 Some later models are equipped with a hydraulic clutch, which means that the clutch fluid reservoir, located to the left of the brake booster, must be checked regularly to make sure it is full. Clean the top and sides of the reservoir to prevent contamination by dirt as the cover is removed and if fluid must be added, check the cylinder and connections for leaks.

Differential oil

43 Like the manual transmission, the differential has an inspection and fill plug which must be removed to check the oil level. If the vehicle is raised to gain access to the plug, be sure to support it securely on jackstands — do not crawl under a vehicle supported only by a jack.

44 Remove the plug, which is located either in the cover plate or on the side of the differential carrier, and use your little finger to reach inside the axle housing to feel the oil level. It should be near the bottom of the plug hole (see illustration).

45 If it isn't, add the recommended oil through the plug hole with a syringe or squeeze bottle.

46 Install and tighten the plug securely and check for leaks after the first few miles of driving.

Power steering fluid

47 Unlike manual steering, the power steering system relies on fluid which may, over a period of time, require replenishing.

48 The reservoir for the power steering pump is located at the front of the engine.

49 During this check, the front wheels should be pointed straight ahead and the engine should be off.

50 Use a clean rag to wipe off the reservoir cap and the area around the cap. This will help prevent foreign matter from entering the reservoir during the check.

51 Make sure the engine is at normal operating temperature.

52 Remove the dipstick (see illustration), wipe it off with a clean rag, reinsert it, then withdraw it and note the fluid level. The level should be between the Full cold (or Add) and Full Hot marks (see illustration).

53 If additional fluid is required, pour the specified type directly into the reservoir, using a funnel to prevent spills.

54 If the reservoir requires frequent fluid additions, all power steering hoses, hose connections, the power steering pump and the steering box should be carefully checked for leaks.

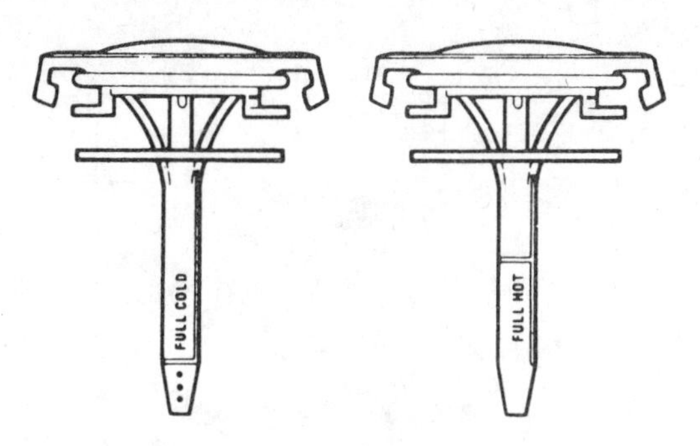

5.52b The power steering dipstick will typically be marked as shown here

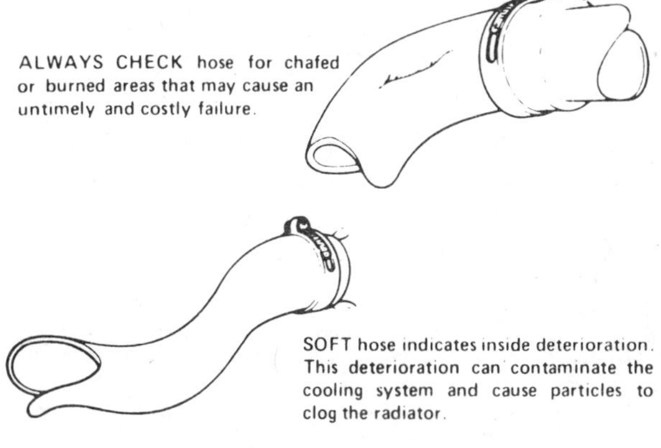

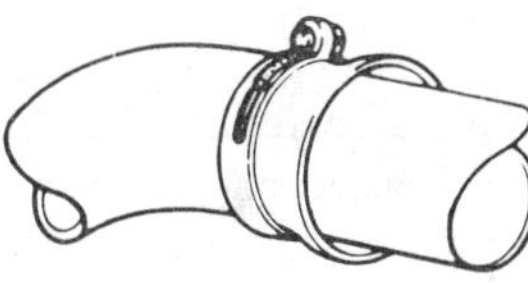

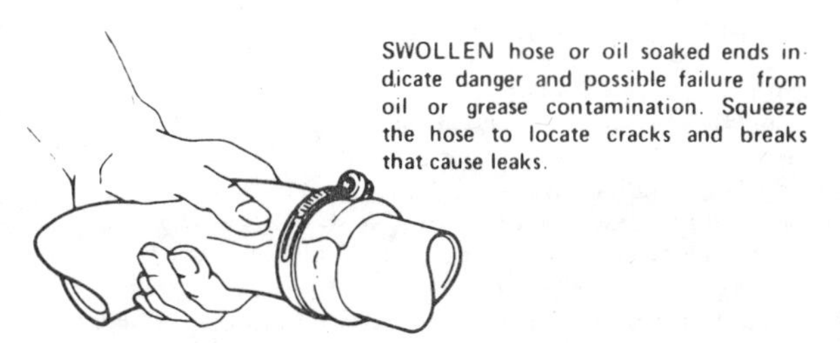

6.4 Hoses, like drivebelts, have a habit of failing at the worst possible time (to prevent the surprise of a blown radiator or heater hose, inspect them as shown here)

6 Cooling system check

1 Many major engine failures can be attributed to a faulty cooling system. If the vehicle is equipped with an automatic transmission, the cooling system also plays an important role in prolonging transmission life.

2 The cooling system should be checked with the engine cold. Do this before the vehicle is driven for the day or after the engine has been shut off for at least 3 hours.

3 Remove the radiator cap and thoroughly clean the cap, inside and out, with clean water. Also clean the filler neck on the radiator. All traces of corrosion should be removed.

4 Carefully check the upper and lower radiator hoses along with the smaller diameter heater hoses. Inspect each hose along its entire length, replacing any hose which is cracked, swollen or shows signs of

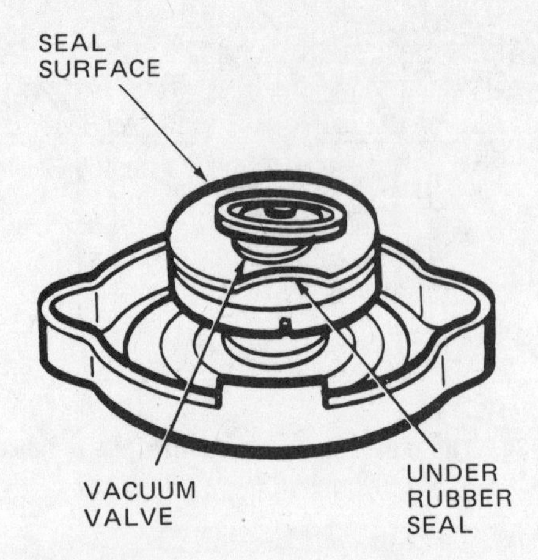

6.7 The radiator cap components should be cleaned and inspected at regular intervals

deterioration. Cracks may become more apparent if the hose is squeezed (see illustration).
5 Make sure that all hose connections are tight. A leak in the cooling system will usually show up as white or rust colored deposits on the areas adjoining the leak.
6 Use compressed air or a soft brush to remove bugs, leaves, etc. from the front of the radiator or air-conditioning condenser. Be careful not to damage the delicate cooling fins or cut yourself on them.
7 Check the radiator cap seal and spring for deterioration (see illustration).
8 Have the cap and system pressure tested. If you do not have a pressure tester, most gas stations and repair shops will do this for a minimal charge.

7 Exhaust system check

1 With the engine cold (at least three hours after the vehicle has been driven), check the complete exhaust system from its starting point at the engine to the end of the tailpipe. This should be done on a hoist where unrestricted access is available.
2 Check the pipes and connections for signs of leakage and corrosion, indicating a potential failure. Make sure that all brackets and hangers are in good condition and tight.
3 Inspect the underside of the body for holes, corrosion, open seams, etc., which may allow exhaust gases to enter the passenger compartment. Seal all body openings with silicone or body putty.
4 Rattles and other noises can often be traced to the exhaust system, especially the mounts and hangers. Try to move the pipes, muffler and catalytic converter (if so equipped). If the components can come in contact with the body or suspension parts, secure the exhaust system with new mounts.
5 Check the running condition of the engine by inspecting inside the end of the tailpipe. The exhaust deposits here are an indication of engine state-of-tune. If the pipe is black and sooty or coated with white deposits, the engine is in need of a tune-up, including a thorough carburetor inspection and adjustment.

8 Exhaust heat control valve check

Refer to illustration 8.2

1 The exhaust heat control valve, which is mounted between the exhaust manifold and one branch of the exhaust pipe, is controlled by a bimetal spring on early models and a vacuum diaphragm on later models. The valve routes hot exhaust gases to the intake manifold heat riser during cold engine operation, helping to eliminate condensation of fuel on the cold surfaces of the intake tract and providing better evaporation of the fuel/air mixture. The net result is better driveability and faster warm-up.

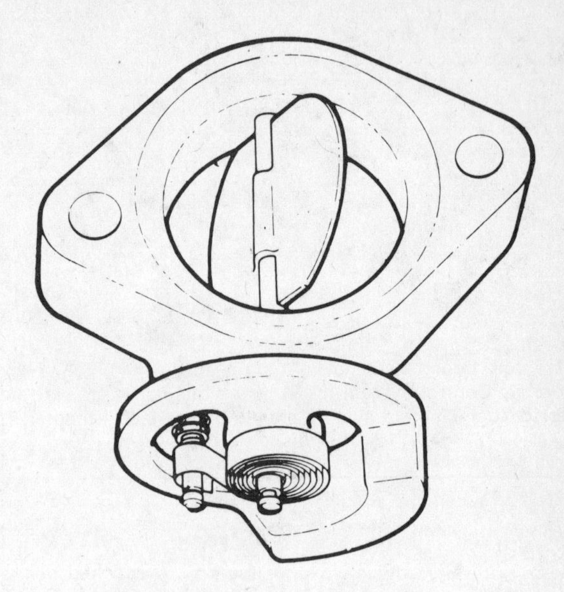

8.2 The exhaust heat control valve (bimetal spring type shown) should move freely by hand

Bimetal thermostat valve

2 To check the operation of the valve, first make sure the spring is hooked on the stop pin (see illustration). Start the engine (it must be completely cold); see if the valve closes and slowly starts to open as the engine warms.
3 If the valve does not operate as described, check the valve for freedom of movement with your hand – careful, it's hot! If the valve is stuck it can sometimes be loosened with penetrating oil. It should be lubricated periodically with a dry film lubricant.

Vacuum operated valve

4 Check the valve for freedom of movement with your hand. If it is stuck, it may be loosened with penetrating oil.
5 Start the engine and allow it to reach normal operating temperature. Disconnect the vacuum hose from the valve and verify that vacuum is present at the hose opening. If not, check the vacuum hose and source.

All models

6 If the valve is stuck and cannot be loosened it should be replaced with a new one, since it can cause overheating if stuck shut and prevent quick warm-ups if stuck open.

9 Fuel system check

Warning: *There are certain precautions to take when inspecting or servicing the fuel system components. Work in a well ventilated area and do not allow open flames (cigarettes, appliance pilot lights, etc.) to get near the work area. Mop up spills immediately and do not store fuel soaked rags where they could ignite.*

1 If your vehicle is equipped with fuel injection, refer to the fuel injection pressure relief procedure (Chapter 4) before servicing any component of the fuel system.
2 The fuel system is under a small amount of pressure, so if any fuel lines are disconnected for servicing, be prepared to catch the fuel as it spurts out. Plug all disconnected fuel lines immediately after disconnection to prevent the tank from emptying.
3 The fuel system is most easily checked with the vehicle raised on a hoist so the components underneath the vehicle are readily visible and accessible.
4 If the smell of gasoline is noticed while driving or after the vehicle has been in the sun, the system should be thoroughly inspected immediately.
5 Remove the gas filler cap and check for damage, corrosion or a

broken sealing imprint on the gasket. Replace the cap with a new one if necessary.
6 With the vehicle raised, inspect the gas tank and filler neck for punctures, cracks and other damage. The connection between the filler neck and the tank is especially critical. Sometimes a rubber filler neck will leak due to loose clamps or deteriorated rubber, problems a home mechanic can usually rectify. **Warning:** *Do not, under any circumstances, try to repair a fuel tank yourself (except rubber components). A welding torch or any open flame can easily cause the fuel vapors to explode if the proper precautions are not taken.*
7 Carefully check all rubber hoses and metal lines leading away from the fuel tank. Check for loose connections, deteriorated hoses, crimped lines and other damage. Follow the lines to the front of the vehicle, carefully inspecting them all the way. Repair or replace damaged sections as necessary.
8 If a fuel odor is still evident after the inspection, check the Evaporative Control System (ECS). If the fuel odor persists, the carburetor or fuel injection system should be serviced (Chapter 4).

10 Carburetor choke check

1 The choke operates only when the engine is cold, so this check should be performed before the engine has been started for the day.
2 Open the hood and remove the top plate of the air cleaner assembly. It is usually held in place by a wing nut. Place the top plate and nut aside, out of the way of moving engine components.
3 Look at the top of the carburetor at the center of the air cleaner housing. You will notice a flat plate in the carburetor opening. On 4-barrel carburetors, the plate covers only the front two barrels of the carburetor. **Caution:** *Be careful not to drop anything down into the carburetor when the top of the air cleaner is off.*
4 Have an assistant press the accelerator pedal to the floor. The plate should close completely. Have the assistant start the engine while you watch the plate in the carburetor. **Warning:** *Do not position your face directly over the carburetor — the engine could backfire and cause serious burns.* When the engine starts, the choke plate should open slightly.
5 Allow the engine to continue running at idle speed. As the engine warms to operating temperature the plate should slowly open, allowing more air to enter through the top of the carburetor. On some vehicles the plate will not open all the way unless the throttle is quickly opened more than half-way and released. If the plate does not seem to be moving, depress the accelerator quickly to see if it releases the choke linkage.
6 After a few minutes, the choke plate should be all the way open to the vertical position.
7 You will notice that engine speed corresponds with the plate opening. With the plate completely closed the engine should run at a fast idle. As the plate opens, the engine speed will decrease.
8 If a malfunction is detected during the above checks, refer to Chapter 4 for specific information related to adjusting and servicing choke components.

11 Thermo-controlled air cleaner check

Refer to illustration 11.5

1 The thermostatically controlled air cleaner draws air to the carburetor from two locations, depending upon engine temperature.
2 This is a visual check. If access is limited, a small mirror may have to be used.
3 Open the hood and locate the flapper door inside the air cleaner assembly. It will be located inside the long snorkel of the air cleaner housing.
4 If there is a flexible air duct attached to the end of the snorkel, leading to an area behind the grille, disconnect it at the snorkel. This will enable you to look through the end of the snorkel and see the flapper door.
5 The check should be done when the engine and outside air are cold. Start the engine and look through the snorkel at the door, which should move to a closed position (see illustration). With the door closed, air cannot enter through the end of the snorkel, but instead enters the air cleaner through a flexible duct attached to the exhaust manifold.

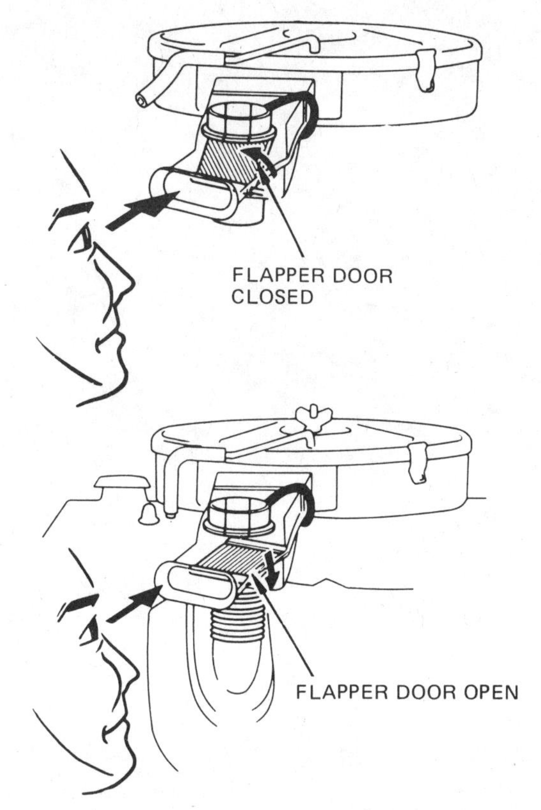

11.5 This is what you should see as you look into the end of the air cleaner snorkel at the flapper door (at the top, the flapper door is closed, as it should be when the engine is cold; after the engine has warmed up, the flapper door should open as shown at the bottom)

6 As the engine warms to operating temperature, the flapper door should open to allow air through the snorkel end. Depending upon ambient temperature, this may take 10 to 15 minutes. To speed up the check you can reconnect the snorkel air duct, drive the vehicle and then check to see if the door is completely open.
7 If the thermo-controlled air cleaner is not operating properly see Chapter 6 for more information.

12 Evaporative Control System (ECS) check

Note: *Refer to Chapter 6 for illustrations showing the locations of the ECS system components.*

1 The evaporative emissions system consists of a charcoal canister, the lines connecting the canister to the air cleaner and to the fuel tank, and the fuel tank filler can.
2 Inspect the fuel filler cap and make sure the gasket sealing the cap is in good condition. It should not be cracked, broken or indicate that leakage has occurred.
3 Inspect the lines leading to the charcoal canister from the fuel tank. They should be in good shape and the rubber should not be cracked, checked or leaking.
4 Check all of the clamps and make sure they are sealing the hoses. Check the carbon filled canister for signs of leakage, overfilling and damage. In most cases, a carbon canister will last the lifetime of the vehicle. However, certain situations will require replacement. If the carbon canister is damaged or has been leaking, replace it with a new one.
5 Check all the lines leading from the charcoal canister to the air cleaner. In some cases there will be two lines, one leading from the carburetor float bowl (to vent it) and one line leading from the canister to the air cleaner for burning of the accumulated vapors.
6 Replace any lines in questionable condition and exercise the same precautions necessary when dealing with fuel lines and the fuel filter.

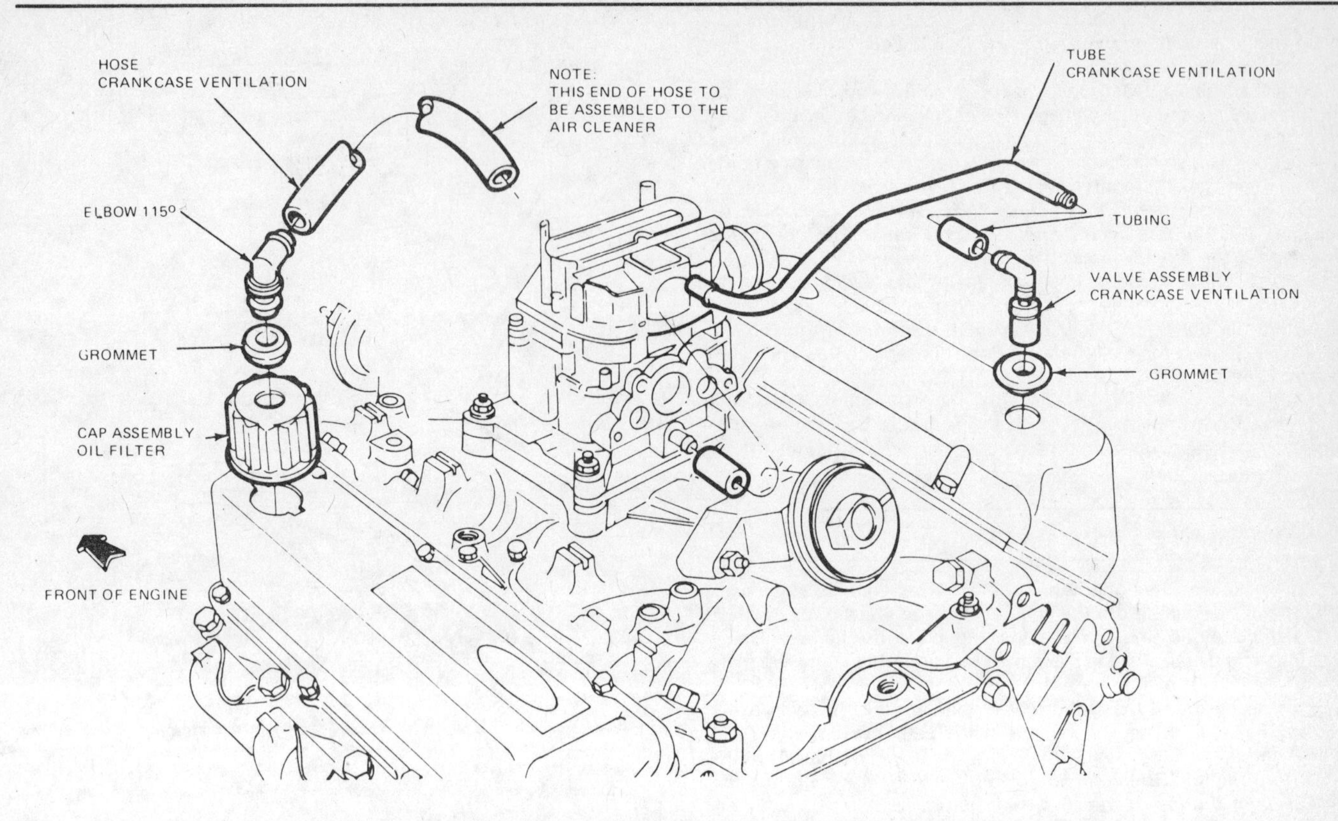

14.1 A typical Positive Crankcase Ventilation (PCV) system looks like this (V8 engine shown)

13 Exhaust Gas Recirculation (EGR) valve check

1 The EGR valve is located on the intake manifold, adjacent to the carburetor or fuel injection unit. Most of the time, when a problem develops in this system, it is due to a stuck or corroded EGR valve.
2 With the engine cold to prevent burns, reach under the EGR valve and manually push on the diaphragm. Using moderate pressure, you should be able to move the diaphragm within the housing.
3 If the diaphragm does not move or moves only with much effort, replace the EGR valve with a new one. If in doubt about the condition of the valve, compare the free movement of your EGR valve with a new valve.
4 Refer to Chapter 6 for more information on the EGR system.

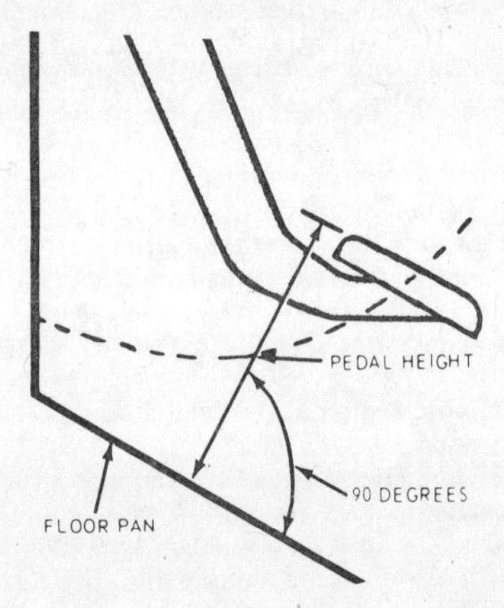

16.2 Measuring clutch pedal height

14 Positive Crankcase Ventilation (PCV) system check and valve replacement

Refer to illustration 14.1

Check

Note: *The following tests are to be performed with the engine idling and at normal operating temperature.*

1 Remove the PCV valve from the grommet located in the rocker arm cover (see illustration). Make sure the connection to the PCV valve at the inlet side remains intact.
2 Check for strong suction, which will be accompanied by a hissing noise at the valve. You can place a finger over the valve inlet to feel the suction. At the same time your finger is blocking the valve, any vacuum leaks in the connections and hose should become apparent.
3 Install the PCV valve back in the grommet.
4 Loosely plug the air inlet hose after removing it from the air cleaner connection. Place a small piece of stiff paper over the opening to block it. After approximately one minute, the paper should be held against the hose opening by strong suction. **Note:** *The following tests will be made with the engine shut off.*
5 Remove the PCV valve from the grommet and shake it. It should make a clicking, metallic noise, which indicates the valve is not stuck.
6 Check the hose leading to the PCV valve, as well as the air inlet connections, PCV valve grommet and air inlet gasket at the oil filler cap (if so equipped). If any loose connections are found, tighten the connections or replace the clamps.
7 If the hoses are damaged or leaking, replace the hoses.

Valve replacement

8 Replace the PCV valve if it fails any of the above tests. Make sure that the replacement valve is identical to the original equipment. An exception to this rule occurs after high mileage when a special high flow PCV valve may be installed.
9 Remove the valve from the grommet and remove the clamp securing the valve to the hose.
10 Insert the PCV valve into the grommet in the rocker arm cover.
11 Connect the hose linking the crankcase or manifold to the top of the PCV valve. Make sure the hose is not cracked or leaking. Install the clamp over the hose end.
12 Start the engine and make sure that the PCV system functions as described above.
13 On V8 engines, the PCV valve is attached to the crankcase

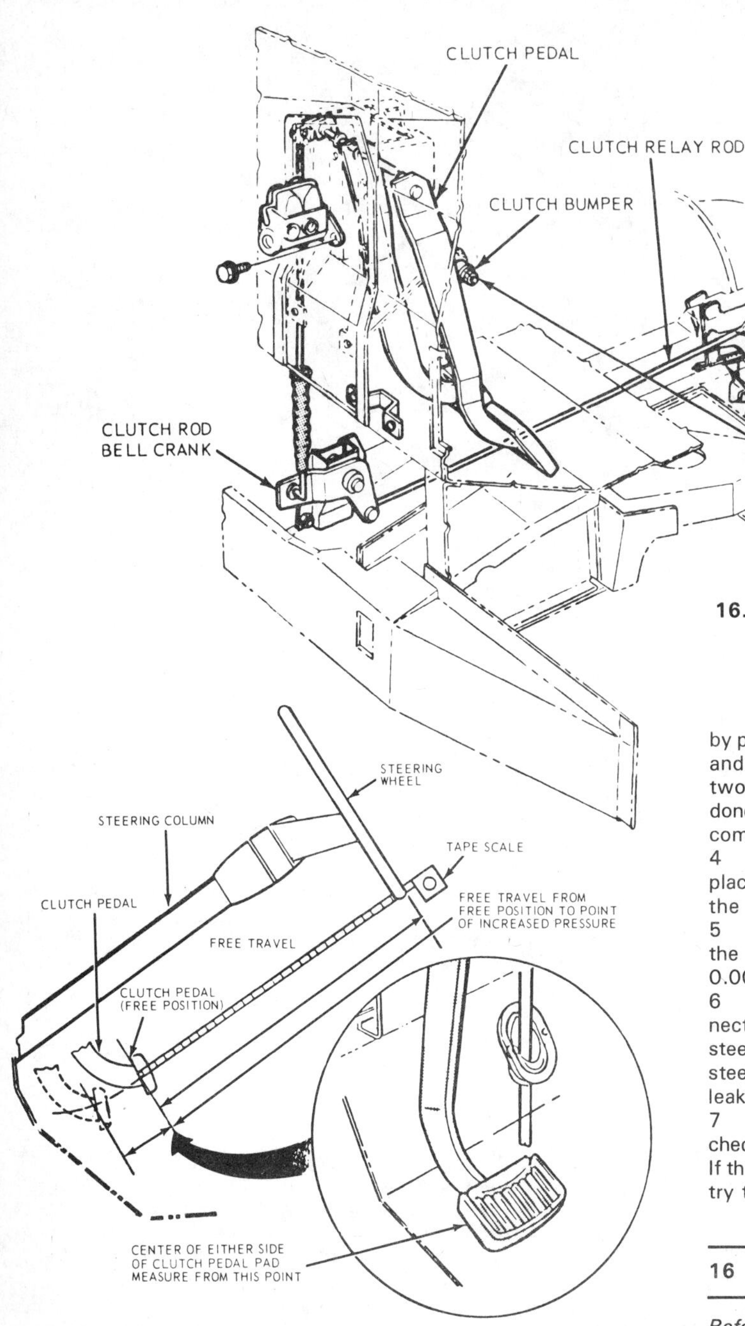

16.3 Typical clutch linkage, indicating pedal height adjusting points

16.5 Checking the clutch pedal free play

breather/oil filler cap. Be sure to remove, clean (with solvent) and replace the cap whenever PCV valve maintenance is required.

14 Replace the crankcase ventilation filter in the air cleaner housing. To do this, remove the air cleaner top plate, lift out the small, rectangular filter and replace it with a new one (see Section 19 for more information).

15 Suspension and steering check

1 Whenever the front of the vehicle is raised for service it is a good idea to visually check the suspension and steering components for wear.

2 Indications of a fault in these systems include excessive play in the steering wheel before the front wheels react, excessive sway around corners, body movement over rough roads and binding at some point as the steering wheel is turned.

3 Before the vehicle is raised for inspection, test the shock absorbers by pushing down to rock the vehicle at each corner. If you push down and the vehicle does not come back to a level position within one or two bounces, the shocks are worn and must be replaced. As this is done, listen for squeaks and strange noises coming from the suspension components.

4 Raise the front of the vehicle and support it firmly on jackstands placed under the frame rails. Make sure the vehicle cannot fall from the stands.

5 Grab the top and bottom of the front tire with your hands and rock the tire/wheel on the spindle. If there is movement of more than 0.005-inch, the wheel bearings should be serviced.

6 Crawl under the vehicle and check for loose bolts, broken or disconnected parts and deteriorated rubber bushings on all suspension and steering components. Look for grease or fluid leaking from around the steering box. Check the power steering hoses and connections for leaks. Check the balljoints for wear.

7 Have an assistant turn the steering wheel from side-to-side and check the steering components for free movement, chafing and binding. If the steering does not react with the movement of the steering wheel, try to determine where the slack is located.

16 Clutch pedal free play check and adjustment

Refer to illustrations 16.2, 16.3, 16.5 and 16.6

1 Econoline vans are equipped with one of two types of clutch release mechanisms. The first and most widely used is a mechanical system which requires periodic adjustment. The second type, used on late model vehicles, is hydraulic and requires no maintenance except to make sure the fluid reservoir is full.

2 To adjust the mechanical system, first check the distance between the center of the clutch pedal and the floor with the clutch in the engaged position (pedal all the way up). The distance is known as pedal height and should be within the specified range (see illustration).

3 If the clutch pedal height is outside the limits, loosen the nut on the clutch pedal bumper under the instrument panel and rotate the eccentric stop (or bumper) until the height of the pedal is within the specified limits. Retighten the locknut (see illustration).

4 Check the clutch pedal free play. Slowly depress the clutch pedal by hand until firm resistance is felt. This indicates that the clutch release bearing is in contact with the pressure plate release levers.

5 The amount the pedal moves down before resistance is felt should be within the specified limit. It can be checked by holding a tape measure alongside the pedal as it is pushed down and then released (see illustration).

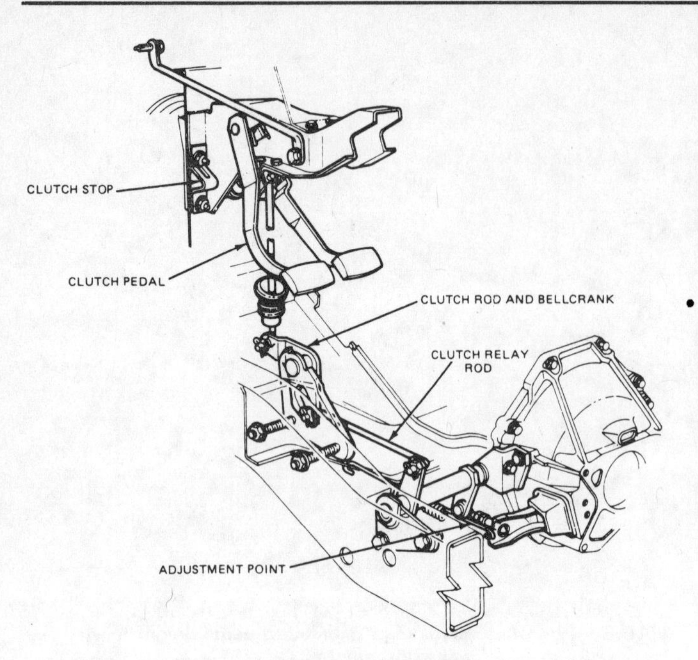

16.6 Typical clutch linkage, indicating clutch pedal free play adjusting point

6 If the amount of pedal free play is outside the specified limits, raise the vehicle, support it on jackstands and locate the adjustment rod on the left side of the clutch housing (see illustration).
7 Loosen the locknut on the release rod and rotate the clutch lever end of the rod in or out as necessary until the pedal free play is as specified. Hold the release rod firmly and tighten the locknut.

17 Brake check

Refer to illustrations 17.6, 17.18, 17.24a, 17.24b, 17.26, and 17.41

1 The brakes should be inspected at the specified intervals, as well as every time the wheels are removed, or whenever a problem is suspected. Indications of a potential problem in the braking system include: the vehicle pulls to one side when the brake pedal is depressed; noises coming from the brakes when they are applied; excessive brake pedal travel; pulsating pedal; and leakage of fluid (usually seen on the inside of the tire or wheel).
2 Nearly all vehicles covered by this manual are equipped with disc brakes at the front and drum brakes at the rear. However, if your vehicle is equipped with drum brakes at the front, follow the basic procedures for the rear drum brakes.

Disc brakes — inspection

3 Disc brakes can be visually checked without removing any parts except the wheels.
4 Most later model vehicles come equipped with a wear sensor attached to the inner pad. This is a small, bent piece of metal which is visible from the inboard side of the brake caliper. When the pads wear to the danger limit, the metal sensor rubs against the disc and makes a screeching sound.
5 Raise the vehicle and place it securely on jackstands. Remove the front wheels. Now visible is the disc brake caliper, which contains the pads. There is an outer brake pad and an inner pad. Both should be inspected.
6 Check the pad thickness by looking through the oval shaped inspection hole in the caliper (see illustration). If the pad lining material is 1/8-inch or less in thickness, the pads should be replaced. Keep in mind that the pad material is riveted or bonded to a metal shoe, and the metal portion is not included in this measurement.
7 Since it may be difficult to measure the exact thickness of the remaining pad lining material, if you are in doubt as to the pad thickness they should be removed for a more thorough check. Refer to Chapter 9.

17.6 The brake pad lining (arrow) can be checked by looking through the caliper inspection hole (remember, the measurement does not include the metal backing plate)

8 Check the disc for score marks, gouges and burned spots. If these conditions exist, the hub/disc assembly should be removed for servicing. If the disc is worn or damaged,it can be resurfaced by an automotive machine shop.

Hydraulic lines and parking brake cables — inspection

9 Before installing the wheels, check for leakage around the brake hose connections at the caliper. Also check the hoses for cracks, chafing, leaks and deterioration.
10 Replace the hose, lines or fittings if leakage is evident.
11 Inspect the entire system. Trace the hydraulic lines from the master cylinder to the brake equalizing chamber and then out to each individual wheel. Pay particular attention to the rubber coated brake hoses that lead to the front calipers and to the flexible rubber line that connects the brake line at the rear of the vehicle's frame to the T-fitting located on the rear axle housing.
12 Carefully check the parking brake system (cables, linkage and connecting points). If the cables are frayed or damaged, replace the necessary parts.

Drum brakes — inspection and cleaning

Note: *If your vehicle is equipped with full-floating rear axles (identified by 8-lug wheels and a protruding hub), a special large deep socket will be required to perform the following service procedure. Also, new lockwashers for the axle bolts, a new gasket and a new bearing adjusting nut locking wedge will be needed, depending on the vehicle model year.*

13 The brake drum must be removed in order to inspect the linings and hardware.
14 Loosen the lug nuts approximately 1/2-turn. On models with full-floating axle hubs, loosen the axle retaining bolts in the center hub.
15 Raise the vehicle and support it securely on jackstands.
16 Remove the lug nuts and the wheel.
17 On vehicles equipped with semi-floating axles (5-lug wheels and no protruding center hub) unthread the spring clips from the studs and pull the drum off the axle flange. If the drum will not come off easily, check to make sure the parking brake is fully released. If the drum still won't come off, back off the brake shoe adjustment. **Note:** *Steps 18 through 24 and 32 through 41 apply only to 1975 and later vehicles with full-floating rear axles. For 1974 and earlier models, refer to Chapter 9 (Section 11) to remove and install the rear brake drums and adjust the bearing preload.*
18 Heavy-duty vehicles equipped with full-floating rear axles require removal of the axle. Remove the previously loosened axle retaining bolts and washers and remove the axle and gasket (see illustration).
19 Carefully pry the locking wedge out of the adjusting nut with a screwdriver. **Caution:** *This must be done before the bearing adjusting nut is removed or turned.*
20 Remove the large bearing adjusting nut with the appropriate size

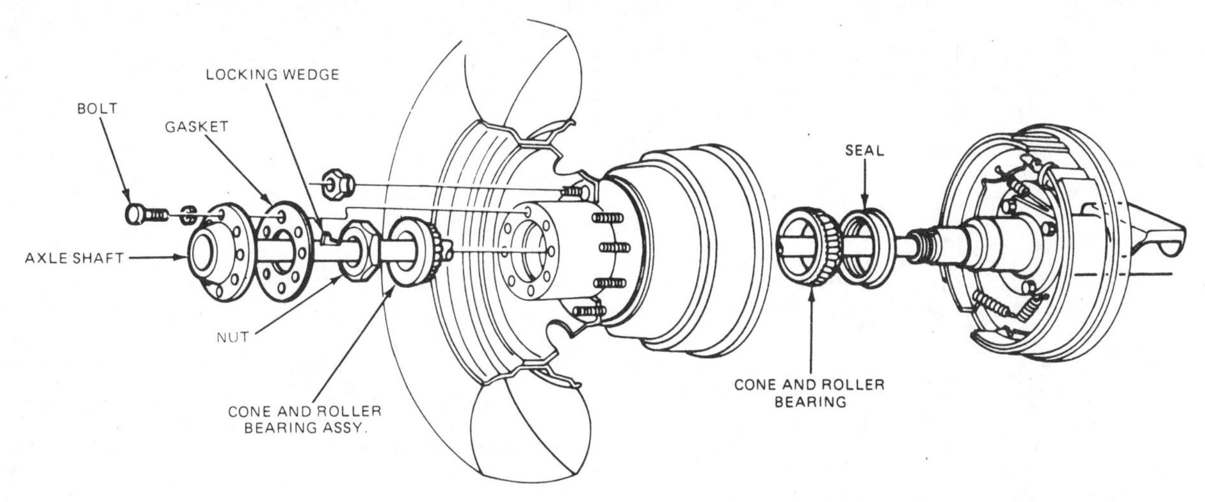

17.18 Full-floating rear hub assembly components — exploded view (1975 and later models)

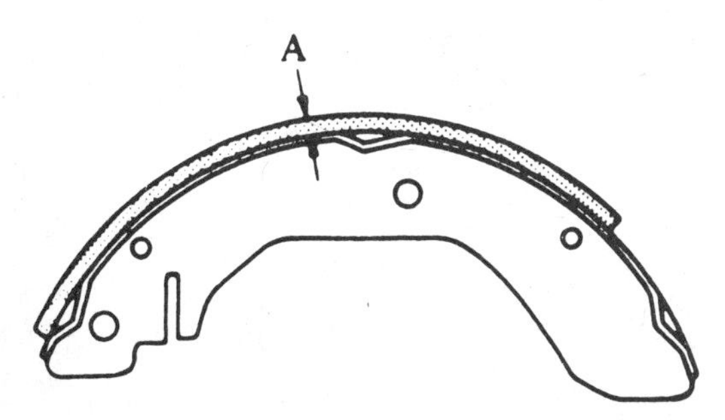

17.24a The lining thickness of the rear brake shoes (A) is measured from the surface of the lining to the metal shoe

deep socket. **Caution:** *Never use a chisel to remove the nut.*

21 Pull the drum off the spindle approximately two to three inches and then push it back on. This will pull the outer wheel bearing out onto the spindle for easier removal. See Chapter 9 for instructions on backing off the brake adjustment for drum removal if it won't pull off easily.

22 Remove the brake drum. Notice that the inner wheel bearing will be retained in the drum by the inner wheel bearing seal. Use caution not to damage the seal.

23 With the drum removed, carefully vacuum or brush away any accumulations of brake lining material and dust. Do not blow this material out with compressed air — it contains asbestos and is hazardous to breathe.

24 Check the thickness of the lining material (see illustrations). If it is less than 1/8-inch thick, the brake shoes should be replaced with new ones. The shoes should also be replaced if they are cracked, glazed (shiny surface) or wet from brake fluid or oil.

25 Check the brake return springs, parking brake cable (rear brakes) and self adjusting brake mechanisms for condition and correct installation.

26 Carefully check the brake components for signs of fluid leakage. Use your finger to carefully pry back the lip of the rubber wheel cylinder cups. These are located at the top of the brake assembly (see illustration). Any leakage at these cups is an indication that the wheel cylinders should be overhauled immediately (Chapter 9). Also check the connections at the rear of the brake backing plate for any signs of leakage.

27 If the wheel cylinders are dry and there are signs of grease or oil in the brake assembly area, the axle seal is defective and should be replaced. Grease or oil will ruin asbestos and rubber parts, so any accumulation of grease or oil necessitates replacement of the brake shoes and/or rubber parts affected by it.

28 Wipe the inside of the drum with a clean rag and brake cleaning solvent or denatured alcohol. Again, be careful not to inhale the asbestos dust.

29 Check the inside of the drum for cracks, score marks, deep grooves and hard spots, which will appear as small discolored spots. If these imperfections cannot be removed with fine emery cloth and light rub-

17.24b If the lining material is riveted to the shoes, the measurement is taken from the top of the rivet heads (arrows) to the surface of the lining material

17.26 During the brake inspection, the wheel cylinders should be carefully checked for leaks in the areas indicated by the arrows

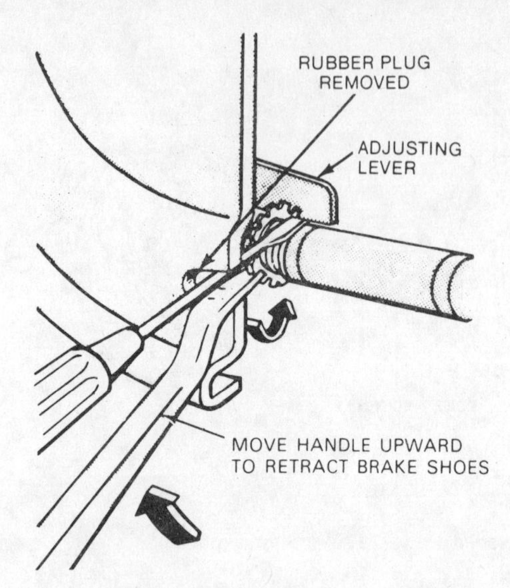

17.41 To adjust the brakes, move the adjusting lever aside with a screwdriver and turn the star wheel with a brake adjusting tool

bing, the drum must be taken to an automotive machine shop, parts house or brake specialist with the equipment required to machine the drums.

30 If, after the inspection and cleaning process, all parts are in good working condition, reinstall the brake drum, retaining clips and wheel (semi-floating axle only).

31 On 1975 and later full-floating axle assemblies only, the following Steps (32 through 41) should be followed.

32 Install the drum after carefully inspecting the hub seal and wear surface of the spindle. Make sure the inner wheel bearing has plenty of lubricant and the bearing area is free of dust. If the bearing is dirty or dry the seal will have to be removed with a seal removing tool or two screwdrivers. Discard the seal after comparing it to its replacement. Remove the bearing, clean it along with the housing, and repack it with the recommended lubricant. Install the bearing and seal in the drum.

33 As a temporary means of protection, wrap the threads of the spindle with tape.

34 Very carefully slide the complete hub/drum assembly over the spindle, taking care not to damage the inner seal. After the hub/drum assembly is positioned on the spindle, remove the tape.

35 Install the outer bearing, making sure that it is clean and packed with the correct lubricant, and start the large adjusting nut by hand. While rotating the drum assembly, tighten the nut with the deep socket to the specified torque, then back of the nut the specified amount.

36 Position the locking wedge in the keyway slot and carefully tap it into place until it is seated. **Note:** *It must not be bottomed against the shoulder of the nut when it is seated. If it is, a new wedge must be installed. The adjusting nut and locking wedge can be used over again, providing the locking wedge cuts a new groove in the nylon retainer material after the nut has been loosened the specified amount. The wedge must not be pressed into an existing groove.*

37 Make sure that the drum/hub assembly rotates freely. **Caution:** *Under no circumstances should the bearings be preloaded. Due to the high capacity weight rating of this type of axle assembly, the installation procedure must be adhered to strictly. Damage to the bearings and spindle can result from improper installation.*

38 Install the axle along with a new gasket. Install the axle retaining bolts and tighten them to the specified torque (use new lockwashers on the bolts).

39 Check the end play of the hub and drum assembly with a dial indicator. If you do not own one, it can be rented from most rental outlets. The correct end play is 0.001-to-0.010-inch. If not correct, repeat the previous steps, beginning with a loose bearing.

Drum brake adjustment

40 Adjust the brakes (if they have been replaced or had to be retracted for removal purposes) by removing the rubber plug at the rear of the brake backing plate.

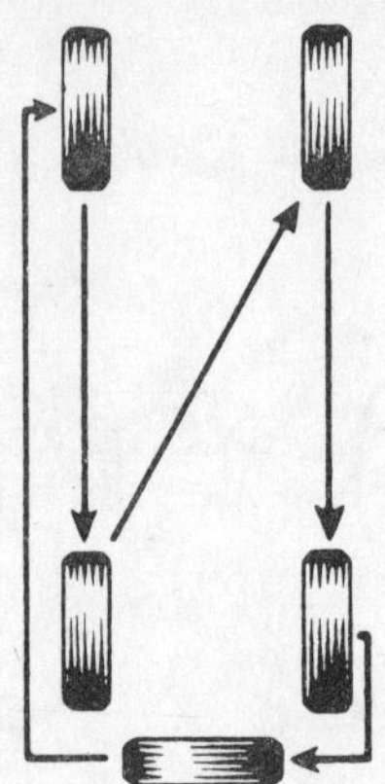
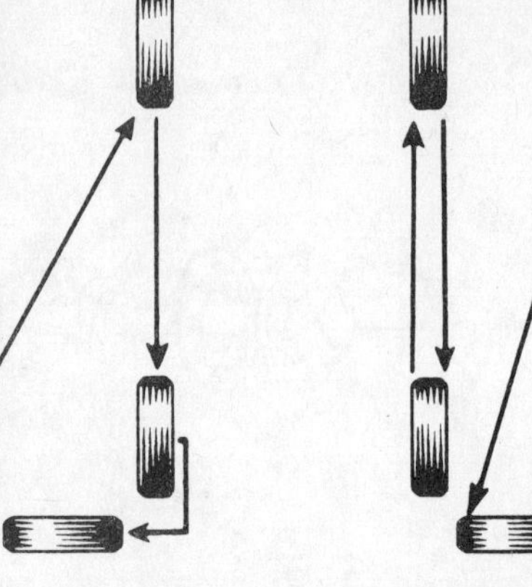

18.1 Tire rotation diagram

41 Use a brake adjustment tool inserted through the slot in the backing plate to rotate the star wheel adjuster until the brakes drag against the drums (see illustration). You will have to keep rotating the brake drum while you are performing this operation.

42 Loosen the star wheel adjuster to a point where the dragging just stops. If the drum drags heavily in one spot it is egg-shaped and must be resurfaced or replaced for proper brake operation.

43 Install the wheel and tighten the lug nuts. Lower the vehicle to the ground and pump the brake pedal several times to verify correct brake operation and "feel" before attempting to drive the vehicle.

18 Tire rotation

Refer to illustration 18.1

1 The tires should be rotated at the specified intervals and whenever uneven wear is noticed (see illustration). Since the vehicle will be raised and the tires removed, check the brakes and wheel bearings.

2 Refer to the information in *Jacking and towing* at the front of this manual for the proper procedures to follow when raising the vehicle and changing a tire. However, if the brakes are to be checked, do not apply the parking brake as stated. Make sure the tires are blocked to prevent the vehicle from rolling.

3 Preferably, the entire vehicle should be raised at the same time. This can be done on a hoist or by jacking up each corner and then lowering the vehicle onto jackstands placed under the frame rails. Always use four jackstands and make sure the vehicle is firmly supported.

4 After rotation, check and adjust the the tire pressures as necessary and be sure to check the lug nut tightness.

19 Air filter and PCV filter check and replacement

Refer to illustrations 19.8 and 19.14

1 At the specified intervals the air filter and PCV filter should be replaced with new ones. A thorough program of preventative maintenance would call for the two filters to be inspected between changes.

Air filter

Carburetor equipped vehicles

2 The air filter is located inside the air cleaner housing on the top of the engine. The filter is generally replaced by removing the wing nut at the top of the air cleaner assembly and lifting off the top plate.

3 While the top plate is off, be careful not to drop anything down into the carburetor.

4 Lift the air filter element out of the housing.

5 Wipe out the inside of the housing with a clean rag.

6 Place the old filter (if in good condition) or the new filter (if the specified interval has elapsed) into the air cleaner housing. Make sure it seats properly in the bottom of the housing.

7 Reinstall the top plate and tighten the wing nut.

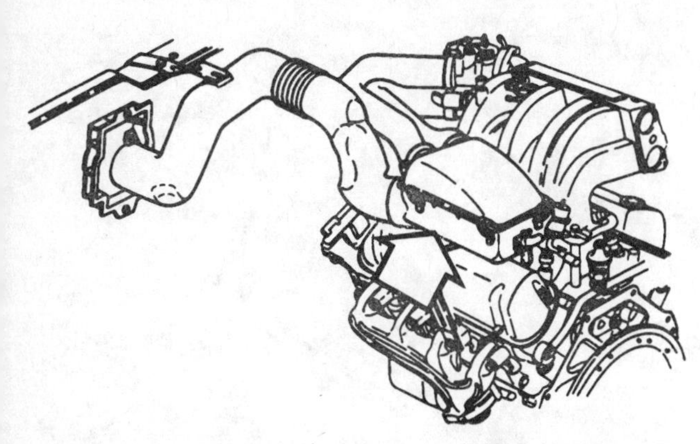

19.8 On EFI equipped vehicles, the air filter is located in a housing (arrow) adjacent to the intake manifold

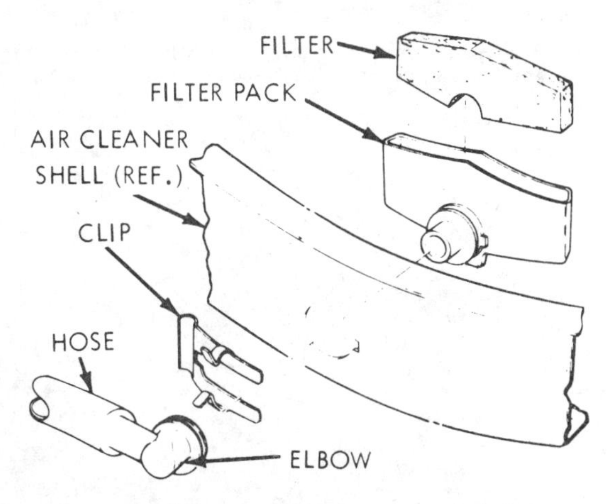

19.14 PCV filter components – exploded view

EFI equipped vehicles

8 Release the four clips securing the cover to the housing, then lift off the cover to expose the filter element (see illustration).

9 Check the inside surfaces of the cover and housing for evidence of dust leakage past the filter element due to damaged seals, incorrect filter installation or inadequate clip tension.

10 Lift out the filter element and clean the sealing surfaces on the housing and cover.

11 Install the new element. Make sure the side marked TOP is facing up.

12 Position the cover and secure it in place with the clips.

PCV filter

13 The PCV filter is also located inside the air cleaner housing. Remove the top plate and air filter as described previously, then locate the PCV filter on the side of the housing.

14 Detach the hose and remove the retaining clip on the outside of the housing, then remove the PCV filter assembly (see illustration).

15 Install a new PCV filter, then reinstall the retaining clip, air filter, top plate and any hoses which were disconnected.

20 Battery check and maintenance

Refer to illustrations 20.1, 20.5a, 20.5b, 20.5c and 20.5d

Warning: *Certain precautions must be followed when checking or servicing the battery. Hydrogen gas, which is highly flammable, is always present in the battery cells, so keep lighted tobacco and all other open flames or sparks away from the battery. The electrolyte inside the battery is actually dilute sulfuric acid, which can be hazardous to your skin and cause injury if splashed in the eyes. It will also ruin clothes and painted surfaces.*

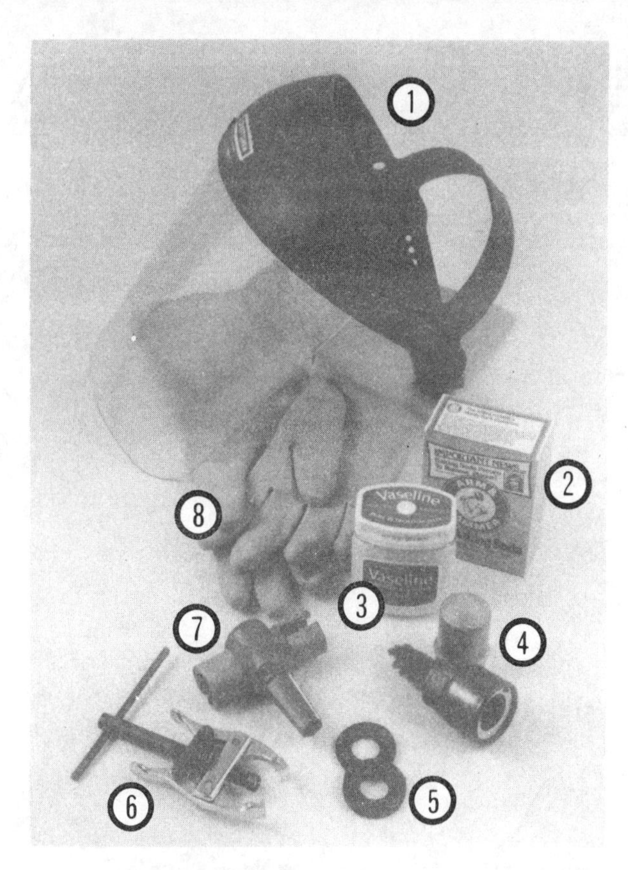

20.1 Tools and materials required for battery maintenance

1 ***Face shield/safety goggles*** *– When removing corrosion with a brush, the acidic particles can easily fly up into your eyes*
2 ***Baking soda*** *– A solution of baking soda and water can be used to neutralize corrosion*
3 ***Petroleum jelly*** *– A layer of this on the battery posts will help prevent corrosion*
4 ***Battery post/cable cleaner*** *– This wire brush cleaning tool will remove all traces of corrosion from the battery posts and cable clamps*
5 ***Treated felt washers*** *– Placing one of these on each post, directly under the cable clamps, will help prevent corrosion*
6 ***Puller*** *– Sometimes the cable clamps are very difficult to pull off the posts, even after the nut/bolt has been completely loosened. This tool pulls the clamp straight up and off the post without damage*
7 ***Battery post/cable cleaner*** *– Here is another cleaning tool which is a slightly different version of number 4 above, but it does the same thing*
8 ***Rubber gloves*** *– Another safety item to consider when servicing the battery; remember that's acid inside the battery!*

1 Tools and materials required for battery maintenance include eye and hand protection, baking soda, petroleum jelly, a battery cable puller and cable/terminal post cleaning tools (see illustration).

Checking

2 Check the battery case for cracks and evidence of leakage.

3 To check the electrolyte level in the battery, refer to Section 5. **Note:** *Many models are equipped with maintenance-free batteries which have no provision for adding water. Some models may be equipped with translucent batteries so the electrolyte level can be observed without removing any cell caps. On these batteries, the level should be between the upper and lower marks.*

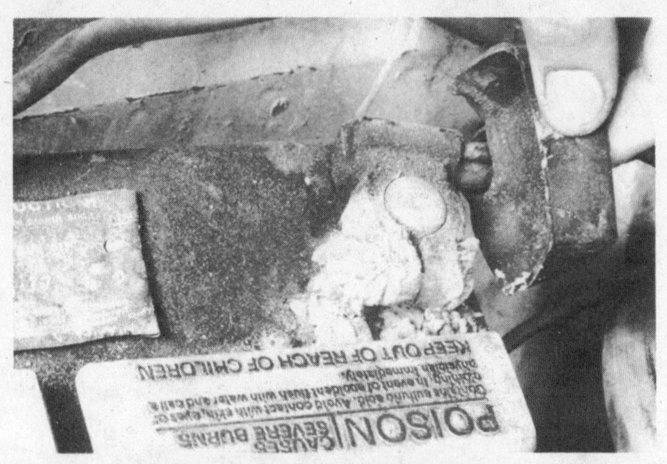

20.5a Battery terminal corrosion usually appears as light, fluffy powder

20.5b Removing a cable from the battery post (always remove the ground cable first and hook it up last)

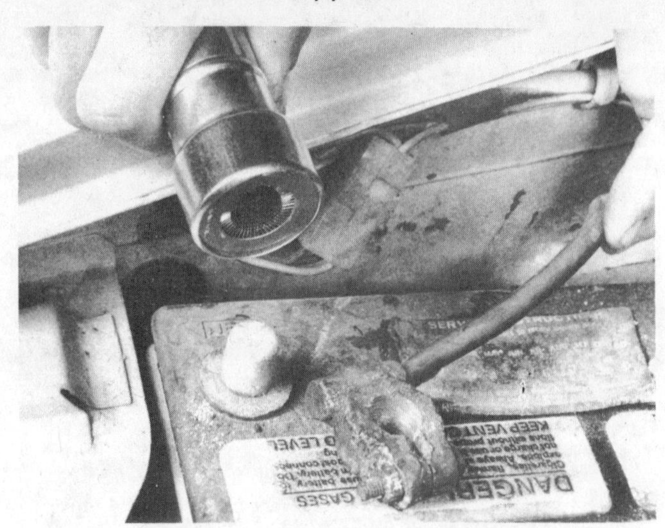

20.5c Cleaning a battery post with the special tool

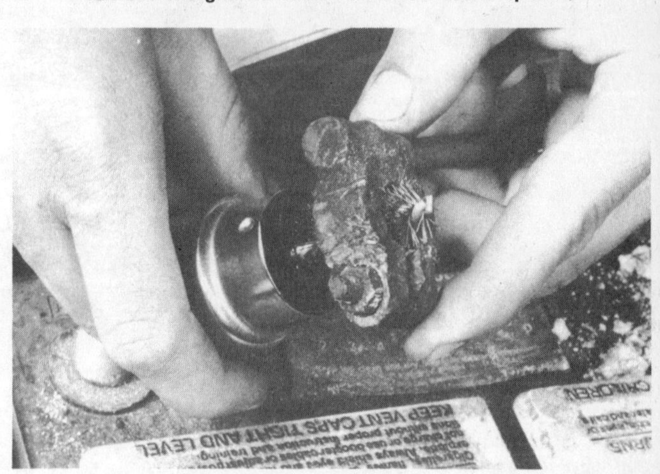

20.5d Cleaning the battery cable clamp with the special tool

4 Periodically have the specific gravity of the electrolyte checked with a hydrometer. This is especially important during cold weather. If the reading is below the specified range, the battery should be recharged. Maintenance-free batteries have a built-in hydrometer which indicates the battery state-of-charge. This check can usually be done by a service station for a minimal charge.

5 Check the tightness of the battery cable clamps to ensure good electrical connections. If corrosion is evident, remove the cables from the battery terminals (a puller may be required), clean them with a battery terminal brush, then reinstall them. Corrosion can be kept to a minimum by applying a layer of petroleum jelly or grease to the terminals and cable clamps after they are assembled (see illustrations).

6 Check the entire length of each battery cable for corrosion, cracks and frayed conductors. Replace the cables with new ones if they are damaged (Chapter 5).

7 Make sure that the rubber protector over the positive terminal is not torn or missing. It should completely cover the terminal.

8 Make sure that the battery is securely mounted, but do not over-tighten the clamp bolts.

9 The battery case and caps should be kept clean and dry. If corrosion is evident, clean the battery as described below.

10 If the vehicle is not being used for an extended period, disconnect the battery cables and have the battery charged approximately every six weeks.

Cleaning

11 Corrosion on the battery hold-down components and inner fender panel can be removed by washing with a solution of water and baking soda. Once the area has been thoroughly cleaned, rinse it with clear water.

12 Corrosion on the battery case and terminals can also be removed with a solution of water and baking soda and a stiff brush. Be careful that none of the solution is splashed into your eyes or onto your skin (wear protective gloves). Do not allow any of the baking soda and water solution to get into the battery cells. Rinse the battery thoroughly once it is clean.

13 Metal parts of the vehicle which have been damaged by spilled battery acid should be painted with a zinc-based primer and paint. Do this only after the area has been thoroughly cleaned and dried.

Charging

14 As mentioned before, if the battery's specific gravity is below the specified level, the battery must be recharged.

15 If the battery is to remain in the vehicle during charging, disconnect the cables from the battery to prevent damage to the electrical system.

16 When batteries are being charged, hydrogen gas, which is very explosive and flammable, is produced. Do not smoke or allow open flames near a charging or a recently charged battery. Also, do not plug in the battery charger until the connections have been made at the battery posts.

17 The average time necessary to charge a battery at the normal rate is from 12 to 16 hours. Always charge the battery slowly. A quick charge or boost charge is hard on a battery and will shorten its life. Use a battery charger that is rated at no more than 1/10 the amp/hour rating of the battery.

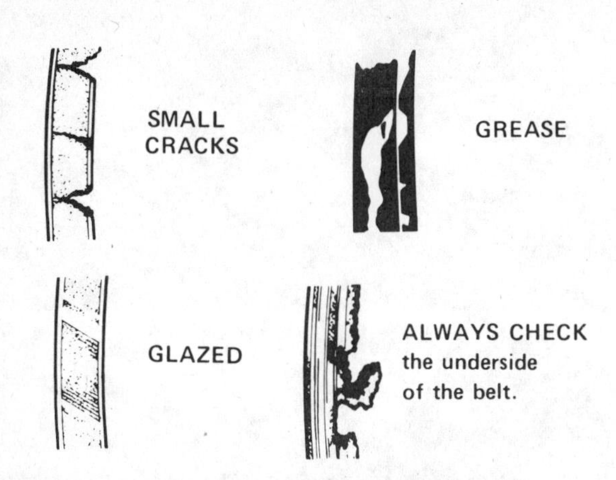

21.3 Here are some of the more common problems associated with drivebelts (check the belts very carefully to prevent an untimely breakdown)

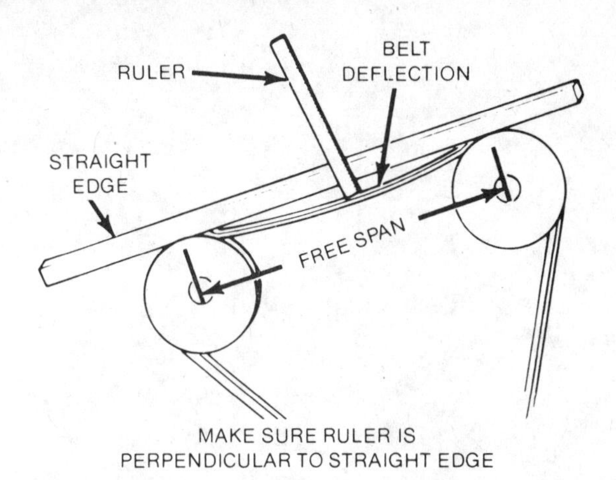

21.4 Measuring drivebelt deflection with a straightedge and ruler

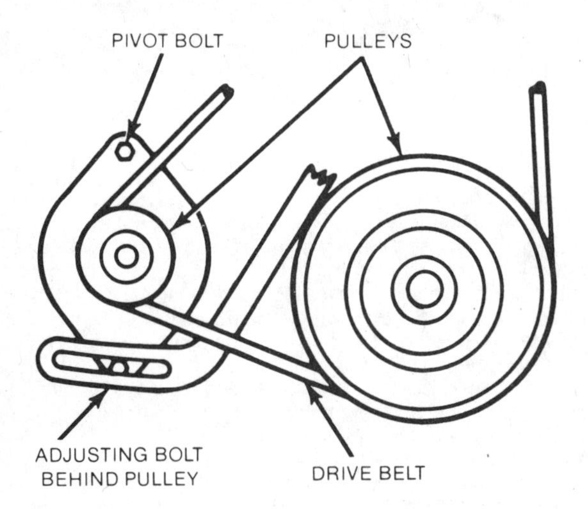

21.6 Belt-driven components have an adjusting bolt and a pivot bolt — both must be loosened when adjusting belt tension

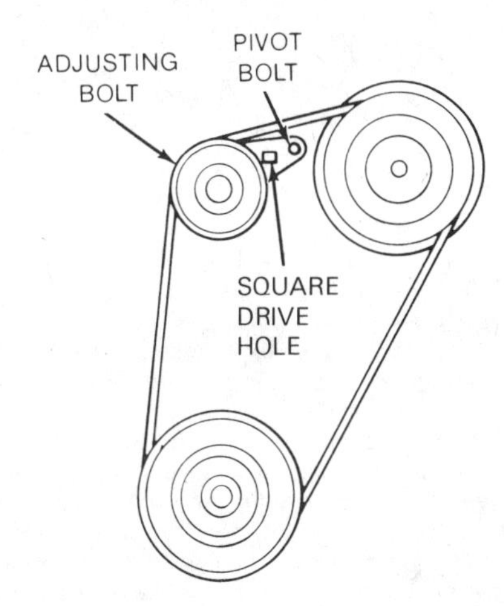

21.8 Idler pulleys often have a square hole, used to accept a breaker bar to lever the pulley and tension the belt

18 Remove all of the cell caps and cover the holes with a clean cloth to prevent the spattering of electrolyte. Hook the battery charger leads to the battery posts (positive to positive, negative to negative), then plug in the charger. Make sure it is set at 12 volts if it has a selector switch.

19 Watch the battery closely during charging to make sure that it does not overheat.

20 The battery can be considered fully charged when it is gassing freely and there is no increase in specific gravity during three successive readings taken at hourly intervals. Overheating of the battery during charging at normal charging rates, excessive gassing and continual low specific gravity readings are indications that the battery should be replaced with a new one.

21 Drivebelt check and adjustment

Refer to illustrations 21.3, 21.4, 21.6 and 21.8

1 The drivebelts, or V-belts as they are sometimes called, are located at the front of the engine and play an important role in the overall operation of the vehicle and its components. Due to their function and material make-up, the belts are prone to failure after a period of time and should be inspected and adjusted periodically to prevent major engine damage.

2 The number of belts used on a particular vehicle depends on the accessories installed. Drivebelts are used to turn the alternator, power steering pump, water pump and air-conditioning compressor. Depending on the pulley arrangement, a single belt may be used to drive more than one of these components.

3 With the engine off, open the hood and locate the various belts at the front of the engine. Using your fingers (and a flashlight, if necessary), move along the belts checking for cracks and separation of the belt plies. Also check for fraying and glazing, which gives the belt a shiny appearance. Both sides of the belt should be inspected, which means you will have to twist the belt to check the underside (see illustration).

4 The tension of each belt is checked by laying a straightedge, such as a wooden stick, across the longest free span (distance between two pulleys) of the drivebelt to be measured. Apply pressure with your hand and see how much the belt moves down (deflects). A rule of thumb is that if the distance from pulley center-to-pulley center is between 7 and 11 inches, the belt should deflect 1/4-inch. If the belt is longer and travels between pulleys spaced 12 to 16 inches apart, the belt should deflect 1/2-inch (see illustration).

5 If it is necessary to adjust the belt tension, either to make the belt tighter or looser, it is done by moving the belt driven component on a bracket or by moving an idler pulley.

6 For each component there will be an adjustment or strap bolt and a pivot bolt (see illustration). Both bolts must be loosened slightly to enable you to move the component.

7 After the bolts have been loosened, move the component away from the engine to tighten the belt or toward the engine to loosen the belt. Hold the component in position and check the belt tension. If it is correct, tighten the bolts until just snug, then recheck the tension. If it is correct, tighten the bolts.

8 It will often be necessary to use some sort of pry bar to move the component while the belt is adjusted. If this must be done to gain the

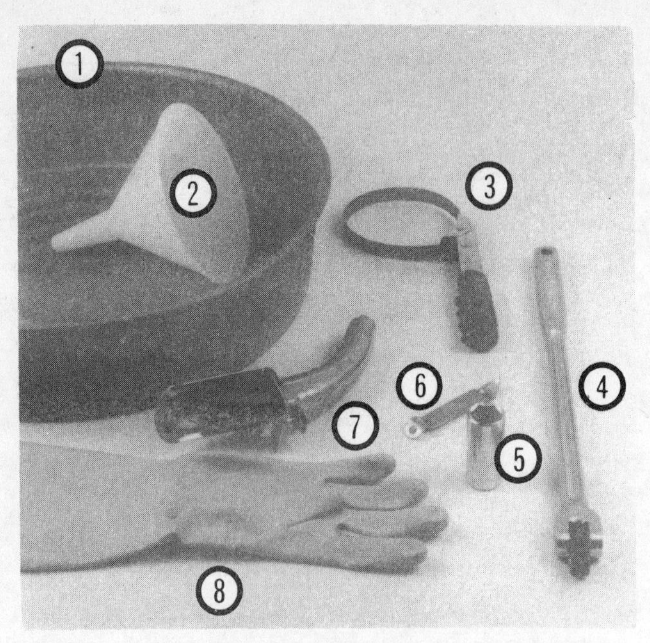

22.3 These tools are required when changing the engine oil and filter

1 **Drain pan** — *It should be fairly shallow in depth, but wide in order to prevent spills*
2 **Funnel** — *To prevent spills when adding oil to the engine (particularly the six cylinder)*
3 **Filter wrench** — *Shown is a metal band-type wrench, but other types will work as well*
4 **Breaker bar** — *Sometimes the oil drain plug is pretty tight and a long breaker bar is needed to loosen it*
5 **Socket** — *To be used with the breaker bar or a ratchet (must be the correct size to fit the drain plug)*
6 **Can opener** — *Used to open the new oil cans*
7 **Oil spout** — *Can be used in place of the funnel when adding oil to the engine (particularly a V8)*
8 **Rubber gloves** — *When removing the drain plug and filter it is inevitable that you will get oil on your hands (the gloves will prevent burns)*

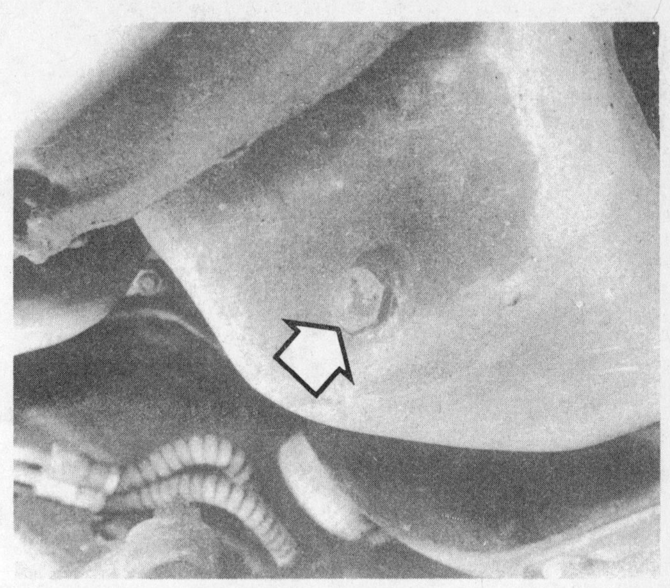

22.9 The engine oil drain plug is located near the bottom of the oil pan

proper leverage, be very careful not to damage the component being moved or the part being pried against. If an idler pulley is used to tension the belt, it may have a square hole in it to accept a breaker bar, which can be used to lever the idler pulley and tension the belt (see illustration).

22 Engine oil and filter change

Refer to illustrations 22.3, 22.9, 22.13a, 22.13b, 22.14 and 22.19

1 Frequent oil changes may be the best form of preventative maintenance available to the home mechanic. When engine oil ages, it becomes diluted and contaminated, leading to premature engine wear.
2 Although some sources recommend oil filter changes every other oil change, we feel that the minimal cost of an oil filter and the relative ease with which it is installed dictate that a new filter be used whenever the oil is changed.
3 The tools necessary for an oil and filter change are a wrench to fit the drain plug at the bottom of the oil pan, an oil filter wrench to remove the old filter, a container with at least a six quart capacity to drain the old oil into and a funnel or oil can spout to help pour fresh oil into the engine (see illustration).
4 You should have plenty of clean rags and newspapers handy to mop up any spills. Access to the underside of the vehicle is greatly improved if the vehicle can be lifted on a hoist, driven onto ramps or supported by jackstands. **Warning:** *Do not work under a vehicle which is supported only by a bumper, hydraulic or scissors jack.*
5 If this is your first oil change on the vehicle, it is recommended that you crawl underneath and familiarize yourself with the locations of the oil drain plug and the oil filter. The engine and exhaust components will be warm during the actual work, so it is a good idea to figure out any potential problems before the engine is started.
6 Allow the engine to warm to normal operating temperature. If new oil or any tools are needed, use this warm-up time to gather everything necessary for the job.
7 With the engine oil warm (warm engine oil will drain better and more built-up sludge will be removed with the oil), raise and support the vehicle on jackstands.
8 Move all necessary tools, rags and newspapers under the vehicle. Position the drain pan under the drain plug. Keep in mind that the oil will initially flow from the engine with some force, so locate the pan accordingly.
9 Being careful not to touch any of the hot exhaust components, use the wrench to remove the drain plug near the bottom of the oil pan (see illustration). Depending on how hot the oil has become, you may want to wear gloves while unscrewing the plug the final few turns.
10 Allow the old oil to drain into the pan. It may be necessary to move the pan as the oil flow slows to a trickle.
11 After all the oil has drained, wipe off the drain plug with a clean rag. Small metal particles may cling to the plug and would immediately contaminate the new oil.
12 Clean the area around the drain plug opening and reinstall the plug. Tighten it securely with the wrench. If a torque wrench is available, use it to tighten the plug.
13 Move the drain pan into position under the oil filter (see illustrations).
14 Use the filter wrench to loosen the oil filter. Chain or metal band-type filter wrenches may distort the filter canister, but this is of no concern as the filter will be discarded. On early model vehicles with a V8 engine and power steering, the oil filter may be difficult to remove. Turn the front wheels to the right as far as possible, then unscrew the filter and slide it to the rear to remove it (see illustration).
15 Sometimes the oil filter is on so tight it cannot be loosened, or it is positioned in an area which is inaccessible with a filter wrench. As a last resort, you can punch a metal bar or long screwdriver directly through the side of the canister and use it as a T-bar to turn the filter. If you do so, be prepared for oil to spurt out of the canister as it is punctured.
16 Completely unscrew the old filter. Be careful, it is full of oil. Empty the oil inside the filter into the drain pan.
17 Compare the old filter with the new one to make sure they are the same type.
18 Use a clean rag to remove all oil, dirt and sludge from the area where the oil filter mounts to the engine. Check the old filter to make sure

22.13a Oil filter location on a V8 engine

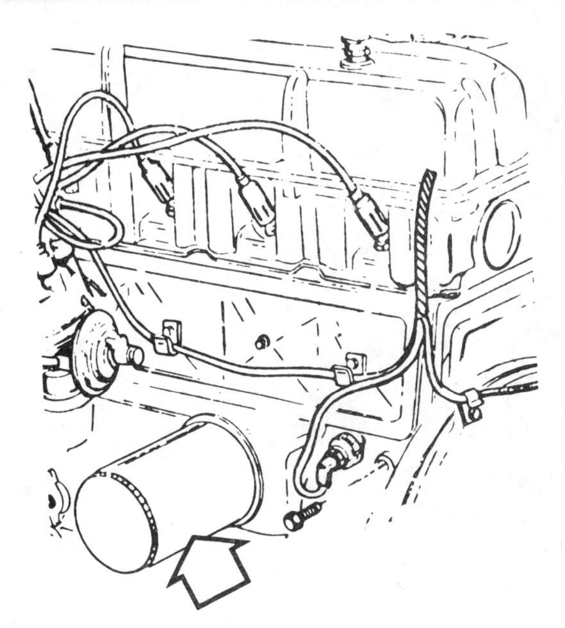

22.13b Oil filter location on a six cylinder engine

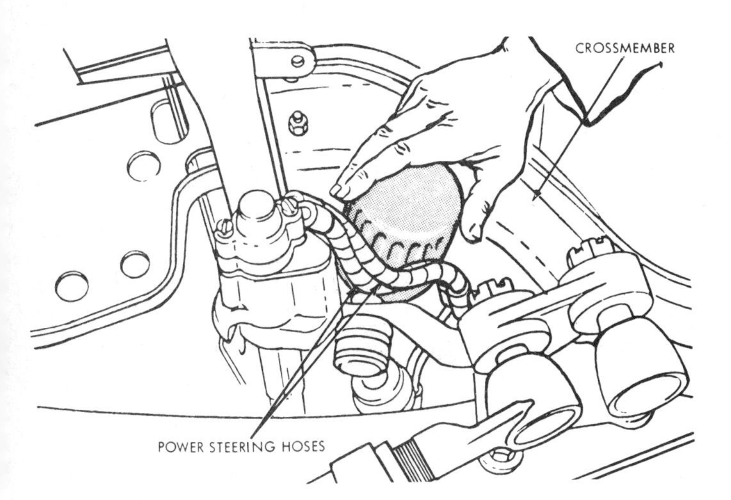

22.14 Some V8 models with power steering require that the oil filter be unscrewed and slid to the rear to clear the power steering hoses

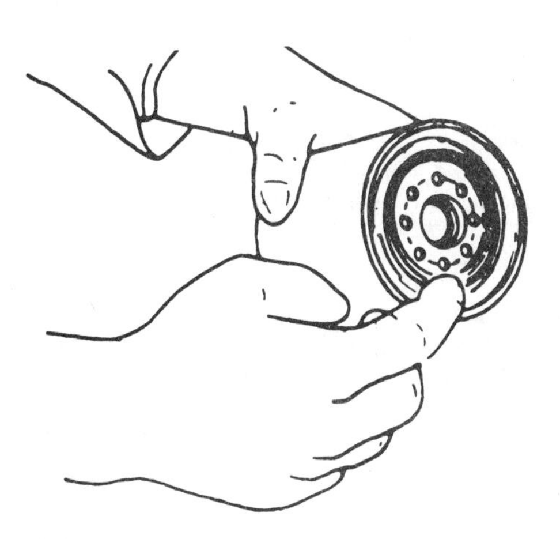

22.19 Before installing the new oil filter, spread a thin coat of fresh oil on the rubber gasket

the rubber gasket is not stuck to the engine mounting surface. If the gasket is stuck to the engine (use a flashlight to check if necessary), remove it.

19 Open one of the cans of new oil and apply a light coat of oil to the rubber gasket of the new filter (see illustration).

20 Attach the filter to the engine, following the tightening directions printed on the filter canister or packing box. Most filter manufacturers recommend against using a filter wrench due to the possibility of over-tightening and damage to the gasket.

21 Remove all tools, rags, etc. from under the vehicle, being careful not to spill the oil in the drain pan, then lower the vehicle.

22 Move to the engine compartment and locate the oil filler cap on the engine. In most cases there will be a screw-off cap on the rocker arm cover. The cap will most likely be labeled *Engine Oil* or *Oil*.

23 If an oil can spout is used, push the spout into the top of the oil can and pour the fresh oil into the filler opening. A funnel may also be used.

24 Pour about four quarts of fresh oil into the engine. Wait a few minutes to allow the oil to drain into the pan, then check the level on the oil dipstick (see Section 5 if necessary). If the oil level is between the upper and lower marks, start the engine and allow the new oil to circulate.

25 Run the engine for only about a minute and then shut it off. Immediately look under the vehicle and check for leaks at the oil pan drain plug and around the oil filter. If either one is leaking, tighten it with a bit more force.

26 With the new oil circulated and the filter now completely full, recheck the level on the dipstick and add enough oil to bring the level to the upper mark on the dipstick.

27 During the first few trips after an oil change, make it a point to check frequently for leaks and proper oil level.

28 The old oil drained from the engine cannot be reused in its present state and should be disposed of. Oil reclamation centers, auto repair shops and gas stations will normally accept the oil, which can be refined and used again. After the oil has cooled it can be drained into a suitable container (capped plastic jugs, topped bottles, milk cartons, etc.) for transport to one of these disposal sites.

23 Fuel filter replacement

Refer to illustrations 23.1, 23.5, 23.9, 23.10 and 23.22

Warning: *Gasoline is extremely flammable so extra safety precautions must be observed when working on any part of the fuel system. Do not smoke and do not allow bare light bulbs or open flames near the vehicle. Also, do not perform this maintenance procedure in a garage where a natural gas type appliance, such as a water heater or clothes dryer, with a pilot light is installed.*

Carburetor equipped engines

1 On these models, the fuel filter is usually located at the fuel inlet fitting on the carburetor (see illustration). On some models an in-line

23.1 The fuel filter is usually attached to the carburetor as shown here

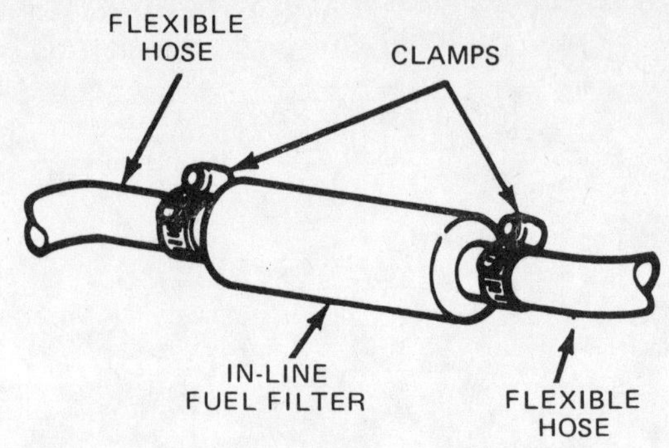

23.5 Typical in-line fuel filter

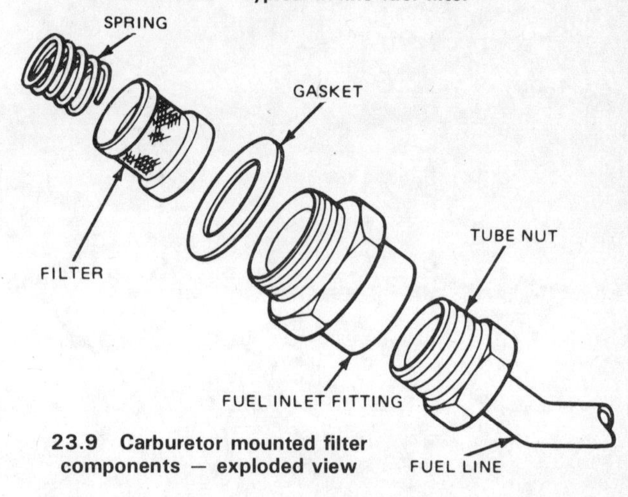

23.9 Carburetor mounted filter components — exploded view

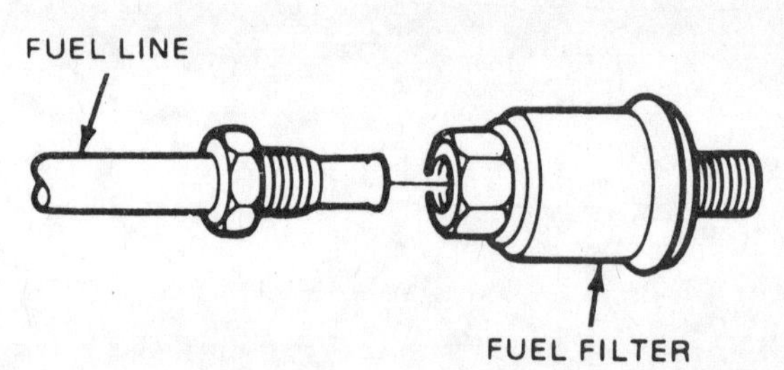

23.10 Some models have a fuel filter that threads into the carburetor inlet — it is replaced as an assembly

filter, located between the fuel pump and carburetor, may be installed.
2 The job should be done with the engine cold, after sitting at least three hours. To prevent any possibility of sparks, disconnect the negative (–) battery cable from the battery and position it away from the battery post.
3 Obtain the replacement filter (make sure it is for your specific vehicle and engine) and some clean rags.
4 Remove the air cleaner assembly. If vacuum hoses must be disconnected, be sure to note their positions and/or tag them to ensure that they are reinstalled correctly.

In-line filter

5 If your filter is located in the line between the fuel pump and carburetor, release the clamps on the rubber hoses and carefully remove the filter from the line (see illustration). Use caution in this step, as the gasoline will be under slight pressure and may spurt out.
6 After removing the filter, drain and discard it. Replace the filter with an exact duplicate and install it with new rubber connecting hoses, which should come with the new filter. Also replace the clamps with new ones.
7 Tighten the clamps, start the engine and check for leaks.

Carburetor inlet filter

8 Place some rags under the fuel filter and fittings to catch any gasoline that is spilled as the fittings are loosened.
9 Hold the large fuel inlet fitting at the carburetor with a wrench as the fuel line tube nut is loosened. A flare nut wrench should be used on the line fitting, if available. Make sure the fitting next to the carburetor is held securely while the fuel line is disconnected (see illustration).
10 Some filters at the carburetor are integral with the inlet connection and may have a rubber hose or metal line attached. If this type is being replaced, remove the hose or line, unscrew the filter/fitting and screw in a new filter assembly. Reattach the rubber hose and tighten the clamp or install the metal line (see illustration).
11 After the fuel line is disconnected, move it aside for better access to the inlet fitting. Do not crimp the fuel line.
12 Unscrew the fuel inlet fitting, which was previously held steady. As this fitting is drawn away from the carburetor body, be careful not to lose the thin gasket on the nut or the spring located behind the fuel filter. Pay close attention to how the filter is installed.
13 Compare the old filter with the new one to make sure they are the same length and design.
14 Reinstall the spring in the carburetor body (if equipped).
15 Place the new filter in position. It will have a rubber gasket and check valve at one end, which should point away from the carburetor.
16 Install a new gasket on the fuel inlet fitting (a gasket is usually supplied with the new filter) and start the fitting. Make sure it is not cross-threaded, then tighten it securely. If a torque wrench is available, tighten the fitting to 7 to 10 ft-lbs (10 to 14 Nm). Do not overtighten it, as the hole can strip easily, causing fuel leaks.
17 Hold the fuel inlet fitting securely with a wrench while the fuel line is connected. Again, be careful not to cross-thread the tube nut, and tighten it securely.
18 Plug the vacuum hose which leads to the air cleaner vacuum motor so the engine can be run.
19 Start the engine and check carefully for leaks. If the fuel line connector leaks, disassemble it and check for stripped or damaged threads. If the fuel line tube nut has stripped threads, remove the entire line and have a repair shop install a new fitting. If the threads look all right, purchase some thread sealing tape and wrap the threads with it. Reinstall and tighten it securely. Inlet repair kits are available at most auto parts stores to overcome leaking at the fuel inlet filter nut.
20 Reinstall the air cleaner assembly, connecting the hoses in their original positions.

Fuel injected engines

21 Fuel is filtered at two different points on fuel injected models. All filters used are of one piece construction and cannot be cleaned. Replace the filter if it becomes clogged. Fuel is first filtered inside the tank at the fuel pump inlet. Servicing of this filter requires removal of the fuel pump.
22 Fuel is again filtered at a reservoir-type filter found downstream from the fuel tank. This filter is attached to the left frame rail (see illustration). **Note:** *This fuel filter is designed to last the life of the vehicle*

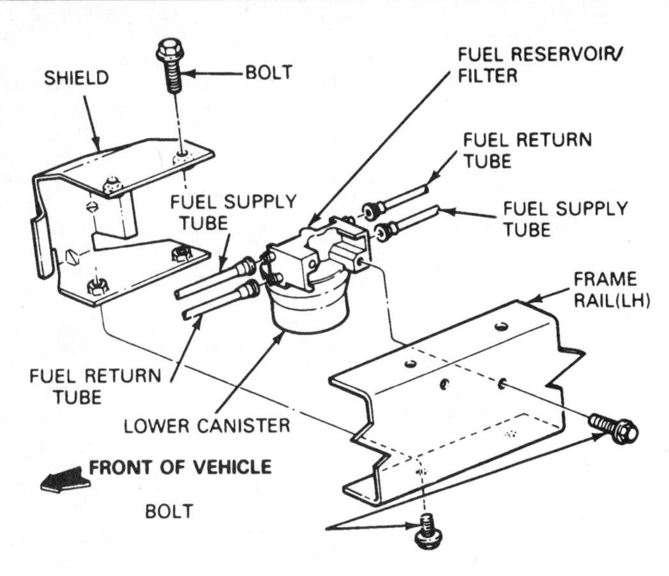

23.22 Fuel-injected models have a filter attached to the frame rail

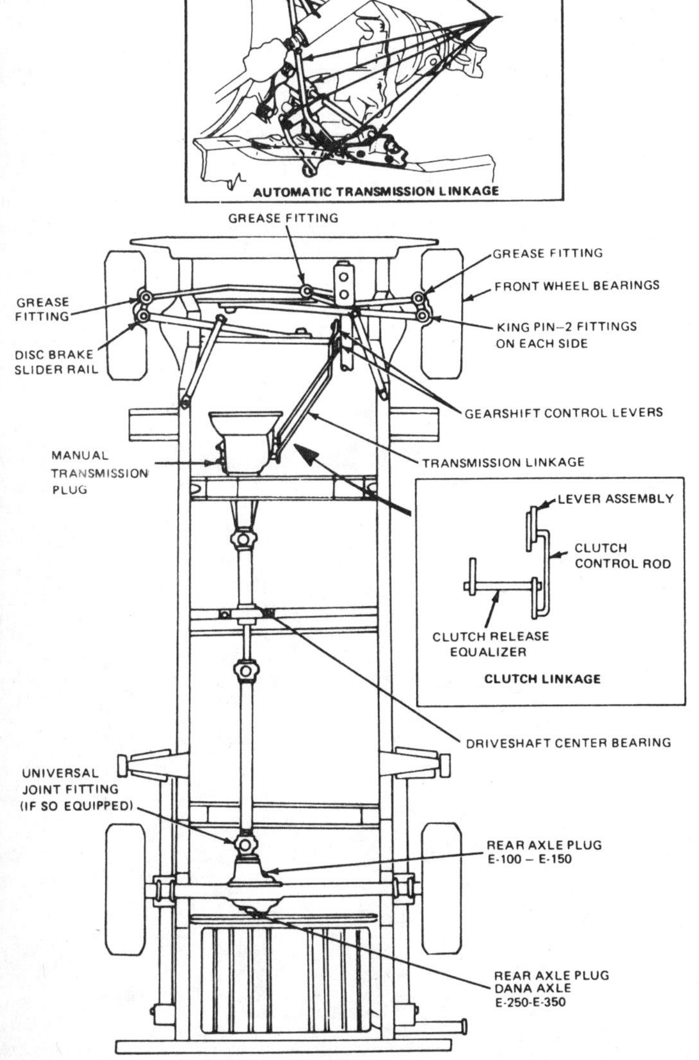

24.2 Chassis component lubrication diagram

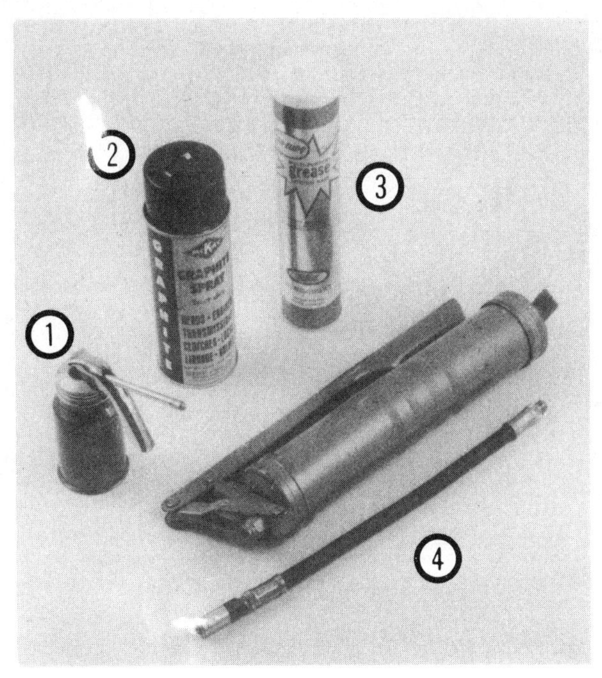

24.1 Materials required for chassis and body lubrication

1 **Engine oil** – *Light engine oil in a can like this can be used for door and hood hinges*
2 **Graphite spray** – *Used to lubricate lock cylinders*
3 **Grease** – *Grease, in a variety of types and weights, is available for use in a grease gun. Check the Specifications for your requirements*
4 **Grease gun** – *A common grease gun, shown here with a detachable hose and nozzle, is needed for chassis lubrication. After use, clean it thoroughly!*

and should not need servicing. Before removing this filter, depressurize the fuel system as described in Chapter 4 and loosen the fuel tank cap. The front of the vehicle should be raised to prevent loss of fuel from the tank.
23 Remove the shield (if equipped) from the frame rail.
24 Remove the filter canister with a strap-type oil filter wrench and slide the filter and canister out of the frame rail. **Note:** *It may be necessary to remove the reservoir mounting bolts and pull the reservoir away from the frame rail to remove the canister.* The filter will be full of fuel. Empty and discard the filter.
25 Install the new filter in the canister and place a new O-ring in the groove in the canister.
26 Reinstall the canister/filter assembly, making sure the O-ring stays in place.
27 Tighten the canister one-sixth of a turn past initial O-ring compression.
28 Start the engine and check the canister for leaks.
29 Replace the shield (if equipped).

24 Chassis, body and driveline lubrication

Refer to illustrations 24.1, 24.2 and 24.6

1 A grease gun and a cartridge filled with the proper grease (see *Recommended lubricants and fluids*), graphite spray and an oil can filled with engine oil will be required to lubricate chassis, body and driveline components (see illustration). In some cases, plugs may be installed in place of grease fittings, in which case grease fittings will have to be purchased and installed.
2 Using the accompanying diagram (see illustration), which indicates where the various grease fittings are located, look under the vehicle to find the components and determine if grease fittings or plugs are installed. If there are plugs, remove them with a wrench and buy grease fittings, which will thread into the component. A Ford dealer or auto

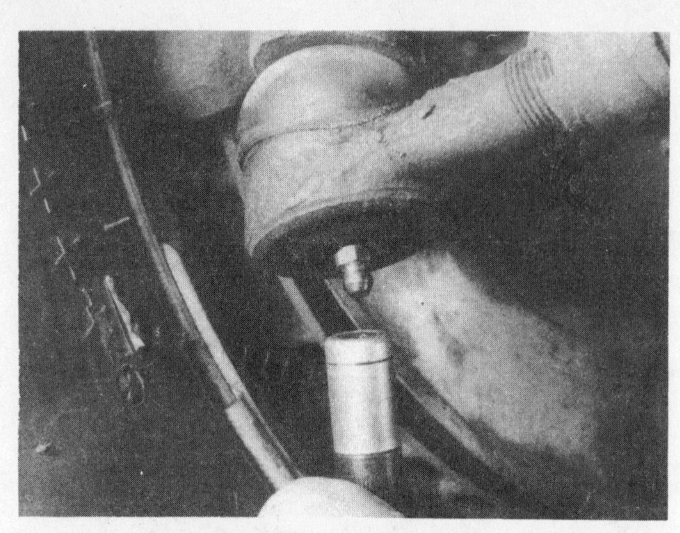

24.6 Clean each grease fitting nipple with a rag before attaching the grease gun and pumping in grease

parts store will be able to supply replacement fittings. Straight, as well as angled, fittings are available.

3 For easier access under the vehicle, raise it with a jack and place jackstands under the frame. Make sure the vehicle is securely supported by the stands.

4 Before proceeding, force a little grease out of the nozzle to remove any dirt from the end of the grease gun. Wipe the nozzle clean with a rag.

5 With the grease gun, plenty of clean rags and the diagram, crawl under the vehicle and begin lubricating the components.

6 Wipe the grease fitting nipple clean and push the nozzle firmly over it (see illustration). Squeeze the trigger on the grease gun to force grease into the component. **Caution:** *The control arm balljoints should be lubricated until the rubber reservoir is firm to the touch. Do not pump too much grease into these fittings as it could rupture the reservoir.* For all other suspension and steering fittings, continue pumping grease into the nipple until it seeps out of the joint between the two components. If the grease escapes around the grease gun nozzle, the nipple is clogged or the nozzle is not seated on the fitting. Resecure the gun nozzle to the fitting and try again. If necessary, replace the fitting.

7 Wipe the excess grease from the components and the fitting. Follow the procedure for all remaining fittings.

8 While you are under the vehicle, clean and lubricate the parking brake cable, along with the cable guides and levers. This can be done by smearing some of the chassis grease onto the cable and its related parts.

9 Place a few drops of polyethylene grease (motor oil will work also) on the transmission shift linkage rods and swivels. Lubricate the clutch linkage components at the cross-shaft grease fitting.

10 Use the grease gun to lubricate the driveshaft U-joints if they are equipped with grease fittings.

11 Lower the vehicle to the ground for the body lubrication process.

12 Open the hood and smear a little chassis grease on the hood latch mechanism. If the hood has an inside release, have an assistant pull the release knob from inside the vehicle as you lubricate the cable at the latch.

13 Lubricate all the hinges (door, hood, etc.) with a few drops of light engine oil to keep them in proper working order.

14 The key lock cylinders can be lubricated with spray-on graphite, which is available at auto parts stores.

25 Cooling system servicing (draining, flushing and refilling)

Refer to illustration 25.6

1 On a periodic basis, the cooling system should be drained, flushed and refilled to replenish the antifreeze mixture and prevent formation of rust and corrosion, which can impair the performance of the cooling system and ultimately cause engine damage.

2 At the same time the cooling system is serviced, all hoses and the radiator cap should be inspected and replaced if defective.

3 Since antifreeze is a corrosive and poisonous solution, be careful not to spill any of the coolant mixture on the paint or your skin. If this happens, rinse immediately with plenty of clean water. Also, consult your local authorities about the dumping of antifreeze before draining the cooling system. In many areas reclamation centers have been set up to collect automobile oil and drained antifreeze/water mixtures, rather than allowing them to be added to the sewage system.

4 With the engine cold, remove the radiator cap.

5 Move a large container under the radiator to catch the coolant as it is drained.

6 Drain the radiator. Most models are equipped with a drain plug at the bottom (see illustration). If this drain has excessive corrosion and cannot be opened easily, or if the radiator is not equipped with a drain, disconnect the lower radiator hose to allow the coolant to drain. Be careful that none of the coolant is splashed onto your skin or into your eyes.

7 If accessible, remove the engine coolant drain plugs.

8 Disconnect the hose from the coolant reservoir and remove the reservoir. Flush it out with clean water.

9 Place a garden hose in the radiator filler neck and flush the system until the water runs clear at all drain points.

10 In severe cases of contamination or clogging of the radiator, remove it (see Chapter 3) and reverse flush it. This involves inserting the hose in the bottom radiator outlet to allow the clean water to run against the normal flow, draining through the top. A radiator repair shop should be consulted if further cleaning or repair is necessary.

11 When the coolant is regularly drained and the system refilled with the correct antifreeze/water mixture, there should be no need to use chemical cleaners or descalers.

12 To refill the system, reconnect the radiator hoses and install the drain plugs securely in the engine. Special thread sealing tape (available

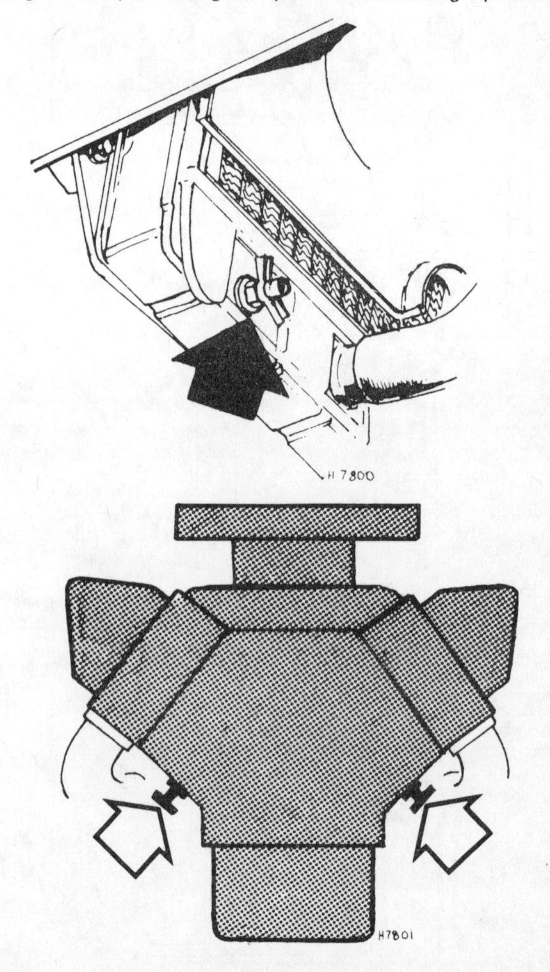

25.6 Most models have both a radiator drain (top) and one or two engine drain plugs (bottom) (V8 engine shown)

at auto parts stores) should be used on the drain plugs. Install the reservoir and the overflow hose, where applicable.
13 Fill the radiator to the base of the filler neck and then add more coolant to the reservoir until it is about half full.
14 Run the engine until normal operating temperature is reached and, with the engine idling, add coolant up to the specified level (see Section 5). Install the radiator cap so that the arrows are in alignment with the overflow hose. Install the reservoir cap.
15 Always use a mixture of antifreeze and water in the proportion called for on the antifreeze container or in your owner's manual. Chapter 3 also contains information on antifreeze mixtures.
16 Keep a close watch on the coolant level and the various cooling system hoses during the first few miles of driving. Tighten the hose clamps and add more coolant as necessary.

26 Underhood hose check and replacement

Caution: *Replacement of air conditioning hoses must be left to a dealer or air conditioning specialist who has the proper equipment to depressurize the system safely. Never remove air conditioning components or hoses until the system has been depressurized.*

Vacuum hoses

1 High temperatures under the hood can cause the deterioration of the rubber and plastic hoses used for engine, accessory and emission systems operation.
2 Periodic inspection should be made for cracks, loose clamps, material hardening and leaks.
3 Some, but not all, vacuum hoses use clamps to secure the hoses to fittings. Where clamps are used, check to be sure they haven't lost their tension, allowing the hose to leak. Where clamps are not used, make sure the hose has not expanded and/or hardened where it slips over the fitting, allowing it to leak.
4 It is quite common for vacuum hoses, especially those in the emissions system, to be color coded or identified by colored stripes molded into the hose. Various systems require hoses with different wall thicknesses, collapse resistance and temperature resistance. When replacing hoses be sure to use the same hose material on the new hose.
5 Often the only effective way to check a hose is to remove it completely from the vehicle. Where more than one hose is removed, be sure to label the hoses and their attaching points to insure proper reattachment.
6 When checking vacuum hoses, be sure to include any plastic T-fittings in the check. Check the fittings for cracks and the hose where it fits over the fitting for enlargement, which could cause leakage.
7 A small piece of vacuum hose (1/4-inch inside diameter) can be used as a stethoscope to detect vacuum leaks. Hold one end of the hose to your ear and probe around vacuum hoses and fittings, listening for the "hissing" sound characteristic of a vacuum leak. **Caution:** *When probing with the vacuum hose stethoscope, be careful not to allow your body or the hose to come into contact with moving engine components such as drivebelts, the cooling fan, etc.*

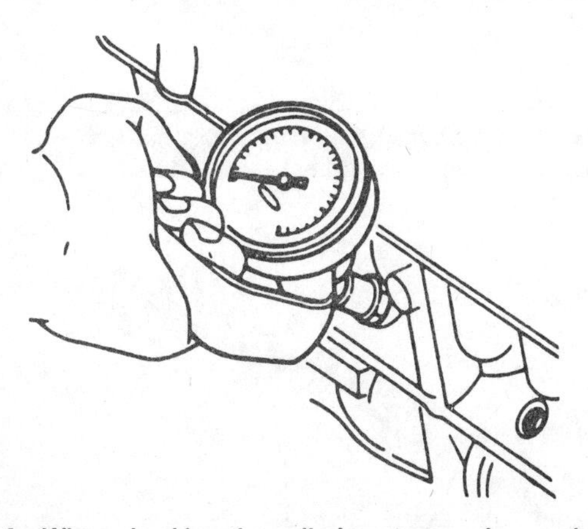

28.4 When checking the cylinder compression, make sure the gauge is firmly seated in the plug hole or the reading will be false

Fuel hose

8 **Warning:** *There are certain precautions which must be taken when inspecting or servicing fuel system components. Work in a well ventilated area and do not allow open flames (cigarettes, appliance pilot lights, etc.) or bare light bulbs near the work area. Mop up any spills immediately and do not store fuel soaked rags where they could ignite.*
9 The fuel lines are usually under a small amount of pressure, so if any fuel lines are to be disconnected be prepared to catch fuel spillage.
10 Check all rubber fuel lines for deterioration and chafing. Check especially for cracking in areas where the hose bends and just before clamping points, such as where a hose attaches to the fuel pump, fuel filter and carburetor or fuel injection unit.
11 High quality fuel line, usually identified by the word *Fluroelastomer* printed on the hose, should be used for fuel line replacement. Under no circumstances should unreinforced vacuum line, clear plastic tubing or water hose be used for fuel line replacement.
12 Spring-type clamps are commonly used on fuel lines. These clamps often lose their tension over a period of time, and can be "sprung" during the removal process. Therefore it is recommended that all spring-type clamps be replaced with screw clamps whenever a hose is replaced.

Metal lines

13 Sections of metal line are often used for fuel line between the fuel pump and carburetor or fuel injection unit. Check carefully to be sure the line has not been bent and crimped and that cracks have not started in the line in the area of bends.
14 If a section of metal fuel line must be replaced, only seamless steel tubing should be used, since copper and aluminum tubing do not have the strength necessary to withstand normal engine operating vibration.
15 Check the metal brake lines where they enter the master cylinder and brake proportioning unit (if used) for cracks in the lines or loose fittings. Any sign of brake fluid leakage calls for an immediate thorough inspection of the brake system.

27 Valve adjustment

All Ford truck light duty engines are equipped with hydraulic valve lifters. Under normal circumstances these lifters maintain the correct clearance for the valve train. Therefore, no adjustment is necessary.

In some instances, especially when repeated valve jobs have been performed on an engine, the valve clearance will have to be manually set. In these cases, a special procedure is required to bleed down the hydraulic lifter, check the clearance and replace the pushrod with a longer or shorter service unit. See Chapter 2 for the correct procedure to install a different size pushrod for correct valve adjustment.

28 Compression check

Refer to illustration 28.4

1 A compression check will tell you what mechanical condition the upper end (pistons, rings, valves, etc.) of your engine is in. Specifically, it can tell you if the compression is down due to leakage caused by worn piston rings, defective valves and seats or a blown head gasket.
2 Begin by cleaning the area around the spark plugs before you remove them. This will keep dirt from falling into the cylinders while you are performing the compression test.
3 Disconnect the ignition switch feed wire at the distributor. This is the pink wire coming from the ignition coil. Block the throttle and choke valves open.
4 With the compression gauge in the spark plug hole (see illustration), crank the engine over at least four compression strokes and observe the gauge (the compression should build up quickly). Low compression on the first stroke, which does not build up during successive strokes, indicates leaking valves or a blown head gasket (a cracked head could also be the cause). Record the highest gauge reading obtained.
5 Repeat the procedure for the remaining cylinders and compare the results to the Specifications.
6 Pour a couple of teaspoons of engine oil (a squirt can works great for this) into each cylinder, through the spark plug hole, and repeat the test.
7 If the compression increases after after the oil is added, the piston

rings are definitely worn. If the compression does not increase significantly, the leakage is occurring at the valves or head gasket. Leakage past the valves may be caused by burned valve seats or faces, warped, cracked or bent valves or incorrectly adjusted valves.

8 If two adjacent cylinders have equally low compression, there is a strong possibility that the head gasket between them is blown. The appearance of coolant in the combustion chambers or the crankcase would verify this condition.

9 If the compression is unusually high, the combustion chambers are probably coated with carbon deposits. If that is the case, the cylinder head(s) should be removed and decarbonized.

10 If compression is way down or varies greatly between cylinders, it would be a good idea to have a leak-down test performed by a reputable automotive repair shop. This test will pinpoint exactly where the leakage is occurring and how severe it is.

29 Spark plug replacement

Refer to illustrations 29.2, 29.7 and 29.11

1 The spark plugs are located on both sides of V8 engines and on one side of inline engines. They may or may not be easily accessible for removal. If the vehicle is equipped with air conditioning or power steering, some of the plugs may be tricky to remove. Special extension or swivel tools may be necessary. Check to see if special tools will be needed.

2 In most cases, tools necessary for a spark plug replacement include a plug wrench or spark plug socket, which fits onto a ratchet wrench (this special socket will be insulated inside to protect the plug) and a wire-type feeler gauge to check and adjust the spark plug gap (see illustration). Also, a special spark plug wire removal tool is available for separating the wires from the spark plugs. To simplify installation, obtain a piece of 3/16-inch inside diameter rubber hose, 8-to-12 inches in length, to use when starting the plugs into the head.

3 The best procedure to follow when replacing the spark plugs is to purchase the new spark plugs beforehand, adjust them to the proper gap, and then replace the plugs one at a time. When buying the new spark plugs be sure to obtain the correct type for your specific engine. This information can be found on the Emissions Control Information label located under the hood or in the owner's manual. If differences exist between these sources, purchase the spark plug type specified on the Emissions Control label because the information was printed for your specific engine.

4 Allow the engine to cool completely before attempting to remove the plugs. During this time, each of the new spark plugs can be inspected for defects and the gaps can be checked.

5 The gap is checked by inserting the proper thickness gauge between the electrodes at the tip of the plug. The gap between the electrodes should be the same as that given on the Emissions Control label. The wire should just touch each of the electrodes. If the gap is incorrect, use the notched adjuster on the feeler gauge body to bend the curved side electrode slightly until the proper gap is achieved. If the side electrode is not exactly over the center electrode, use the notched adjuster to align the two.

6 Remove the engine cover to gain access to the plugs.

7 With the engine cool, remove the spark plug wire from one spark plug. Do this by grabbing the boot at the end of the wire, not the wire itself. Sometimes it is necessary to use a twisting motion while the boot and plug wire are pulled free. Using a plug wire removal tool is the easiest and safest method (see illustration).

8 If compressed air is available, use it to blow any dirt or foreign material away from the spark plug area. A common bicycle pump will also work. The idea is to eliminate the possibility of material falling into the cylinder as the spark plug is removed.

9 Place the spark plug wrench or socket over the plug and remove it from the engine by turning in a counterclockwise direction.

10 Compare the spark plug with those shown in the accompanying photos to get an indication of the overall running condition of the engine.

11 Due to the angle at which the spark plugs must be installed on most engines, installation will be simplified by inserting the plug wire terminal of the new spark plug into the 3/16-inch rubber hose, mentioned previously, before it is installed in the cylinder head (see illustration). This procedure serves two purposes: the rubber hose gives you flexibility for establishing the proper angle of plug insertion in the head and, should the threads be improperly lined up, the rubber hose will merely slip on the spark plug terminal when it meets resistance, preventing damage to the cylinder head threads.

12 After installing the plug to the limit of the hose grip, tighten it with the socket. It is a good idea to use a torque wrench for this to ensure that the plug is seated correctly. The correct torque figure is included in the Specifications.

13 Before pushing the spark plug wire onto the end of the plug, in-

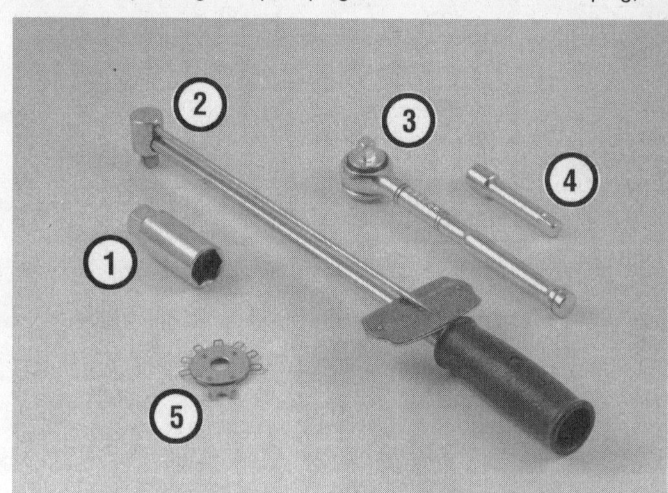

29.2 Tools required for changing spark plugs

1 ***Spark plug socket*** *- This will have special padding inside to protect the spark plug's porcelain insulator*
2 ***Torque wrench*** *- Although not mandatory, using this tool is the best way to ensure the plugs are tightened properly*
3 ***Ratchet*** *- Standard hand tool to fit the spark plug socket*
4 ***Extension*** *- Depending on model and accessories, you may need special extensions and universal joints to reach one or more of the plugs*
5 ***Spark plug gap gauge*** *- This gauge for checking the gap comes in a variety of styles. Make sure the gap for your engine is included*

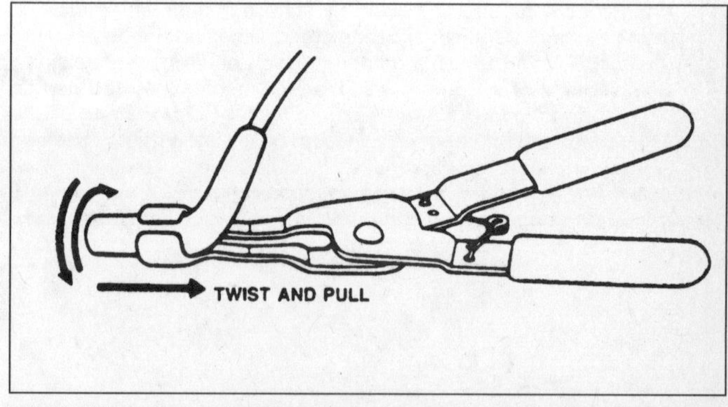

29.7 When removing the spark plug wires, pull only on the boot and twist it back-and-forth

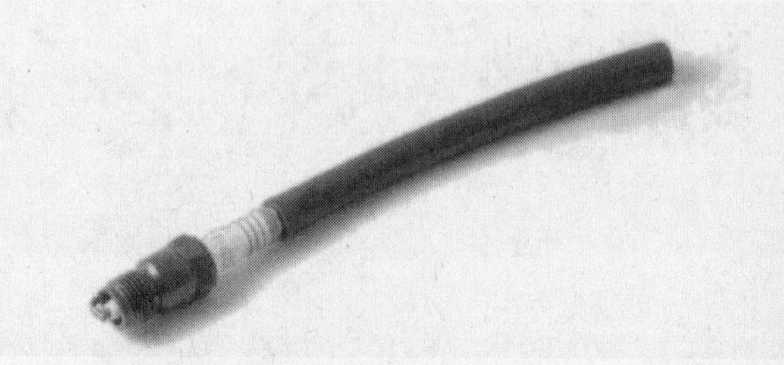

29.11 A length of 3/16-inch ID rubber hose will save time and prevent damaged threads when installing the spark plugs

spect it following the procedures outlined elsewhere in this Chapter.
14 Attach the plug wire to the new spark plug, again using a twisting motion on the boot until it is firmly seated on the spark plug. Make sure the wire is routed away from the exhaust manifold.
15 Follow the above procedure for the remaining spark plugs, replacing them one at a time to prevent mixing up the spark plug wires.

30 Spark plug wire check and replacement

1 The spark plug wires should be checked whenever new spark plugs are installed in the engine.
2 The wires should be inspected one at a time to prevent mixing up the order, which is essential for proper engine operation. Each original plug wire is numbered to help identify its location. If a number is illegible, a piece of tape can be marked with the correct number and wrapped around the plug wire.

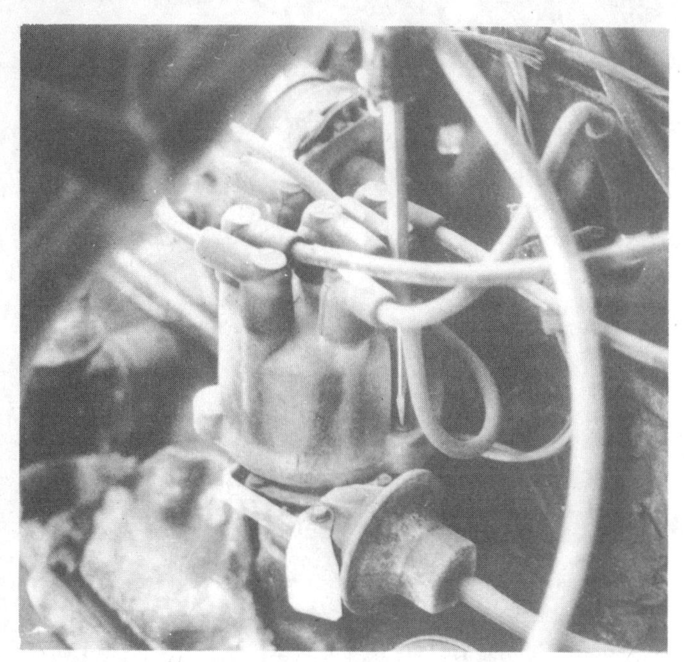
31.1a Use a screwdriver to detach the distributor cap

3 Disconnect the plug wire from the spark plug. A removal tool can be used for this purpose or you can grab the rubber boot, twist slightly and pull the wire free. Do not pull on the wire itself, only on the rubber boot.
4 Check inside the boot for corrosion, which will look like a white crusty powder. Some vehicles use a conductive white grease, which should not be mistaken for corrosion. **Note:** *When any spark plug wire on an electronic ignition system is detached from a spark plug, distributor or coil terminal, silicone grease (Ford No. D7AZ-19A331-A or equivalent electronic application grease) should be applied to the interior surface of the boot before reattaching the wire to the component.*
5 Push the wire and boot back onto the end of the spark plug. It should be a tight fit on the plug end. If not, remove the wire and use pliers to carefully crimp the metal connector inside the wire boot until the fit is snug.
6 Using a clean rag, wipe the entire length of the wire to remove built-up dirt and grease. Once the wire is clean, check for burns, cracks and other damage. Do not bend the wire sharply, since the conductor might break.
7 Disconnect the wire from the distributor. Again, pull only on the rubber boot. Check for corrosion and a tight fit in the same manner as the spark plug end. Replace the wire in the distributor.
8 Check the remaining spark plug wires, making sure they are securely fastened at the distributor and spark plug when the check is complete.
9 If new spark plug wires are required, purchase a set for your specific engine model. Wire sets are available pre-cut, with the rubber boots already installed. Remove and replace the wires one at a time to avoid mix-ups in the firing order.

1

31 Distributor cap and rotor check and replacement

Refer to illustrations 31.1a, 31.1b and 31.1c

1 A screwdriver is used to remove the cap from the distributor (see illustration). Check the distributor cap and rotor for wear. Look for cracks, carbon tracks and worn, burned or loose contacts (see illustrations). Replace the cap and rotor with new parts if defects are found. It is common practice to install a new cap and rotor whenever new spark plug wires are installed.
2 When installing a new cap, remove the wires from the old cap one at a time and attach them to the new cap in the exact same location — do not simultaneously remove all the wires from the old cap or firing order mix-ups may occur.

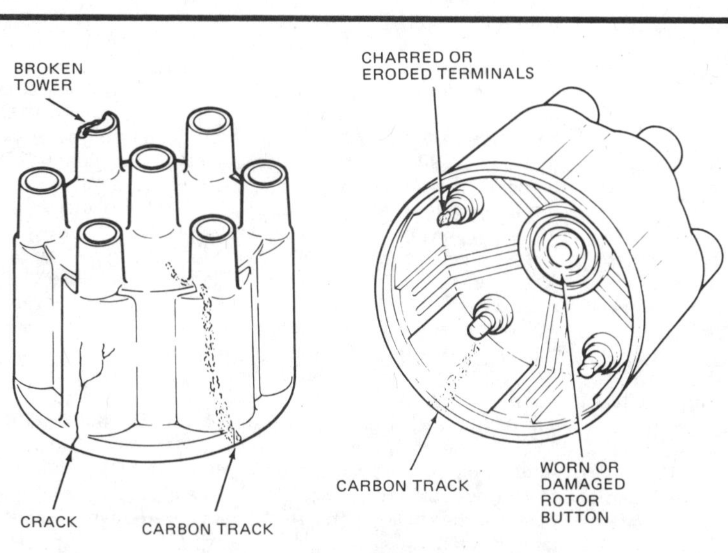

31.1b This illustration shows some of the common defects to look for when inspecting the distributor cap (if in doubt, buy a new one)

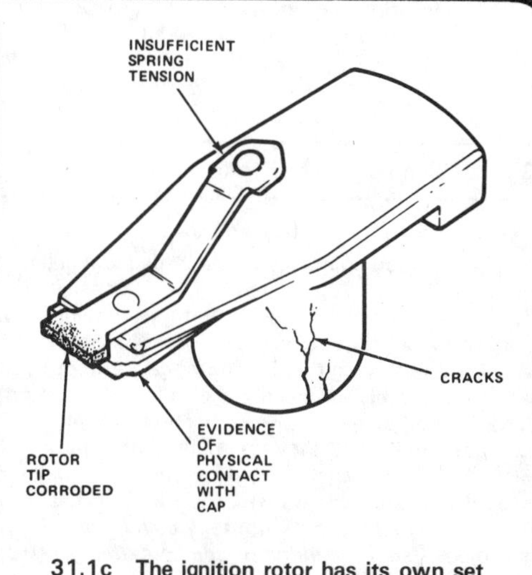

31.1c The ignition rotor has its own set of defects to look for — most notably wear and corrosion at the rotor tip (if in doubt, replace it with a new one)

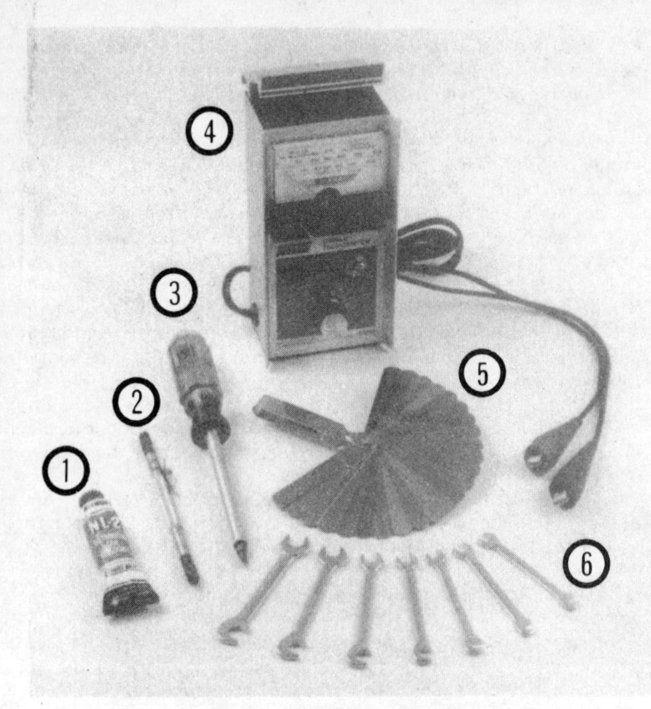

32.1 Tools and materials needed for contact point replacement and dwell angle adjustment

1 **Distributor cam lube** – *Sometimes this special lubricant comes with the new points; however, its a good idea to buy a tube and have it on hand*
2 **Screw starter** – *This tool has special claws which hold the screw securely as it is started, which helps prevent accidental dropping of the screw*
3 **Magnetic screwdriver** – *Serves the same purpose as 2 above. If you do not have one of these special screwdrivers, you risk dropping the point mounting screws down into the distributor body*
4 **Dwell meter** – *A dwell meter is the only accurate way to determine the point setting (gap). Connect the meter according to the instructions supplied with it*
5 **Blade-type feeler gauges** – *These are required to set the initial point gap (space between the points when they are open)*
6 **Ignition wrenches** – *These special wrenches are made to work within the tight confines of the distributor. Specifically, they are needed to loosen the nut/bolt which secures the leads to the points*

32 Ignition points check and replacement

Refer to illustrations 32.1 and 32.2

1 The ignition points must be replaced at regular intervals on vehicles not equipped with electronic ignition. Occasionally, the rubbing block on the points will wear sufficiently to require readjustment. Several special tools are required to replace the points (see illustration).

2 After removing the distributor cap and rotor, the points and condenser assembly are plainly visible (see illustration). The points may be examined by gently prying them open to reveal the condition of the contact surfaces. If they are rough, pitted or dirty they should be replaced. **Caution:** *The following procedure requires the removal and installation of small screws which can easily fall down into the distributor. To retrieve them, the distributor would have to be removed and disassembled. Use a magnetic or spring loaded screwdriver and exercise caution.*

3 To detach the condenser, which should be replaced along with the points, remove the screw that secures the condenser to the point plate. Loosen the nut or screw retaining the condenser lead and the primary lead to the point assembly. Remove the condenser and mounting bracket.

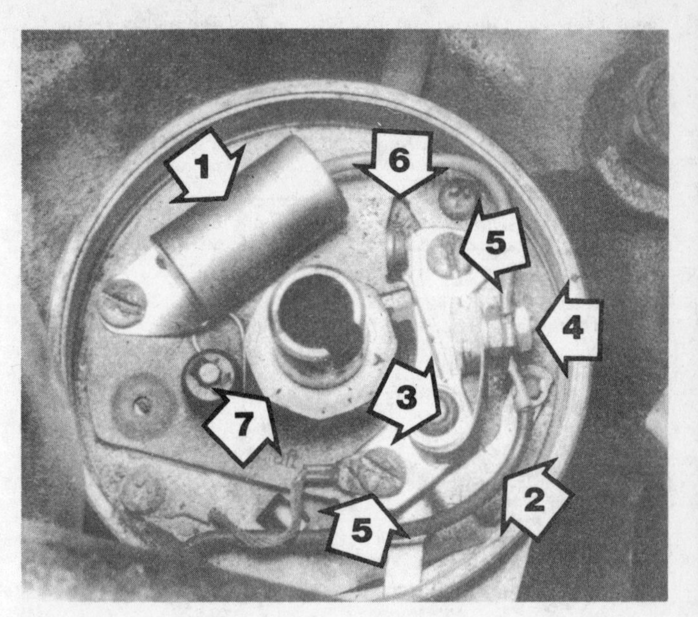

32.2 The Ignition points and related components are accessible after removing the distributor cap and rotor

1 Condenser
2 Primary lead
3 Point assembly
4 Primary/condenser lead connection
5 Mounting screw
6 Adjusting slot
7 Point (distributor) cam

4 Remove the two point assembly mounting screws.

5 Attach the new point assembly to the distributor plate with the two mounting screws. Notice that the ground lead is attached to the rear screw.

6 Install the new condenser and tighten the mounting screw.

7 Attach the primary ignition lead and the condenser lead to the point assembly . Make sure that the forked connectors for the primary ignition and condenser leads do not touch the distributor plate or any other grounded surface.

8 Two adjusting methods are available for setting the points. The first and most effective method involves an instrument called a dwell meter.

9 Connect one dwell meter lead to the primary ignition terminal of the point assembly or to the distributor terminal of the coil. Connect the other lead of the dwell meter to an engine ground. Some dwell meters may have different connecting instructions, so always follow the instrument manufacturer's directions.

10 Have an assistant crank the engine over or use a remote starter and crank the engine with the ignition switch turned to On.

11 Observe the dwell reading on the meter and compare it to the Specifications found at the front of this Chapter or on the engine tune-up decal. If the dwell reading is incorrect, adjust it by first loosening the two point assembly mounting screws a small amount.

12 Move the point assembly plate by inserting a screwdriver into the slot provided and turning the screwdriver. Closing the gap on the points will increase the dwell reading, while opening the gap will decrease the dwell.

13 Tighten the screws after the correct reading is obtained and recheck the setting before reinstalling the rotor and cap.

14 If a dwell meter is unavailable, a feeler gauge can be used to set the points.

15 Have a helper crank the engine in short bursts until the rubbing block of the points rests on a high point of the cam assembly. The rubbing block must be exactly on the apex of one of the cam lobes for correct point adjustment. It may be necessary to rotate the front pulley of the engine with a socket and breaker bar to position the cam lobe exactly.

16 Measure the gap between the contact points with the correct size feeler gauge. If the gap is incorrect, loosen the two mounting screws and move the point assembly until the correct gap is achieved.

17 Retighten the screws and recheck the gap one more time before installing the distributor cap and rotor.

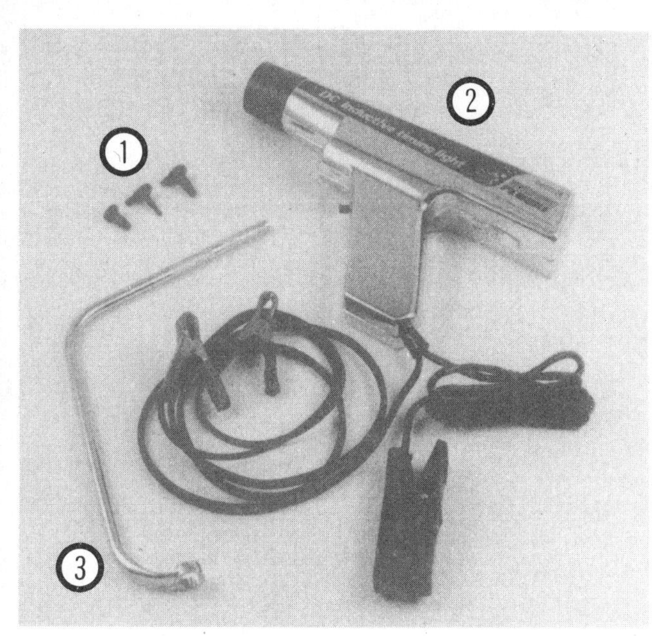

33.1 Tools needed to check and adjust the ignition timing

1 **Vacuum plugs** — *Vacuum hoses will, in most cases, have to be disconnected and plugged. Molded plugs in various shapes and sizes are available for this*
2 **Inductive pick-up timing light** — *Flashes a bright concentrated beam of light when the number one spark plug fires. Connect the leads according to the instructions supplied with the light*
3 **Distributor wrench** — *On some models, the hold-down bolt for the distributor is difficult to reach and turn with conventional wrenches or sockets. A special wrench like this must be used*

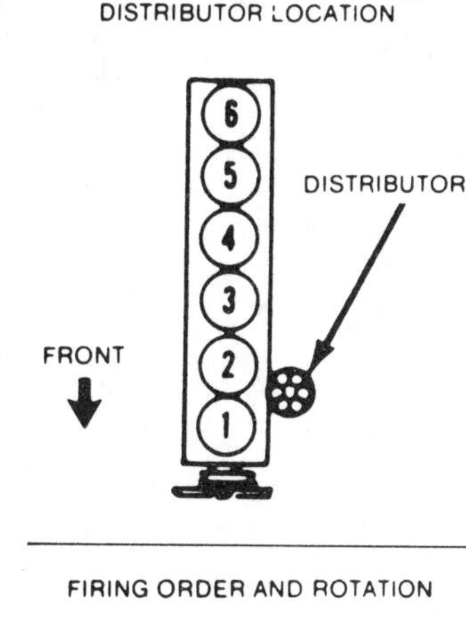

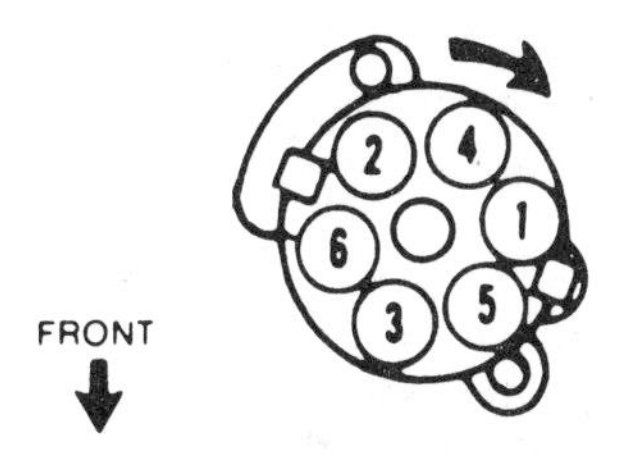

33.2a Cylinder numbers and firing order for six cylinder engines

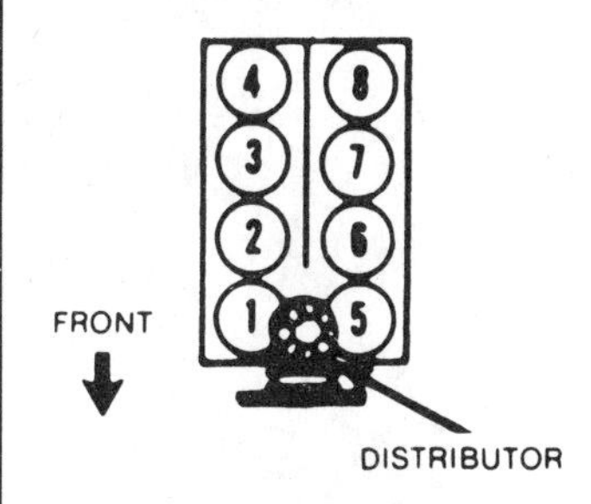

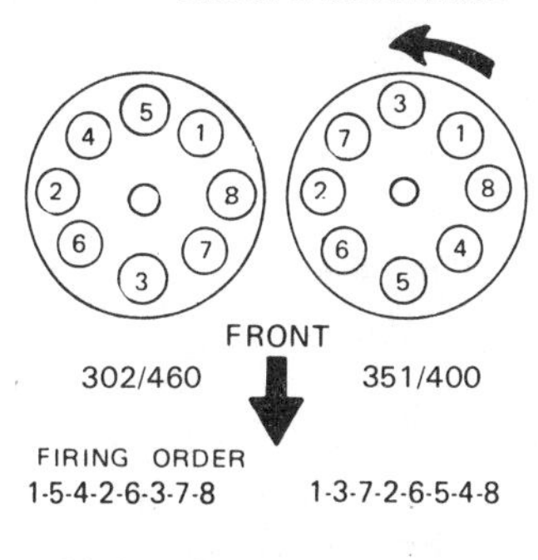

33.2b Cylinder numbers and firing order for V8 engines

33.5a Point the flashing timing light at the timing marks with the engine idling at the specified speed (be careful not to get the timing light leads or your hands tangled in the fan or drivebelts)

33 Ignition timing check and adjustment

Refer to illustrations 33.1, 33.2a, 33.2b, 33.5a, 33.5b, 33.5c and 33.7

Note: *Ignition timing on Duraspark III ignition systems is controlled by an Electronic Engine Control (EEC) and is not adjustable. The information which follows is applicable only to vehicles equipped with other types of ignition systems. The procedure for checking and adjusting ignition timing requires the use of a stroboscopic timing light, available at auto parts stores and rental yards. Timing an engine "by ear" and static timing procedures are not acceptable with today's tight emissions controls and should only be used to initially start and run an engine after it has been disassembled or the distributor has been removed.*

1 Several tools, including an inductive pick-up timing light, a wrench for the distributor hold-down bolt and vacuum line plugs, will be needed for this procedure (see illustration).

2 Start the engine and bring it to normal operating temperature, then shut it off and connect the timing light according to the manufacturer's instructions. The leads are normally connected to the battery terminals and the number one spark plug wire (see illustrations). At the same time, refer to the Emissions Control Information label (located inside the engine compartment) for the recommended ignition timing specifications. Make sure the wiring leads for the timing light are not contacting the exhaust manifold and that they are routed away from any moving parts such as the cooling fan.

3 Remove and plug any vacuum hoses connected to the distributor (if specified on the Emissions Control Information label). Use a golf tee, pencil or bolt to plug any disconnected vacuum lines if the rubber plugs mentioned above are not available. **Warning:** *Make sure that you remove dangling articles of clothing such as a tie or jewelry, which could be caught in the moving components of the engine.*

4 Start the engine and make sure that it is idling at the proper speed according to the Emissions Control Information label.

5 Aim the timing light at the timing marks (see illustration). If the

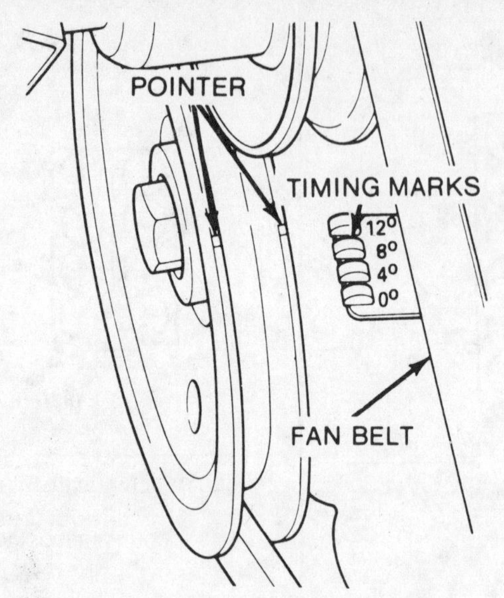

33.5b Six cylinder timing marks look like this – the pointer is a groove in the crankshaft pulley

timing marks do not show up clearly, stop the engine and clean the timing pointer and crankshaft pulley with solvent and a rag (see illustrations).

6 Compare the position of the timing marks to the timing specifications for your vehicle. If the specified mark lines up with the pointer, the ignition timing is correct and no adjustment is necessary. If the timing is incorrect, turn off the engine.

7 Loosen the distributor hold-down bolt with a wrench (a special distributor bolt tool may be required) (see illustration).

8 Restart the engine and, using the timing light, rotate the distributor very slowly until the marks are aligned.

9 Tighten the distributor hold-down bolt.

10 Restart the engine and check the timing to make sure it has not changed while tightening the distributor bolt.

11 Recheck the idle speed to make sure it has not changed significantly. If the idle speed has changed, reset the idle speed and recheck the timing, as timing will vary with engine rpm.

12 Reconnect all vacuum hoses.

34 Intake manifold bolt torque check

At the specified intervals the intake manifold bolts should be checked and retightened to the specified torque to prevent vacuum leaks which would upset the fuel/air mixture. The engine must be completely cool when this is done. A torque wrench, sockets and various extensions, as well as a universal joint and possibly some crowfoot wrenches, will be needed. Refer to Chapter 2 for illustrations of the sequence to follow when tightening the bolts and be sure to use the torque specifications listed there as well.

35 Idle speed and fuel/air mixture check and adjustment

1 Anti-pollution laws have dictated that strict tune-up rules be applied to light duty vehicles, including vans. Idle speed and idle mixtures are covered under these regulations and strict adherence to the correct method of adjustment is required. Due to the wide range of vehicles, models and power train combinations covered in this book, it is impractical to describe every method of idle speed and mixture adjustment. Additionally, certain areas require that these adjustments be made only by qualified technicians using specialized exhaust sampling equipment.

2 If you feel that you are qualified to make these adjustments, please refer to Chapter 4 for the correct procedure for your vehicle. The specifications for your particular vehicle should be clearly displayed on a decal or metal plate inside the engine compartment. The operation

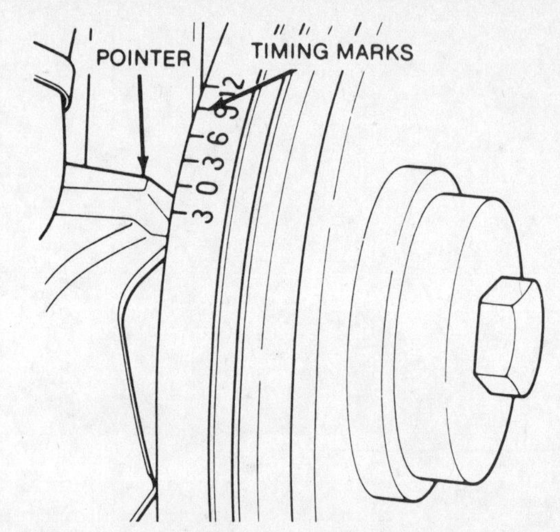

33.5c If you have a V8 engine, the timing marks and pointer will typically look like this (some V8 marks are only visible from the underside of the vehicle)

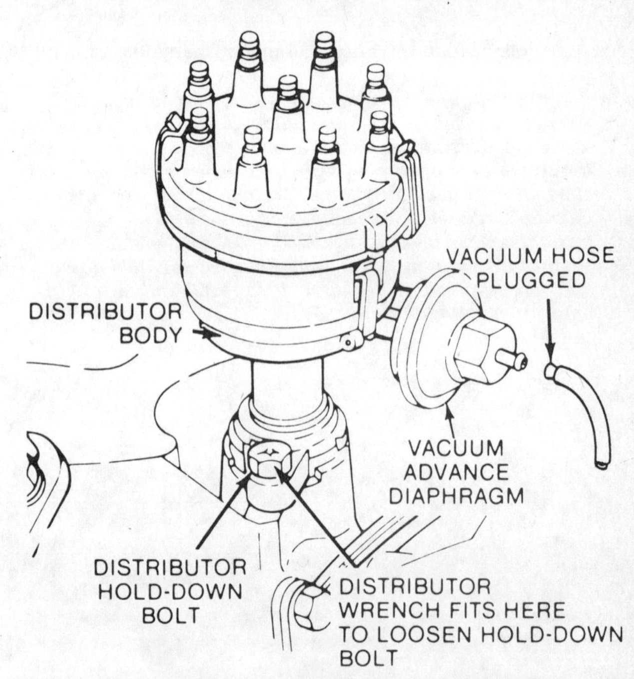

33.7 The hold-down bolt is located at the base of the distributor

of various devices such as throttle positioners, dash pots, solenoids and the connecting and disconnecting of vacuum lines to them, are all variables which must be controlled while making idle speed adjustments. Do not, under any circumstances, attempt to make any adjustments on carburetors or attempt to disable the limiter devices on the idle mixture screws in an effort to modify your vehicle's idle or performance characteristics. Following the factory designated procedures is the best way to guarantee a satisfactory performing engine and comply with federal and state pollution control laws.

36 Differential oil change

Note: *Carefully read through this Section before undertaking this procedure. You will need to purchase the correct type and amount of differential lubricant before draining the old oil out of the vehicle. In some cases you will also need a differential cover gasket and an additive.*

1 The vehicle should be driven for several minutes before draining

Common spark plug conditions

NORMAL

Symptoms: Brown to grayish-tan color and slight electrode wear. Correct heat range for engine and operating conditions.
Recommendation: When new spark plugs are installed, replace with plugs of the same heat range.

WORN

Symptoms: Rounded electrodes with a small amount of deposits on the firing end. Normal color. Causes hard starting in damp or cold weather and poor fuel economy.
Recommendation: Plugs have been left in the engine too long. Replace with new plugs of the same heat range. Follow the recommended maintenance schedule.

CARBON DEPOSITS

Symptoms: Dry sooty deposits indicate a rich mixture or weak ignition. Causes misfiring, hard starting and hesitation.
Recommendation: Make sure the plug has the correct heat range. Check for a clogged air filter or problem in the fuel system or engine management system. Also check for ignition system problems.

ASH DEPOSITS

Symptoms: Light brown deposits encrusted on the side or center electrodes or both. Derived from oil and/or fuel additives. Excessive amounts may mask the spark, causing misfiring and hesitation during acceleration.
Recommendation: If excessive deposits accumulate over a short time or low mileage, install new valve guide seals to prevent seepage of oil into the combustion chambers. Also try changing gasoline brands.

OIL DEPOSITS

Symptoms: Oily coating caused by poor oil control. Oil is leaking past worn valve guides or piston rings into the combustion chamber. Causes hard starting, misfiring and hesitation.
Recommendation: Correct the mechanical condition with necessary repairs and install new plugs.

GAP BRIDGING

Symptoms: Combustion deposits lodge between the electrodes. Heavy deposits accumulate and bridge the electrode gap. The plug ceases to fire, resulting in a dead cylinder.
Recommendation: Locate the faulty plug and remove the deposits from between the electrodes.

TOO HOT

Symptoms: Blistered, white insulator, eroded electrode and absence of deposits. Results in shortened plug life.
Recommendation: Check for the correct plug heat range, over-advanced ignition timing, lean fuel mixture, intake manifold vacuum leaks, sticking valves and insufficient engine cooling.

PREIGNITION

Symptoms: Melted electrodes. Insulators are white, but may be dirty due to misfiring or flying debris in the combustion chamber. Can lead to engine damage.
Recommendation: Check for the correct plug heat range, over-advanced ignition timing, lean fuel mixture, insufficient engine cooling and lack of lubrication.

HIGH SPEED GLAZING

Symptoms: Insulator has yellowish, glazed appearance. Indicates that combustion chamber temperatures have risen suddenly during hard acceleration. Normal deposits melt to form a conductive coating. Causes misfiring at high speeds.
Recommendation: Install new plugs. Consider using a colder plug if driving habits warrant.

DETONATION

Symptoms: Insulators may be cracked or chipped. Improper gap setting techniques can also result in a fractured insulator tip. Can lead to piston damage.
Recommendation: Make sure the fuel anti-knock values meet engine requirements. Use care when setting the gaps on new plugs. Avoid lugging the engine.

MECHANICAL DAMAGE

Symptoms: May be caused by a foreign object in the combustion chamber or the piston striking an incorrect reach (too long) plug. Causes a dead cylinder and could result in piston damage.
Recommendation: Repair the mechanical damage. Remove the foreign object from the engine and/or install the correct reach plug.

the differential oil. This will warm the oil and ensure complete drainage.
2 Move a drain pan, rags, newspapers and tools under the vehicle. With the drain pan under the differential, remove the drain plug from the bottom of the housing. On Dana differentials no drain plug is provided, which means that the differential cover must be removed to drain the fluid. Remove the inspection/fill plug to vent the differential.
3 After the oil has completely drained, wipe the area around the drain hole with a clean rag and install the drain plug. Reinstall the differential cover with a new gasket on Dana differentials.
4 Fill the housing (through the inspection hole) with the recommended lubricant until the level is even with the bottom of the hole. Check the tag on the driver's door post or the tag attached to the differential to determine if your vehicle is equipped with a locking differential. These differentials require an additional additive to supplement the normal differential lubricant (see *Recommended lubricants* at the beginning of this Chapter). Add the prescribed amount of the additive at this time. Install the inspection plug after cleaning it and the threads in the case or cover.
5 After driving the vehicle check for leaks at the drain and inspection plugs.
6 When the job is complete check for metal particles or chips in the drain oil, which indicate that the differential should be thoroughly inspected and repaired (see Chapter 8 for more information).

37 Manual transmission oil change

1 At the specified time intervals the transmission lubricant should be changed to ensure trouble free transmission operation. Before proceeding, purchase the specified transmission oil.
2 Tools necessary for this job include jackstands to support the vehicle in a raised position, a wrench to remove the drain plug, a drain pan capable of holding at least four quarts, newspapers and clean rags.
3 The oil should be drained immediately after the vehicle has been driven. This will remove any contaminants better than if the oil were cold. Because of this, it may be wise to wear protective gloves while removing the drain plug.
4 After the vehicle has been driven to warm up the oil, raise it and place it on jackstands for access underneath. Make sure it is firmly supported and as level as possible.
5 Move the necessary equipment under the vehicle, being careful not to touch any of the hot exhaust components.
6 Place the drain pan under the transmission and remove the drain plug. Be careful not to burn yourself on the oil.
7 Allow the oil to drain completely, then reinstall the plug and tighten it securely.
8 Clean the area around the filler plug (located on the side of the transmission), then remove the plug.
9 Using a syringe or squeeze bottle, fill the transmission to the bottom of the filler hole with the correct grade of oil.
10 Reinstall the filler plug and tighten it securely.
11 Lower the vehicle and test drive it, then check for leaks around the plugs.
12 Pour the old oil into capped jugs (old antifreeze containers, milk cartons, etc.) and dispose of it at a service station or reclamation center.

38 Automatic transmission fluid change

Refer to illustration 38.10

1 At the specified intervals the transmission fluid should be changed and the filter replaced. Since there is no drain plug, the transmission oil pan must be removed from the bottom of the transmission to drain the fluid. On some models of the C4 transmission the fluid can be drained by disconnecting the filler tube from the transmission pan. We recommend, however, that the pan also be removed to make thorough cleaning possible.
2 Before beginning work, purchase the specified transmission fluid (see *Recommended lubricants and fluids* at the front of this Chapter) and a new filter. The necessary gaskets should be included with the filter. If not, purchase an oil pan gasket and a strainer-to-valve body gasket.
3 Other items necessary for this job include jackstands to support the vehicle in a raised position, a wrench to remove the oil pan bolts, a standard screwdriver, a drain pan capable of holding at least five quarts, newspapers and clean rags.

38.10 The automatic transmission filter is held in place with several screws (arrows)

4 The fluid should be drained immediately after the vehicle has been driven. This will remove any built-up sediment better than if the fluid were cold. Because of this, it may be wise to wear protective gloves. Fluid temperature can exceed 350 °F in a hot transmission.
5 After the vehicle has been driven to warm up the fluid, raise it and place it on jackstands for access underneath.
6 Move the necessary equipment under the vehicle, being careful not to touch any of the hot exhaust components.
7 Place the drain pan under the transmission oil pan and loosen, but do not remove, the bolts at one end of the pan.
8 Moving around the pan, loosen all the bolts a little at a time. Be sure the drain pan is in position, as fluid will begin dripping out. Continue in this manner until all of the bolts are removed, except for one at each of the corners.
9 Remove the two bolts at the back of the pan and allow the pan to tilt down, emptying the fluid into the drain pan. If the pan sticks to the transmission body, use a screwdriver to carefully pry it free. Be careful not to scratch the gasket mating surface. While supporting the pan, remove the remaining bolts and detach the pan. Pour the remaining fluid into the drain pan. As this is done, check the fluid for metal particles, which may be an indication of transmission damage.
10 Now visible on the bottom of the transmission is the filter/strainer, held in place by several screws (see illustration).
11 Remove the screws, filter and gasket.
12 Thoroughly clean the transmission oil pan with solvent. Check for metal particles and foreign material. Dry it with compressed air, if available, or a lint-free cloth. It is important that all remaining gasket material be removed from the oil pan mounting flange. Use a gasket scraper or putty knife for this.
13 Clean the filter mounting surface on the valve body. Again, this surface should be smooth and free of any old gasket material.
14 Clean the screen assembly with solvent, then dry it thoroughly (use compressed air if available). Paper or felt-type filters should be replaced with a new one.
15 Place the new or cleaned filter in position, with a new gasket between it and the transmission valve body. Install the mounting screws and tighten them securely.
16 Apply a narrow bead of gasket sealant around the pan mounting surface, with the sealant to the inside of the bolt holes. Press the new gasket into place on the pan, making sure all the holes line up with the holes in the pan.
17 Position the pan against the bottom of the transmission and install the mounting bolts. Tighten the bolts in a criss-cross pattern. Using a torque wrench, tighten the bolts to the specified torque.
18 Lower the vehicle.
19 Open the hood and remove the transmission fluid dipstick.
20 It is best to add a little fluid at a time, continually checking the level with the dipstick. Allow the fluid time to drain into the pan. Add

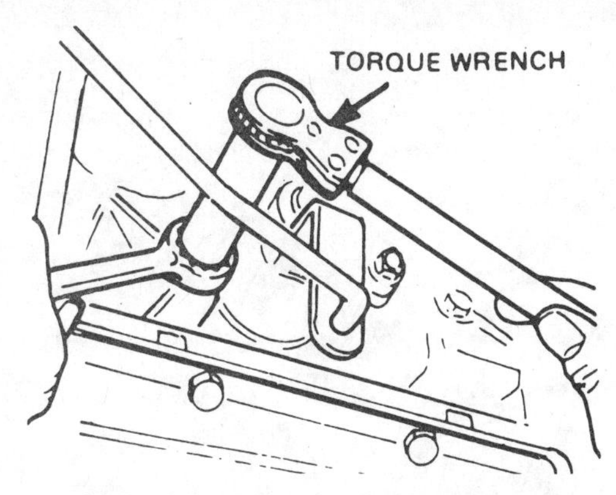

39.2 Adjusting the automatic transmission intermediate band

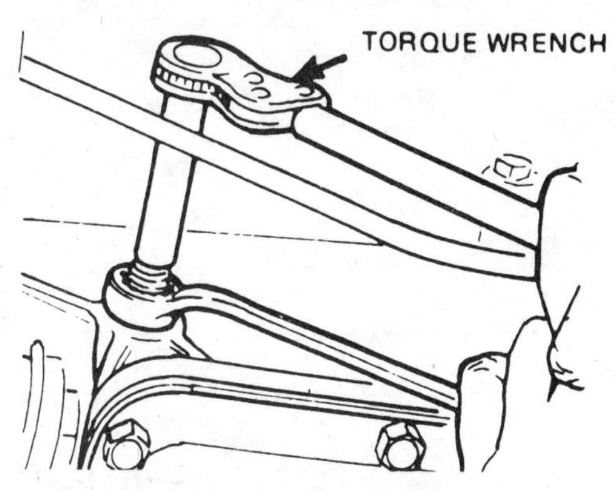

39.4 Adjusting the automatic transmission Low/Reverse band

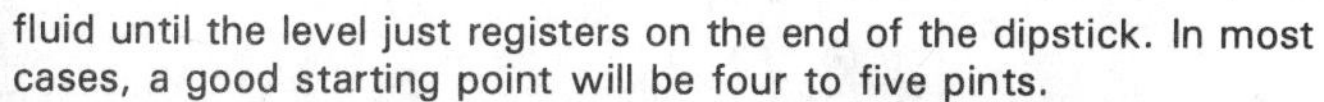

fluid until the level just registers on the end of the dipstick. In most cases, a good starting point will be four to five pints.

21 With the selector lever in Park, apply the parking brake and start the engine without depressing the accelerator pedal (if possible). Do not race the engine at high speed; run at slow idle only.

22 Depress the brake pedal and shift the transmission through each gear. Place the selector back in Park and check the level on the dipstick with the engine still idling. Look under the vehicle for leaks around the transmission oil pan. Add fluid as described in Section 5.

23 Push the dipstick firmly back into its tube and drive the vehicle to reach normal operating temperature. Park the vehicle on a level surface and check the fluid level on the dipstick with the engine idling and the transmission in Park. The level should now be at the Full mark on the dipstick. If not, add more fluid to bring the level up to this point. Do not overfill.

39 Automatic transmission band adjustment

Refer to illustrations 39.2 and 39.4

Intermediate band (C4, C5, C6)

1 The intermediate or front band is used to hold the sun gear stationary to produce second gear. If it is not correctly adjusted there will be noticeable slip during the first-to-second gear shift or on the downshift from high to second gear. The first symptoms of these problems will be very sluggish shifts instead of the usual crisp action.

2 To adjust the intermediate band, loosen, remove and discard the locknut on the band adjustment screw, located on the left side of the case. Tighten the adjusting screw to 10 ft-lbs, then loosen it exactly 1-3/4 turns (C4), 4-1/4 turns (C5) or 1-1/2 turns (C6). Install a new locknut and tighten it to 35 to 45 ft-lbs while holding the adjustment screw to keep it from turning (see illustration).

Low/Reverse band

3 The low and reverse band is operational when the selector lever is placed in the Low or Reverse positions. If it is not correctly adjusted there will be no drive with the selector lever in Reverse.

4 To adjust this band, remove the adjusting screw locknut from the screw (located on the right side of the case) and discard it. Tighten the adjusting screw to 10 ft-lbs, then loosen it exactly three turns. Install a new locknut and tighten it to 35 to 45 ft-lbs while holding the adjusting screw to keep it from turning (see illustration).

40 Wheel bearing check and repack

Refer to illustrations 40.1, 40.2, 40.6, 40.7, 40.8a, 40.8b, 40.9, 40.11 and 40.15

1 In most cases the front wheel bearings will not need servicing until the brake pads are changed. However, the bearings should be checked whenever the front wheels are raised for any reason. Several items, including a torque wrench and special grease, are required for this procedure (see illustration).

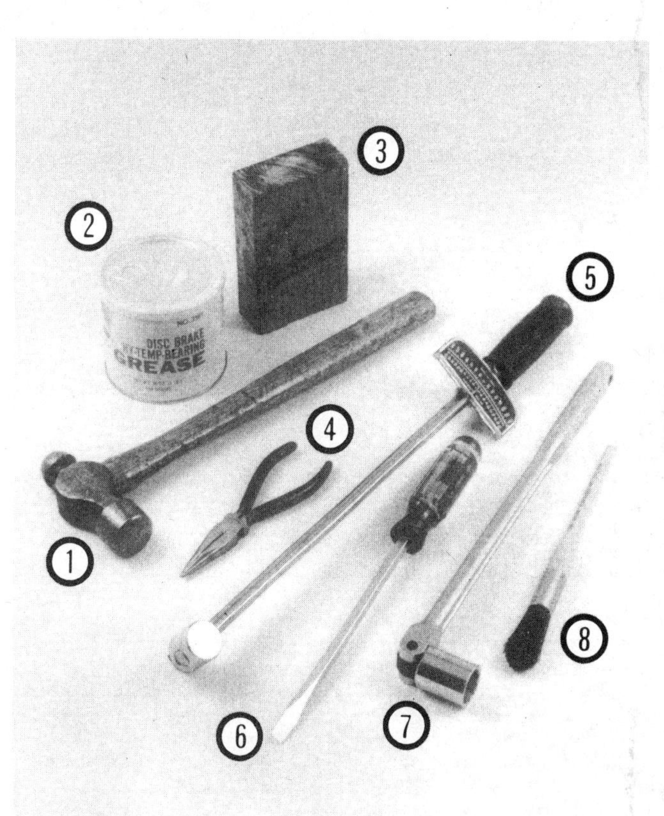

40.1 Tools and materials needed for front wheel bearing maintenance

1 **Hammer** — *A common hammer will do just fine*
2 **Grease** — *High-temperature grease which is formulated specially for front wheel bearings should be used*
3 **Wood block** — *If you have a scrap piece of 2x4, it can be used to drive the new seal into the hub*
4 **Needle nose pliers** — *Used to straighten and remove the cotter pin in the spindle*
5 **Torque wrench** — *This is very important in this procedure; if the bearing is too tight, the wheel won't turn freely — if it is too loose, the wheel will 'wobble' on the spindle. Either way, it could mean extensive damage*
6 **Screwdriver** — *Used to remove the seal from the hub (a long screwdriver would be preferred)*
7 **Socket/breaker bar** — *Needed to loosen the nut on the spindle if it is extremely tight*
8 **Brush** — *Together with some clean solvent, this will be used to remove old grease from the hub and spindle*

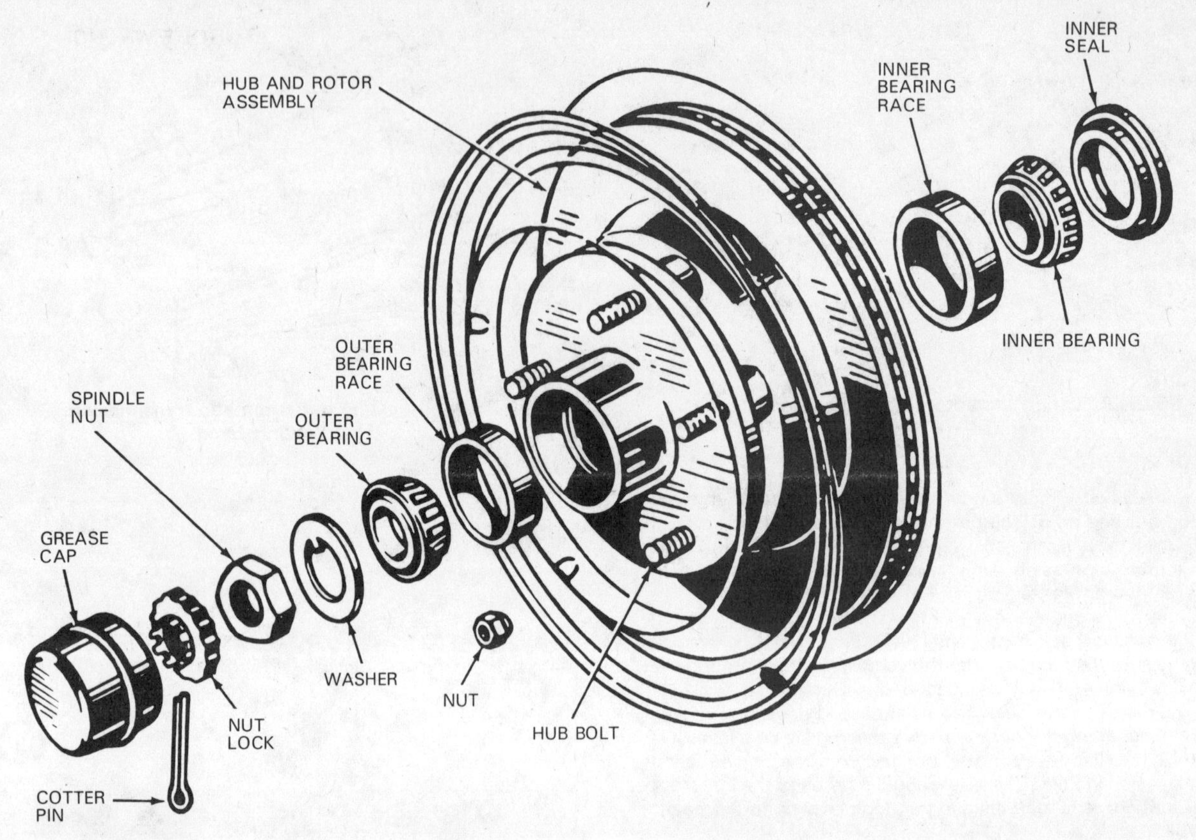

40.2 Exploded view of the front wheel bearing assembly (make sure everything goes back together in this order)

40.6 Removing the hub grease cap with a water pump pliers (it can be pried out with a screwdriver if necessary)

40.7 Remove the cotter pin, . . .

2 With the vehicle securely supported on jackstands, spin the wheel and check for noise, rolling resistance and free play. Now grasp the top of the tire with one hand and the bottom of the tire with the other. Move the tire in and out on the spindle. If it moves more than 0.005-inch, the bearings should be checked, then repacked with grease or replaced if necessary (see illustration).

3 To remove the bearings for replacement or repacking, begin by removing the hub cap and wheel.

4 Remove the brake caliper as described in Chapter 9.

5 Use wire to hang the caliper out of the way. Be careful not to kink or damage the brake hose.

6 Pry the grease cap out of the hub with a screwdriver or large pliers. This cap is located at the center of the hub (see illustration).

7 Use needle nose pliers to straighten the bent ends of the cotter pin and then pull the cotter pin out of the locking nut (see illustration). Discard the cotter pin, as a new one should be used during reassembly.

8 Remove the nut lock, spindle nut and washer from the end of the spindle (see illustrations)

9 Pull the hub assembly out slightly and then push it back into its original position. This should force the outer bearing off the spindle enough so that it can be removed with your fingers (see illustration). Remove the outer bearing, noting how it is installed on the end of the spindle.

10 Pull the hub assembly off the spindle.

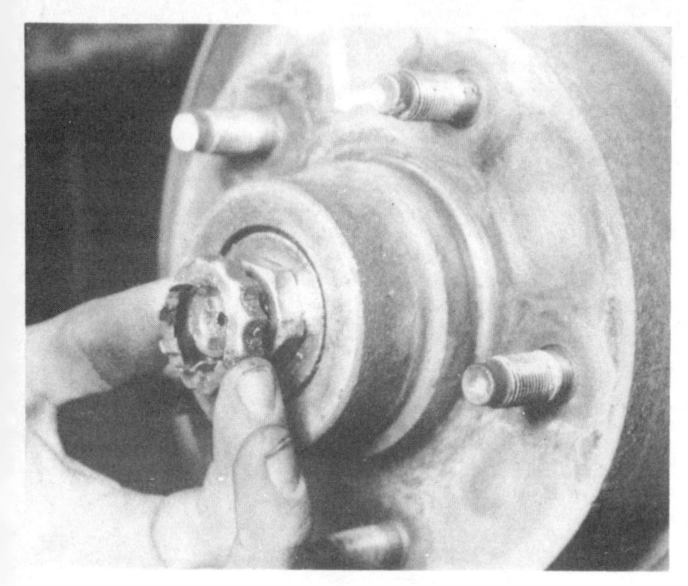

40.8a . . . the nut lock . . .

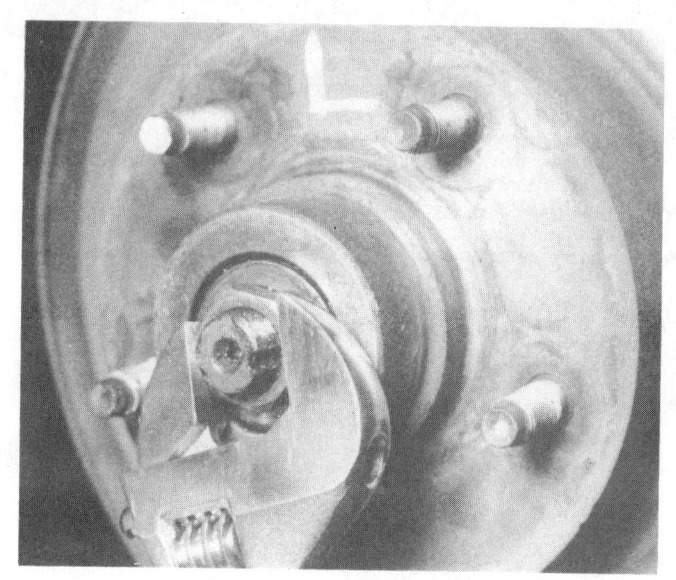

40.8b . . . and the spindle nut, then withdraw the hub to dislodge the outer bearing and washer

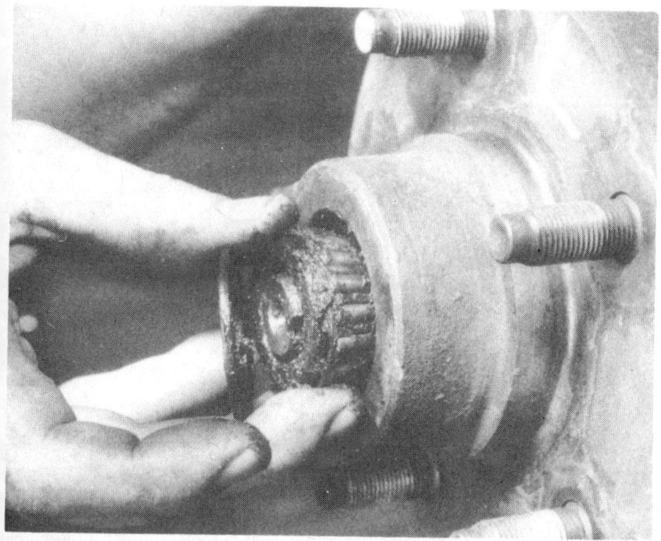

40.9 Push the hub back onto the spindle and remove the washer and bearing

40.11 Use a hammer and screwdriver to remove the seal from the rear of the hub

11 On the rear side of the hub, use a hammer and screwdriver to tap or pry out the inner seal (see illustration). As this is done, note the direction in which the seal is installed.

12 The inner bearing can now be removed from the hub, again noting how it is installed.

13 Use clean solvent to remove all traces of the old grease from the bearing, hub and spindle. A small brush may prove useful. Make sure no bristles from the brush embed themselves inside the bearing rollers. Allow the parts to air dry.

14 Carefully inspect the bearings for cracks, heat discoloration, scored rollers, etc. Check the bearing races inside the hub for cracks, scoring and uneven surfaces. If the bearing races are in need of replacement, take the hub to a repair shop that can press the new races into place.

15 Use an approved high temperature front wheel bearing grease to pack the bearings. Work the grease into the bearings, forcing it between the rollers, cone and cage (see illustration).

16 Apply a thin coat of grease to the spindle at the outer bearing seat, inner bearing seat, shoulder and seal seat.

17 Put a small quantity of grease inboard of each of the bearing races inside the hub. Using your fingers, form a dam at these points to provide extra grease availability and to keep thinned grease from flowing out of the bearing.

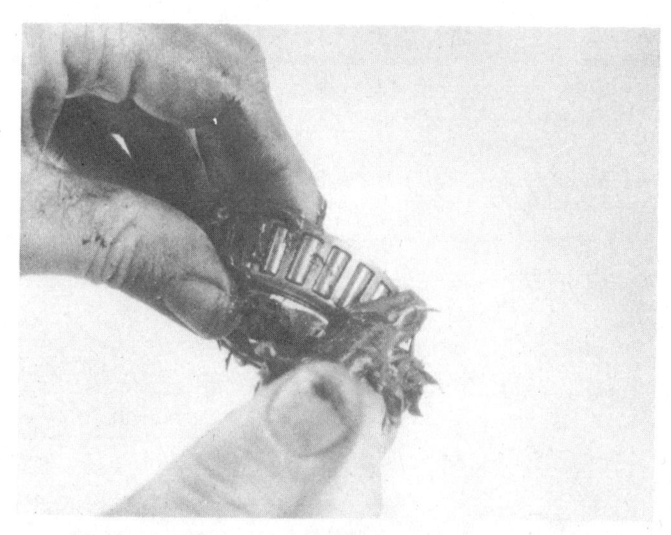

40.15 Pack the grease into each bearing until it is full

18 Place the grease packed inner bearing into the rear of the hub and put a little more grease outboard of the bearing.

19 Place a new seal over the inner bearing and tap the seal with a flat piece of hardwood and a hammer until it is flush with the hub.

20 Carefully place the hub assembly onto the spindle and push the grease packed outer bearing into position.

21 Install the washer and spindle nut. Tighten the nut only slightly (22 to 25 ft-lbs of torque).

22 Spin the hub in a forward direction to seat the bearings and remove any grease which would cause excessive bearing play later.

23 Put a little grease outboard of the outer bearing to provide extra grease availability.

24 Check that the spindle nut is still tight (22 to 25 ft-lbs).

25 Loosen the spindle nut exactly 1/8-turn.

26 Using your hand only (not a wrench) tighten the nut until it is snug. Install a new cotter pin through the hole in the spindle and the spindle nut. If the nuts slits do not line up with the hole in the spindle, loosen the nut slightly until they do. From the hand tight position the nut should not be loosened any more than one-half flat to install the cotter pin.

27 Bend the ends of the new cotter pin until they are flat against the nut. Cut off any extra length which could interfere with the dust cap.

28 Install the dust cap, tapping it into place with a rubber mallet.

29 Reinstall the brake caliper as described in Chapter 9.

30 Install the tire/wheel assembly and tighten the lug nuts.

31 Grab the top and bottom of the tire and check the bearings in the same manner as described at the beginning of this Section.

32 Lower the vehicle to the ground and completely tighten the lug nuts. Install the hub cap, using a rubber mallet to seat it.

Chapter 2 Part A Six cylinder inline engines

Refer to Chapter 13 for Specifications and information on 1987 and later models

2A

Contents

Specifications

Note: *Additional specifications can be found in Chapter 2, Part C*

Camshaft and drive gears

Camshaft end play	0.009 in maximum
Camshaft gear-to-crankshaft gear backlash	0.002 to 0.004 in
Camshaft gear runout limit (assembled)	
thru 1976	0.006 in TIR
1977 and later	0.005 in TIR
Crankshaft gear runout limit (assembled)	
thru 1976	0.003 in TIR
1977 and later	0.005 in TIR

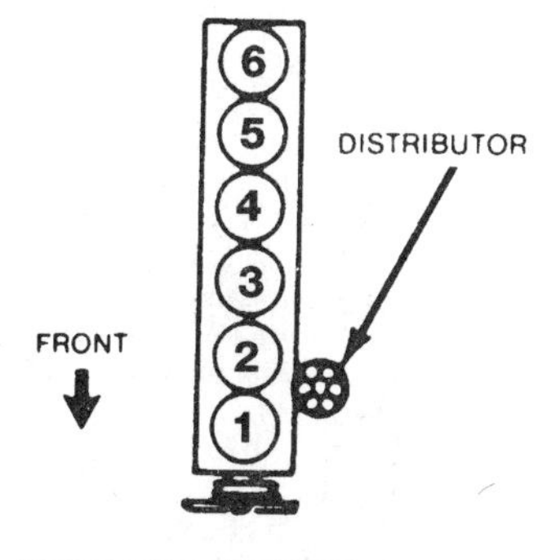

Oil pump

Outer race-to-housing clearance	0.001 to 0.013 in
Rotor assembly end clearance	0.004 in maximum
Driveshaft-to-housing bearing clearance	0.0015 to 0.0030 in
Relief spring tension	20.6 to 22.6 lbs. @ 2.49 in
Relief valve clearance	0.0015 to 0.0030 in

Torque specifications	**Ft-lbs**
Camshaft thrust plate-to-block bolts	9 to 12
Timing gear cover bolts	12 to 18
Cylinder head bolts	
thru 1973	70 to 75
1974 thru 1979	70 to 85
1980 and later	
step 1	55
step 2	65
step 3	85
Damper-to-crankshaft bolt	130 to 150
Fuel pump-to-block bolts	12 to 18
Flywheel/driveplate-to-crankshaft bolts	75 to 85
Exhaust manifold-to-cylinder head bolts	
thru 1976	23 to 28
1977 thru 1981	28 to 33
1982 and later	22 to 32
Intake-to-exhaust manifold bolts	28 to 32
Intake manifold-to-cylinder head bolts	
thru 1976	23 to 28
1977 and later	22 to 32

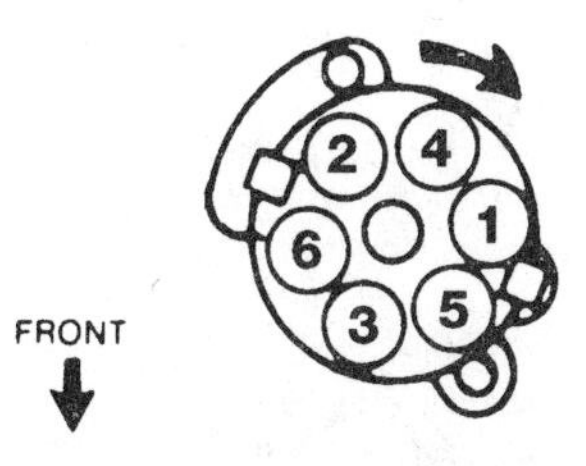

Cylinder numbers and firing order for six cylinder engines

Torque specifications (continued)	Ft-lbs
Oil filter insert-to-block/adapter bolts	
thru 1984	20 to 30
1985 and later	15 to 35
Oil filter adapter-to-block bolts	40 to 50
Oil pan-to-block bolts	10 to 12
Oil pickup tube-to-pump bolts	10 to 15
Oil pump-to-block bolts	10 to 15
Oil pump cover plate bolts	
thru 1973	9 to 15
1974 and later	6 to 14
Pulley-to-damper bolts	35 to 50
Rocker arm cover bolts	4 to 7
Pushrod cover bolts	15 to 20 in-lbs
Alternator bracket-to-block bolt	30 to 45
Alternator adjusting arm-to-block bolt	19 to 27
Alternator adjusting arm-to-alternator bolt	24 to 40
Thermactor pump bracket-to-block bolt	22 to 32
Thermactor pump pivot bolt	30 to 35
Thermactor pump adjusting arm-to-pump bolt	22 to 32
Thermactor pump pulley-to-pump hub bolt	12 to 18
Rocker arm stud nut (1979 on)	17 to 23
Rocker arm stud nut breakaway torque	4.5 to 15

1 General information

This Part of Chapter 2 is devoted to repair procedures for inline six cylinder engines, as well as the removal and installation steps. All information concerning engine block and cylinder head servicing can be found in Part C of this Chapter.

Many of the repair procedures included in this Part are based on the assumption that the engine is still installed in the vehicle. Therefore, if this information is being used during a complete engine overhaul, with the engine already out of the vehicle and on a stand, many of the steps included here will not apply.

The specifications included in this Part of Chapter 2 apply only to the engines and procedures found here. For specifications related to V8 engines, refer to Part B. The specifications necessary for rebuilding the block and cylinder head are included in Part C.

Three versions of the Ford inline six cylinder engine were installed in vans over the years covered by this manual. The 240 cubic inch engine was offered through production year 1975. The 300 cubic inch light duty and heavy duty engines were available in all vehicles (heavy duty engines were not available in California).

All six cylinder engines have a block made of special high grade cast iron. The crankshaft is supported by seven main bearings. The 240 and 300 cubic inch light duty engines have a nodular iron crankshaft, while the heavy duty engine has a forged alloy steel crankshaft. Pistons are made of an aluminum alloy with integral, cast in steel struts. The rocker arms are ball pivot stud mounted with positive stop studs used from 1978 on. Lifters are hydraulic and self adjusting. Timing gears are helical cut. The heavy duty 300 cubic inch engines have exhaust valve rotators, heavy duty pistons, special valves and a larger radiator.

2 Valve spring, retainer and seal – replacement (in vehicle)

Refer to illustration 2.10

Note: *Broken valve springs and retainers or defective valve stem seals can be replaced without removing the cylinder head on engines that have no damage to the valves or valve seats. Two special tools and a compressed air source are required to perform this operation, so read through this Section carefully and rent or buy the tools before beginning the job.*

1 Remove the engine cover and air cleaner.
2 Remove the accelerator cable return spring.
3 Remove the accelerator cable linkage at the carburetor.
4 Disconnect the choke cable at the carburetor.
5 Detach the PCV valve from the rocker arm cover and remove the rocker arm cover.
6 Remove the spark plug from the cylinder which has the bad component.
7 Crank the engine until the piston in the cylinder with the bad component is at top dead center on the compression stroke (refer to Section 23 in Chapter 2C). Note that the distributor should be marked opposite the appropriate terminal – not necessarily number one.
8 Install a special air line adapter which will screw into the spark plug hole and connect to a compressed air source. Remove the rocker arm stud nut, fulcrum seat, rocker arm and pushrod.
9 Apply compressed air to the cylinder. If the engine rotates until the piston is at bottom dead center, be very careful that you do not drop the valve into the cylinder, as it will fall all the way in. Do not release the air pressure or the valve will drop through the guide.
10 Compress the valve spring with the special tool designed for this purpose (see illustration).
11 Remove the keepers, spring retainer and valve spring, then remove the valve stem seal. **Note:** *If air pressure fails to hold the valve in the closed position during this operation, there is apparently damage to the seat or valve. If this condition exists, remove the cylinder head for further repair operations.*
12 If air pressure has forced the piston to the bottom of the cylinder,

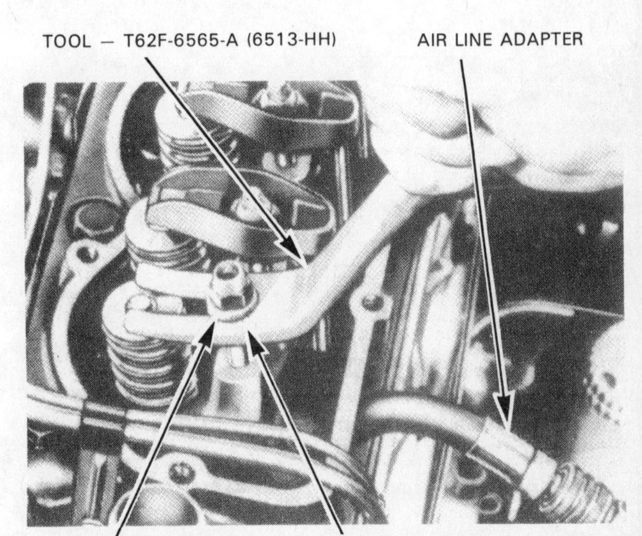

2.10 The valve spring can be compressed with a special tool to remove the keepers when servicing the seals with the cylinder head installed (note the air line adapter that allows the cylinder to be pressurized from an outside source)

wrap a rubber band, tape or string around the top of the valve stem so that the valve will not fall through into the cylinder if it is dropped. Release the air pressure.

13 Inspect the valve stem for damage. Rotate the valve in its guide and check the valve stem tip for eccentric movement, which would indicate a bent valve.

14 Move the valve up and down through its normal travel and check that the valve guide and stem do not bind. If the valve stem binds, either the valve is bent or the guide is damaged and the head will have to be removed for repair.

15 Reapply air pressure to the cylinder to retain the valve in the closed position.

16 Lubricate the valve stem with engine oil and install a new valve stem seat.

17 Install the spring in position over the valve. Make sure that the closed coil end of the spring is positioned next to the cylinder head.

18 Install the valve spring retainer. Compress the valve spring, using the valve spring compressor, and install the valve spring keepers. Remove the valve spring compressor and make sure that the valve spring keepers are seated.

19 Apply engine oil to both ends of the pushrod.

20 Install the pushrod.

21 Apply engine oil to the tip of the valve stem.

22 Apply engine oil to the fulcrum seats and socket.

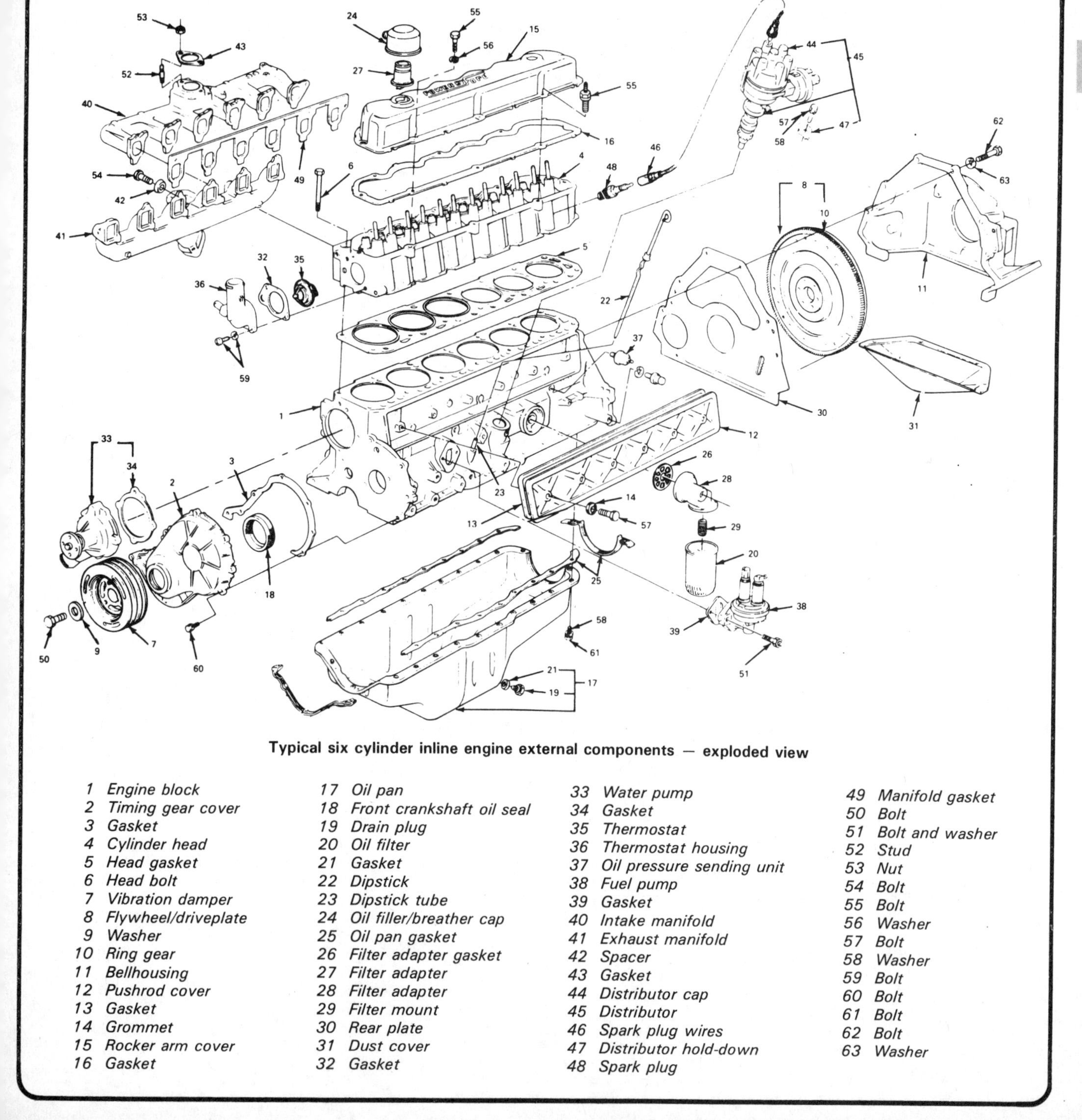

Typical six cylinder inline engine external components — exploded view

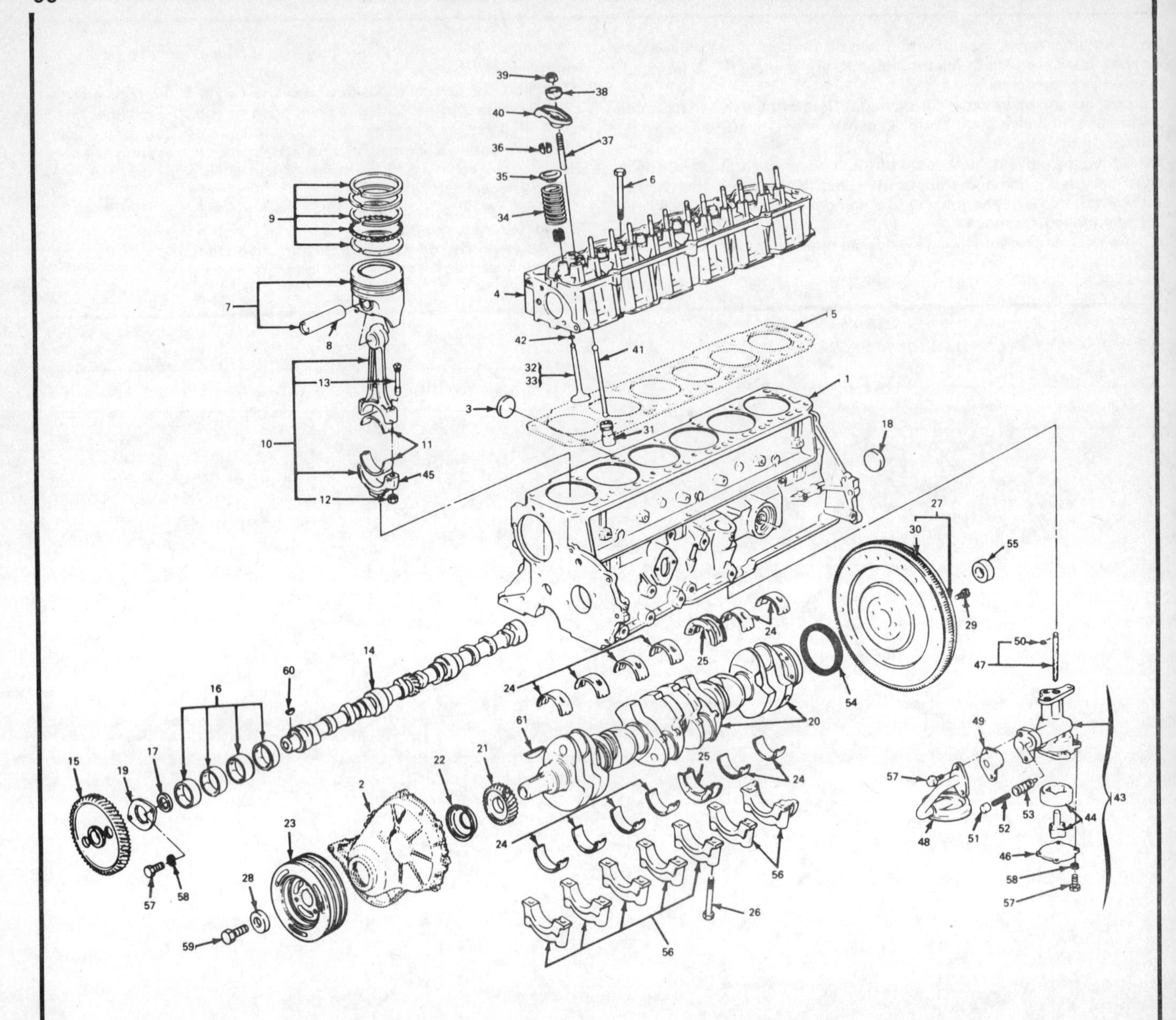

Typical six cylinder inline engine internal components — exploded view

1 *Engine block*
2 *Timing gear cover*
3 *Soft plugs*
4 *Cylinder head*
5 *Head gasket*
6 *Head bolt*
7 *Piston assembly*
8 *Piston pin*
9 *Piston ring set*
10 *Connecting rod*
11 *Bearing inserts*
12 *Nut*
13 *Bolt*
14 *Camshaft*
15 *Gear*
16 *Cam bearing*
17 *Spacer*
18 *Plug*
19 *Thrust plate*
20 *Crankshaft*
21 *Gear*
22 *Oil slinger*
23 *Vibration damper*
24 *Main bearing inserts*
25 *Thrust bearing*
26 *Main bearing cap bolt*
27 *Flywheel/driveplate*
28 *Washer*
29 *Bolt*
30 *Ring gear*
31 *Lifter*
32 *Exhaust valve*
33 *Intake valve*
34 *Valve spring*
35 *Retainer*
36 *Keepers*
37 *Rocker arm stud*
38 *Fulcrum seat*
39 *Stud nut*
40 *Rocker arm*
41 *Pushrod*
42 *Valve stem seal*
43 *Oil pump*
44 *Oil pump rotor and driveshaft*
45 *Connecting rod cap*
46 *Oil pump cover plate*
47 *Oil pump intermediate shaft*
48 *Oil pickup tube*
49 *Gasket*
50 *Snap-ring*
51 *Plug*
52 *Relief valve spring*
53 *Plunger*
54 *Rear crankshaft oil seal*
55 *Clutch pilot bearing*
56 *Main bearing cap*
57 *Bolt*
58 *Washer*
59 *Bolt*
60 *Woodruff key*
61 *Woodruff key*

23 Install the rocker arm, fulcrum seat and stud nut. Adjust the valve clearance following the procedure in Chapter 2, Part C. No adjustment is necessary on 1979 through 1986 models; tighten the rocker arm stud nut to the specified torque.
24 Remove the air source and the adapter from the spark plug hole.
25 Install the spark plug and connect the spark plug wire.
26 Install a new rocker arm cover gasket and attach the rocker arm cover to the engine.
27 Connect the accelerator cable to the carburetor.
28 Install the accelerator cable return spring. Connect the choke cable to the carburetor.
29 Install the PCV valve in the rocker arm cover and make sure that the line connected to the PCV valve is positioned correctly on both ends.
30 Install the air cleaner.
31 Start and run the engine, making sure that there are no oil leaks and that there are no unusual sounds coming from the valve assembly.

3 Manifolds – removal and installation

Refer to illustration 3.14

Removal

1 Remove the engine cover, air cleaner and related air cleaner components.
2 Disconnect the choke cable at the carburetor.
3 Disconnect the accelerator cable or linkage at the carburetor. Remove the accelerator return spring.
4 Remove the kickdown rod return spring and the kickdown shaft from its linkage at the carburetor.
5 Disconnect the fuel inlet line from the carburetor.
6 Disconnect the vacuum advance line for the distributor from the carburetor.
7 Disconnect the exhaust pipe from the exhaust manifold and support it out of the way.
8 Disconnect the power brake booster vacuum line from the intake manifold.
9 Remove the ten bolts and three nuts retaining the manifolds to the cylinder head. Separate the manifold assemblies from the engine.
10 Remove and scrape the gaskets from the mating surface of the manifolds. If the manifolds are to be replaced or changed, remove the nuts connecting the intake manifold to the exhaust manifold.

Installation

11 Install new studs in the exhaust manifold for the exhaust pipe.
12 If the intake and the exhaust manifolds have been separated, coat the mating surfaces with graphite grease. Place the exhaust manifold over the studs on the intake manifold. Connect the two with the lockwashers and nuts. Tighten the nuts finger tight.
13 Install a new intake manifold-to-head gasket.
14 Coat the mating surfaces lightly with graphite grease. Place the manifold assemblies against the mating surface of the cylinder head, making sure that the gaskets are positioned correctly. Install the washers, bolts and nuts finger tight and make sure everything is aligned correctly. Tighten the nuts and bolts to the specified torque in order (see illustration). Tighten the exhaust-to-intake manifold nuts to the specified torque (if they were removed).
15 Attach a new gasket to the exhaust pipe and fasten the pipe to the exhaust manifold.
16 Attach the crankcase vent hose to the intake manifold inlet tube and tighten the hose clamp.
17 Attach the fuel line to the carburetor.
18 Install the distributor vacuum hose on the fitting at the carburetor.
19 Connect the accelerator cable to the carburetor and install the return spring. Connect the choke cable to the carburetor and adjust the choke.
20 Install the bellcrank assembly and kickdown rod return spring.
21 Adjust the transmission control kickdown linkage as described in Chapter 7.
22 Install the air cleaner. Readjust the engine idle speed and idle fuel mixture as described in Chapter 4.

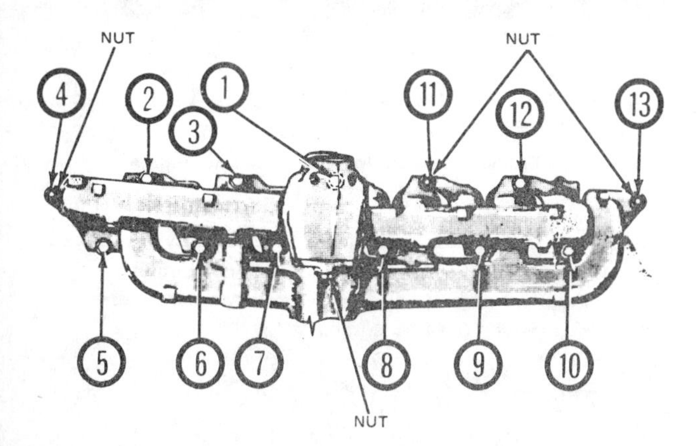

3.14 Manifold bolt/nut tightening sequence

4 Flywheel/driveplate – removal and installation

Note: *These instructions are valid only if the engine has been removed from the vehicle. If the engine is still in the vehicle, the transmission, bellhousing and clutch assembly must be removed to expose the flywheel (refer to Chapter 7). Automatic transmission equipped vehicles are equipped with a driveplate rather than a flywheel. It can be separated from the crankshaft by removing the mounting bolts. If the ring gear teeth are worn or damaged or if the driveplate is damaged, replace it with a new one and tighten the mounting bolts to the specified torque.*

1 Mark the flywheel and the crankshaft end with a center punch to ensure installation in the same relative position.
2 To keep the crankshaft from turning, wedge a large screwdriver or pry bar between the ring gear teeth and the engine block. It must be positioned so that as the crankshaft moves, the tool bears against the block. Make sure that the tool is not pushed against the oil pan. Another method of preventing crankshaft rotation involves holding the front pulley retaining bolt with a large wrench or socket and breaker bar. This method may require a helper. **Warning:** *Support the flywheel before performing the following step as it could fall off and be damaged or cause injury.*
3 Remove the flywheel mounting bolts from the crankshaft flange.
4 Remove the flywheel from the crankshaft by pulling straight back on it.
5 Inspect the flywheel clutch disc mating surface for scoring, heat marks, cracks and warpage. If any of these conditions exist, the flywheel should be taken to an automotive machine shop to be resurfaced (or replaced with a new one). If the flywheel is cracked it must be replaced with a new one.
6 Installation is the reverse of removal. The mounting bolts should be coated with a thread locking compound and tightened in a crisscross pattern to the specified torque.

5 Rocker arm and pushrod covers – removal and installation

Note: *On some engines the accelerator cable and bracket must be removed to provide room to lift off the rocker arm cover.*

Removal

1 Remove the engine cover and air cleaner then detach the positive crankcase ventilation hose from the top of the rocker arm cover.
2 Disconnect the air vent tube from the oil filler cap and remove the filler cap and tube from the rocker arm cover.
3 Disconnect the fuel supply hose at the fuel pump and at the carburetor. Remove the fuel supply hose.
4 Remove the distributor from the engine as described in Chapter 5.
5 Remove the ignition coil and bracket from the side of the engine.
6 Remove the bolts holding the rocker arm cover to the cylinder head.
7 Detach the rocker arm cover from the cylinder head and clean the old gasket from the mating surfaces.
8 Remove the bolts from the pushrod cover at the left side of the engine.
9 Detach the pushrod cover from the side of the engine and clean the old gasket from the cover and engine block.

Installation

10 Attach a new gasket to the pushrod cover. RTV-type gasket sealant will keep it positioned correctly.
11 Place the pushrod cover on the engine and tighten the bolts evenly and securely.
12 Install a new rocker arm cover gasket in the rocker arm cover. RTV-type gasket sealant will hold it in place.
13 Install the rocker arm cover on the cylinder head, making sure that the bolt holes line up.
14 Install the rocker arm cover bolts and tighten them to the specified torque.
15 Install the components listed in Steps 1 through 5 in reverse order.
16 After the engine has been started, run it until it reaches normal operating temperature. Check the pushrod cover and rocker arm cover for leaks.

6 Rocker arms and pushrods — removal, inspection and installation

1 Remove the rocker arm cover as described in the previous Section.
2 Remove the rocker arm stud nut, fulcrum seat and rocker arm from each cylinder. Keep them in order or mark them if they are to be reinstalled so they can be replaced in their original positions.
3 Inspect each rocker arm for signs of excessive wear, galling and damage.
4 Make sure the oil hole at the pushrod end of the rocker arm is open.
5 Inspect each rocker arm fulcrum for galling and check for wear on the face. If any of these conditions exist, replace the rocker arms and fulcrums as an assembly.
6 To remove the pushrods, pull them straight up out of the lifter pocket and through the cylinder head.
7 Inspect the pushrods to see if they are bent, cracked or excessively worn. If any of these conditions exist, replace the pushrods with new ones.
8 Apply engine oil or assembly lube to the top of each valve stem and the pushrod guides in the cylinder head.
9 Apply lubricant to the rocker arm fulcrum seats and the fulcrum seat sockets in the rocker arms.
10 Install each pushrod (with lubricant applied to both ends).
11 Install each rocker arm, fulcrum seat and stud nut and tighten the stud nuts to the proper torque.
12 Replace the rocker arm cover and gasket as described in the previous Section.

7 Cylinder head — removal and installation

Refer to illustration 7.24

Removal

Note: *If the engine has been removed from the vehicle, you may skip Steps 1 through 14 and begin with Step 15. If the engine is in the vehicle, open the hood, remove the engine cover and cover the seats to avoid damage.*

1 Drain the cooling system (see Chapter 1).
2 Remove the air cleaner and hoses leading to the air cleaner.
3 Disconnect the PCV valve from the rocker arm cover and remove the valve.
4 Disconnect the vent hose from the intake manifold inlet tube and remove it.
5 Disconnect the carburetor fuel inlet line leading from the fuel pump.
6 Disconnect the distributor vacuum line at the carburetor.
7 Disconnect the choke cable at the carburetor and position the choke cable and housing out of the way. Secure the cable and housing to the firewall or fender well.
8 Disconnect the accelerator cable or accelerator linkage from the carburetor. Remove the accelerator return spring.
9 Disconnect the kickdown link at the carburetor (vehicles with an automatic transmission).
10 Disconnect the upper radiator hose from the thermostat outlet.
11 Remove the heater hose at the coolant outlet elbow.
12 Remove the nuts securing the exhaust pipe to the exhaust manifold and support the pipe out of the way.
13 Mark the wires leading to the coil and disconnect them. Remove the coil bracket mounting bolt. Secure the coil and bracket out of the way.
14 Remove the rocker arm cover.
15 Loosen the rocker arm stud nuts so that the rocker arms can be rotated to one side.
16 Remove the pushrods in sequence and label them so they can be installed in their original locations. A numbered box or rack will keep them properly organized.
17 Disconnect the spark plug wires at the spark plugs.
18 Remove the cylinder head bolts. If you have an engine hoist or similar device handy, attach eyelet bolts at the two ends of the cylinder head in the holes provided and lift the cylinder head off of the engine block. If equipment of this nature is not available, use a helper and pry the cylinder head up off of the engine block. **Caution:** *Do not wedge any tools between the cylinder head and block gasket mating surfaces.*
19 Clean all old gasket material or sealant from the head gasket mating surface. Be careful not to scratch the sealing surface.
20 Clean the threads on all cylinder head bolts very thoroughly.

Installation

21 Make sure that the cylinder head and engine block mating surfaces are clean, flat and prepared for the new cylinder head gasket. Clean the exhaust manifold and exhaust pipe gasket surfaces.
22 Position the gasket over the dowel pins on the engine block, making sure that it is facing the right direction and that the correct surface is exposed. Gaskets are often marked *front* and *this side up* to aid the installer.
23 Using the previously installed lifting hooks (or two people), carefully lower the cylinder head into place on the block. Take care not to move the head sideways or to scrape it across the surface, as it can dislodge the gasket and damage the mating surfaces.
24 Coat the cylinder head bolts with light engine oil and thread the bolts into the block. Tighten the bolts following the cylinder head bolt tightening sequence (see illustration). Work up to the final torque in three steps to avoid warping the head.
25 Apply a coat of engine oil to the rocker arm fulcrum seats and the sockets in the rocker arms.
26 Install the pushrods, the rocker arms, the rocker arm fulcrum seats and the retaining nuts.
27 Install the rocker arm cover. The remaining steps are the reverse of the removal procedure.
28 Start the engine and allow it to reach operating temperature. Shut it off, allow it to cool and retorque the head bolts.

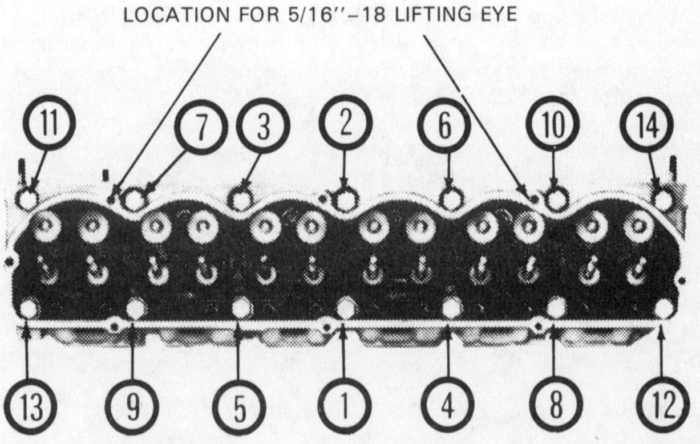

LOCATION FOR 5/16"–18 LIFTING EYE

11 7 3 2 6 10 14

13 9 5 1 4 8 12

7.24 Cylinder head bolt tightening sequence

8 Timing cover and gears — removal and installation

Refer to illustrations 8.14, 8.15 and 8.19

Note: *The following procedure requires a gear puller and gear installation tools available from Ford dealers.*

Removal

1 Drain the cooling system.
2 Remove the fan shroud and the radiator (Chapter 3).

8.14 Before removing the timing gears, the marks must be aligned (arrows)

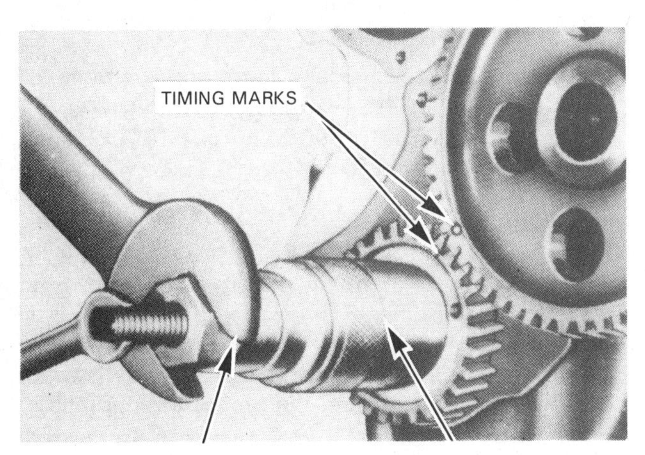

8.19 The timing marks must be aligned when installing the crankshaft gear (note the special tool being used to push the gear into place on the shaft)

3 Remove the alternator adjustment bolt.
4 Loosen the drivebelt and swing the adjusting arm up out of the way.
5 Remove the fan, drivebelts, fan spacer and pulley.
6 Remove the large bolt and washer from the crankshaft nose. It may be necessary to prevent the crankshaft from rotating by putting the transmission in gear (if the engine is still in the vehicle) or by holding the flywheel or crankshaft flange with a suitable tool if the engine is out of the vehicle.
7 Remove the vibration damper with a puller.
8 Remove the front oil pan bolts.
9 Remove the timing cover bolts.
10 Remove the cover and scrape the old gasket from the mating surfaces of the cover and the engine block.
11 Remove the crankshaft oil seal by pushing it out of the front cover with a large drift. Be careful not to damage the front cover while performing this operation.
12 Remove any chemical sealants from the seal bore of the cover. Check the bore carefully for anything that would prevent the new seal from seating properly in the cover.
13 Before removing the gears, the camshaft end play, timing gear backlash and timing gear runout should be checked as described in Section 9.
14 Turn the crankshaft or camshaft until the timing marks of the gears are aligned (see illustration). **Caution:** *Do not turn either the crankshaft or camshaft while the gears are removed. Serious internal engine damage can result from rotating either one independent of the other.*

8.15 A puller must be used to detach the camshaft gear from the camshaft

2A

15 Using a gear puller, remove the gear from the camshaft (see illustration).
16 Using a puller, remove the timing gear from the crankshaft.

Installation

17 Make sure that the camshaft end play, timing gear backlash and timing gear runout are within the Specifications. Do not install the camshaft gear until all of these camshaft related tolerances are correct.
18 Align the key spacer and thrust plate before installing the camshaft drive gear onto the camshaft. Make sure that the timing marks are properly aligned.
19 Install the crankshaft gear using the special drive tool (see illustration). A substitute special tool can be fashioned using a bolt and nut with a thread that matches the thread of the vibration damper bolt. Use a bolt approximately 2-1/2 inches long. Place the crankshaft gear on the crankshaft. Position the damper on the crankshaft. Thread the bolt (with the nut run all the way up the bolt) into the crankshaft nose. After the bolt is threaded into the crankshaft as far as possible, use the nut to push the damper onto the crankshaft. Make sure the damper is correctly aligned with the key on the crankshaft and check the internal bore of the damper as well as the outside surface of the crankshaft if resistance is felt. The damper will push the crankshaft gear into position. Remove the damper with a puller. A large deep socket (if you have access to one) can also be used to force the gear onto the crankshaft and will save having to pull the damper back off.
20 Install the crankshaft oil slinger in front of the crankshaft drive gear. Note that the cupped side faces away from the engine.
21 Coat the outside edge of the new crankshaft oil seal with grease and install the seal in the cover using an appropriate drive tool. Make sure the seal is seated completely in the bore.
22 If the oil pan is still on the engine, cut the old front oil pan seal flush at the cylinder block-to-pan junction. Remove the old seal.
23 Clean all gasket surfaces on the camshaft cover, block and oil pan.
24 If the oil pan is in place, cut and install a new pan seal so that it is flush with the engine block-to-oil pan junction.
25 Align the pan seal locating tabs with the holes in the oil pan. Make sure that the seal tabs pull all the way through so that the seal is completely seated. Apply RTV-type gasket sealant to the block and pan mating surfaces, particularly to the corner junctions of the block, oil pan and cover.
26 Position the cover over the end of the crankshaft and onto the cylinder block. Start the cover and pan bolts by hand.
27 Slide an alignment tool over the crankshaft to make sure that the cover is located correctly before tightening the bolts. If no alignment tool is available, locate the vibration damper over the nose of the crankshaft to center the seal. If the seal is not centered, the high spots will cause oil leakage around the vibration damper.
28 Tighten the cover and oil pan bolts to the correct torque.
29 Make sure that the oil pan bolts are tightened first to compress the pan seal so that the alignment of the cover is retained.
30 Lubricate the nose of the crankshaft, the inner hub of the vibration damper and the seal surface with engine oil.
31 Align the damper keyway with the key on the crankshaft and install

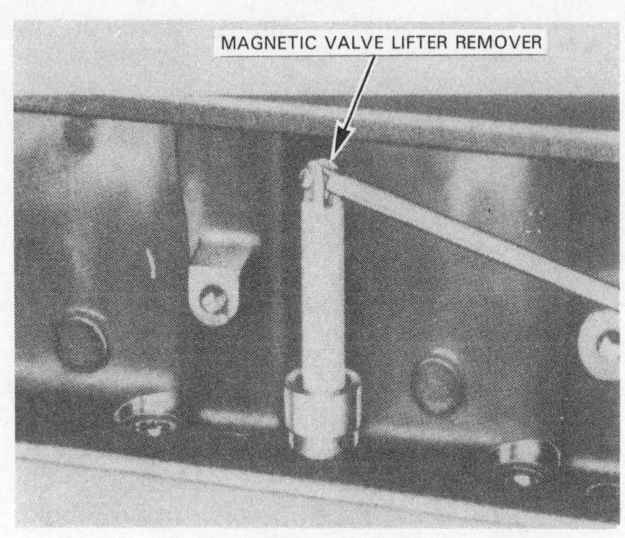

9.11 The lifters can be removed with a special magnetic tool

9.19 Checking camshaft end play with a dial indicator

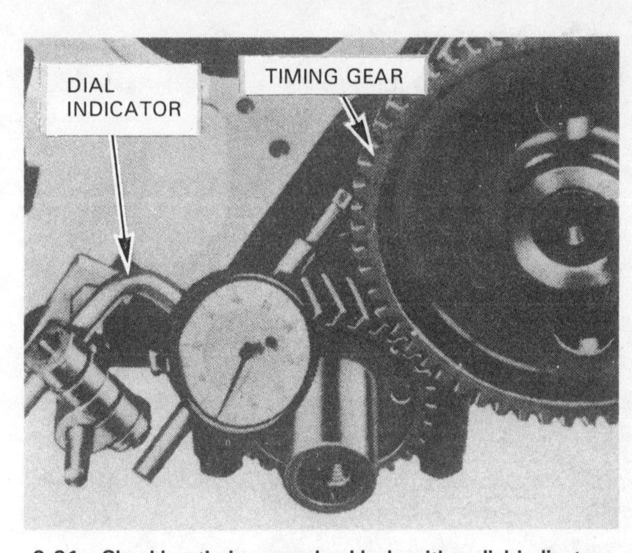

9.21 Checking timing gear backlash with a dial indicator

the damper.
32 Install the damper bolt and washer and tighten the bolt to the proper torque.
33 Install the alternator adjusting arm.
34 Install the pulleys, drivebelts, spacer and fan.
35 Adjust all drivebelts.
36 Install the fan shroud, radiator and hoses.
37 Fill the cooling system.
38 Fill the engine with oil if the oil has been drained.
39 Start and operate the engine at a fast idle and check for leaks of any type.

9 Camshaft and lifters – removal and installation

Refer to illustrations 9.11, 9.19, 9.21, 9.27 and 9.42

Note: *If the engine is in the vehicle, open the hood, remove the engine cover and cover the seats to prevent damage.*

Removal

1 Remove the air cleaner.
2 Remove the PCV valve from the rocker arm cover.
3 Disconnect the choke cable at the carburetor.
4 Disconnect the throttle cable at the carburetor. Remove the accelerator return spring.
5 Remove the coil bracket bolt and swing the coil out of the way. Support it securely and do not let it hang by the wires.
6 Remove the rocker arm cover.
7 Disconnect the spark plug wires from the spark plugs and the coil wire from the coil.
8 Remove the distributor cap and spark plug wire assembly.
9 Remove the pushrod cover from the side of the block.
10 Loosen the rocker arm stud nuts until the rocker arms are free. Turn the rocker arms to the side and remove the pushrods. Make sure that the pushrods are numbered or marked so that they can be installed in the same locations.
11 Remove the valve lifters with a special magnetic valve lifter tool (see illustration).
12 Drain the cooling system and the oil from the crankcase.
13 Remove the radiator.
14 Remove the front timing cover.
15 Remove the oil pan and oil pump.
16 Disconnect the fuel lines at the fuel pump. Remove the fuel pump mounting bolts and secure the fuel pump out of the way.
17 Disconnect the vacuum line at the distributor and remove the vacuum line from the carburetor. **Note:** *The following checking procedures require a magnetic base dial indicator.*
18 Check the camshaft end play by pushing the camshaft all the way to the rear.
19 Install a dial indicator so that the indicator stem is on the camshaft sprocket mounting bolt. Zero the dial indicator in this position (see illustration).
20 Using a large screwdriver between the camshaft gear and the block, pull the camshaft forward and release it. The reading on the dial indicator will give you the end play measurement. Compare it to the Specifications. If the end play is excessive, check the spacer for correct installation. If the spacer is correctly installed and the end play is excessive, replace the thrust plate with a new one.
21 Check the timing gear backlash by installing a dial indicator on the engine block and positioning the stem against the timing gear (see illustration).
22 Zero the pointer on the dial indicator.
23 While holding the crankshaft still, move the camshaft timing gear until it takes up the slack in the gear train.
24 Read the dial indicator to obtain the gear backlash.
25 Compare the results to the Specifications.
26 If the backlash is excessive, replace the timing gear and the crankshaft gear with new ones.
27 To check the timing gear runout, install a dial indicator on the engine block with the stem touching the timing gear face (see illustration).
28 Hold the camshaft gear against the camshaft thrust plate and zero the indicator.
29 Rotate the crankshaft to turn the camshaft while holding the camshaft gear against the thrust plate.
30 Rotate through one complete revolution of the camshaft. Observe the reading on the dial indicator during this procedure.
31 If the runout exceeds the Specifications, remove the camshaft gear and check for foreign objects or burrs between the camshaft and gear flanges. If this condition does not exist and the runout is excessive,

9.27 Checking timing gear runout with a dial indicator

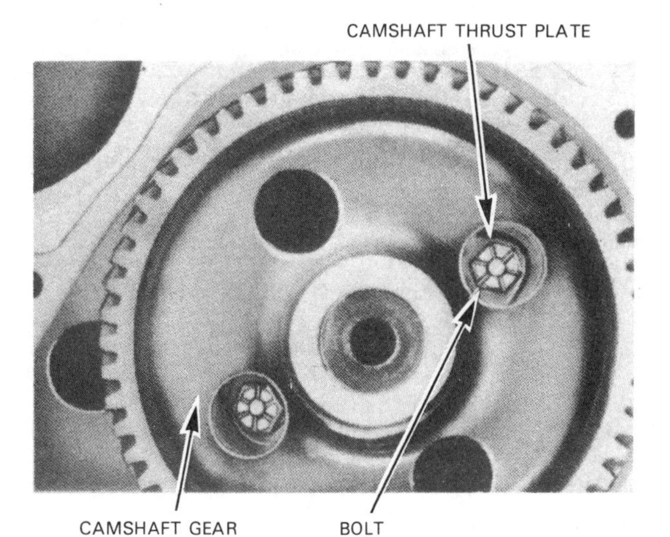

9.42 The camshaft thrust plate bolts can be removed/installed by inserting a socket through the holes in the gear

2A

the gear must be replaced with a new one. Use a similar procedure to check the crankshaft gear runout. Make sure that the crankshaft is situated against one end of the thrust bearing. This will prevent you from obtaining a crankshaft end play measurement as opposed to the actual runout of the crankshaft gear.
32 Turn the crankshaft until the timing marks are directly adjacent to each other.
33 Remove the camshaft thrust plate bolts.
34 Withdraw the camshaft from the engine block, being careful that the lobes do not catch on the camshaft bearings (they can scrape and damage them easily).
35 Remove the camshaft from the gear using a hydraulic press. This procedure will have to be completed by an automotive machine shop.
36 Remove the key, the thrust plate and the spacer.
37 Camshaft and lifter inspection procedures are included in Chapter 2, Part C.

Installation

38 Make sure that the camshaft end play, timing gear backlash and/or timing gear runout are within the Specifications before installing the camshaft.
39 Oil the camshaft bearing journals and apply engine assembly lubricant to all of the lobes.
40 Install the key, spacer and thrust plate on the front of the camshaft. Install the gear on the camshaft using the special tool. An alternative is to use a bolt that will fit the threaded hole in the end of the camshaft. Put a nut and large flat washer on the bolt. Thread the bolt into the camshaft with the gear in place. Hold the bolt stationary and turn the nut down the bolt to push the gear into place on the camshaft. Remove the bolt and nut combination after the cam gear is in place.
41 Install the camshaft in the engine. Be careful not to nick or damage the camshaft bearings. Install the camshaft bolts, making sure the camshaft gear is aligned with the crankshaft gear.
42 Tighten the camshaft thrust plate bolts to the specified torque, using the access holes provided in the camshaft gear (see illustration).
43 Install the timing cover as described in Section 8.
44 Do not turn the crankshaft until the distributor is installed, as the timing marks must remain aligned.
45 Install the oil pump and oil pan.
46 Lubricate the bottom of the lifters with engine assembly lube and install them in their original locations. Install the pushrods, align the rocker arms and tighten the nuts. Install the rocker arm and pushrod covers and hook up the throttle cable.
47 Install the distributor as described in Chapter 5.
48 Install the fuel pump as described in Chapter 4.
49 Install the vacuum line connecting the distributor to the carburetor.
50 Connect the fuel lines to the fuel pump.
51 Fill the crankcase with oil.
52 Install the radiator and fill the cooling system with the specified coolant.
53 Start the engine and check for oil, coolant and fuel leaks.

10 Oil pan — removal and installation

Refer to illustrations 10.17 and 10.18

Note: *The following procedure is based on the assumption that the engine is in place in the vehicle. If it has been removed, merely unbolt the oil pan and detach it from the block. Many of the steps outlined below apply only to later model vehicles. If you are working on an early model vehicle, modify the procedure as required.*

Removal

1 If the vehicle is equipped with air conditioning, have the system discharged so the compressor can be removed and positioned out of the way. Do not attempt to do this yourself as serious injury can result.
2 Remove the engine cover and cover the seats to prevent damage, then remove the air cleaner and carburetor (refer to Chapter 4 if necessary).
3 Remove the EGR valve. If the vehicle is an E-250HD or E-350, disconnect the Thermactor check valve inlet hose and remove the valve (see Chapter 6).
4 Remove the upper radiator hose, then unbolt the fan shroud and position it over the fan.
5 If the vehicle is equipped with an automatic transmission, unbolt and remove the filler tube.
6 Remove the exhaust manifold-to-pipe nuts.
7 Raise the vehicle and support it securely on jackstands, then disconnect and plug the fuel inlet hose at the pump.
8 Remove the alternator heat shield and the front engine support nuts.
9 Remove the power steering return line clip located in front of the number one crossmember.
10 Disconnect the lower radiator hose and the automatic transmission cooler lines.
11 Remove the starter (Chapter 5).
12 Raise the engine with a jack and place three one-inch thick wood blocks under the engine mounts.
13 Remove the dipstick tube from the oil pan.
14 Remove the oil pan bolts, then detach the pan from the block. Do not pry between the block and pan or damage to the sealing surfaces may result.
15 Reach in and unbolt the oil pump pickup tube from the pump, then detach the pan and tube from the engine.
16 Clean the oil pan and engine block gasket mating surfaces and make sure the bolt holes in the block are clear. Clean the pickup tube and screen assembly with solvent and blow dry with compressed air.

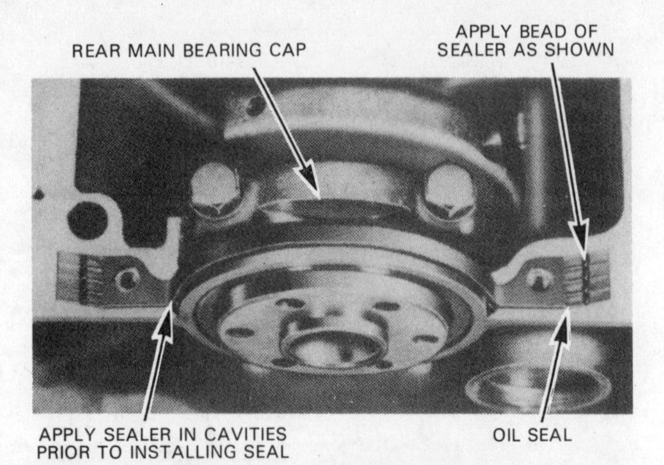

10.17 Rear oil pan seal installation details

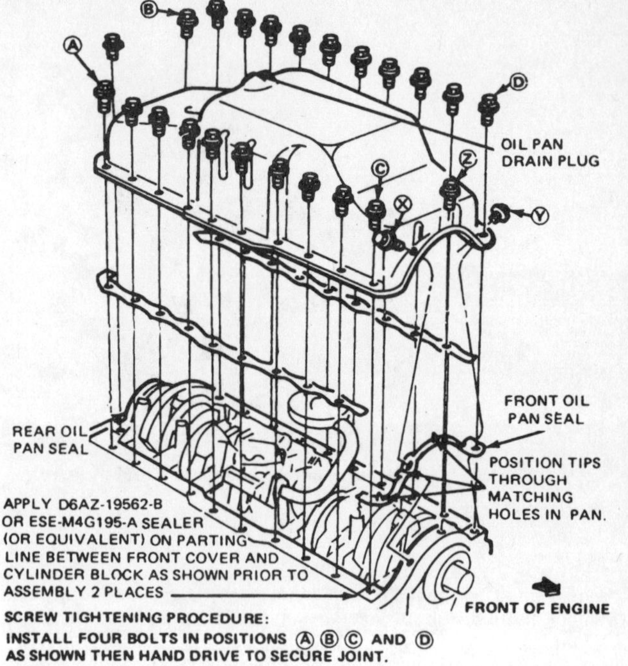

10.18 Front oil pan seal and gasket installation details

Installation

17 Remove the old rear main bearing cap-to-pan seal and install a new one. Apply RTV-type sealant to the seal cavity where the main bearing cap mates with the block and at the point where the oil pan gasket mates with the seal (see illustration).
18 Attach the new gaskets and front seal to the oil pan with RTV-type sealant or contact cement. The front seal has molded tips that can be pushed through the holes in the pan to hold it in place (see illustration). Position the oil pump pickup tube in the pan.
19 Reattach the pan and tube assembly to the engine and install the pickup tube mounting bolts. Be sure to tighten them securely.
20 Install the oil pan mounting bolts and tighten them in three steps to the specified torque. Start at the center of the pan and work out toward the ends in a circular pattern.
21 The remaining steps are the reverse of removal.

11 Oil pump – removal and installation

1 Remove the oil pan as described in Section 10.
2 Remove the bolts retaining the oil pump to the block.
3 Detach the oil pump assembly.
4 Clean the mating surfaces of the oil pump and the block.
5 Before installation, prime the pump by filling the inlet opening with oil and rotating the pump shaft until the oil spurts out of the outlet.
6 Attach the oil pump to the engine block using the two retaining bolts.
7 Tighten the bolts to the specified torque.
8 Install the oil pan by referring to Section 10.

12 Oil pump – disassembly, inspection and reassembly

Refer to illustrations 12.6 and 12.7

1 Remove the two bolts securing the pickup tube to the oil pump, then detach the pickup.
2 Clean the oil pump with solvent and dry it thoroughly with compressed air.
3 Remove the oil pump housing cover. It is retained by four bolts.
4 Use a brush to clean the inside of the pump housing and the pressure relief valve chamber. Make sure that the interior of the oil pump is clean.
5 Visually check the inside of the pump housing and the outer race and rotor for excessive wear, scoring and other damage. Check the mating surface of the pump cover for wear, grooves and damage. If any of these conditions exist, replace the pump with a new one.
6 Measure the outer race-to-housing clearance with a feeler gauge (see illustration).
7 Using a straightedge and feeler gauge, measure the end plate-to-rotor assembly clearance (see illustration).
8 Check the driveshaft-to-housing bearing clearance by measuring the inside diameter of the housing bearing and subtracting that figure from the outside diameter of the driveshaft.

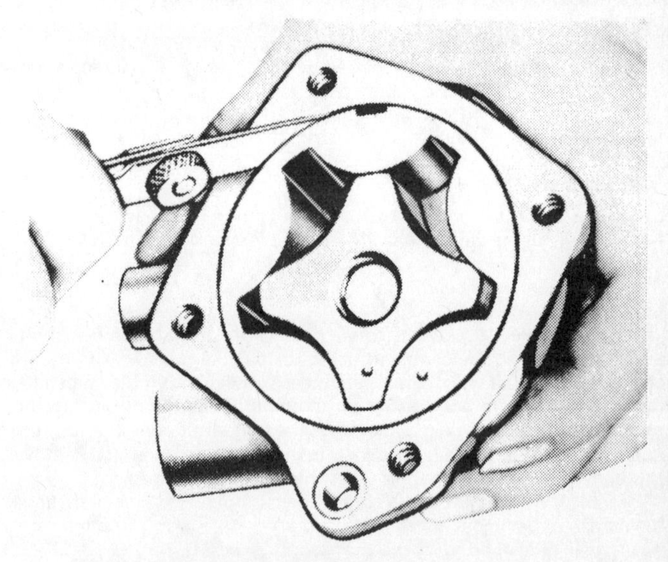

12.6 Checking the oil pump outer race-to-housing clearance with a feeler gauge

9 If any components fail the checks mentioned, replace the entire oil pump, as the components are not available as separate parts.
10 Inspect the relief valve spring for wear and a collapsed condition.
11 Check the relief valve piston for scoring, damage and free operation within its bore.
12 If the relief valve fails any of the above tests, replace the entire relief valve assembly with a new one.
13 Install the rotor, outer housing and race in the oil pump.
14 Install the rotor housing cover and the four retaining bolts and tighten them to the specified torque.
15 Attach the pickup tube to the oil pump body using a new gasket. Tighten the bolts to the specified torque.

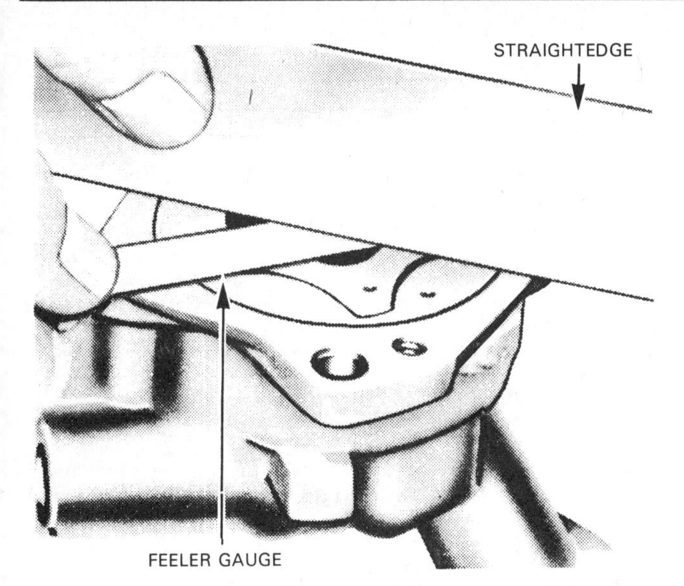

12.7 Checking the oil pump end plate-to-rotor clearance with a straightedge and feeler gauge

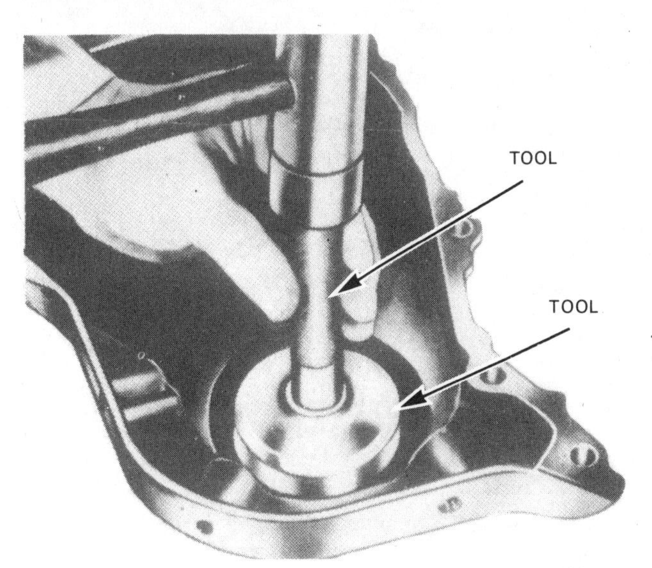

13.6 Installing a new front crankshaft oil seal in the timing gear cover with the special tool

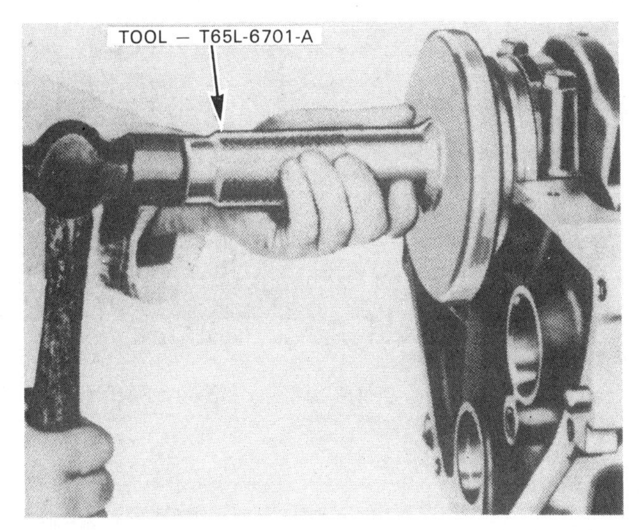

13.21 Installing a new rear crankshaft oil seal with the special drift-type tool

13 Crankshaft oil seals — replacement

Refer to illustrations 13.6 and 13.21

Front crankshaft seal

Note: *The following operation requires a special tool for proper seal installation.*

1 Drain the cooling system and crankcase.
2 Remove the radiator.
3 Remove the timing gear cover (Section 8).
4 Drive out the oil seal with a drift punch.
5 Clean out the recess in the cover.
6 Coat the outer edge of the new seal with grease and install it using the special tool designed for this operation (see illustration). As an alternative, a large piece of pipe can be used to push the new seal in. However, use extreme caution as the seal can be damaged easily with this method. Drive in the seal until it is completely seated in the recess. Make sure that the spring is properly positioned within the seal.
7 Install the timing cover on the engine as described in Section 8.

Rear crankshaft seal

Note: *If rear crankshaft oil seal replacement is the only operation being performed, it can be accomplished with the engine in the vehicle. If, however, the oil seal is being replaced along with the rear main bearing, the engine must be removed.*

8 Disconnect the negative cable from the battery, then remove the starter.
9 Remove the transmission (Chapter 7).
10 On manual transmission equipped vehicles, remove the pressure plate, disc and clutch assembly. On automatic transmission equipped vehicles, remove the driveplate.
11 Remove the flywheel bolts and detach the flywheel and engine rear cover plate.
12 Use an awl to punch two holes in the rear crankshaft oil seal.
13 Punch the holes on opposite sides of the crankshaft journal, just above the bearing cap-to-engine block junction.
14 Thread a sheet metal screw into each punched hole.
15 Use two large screwdrivers or small pry bars and pry against both screws at the same time to remove the seal. A block or blocks of wood placed against the engine will provide additional leverage.
16 Be careful when performing this operation that you do not damage the crankshaft journal.
17 Clean the oil seal recess in the rear of the engine block and the main bearing cap.
18 Inspect, clean and polish the oil seal contact surface of the crankshaft.
19 Coat the new oil seal with a light film of engine oil.
20 Coat the crankshaft with a light film of engine oil.
21 Start the seal into the cavity in the back of the engine with the seal lip facing forward and install it with the special drive tool (see illustration). Make sure that the tool stays in alignment with the crankshaft until the tool contacts the block. See Step 6 for seal installation alternatives.
22 Make sure that the seal has been installed correctly after removing the tool.
23 Install the engine rear cover plate.
24 Attach the flywheel (or driveplate) to the crankshaft.
25 Install the clutch assembly.
26 Install the transmission as described in Chapter 7.
27 Install the starter and hook up the battery cable.

14 Engine — removal and installation

Note: *Due to the wide range of model years covered by this manual, the following instructions are general in nature and may also cover some steps not applicable to your particular vehicle. On 1973 through 1978 models, the transmission must be removed first (see Chapter 7) in order to facilitate engine removal. Be sure to support the engine with a hoist as the transmission is removed.*

1 Open the hood and doors and install seat covers to protect the upholstery, then remove the engine cover. On 1973 through 1978

models, remove the right seat.
2 Drain the engine oil and coolant, remove the air cleaner and disconnect the battery cables from the battery.
3 Remove the front bumper, followed by the grille and lower gravel deflector as an assembly.
4 Refer to Chapter 3 and remove the radiator and shroud. Remove the fan and water pump pulley.
5 Disconnect the heater hoses from the engine.
6 Refer to Chapter 5 and remove the alternator.
7 Refer to Chapter 10 and remove the power steering pump and bracket.
8 Disconnect the fuel line at the fuel pump and plug it.
9 Disconnect the distributor and sending unit wires from the engine or the EEC harness from all sensors.
10 Disconnect the brake booster hose from the engine.
11 Disconnect the accelerator cable and remove the bracket from the engine. Disconnect the choke cable (early models) and remove the throttle return spring.
12 Disconnect the automatic transmission kickdown linkage at the bellcrank.
13 Remove the exhaust manifold heat deflector, then remove the nuts and separate the pipe from the manifold.
14 Disconnect both ends of the transmission vacuum line from the intake manifold and junction.
15 Remove the upper engine-to-transmission bolts.
16 Remove the automatic transmission dipstick tube support bolt from the intake manifold.
17 Raise the vehicle and support it securely with jackstands.
18 Refer to Chapter 5 and remove the starter.
19 Remove the flywheel inspection cover or torque converter drain plug cover, then remove the torque converter-to-driveplate bolts. The crankshaft will have to be turned to bring each bolt into view.
20 Remove the front engine support nuts. Remove the oil filter.
21 Support the transmission, then remove the remaining engine-to-transmission bolts.
22 On manual transmission equipped vehicles, disconnect the clutch return spring.
23 On automatic transmission equipped vehicles, secure the torque converter in the housing and detach the oil cooler lines from the clips on the engine.
24 Attach a chain and hoist to the engine, then move it forward to separate it from the transmission and remove it from the engine compartment.
25 Installation is the reverse of removal.

Chapter 2 Part B V8 engines

Refer to Chapter 13 for Specifications and information on 1987 and later models

Contents

2B

Specifications

Note: *Additional specifications can be found in Chapter 2, Part C*

Oil pump

Outer race-to-housing clearance	
302 (all) and 351W (thru 1979)	0.001 to 0.013 in
351W (1980 and later)	0.001 to 0.003 in
351M, 400 and 460	0.001 to 0.013 in
Rotor assembly end clearance limit	0.004 in
Inner rotor-to-outer rotor tip clearance	
351M, 400 and 460	0.012 in
Driveshaft-to-housing clearance	0.0015 to 0.0030 in
Relief spring tension (lbs @ specified length)	
302 (thru 1979)	10.6 to 12.2 lbs @ 1.704 in
302 (1980 and later)	10.6 to 12.2 lbs @ 1.740 in
351W	18.2 to 20.2 lbs @ 2.490 in
351M, 400 and 460	20.6 to 22.6 lbs @ 2.490 in
Relief valve clearance	0.0015 to 0.0030 in

Torque specifications	**Ft-lbs**
Camshaft sprocket bolt	40 to 45
Camshaft thrust plate bolts	9 to 12
Connecting rod nuts/main bearing cap bolts	See Chapter 2, Part C
Timing cover bolts	
460	15 to 21
all others	12 to 18
Cylinder head bolts	
302	
1st step	50
2nd step	60
3rd step	65 to 72
351W	
1st step	85
2nd step	95
3rd step	105 to 112
351M and 400	
1st step	75
2nd step	95 to 105
460	
1st step	80
2nd step	110
3rd step	130 to 140

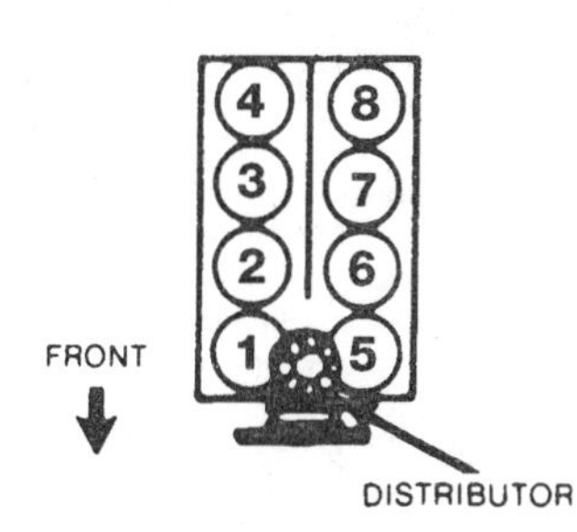

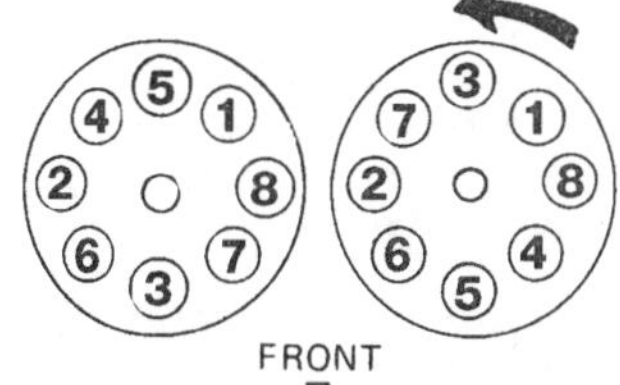

Cylinder numbers and firing order for V8 engines

Torque specifications (continued)	Ft-lbs
Vibration damper bolt	70 to 90
Flywheel/driveplate bolts	75 to 85
Intake manifold bolts	
302 and 351W	
thru 1974	17 to 25
1975 and later	23 to 25
351M and 400	
5/16-inch bolts	17 to 25
3/8-inch bolts	22 to 32
460	25 to 32
Exhaust manifold bolts	
302, 351W, 351M and 400	
thru 1974	12 to 16
1975 and later	18 to 24
460	28 to 33
Oil pan bolts	
1/4–20 bolts	7 to 9
5/16–18 bolts	9 to 11
Oil pump bolts	22 to 25
Rocker arm stud/bolt-to-cylinder head	18 to 25
Rocker arm stud nut	
302 and 351W	17 to 23
351M, 400 and 460	18 to 25
Rocker arm cover bolts	3 to 5

1 General information

This Part of Chapter 2 is devoted to repair procedures for V8 engines, as well as the removal and installation steps. All information concerning engine block and cylinder head servicing can be found in Part C of this Chapter.

Many of the repair procedures included in this part are based on the assumption that the engine is still installed in the vehicle. Therefore, if this information is being used during a complete engine overhaul, with the engine already out of the vehicle and on a stand, many of the steps included here will not apply.

The specifications included in this Part of Chapter 2 apply only to the engines and procedures found here. For six cylinder engine specifications, refer to part A. The specifications necessary for rebuilding the block and cylinder heads are included in Part C.

Several V8's of various displacements were used during the model years covered by this manual. All of them are gasoline engines with overhead valves actuated by hydraulic lifters and have five main bearing crankshafts.

Although all the V8's are of the same fundamental design, they are divided into three families. The 302 and 351W engines constitute one family. The 351M and 400 engines belong to the second family, while the final family, known as big blocks, contains the largest displacement engine available in Ford vans, the 460 cubic inch engine.

2 Valve spring, retainer and seal — replacement on vehicle

Note: *Broken valve springs and retainers or defective valve stem seals can be replaced without removing the cylinder head on engines that have no damage to the valves or valve seats. Two special tools and a compressed air source are required to perform this operation, so read through this Section carefully and rent or buy the tools before beginning the job. Refer to the illustrations in Chapter 2, Part A.*

1 Open the hood, remove the engine cover and cover the seats to prevent damage to the upholstery. Remove the air cleaner and related components, as well as anything restricting access to the rocker arm cover(s).
2 Remove the accelerator cable return spring.
3 Remove the accelerator cable linkage at the carburetor.
4 Disconnect the choke cable at the carburetor.
5 Remove the PCV valve from the rocker arm cover and remove the rocker arm cover(s).
6 Remove the spark plug from the cylinder which has the bad component.
7 Turn the crankshaft until the piston in the cylinder with the bad component is almost at top dead center on the compression stroke.
8 Install an air line adapter which screws into the spark plug hole and connects to a compressed air source. Remove the rocker arm stud nut or bolt, oil deflector, fulcrum, rocker arm and pushrod.
9 Apply compressed air to the cylinder.
10 Compress the valve spring with the valve spring tool.
11 Remove the keepers, spring retainer and valve spring, then remove the valve stem seal. **Note:** *If air pressure fails to hold the valve in the closed position during this operation, the valve face or seat is probably damaged. If this condition exists, remove the cylinder head for additional repair operations.*
12 Wrap a rubber band or tape around the valve stem so that the valve will not fall into the combustion chamber. Release the air pressure.
13 Inspect the valve stem for damage. Rotate the valve in the guide and check the valve stem tip for eccentric movement, which would indicate a bent valve.
14 Move the valve up-and-down in the guide and make sure it does not bind. If the valve stem binds, either the valve is bent or the guide is damaged. In either case, the head will have to be removed for repair.
15 Reapply air pressure to the cylinder to retain the valve in the closed position.
16 Lubricate the valve stem with engine oil and install a **new** valve stem seal.
17 Install the spring in position over the valve. Make sure that the closed coil end of the spring is positioned against the cylinder head.
18 Install the valve spring retainer. Compress the valve spring and install the valve spring keepers. Remove the spring tool and make sure that the keepers are seated.
19 Apply engine oil to both ends of the pushrod.
20 Install the pushrod.
21 Apply engine oil to the tip of the valve stem.
22 Apply engine oil to the fulcrum seat and socket.
23 Install the rocker arm, fulcrum, oil deflector and stud nut or bolt. Adjust the valve clearance following the procedure in Chapter 2C.
24 Remove the air source and the adapter from the spark plug hole.
25 Install the spark plug and connect the spark plug wire.
26 Install a new rocker arm cover gasket and attach the rocker arm cover to the engine.
27 Connect the accelerator cable to the carburetor.
28 Install the accelerator cable return spring. Connect the choke cable to the carburetor.
29 Install the PCV valve in the rocker arm cover and make sure that the line connected to the PCV valve is positioned correctly at both ends.
30 Install the air cleaner.
31 Start and run the engine, then check for oil leaks and unusual sounds coming from the valve assembly area.

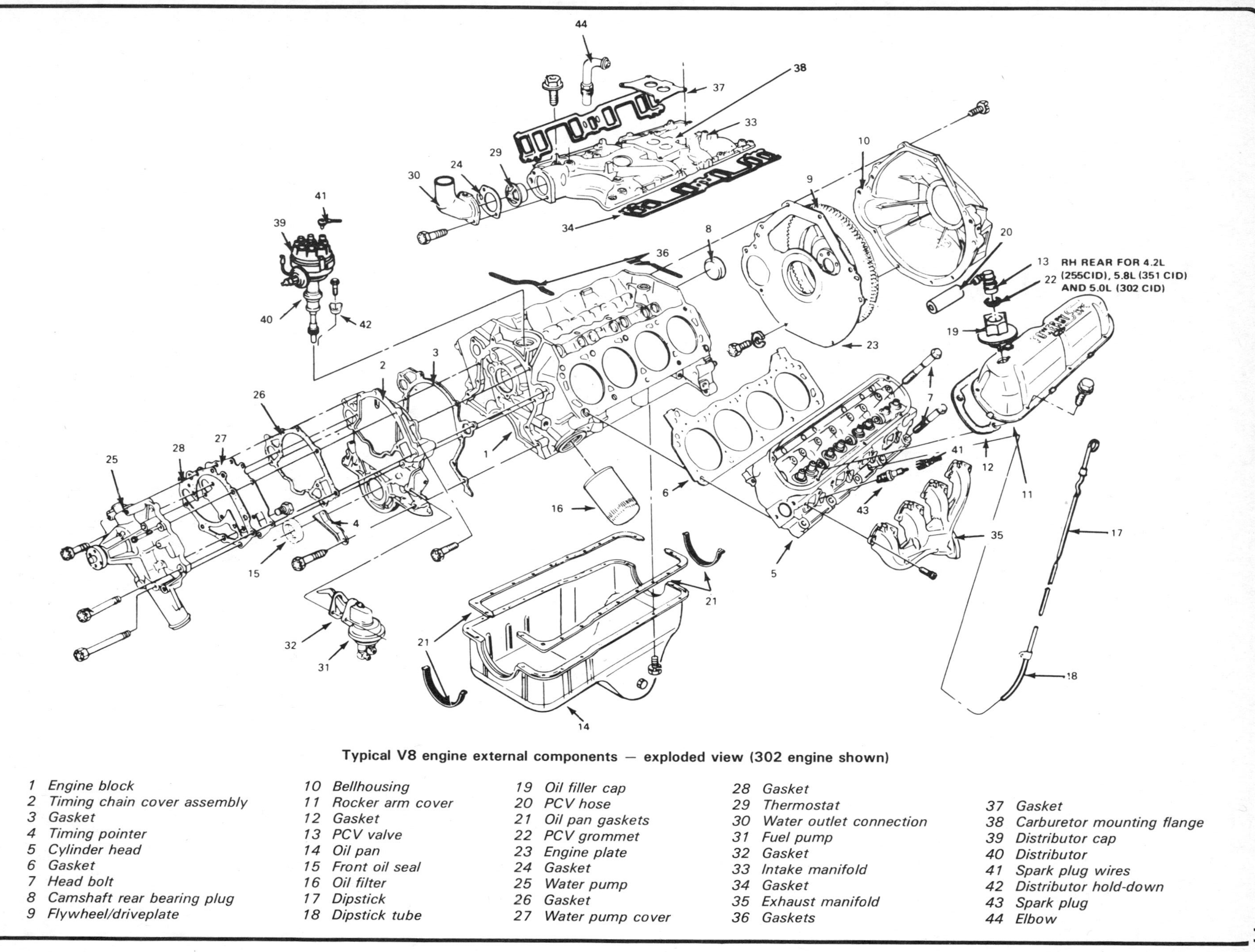

Typical V8 engine external components — exploded view (302 engine shown)

1 Engine block
2 Timing chain cover assembly
3 Gasket
4 Timing pointer
5 Cylinder head
6 Gasket
7 Head bolt
8 Camshaft rear bearing plug
9 Flywheel/driveplate
10 Bellhousing
11 Rocker arm cover
12 Gasket
13 PCV valve
14 Oil pan
15 Front oil seal
16 Oil filter
17 Dipstick
18 Dipstick tube
19 Oil filler cap
20 PCV hose
21 Oil pan gaskets
22 PCV grommet
23 Engine plate
24 Gasket
25 Water pump
26 Gasket
27 Water pump cover
28 Gasket
29 Thermostat
30 Water outlet connection
31 Fuel pump
32 Gasket
33 Intake manifold
34 Gasket
35 Exhaust manifold
36 Gaskets
37 Gasket
38 Carburetor mounting flange
39 Distributor cap
40 Distributor
41 Spark plug wires
42 Distributor hold-down
43 Spark plug
44 Elbow

3 Manifolds – removal and installation

Refer to illustrations 3.5, 3.9, 3.14, 3.19, 3.27a, 3.27b, 3.27c, 3.31, 3.32, 3.34, 3.36a, 3.36b and 3.36c

Intake manifold – removal

Note: *Due to the weight and bulk of the large displacement engine intake manifold, it is recommended that an engine hoist or puller coupled with lifting hooks (available from hardware stores) be used to lift the manifold from the engine. They can be removed by hand – however, the aid of a helper is suggested to avoid damaging engine components or causing personal injury. Due to the need to remove EFI components to gain access to the manifold, the intake manifold removal and installation procedure for EFI engines is in Chapter 4.*

1 Drain the cooling system (see Chapter 1).

2 Remove the engine cover and cover the seats to avoid damage to the upholstery.

3 Carefully mark and disconnect all emissions control hose connections at the air cleaner. Remove the air cleaner assembly.

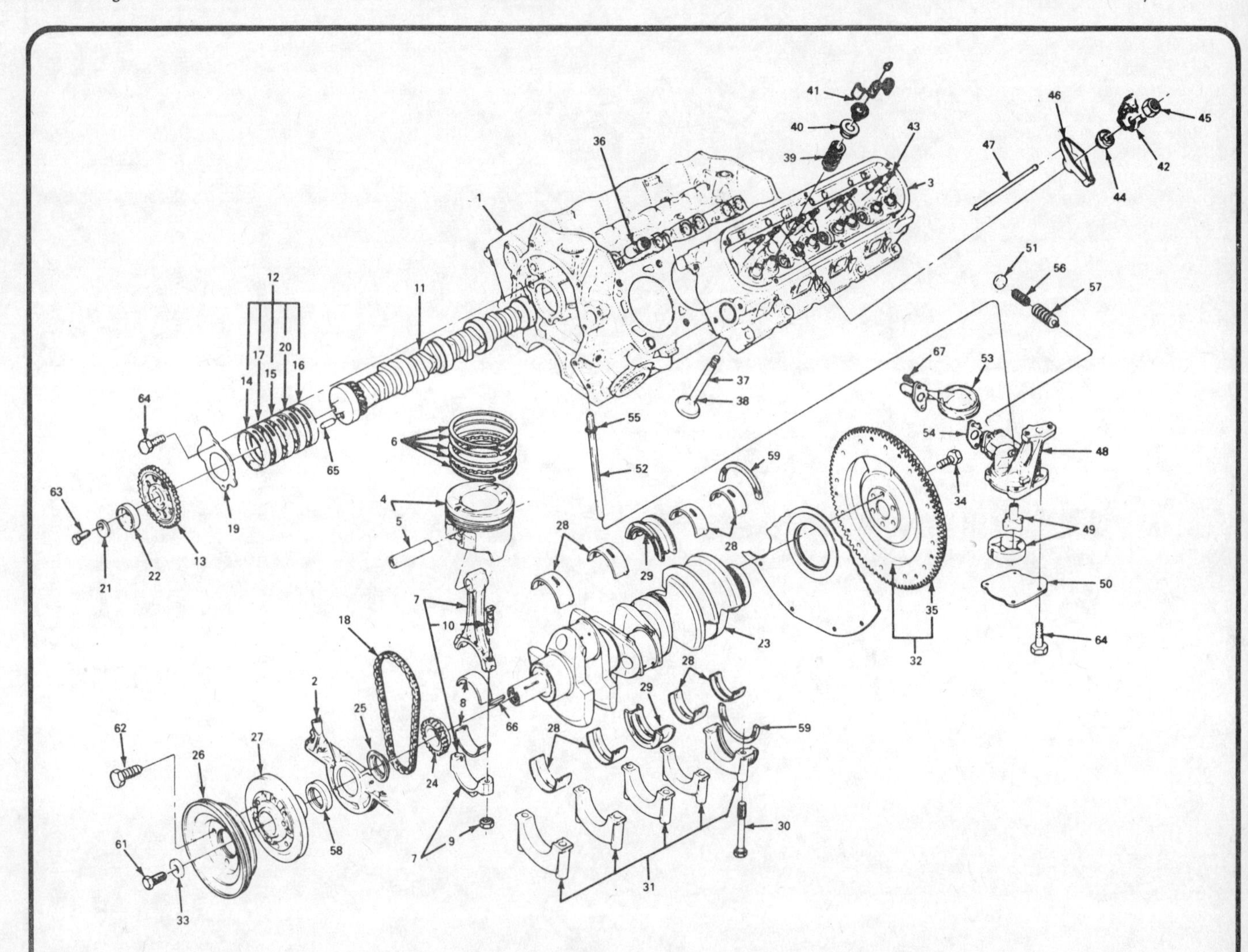

Typical V8 engine internal components – exploded view (302 engine shown)

1 Engine block
2 Timing cover
3 Cylinder head
4 Piston assembly
5 Piston pin
6 Piston ring set
7 Connecting rod
8 Connecting rod bearing inserts
9 Nut
10 Bolt
11 Camshaft
12 Camshaft bearings
13 Camshaft sprocket
14 Cam bearing no. 1
15 Cam bearing no. 3
16 Cam bearing no. 5
17 Cam bearing no. 2
18 Timing chain
19 Thrust plate
20 Cam bearing no. 4
21 Washer
22 Fuel pump eccentric
23 Crankshaft
24 Crankshaft sprocket
25 Oil slinger
26 Pulley
27 Vibration damper
28 Main bearing
29 Thrust bearing
30 Main bearing cap bolt
31 Main bearing cap
32 Flywheel/driveplate
33 Washer
34 Bolt
35 Ring gear
36 Lifter
37 Exhaust valve
38 Intake valve
39 Valve spring
40 Retainer
41 Keepers
42 Baffle
43 Rocker arm stud
44 Fulcrum seat
45 Nut
46 Rocker arm
47 Pushrod
48 Oil pump
49 Oil pump driveshaft
50 Oil pump cover plate
51 Relief valve plug
52 Intermediate shaft
53 Pick-up tube
54 Gasket
55 Ring
56 Relief valve spring
57 Relief valve plunger
58 Front oil seal
59 Rear oil seal
60 Rear plate
61 Bolt
62 Bolt
63 Bolt
64 Bolt
65 Dowel pin
66 Woodruff key
67 Bolt

4 Disconnect the upper radiator hose from the thermostat housing and remove the thermostat housing.
5 Disconnect the heater hose and loosen the water pump bypass hose clamp at the intake manifold connections (see illustration).
6 On 460 engines, remove the rocker arm covers.
7 Disconnect the spark plug wires at the spark plugs.
8 Unclip and remove the distributor cap together with the spark plug wires.
9 Disconnect the primary wire leads at the coil and mark them so they can be reinstalled correctly. Remove the coil mounting bracket bolt and detach the coil (see illustration).
10 Remove the fuel inlet line from the carburetor and position it out of the way. If the line is steel, it may be necessary to disconnect it or loosen it at the fuel pump.
11 Disconnect the vacuum advance hose(s) from the distributor and mark them to ensure correct reinstallation.
12 Remove the distributor as described in Chapter 5.
13 As the distributor is removed, make sure the distributor-to-oil pump driveshaft remains in the engine and connected to the oil pump. Plug the hole with a clean rag to prevent debris from entering the engine.
14 Remove the wire from the coolant temperature sending unit (see illustration).
15 Remove the wires from any sensors mounted in the intake manifold or the thermostat housing.
16 Remove the throttle cable or linkage at the carburetor (see Chapter 4).
17 If the vehicle is equipped with an automatic transmission, remove the kickdown cable at the carburetor.
18 If the vehicle is equipped with a cruise control unit, remove the pull cable and activating unit from the intake manifold.
19 Remove the vacuum hose leading to the power brake booster (see illustration). Secure this hose to the firewall, out of the way.
20 Remove the carburetor from the intake manifold if it is to be serviced separately.
21 Remove any wiring leading to components located on the intake manifold. Locate the wires where they won't be damaged.
22 Disconnect the crankcase vent hose leading from the manifold to the rocker arm cover.
23 Remove the bolts retaining the intake manifold to the cylinder heads.
24 Attach the lifting hooks at opposite corners and lift the intake manifold from the engine. It may be necessary to pry the manifold away from the cylinder heads. Use caution to avoid damaging the mating surfaces.

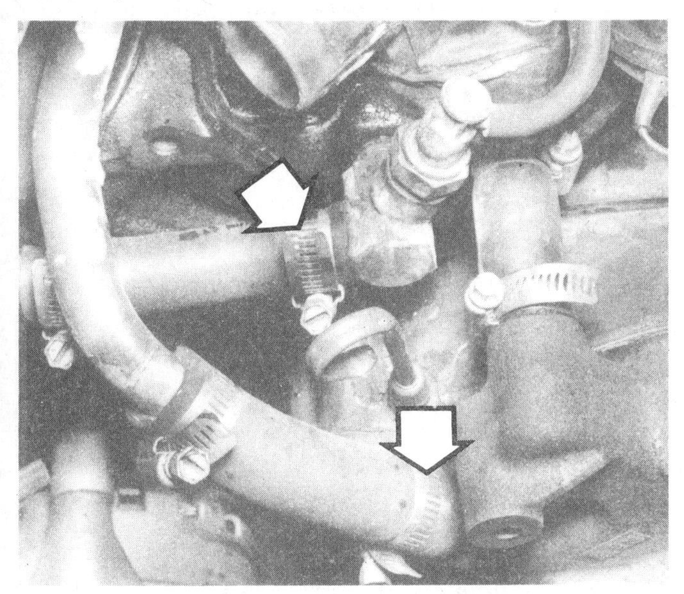

3.5 The heater hose and water pump bypass hose clamp (arrows) must be removed to detach the intake manifold

3.9 The ignition coil is held in a bracket that is bolted to the intake manifold on most engines

3.14 Detach the wire from the coolant temperature sending unit (arrow) . . .

3.19 . . . and remove the power brake booster vacuum hose from the manifold fitting

3.27a Carefully scrape all traces of the old gasket from the head and manifold mating surfaces

3.27b The bolt hole threads must be clean and dry to ensure accurate torque readings when the manifold is installed

3.27c Clean the bolt holes with compressed air, but be very careful — wear safety goggles!

25 Remove the end gaskets from the top of the engine block.
26 Remove the oil gallery splash pan from the engine (if so equipped).
27 Clean the mating surfaces of the intake manifold and cylinder heads. Be careful not to allow any material to fall into the intake ports (see illustration). Use a tap of the correct size to chase the threads in the bolt holes, then use compressed air to remove the debris from the holes (see illustrations). Wear safety goggles to protect your eyes when using the compressed air.

Intake manifold — disassembly and reassembly

302, 351 and 400 engines

28 If the manifold assembly is to be disassembled, identify all vacuum hoses before disconnecting them. Remove the coolant outlet housing and the thermostat. Remove the ignition coil, coolant temperature sending unit, carburetor (if not previously removed), spacer (on EEC engines remove the EGR cooler and related parts), gasket, vacuum fitting, accelerator return spring bracket and choke cable bracket.

460 engine

29 If the manifold assembly is to be disassembled, identify all vacuum hoses before disconnecting them. Remove the coolant outlet housing, gasket and thermostat. Remove the automatic choke jet tubes, carburetor (if not previously removed), spacer and gaskets. Remove the thermostatic choke heater tube, the engine temperature sending unit and the EGR valve and gasket. Discard all gaskets.

Intake manifold — installation

30 If the manifold was disassembled, reassemble it by reversing the disassembly procedure. When installing the temperature sending unit, coat the threads with electrically conductive sealant and coat the thermostat gasket with water resistant sealant.

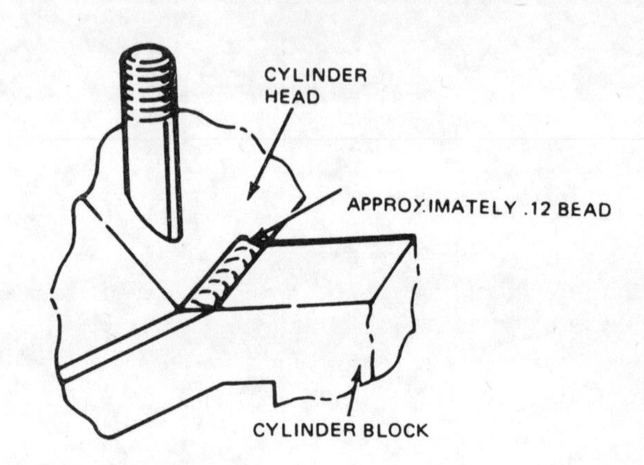

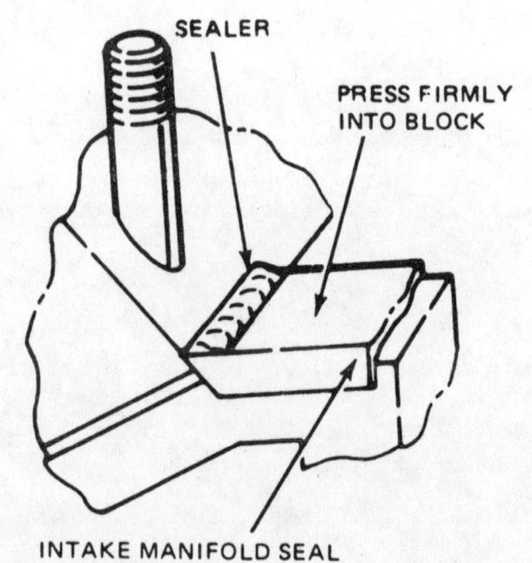

3.31 Apply RTV-type sealant to the block-to-head junctions and the intake manifold end seal-to-head junctions as shown here

31 Apply RTV-type gasket sealant to the mating points at the junctions of the cylinder heads and engine block (see illustration). **Note:** *Do not apply sealant to the waffle section of the end seals on 351W and 400 engines as the sealant will rupture the seals.*
32 Position new seals on the engine block and press the seal locating tabs into the holes in the mating surface (see illustration). This is a very critical step, as correct intake manifold sealing depends on the correct installation of these gaskets.
33 Apply RTV-type gasket sealant to the ends of the intake manifold seal (see illustration 3.31). For 351W and 400 engines, see the note in Step 31.

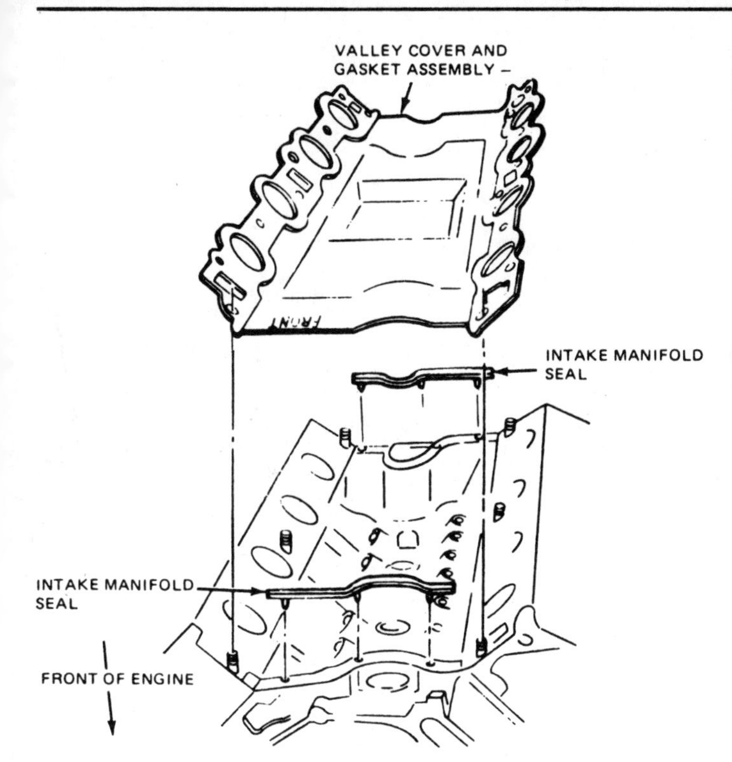

3.32 Be sure the end seal locating tabs and pegs (if used) are seated before lowering the manifold gasket into place

3.34 Align the new intake manifold gaskets by seating them over the dowel pins or studs in the cylinder heads (arrows)

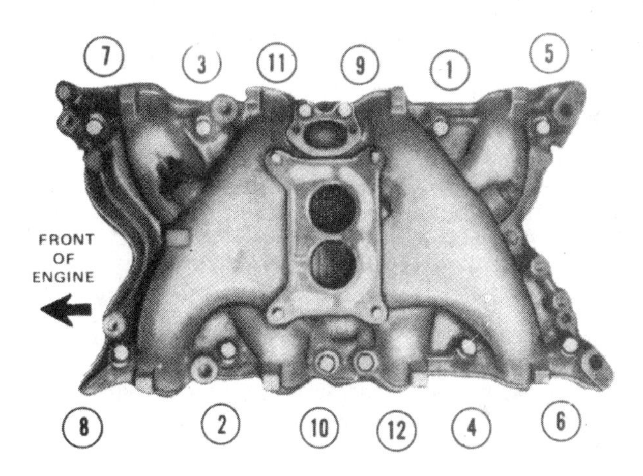

3.36b 351M and 400 engine intake manifold bolt tightening sequence

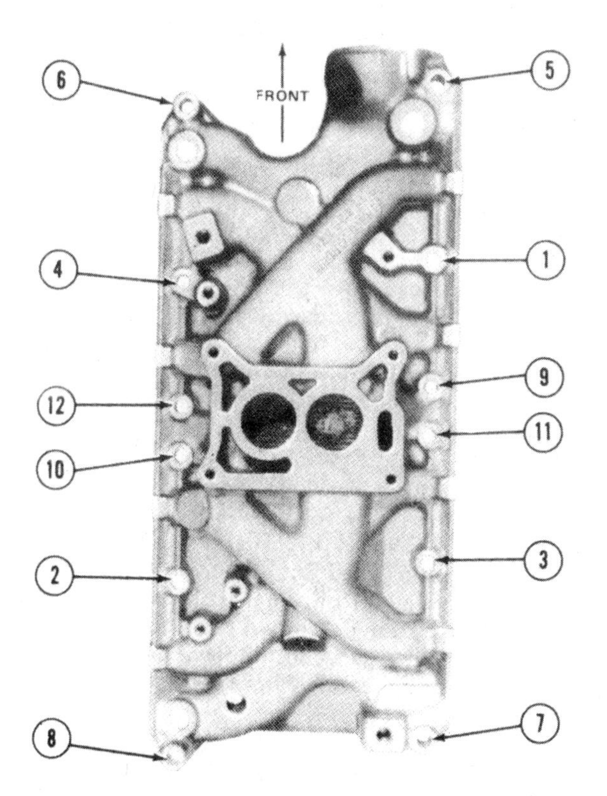

3.36a 302 and 351W engine intake manifold bolt tightening sequence

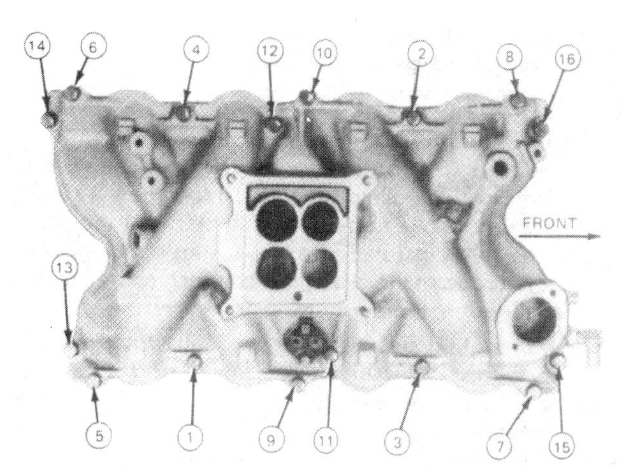

3.36c 460 engine intake manifold bolt tightening sequence

34 Position the intake manifold gasket on the block and cylinder heads with the alignment notches fitting over the dowels on the block. Be sure that all of the holes in the gasket are aligned with the corresponding holes in the cylinder heads (see illustration).

35 Lower the intake manifold into position with a lift or an assistant, being careful not to disturb the gaskets. After the manifold is in place, run a finger around the seal area to make sure that the seals are in place. If the seals are not in place, remove the manifold and reposition the seals.

36 Install the intake manifold retaining bolts finger tight. Tighten the intake manifold bolts in the recommended sequence (see illustrations). Note that different engines have different bolt tightening sequences. Work up to the final torque in three steps to avoid warping the manifold.

4.9 The rocker arm cover is held in place with several bolts (arrows)

4.11 Being careful not to damage the mating surface of the head, carefully remove the rocker arm cover gasket with a gasket scraper or putty knife

37 Install the remaining components in the reverse order of removal.
38 Start and run the engine and allow it to reach operating temperature. After it has reached operating temperature, check carefully for oil and coolant leaks.
39 Shut the engine off and retighten the manifold bolts while the engine is still warm.

Exhaust manifolds — removal

40 If the right exhaust manifold is being removed, remove the air cleaner, intake duct and heat stove.
41 If the left exhaust manifold is being removed, remove the engine oil filter on 351M and 400 models. Remove the speed control bracket (if so equipped), and, on all engines except the 460, remove the oil dipstick and tube assembly.
42 On vehicles equipped with a column selector and automatic transmission, disconnect the cross lever shaft for the automatic transmission selector to provide the clearance necessary to remove the manifold.
43 Remove the bolts holding the exhaust pipe to the exhaust manifolds.
44 Remove the spark plug heat shields (if so equipped).
45 If so equipped, flatten the tabs on the lock plates used to secure the exhaust manifold bolts. Remove the exhaust manifold mounting bolts. **Note:** *Keep track of the location of each bolt, since some of them may be shoulder stud bolts used to attach other components to the engine — they must be reinstalled in their original locations.*
46 Detach the manifold from the head.
47 Clean the mating surfaces of the manifold and the cylinder head.
48 Clean the mounting flange of the exhaust manifold and the exhaust pipe.

Exhaust manifolds — installation

49 Apply graphite grease to the mating surface of the exhaust manifold.
50 Position the exhaust manifold on the head and install the bolts. Tighten the bolts to the specified torque in three steps, working from the center to the ends. If so equipped, bend the tabs on the lock plates to keep the bolts from loosening.
51 If so equipped, install the spark plug heat shields.
52 Install the gasket or spacer between the exhaust pipe and the exhaust manifold outlet.
53 Connect the exhaust pipe to the exhaust manifold using new nuts.
54 Tighten the nuts, making sure that the exhaust pipe is situated squarely in the exhaust manifold outlet.
55 Install the oil filter if the left exhaust manifold was removed (351M or 400 engine).
56 If removed, install the oil dipstick assembly.
57 If the vehicle is equipped with a column shifter, install the automatic transmission selector cross shaft at the chassis and the engine block.
58 If the right exhaust manifold was replaced, install the air cleaner heat stove, air cleaner and intake duct.
59 Start the engine and check for exhaust leaks.

4 Rocker arm covers — removal and installation

Refer to illustrations 4.9 and 4.11

1 Raise the hood and remove the engine cover, then cover the seats to avoid damage to the upholstery. Remove the air cleaner and intake duct assembly.
2 Remove the crankcase ventilation hoses and lines (where applicable). Make sure that all lines and hoses have been removed from the rocker arm covers and position them out of the way.
3 Remove the PCV valve from the oil filler cap or rocker arm cover.
4 Some 302 engines are equipped with an electric solenoid and vacuum line that must be removed.
5 Remove the vacuum solenoid mounted on the left rocker arm cover (if so equipped).
6 Disconnect the spark plug wires. Mark them so they can be installed in their original locations.
7 Remove the spark plug wires clipped to the rocker arm covers and position them out of the way.
8 Remove the clips retaining the wiring looms running along the left rocker arm cover.
9 Remove the rocker arm cover retaining bolts (see illustration).
10 Detach the rocker arm cover(s) from the engine.
11 Remove all old gasket material and sealant from the rocker arm cover and cylinder head surfaces (see illustration).
12 Make sure the gasket surfaces of the rocker arm covers are flat and smooth, particularly around the bolt holes. Use a hammer and a block of wood to flatten them if they are deformed.
13 Attach a new rocker arm cover gasket to the cover. Notice that there are tabs provided in the cover to retain the gasket. It may be necessary to apply RTV-type gasket sealant to the corners of the cover to retain the gasket there.
14 Position the rocker arm cover on the cylinder head, making sure that the bolt holes are aligned correctly.
15 Install the rocker arm cover retaining bolts finger tight.
16 Working from the center of the rocker arm cover out, tighten the bolts to the specified torque. **Caution:** *Do not overtighten the bolts or the covers will warp and the gaskets will be pushed out of position, causing leaks.*
17 The remainder of the installation procedure is the reverse of removal.
18 Start the engine and run it until it reaches normal operating temperature, then make sure that there are no leaks.

5.3a Loosen the nut or bolt (arrow) and pivot the rocker arm to the side to remove the pushrod

5.3b A perforated cardboard box can be used to store the pushrods to ensure installation in their original locations

5 Rocker arms and pushrods — removal, inspection and installation

Refer to illustrations 5.3a and 5.3b

Removal

1 Remove the rocker arm covers as described in Section 4.
2 Remove the rocker arm fulcrum bolts or nuts.
3 Remove the oil deflectors (351M, 400 and 460 engines only), fulcrums, fulcrum guides (302 and 351W engines only) and the rocker arms. If only the pushrods are being removed, loosen the fulcrum bolts/nuts and rotate the rocker arms out of the way of the pushrods (see illustration). Keep the pushrods organized to ensure that they are reinstalled in their original locations (see illustration).

Inspection

4 Check the rocker arms for excessive wear, cracks and other damage, especially where the pushrods and valve stems contact the rocker arm faces.
5 Make sure the hole at the pushrod end of the rocker arm is open.
6 Check the rocker arm fulcrum contact area for wear and galling. If the rocker arms are worn or damaged, replace them with new ones and use new fulcrum seats as well.
7 Inspect the pushrods for cracks and excessive wear at the ends. Roll each pushrod across a flat surface to see if they are bent.

Installation

8 Apply engine oil or assembly lube to the top of the valve stem and the pushrod guide in the cylinder head.
9 Apply engine oil to the rocker arm fulcrum seat and the fulcrum seat socket in the rocker arm.
10 Install the pushrods.
11 Install the fulcrum guides (302 and 351W engines only), rocker arms, fulcrums, oil deflector (351M and 400 engines only) and fulcrum bolts or nuts.
12 Tighten the fulcrum bolts or nuts to the specified torque.
13 Replace the rocker arm covers and gaskets as described in Section 4.
14 Start the engine and check for roughness and/or noise. Refer to Section 23 of Chapter 2C for any corrections that may have to be made for either a rough running or excessively noisy engine.

6 Cylinder heads — removal and installation

Refer to illustrations 6.12, 6.15 and 6.19

Removal

1 Remove the intake manifold and carburetor as an assembly as described previously in this Chapter.
2 Remove the rocker arm covers.

6.12 The hoist chain can be attached to the head with bolts and washers threaded into the rocker arm bolt holes

3 If the left cylinder head is being removed on a vehicle equipped with factory air conditioning, detach the compressor and support it securely at the side of the engine compartment. **Warning:** *Do not disconnect the air conditioning hoses, as serious injury or damage to the system will result.*
4 If the left cylinder head is being removed and the vehicle is equipped with power steering, remove the power steering bracket retaining bolt from the left cylinder head.
5 Position the power steering pump out of the way so it will not leak fluid.
6 If the right cylinder head is being removed, remove the alternator mounting bracket through-bolt.
7 Remove the air cleaner inlet tube (if so equipped) from the right cylinder head.
8 Remove the ground wire connected to the rear of the cylinder head.
9 Remove the exhaust manifold mounting bolts.
10 Remove the pushrods as described in Section 5. The rocker arms can remain in place on each head. Be sure to mark the pushrods, as they must be reinstalled in their original locations.
11 Loosen the cylinder head bolts by reversing the order shown in the tightening sequence diagram (illustration 6.19), then remove the bolts from the heads. Keep them in order so they can be installed in their original locations.
12 Using a hoist (or an assistant), carefully remove the cylinder head from the block, using care to avoid damaging the gasket mating surfaces (see illustration).
13 For cylinder head inspection procedures, see Chapter 2, Part C.

6.15 Be certain that the cylinder head gaskets are correctly positioned right side up (note the FRONT mark on the type shown here)

Installation

14 Make sure that the cylinder head and engine block mating surfaces are clean and flat.
15 Position a new head gasket over the dowel pins on the block. Make sure that the head gasket is facing the right direction and that the correct surface is exposed. Gaskets are sometimes marked *FRONT* and *TOP* (see illustration).
16 Using a hoist (or an assistant), carefully lower the cylinder head into place on the block. Take care not to move the head sideways or scrape it across the block as it can dislodge the gasket and/or damage the gasket surfaces.
17 Coat the cylinder head bolts with light engine oil and thread them into the block.
18 Tighten the bolts finger tight.
19 Tighten the bolts in the prescribed sequence (see illustration). Work up to the final torque in three steps, going through the pattern completely on each step.
20 Apply engine oil to the rocker arm fulcrum seats and sockets.
21 Install the pushrods and the rocker arm cover(s).
22 The remainder of the installation procedure is the reverse of removal.
23 Start and run the engine and check carefully for leaks and unusual noises.

7 Timing cover and chain – removal and installation

Refer to illustrations 7.14, 7.21, 7.23, 7.27, 7.30, 7.32, 7.39 and 7.43

Removal

1 Drain the cooling system.
2 Remove the screws attaching the radiator shroud to the radiator.
3 Remove the bolts attaching the fan to the water pump shaft.
4 Remove the fan and the radiator shroud.
5 Disconnect the upper radiator hose at the thermostat housing.
6 Disconnect the lower radiator hose at the water pump outlet.
7 Disconnect the transmission oil cooler lines at the radiator (if so equipped).
8 Loosen the alternator mounting and adjusting bolts to relieve the tension on the drivebelt. If the vehicle is equipped with air conditioning, loosen the air conditioning idler pulley.
9 Remove the air pump (if so equipped).
10 Remove the drivebelts and the water pump pulley.
11 Remove the bolt attaching the air conditioning compressor support,

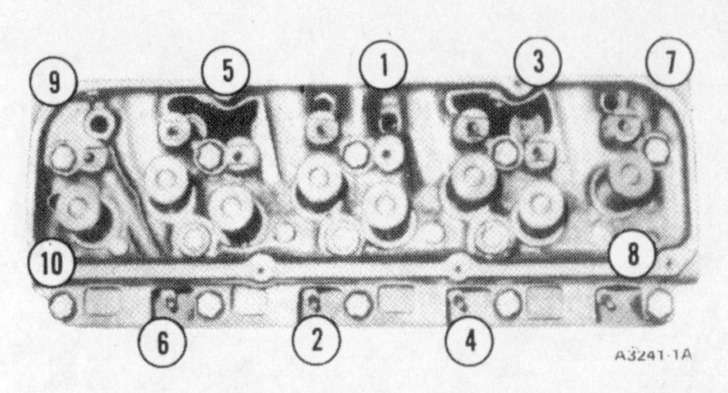

6.19 Cylinder head bolt tightening sequence

water pump and compressor. Remove the compressor support. **Warning:** *Do not loosen or remove the air conditioning hoses, as serious personal injury or damage to the system could result.*
12 Remove the bolts and washers retaining the crankshaft pulley to the vibration damper, then remove the pulley.
13 Remove the large bolt and washer retaining the damper to the crankshaft.
14 Remove the vibration damper from the crankshaft with a puller (see illustration).
15 Remove the bypass hose from the top of the water pump.
16 Disconnect the heater return hose or tube at the water pump.
17 Remove and plug the fuel line at the fuel pump.
18 Disconnect the fuel line at the carburetor.
19 Remove the fuel feed line from the fuel pump.
20 Remove the fuel pump (Chapter 4).
21 Remove the bolts holding the timing cover to the engine block and those holding the oil pan to the timing cover (see illustration).
22 Remove the timing cover and water pump as an assembly.
23 Use a knife or similar tool to cut the oil pan seal flush with the engine block mating surface (see illustration).
24 Remove the timing cover gasket and oil pan seal.
25 If the cover is being replaced with a new one, remove the water pump from the timing cover.
26 Check the timing chain deflection by rotating the crankshaft in a counterclockwise direction to take up the slack on the left side of the chain.
27 Establish a reference point on the block and use a ruler to check the distance from the reference point to the left side of the chain (see illustration).
28 Rotate the crankshaft in the opposite direction to take up the slack on the opposite (or right) side of the chain.
29 Force the left side of the chain out and measure the distance between the reference point and the chain. This will give you the deflection. If the deflection exceeds 1/2-inch, the timing chain and sprockets should be replaced with new parts.
30 If the timing chain and sprockets are being removed, turn the engine until the timing marks are aligned (see illustration).
31 Remove the camshaft sprocket retaining bolt, washer, fuel pump eccentric (two piece on 460 engines) and the front oil slinger (if present) from the crankshaft.
32 Slide the timing chain and sprockets forward and off of the camshaft and crankshaft as an assembly (see illustration).

Installation

33 Assemble the timing chain and sprockets so the timing marks are in alignment.
34 Install the chain and sprockets onto the camshaft and crankshaft as an assembly. Make sure that the timing marks remain in proper alignment during the installation procedure.
35 Install the oil slinger (if so equipped) over the nose of the crankshaft.
36 Install the fuel pump eccentric, camshaft sprocket retaining bolt and washer. Tighten the bolt to the specified torque. Lubricate the timing chain and gears with engine oil.
37 Clean the old gasket material from the gasket mating surface on the oil pan.
38 Coat the gasket surface of the oil pan with RTV-type gasket sealant. Cut and position the required sections of new front cover-to-oil pan seal on the oil pan.

7.14 Remove the vibration damper with a puller

7.21 When removing the timing chain cover, remove only the oil pan bolts that thread into the cover — leave the rest in place

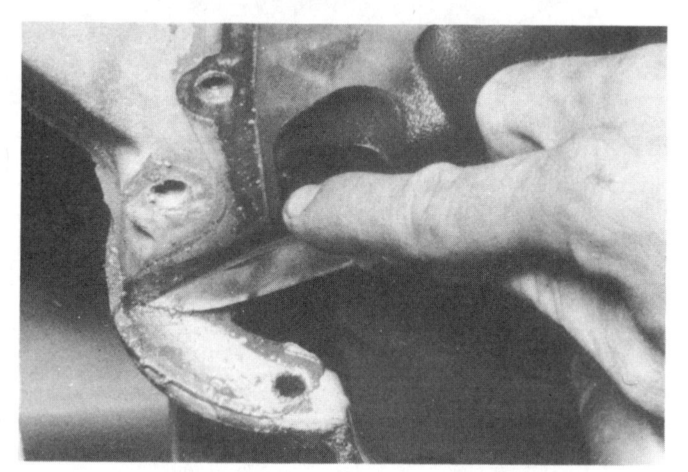

7.23 The timing chain cover-to-oil pan gasket must be cut at the block-to-pan junction with a knife

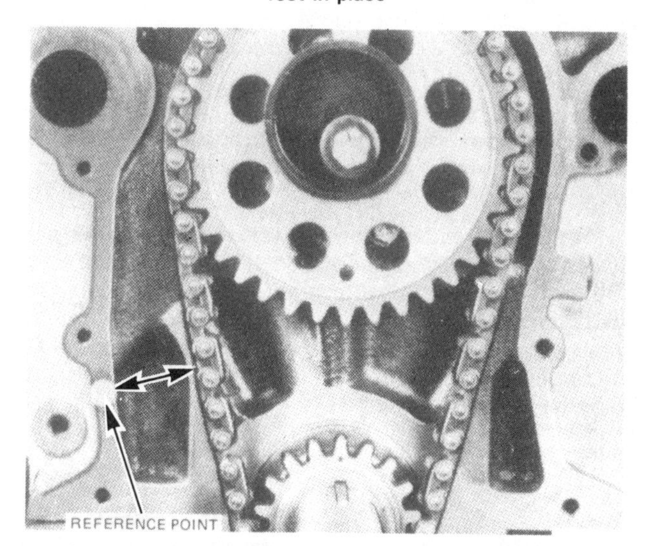

7.27 Mark a reference point on the block, then measure the distance from the point to the chain (push the chain in and take another measurement to determine timing chain deflection)

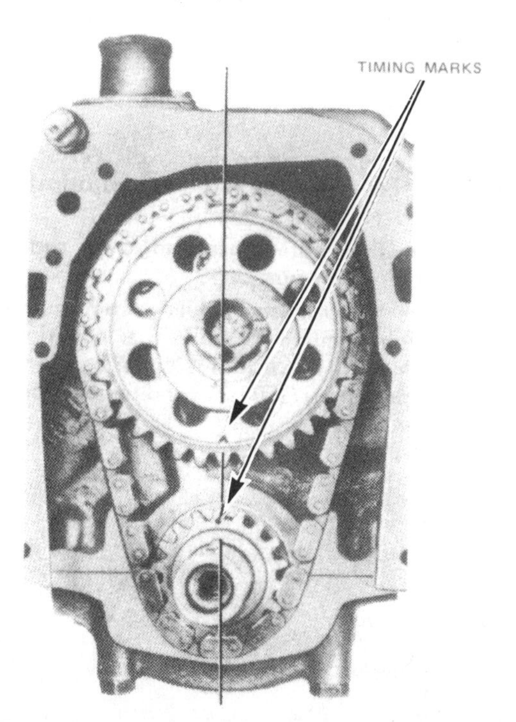

7.30 The camshaft and crankshaft timing marks must be aligned as shown here (make sure the marks are facing each other)

7.32 With the timing marks aligned, the chain and sprockets can be removed as an assembly

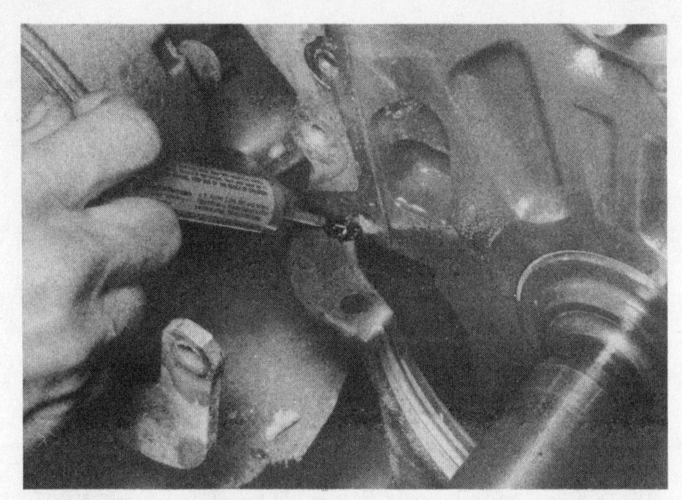

7.39 Before positioning the new oil pan gasket sections, apply a bead of RTV-type sealant at the junction of the oil pan and block as shown here

39 Apply sealant at the corners of the mating surfaces (see illustration).
40 Coat the gasket surfaces of the cover with sealant and install a new gasket. Coat the mating surface on the block with sealant.
41 Position the timing cover on the block after installing a new front crankshaft seal (Section 9). Coat the seal lip with grease.
42 Use care when installing the cover to avoid damaging the front crankshaft seal.
43 Install the cover alignment tool to position the cover properly (see illustration). If no alignment tool is available, use the vibration damper to position the seal. It may be necessary to force the cover down slightly to compress the oil pan seal. This can be done by inserting a punch into the bolt holes.
44 Coat the threads of the retaining bolts with RTV-type sealant and install them.
45 While holding the cover in alignment, tighten the oil pan-to-front cover retaining bolts.
46 Remove the alignment punch.
47 Tighten the bolts holding the cover to the engine block and remove the alignment tool or vibration damper.
48 Apply a thin coat of grease to the vibration damper seal contact surface.
49 Install the crankshaft spacer.
50 Install the Woodruff key in the crankshaft and slide the vibration damper into position.
51 Install the crankshaft damper retaining bolt and washer. Tighten the bolt to the specified torque.
52 This bolt may be used to push the damper onto the crankshaft if the proper installation tool is not available.
53 Attach the crankshaft pulley to the damper and install the pulley retaining bolts.
54 Install the fuel pump with a new gasket.
55 Connect the fuel lines to the fuel pump.
56 The remainder of the installation procedure is the reverse of removal. Make sure that all bolts are tightened securely.
57 If any coolant entered the oil pan when separating the timing chain cover from the block, the crankcase oil should be changed and the oil filter replaced.
58 Start and run the engine at a fast idle and check for coolant and oil leaks.
59 Check the engine idle speed and ignition timing.

8 Camshaft and lifters — removal and installation

Refer to illustrations 8.6 and 8.17

Note: *Refer to Chapter 2, Part C, Section 15 and check the camshaft lobe lift as described there before removing the camshaft.*

Removal

1 Open the hood and remove the engine cover, then cover the seats to avoid damage to the upholstery. After marking the hoses to simplify

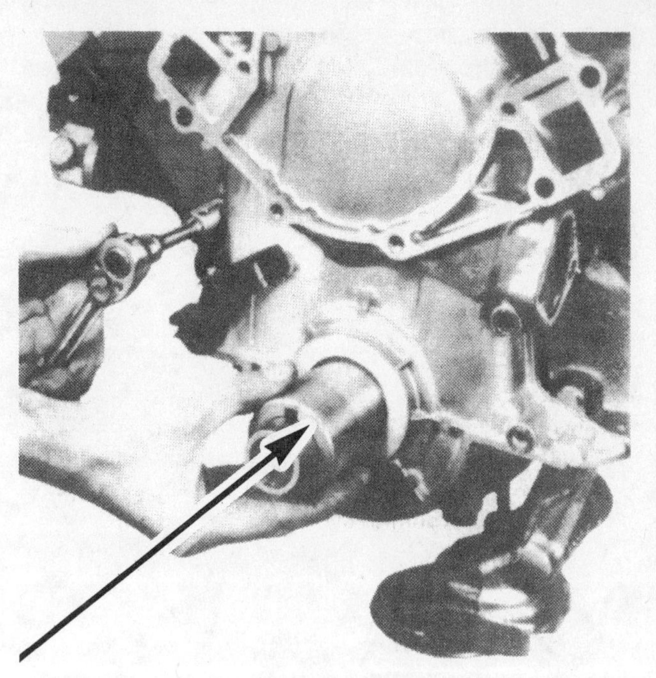

7.43 Use the special tool (shown here) or the vibration damper to align the cover over the crankshaft before tightening the bolts

installation and disconnecting them, remove the air cleaner.
2 Remove the intake manifold (see Section 3).
3 Remove the rocker arm covers (see Section 4).
4 Loosen the rocker arm nuts/bolts and rotate the rocker arms to the side.
5 Mark the pushrods if they are to be reused and remove them from the engine.
6 Remove the valve lifters from the engine using a special tool designed for this purpose (see illustration). Sometimes they can be removed with a magnet if there is no varnish build-up or wear. If they are stuck in their bores you will have to obtain the special tool designed for grasping lifters internally and work them out.
7 Remove the timing cover, chain and sprockets (see Section 7).
8 Remove the grille, radiator and air conditioning condenser (see Chapter 3).
9 Remove the fuel pump.
10 Remove the bolts securing the camshaft thrust plate to the engine block.
11 Slowly withdraw the camshaft from the engine, being careful not to nick, scrape or otherwise damage the bearings with the cam lobes.
12 See Chapter 2, Part C for camshaft and lifter inspection procedures.

Installation

13 Lubricate the camshaft journals with engine oil and apply engine assembly lube to the cam lobes.
14 Slide the camshaft into position, being careful not to scrape or nick the bearings.
15 Install the camshaft thrust plate. Tighten the thrust plate retaining bolts to the specified torque.
16 Check the camshaft end play by pushing it toward the rear of the engine.
17 Install a dial indicator so that the stem is on the camshaft sprocket retaining bolt (see illustration). Zero the dial indicator.
18 Position a large screwdriver between the camshaft gear and the block. **Caution:** *Do not pry against aluminum or nylon camshaft sprockets when any type of valve train load is on the camshaft, as damage to the sprocket can result.*
19 Force the camshaft forward with the screwdriver and release it.
20 Compare the dial indicator reading to the Specifications in Part C of this Chapter.
21 If the end play is excessive, check the spacer for correct installation before it is removed. If the spacer is installed correctly, replace the thrust plate. Notice that the thrust plate has a groove on it, which should face **IN** on all engines.

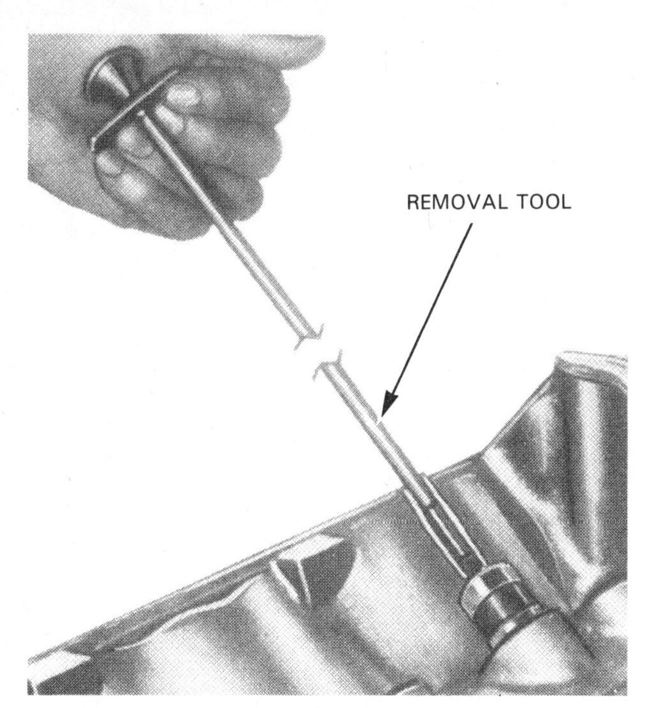

8.6 A special tool may be required to remove the lifters from the bores

8.17 Checking camshaft end play with a dial indicator

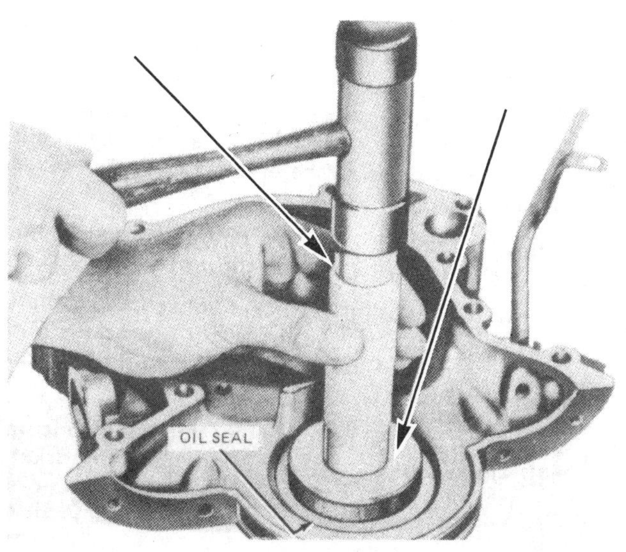

9.4 On some engines, where the timing chain cover must be removed, a special tool is used to install and seat the front oil seal (a large drift can be used if care is exercised)

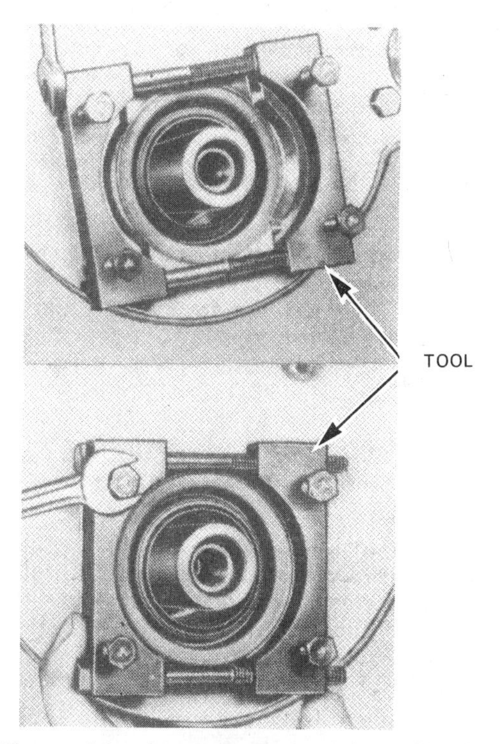

9.5 The special tool is clamped around the seal (top), then the bolts are tightened to pry the seal from the timing cover (bottom)

22 Check the timing chain deflection as described in Section 7.

23 Install the hydraulic valve lifters in their original bores if the old ones are being used. Make sure they are coated with engine assembly lube. Never use old lifters with a new camshaft or new lifters with an old camshaft.

24 The remainder of the installation procedure is the reverse of removal.

9 Crankshaft oil seals – replacement

Refer to illustrations 9.4, 9.5, 9.7, 9.14 and 9.17

Note: *Some early model 302, 351 and 400 engines require removal of the timing chain cover to replace the oil seal. If you are working on one of these engines, follow the seal replacement procedure for the 460 engine outlined in Steps 1 through 4.*

Front seal

460 engine

1 On 460 engines, remove the timing chain cover (see Section 7).

2 Tap the front crankshaft seal out with a drift punch and hammer. **Note:** *The front oil seal should be replaced with a new one whenever the cover is removed.*

3 Clean the groove in the front cover.

4 Install the new front oil seal using a special drive tool (see illustration), or a large drift if the drive tool is not available. If you are using a large drift, be very careful not to damage the seal. Make sure that the spring remains positioned inside the seal.

All other engines

5 Remove the front seal using the special puller available from Ford (see illustration). The cover does not have to be removed from the engine.

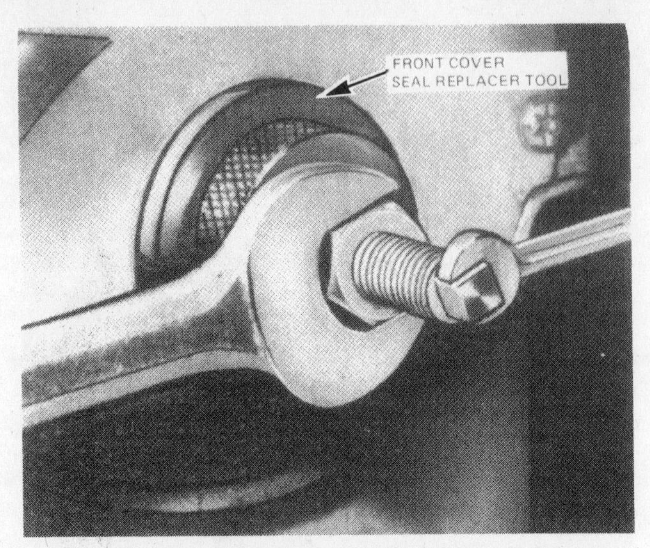

9.7 On engines where the timing chain cover does not require removal, a special tool is used to push the seal into the bore

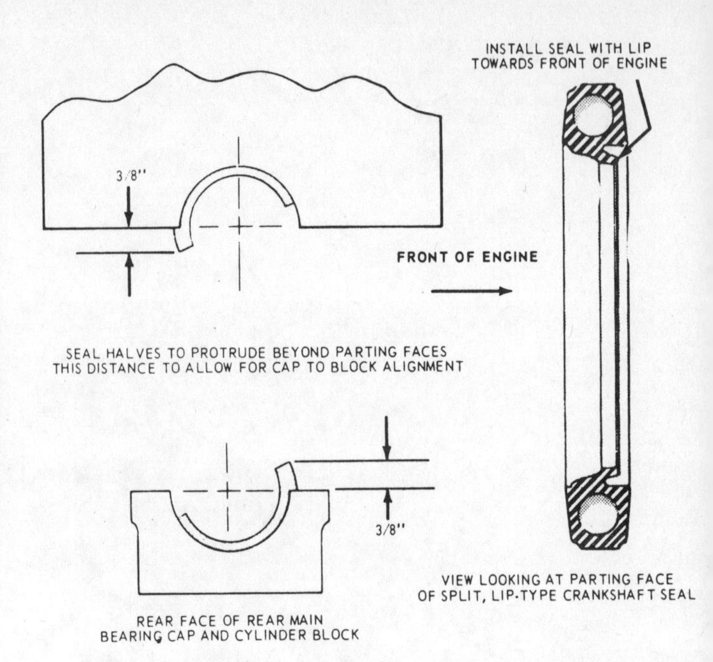

9.14 The split-type rear oil seal must protrude from the block and cap as shown (make sure the lip faces the front of the engine and remove the pin if one was present)

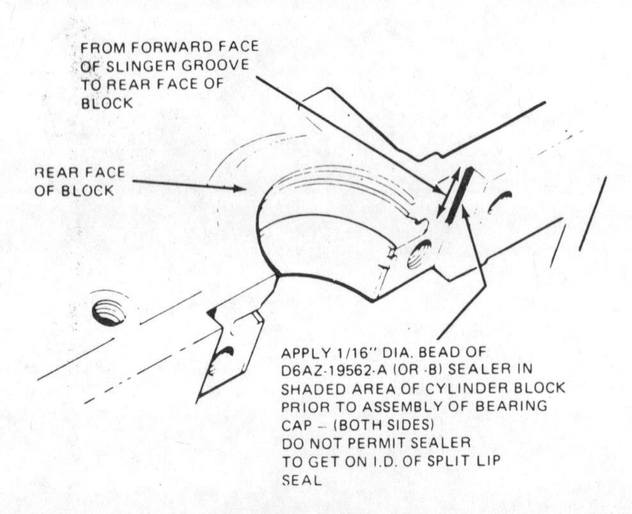

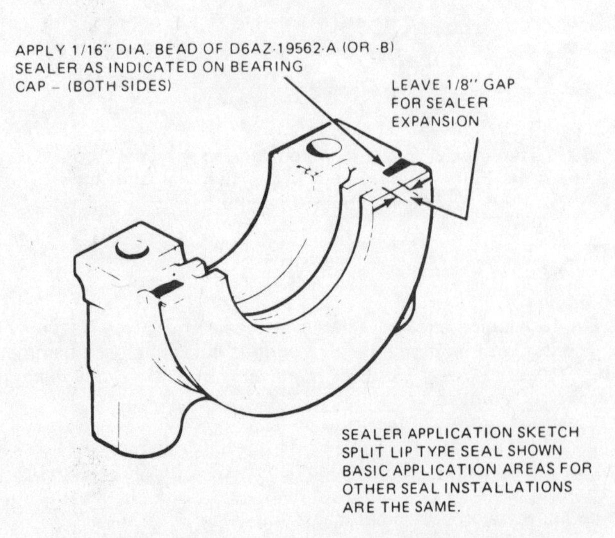

9.17 Apply RTV-type sealant to the block and cap as shown here before installing the cap (don't get sealer on the seal ends or the bearing surfaces)

6 Clean the crankshaft front seal groove.

7 Install the new seal using the special drive tool available from Ford (see illustration). A bolt or threaded rod, nut and very large washer may work as well.

Rear Seal – two-piece

Engine in vehicle

8 The oil pan and oil pump must be removed to gain access to the seal, but the crankshaft can remain in place.

9 Loosen the main bearing cap bolts slightly to allow the crankshaft to drop no more than 1/32-inch.

10 Remove the rear main bearing cap and detach the oil seal from the cap. If so equipped, remove the oil seal locating pin from the cap. The pin is not used with the split-type seal that will be installed. To remove the portion of the rear main seal housed in the block, thread a small sheet metal screw into one end of the seal and pull on the screw to rotate the seal out of the groove. Exercise extreme care during this procedure to prevent scratching or damaging the crankshaft seal surface.

11 Carefully clean the seal grooves in the cap and block with a brush dipped in solvent.

12 Inspect the bearing cap and engine block mating surfaces and seal grooves for nicks, burrs and scratches. Remove any defects with a fine file or deburring tool.

13 Dip both new seal halves in clean engine oil.

14 Push one seal section into the block groove with the lip facing the **front** of the engine. Leave one end protruding from the block 3/8-inch and make sure it is completely seated (see illustration). Make sure no rubber has been shaved from the outer edge of the seal by the edge of the groove and wipe any engine oil off the cap-to-block mating surfaces.

15 Install the remaining seal half in the rear main bearing cap. In this case, leave the opposite end of the seal protruding from the cap the same distance the block seal is protruding from the block. Make sure the seal lip faces the **front** of the engine.

16 Tighten all but the rear main bearing cap bolts to the specified torque.

17 Apply a thin, even film of RTV-type gasket sealant to the areas of the cap and block indicated in the accompanying illustration. **Caution:** *Do not get any sealant on the bearing face, crankshaft journal or seal lip. Lubricate the seal lips with moly-base grease or engine assembly lube.*

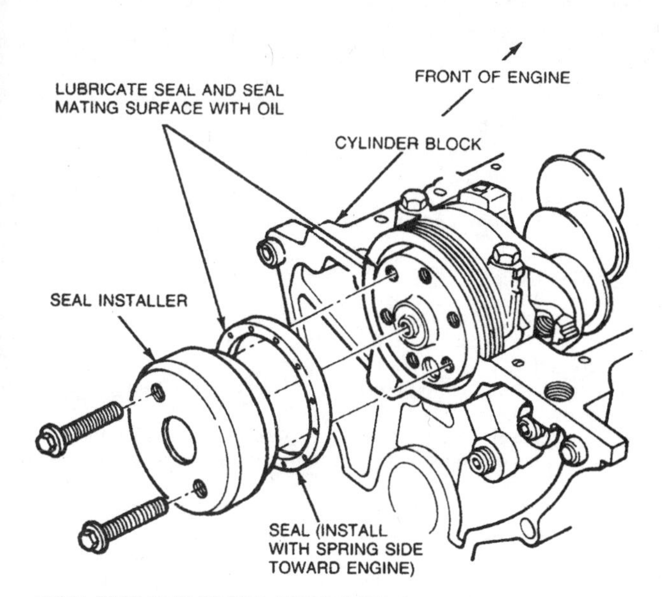

NOTE: REAR FACE OF SEAL MUST BE WITHIN 0.127mm (0.005-INCH) OF THE REAR FACE OF THE BLOCK

Fig. 9.29 One piece rear seal installation

18 Carefully install the rear bearing cap, making sure the protruding seal ends enter the grooves and that no sealant gets on the seals. Install the bolts and tighten them to the specified torque.
19 Install the oil pan and oil pump.
20 The remainder of the installation procedure is the reverse of removal.

Engine out of vehicle

21 When the engine is being rebuilt (and the crankshaft has been removed), the seal installation procedure is slightly different, since the seal halves are installed in the block and cap before the crankshaft is installed.
22 On some engines, a pin is used to locate the seal section in the cap groove. Be sure to remove the pin before installing the new seal (it does not require the pin).
23 Follow the procedure described above, but note that the seal section installed in the block does not have to be pushed into place from one end. Since the crankshaft is not in the way, the seal can be installed in the block groove in the same manner as the cap groove.
24 After the seal has been positioned in the block, the crankshaft can be laid in place (Chapter 2, Part C) and the main bearing caps installed.

Rear Seal – one-piece

25 On models with a one piece rear seal, the transmission and flywheel or flexplate must be removed to gain access to the seal. Refers to Chapters 7 and 2B for these procedures.
26 Using a sharp awl, punch a hole into the seal metal surface between the seal lip and the engine block. Screw in the threaded end of a slide hammer, dent puller, or equivalent tool. Using the puller remove the seal. **Caution:** *Do not scratch or damage the crankshaft or oil sealing surface.*
27 Clean the crankshaft and oil seal recess in the cylinder block and main bearing cap. Inspect the crankshaft to seal contact surface for scratches or nicks that could damage the new seal lip and cause oil leaks.
28 Coat the new seal and the crankshaft with a light coat of engine oil.
29 Start the new seal into the recess with the seal lip facing inward and install with an appropriate tool (see illustration). Either Ford tool number T65P-6701-A or T82L-6701-A may be used or you may be able to fashion your own tool from a large section of plastic or metal pipe. Press the seal squarely into the recess bore until the tool contacts the engine block surface. Remove the tool and inspect the seal to ensure it was not damaged during installation.
30 Install the flywheel and transmission (refer to Chapters 7 and 2B).

10 Engine — removal and installation

Refer to illustration 10.32

Removal — thru 1973

1 Remove the engine cover.
2 Remove the bolts securing the right seat to the floor and lift the seat out of the vehicle.
3 Refer to Chapter 1 if necessary and drain the cooling system.
4 Refer to Chapter 4 if necessary and remove the air cleaner and intake duct assembly, disconnecting the crankcase ventilation hose at the same time.
5 Disconnect the battery and alternator ground cables from the cylinder block.
6 Detach the oil filler tube from the dash panel bracket and rocker cover and remove it.
7 Detach the upper and lower coolant hoses from the radiator and, on automatic transmission equipped vehicles, disconnect the hoses from the oil cooler. Plug the ends of the hoses.
8 Remove the radiator bolts and lift out the radiator.
9 Disconnect the heater hoses from the engine and tie them out of the way.
10 Remove the cooling fan bolts and the fan, spacer, pulley and drivebelt.
11 Disconnect the accelerator linkage from the shaft on the left cylinder head. Detach the kickdown rod from the carburetor and vacuum line from the intake manifold (automatic transmission models only).
12 Unclip the electrical wiring harness from the left rocker arm cover and tie it out of the way.
13 Remove the upper nut securing the right exhaust manifold to the exhaust pipe.
14 Raise the front of the vehicle and support it securely on jackstands.
15 Drain the engine oil into a container and unscrew the oil filter.
16 Disconnect the fuel line from the fuel pump.
17 Disconnect the oil dipstick tube bracket from the exhaust manifold and oil pan.
18 On models equipped with a manual transmission, remove the bolts securing the equalizer arm bracket to the engine block and clutch housing. Disconnect the clutch linkage and remove the bracket and springs.
19 Disconnect the cable from the starter. Remove the bolts and detach the starter motor.
20 On vehicles with a manual transmission, disconnect the driveshaft from the rear axle and carefully pull the driveshaft from the rear of the transmission (refer to Chapter 7 if necessary).
21 On vehicles with an automatic transmission, disconnect the driveshaft at the companion flange.
22 Disconnect the speedometer cable and shift linkage from the transmission.
23 Place a jack under the transmission and raise it sufficiently to support the weight of the transmission. Remove the bolts securing the crossmember to the chassis. Lower the jack slightly and remove the nut and bolt securing the engine rear support to the crossmember. Remove the crossmember.
24 On vehicles with a manual transmission, remove the bolts securing the transmission to the clutch housing, withdraw the transmission and lower it to the ground.
25 If an automatic transmission is installed, remove the lower front cover plate from the torque converter housing, remove the dipstick tube and drain the transmission fluid into a container. Remove the nuts securing the converter to the driveplate and disconnect the oil cooler and vacuum lines from the transmission. Remove the bolts retaining the transmission to the rear of the engine and withdraw the transmission to the rear and down.
26 Disconnect the front exhaust pipes from the exhaust manifolds.
27 Remove the nuts securing the two front engine mounts to the support frame.
28 Remove the bolt securing the bellcrank assembly to the side of the engine block and tie the bellcrank out of the way.
29 Lower the front of the vehicle to the ground.
30 Remove the bolts securing the alternator and adjusting arm to the engine and position it out of the way.
31 Remove the stud from the carburetor air inlet and disconnect the fuel hose from the fuel pump.

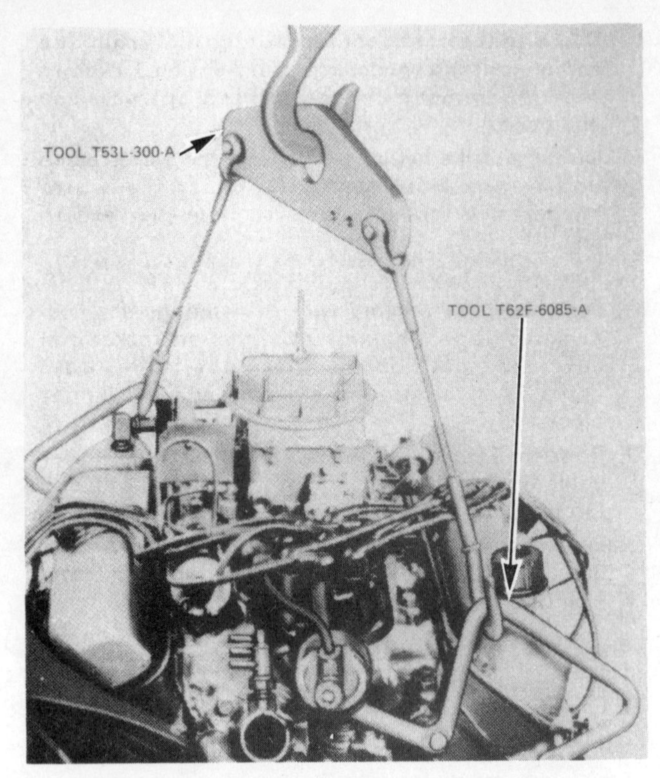

10.32 A sling or chain arrangement must be used to remove the engine from the engine compartment

32 Attach a sling and hoist to the engine (see illustration). Carefully lift the engine up and to the rear and withdraw it through the right door opening.

Removal — 1974 and later

33 Remove the engine cover.
34 Open the hood and mark the position of the hinges with a pencil before removing the retaining bolts and lifting away the hood.
35 Disconnect the battery cables.
36 Refer to Chapter 1 if necessary and drain the cooling system.
37 Remove the radiator grille and gravel deflector assembly.
38 Remove the upper grille support bracket, hood lock support bracket and condenser upper mounting brackets where used.
39 If air conditioning is installed have the system discharged by an air conditioning specialist (do not attempt to do it yourself) and remove the condenser.
40 Remove the accelerator cable bracket assembly.
41 Disconnect the upper and lower hoses from the radiator and the two heater hoses from the engine. If the vehicle is equipped with an automatic transmission, disconnect the hoses from the transmission cooler. Plug the ends to prevent contamination by dirt. If an engine oil cooler is installed, disconnect the lines at the filter adapter. Do not attempt to loosen the lines at the oil cooler.
42 Remove the fan shroud and fan assembly. Undo the retaining bolts and lift the radiator out of the vehicle.
43 Loosen the alternator adjusting bolts and pivot it in towards the engine. Remove the wires from the rear of the alternator.
44 Remove the air cleaner assembly complete with the inlet duct and valve (refer to Chapter 4 if necessary).
45 Remove the exhaust manifold shroud and the flexible tube from the manifold.
46 Disconnect the throttle linkage from the carburetor and remove the accelerator cable bracket from the engine.
47 If an automatic transmission is installed, disconnect the shift rod from the carburetor.
48 Disconnect the fuel lines and hoses, choke and vacuum lines from the carburetor. Remove the retaining nuts and lift off the carburetor and spacer.
49 Raise the front of the vehicle and support it securely on jackstands.
50 Drain the engine oil into a container and unscrew the oil filter.
51 Detach the front exhaust pipes from the engine manifolds.
52 Remove the transmission filler tube retaining bracket from the right cylinder head.
53 Remove the nuts and bolts securing the two front engine mounts to the supports.
54 Disconnect the lead from the starter motor and remove the motor.
55 On vehicles equipped with an automatic transmission, remove the cover plate and unscrew the bolts securing the converter to the driveplate. Remove the three bolts retaining the adapter plate to the converter housing.
56 Remove the four bolts securing the converter housing to the engine and the bolt attaching the ground cable to the engine block.
57 In the case of vehicles equipped with a manual transmission, remove all the bolts securing the transmission to the engine.
58 Lower the front of the vehicle to the ground and place a jack beneath the transmission to support its weight.
59 If power steering is installed, loosen the pump bracket bolts and remove the drivebelt. Unscrew the bolts from the front of the power steering support bracket.
60 Disconnect the vacuum lines from the rear of the intake manifold.
61 Unclip the wiring harness from the engine and position it out of the way.
62 Remove the speed control servo and the accelerator cable bracket from the intake manifold and position them out of the way.
63 If air conditioning is installed, disconnect the clutch lead wire from the compressor.
64 Make sure the transmission is securely supported and, on automatic transmission equipped models, remove the two upper converter housing-to-engine bolts.
65 Attach a lifting sling or chain to the engine and then carefully lift the engine forward to clear the transmission input shaft. Finally, withdraw the engine from the front of the vehicle.

Installation

66 Using the same sling or chain used for engine removal, raise the engine and position it in the engine compartment.
67 Lower the engine, keeping all lines, hoses, wires and cables well clear of the sides. It is a good idea to have a second person guiding the engine while it is being lowered.
68 The tricky part is mating the engine to the transmission, which involves locating the transmission input shaft in the clutch and flywheel. Provided that the clutch plate has been centered correctly as described in Chapter 8, there should be little difficulty. Grease the splines of the transmission input shaft first. It may be necessary to rock the engine from side-to-side in order to get it into place. Do not, under any circumstances, strain the transmission input shaft. This could occur if the shaft was not completely seated and the engine was raised or lowered more than the amount required for very slight adjustments of position.
69 As soon as the engine is seated against the transmission, install the bolts holding the two together.
70 Lower the engine onto the mounting brackets at the front and install and tighten the nuts and washers.
71 Refill the engine with fresh oil and coolant after it is installed.

Chapter 2 Part C
General engine overhaul procedures

Contents

Specifications

240 and 300 CID six cylinder engines

General

Bore and stroke	
240	4.00 x 3.18 in
300	4.00 x 3.98 in
Oil pressure (at 2000 rpm — normal operating temperature)	
thru 1972	35 to 60 psi
1973 and later	40 to 60

Engine block

Cylinder bore	
diameter	
thru 1973	4.000 to 4.0036 in
1974 and later	4.000 to 4.0048 in
taper limit	0.010 in
out-of-round limit	0.005 in
Deck warpage limit	0.003 in per 6 in or 0.006 in overall

Pistons and rings

Piston	
diameter (thru 1979)	
coded red	3.9984 to 3.9990 in
coded blue	3.9996 to 4.0002 in
diameter (1980 and later)	
coded red	3.9982 to 3.9988 in
coded blue	3.9994 to 4.0000 in
Piston-to-cylinder bore clearance	
standard	0.0014 in
service limit	
thru 1984	0.0022 in
1985 and later	0.0018 in

240 and 300 CID six cylinder engines (continued)

Piston ring-to-groove side clearance	
standard	
top ring	
thru 1976	0.002 to 0.004 in
1977 and later	0.0019 to 0.0036 in
2nd ring	
240	0.0025 to 0.004 in
300	0.002 to 0.004 in
oil ring	Snug fit in groove
service limit	
240	0.006 in
300	0.002 in maximum increase in clearance
Piston ring end gap	
top ring	0.010 to 0.020 in
2nd ring	0.010 to 0.020 in
oil ring	0.015 to 0.055 in

Crankshaft and flywheel

Main journal	
diameter	2.3982 to 2.3990 in
taper limit	
thru 1973	0.0003 in per inch
1974 and later	0.0005 in per inch
out-of-round limit	
thru 1973	0.0004 in
1974 and later	0.0006 in
runout limit	0.002 in
Main bearing oil clearance	
standard	
240	0.0005 to 0.0015 in
300	0.0008 to 0.0015 in
service limit	
240	0.0022 in
300	
light duty	0.0026 in
heavy duty	0.0028 in
Connecting rod journal	
diameter	2.1228 to 2.1236 in
taper limit	
thru 1973	0.0003 in per inch
1974 and later	0.0006 in per inch
out-of-round limit	
thru 1973	0.0004 in
1974 and later	0.0006 in
Connecting rod bearing oil clearance	
standard	0.0008 to 0.0015 in
service limit	
1973 thru 1976 300 CID only	0.0027 in
all others	0.0024 in
Connecting rod side clearance	
standard	0.006 to 0.013 in
service limit	0.018 in
Crankshaft end play	
standard	0.004 to 0.008 in
service limit	0.012 in
Flywheel clutch face runout limit	0.010 in
Flywheel ring gear lateral runout limit	
manual transmission	
thru 1974	0.045 in
1975 and later	0.040 in
automatic transmission	
thru 1974	0.040 in
1975 and later	0.060 in

Camshaft

Bearing journal	
diameter	2.017 to 2.018 in
runout limit	0.008 in TIR
out-of-round limit	0.0005 in
Bearing oil clearance	
standard	0.001 to 0.003 in
service limit	0.006 in
Lobe lift	0.2490 in

Cylinder head and valve train

Head warpage limit	0.006 in per 6 in or 0.007 in overall
Valve seat angle	45°
Valve seat width	
intake	0.060 to 0.080 in
exhaust	0.070 to 0.090 in
Valve seat runout limit	
thru 1973	0.0015 in TIR
1974 and later	0.002 in TIR
Valve face angle	44°
Valve face runout limit	0.002 in TIR
Minimum valve margin width	1/32 in
Valve stem diameter — standard	
intake	0.3416 to 0.3423 in
exhaust	0.3416 to 0.3423 in
Valve guide diameter	
intake	0.3433 to 0.3443 in
exhaust	0.3433 to 0.3443 in
Valve stem-to-guide clearance	
intake	
standard	0.0010 to 0.0027 in
service limit	0.0055 in
exhaust	
standard	0.0010 to 0.0027 in
service limit	0.0055 in
Valve spring free length (approx.)	
intake	
thru 1983	1.99 in
1984	1.97 in
1985 and later	1.96 in
exhaust	
thru 1984	1.87 in
1985 and later	1.78 in
Valve spring installed height	
intake	
thru 1984	1.69 to 1.72 in
1985 and later	1.61 to 1.67 in
exhaust	
thru 1984	1.56 to 1.59 in
1985 and later	1.44 to 1.50 in
Valve spring out-of-square limit	0.078 (5/64) in
Collapsed lifter gap (valve clearance)	
240	
thru 1973	
allowable	0.074 to 0.174 in
desired	0.124 in
1974 and later	
allowable	0.100 to 0.200 in
desired	0.100 to 0.150 in
300	
thru 1973	
allowable	0.074 to 0.174 in
desired	0.124 in
1974 and later	
allowable	0.100 to 0.200 in
desired	0.125 to 0.175 in
Lifter diameter	0.8740 to 0.8745 in
Lifter bore diameter	0.8752 to 0.8767 in
Lifter-to-bore clearance	
standard	0.0007 to 0.0027 in
service limit	0.005 in
Pushrod runout limit	0.015 in

Oil pump

Outer race-to-housing clearance	0.001 to 0.013 in
Rotor assembly end clearance	0.004 in maximum
Driveshaft-to-housing bearing clearance	0.0015 to 0.0030 in
Relief spring tension	20.6 to 22.6 lbs @ 2.49 in
Relief valve clearance	0.0015 to 0.0030 in

Torque specifications*	**Ft-lbs**
Main bearing cap bolts	60 to 70
Connecting rod cap nuts	40 to 45
Rocker arm stud nut breakaway torque	4.5 to 15
Rocker arm stud nut (1979 on)	17 to 23

* *Additional torque specifications can be found in Chapter 2, Part A*

302 and 351W CID V8 engines

General

Bore and stroke	
302	4.00 x 3.00 in
351W	4.00 x 3.50 in
Oil pressure (at 2000 rpm — normal operating temperature)	
302	
thru 1972	35 to 60 psi
1973 and later	40 to 60 psi
351W	40 to 65 psi

Engine block

Cylinder bore	
diameter	
302	
thru 1973	4.0004 to 4.0036 in
1974 and later	4.0004 to 4.0052 in
351W	4.0000 to 4.0048 in
taper limit	0.0010 in
out-of-round limit	0.005 in
Deck warpage limit	0.003 in per 6 in or 0.006 in overall

Pistons and rings

Piston diameter	
302 (coded red)	3.9984 to 3.9990 in
302 (coded blue)	3.9996 to 4.0002 in
351W (coded red)	3.9978 to 3.9984 in
351W (coded blue)	3.9990 to 3.9996 in
Piston-to-bore clearance — selective fit	
1977 thru 1979 351W only	0.0022 to 0.0030 in
all others	0.0018 to 0.0026 in
Piston ring-to-groove clearance	
top compression ring	
thru 1976	0.002 to 0.004 in
1977 thru 1981	0.0019 to 0.0036 in
bottom compression ring	0.002 to 0.004 in
oil control ring	Snug fit in groove
service limit	0.002 in maximum increase in clearance
Piston ring end gap	
top compression ring	0.010 to 0.020 in
bottom compression ring	0.010 to 0.020 in
oil control ring	
thru 1978 and 1985 and later	0.015 to 0.055 in
1979	0.015 to 0.035 in
1980 thru 1984	0.010 to 0.035 in

Crankshaft and flywheel

Main journal	
diameter	
302	2.2482 to 2.2490 in
351W	2.9994 to 3.0002 in
taper limit — maximum per inch	
302	
thru 1974	0.0003 in
1975, 1976 and 1980 thru 1983	0.0006 in
1977 thru 1979 and 1984 thru 1986	0.0005 in
351W	0.0005 in
out-of-round limit	
302	
thru 1973	0.0004 in
1974 and later	0.0006 in
351W	0.0006 in
runout limit	
thru 1973	0.004 in
1974 and later	0.005 in
Main bearing oil clearance	
standard	
302	
No. 1 bearing	0.0001 to 0.0015 in
all others	0.0005 to 0.0015 in
351W (thru 1978)	
No. 1 bearing	0.0005 to 0.0015 in
all others	0.0008 to 0.0015 in
351W (1979 and later — all bearings)	0.0008 to 0.0015 in

Item	Specification
service limit	
302	
No. 1 bearing	0.002 in
all others	0.0024 in
351W (thru 1978)	
No. 1 bearing	0.0005 to 0.0024 in
all others	0.0008 to 0.0026 in
351W (1979 and later — all bearings)	0.0008 to 0.0026 in
Connecting rod journal	
diameter	
302	2.1228 to 2.1236 in
351W	2.3103 to 2.3111 in
taper limit — maximum per inch	
thru 1974	0.0004 in
1975 and later	0.0006 in
out-of-round limit	
thru 1974	0.0004 in
1975 and later	0.0006 in
Connecting rod bearing oil clearance (302)	
standard	0.0008 to 0.0015 in
service limit	
thru 1976	0.0026 in
1977 and later	0.0024 in
Connecting rod bearing oil clearance (351W)	
standard	0.0008 to 0.0015 in
service limit	
thru 1979	0.0026 in
1980 and later	0.0025 in
Connecting rod side clearance	
standard	0.010 to 0.020 in
service limit	0.023 in
Crankshaft end play	
standard	0.004 to 0.008 in
service limit	0.012 in
Flywheel clutch face runout limit	0.010 in
Flywheel ring gear lateral runout limit	
manual transmission	
thru 1974	0.040 in
1975 and later	0.030 in
automatic transmission	0.060 in

2C

Camshaft

Item	Specification
Bearing journal diameter	
No. 1	2.0805 to 2.0815 in
No. 2	2.0655 to 2.0665 in
No. 3	2.0505 to 2.0515 in
No. 4	2.0355 to 2.0365 in
No. 5	2.0205 to 2.0215 in
Bearing oil clearance	
standard	0.001 to 0.003 in
service limit	0.006 in
Front bearing location	0.005 to 0.020 in below the front face of the engine block
Lobe lift (302)	
intake	
thru 1976	0.2303 in
1977 and later	0.2375 in
exhaust	
thru 1976	0.2375 in
1977 thru 1981	0.2470 in
1982 and later	0.2474 in
Lobe lift (351W)	
intake	0.2600 in
exhaust	0.2600 in
Maximum allowable lift loss	0.005 in
End play	
standard	0.001 to 0.007 in
service limit	0.009 in
Timing chain deflection limit	0.500 in
Crankshaft sprocket runout limit (assembled)	
thru 1975	0.006 in
1976 thru 1979	0.005 in
Camshaft sprocket runout limit	0.005 in

Cylinder heads and valve train

Item	Specification
Head warpage limit	0.003 in per 6 in or 0.006 in total
Valve seat angle	45°

302 and 351W CID V8 engines (continued)

Valve seat width	0.060 to 0.080 in
Valve seat runout limit	0.002 in
Valve face angle	44°
Valve face runout limit	0.002 in
Minimum valve margin width	1/32 in
Valve stem diameter	
intake	0.3416 to 0.3423 in
exhaust	0.3411 to 0.3418 in
Valve guide diameter	0.3433 to 0.3443 in
Valve stem-to-guide clearance	
intake	
standard	0.0010 to 0.0027 in
service limit	0.0055 in
exhaust	
standard	0.0015 to 0.0032 in
service limit	0.0055 in
Valve spring free length (302)	
intake	
thru 1978	1.94 in
1979 and later	2.04 in
exhaust	
thru 1973	1.94 in
1974 and later	1.85 in
Valve spring free length (351W)	
intake	
thru 1978	2.06 in
1979 and later	2.04 in
exhaust	
thru 1976	2.12 in
1977 and 1978	1.87 in
1979 and later	1.85 in
Valve spring installed height (302)	
intake	
thru 1978 and 1981 thru 1986	1.67 to 1.70 in
1979 and 1980	1.67 to 1.80 in
exhaust	
thru 1973	1.67 to 1.70 in
1974 thru 1978	1.59 to 1.61 in
1979 and later	1.58 to 1.61 in
Valve spring installed height (351W)	
intake	
thru 1978	1.76 to 1.81 in
1979 and later	1.76 to 1.80 in
exhaust	
thru 1976	1.81 to 1.84 in
1977 and 1978	1.59 to 1.61 in
1979 and later	1.58 to 1.61 in
Valve spring out-of-square limit	(0.078) 5/64 in
Collapsed lifter gap (valve clearance)	
302	
allowable	
thru 1976	0.090 to 0.190 in
1977 and later	0.071 to 0.193 in
desired	
thru 1974	0.090 to 0.140 in
1975 and 1976	0.115 to 0.165 in
1977 and later	0.096 to 0.165 in
351W	
allowable	
thru 1976	0.106 to 0.206 in
1977 and later	0.098 to 0.198 in
desired	
thru 1976	0.131 to 0.181 in
1977 and later	0.123 to 0.173 in
Lifter diameter	0.8740 to 0.8745 in
Lifter bore diameter	0.8752 to 0.8767 in
Lifter-to-bore clearance	
standard	0.0007 to 0.0027 in
service limit	0.005 in
Pushrod runout limit	0.015 in

Oil pump

Outer race-to-housing clearance	
302 (all) and 351W (thru 1979)	0.001 to 0.013 in
351W (1980 and later)	0.001 to 0.003 in

Rotor assembly end clearance limit	0.004 in
Driveshaft-to-housing clearance	0.0015 to 0.0030 in
Relief spring tension (lbs @ specified length)	
302 (thru 1979)	10.6 to 12.2 lbs @ 1.704 in
302 (1980 and later)	10.6 to 12.2 lbs @ 1.740 in
351W	18.2 to 20.2 @ 2.490 in
Relief valve clearance	0.0015 to 0.0030 in

Torque specifications*	**Ft-lbs**
Connecting rod nuts	
302	19 to 24
351W	40 to 45
Main bearing cap bolts	
302	60 to 70
351W	95 to 105

* *Additional torque specifications can be found in Chapter 2, Part B*

351M and 400 CID V8 engines

General

Bore and stroke	
351M	4.00 x 3.50 in
400	4.00 x 4.00 in
Oil pressure (at 2000 rpm — normal operating temperature)	50 to 75 psi

Engine block

Cylinder bore diameter	4.0000 to 4.0048 in
Taper limit	0.010 in
Out-of-round limit	0.005 in
Deck warpage limit	0.003 in per 6 in or 0.006 in overall

Pistons and rings

Piston Diameter	
coded red	3.9982 to 3.9988 in
coded blue	3.9994 to 4.0000 in
Piston-to-cylinder bore clearance — selective fit	0.0014 to 0.0022 in
Piston ring-to-groove clearance	
top compression ring	
1982 400 only	0.0030 to 0.0040 in
all others	0.0019 to 0.0036 in
bottom compression ring	
1982 400 only	0.0030 to 0.0040 in
all others	0.002 to 0.004 in
oil control ring	Snug fit in groove
service limit	0.002 in maximum increase in clearance
Piston ring end gap	
top compression ring	0.010 to 0.020 in
bottom compression ring	0.010 to 0.020 in
oil control ring	
Thru 1978	0.015 to 0.055 in
1979 and later	0.010 to 0.035 in

Crankshaft and flywheel

Main journal	
diameter	2.9994 to 3.0002 in
taper limit — maximum per inch	0.0005 in
out-of-round limit	0.0006 in
runout limit	0.005 in
Main bearing oil clearance	
desired	0.0008 to 0.0015 in
allowable	0.0008 to 0.0026 in
Connecting rod journal	
diameter	2.3103 to 2.3111 in
taper limit — maximum per inch	0.0006 in
out-of-round limit	0.0006 in
Connecting rod bearing oil clearance	
desired	0.0008 to 0.0015 in
allowable	0.0008 to 0.0025 in
Connecting rod side clearance	
standard	0.010 to 0.020 in
service limit	0.023 in
Crankshaft end play	
standard	0.004 to 0.008 in
service limit	0.012 in

351M and 400 CID V8 engines (continued)

Camshaft

Bearing journal diameter	
No. 1	2.1238 to 2.1248 in
No. 2	2.0655 to 2.0665 in
No. 3	2.0505 to 2.0515 in
No. 4	2.0355 to 2.0365 in
No. 5	2.0205 to 2.0215 in
Bearing oil clearance	
standard	0.001 to 0.003 in
service limit	0.006 in
Front bearing location	0.040 to 0.060 in (1982 400 – 0.060 in) below the front face of the engine block
Lobe lift (thru 1978)	
351M	
intake	0.235 in
exhaust	0.235 in
400	
intake	0.2474 in
exhaust	0.250 in
Lobe lift (1979 and later – all engines)	
intake	0.250 in
exhaust	0.250 in
Maximum allowable lift loss	0.005 in
End play	
standard	0.001 to 0.006 in
service limit	0.009 in
Timing chain defection limit	0.500 in

Cylinder heads and valve train

Head warpage limit	0.003 in per 6 in or 0.006 in overall
Valve seat angle	45°
Valve seat width	
intake	0.060 to 0.080 in
exhaust	0.070 to 0.090 in
Valve seat runout limit	0.002 in
Valve face angle	44°
Valve face runout limit	0.002 in
Minimum valve margin width	1/32 in
Valve stem diameter	
intake	0.3416 to 0.3423 in
exhaust	0.3411 to 0.3418 in
Valve guide diameter	0.3433 to 0.3443 in
Valve stem-to-guide clearance	
intake	
standard	0.0010 to 0.0027 in
service limit	0.005 in
exhaust	
standard	0.0015 to 0.0032 in
service limit	0.005 in
Valve spring free length	
intake (all) and exhaust (thru 1978)	2.06 in
exhaust (1979 on)	1.93 in
Valve spring installed height	
intake	1.81 to 1.84 in
exhaust	
thru 1978	1.81 to 1.84 in
1979 and later	1.69 to 1.72 in
Valve spring out-of-square limit	0.078 (5/64) in
Collapsed lifter gap (valve clearance)	
allowable	
1982 400 only	0.200 in
all others	0.100 to 0.200 in
desired	
1982 400 only	0.175 in
all others	0.125 to 0.175 in
Lifter diameter	0.8740 to 0.8745 in
Lifter bore diameter	0.8752 to 0.8767 in
Lifter-to-bore clearance	
standard	0.0007 to 0.0027 in
service limit	0.005 in
Pushrod runout limit	0.015 in

Oil pump

Outer race-to-housing clearance	0.001 to 0.013 in

Rotor assembly end clearance limit	0.004 in
Inner rotor-to-outer rotor tip clearance	0.012 in
Driveshaft-to-housing clearance	0.0015 to 0.0030 in
Relief spring tension (lbs @ desired length)	20.6 to 22.6 @ 2.49 in
Relief valve clearance	0.0015 to 0.0030 in

Torque specifications*	**Ft-lbs**
Connecting rod nuts	40 to 45
Main bearing cap bolts	95 to 105

* *Additional torque specifications can be found in Chapter 2, Part B*

460 CID V8 engine

General

Bore and stroke	4.360 x 3.850 in
Oil pressure (at 2000 rpm — normal operating temperature)	40 to 65 psi

Engine block

Cylinder bore diameter	4.3600 to 4.3636 in
Taper limit	0.010 in
Out-of-round	
maximum	0.0015 in
service limit	0.005 in
Deck warpage limit	0.003 in per 6 in or 0.006 in overall

Pistons and rings

Piston Diameter	
coded red	4.3585 to 4.3591 in
coded blue	4.3597 to 4.3603 in
Piston to bore clearance — selective fit	0.0022 to 0.0030 in
Piston ring-to-groove clearance	
compression rings (both)	
thru 1976	0.002 to 0.004 in
1977, 1978 and 1982 and later	0.0025 to 0.0045 in
1979 thru 1981	
top ring	0.0019 to 0.0036 in
second ring	0.002 to 0.004 in
oil control ring	Snug fit in groove
service limit	0.002 in maximum increase in clearance
Piston ring end gap	
compression rings (both)	0.010 to 0.020 in
oil control ring	
thru 1976	0.015 to 0.055 in
1977 and 1978	0.010 to 0.030 in
1979 and later	0.010 to 0.035 in

Crankshaft and flywheel

Main journal	
diameter	2.9994 to 3.0002 in
taper limit — maximum per inch	0.0005 in
out-of-round limit	0.0006 in
runout limit	0.005 in
Main bearing oil clearance	
thru 1974	
standard	
No. 1 bearing	0.0004 to 0.0015 in
all others	0.0012 to 0.0015 in
service limit	
No. 1 bearing	0.002 in
all others	0.0028 in
1975 thru 1979	
standard	
No. 1 bearing	0.0008 to 0.0015 in
all others	0.0008 to 0.0026 in
service limit	
No. 1 bearing	0.002 in
all others	0.0026 in
1980 on	
desired	0.0008 to 0.0015 in
allowable	0.0008 to 0.0026 in

460 CID V8 engine (continued)

Connecting rod journal	
diameter	2.4992 to 2.5000 in
taper limit — maximum per inch	0.0006 in
out-of-round limit	0.0006 in
Connecting rod bearing oil clearance	
desired	0.0008 to 0.0015 in
allowable	0.0008 to 0.0025 in
Connecting rod side clearance	
standard	0.010 to 0.020 in
service limit	0.023 in
Crankshaft end play	
standard	0.004 to 0.008 in
service limit	0.012 in

Camshaft

Bearing journal diameter (all)	2.1238 to 2.1248 in
Bearing oil clearance	
standard	0.001 to 0.003 in
service limit	0.006 in
Front bearing location	0.040 to 0.060 in below the front face of the engine block
Lobe lift	
intake	
thru 1978	0.2530 in
1979 only	0.2526 in
1980 and later	0.2520 in
exhaust	0.2780 in
Maximum allowable lift loss	0.005 in
End play	
standard	
thru 1977	0.003 to 0.007 in
1978 and later	0.001 to 0.006 in
service limit	0.009 in
Timing chain deflection limit	0.500 in
Timing sprocket runout limit (assembled)	0.005 in

Cylinder heads and valve train

Head warpage limit	0.003 in per 6 in or 0.006 in overall
Valve seat angle	45°
Valve seat width	0.060 to 0.080 in
Valve seat runout limit	
thru 1978	0.0015 in
1979 and later	0.002 in
Valve face angle	44°
Valve face runout limit	0.002 in
Minimum valve margin width	1/32 in
Valve stem diameter	0.3416 to 0.3423 in
Valve guide diameter	0.3433 to 0.3443 in
Valve stem-to-guide clearance	
standard	0.0010 to 0.0027 in
service limit	0.0055 in
Valve spring free length	
thru 1976	2.03 in
1977 and later	2.06 in
Valve spring installed height	1.80 to 1.83 in
Valve spring out-of-square limit	0.078 (5/64) in
Collapsed lifter gap (valve clearance)	
allowable	0.075 to 0.175 in
desired	
thru 1974	0.075 to 0.125 in
1975 and later	0.100 to 0.150 in
Lifter diameter	0.8740 to 0.8745 in
Lifter bore diameter	0.8752 to 0.8767 in
Lifter-to-bore clearance	
standard	0.0007 to 0.0027 in
service limit	0.005 in
Pushrod runout limit	0.015 in

Oil pump

Outer race-to-housing clearance	0.001 to 0.013 in
Rotor assembly end clearance limit	0.004 in
Driveshaft-to-housing clearance	0.0015 to 0.0030 in
Relief spring tension (lbs @ specified length)	20.6 to 22.6 @ 2.49 in
Relief valve clearance	0.0015 to 0.0030 in

Torque specifications*	Ft-lbs
Connecting rod nuts	
thru 1981	40 to 45
1982 and later	45 to 50
Main bearing cap bolts	95 to 105

* *Additional torque specifications can be found in Chapter 2, Part B*

1 General information

Included in this portion of Chapter 2 are the general overhaul procedures for the cylinder head(s) and internal engine components. The information ranges from advice concerning preparation for an overhaul and the purchase of replacement parts to detailed, step-by-step procedures covering removal and installation of internal engine components and the inspection of parts.

The following Sections have been written based on the assumption that the engine has been removed from the vehicle. For information concerning in vehicle engine repair, as well as removal and installation of the external components necessary for the overhaul, see Section 2 in this Part and the portion of Chapter 2 applicable to your particular engine.

The specifications included here in Part C are only those necessary for the inspection and overhaul procedures which follow. Refer to the appropriate Part of Chapter 2 for additional specifications related to the various engines covered in this manual.

2 Repair operations possible with the engine in the vehicle

Many major repair operations can be accomplished without removing the engine from the vehicle.

If possible, clean the engine compartment and the exterior of the engine with some type of pressure washer before any work is done. A clean engine will make the job easier and will help keep dirt out of the internal areas of the engine.

Remove the engine cover and open the hood to provide as much working room as possible. Cover the front of the vehicle and the seats to prevent damage to the upholstery and painted surfaces.

If oil or coolant leaks develop, indicating a need for gasket or seal replacement, the repairs can generally be made with the engine in the vehicle. The oil pan gasket, the cylinder head gasket(s), intake and exhaust manifold gaskets, timing cover gaskets and the crankshaft oil seals are usually accessible with the engine in place.

Exterior engine components, such as the water pump, the starter motor, the alternator, the distributor, the fuel pump and the carburetor or EFI components, as well as the intake and exhaust manifolds, can be removed for repair with the engine in place.

Since the cylinder head(s) can be removed without pulling the engine, valve component servicing can also be accomplished with the engine in the vehicle.

Replacement of, repairs to or inspection of the timing sprockets and chain or gears and the oil pump are all possible with the engine in place.

In extreme cases caused by a lack of necessary equipment, repair or replacement of piston rings, pistons, connecting rods and rod bearings is possible with the engine in the vehicle. However, this practice is not recommended because of the cleaning and preparation work that must be done to the components involved.

3 Engine overhaul — general information

It is not always easy to determine when, or if, an engine should be completely overhauled, as a number of factors must be considered.

High mileage is not necessarily an indication that an overhaul is needed, while low mileage does not preclude the need for an overhaul. Frequency of servicing is the single most important consideration. An engine that has had regular and frequent oil and filter changes, as well as other required maintenance, will most likely give many thousands of miles of reliable service. Conversely, a neglected engine may require an overhaul very early in its life.

Excessive oil consumption is an indication that piston rings and/or valve guides are in need of attention (make sure that oil leaks are not responsible before deciding that the rings and guides are bad). Check the cylinder compression or have a leakdown test performed by an experienced mechanic to determine the extent of the work required.

If the engine is making obvious knocking or rumbling noises, the connecting rod and/or main bearings are probably at fault. Check the oil pressure with a gauge installed in place of the oil pressure sending unit. If the pressure is extremely low, the bearings and/or oil pump are probably worn out.

Loss of power, rough running, excessive valve train noise and high fuel consumption may also point to the need for an overhaul, especially if they are all present at the same time. If a complete tune-up does not remedy the situation, major mechanical work is the only solution.

An engine overhaul generally involves restoring the internal parts to the specifications of a new engine. During an overhaul the piston rings are replaced and the cylinder walls are reconditioned by boring and/or honing. If the cylinders are bored new pistons will be required. The main and connecting rod bearings are replaced with new ones and, if necessary, the crankshaft may be ground undersize to restore the journals. The valves are generally serviced as well, since they are usually in less-than-perfect condition at this point. While the engine is being overhauled, other components, such as the carburetor, distributor, starter and alternator can be rebuilt. The end result is a like new engine that will give many trouble free miles.

Before beginning the engine overhaul, read through the entire procedure to familiarize yourself with the scope and requirements of the job. Overhauling an engine is not difficult, but it is time consuming. Plan on the vehicle being tied up for a minimum of two weeks, especially if parts must be taken to an automotive machine shop for repair or reconditioning. Check on availability of parts and make sure that any necessary special tools and equipment are obtained in advance. Most work can be done with hand tools, although a number of precision measuring tools are required for inspecting parts to determine if they must be replaced. Often an automotive machine shop will handle the inspection of parts and offer advice concerning reconditioning and replacement. **Note:** *Always wait until the engine has been completely disassembled and all components have been inspected before deciding what service and repair operations must be performed by a machine shop.*

Since the condition of the block is the major factor to consider when determining whether to overhaul the original engine or buy a rebuilt one, never purchase parts or have machine work done on other components until the block has been thoroughly inspected.

As a final note, to ensure maximum life and minimum trouble from a rebuilt engine, everything must be assembled with care in a spotlessly clean environment.

4 Engine rebuilding alternatives

The home mechanic has a number of options available when performing an engine overhaul. The decision to replace the engine block, piston/connecting rod assemblies and crankshaft depends on several factors, with the primary consideration being the condition of the block. Other considerations are cost, access to machine shop facilities, parts availability, time required to complete the project and experience.

Some of the rebuilding alternatives include:

Individual parts — If the inspection procedures reveal that the engine block and most engine components are in reusable condition, purchasing individual parts may be the most economical alternative. The block, crankshaft and piston/connecting rod assemblies should all be inspected carefully. Even if the block shows little wear, the cylinder bores should be honed.

Master kit — This rebuild package usually contains a reground crankshaft and a set of pistons and connecting rods. The pistons will already be installed on the connecting rods. These kits are commonly available for standard cylinder bores, as well as for engine blocks which have been bored to a regular oversize.

Short block – A short block consists of an engine block with a crankshaft and piston/connecting rod assemblies already installed. All new bearings are incorporated and all clearances will be correct. Your head(s), valve train components, and external parts are then bolted to the short block.

Long block – A long block consists of a short block plus an oil pump, oil pan, cylinder head(s), rocker arm cover(s), valve train components and timing chain/sprockets or gears. All components are installed with new bearings, seals and gaskets incorporated throughout. The installation of manifolds and external parts is all that is necessary.

Give careful thought to which alternative is best for you and discuss the situation with local automotive machine shops and dealer parts and service departments before ordering or purchasing replacement parts.

5 Engine removal – methods and precautions

If it has been decided that an engine must be removed for overhaul or major repair work, certain preliminary steps should be taken.

Locating a suitable work area is extremely important. A shop is, of course, the most desirable place to work. Adequate work space, along with storage space for the vehicle, is very important. If a shop or garage is not available, at the very least a flat, level, clean work surface made of concrete or asphalt is required.

Cleaning of the engine compartment and engine prior to removal will help you keep tools clean and organized.

An engine hoist will be necessary. Make sure that the equipment is rated in excess of the combined weight of the engine and its accessories. Safety is of primary importance, considering the potential hazards involved in removing the engine from the vehicle.

If the engine is being removed by a novice, a helper should be available. Advice and aid from someone more experienced would also be helpful. There are many instances when one person cannot simultaneously perform all of the operations required when removing the engine from the vehicle.

Plan the operation ahead of time. Arrange for all of the tools and equipment you will need prior to beginning the job. Some of the equipment necessary to perform engine removal and installation safely and with relative ease are, in addition to an engine hoist, a heavy duty floor jack, complete sets of wrenches and sockets as described in the front of this manual, wooden blocks and plenty of rags and cleaning solvent for mopping up the inevitable spills. If the hoist is to be rented, make sure that you arrange for it in advance and perform in advance all of the operations possible without it. This will save you money and time.

Plan for the vehicle to be out of use for a considerable amount of time. A machine shop will be required to perform some of the work which the home mechanic cannot accomplish due to a lack of special equipment. These shops often have a busy schedule, so it would be a good idea to consult them before removing the engine in order to accurately estimate the amount of time required to rebuild or repair components that may need work.

Always use extreme caution when removing and installing the engine. Serious injury can result from careless actions. Plan ahead. Take your time and a job of this nature, although major, can be accomplished successfully.

6 Engine overhaul – disassembly sequence

1 It is much easier to disassemble and work on an engine if it is mounted on a portable engine stand. These stands can often be rented for a reasonable fee from an equipment rental yard. Before the engine is mounted on a stand, the flywheel or driveplate should be removed from the engine.

2 If a stand is not available it is possible to disassemble the engine with it blocked up on a sturdy workbench or on the floor. Be extra careful not to tip or drop the engine when working without a stand.

3 If you are going to obtain a rebuilt engine, all external components must come off your engine to be transferred to the replacement engine, just as they will if you are doing a complete engine overhaul yourself. These include:

Alternator and brackets
Accessory drivebelts and pulleys
Emissions control components
Distributor, coil, spark plug wires and spark plugs
Thermostat and housing
Water pump and hoses
Carburetor or fuel injection components and lines
Intake and exhaust manifolds
Oil filter
Fuel pump
Engine mounts
Clutch
Flywheel/driveplate

Note: *When removing the external components from the engine, pay close attention to details that may be helpful or important during installation. Note the installed position of gaskets, seals, spacers, pins, washers, bolts and other small items.*

4 If you are obtaining a short block, which consists of the engine block, crankshaft, pistons and connecting rods all assembled, then the cylinder head(s), oil pan and oil pump will have to be removed as well. See *Engine rebuilding alternatives* for additional information regarding the different possibilities to be considered.

5 If you are planning a complete overhaul, the engine must be disassembled and the internal components removed in the following order:

Rocker arm cover(s)
Exhaust and intake manifolds
Rocker arms and pushrods
Valve lifters
Cylinder head(s)
Front pulley and vibration damper assembly
Timing cover, chain and gears
Camshaft
Oil pan
Oil pump
Piston/connecting rod assemblies
Crankshaft and bearings

6 Before beginning the disassembly and overhaul procedures, make sure the following items are available:

Common hand tools
Small cardboard boxes or plastic bags for storing parts
Gasket scraper
Ridge reamer
Vibration damper puller
Micrometers and/or dial caliper
Telescoping gauges
Dial indicator set
Valve spring compressor
Cylinder surfacing hone
Piston ring groove cleaning tool
Electric drill motor
Tap and die set
Wire brushes
Cleaning solvent

7 Cylinder head – disassembly

Refer to illustrations 7.2, 7.3a and 7.3b

Note: *New and rebuilt cylinder heads are commonly available for most engines at dealerships and auto parts stores. Due to the specialized tools necessary for some disassembly and inspection procedures, it may be more practical and economical for the home mechanic to purchase a replacement head or heads rather than taking the time to disassemble, inspect and recondition the original(s).*

1 Cylinder head disassembly involves removal and disassembly of the intake and exhaust valves and their related components. If they are still in place, remove the nuts or bolts, baffles, fulcrum seats and rocker arms or the rocker arm shaft assemblies from the cylinder head. Label the parts or store them separately so they can be reinstalled in their original locations.

2 Before the valves are removed, arrange to label and store them, along with their related components, so they can be kept separate and

7.2 A small plastic bag, with an appropriate label, can be used to store the valve train components so they can be kept together and reinstalled in the correct guide

reinstalled in the same valve guides they are removed from (see illustration).

3 Compress the valve spring on the first valve with a spring compressor and remove the keepers (see illustration). Carefully release the valve spring compressor and remove the retainer, the shield or rotator (if so equipped), the springs, the valve guide seal, any spring shims and the valve from the head. If the valve binds in the guide (won't pull through), push it back into the head and deburr the area around the keeper groove and the end of the valve stem with a fine file or whetstone (see illustration).

4 Repeat the procedure for the remaining valves. Remember to keep all the parts for each valve together so they can be reinstalled in the same locations.

5 Once the valves have been removed and safely stored, the head should be thoroughly cleaned and inspected. If a complete engine overhaul is being done, finish the engine disassembly procedures before beginning the cylinder head cleaning and inspection process.

8 Cylinder head — cleaning and inspection

Refer to illustrations 8.12, 8.19, 8.28a and 8.28b

1 Thorough cleaning of the cylinder head and related valve train components, followed by a detailed inspection, will enable you to decide how much valve service work must be done during the engine overhaul.

Cleaning

2 Scrape away all traces of old gasket material and sealing compound from the head gasket, intake manifold and exhaust manifold sealing surfaces.

3 Remove any built-up scale from the coolant passages.

4 Run a stiff wire brush through the oil holes to remove any deposits that may have formed in them.

5 Run an appropriate size tap into each threaded hole to remove any corrosion and thread sealant that may be present. If compressed air is available, use it to clear the holes of debris produced by this operation.

6 Clean the exhaust and intake manifold stud threads in a similar manner with an appropriate size die. Clean the rocker arm pivot bolt or stud threads with a wire brush.

7 Clean the cylinder head with solvent and dry it thoroughly. Compressed air will speed the drying process and ensure that all holes and recessed areas are clean. **Note:** *Decarbonizing chemicals are available and may prove very useful when cleaning cylinder heads and valve train components. They are very caustic and should be used with caution. Be sure to follow the instructions on the container.*

8 Clean the rocker arms, fulcrum seats or rocker arm shafts and

7.3a Use a valve spring compressor to compress the spring, then remove the keepers from the valve stem

7.3b If the valve won't pull through the guide, deburr the edge of the stem end and the area around the top of the keeper groove with a file

pushrods with solvent and dry them thoroughly. Compressed air will speed the drying process and can be used to clean out the oil passages.

9 Clean all the valve springs, keepers, retainers, shields and spring shims with solvent and dry them thoroughly. Do the components from one valve at a time to avoid mixing up the parts.

10 Scrape off any heavy deposits that may have formed on the valves, then use a motorized wire brush to remove deposits from the valve heads and stems. Again, make sure the valves do not get mixed up.

Inspection

Cylinder head

11 Inspect the head very carefully for cracks, evidence of coolant leakage and other damage. If cracks are found, a new cylinder head should be obtained.

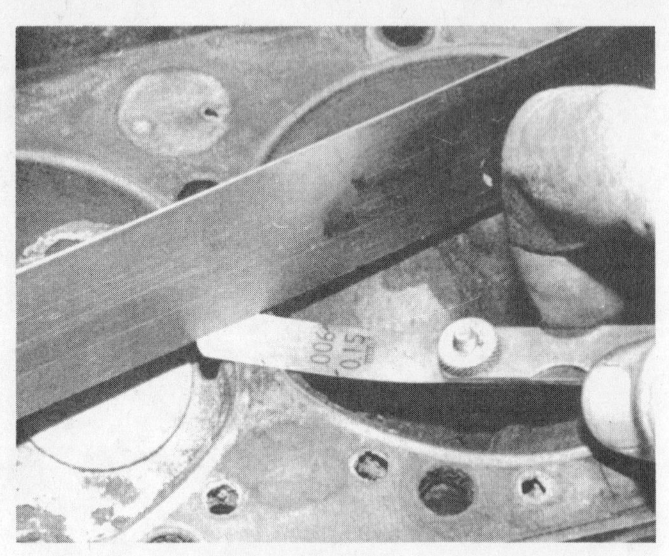

8.12 Check the cylinder head gasket surface for warpage by trying to slip a feeler gauge under the straightedge (see the Specifications for the maximum warpage allowed and use a feeler gauge of that thickness)

12 Using a straightedge and feeler gauge, check the head gasket mating surface for warpage (see illustration). If the warpage exceeds the specified limit, the head must be resurfaced at a machine shop.
13 Examine the valve seats in each of the combustion chambers. If they are pitted, cracked or burned, the head will require valve service that is beyond the scope of the home mechanic.
14 A preliminary examination of the valve guides will indicate whether the heads can be reassembled with new seals, or whether guide reconditioning or replacement is necessary.
15 A complete examination of the valve guides is a job that should be performed by an automotive machine shop, where precision measuring equipment is available. In addition to checking the valve stem-to-guide clearance, the shop will be able to check for a bellmouth wear pattern in the guide.
16 Before taking the head(s) to a shop for a valve guide check, you can make a preliminary check with a dial indicator.
17 Remove the valve assembly (valve, valve spring, spring shims, retainer and keepers) from the head.
18 Reinsert the valve into the guide. If you have removed all the valves from the head, make sure the valve is the one that was originally installed in that guide. Don't mix up valve components.
19 Mount the dial indicator on the head so that the plunger rests against the valve stem (see illustration), as close to the top of the guide boss as possible.
20 Raise the valve until the head is between 1/16 and 1/8-inch off the seat.
21 Rock the valve stem back-and-forth in the guide, in line with the plunger of the dial indicator, and note the amount of lateral movement shown on the indicator.
22 Check the specifications for the valve stem-to-guide clearance for your engine. The specifications will give a minimum/maximum or a service limit figure. If the movement of the valve stem is more than the maximum or service limit figure given, the valve guides should be checked by a shop. If movement is less than the maximum or service limit figure, only new seals will be required.
23 Even if the valve guides are in acceptable shape, if the valve seats and faces are going to be reground by a shop, the guides should be checked and, if necessary, reconditioned.

Rocker arm components

24 Check the rocker arm faces (that contact the pushrod ends and valve stems) for pits, wear and rough spots. Check the pivot contact areas as well.
25 Inspect the pushrod ends for scuffing and excessive wear. Roll the pushrod on a flat surface, such as a piece of glass, to determine if it is bent.
26 Any damaged or excessively worn parts must be replaced with new ones.

8.19 A dial indicator can be used to determine the valve stem-to-guide clearance (move the valve stem as indicated by the arrows)

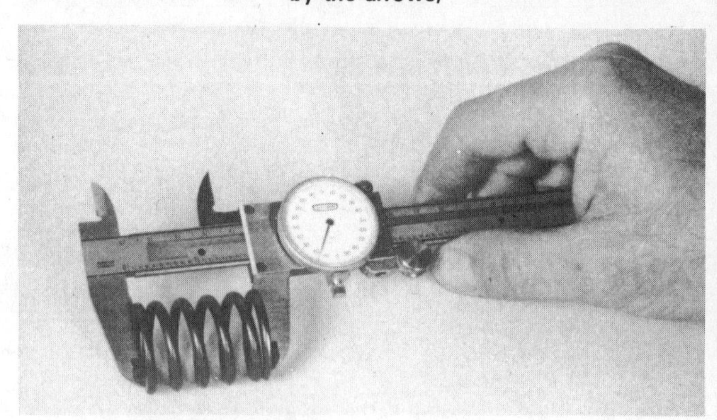

8.28a Measure the free length of each valve spring with a dial or vernier caliper

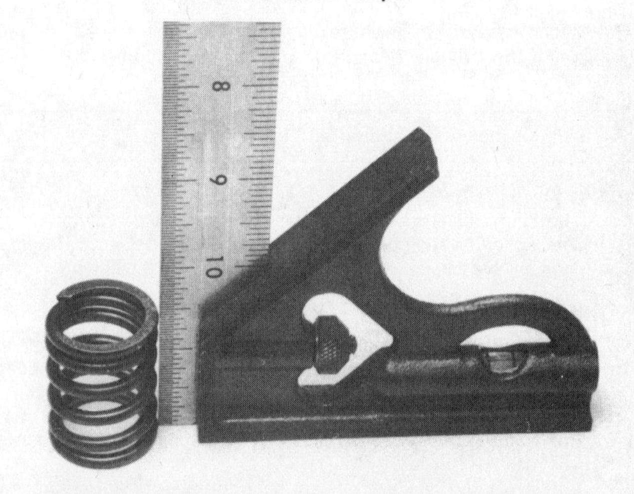

8.28b Check each valve spring for squareness

Valves

27 Carefully inspect each valve face for cracks, pits and burned spots. Check the valve stem and neck for cracks. Rotate the valve and check for any obvious indication that it is bent. Check the end of the stem for pits and excessive wear. The presence of any of these conditions indicates the need for valve service by a repair shop.

Valve components

28 Check each valve spring for wear on the ends and pits. Measure the free length (see illustration) and compare it to the Specifications.

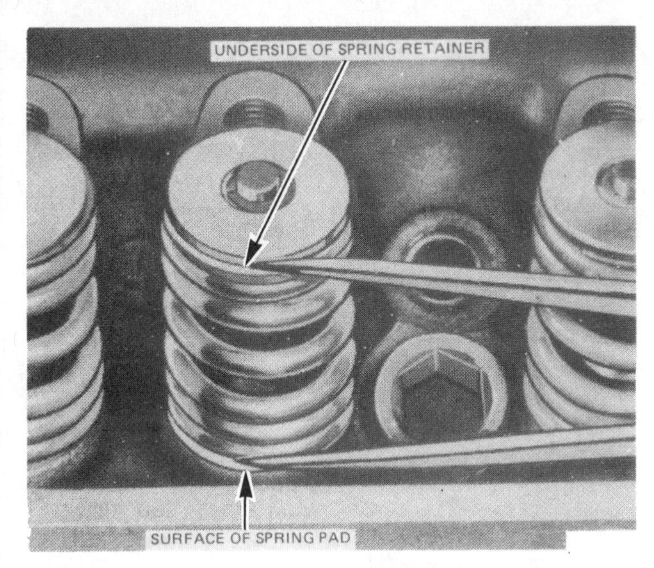

10.4 Be sure to check the valve spring installed height (adjust dividers to match the distance from the spring seat or shim to the underside of the retainer, then transfer the dividers to a ruler to determine the height in inches)

11.1 A ridge reamer is required to remove the ridge from the top of the cylinder before removing the pistons

Any springs that are shorter than specified have sagged and should not be reused. Stand the spring on a flat surface and check it for squareness (see illustration). The springs should be checked for pressure with a special fixture (take the springs to an automotive machine shop for this check). Weak springs will cause the engine to run poorly, so don't ignore this check.
29 Check the spring retainers or rotators (if equipped) and keepers for obvious wear and cracks. Any questionable parts should be replaced with new ones, as extensive damage will occur in the event of failure during engine operation.
30 If the inspection process indicates that the valve components are in generally poor condition or worn beyond the limits specified, which is usually the case in an engine that is being overhauled, reassemble the valves in the cylinder head and refer to Section 9 for valve servicing recommendations.
31 If the inspection turns up no excessively worn parts, and if the valve faces and seats are in good condition, the valve train components can be reinstalled in the cylinder head without major servicing. Refer to the appropriate Section for cylinder head reassembly procedures.

9 Valves — servicing

1 Because of the complex nature of the job and the special tools and equipment needed, servicing of the valves, the valve seats and the valve guides (commonly known as a valve job) is best left to a professional.
2 The home mechanic can remove and disassemble the head, do the initial cleaning and inspection, then reassemble and deliver the head to a dealer service department or a reputable automotive machine shop for the actual valve service work.
3 The shop will remove the valves and springs, recondition or replace the valves and valve seats, recondition the valve guides, check and replace the valve springs, spring retainers and keepers (as necessary), replace the valve seals with new ones, reassemble the valve components and make sure the installed spring height is correct. The cylinder head gasket surface will also be resurfaced if it is warped.
4 After the valve job has been performed by a professional, the head will be in like new condition. When the head is returned, be sure to clean it very thoroughly before installation on the engine to remove any metal particles and abrasive grit that may still be present from the valve service or head resurfacing operations. Use compressed air, if available, to blow out all the oil holes and passages.

10 Cylinder head — reassembly

Refer to illustration 10.4

1 If the head was sent out for valve servicing, the valves and related components will already be in place.
2 If the head must be reassembled, lay all the spring seats or shims, if equipped, in place over the guides they were removed from, then lubricate and install one of the valves. Use engine assembly lube or moly-base grease on the valve stem. Slip the new oil seal over the valve stem and seat it on the guide. **Note:** *Some later model V8 engines are equipped with seals that must be carefully driven onto the guide with a deep socket and small hammer.*
3 Install the valve springs, retainer, shield or rotator (if equipped) and the keepers. Note that the closed coil end of the spring must be installed closest to the head. When compressing the springs, do not let the retainers contact the valve guide seals. Make certain that the keepers are securely locked in the retaining grooves.
4 Double-check the installed valve spring height (if it was correct before disassembly, it should still be within the specified limits). Measure from the top of the shim (if used) or the spring seat to the top of the spring (see illustration). If the height is not as specified, shims can be added under the spring to decrease the height. Do not, under any circumstances, decrease the spring height under the specified minimum.
5 Install the pushrods and rocker arms and tighten the bolts/nuts to the specified torque (where applicable). Be sure to lubricate the fulcrums or shafts with moly-base grease or engine assembly lube.

11 Piston/connecting rod assembly — removal

Refer to illustrations 11.1, 11.5, 11.6 and 11.7

Note: *Prior to removing the piston/connecting rod assemblies, remove the cylinder head(s), the oil pan and the oil pump by referring to the appropriate Part of Chapter 2.*

1 Using a ridge reamer, completely remove the ridge at the top of each cylinder, following the manufacturer's instructions provided with the ridge reaming tool (see illustration). Failure to remove the ridge before attempting to remove the piston/connecting rod assembly will result in broken pistons.
2 After the cylinder ridges have been removed, turn the engine upside-down on the engine stand, with the crankshaft at the top.
3 Before the connecting rods are removed, check the side clearance as follows. Mount a dial indicator with its stem in line with the crankshaft and touching the side of the number one connecting rod cap.
4 Push the connecting rod backward, as far as possible, and zero the dial indicator. Push the connecting rod all the way to the front and check the reading on the dial indicator. The distance that it moves is the side clearance. If the side clearance exceeds the service limit, a

11.5 Check the connecting rod side clearance with a feeler gauge as shown

11.6 If your engine does not have numbers stamped or cast into the rod bearing caps, mark them with a center punch before removing the nuts – during reassembly, each rod cap must be matched to the rod it was removed from

11.7 To prevent damage to the crankshaft journals and cylinder walls, slip sections of hose over the rod bolts before removing the pistons

12.1a If the main bearing caps are not identified with numbers and arrows to indicate position and direction, mark them with a center punch

new connecting rod will be required. Repeat the procedure for the remaining connecting rods.

5 An alternative method is to slip feeler gauge blades between the connecting rod and the crankshaft throw until the play is removed (see illustration). The side clearance is equal to the thickness of the feeler gauge(s).

6 Check the connecting rods and connecting rod caps for identification marks. If they are not plainly marked, identify each rod and cap, using a small punch to make the appropriate number of indentations to indicate the cylinder it is associated with (see illustration).

7 Loosen each of the connecting rod cap nuts approximately 1/2-turn. Remove the number one connecting rod cap and bearing insert. Do not drop the bearing insert out of the cap. Slip a short length of plastic or rubber hose over each connecting rod bolt to protect the crankshaft journal and cylinder wall when the piston is removed (see illustration). Push the connecting rod/piston assembly out through the top of the engine. Use a wooden hammer handle to push on the upper bearing insert in the connecting rod. If resistance is felt, double-check to make sure that all of the ridge was removed from the cylinder.

8 Repeat the procedure for the remaining pistons. After removal, reassemble the connecting rod caps and bearing inserts in their respective connecting rods and install the cap nuts finger tight. Leaving the old bearing inserts in place until reassembly will help prevent the connecting rod bearing surfaces from being accidentally nicked or gouged.

12 Crankshaft – removal

Refer to illustrations 12.1a and 12.1b

Note: *The crankshaft can be removed only after the engine has been removed from the vehicle. It is assumed that the flywheel/driveplate, vibration damper, timing chain and sprocket, oil pan, oil pump and pick-up tube assembly and the piston/connecting rod assemblies have already been removed.*

1 Loosen each of the main bearing cap bolts 1/4-turn at a time, until they can be removed by hand. Check the main bearing caps to see if they are marked as to their locations. They are usually numbered consecutively from the front of the engine to the rear. If they are not,

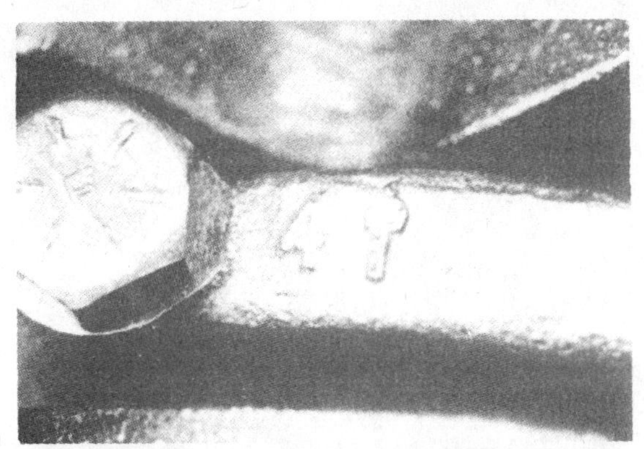

12.1b Typical main bearing cap identification marks

13.1b Use pliers to lever each soft plug through the hole in the block

mark them with number stamping dies or a center punch (see illustration). Most main bearing caps have a cast in arrow, which points to the front of the engine, and a number, indicating its position on the block (see illustration).

2 Gently tap the caps with a soft face hammer, then separate them from the engine block. If necessary, use the main bearing cap bolts as levers to remove the caps. Try not to drop the bearing insert if it comes out with the cap.

3 Carefully lift the crankshaft out of the engine. It is a good idea to have an assistant available, since the crankshaft is quite heavy. With the bearing inserts in place in the engine block and in the main bearing caps, return the caps to their respective locations on the engine block and tighten the bolts finger tight.

13 Engine block – cleaning

Refer to illustrations 13.1a, 13.1b and 13.10

1 Remove the soft plugs from the engine block. Knock the plugs into the block using a hammer and punch, then grasp them with large pliers and pull them back through the holes (see illustrations).

2 Using a gasket scraper, remove all traces of gasket material from the engine block. Be very careful not to nick or gouge the gasket sealing surfaces.

3 Remove the main bearing caps and separate the bearing inserts from the caps and the engine block. Tag the bearings according to which cylinder they were removed from and whether they were in the cap or the block, then set them aside.

4 Remove the threaded oil gallery plugs from the front and back of the block.

5 If the engine is extremely dirty it should be taken to an automotive machine shop to be hot tanked. Any bearings left in the block, such as the camshaft bearings, will be damaged by the cleaning process, so plan on having new ones installed while the block is at the machine shop.

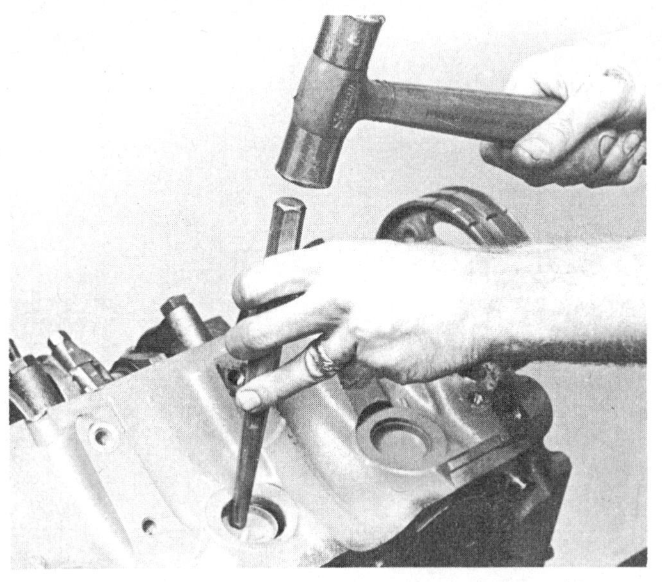

13.1a A hammer and large punch can be used to drive the soft plugs into the block

2C

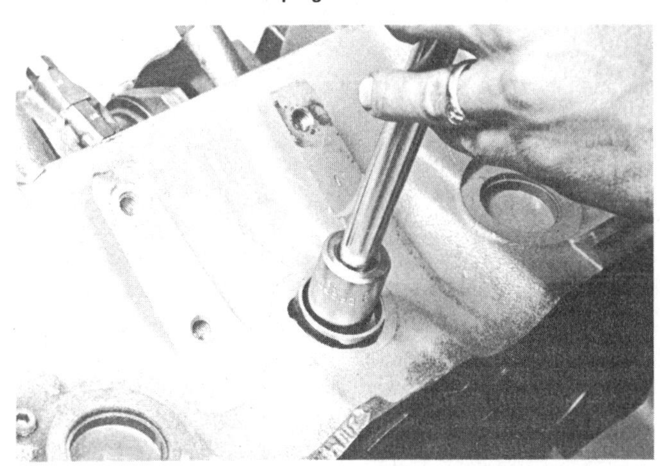

13.10 A large socket on an extension can be used to drive the new soft plugs into their bores

6 After the block is returned, clean all oil holes and oil galleries one more time. Brushes for cleaning oil holes and galleries are available at most auto parts stores. Flush the passages with warm water until the water runs clear, dry the block thoroughly and wipe all machined surfaces with a light, rust preventative oil. If you have access to compressed air, use it to speed the drying process and to blow out all the oil holes and galleries.

7 If the block is not extremely dirty or sludged up, you can do an adequate cleaning job with warm soapy water and a stiff brush. Take plenty of time and do a thorough job. Regardless of the cleaning method used, be very sure to thoroughly clean all oil holes and galleries, dry the block completely and coat all machined surfaces with light oil.

8 The threaded holes in the block must be clean to ensure accurate torque readings during reassembly. Run the proper size tap into each of the holes to remove any rust, corrosion or thread sealant and to restore damaged threads. If possible, use compressed air to clear the holes of debris produced by this operation. Now is a good time to thoroughly clean the threads on the head bolts and the main bearing cap bolts as well.

9 Reinstall the main bearing caps and tighten the bolts finger tight.

10 After coating the sealing surfaces of the new soft plugs with gasket sealer, install them in the engine block (see illustration). Make sure they

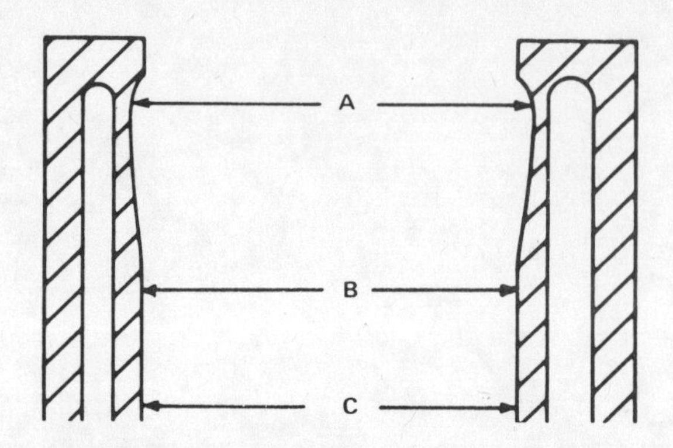

14.4a Measure the diameter of each cylinder just under the wear ridge (A), at the center (B), and at the bottom (C)

14.4b A telescoping gauge (snap-gauge) can be used to determine the cylinder bore diameter

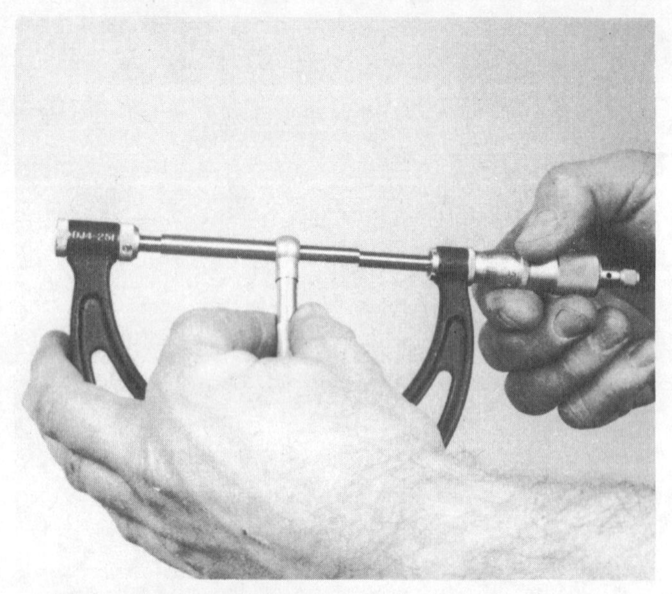

14.4c The gauge is then measured with a micrometer to determine the bore size

14.7a Honing a cylinder with a 'bottle brush' hone

14.7b Honing a cylinder with a stone-type hone

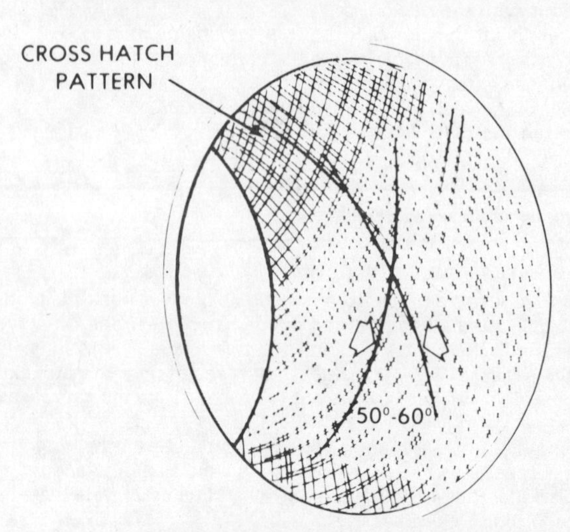

14.7c The cylinder hone should leave a smooth, crosshatch pattern with the lines intersecting at 60-degree angle

re driven in straight and seated properly or leakage could result. Special ools are available for this purpose, but equally good results can be btained using a socket with an outside diameter that will just slip in- o the soft plug and a hammer.

1 If the engine is not going to be reassembled right away, cover it vith a large plastic trash bag to keep it clean.

4 Engine block — inspection

Refer to illustrations 14.4a, 14.4b, 14.4c, 14.7a, 14.7b and 14.7c

Thoroughly clean the engine block as described in Section 13 and double-check to make sure that the ridge at the top of each cylinder has been completely removed.

2 Visually check the block for cracks, rust and corrosion. Check for stripped threads in the threaded holes.

3 Check the cylinder bores for scuffing and scoring.

4 Using the appropriate precision measuring tools, measure each cylinder's diameter at the top (just under the ridge), center and bottom of the cylinder bore, parallel to the crankshaft axis (see illustrations). Next, measure each cylinder's diameter at the same three locations across the crankshaft axis. Compare the results to the Specifications. f the cylinder walls are badly scuffed or scored, or if they are out-of-round or tapered beyond the limits given in the Specifications, have the engine block rebored and honed at an automotive machine shop. f a rebore is done, oversize pistons and rings will be required.

5 If the cylinders are in reasonably good condition and not worn to the outside of the limits, they do not have to be rebored. Honing is all that is necessary.

6 Before honing the cylinders, install the main bearing caps (without the bearings) and tighten the bolts to the specified torque.

7 To hone the cylinders you will need a hone. Two types are available, the traditional stone type and the "bottle brush" or flex hone type. Both will do a good job. For the less experienced mechanic the "bottle brush" type will probably be easier to use. You will also need plenty of light oil or honing oil, some rags and an electric drill motor. Mount the hone in the drill motor, compress the stones and slip the hone into the first cylinder (see illustrations). Lubricate the cylinder thoroughly, turn on the drill and move the hone up-and-down in the cylinder at a pace which will produce a crosshatch pattern on the cylinder walls with the crosshatch lines intersecting at approximately a 60° angle (see illustration). Be sure to use plenty of lubricant and do not take off any more material than is absolutely necessary to produce the desired finish. Do not withdraw the hone from the cylinder while it s running. Instead, shut off the drill and continue moving the hone up-and-down in the cylinder until it comes to a complete stop, then withdraw the hone. Wipe the oil out of the cylinder and repeat the procedure for the remaining cylinders. Remember, do not remove too much material from the cylinder wall. If you do not have the tools or do not desire to perform the honing operation, most automotive machine shops will do it for a reasonable fee.

8 After the honing job is complete, chamfer the top edges of the cylinder bores with a small file so the rings will not catch as the pistons are installed.

9 The entire engine block must be thoroughly washed again with warm, soapy water to remove all traces of the abrasive grit produced during the honing operation. The cylinder bores must be cleaned until a white cloth can be wiped through them and not pick up any traces of oil or debris.

10 Be sure to run a brush through all oil holes and galleries and flush them with running water. After rinsing, dry the block and apply a coat of light rust preventative oil to all machined surfaces. Wrap the block in a plastic trash bag to keep it clean and set it aside until reassembly.

15 Camshaft, lifters and bearings — inspection and bearing replacement

Refer to illustrations 15.3, 15.8 and 15.11

Camshaft

1 Camshaft wear most often shows up as loss of lobe lift, which can be checked before the engine is disassembled.

2 Remove the rocker arm cover(s), then remove the nuts and separate the rocker arms and fulcrum seats or the rocker arm shaft(s) from the cylinder head(s).

3 Beginning with the number one cylinder, mount a dial indicator with the stem resting on the end of, and directly in line with, the first valve pushrod (see illustration).

4 Rotate the crankshaft very slowly in the direction of normal rotation until the lifter is on the heel of the cam lobe. At this point the pushrod will be at its lowest position.

5 Zero the dial indicator, then very slowly rotate the crankshaft until the pushrod (cam lobe) is at its highest position. Note and record the reading on the dial indicator, then compare it to the lobe lift specifications. Repeat the procedure for each of the remaining valves.

6 If the lobe lift measurements are not as specified, a new camshaft should be installed.

7 After the camshaft has been removed from the engine, cleaned with solvent and dried, inspect the bearing journals for uneven wear, pitting and evidence of seizure. If the journals are damaged, the bearing inserts in the block are probably damaged as well. Both the camshaft and bearings will have to be replaced.

8 Measure the bearing journals with a micrometer (see illustration) to determine if they are excessively worn or out-of-round. If there is any question regarding camshaft and/or bearing wear have the cam bearings measured by a machine shop to check for proper oil clearance.

2C

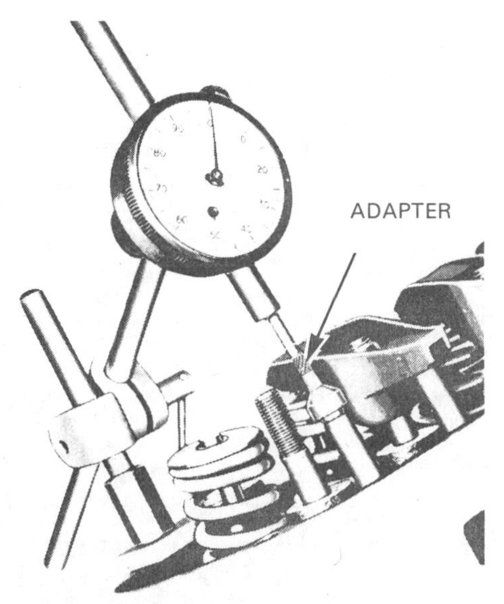

15.3 A dial indicator can be mounted as shown to check the camshaft lobe lift

15.8 The camshaft bearing journal diameter is checked to pinpoint excessive wear and out-of-round conditions

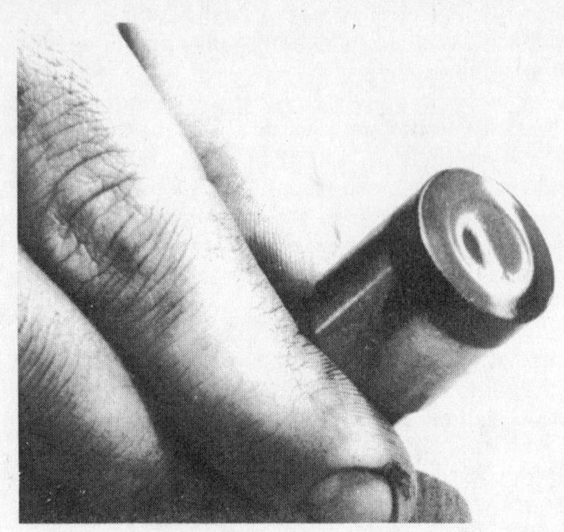

15.11 If the bottom of any lifter is worn concave, scratched or galled, replace the entire set with new lifters

9 Check the camshaft lobes for heat discoloration, score marks, chipped areas, pitting and uneven wear. If the lobes are in good condition and if the lobe lift measurements are as specified, the camshaft can be reused.

Lifters

10 Clean the lifters with solvent and dry them thoroughly without mixing them up.
11 Check each lifter wall, pushrod seat and foot for scuffing, score marks and uneven wear. Each lifter foot (the surface that rides on the cam lobe) must be slightly convex, although this can be difficult to determine by eye (the lifter foot curve is the arc of a 60-inch circle). If the base of the lifter is concave (see illustration), the lifters and camshaft must be replaced. If the lifter walls are damaged or worn (which is not very likely), inspect the lifter bores in the engine block as well. If the pushrod seats are worn, check the pushrod ends.
12 If new lifters are being installed, a new camshaft must also be installed. If a new camshaft is installed, then use new lifters as well. Never install used lifters unless the original camshaft is used and the lifters can be installed in their original locations.

Bearing replacement

13 Camshaft bearing replacement requires special tools and expertise that place it outside the scope of the home mechanic. Take the block to an automotive machine shop to ensure that the job is done correctly.

16 Piston/connecting rod assembly — inspection

Refer to illustrations 16.4, 16.10 and 16.11

1 Before the inspection process can be carried out the piston and connecting rod assembly must be cleaned and the piston rings removed from the pistons. **Note:** *Always use new piston rings when the engine is reassembled.*
2 Remove the rings from the pistons. Do not nick or gouge the pistons in the process.
3 Scrape all traces of carbon from the top of each piston. A hand held wire brush or a piece of fine emery cloth can be used once the majority of the deposits have been scraped away. Do not, under any circumstances, use a wire brush mounted in a drill motor to remove deposits from the pistons. The piston material is soft and will be eroded away by the wire brush.
4 Use the special tool commonly available or a piece of broken ring to remove carbon deposits from the ring grooves. Be very careful to remove only the carbon deposits. Do not remove any metal and do not nick or scratch the sides of the ring grooves (see illustration).
5 Once the deposits have been removed, clean the piston/rod assemblies with solvent and dry them thoroughly. Make sure that the oil return holes in the back sides of the ring grooves are clear.
6 If the pistons are not damaged or worn excessively, and if the engine block is not rebored, new pistons will not be necessary. Normal piston wear appears as even vertical wear on the piston thrust surfaces and slight looseness of the top ring in its groove. New piston ring should always be used when an engine is rebuilt.
7 Carefully inspect each piston for cracks around the skirt, at th pin bosses and at the ring lands.
8 Look for scoring and scuffing on the thrust faces of the skirt, hole in the piston crown and burned areas at the edge of the crown. If th skirt is scored or scuffed, the engine may have been suffering fror overheating and/or abnormal combustion, which caused excessivel high operating temperatures. The ignition, cooling and lubricatio systems should be checked thoroughly. A hole in the piston crown an extreme to be sure, is an indication that abnormal combustio (preignition) was occurring. Burned areas at the edge of the pisto crown are usually evidence of spark knock (detonation). If any of th above problems exist, the causes must be corrected or the damag will occur again.
9 Corrosion of the piston (evidenced by pitting) indicates that coolan is leaking into the combustion chamber and/or the crankcase. Again the cause must be corrected or the problem may persist in the rebuil engine.
10 Measure the piston ring side clearance by laying a new ring in eac ring groove and slipping a feeler gauge in between the ring and th

16.4 The piston ring grooves can be cleaned with a piece of broken ring, as shown here, or a special ring groove cleaning tool

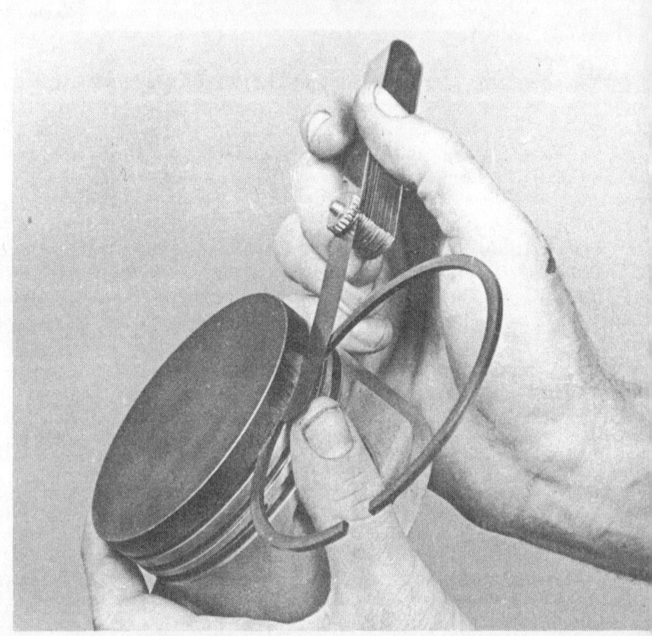

16.10 Check the ring side clearance with a feeler gauge at several points around the groove

16.11 Measure the piston diameter at a 90° angle to the piston pin and in line with it

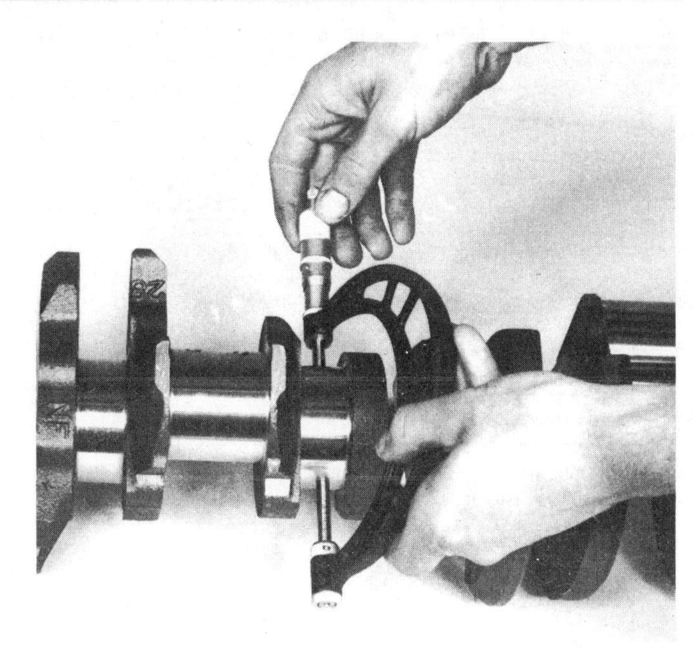

17.2 Measure the diameter of each crankshaft journal at several points to detect taper and out-of-round conditions

edge of the ring groove (see illustration). Check the clearance at three or four locations around each groove. Be sure to use the correct ring for each groove. If the side clearance is greater than specified, new pistons will be required.

11 Check the piston-to-bore clearance by measuring the bore (see Section 14) and the piston diameter. Make sure that the pistons and bores are correctly matched. Measure the piston across the skirt, at a 90° angle to and in line with the piston pin (see illustration). Subtract the piston diameter from the bore diameter to obtain the clearance. If it is greater than specified, the block will have to be rebored and new pistons and rings installed.

12 If the pistons must be removed from the connecting rods, such as when new pistons are installed, or if the piston pins have too much play in them, they should be taken to an automotive machine shop. While they are there have the connecting rods checked for bend and twist. Unless new pistons or connecting rods are to be installed, do not detach the pistons from the connecting rods.

13 Check the connecting rods for cracks and other damage. Remove the rod caps, lift out the old bearing inserts, wipe the rod and cap bearing surfaces clean and inspect them for nicks, gouges and scratches. After checking the rods, replace the old bearings, slip the caps into place and tighten the nuts finger tight.

17 Crankshaft — inspection

Refer to illustration 17.2

1 Clean the crankshaft with solvent and dry it thoroughly. Be sure to clean the oil holes with a stiff brush and flush them with solvent. Check the main and connecting rod bearing journals for uneven wear, score marks, pitting and cracks. Check the remainder of the crankshaft for cracks and other damage.

2 Using a micrometer, measure the diameter of the main and connecting rod journals (see illustration) and compare the results to the Specifications. By measuring the diameter at a number of points around the journal's circumference you will be able to determine whether or not the journal is out-of-round. Take measurements at each end of the journal, near the crank counterweights, to determine whether the journal is tapered.

3 If the crankshaft journals are damaged, tapered, out-of-round or worn beyond the limits given in the Specifications, have the crankshaft reground by an automotive machine shop. Be sure to use the correct oversize bearing inserts if the crankshaft is reconditioned.

18 Main and connecting rod bearings — inspection

1 Even though the main and connecting rod bearings should be replaced with new ones during the engine overhaul, the old bearings should be retained for close examination. They will often reveal valuable information about the condition of the engine.

2 Bearing failure occurs mainly because of lack of lubrication, the presence of dirt or other foreign particles, overloading the engine and corrosion. Regardless of the cause of bearing failure, it must be corrected before the engine is reassembled to prevent it from happening again.

3 When examining the bearings, remove them from the engine block, the main bearing caps, the connecting rods and the rod caps and lay them out on a clean surface in the same positions as their location in the engine. This will enable you to match any bearing problems with the corresponding crankshaft journal.

4 Dirt and other foreign particles get into the engine in a variety of ways. They may be left in the engine during assembly, or they may pass through filters or breathers. They may get into the oil, and from there into the bearings. Metal particles from machining operations and normal engine wear are often present. Abrasives are sometimes left in engine components after reconditioning, especially when parts are not thoroughly cleaned using the proper cleaning methods. Whatever the source, these foreign objects often end up embedded in the soft bearing material and are easily recognized. Large particles will not embed in the bearing and will score or gouge the bearing and shaft. The best prevention for this cause of bearing failure is to clean all parts thoroughly and keep everything spotlessly clean during engine assembly. Frequent and regular engine oil and filter changes are also recommended.

5 Lack of lubrication (or lubrication breakdown) has a number of interrelated causes. Excessive heat, which thins the oil, overloading, which squeezes the oil from the bearing face, and oil leakage or throwoff from excessive bearing clearances, worn oil pump or high engine speeds all contribute to lubrication breakdown. Blocked oil passages, which usually are the result of misaligned oil holes in a bearing shell, will also oil starve a bearing and destroy it. When lack of lubrication is the cause of bearing failure the bearing material is wiped or extruded from the steel backing of the bearing. Temperatures may increase to the point where the steel backing turns blue from overheating.

6 Driving habits can have a definite effect on bearing life. Full throttle, low speed operation (lugging the engine) puts very high loads on bear-

ings, which tends to squeeze out the oil film. These loads cause the bearings to flex, which produces fine cracks in the bearing face (fatigue failure). Eventually the bearing material will loosen in pieces and tear away from the steel backing.

7 Short trip driving leads to corrosion of bearings because insufficient engine heat is produced to drive off the condensed water and corrosive gases. These products collect in the engine oil, forming acid and sludge. As the oil is carried to the engine bearings, the acid attacks and corrodes the bearing material.

8 Incorrect bearing installation during engine assembly will lead to bearing failure as well. Tight fitting bearings leave insufficient bearing oil clearance and will result in oil starvation. Dirt or foreign particles trapped behind a bearing insert result in high spots on the bearing, which can lead to failure.

19 Engine overhaul — reassembly sequence

1 Before beginning engine reassembly, make sure you have all the necessary new parts, gaskets and seals as well as the following items on hand:

Common hand tools
1/2-inch drive torque wrench
Piston ring installation tool
Piston ring compressor
Lengths of rubber to fit over connecting rod bolts
Plastigage
Feeler gauges
Fine-tooth file
Engine oil
Engine assembly lube or moly-base grease
RTV-type gasket sealant
Anaerobic-type gasket sealant
Thread locking compound

2 In order to save time and avoid problems, engine reassembly should be done in the following order:

Piston rings
Rear main oil seal
Crankshaft and main bearings
Piston/connecting rod assemblies
Oil pump
Oil pan
Camshaft
Timing chain/sprockets or gears
Timing chain/gear cover
Valve lifters
Cylinder head and pushrods
Intake and exhaust manifolds
Oil filter
Rocker arm cover(s)
Fuel pump
Water pump
Flywheel/driveplate
Carburetor/fuel injection components
Thermostat and housing
Distributor, coil, spark plug wires and spark plugs
Emissions control components
Alternator

20 Piston rings — installation

Refer to illustrations 20.3a, 20.3b and 20.12

1 Before installing the new piston rings, the ring end gaps must be checked. It is assumed that the piston ring side clearance has been checked and verified correct (Section 16).

2 Lay out the piston/connecting rod assemblies and the new ring sets so the ring sets will be matched with the same piston and cylinder during both end gap measurement and engine assembly.

3 Insert the top (number one) ring into the cylinder and square it up with the cylinder walls by pushing it in with the top of the piston (see illustration). The ring should be near the lower limit of ring travel. To measure the end gap, slip a feeler gauge between the ends of the ring (see illustration). Compare the measurement to the Specifications.

4 If the gap is larger or smaller than specified, double-check to make sure that you have the correct rings before proceeding.

5 If the gap is too small it must be enlarged or the ring ends may come in contact with each other during engine operation, which can cause serious damage to the engine. The end gap can be increased by filing the ring ends very carefully with a fine file. Mount the file in a vise equipped with soft jaws, slip the ring over the file with the ends contacting the file face and slowly move the ring to remove material from the ends. When performing this operation, file only from the outside in.

6 Excess end gap is not critical unless it is greater than 0.040-inch (1 mm). If the ring end gap is greater than 0.040-inch you have the wrong rings.

7 Repeat the procedure for each ring. Remember to keep rings, pistons and cylinders matched up.

8 Once the ring end gaps have been checked the rings can be installed on the pistons.

9 The oil control ring (lowest one on the piston) is installed first. It is composed of three separate components. Slip the spacer/expander into the groove, then install the lower side rail. Do not use a piston ring installation tool on the oil ring side rails, as they may be damaged. Instead, place one end of the side rail into the groove between the spacer/expander and the ring land, hold it firmly in place and slide a finger around the piston while pushing the rail into the groove. Install the upper side rail in the same manner.

10 After the three oil ring components have been installed, check to make sure that both the upper and lower side rails can be turned smoothly in the ring groove.

11 The number two (middle) ring is installed next. It should be stamped with a mark so it can be readily distinguished from the top ring. **Note:** *Always follow the instructions printed on the ring package or*

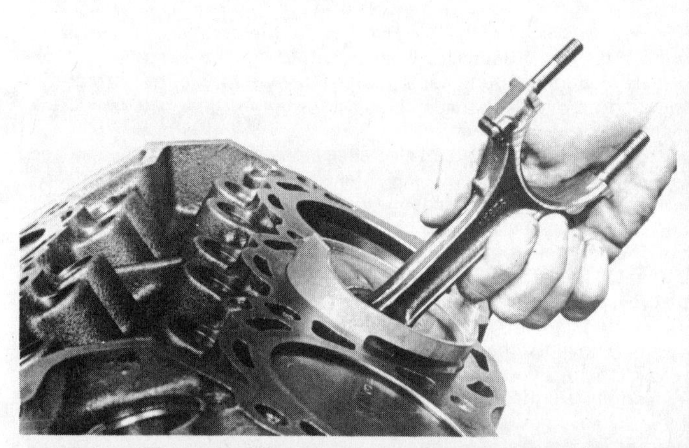

20.3a When checking piston ring end gap, the ring must be square in the cylinder bore — this is done by pushing the ring down with the top of a piston as shown

20.3b With the ring square in the cylinder, measure the end gap with a feeler gauge

box — different manufacturers may require different approaches. Do not mix up the top and middle rings, as they have different cross sections.

12 Use a piston ring installation tool and make sure that the identification mark is facing the top of the piston, then slip the ring into the middle groove on the piston (see illustration). Do not expand the ring any more than is necessary to slide it over the piston.

13 Install the number one (top) ring in the same manner. Make sure the identifying mark is facing up.

14 Repeat the procedure for the remaining pistons and rings.

20.12 Installing the compression rings with a ring expander — the mark (arrow) must face up

21 Crankshaft — installation and main bearing oil clearance check

Refer to illustrations 21.22 and 21.24

1 Crankshaft installation is the first step in engine reassembly. It is assumed at this point that the engine block and crankshaft have been cleaned, inspected and repaired or reconditioned.

2 Remove the main bearing cap bolts and lift out the caps. Lay them out in the proper order to ensure that they are installed correctly.

3 If they are still in place, remove the old bearing inserts from the block and the main bearing caps.

4 Wipe the main bearing surfaces of the block and caps with a clean, lint free cloth. They must be kept absolutely clean. This includes keeping them free of oil. If anything (dust, oil, gasket sealer or anything else), is trapped between the main bearing cap and the bearing insert it will cause a high spot when the crankshaft is installed, which can quickly wipe the babbit material from the bearing. Oil between the cap and insert will also change the bearing clearance and it can also act as a shield to prevent heat from being transferred from the bearing to the block, again destroying the bearing.

5 Clean the back sides of the new main bearing inserts and lay one bearing half in each main bearing saddle in the block. Lay the other bearing half from each bearing set in the corresponding main bearing cap. Make sure the tab on the bearing insert fits into the recess in the block or cap. **Caution:** *Make sure the bearing inserts with the oil holes and grooves are installed in the block — not the caps.* Do not hammer the bearing into place and do not nick or gouge the bearing faces. *No lubrication should be used at this time.*

6 The flanged thrust bearing must be installed in the number three cap and saddle on V8 engines and the number five cap and saddle on six cylinder engines.

7 Clean the faces of the bearings in the block and the crankshaft main bearing journals with a clean, lint free cloth. Check or clean the oil holes in the crankshaft, as any dirt here can go only one way — straight through the new bearings.

8 Once you are certain that the crankshaft is clean, carefully lay it in position (an assistant would be very helpful here) in the main bearings.

9 If you checked the main and rod journal sizes with a micrometer in Section 17, and are sure you have purchased the correct size bearing inserts, the following steps can be skipped. If, however, you were not able to check the journal sizes, or if you want to double-check to make sure you have the right size bearings, you can use Plastigage to check actual bearing clearances.

10 Trim several pieces of the appropriate size Plastigage so they are slightly shorter than the width of the main bearings and place one piece on each crankshaft main bearing journal, parallel with the journal axis. Do not lay them across the oil holes.

11 Clean the faces of the bearings in the caps and install the caps in their respective positions (do not mix them up) with the arrows pointing toward the front of the engine. Do not disturb the Plastigage.

12 Starting with the center main and working out toward the ends, tighten the main bearing cap bolts, in three steps, to the specified torque. *Do not rotate the crankshaft at any time during this operation.*

13 Remove the bolts and carefully lift off the main bearing caps. Keep them in order. Do not disturb the Plastigage or rotate the crankshaft. If any of the main bearing caps are difficult to remove, tap them gently from side-to-side with a soft face hammer to loosen them.

14 Compare the width of the crushed Plastigage on each journal to the scale printed on the Plastigage container to obtain the main bearing oil clearance. Check the Specifications to make sure it is correct.

15 If the clearance is not correct, double-check to make sure you have the right size bearing inserts. Also, make sure that no dirt or oil was between the bearing inserts and the main bearing caps or the block when the clearance was measured.

16 Carefully scrape all traces of the Plastigage material off the main bearing journals and the bearing faces. Do not nick or scratch the bearing faces.

17 Carefully lift the crankshaft out of the engine. Clean the bearing faces in the block, then apply a thin layer of moly-base grease or engine assembly lube to each of the bearing surfaces. Be sure to coat the thrust flange faces as well as the journal face of the thrust bearing.

18 Refer to the appropriate part of Chapter 2 and install the rear main oil seal in the block and bearing cap (V8 engines only). Lubricate the rear main bearing oil seal, where it contacts the crankshaft, with moly-base grease or engine assembly lube.

19 Make sure the crankshaft journals are clean, then lay the crankshaft back in place in the block. Clean the faces of the bearings in the caps, then apply a thin layer of moly-base grease to each of the bearing faces. Install the caps in their respective positions with the arrows on the caps pointing toward the front of the engine. Install the bolts.

20 Tighten all except the thrust bearing cap bolts to the specified torque. Tighten the thrust bearing cap bolts finger tight. Tap the end of the crankshaft, first backward, then forward, using a lead or brass hammer to line up the main bearing and crankshaft thrust surfaces. Retighten all main bearing cap bolts to the specified torque.

21 On all models, rotate the crankshaft a number of times by hand and check for any obvious binding.

22 Check the crankshaft end play. Mount a dial indicator with the stem in line with the crankshaft and just touching one of the crank throws (see illustration).

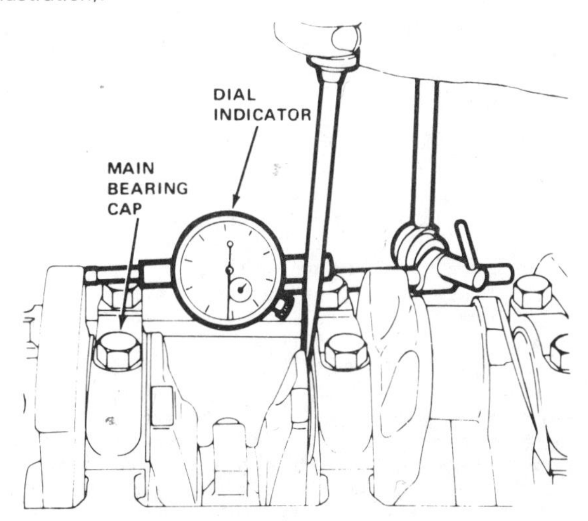

21.22 Checking crankshaft end play with a dial indicator

2C

21.24 Checking crankshaft end play with a feeler gauge

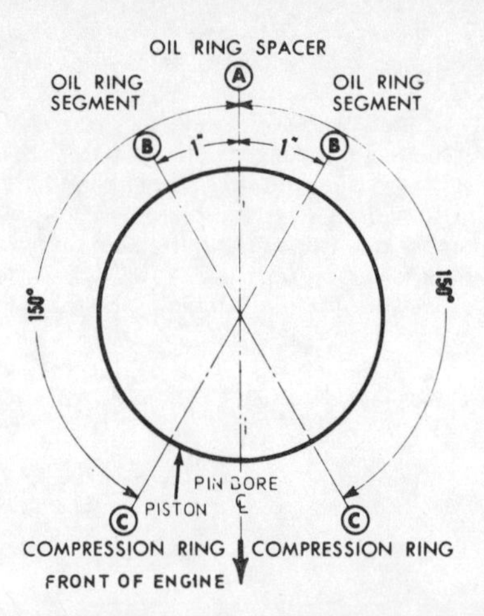

22.5 Before installing the pistons, position the piston ring end gaps as shown here

23 Push the crankshaft all the way to the rear and zero the dial indicator. Pry the crankshaft to the front as far as possible and check the reading on the dial indicator. The distance it moves is the end play. If it is different than specified, check to make sure you have the correct bearings and that the thrust bearing is installed in the correct position.
24 If a dial indicator is not available, a feeler gauge can be used. Gently pry or push the crankshaft all the way to the front of the engine. Slip feeler gauge blades between the crankshaft and the front face of the thrust main bearing to determine the clearance (see illustration).
25 On six cylinder engines, the rear crankshaft oil seal can now be installed (see Chapter 2, Part A).

22 Piston/connecting rod assembly — installation and bearing oil clearance check

Refer to illustrations 22.5, 22.8, 22.10, 22.12 and 22.14

1 Before installing the piston and connecting rods, each cylinder must be perfectly clean, the top edge of each cylinder must be chamfered and the crankshaft must be in place.
2 Remove the connecting rod cap from the end of the number one connecting rod. Remove the old bearing inserts and wipe the bearing surfaces of the connecting rod and cap with a clean, lint free cloth (they must be kept spotlessly clean).
3 Clean the back side of the new upper bearing half, then lay it in place in the connecting rod. Make sure that the tab on the bearing fits into the recess in the rod. Do not hammer the bearing insert into place and be very careful not to nick or gouge the bearing face. Do not lubricate the bearing at this time.
4 Clean the back side of the other bearing half and install it in the rod cap. Again, make sure the tab on the bearing fits into the recess in the cap and do not apply any lubricant. It is critically important that the mating surfaces of the bearing and connecting rod are perfectly clean and oil free when they are assembled.
5 Position the piston ring gaps as shown (see illustration), then slip sections of plastic or rubber hose over the connecting rod cap bolts.
6 Lubricate the piston and rings with clean engine oil and place a ring compressor over the piston. Leave the skirt protruding about 1/4-inch to guide the piston into the cylinder. The rings must be compressed until they are flush with the piston body.
7 Rotate the crankshaft until the number one connecting rod journal is as far from the number one cylinder as possible (bottom dead center) and apply a coat of engine oil to the cylinder wall.
8 With the notch or arrow on top of the piston facing the front of the engine (see illustration), gently place the piston/connecting rod assembly into the number one cylinder bore and rest the bottom edge of the ring compressor on the engine block. Tap the top edge of the ring compressor to make sure it is contacting the block around its entire circumference.

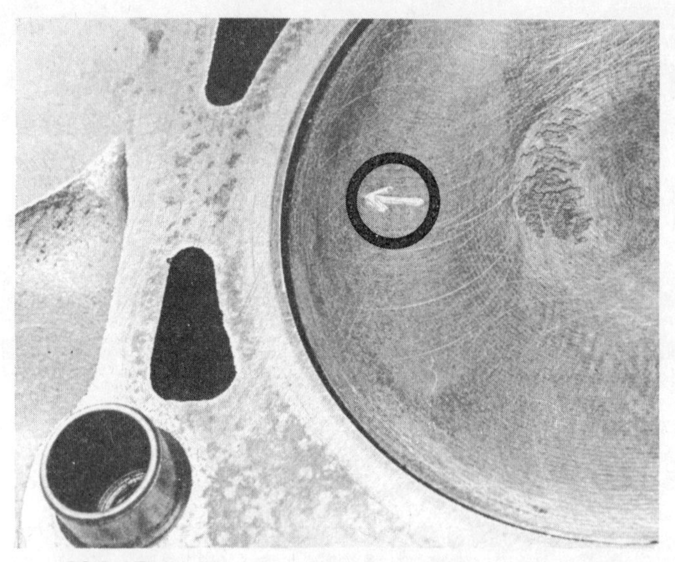

22.8 The arrow or notch in each piston must face the front of the engine as the pistons are installed

9 Clean the number one connecting rod journal on the crankshaft and the bearing faces in the rod.
10 Carefully tap on the top of the piston with the end of a wooden hammer handle (see illustration) while guiding the end of the connecting rod into place on the crankshaft journal. The piston rings may try to pop out of the ring compressor just before entering the cylinder bore, so keep some downward pressure on the ring compressor. Work slowly, and if any resistance is felt as the piston enters the cylinder, stop immediately. Find out what is hanging up and fix it before proceeding. *Do not, for any reason, force the piston into the cylinder, as you will break a ring and/or the piston.*
11 If you checked the rod journal diameter with a micrometer in Section 17, you know what size bearings will be needed. If, however, you either weren't able to check the journal diameter, or you want to double-check to make sure you have the right size bearings, you can use Plastigage to check the bearing clearance.
12 Cut a piece of the appropriate size Plastigage slightly shorter than the width of the connecting rod bearing and lay it in place on the number one connecting rod journal, parallel with the journal axis (it must not cross the oil hole in the journal) (see illustration).
13 Clean the connecting rod cap bearing face, remove the protective hoses from the connecting rod bolts and gently install the rod cap in place. Make sure the mating mark on the cap is on the same side as the mark on the connecting rod. Install the nuts and tighten them to

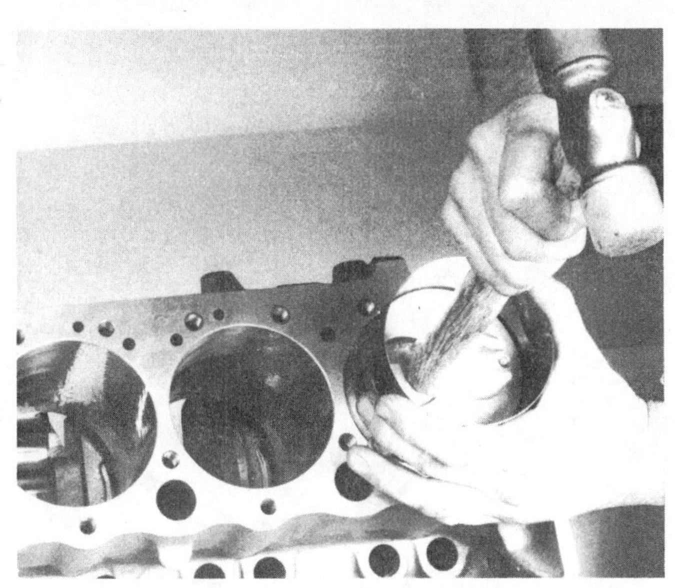

22.10 The piston can be driven (gently) into the cylinder bore with the end of a wooden hammer handle

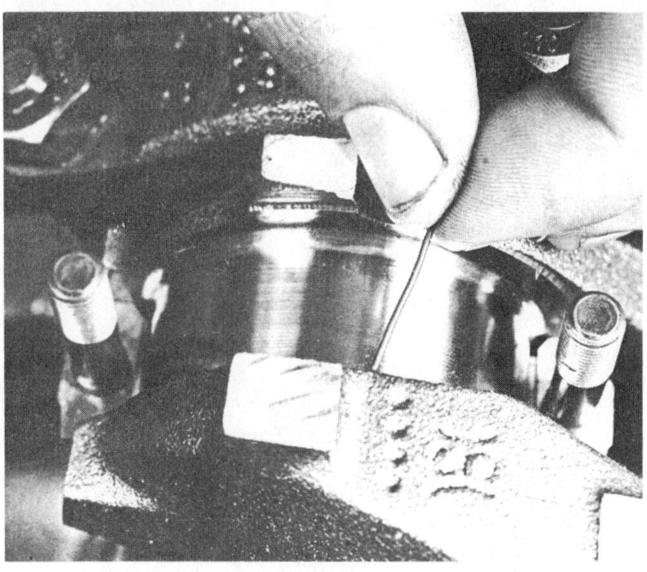

22.12 Lay the Plastigage strips on the bearing journals, parallel to the crankshaft centerline

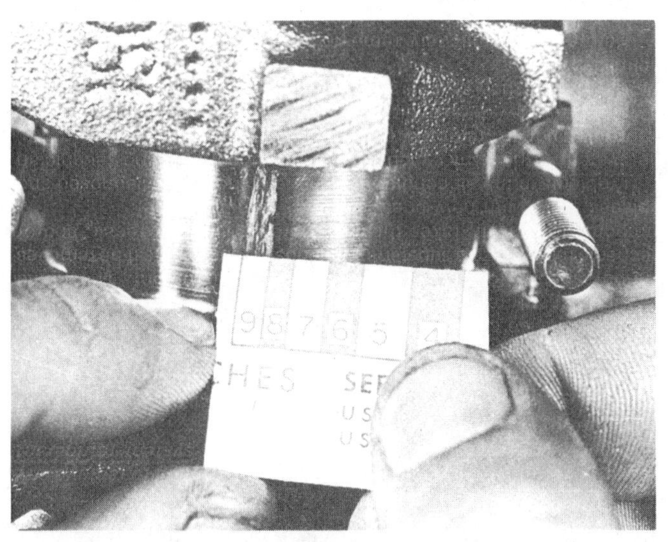

22.14 Measuring the width of the crushed Plastigage to determine the bearing oil clearance

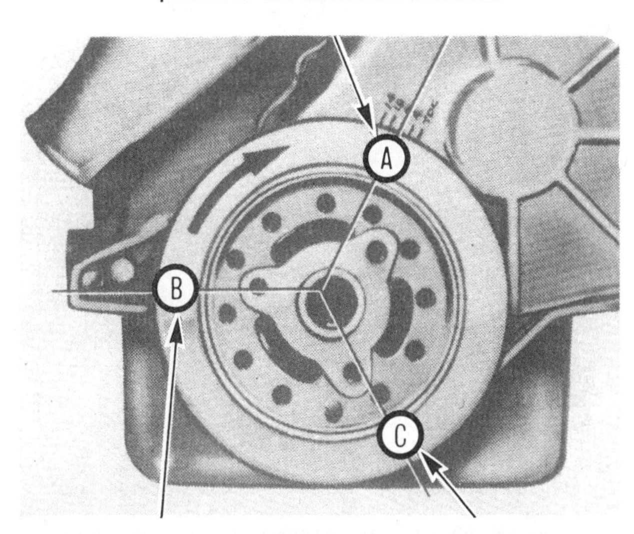

23.2 Mark the six cylinder engine crankshaft pulley at two additional points, 120° from the timing mark (A in this illustration)

the specified torque, working up to it in three steps. *Do not rotate the crankshaft at any time during this operation.*

14 Remove the rod cap, being very careful not to disturb the Plastigage. Compare the width of the crushed Plastigage to the scale printed on the Plastigage container to obtain the oil clearance (see illustration). Compare it to the Specifications to make sure the clearance is correct. If the clearance is not correct, double-check to make sure that you have the correct size bearing inserts.

15 Carefully scrape all traces of the Plastigage material off the rod journal and bearing face. Be very careful not to scratch the bearing — use your fingernail or a piece of hardwood. Make sure the bearing faces are perfectly clean, then apply a layer of moly-base grease or engine assembly lube to both of them. You will have to push the piston up into the cylinder to expose the face of the bearing insert in the connecting rod. Be sure to return the protective hoses to the rod bolts first.

16 Slide the connecting rod back into place on the journal, remove the protective hoses from the rod cap bolts, install the rod cap and tighten the nuts to the specified torque. Again, work up to the torque in three steps.

17 Repeat the entire procedure for the remaining piston and connecting rod assemblies. Keep the back sides of the bearing inserts and the inside of the connecting rod and cap perfectly clean when assembling them. Make sure you have the correct piston for each cylinder and that the notch or arrow on the piston faces the front of the engine when the piston is installed. Use plenty of oil to lubricate the piston before installing the ring compressor. When installing the rod caps for the final time, be sure to lubricate the bearing faces adequately.

18 After the piston and connecting rod assemblies have been properly installed, rotate the crankshaft a number of times by hand and check for any obvious binding.

19 As a final step, the connecting rod end play must be checked. Refer to Section 11 for this procedure. Compare the measured end play to the Specifications.

23 Valve adjustment

Refer to illustrations 23.2, 23.20a, 23.20b, 23.20c, 23.22, 23.23a, 23.23b and 23.23c

Six cylinder engines (1969 through 1978 only)

1 Connect an auxiliary starter switch to the starter solenoid.

2 Mark the vibration damper pulley with two chalk marks spaced 120° apart on either side of the timing mark (see illustration).

3 Position the piston in the number one cylinder at top dead center

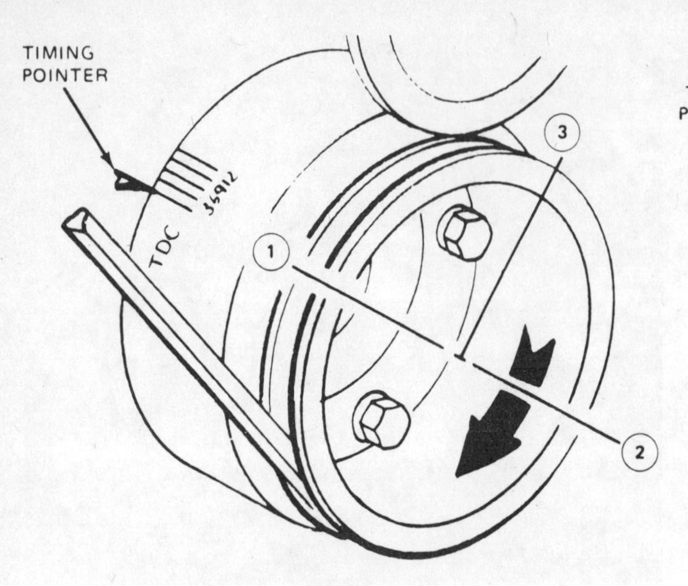

23.20a On all 1969 through 1982 V8 engines, mark the crankshaft pulley at the TDC mark and two additional points (90° and 180° from the TDC mark)

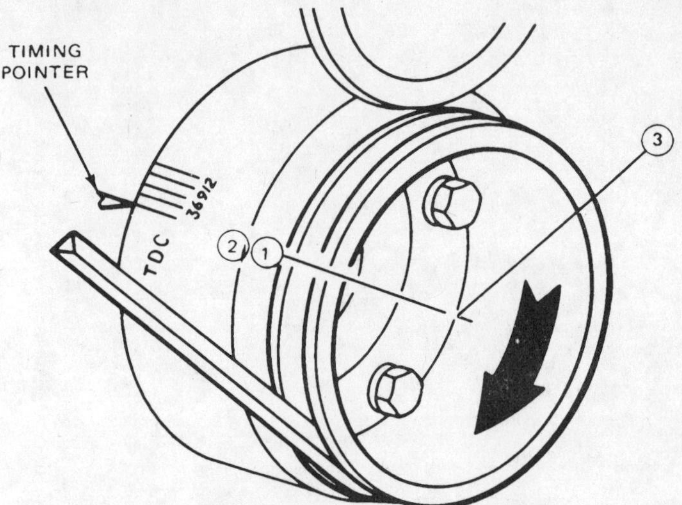

23.20b On 1983 through 1986 302 and 351W V8's, positions 1 and 2 are 360° apart at the TDC mark, while position 3 is 90° clockwise

on the compression stroke. To do this, remove all of the spark plugs from the engine, then locate the number one cylinder spark plug wire and trace it back to the distributor. Make a mark on the distributor body directly below the terminal where the number one spark plug wire attaches to the distributor cap, then remove the cap and wires from the distributor. Slip a wrench or socket over the large bolt at the front of the crankshaft and slowly turn it in a clockwise direction (viewed from the front) until the notch in the crankshaft pulley is aligned with the O on the timing mark tag. At this point the rotor should be pointing at the mark you made on the distributor body. If it is not, turn the crankshaft one complete revolution (360°) in a clockwise direction. If the rotor is now pointing at the mark on the distributor body, then the number one piston is at TDC on the compression stroke.

4 Remove the rocker arm cover.

5 Using a torque wrench, check the breakaway torque required to turn each locknut for the number one cylinder rocker arms counterclockwise. If the breakaway torque is less than specified, replace the locknuts with new ones. If the breakaway torque is still not correct, replace the studs with new ones.

6 Adjust the clearance for the number one cylinder valves by first loosening each rocker arm stud nut until there is clearance between the rocker arm and pushrod. You can keep a finger on the pushrod while loosening the rocker arm retaining nut and feel the point at which the tension is relieved from the pushrod.

7 Tighten the nut slowly while rotating the pushrod between your thumb and finger until the rocker arm just begins to touch the pushrod and valve (zero clearance). Note the exact position of your ratchet or socket handle.

8 Turn the locating nut in exactly *one turn* to place the hydraulic lifter plunger in the desired operating range.

9 Repeat this procedure for the remaining valve.

10 Rotate the engine exactly 120° clockwise (to the chalk mark) and adjust the number five intake and exhaust valves.

11 Rotate the engine 120° farther and adjust the number three intake and exhaust valves.

12 Rotate the engine 120° (back to the top dead center point) and adjust the number six intake and exhaust valves.

13 Again, rotate the engine 120° and adjust the number two intake and exhaust valves.

14 Finally, rotate the engine 120° more and adjust the number four intake and exhaust valves.

15 Attach a new rocker arm cover gasket and install the cover.

16 Connect all components and lines previously removed to obtain access to the rocker arms.

17 Start the engine and allow it to reach normal operating temperature. Make sure that the engine is running smoothly and that there is no roughness, which could be caused by a tight valve. Also listen for valve

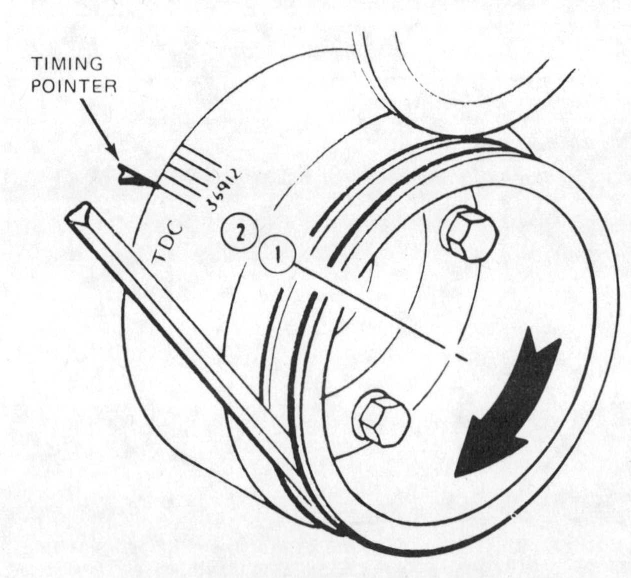

23.20c On 1983 through 1986 460 V8's, the TDC mark is the only reference point needed when adjusting the valves

lifter noise which could be caused by a loose valve. If any of these problems occur, recheck the valve adjustments.

V8 engines (all) and 1979-on six cylinder engines

Note: *Since no provision is made for valve adjustments on these engines, service pushrods are available 0.060-inch shorter and 0.060-inch longer than standard pushrods to provide adjustment capability (although some early 302 V8 engines are equipped with adjustable rocker arms on which the valve clearance is changed by turning the rocker arm nut as needed). Normally these engines do not need any valve adjustments, as the lash is taken care of by the hydraulic lifter. If major engine work is done, such as a valve job, which will alter the relationship between valve train components, the following valve checking procedure is provided. If you have a running engine that has symptoms of valve clearance problems, such as excessive noise from a lifter, check for a defective part. Normally an engine will not reach a point where it needs a valve adjustment unless a component malfunction has occurred. Hydraulic lifter failure or excessive rocker arm wear are two likely examples of component failure.*

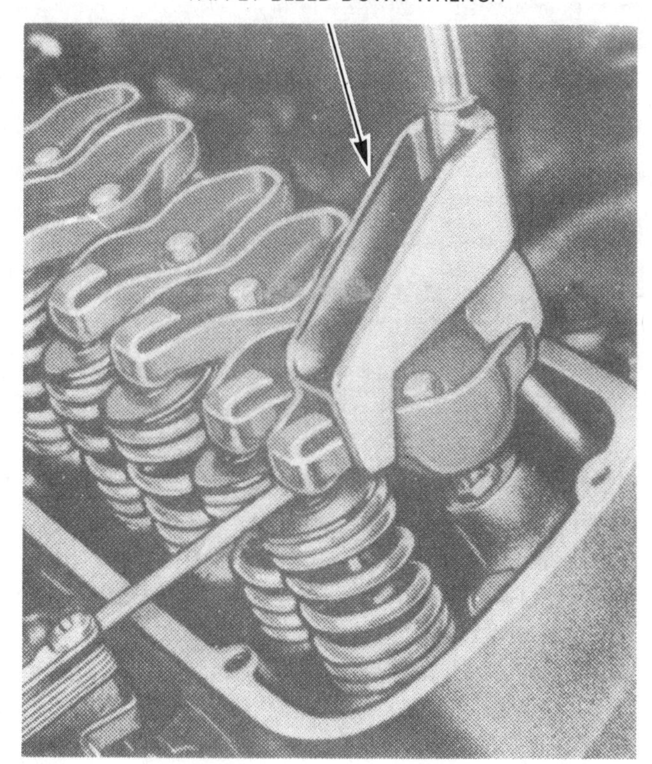

23.22 Checking the valve clearances (note the use of a tappet [lifter] bleed down wrench)

18 Make sure that the lifter is compressed (not pumped up with oil). This is accomplished during engine assembly by installing new lifters or by compressing the lifter and relieving it of all internal oil pressure. If you are compressing a lifter that is pumped up from use, a special tool will be necessary.
19 Position the number one piston at top dead center on the compression stroke (refer to Step 3 above). Note that V8 engines have the ignition timing numbers on the pulley and a pointer on the engine. Finding TDC is accomplished the same way.
20 With the crankshaft in this position, mark the pulley as shown (see illustrations).
21 On six cylinder engines, divide the pulley into 120° increments as described in Step 2, then proceed to Steps 25 and 26 for the valve adjustment procedure.
22 With the number one piston at TDC, check the gaps between the number one cylinder rocker arms and the valve stems with a feeler gauge (see illustration). Compare the results to the Specifications. Make sure the hydraulic lifter is fully compressed. If the clearance is less than specified, install a shorter service pushrod. If the clearance is greater than specified, install a longer service pushrod.
23 Rotate the crankshaft clockwise to position number two (see illustrations). On 1983 through 1986 302, 351W and 460 engines, position 2 is 360° from position 1. Check the valve clearance for the cylinders indicated. Make sure the directions apply to your specific engine type.
24 Rotate the engine clockwise to position number 3 and adjust the corresponding valves. Some engines will have only two positions.
25 For six cylinder engines, adjust both valves for cylinder number one at position A as shown in illustration 23.2. Rotate the crankshaft to position B. Adjust both valves for cylinder number five. Rotate the crankshaft 120° to position C and adjust both valves for cylinder number three.
26 Rotate the crankshaft 120° back to position A and adjust both valves for cylinder number six. Repeat the above procedure and adjust both valves for cylinder number two at position B. Finally, complete the procedure by adjusting both valves for cylinder number four at position C.

POSITION 1

No. 1 Intake — No. 1 Exhaust
No. 7 Intake — No. 5 Exhaust
No. 8 Intake — No. 4 Exhaust

POSITION 2

No. 5 Intake — No. 2 Exhaust
No. 4 Intake — No. 6 Exhaust

POSITION 3

No. 2 Intake — No. 7 Exhaust
No. 3 Intake — No. 3 Exhaust
No. 6 Intake — No. 8 Exhaust

23.23a Valve clearance check/adjustment at each position — 302 V8's (all years) and 460 V8's (through 1982)

POSITION 1

No. 1 Intake — No. 1 Exhaust
No. 4 Intake — No. 3 Exhaust
No. 8 Intake — No. 7 Exhaust

POSITION 2

No. 3 Intake — No. 2 Exhaust
No. 7 Intake — No. 6 Exhaust

POSITION 3

No. 2 Intake — No. 4 Exhaust
No. 5 Intake — No. 5 Exhaust
No. 6 Intake — No. 8 Exhaust

23.23b Valve clearance check/adjustment at each position (all 351 and 400 V8's)

POSITION 1

No. 1 Intake — No. 1 Exhaust
No. 3 Intake — No. 8 Exhaust
No. 7 Intake — No. 5 Exhaust
No. 8 Intake — No. 4 Exhaust

POSITION 2

No. 2 Intake — No. 2 Exhaust
No. 4 Intake — No. 3 Exhaust
No. 5 Intake — No. 6 Exhaust
No. 6 Intake — No. 7 Exhaust

23.23c Valve clearance check/adjustment at each position (1983 through 1986 460 V8's)

24 Initial start-up and break-in after overhaul

1 Once the engine has been installed in the vehicle, double-check the engine oil and coolant levels.
2 With the spark plugs out of the engine and the coil high tension lead grounded to the engine block, crank the engine over until oil pressure registers on the gauge (if so equipped) or until the oil light goes off.
3 Install the spark plugs, hook up the plug wires and the coil high tension lead.
4 Make sure the carburetor choke plate is closed, then start the engine. It may take a few moments for gasoline to reach the carburetor, but the engine should start without a great deal of effort.
5 As soon as the engine starts it should be set at a fast idle to ensure proper oil circulation and allowed to warm up to normal operating temperature. While the engine is warming up make a thorough check for oil and coolant leaks.
6 Shut the engine off and recheck the engine oil and coolant levels. Restart the engine and check the ignition timing and the engine idle speed (refer to Chapter 1). Make any necessary adjustments.
7 Drive the vehicle to an area with minimum traffic, accelerate at full throttle from 30 to 50 mph, then allow the vehicle to slow to 30 mph with the throttle closed. Repeat the procedure 10 or 12 times. This will load the piston rings and cause them to seat properly against the cylinder walls. Check again for oil and coolant leaks.
8 Drive the vehicle gently for the first 500 miles (no sustained high speeds) and keep a constant check on the oil level. It is not unusual for an engine to use oil during the break-in period.
9 At 500 miles change the oil and filter and retorque the cylinder head bolts.
10 For the next few hundred miles drive the vehicle normally. Do not pamper it or abuse it.
11 After 2000 miles, change the oil and filter again and consider the engine fully broken in.

Chapter 3
Cooling, heating and air conditioning systems

Refer to Chapter 13 for information on 1987 and later models

Contents

Specifications

Drivebelt tension	See Chapter 1
Thermostat	
starts to open	170°F
fully open	210°F

Torque specifications	**Ft-lbs**
Water outlet (thermostat housing) bolts	12 to 15
Water pump mounting bolts	15 to 18
Radiator mounting bolts	10 to 15
Fan mounting bolts	12 to 18

1 General information

Ford vans produced during the years covered by this manual have a cooling system consisting of a radiator, a thermostat for temperature control and an impeller-type water pump driven by a belt from the crankshaft pulley.

The cooling fan is mounted on the front of the water pump. Certain vehicles with extra heavy-duty cooling systems incorporate an automatic clutch which disengages the fan at high speeds or when the outside temperature is sufficient to maintain a low radiator temperature. On some models a fan shroud is mounted to the rear of the radiator to increase cooling efficiency.

The cooling system is pressurized with a spring loaded radiator cap. Cooling efficiency is improved by raising the boiling point of the coolant through increased cooling system pressure. If the coolant temperature rises above the cap release point, the extra pressure in the system forces the radiator cap internal spring loaded valve off its seat and exposes the overflow pipe or the coolant recovery reservoir connecting tube to allow the displaced coolant a path of escape.

The cooling system functions as follows: coolant from the radiator circulates up the lower radiator hose to the water pump, where it is pumped into the engine block and around the water passages to cool the engine. The coolant then travels up into the cylinder head(s), around the combustion chambers and valve seats absorbing heat, before it finally passes out through the thermostat. When the engine is running at its correct temperature the coolant flowing from the cylinder head(s) diverges to flow through the intake manifold, vehicle interior heater (when activated) and the radiator.

When the engine is cool (below normal operating temperature), the thermostat valve is closed, preventing the coolant from flowing through the radiator, restricting the flow to the engine. The restriction in the flow of coolant enables the engine to quickly warm to correct operating temperature.

The coolant temperature is monitored by a sensor mounted in either the cylinder head or the intake manifold. The sensor, together with the gauge on the instrument panel, gives a continuous indication of coolant temperature to the driver.

Normal maintenance consists of checking the coolant level at regular intervals, inspecting the hoses and connections for leaks and material deterioration and checking the cooling fan drivebelt tension and condition (refer to Chapter 1 for details).

The heating system utilizes heat produced by the engine to warm the interior of the vehicle by routing coolant through hoses attached to the heater and engine block. The system is controlled through dash mounted levers and switches inside the vehicle.

Air conditioning is an optional accessory, with all components contained in the engine compartment with the exception of the controls, which are also dash mounted inside the vehicle. The system, like the water pump, is crankshaft driven by an accessory drivebelt.

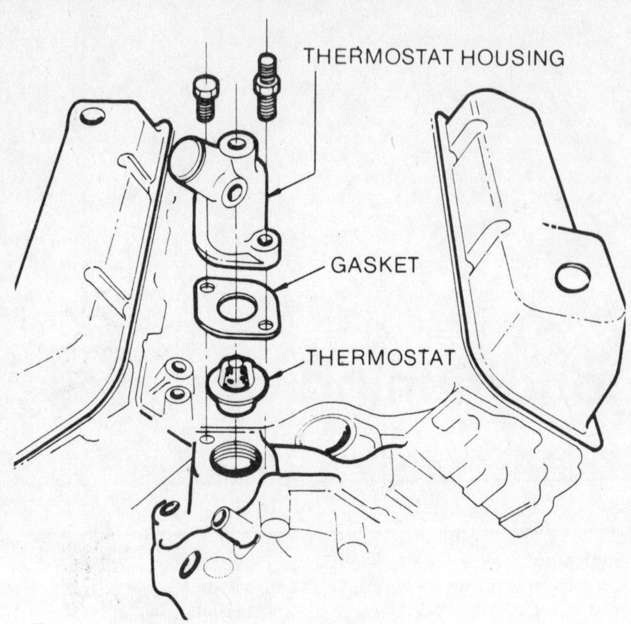

3.4 **Thermostat and related components — exploded view (V8 engine shown)**

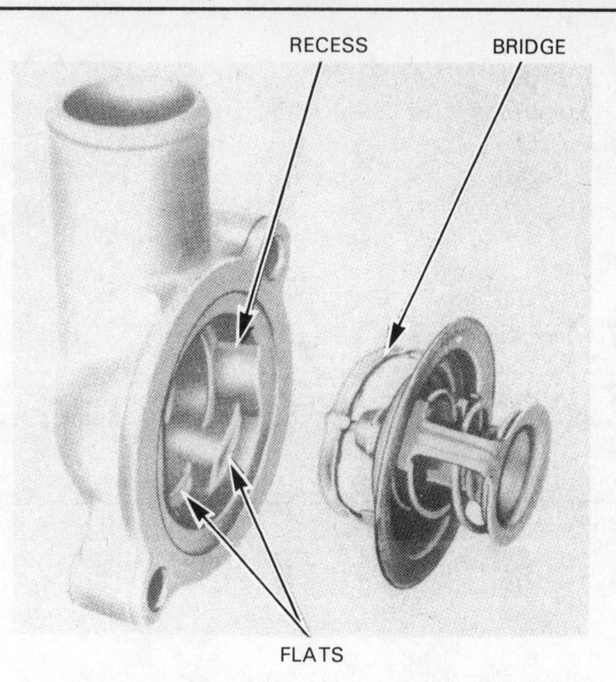

3.13 The thermostat on six cylinder engines is attached to the housing and locked into place by turning it — make sure the bridge faces into the housing elbow

2 Coolant — general information

Warning: *Do not allow antifreeze to come into contact with your skin or painted surfaces of the vehicle. Flush contaminated areas immediately with plenty of water. Antifreeze can be fatal if swallowed by children or pets (they like its sweet taste). Wipe up garage floor and drip pan coolant spills immediately. Keep antifreeze containers covered and repair cooling system leaks as soon as they are discovered.*

1 It is recommended that the cooling system contain a coolant solution of water and ethylene glycol antifreeze which will give protection to at least -20°F at all times. This provides protection against corrosion and increases the coolant boiling point. When handling antifreeze, be careful not to spill it on painted surfaces, since it will cause damage if not removed immediately.

2 The cooling system should be drained, flushed and refilled at the recommended intervals (see Chapter 1). The use of coolant solutions for longer than the specified intervals can cause damage and encourage the formation of rust and scale since the corrosion inhibitors in the antifreeze gradually lose their efficiency.

3 Before adding antifreeze to the system, check all hose connections and look for signs of leakage around the thermostat and water pump.

4 The exact mixture of antifreeze-to-water which you should use depends upon the relative weather conditions. Consult the information provided by the antifreeze manufacturer on the container label. Generally speaking, antifreeze and water are mixed in a 50/50 ratio.

5 To prevent damage to the cooling system when ambient temperatures are below freezing, when adding water or antifreeze, always operate the engine at fast idle for 30 minutes before letting the vehicle stand with the engine off for prolonged periods. This will allow a uniform mixture throughout the cooling system and prevent damage by freezing, providing that sufficient antifreeze is used.

3 Thermostat — removal, inspection and installation

Refer to illustrations 3.4, 3.13 and 3.16

Warning: *The engine must be completely cool before beginning this procedure.*

Removal

1 Drain the radiator so that the coolant level is below the thermostat housing.

Inline six cylinder engines

2 Remove the thermostat housing bolts, then pull it away from the cylinder head enough to provide access to the thermostat. You may have to tap the housing lightly with a soft face hammer to break the gasket seal.

3 Remove the thermostat and gasket, noting how it is installed to ensure correct reinstallation.

V8 engines

4 Follow the upper radiator hose back to the engine to locate the thermostat housing (see illustration).

5 Remove the thermostat housing bolts and detach the housing from the manifold. You may have to remove the bypass hose to provide access to the housing and you may have to tap lightly on the thermostat housing to break the gasket seal.

6 Carefully bend the radiator hose up and remove the thermostat and gasket, noting how it is installed to ensure correct installation.

Inspection

7 Due to the minimal cost of a replacement thermostat, it is usually better to purchase a new unit rather than attempt to check the old one to verify that it is operating correctly. However, the following procedure can be used to detect a faulty thermostat.

8 Heat a pan of water on the kitchen stove until the temperature nears those listed in the Specifications. A candy thermometer can be used to monitor the temperature.

9 Using wire, suspend the thermostat in the hot water. The valve should open approximately 1/4-inch at the specified temperature.

10 If the thermostat does not react to temperature variations as stated above, or if there are any visible defects (corrosion, cracks, etc.) the thermostat should be replaced with a new one.

Installation

Inline six cylinder engines

11 After cleaning the thermostat housing and cylinder head gasket surfaces, coat a new gasket with RTV-type sealant and position the gasket on the cylinder head opening. **Note:** *The gasket must be positioned on the cylinder head before the thermostat is installed.*

12 The thermostat housing contains a locking recess into which the thermostat is turned and locked. Install the thermostat with the bridge section in the housing elbow (facing in).

13 Turn the thermostat clockwise to lock it in position on the flats cast into the elbow (see illustration).

14 Position the housing against the cylinder head and gasket, then install and tighten the bolts to the specified torque.

V8 engines

15 After cleaning the thermostat housing gasket surfaces, coat a new gasket with RTV-type sealant.

16 Install the thermostat in the intake manifold with the copper element toward the engine and the thermostat flange positioned in the recess (see illustration).
17 Position the gasket over the thermostat.
18 Position the thermostat housing against the intake manifold and install and tighten the bolts to the specified torque.
19 Install the bypass hose (if removed) and tighten the hose connections.

All engines
20 Fill the cooling system with the recommended coolant.
21 Start and run the engine until it reaches normal operating temperature, then check the coolant level and look for leaks.

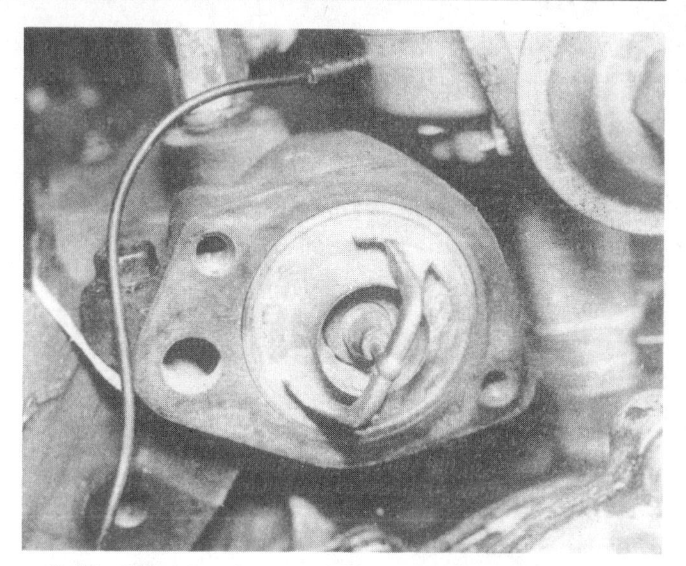

3.16 Make sure the copper element faces into the engine (the bridge will face out as shown here) when installing the thermostat in a V8 engine (302 shown)

4 Radiator — removal, inspection and installation

Refer to illustrations 4.2a and 4.2b

Warning: *The engine must be completely cool before beginning this procedure.*

Removal

1 Drain the cooling system as described in Chapter 1.
2 Remove the lower radiator hose and clamp from the radiator. Be careful not to put excess pressure on the outlet tube, as it can easily be damaged (see illustrations).

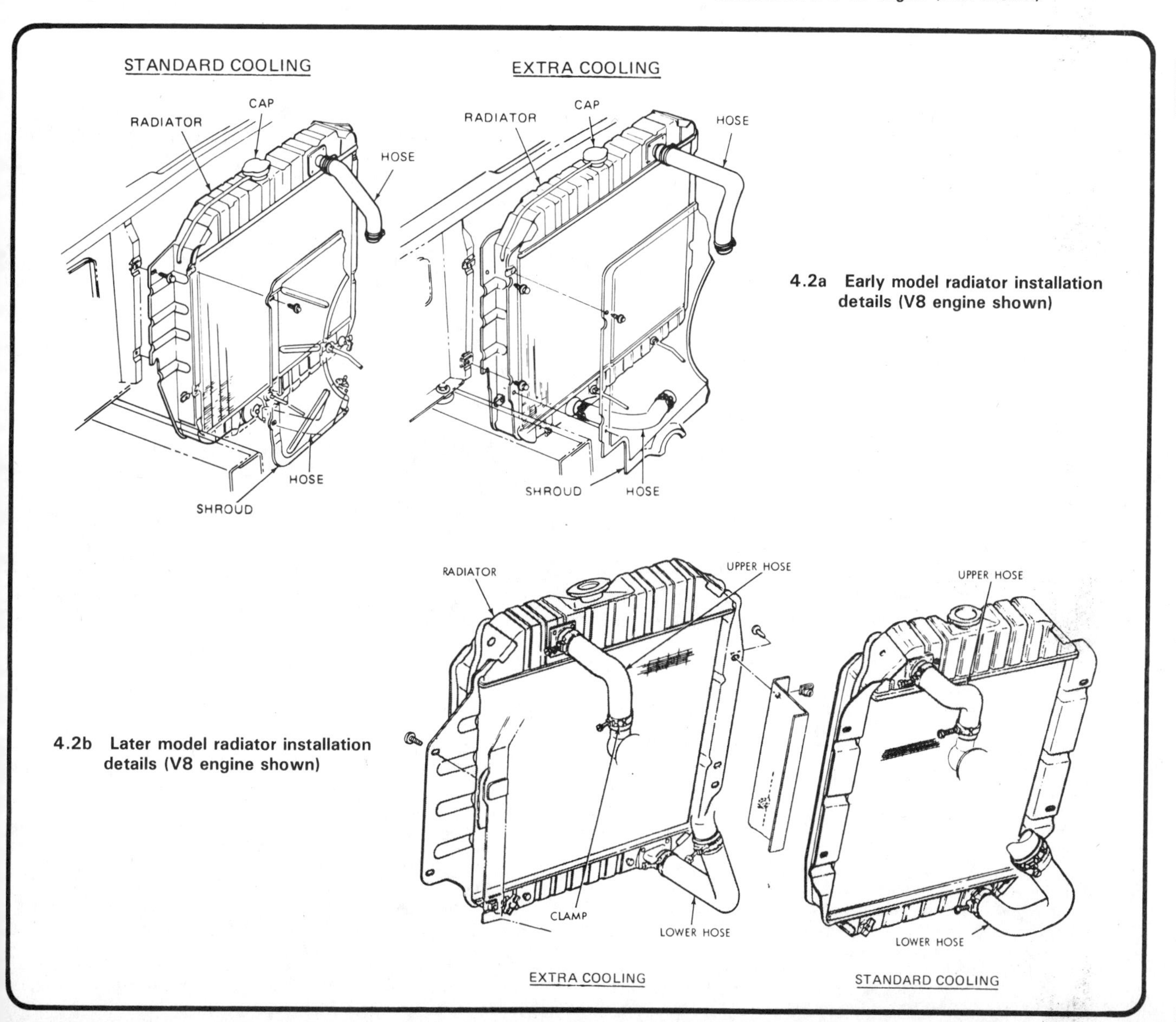

4.2a Early model radiator installation details (V8 engine shown)

4.2b Later model radiator installation details (V8 engine shown)

3 Remove the upper radiator hose and clamp from the radiator top tank.
4 If equipped with an automatic transmission, remove the transmission cooler lines from the bottom of the radiator, taking care not to twist the lines or damage the fittings. It is recommended that you use a flare nut wrench for this job. Plug the ends of the disconnected transmission lines to prevent leakage and stop dirt from entering the system. If so equipped, remove the upper radiator splash shield.
5 Remove the bolts attaching the fan shroud to the radiator support. Place the shroud over the fan, allowing space for radiator removal. Loosen and remove the radiator mounting bolts and lift the radiator out of the engine compartment.

Inspection

6 Carefully check the radiator for signs of leakage, deterioration of the tubes and fins and rust and corrosion (particularly on the inside). Inspect the cooling fins for distortion and damage. In most cases a radiator repair shop should be consulted for repairs. Radiator flushing is covered in Chapter 1.

Installation

7 Installation is the reverse of removal. Be very careful when installing the radiator into the vehicle as the cooling fins, along with the radiator itself, are fragile and can be damaged easily by mishandling or contact with the fan or radiator support. Make sure that the radiator is mounted securely and that all hoses and clamps are in good condition before you connect them to the radiator. Now would be a good time to replace them with new ones (see Chapter 1).
8 After remounting all components related to the radiator, fill it with the recommended coolant as described in Chapter 1.
9 Start the engine and allow it to reach normal operating temperature, then check for leaks.

5 Water pump — checking

1 A defective or failing water pump will usually be noisy or leak coolant.
2 Visually check the water pump for leakage. Pay special attention to the area around the pump seal at the very front of the shaft and the drain hole.
3 The front bearing in the water pump can be checked for roughness and excessive play by removing the drivebelt and grasping the fan by hand to check for movement. Move the fan up-and-down as well as in a circular motion to test for a loose bearing.
4 Visually check the sealing surfaces where the water pump mates to the front cover (or to the engine block on inline six cylinder engines) for signs of leakage.
5 If any of the above conditions are present, the water pump will have to be removed for further checking and/or replacement.

6 Water pump — removal and installation

Refer to illustrations 6.4 and 6.17

Warning: *The engine must be completely cool before beginning this procedure.*

1 Refer to Chapter 1 and drain the cooling system.
2 Remove the air cleaner assembly and air intake duct.
3 Remove the radiator and fan shroud.
4 Remove the fan/clutch retaining bolts from the nose of the water pump and remove the shroud, fan/clutch and spacer (if so equipped) (see illustration).
5 If equipped, loosen the power steering pump mounting bolts. Loosen the power steering pump drivebelt by releasing the adjustment bolt and allowing the power steering pump to move toward the engine.
6 If the power steering pump bracket is retained at the water pump, the power steering pump should be removed and laid to one side to facilitate removal of the bracket.
7 If equipped with air conditioning, do not disconnect any of the hoses or lines. The following procedures can be performed by moving, not disconnecting, the compressor. Loosen the air conditioning compressor top bracket retaining bolts. Remove the bracket on engines that have it secured to the pump. Remove the air conditioner idler arm and assembly.

6.4 The fan clutch is attached to the water pump flange with four bolts (arrow indicates one of the bolts) (V8 engine shown)

6.17 Be sure to remove all of the bolts before attempting to separate the water pump from the engine (V8 engine shown — not all bolts are visible in this case)

8 Remove the air compressor and power steering pump drivebelt.
9 If equipped, remove the air pump pulley hub bolts and remove the bolt and pulley. Remove the air pump pivot bolt, bypass hose and air pump.
10 Loosen the alternator pivot bolt.
11 Remove the retaining bolt and spacer for the alternator.
12 Remove the adjustment arm bolt, pivot bolt and alternator drivebelt.
13 Remove the alternator bracket if it is retained at the water pump.
14 Disconnect the lower radiator hose from the water pump inlet.
15 Disconnect the heater hose from the water pump.
16 Disconnect the bypass hose from the water pump.
17 Remove the water pump retaining bolts (see illustration) and detach the water pump from the front cover or the engine block (depending on engine type). Take note of the installed positions of the various size bolts.
18 Remove the separator plate from the water pump (460 engine only).

7.1 On V8 engines, the coolant temperature sending unit (arrow) is threaded into the intake manifold

19 Remove the gaskets from the mating surfaces of the water pump and from the front cover or engine block.
20 Before installation, remove all gasket material from the water pump, front cover, separator plate mating surfaces and/or engine block. Clean the gasket surfaces with lacquer thinner or acetone.
21 Coat new gaskets with RTV-type sealant and position them on the water pump.
22 Carefully position the water pump on the front cover or engine block.
23 Install the bolts finger tight. Make sure that all gaskets are in place and that the hoses line up without kinking. It may be necessary to transfer some hose fittings from the old pump if you are replacing it with a new pump.
24 Tighten the bolts to the specified torque in a criss-cross pattern.
25 Connect the radiator lower hose and clamp.
26 Connect the heater return hose and clamp.
27 Connect the bypass hose to the water pump.
28 If equipped, install the top air conditioning compressor bracket and idler pulley assembly.
29 Attach the remaining components to the water pump and engine in the reverse order of removal.
30 Adjust the drivebelts and add coolant as described in Chapter 1.
31 Start the engine and make sure there are no leaks. Check the coolant level frequently during the first few weeks of operation to ensure that there are no leaks and that the level in the system is stable.

7 Coolant temperature sending unit — check and replacement

Refer to illustration 7.1

Check

1 The coolant temperature indicating system consists of a sending unit which is screwed into the cylinder head or manifold (see illustration) and a temperature gauge mounted in the instrument panel. When the coolant temperature of the engine is low, the resistance of the sending unit is high and current to the gauge is restricted. This causes the pointer to move only a short distance. As the temperature of the engine increases, the resistance at the sending unit decreases. Current flow then increases and needle movement at the gauge changes. **Caution:** *Do not apply 12-volt current directly to the sending unit terminal at any time, as the voltage will damage the unit.*
2 Start the engine and allow it to run with a thermometer in the radiator neck until a minimum temperature of 180°F is reached.
3 The gauge in the instrument panel should indicate within the normal operating range.

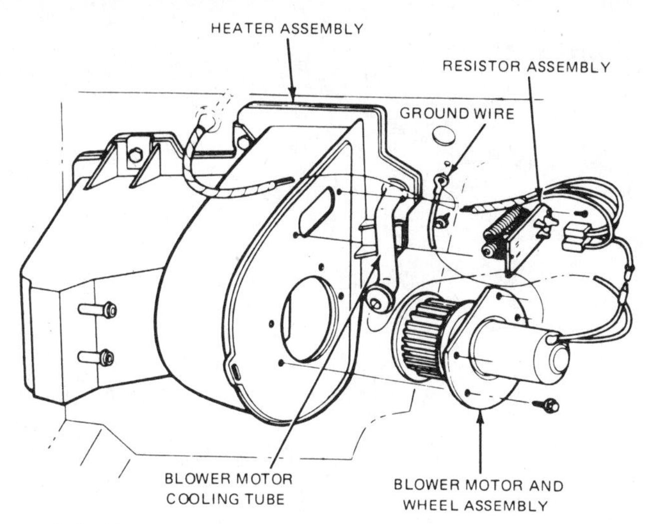

8.2 Heater blower fan and motor assembly — exploded view (later models only)

4 If the gauge does not indicate correctly, disconnect the gauge wire from the terminal at the sending unit.
5 Connect one lead of a 12-volt test light or the positive lead of a voltmeter to the gauge wire that was disconnected.
6 Connect the other test lead to an engine ground.
7 With the ignition switch set to the On or Accessory position, a flashing light or fluctuating voltage indicates that the instrument voltage regulator is operating and the gauge circuit is not grounded.
8 If the light stays on or the voltage reading is steady, the instrument voltage regulator is bad. If no voltage is indicated by the voltmeter or test light, check for an open circuit in the system.
9 If no defects are indicated, yet the gauge does not work properly, replace the sending unit with a new one.

Replacement

Warning: *The engine must be completely cool before beginning this procedure.*

10 Disconnect the cable from the negative battery terminal.
11 Disconnect the temperature sending unit wire at the sending unit.
12 Remove the temperature sending unit from the cylinder head or intake manifold by unscrewing it with the correct size socket.
13 Prepare the new temperature sending unit for installation by applying electrically conductive thread sealing tape or sealer, such as spray-on copper sealer, to the threads.
14 Install the temperature sending unit in the cylinder head or intake manifold.
15 Connect the wire to the sending unit.
16 Connect the cable to the negative battery terminal.
17 Start the engine and verifv that the temperature gauge is operating correctly.

8 Heater blower assembly — removal and installation

Refer to illustration 8.2

Note: *If the heater blower motor fails to operate, it should be checked by an automotive electrical technician. If replacement of the heater blower and/or wheel is indicated, follow the procedure described below. On pre-1974 models, the heater assembly must be removed first (Section 9) and disassembled to gain access to the blower motor and fan.*

1 Disconnect the cable from the negative battery terminal.
2 Working in the engine compartment, disconnect the lead wire(s) from the rear of the blower motor. Remove the screw and detach the ground wire from the dash panel (see illustration).
3 Disconnect the motor cooling tube (if equipped), then remove the screws retaining the blower motor and wheel to the heater case.
4 Remove the blower motor and wheel from the heater case.
5 Remove the blower wheel hub clamp spring and tab lock washer from the motor shaft, then pull the blower wheel from the motor shaft.
6 installation is the reverse of the removal procedure.

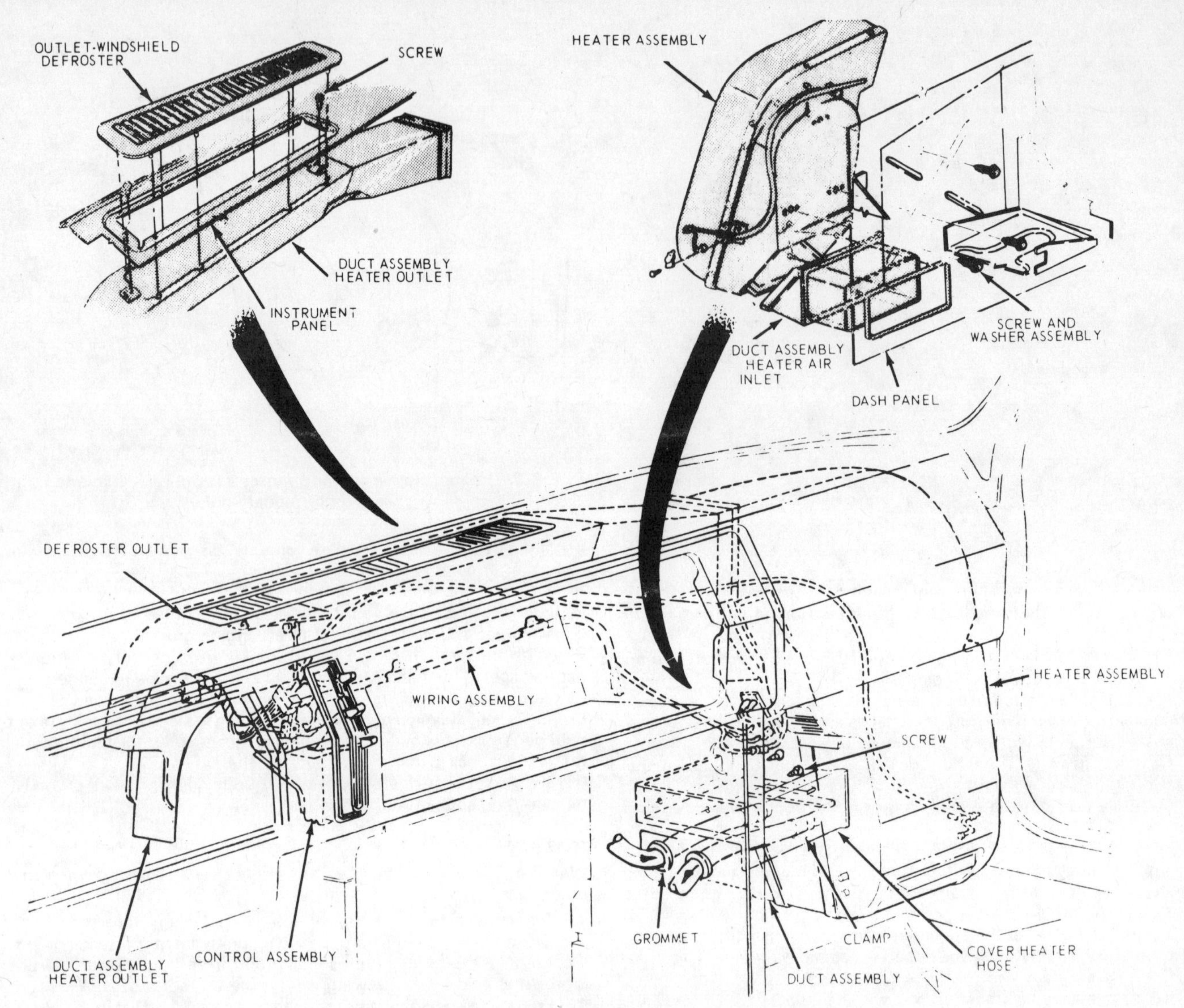

9.6 Early model (pre-1974) heater component layout

9 Heater case and core – removal and installation

Refer to illustrations 9.6 and 9.9

Warning: *The engine must be completely cool before beginning this procedure.*

Removal

1 Disconnect the battery cables (negative first, then positive) and remove the battery.

2 Disconnect the wires from the blower motor resistor and the blower motor.

3 Remove the screw and detach the motor ground wire from the dash panel.

4 Drain the coolant from the radiator into a clean container (see Chapter 1).

5 Disconnect the heater hoses from the heater core on the engine side of the firewall. Remove the plastic strap retaining the hoses to the heater assembly.

6 On pre-1974 models, remove the three heater mounting bolts, move the heater to the side and disconnect the control cable. The heater assembly can now be lifted out (see illustration).

7 On later models, working in the passenger compartment, remove the nuts retaining the left end of the heater case and the right end of the plenum to the dash panel.

8 Remove the nuts retaining the right end of the heater case to the dash panel and remove the heater case assembly from the vehicle.

9 On pre-1974 models, remove the screws and clip, then detach the heater sections and lift out the core. On later models, remove the screws and detach the heater core retainer, then slide the core and seal assembly out of the case (see illustration).

Installation

10 Position the heater core and seal in the heater case.

11 Install the retainer and tighten the screws. On early models, rejoin the heater sections and install the screws and clip.

12 Position the heater case in the dash panel and install the bolts and nuts. Make sure the case seal is positioned correctly over the bolts.

13 On early models, be sure to set the heater control to Off, place the heater in position and pull the air inlet flap closed (to the rear of the vehicle) before connecting the control cable and installing the heater assembly. Make sure the openings in the heater case line up with the defroster and fresh air ducts.

14 Connect the heater hoses to the heater core and tighten the clamps. Install the retaining strap.

15 Connect the wires to the blower motor and blower motor resistor, then attach the ground wire to the dash panel.

16 Fill the radiator as described in Chapter 1.

17 Install the battery and hook up the cables (positive first, then negative).

18 Start the engine and check for leaks.

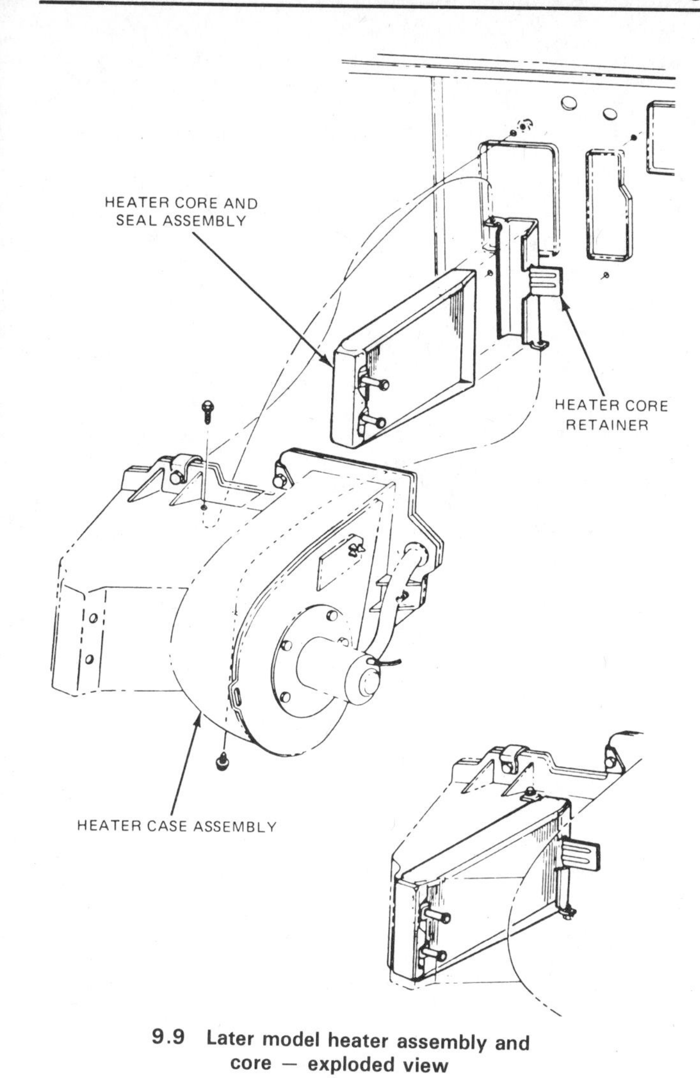

9.9 Later model heater assembly and core — exploded view

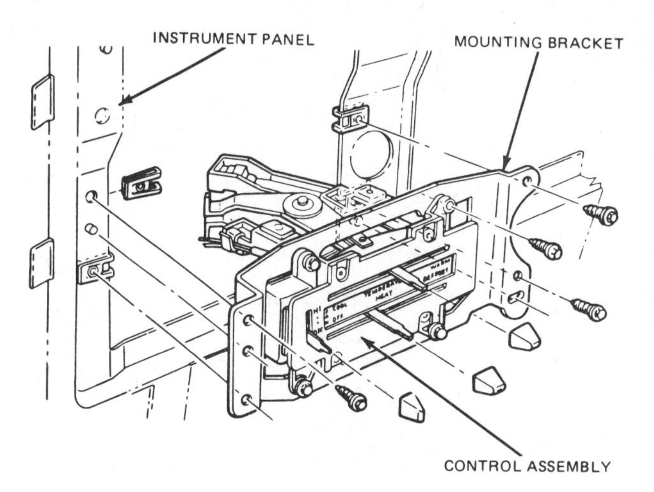

10.2 Later model heater control knobs are press fit on the control levers and the bracket is attached to the instrument panel with several screws

10 Heater control assembly — removal and installation

Refer to illustration 10.2

Note: *This procedure is valid only for 1974 and later model vehicles.*

Removal

1 Disconnect the cable from the negative battery terminal.

2 Remove the knobs from the blower switch and control assembly by placing a small screwdriver between the knob and the control assembly face plate (see illustration). While applying pressure to the rear of the knob, pull the knob off. Repeat for each knob.

3 Remove the applique.

4 Remove the four instrument panel-to-control assembly bracket retaining screws.

5 Disconnect the wiring harness connectors from the blower switch and the light bulb(s).

6 Detach the function and temperature control cables from the control assembly by removing the screws and push nuts.

7 Remove the screws and detach the control assembly from the mounting bracket.

Installation

8 If a new control assembly is being installed, transfer the panel illumination light socket and bulb and the blower switch to the new control panel.

9 Attach the control cables to the control lever arms and install the push nuts. Install the cable clamps and tighten the screws finger tight.

10 The remaining installation steps are the reverse of removal.

11 Be sure to adjust the cables when the job is complete.

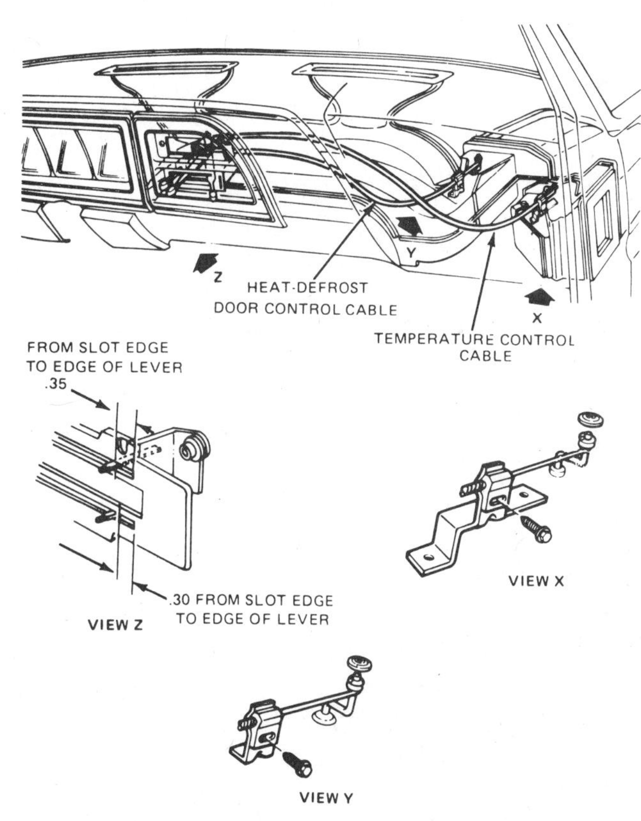

11.1 Control cable adjustment details

11 Heater function and temperature control cables — adjustment

Refer to illustration 11.1

Thru 1983

Function control cable (heat/defrost)

1 Rotate the heat/defrost door arm clockwise as far as possible, then place the function control lever in the Defrost position, allowing a gap of 0.30-inch between the lever and the edge of the slot (see illustration).

2 Secure the cable in the clamp by tightening the screw.

3 Move the lever to Off, then back to Defrost and verify that the gap at the lever is correct.

4 Readjust the cable if necessary.

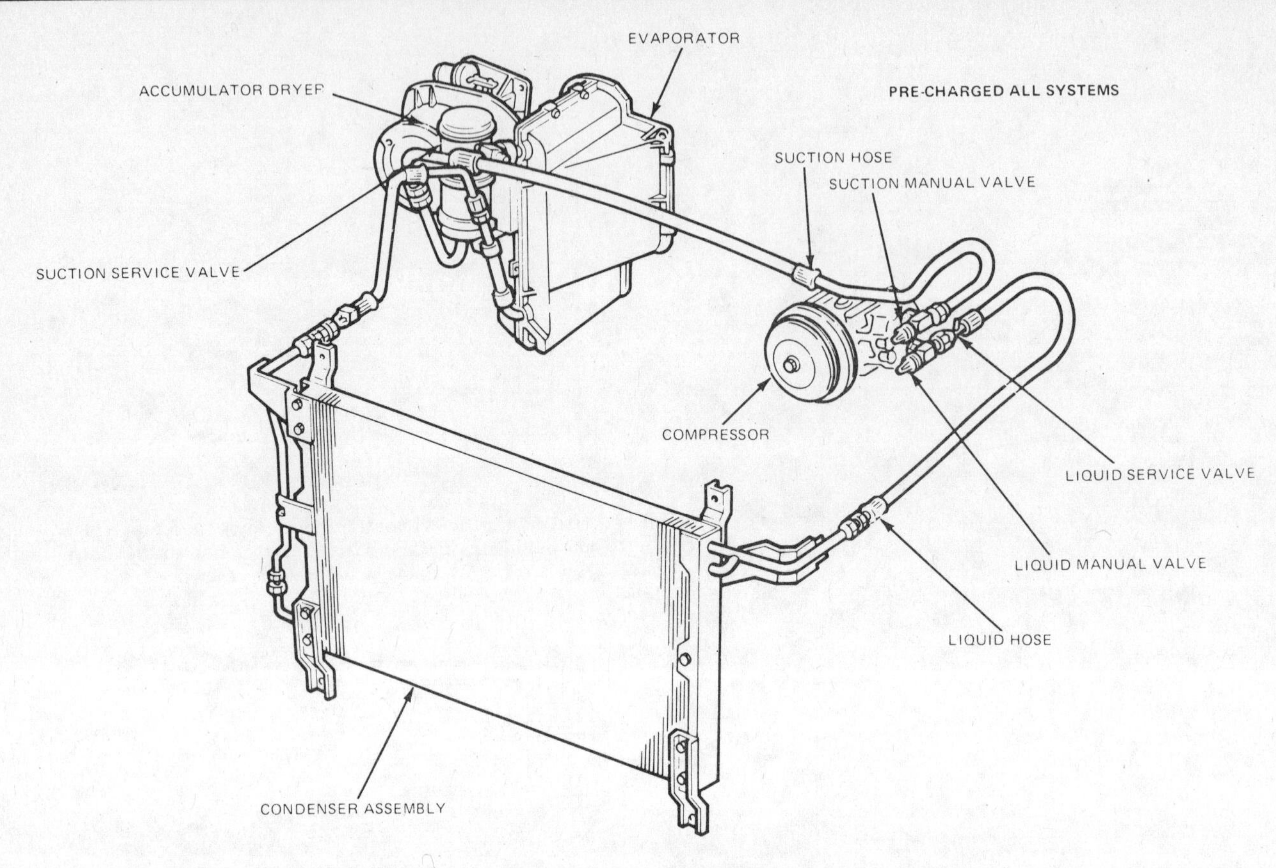

12.2 Typical air conditioning system components

Temperature control cable

5 Rotate the temperature door arm clockwise as far as possible, then place the function control lever in the Warm position, allowing a gap of 0.35-inch between the lever and the edge of the slot.

6 Secure the cable in the clamp by tightening the screw.

7 Move the lever to Cool, then back to Warm and verify that the gap at the lever is correct.

8 Readjust the cable if necessary.

1984 on

Function control cable (heat/defrost)

9 The function cable does not normally require adjustment, but can be checked as follows. Move the function control lever all the way to the left and then all the way to the right. When released, the lever should bounce back slightly from both extremes of travel, indicating that the blend door is adjusted properly. If the control lever moves to either end of the slot without bouncing back, the control cable should be adjusted.

10 Remove the screw attaching the cable to the clamp bracket.

11 Two drill dimples can be seen on the cable mounting flange. If the lever did not bounce back at Defrost, drill an 11/64-inch hole in the dimple closest to the mode door crank. If the lever did not bounce back at Off, drill the hole in the dimple farthest from the mode door crank.

12 Install the cable using the newly drilled hole and tighten the cable clamp screw. Check the cable operation as described in Step 9.

Temperature control cable

13 The cable does not normally require adjustment, but can be checked as follows. Move the temperature control lever all the way to the left and then all the way to the right. When released, the lever should bounce back slightly from both extremes of travel. If the control lever moves to either end of the slot without bouncing back, the control cable should be adjusted.

14 Remove the screw attaching the cable to the clamp bracket.

15 Two drill dimples can be seen on the cable mounting flange. If the lever did not bounce back at Heat, drill an 11/64-inch hole in the dimple closest to the door crank. If the lever did not bounce back at Cool, drill the hole in the dimple farthest from the door crank.

16 Install the cable using the newly drilled hole and tighten the cable clamp screw. Check the cable operation as described in Step 13.

12 Air conditioning system – servicing

Refer to illustration 12.2

Warning: *Before disconnecting any lines or attempting to remove any air conditioning system components, have the system refrigerant evacuated by an air conditioning technician.*

1 Because of the special tools, equipment and skills required to service air conditioning systems, and the differences between the various systems that may be installed on these vehicles, air conditioner servicing cannot be covered in this manual.

2 Component removal, however, can usually be accomplished without special tools and equipment (see illustration). The home mechanic may realize a substantial savings in repair costs if he removes components himself, takes them to a professional for repair, and/or replaces them with new ones (see the Warning above).

3 Problems in the air conditioning system should be diagnosed and the system refrigerant evacuated by an air conditioning technician before component removal/replacement is attempted.

4 Once the new or reconditioned component has been installed, the system should then be charged and checked by an air conditioning technician.

5 Before indiscriminately removing air conditioning system components, get more than one estimate of repair costs from air conditioning service centers. You may find it to be cheaper and less trouble to let the entire operation be performed by someone else.

Chapter 4 Fuel and exhaust systems

Refer to Chapter 13 for Specifications and information on 1987 and later models

Contents

Specifications

Carburetor	See Emissions Control Information label in engine compartment or instructions with rebuild kit
Fuel pump (mechanical type) pressure	
six-cylinder	
thru 1976	4 to 6 psi
1977 on	5 to 7 psi
V8	
thru 1975	5 to 6 psi
1976 on	6 to 8 psi
Volume	
thru 1975	1 pint in 15 seconds (with fuel return line to tank pinched off)
1976 on	1 pint in 20 seconds
Accelerator pump lever clearance (4180C)	0.015 in

Torque specifications	**Ft-lbs (unless otherwise indicated)**
Fuel tank nuts	20 to 25
Carburetor mounting nuts	12 to 15
Fuel pump mounting nuts/bolts	
240 and 300 six-cylinder	12 to 18
302, 351W and 460 V8's	19 to 27
351M and 400 V8's	
bolt	10 to 15
nut	14 to 20
Exhaust system	
bracket-to-crossmember	23 to 32
bracket-to-frame side rail	
5/16-inch bolt	12 to 17
3/8-inch bolt	23 to 32
Exhaust pipe joint U-bolt	25 to 36
Exhaust pipe-to-exhaust manifold	25 to 38
EFI system	
lower intake manifold-to-head bolts	23 to 25
EGR valve-to-upper intake manifold bolts	13 to 19
upper-to-lower intake manifold bolts	15 to 22
throttle body mounting bolts	12 to 18
air bypass valve bolts	71 to 102 in-lbs
throttle position sensor bolts	14 to 16 in-lbs
fuel pressure regulator bolts	27 to 40 in-lbs
fuel injector manifold-to-fuel charging assembly bolts	12 to 15
air cleaner bracket-to-manifold stud	22 to 32
throttle cable bracket-to-manifold bolts	8 to 10

1 General information

The fuel system on most models consists of a fuel tank mounted beneath the rear of the vehicle, a mechanically operated fuel pump and a carburetor. Depending on the year and model, Electronic Fuel Injection (EFI) may be installed and auxiliary fuel tanks may be used in addition to the main tank. A combination of metal and rubber fuel lines are used to connect the main components and, in the case of some vehicles which have auxiliary tanks, there is a fuel flow control valve located between the tanks and the fuel pump.

The carburetor may be a single, dual or four venturi downdraft unit, depending on the engine displacement and year of production. Some carburetors are electronically controlled.

The fuel system (especially the carburetor) is heavily interrelated with the emissions control system on all vehicles produced for sale in the United States. Refer to Chapter 6 for additional information on emissions control systems.

Warning: *Gasoline is extremely flammable, so extra precautions must be taken when working on any part of the fuel system. Do not smoke or allow open flames or bare light bulbs near the work area. Also, do not work in a garage if a natural gas appliance equipped with a pilot light is present.*

2 Carburetor servicing and overhaul — general information

Refer to illustration 2.7

1 A thorough road test and check of carburetor adjustments should be done before any major carburetor service work is done. Specifications for some adjustments are listed on the Vehicle Emissions Control Information label or tune-up decal found in the engine compartment.

2 Some performance complaints directed at the carburetor are actually a result of loose, out-of-adjustment or malfunctioning engine or electrical components. Others develop when vacuum hoses leak, are disconnected or are incorrectly routed. The proper approach to analyzing carburetor problems should include a routine check of the following:

a) Inspect all vacuum hoses and actuators for leaks and correct installation (see Chapter 6).
b) Tighten the intake manifold nuts and carburetor mounting nuts evenly and securely.
c) Perform a cylinder compression test.
d) Clean or replace the spark plugs as necessary.
e) Test the resistance of the spark plug wires.
f) Inspect the ignition primary wires and check the vacuum advance operation. Replace any defective parts.
g) Check the ignition timing according to the instructions listed on the Emissions Control Information label (see Chapter 1).
h) Set the carburetor idle mixture.
i) Check the fuel pump pressure.
j) Inspect the heat control valve in the air cleaner for proper operation.
k) Remove the carburetor air filter element and blow out any dirt with compressed air. If the filter is extremely dirty, replace it with a new one.

3 Carburetor problems usually show up as flooding, hard starting, stalling, severe backfiring, poor acceleration and lack of response to idle mixture screw adjustments. A carburetor that is leaking fuel and/or covered with wet looking deposits definitely needs attention.

4 Diagnosing carburetor problems may require that the engine be started and run with the air cleaner removed. While running the engine without the air cleaner it is possible that it could backfire. A backfiring situation is likely to occur if the carburetor is malfunctioning, but removal of the air cleaner alone can lean the fuel/air mixture enough to produce an engine backfire. Perform this type of testing as quickly as possible and be especially watchful for the potential of backfire and the possibility of starting a fire. **Warning:** *Do not position your face or body directly over the carburetor during inspection and servicing procedures.*

5 Once it is determined that the carburetor is in need of work or an overhaul, several alternatives should be considered. If you are going to attempt to overhaul the carburetor yourself, first obtain a good quality carburetor rebuild kit, which will include all necessary gaskets, internal parts, instructions and a parts list. You will also need carburetor cleaning solvent and some means of blowing out the internal passages of the carburetor with air.

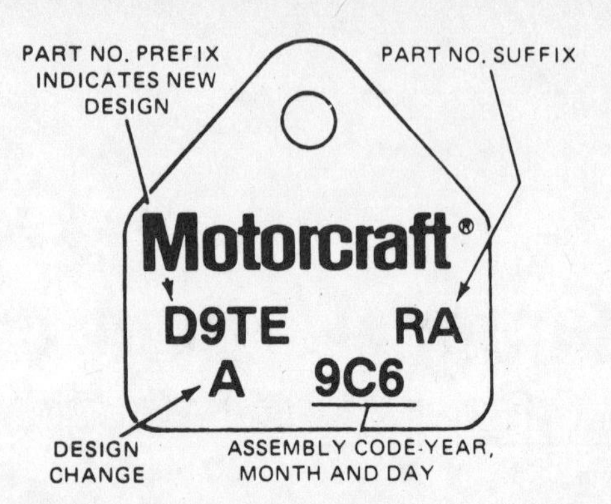

2.7 An identification tag such as this will be attached to the top of the carburetor (copy the numbers off the tag when purchasing a rebuild kit or a new or rebuilt carburetor)

6 Due to the many configurations and variations of carburetors offered on the range of vehicles covered in this manual, it is not feasible for us to do a step-by-step overhaul of each type. The carburetor disassembly and reassembly procedures included in this Chapter are general in nature and apply only to the most widely used carburetors. Quality carburetor overhaul kits contain detailed instructions and illustrations that apply in a more specific manner to each model carburetor.

7 Another alternative is to obtain a new or rebuilt carburetor. These are readily available from dealers and auto parts stores for all engines covered in this manual. The important fact when purchasing one of these units is to make sure the exchange carburetor is identical to the original. Usually a tag is attached to the top of your carburetor (see illustration) which will aid the parts man in determining the exact type of carburetor you have. When obtaining a rebuilt carburetor or a rebuild kit, take time to determine that the kit or carburetor matches your application exactly. Seemingly insignificant differences can make a considerable difference in the overall running condition of your engine.

8 If you choose to overhaul your own carburetor, allow enough time to disassemble the carburetor carefully, soak the necessary parts in the cleaning solvent (usually for at least one-half day or according to the instructions listed on the carburetor cleaner) and reassemble it, which will usually take you much longer than disassembly. When you are disassembling a carburetor, take care to match each part with the illustration in your carburetor kit and lay the parts out in order on a clean work surface to help you reassemble the carburetor. An overhaul by an inexperienced mechanic can result in a vehicle which runs poorly or not at all. To avoid this, use care and patience when disassembling your carburetor so you can reassemble it correctly.

9 When the overhaul is complete, adjustments which may be beyond the ability of the home mechanic may be required, especially on later models. If so, take the vehicle to a Ford dealer or a tune-up shop for final carburetor adjustments which will ensure compliance with emissions regulations and acceptable performance.

3 Carburetor — removal and installation

Warning: *Gasoline is extremely flammable, so extra precautions must be taken when working on any part of the fuel system. Do not smoke or allow open flames or bare light bulbs near the work area. Also, do not work in a garage if a natural gas appliance equipped with a pilot light is present.*

1 Disconnect the negative battery cable from the battery. Remove the hoses connected to the air cleaner. Mark them with coded pieces of tape to simplify reassembly.

2 Remove the air cleaner assembly.

3 Disconnect the fuel line from the carburetor. Plug the end of the line to prevent leakage.

4 Disconnect the electrical leads from the emissions control devices attached to the carburetor. Mark the wires and connections so they can be installed in the proper locations.

5 Remove any vacuum lines attached to the carburetor. Mark them to simplify installation.

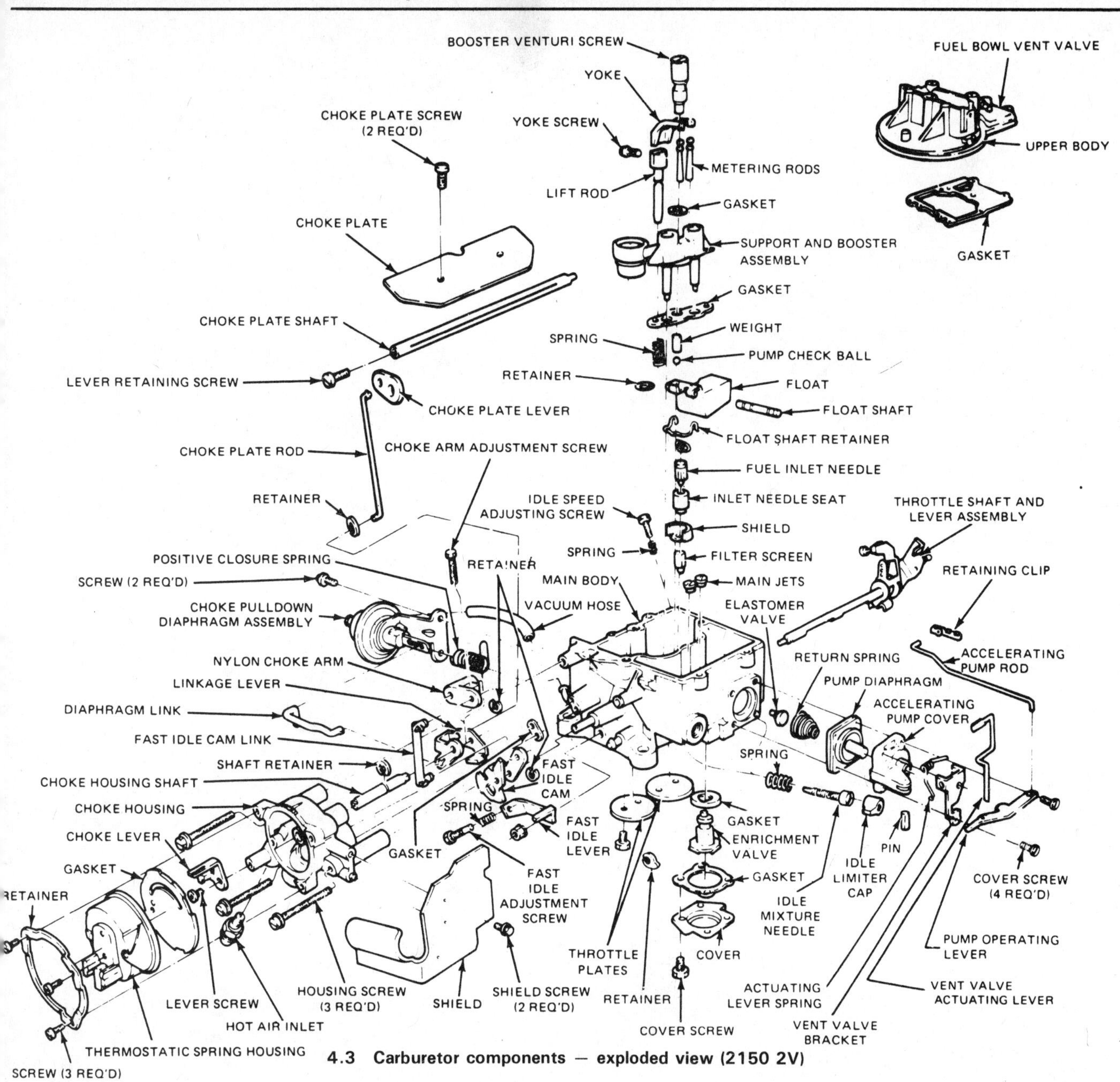

4.3 Carburetor components — exploded view (2150 2V)

6 Disconnect the kickdown lever or cable from the carburetor (if equipped).
7 Disconnect the throttle cable or linkage from the carburetor.
8 Disconnect any under carburetor heater hoses that may be connected to the carburetor.
9 Disconnect any coolant transfer hoses that may be connected to the choke system.
10 Remove the carburetor mounting nuts from the studs in the intake manifold.
11 Lift off the carburetor, spacer plate (if equipped) and gasket(s). Place a piece of cardboard over the intake manifold opening to prevent debris from falling into the engine while the carburetor is off.
12 Before installation, carefully clean the mating surfaces of the intake manifold spacer plate (if equipped) and the base of the carburetor to remove any old gasket material. These surfaces must be perfectly clean and smooth to prevent vacuum leaks.
13 Install a new gasket.
14 Install the carburetor and spacer plate (if equipped) over the studs on the intake manifold.
15 Install the mounting nuts and tighten them to the specified torque. Be careful not to overtighten the nuts as they can warp the base plate of the carburetor.
16 The remaining installation steps are the reverse of removal. Make sure all hoses, wires and cables are returned to their original locations.

4 Carburetor — disassembly and reassembly (2150 2V)

Refer to illustrations 4.3, 4.19 and 4.20

1 Before disassembly, clean the exterior of the carburetor with solvent and wipe it dry with a rag. Select a clean area of the workbench and spread out several layers of newspaper. Obtain several small containers for holding some of the small parts (which could be easily lost). Whenever a part is removed, note how and where it is installed. As each part is removed, place it in order along one edge of the newspaper so that reassembly is made easier.

Disassembly

Upper body

2 Unscrew and remove the fuel filter retainer from the upper body. Recover the filter.
3 Disconnect the choke plate rod at the upper end (see illustration).

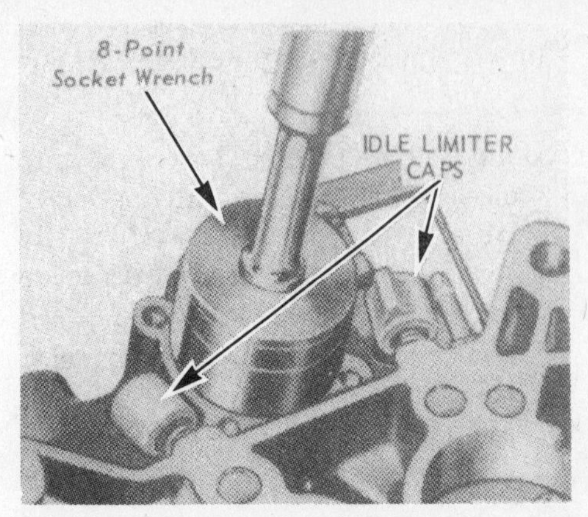

4.19 A large 8-point socket is required to remove the enrichment valve (2150 2V carburetor)

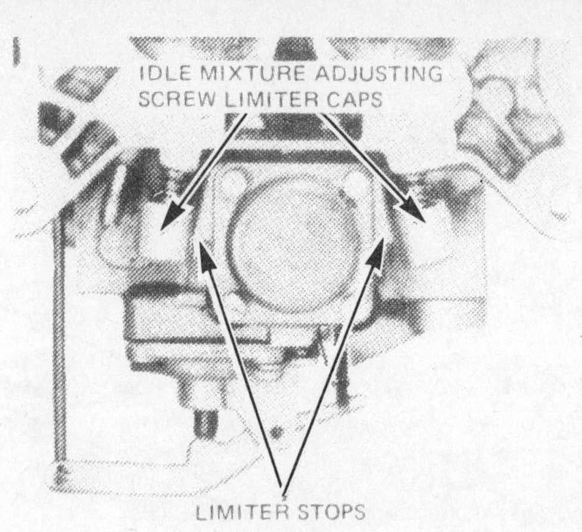

4.20 The idle mixture adjusting screw limiter caps can be pried off the screws

4 Loosen and remove the screws and washers retaining the upper body to the main body. Lift away the upper body and the gasket.
5 Carefully remove the float shaft and lift out the float assembly, followed by the fuel inlet needle valve.
6 Unscrew the inlet needle seat and remove the shield and filter screen.
7 Remove the three enrichment valve vacuum diaphragm cover screws. Remove the washers and diaphragm.

Automatic choke

8 Remove the fast idle cam retainer.
9 Remove the screws and detach the retainer, thermostat housing and gasket.
10 Remove the retaining screws and lift off the choke housing assembly, the fast idle cam and gasket.
11 Remove the choke lever retaining screw and washer. Disconnect the choke plate rod from the linkage lever.
12 Remove the linkage lever and fast idle lever from the choke housing.

Accelerator pump

13 Remove the four pump cover screws and detach the pump cover. Remove the pump diaphragm and return spring.
14 Remove the pump discharge screw assembly, the discharge nozzle and the two gaskets. Remove the two discharge check balls.

Main body

15 If not done prior to this point remove the inlet needle filter screen.
16 Remove the main jets, using the correct size wrench or screwdriver to avoid damaging them.
17 Remove the retaining screw and lift out the booster venturi and metering rod assembly and gasket.
18 Invert the carburetor body, allowing the accelerator pump discharge weight and check ball to drop out into your hand.
19 Remove the enrichment valve cover using the correct type socket wrench (see illustration). Remove the gasket and discard it.
20 Turn the idle mixture needles in, very carefully, counting the number of turns before they bottom. Remove the idle mixture needles and springs. Detach the limiters from the needles/screws (see illustration).
21 If used, remove the anti-stall solenoid and dashpot.
22 If it is necessary to remove the throttle shaft from the main body, carefully mark each throttle plate and associated bore before filing off the flared portion and removing the throttle plate retaining screws.
23 Slide the throttle shaft out of the body and retrieve the bleed actuator.
24 Disassembly is now complete and all parts should be thoroughly cleaned (see Section 2). Remove any sediment in the float chamber and passages. Remove all traces of old gaskets using a sharp knife. When all parts are clean, reassembly can begin.

Reassembly

25 Reassembly of the carburetor is essentially the reverse of the disassembly procedure, but careful attention should be paid to the following points:

Main body

26 Make sure that the idle mixture needles are installed in exactly the same position determined in Step 20, then install a new limiter cap with the stop tab against the rich side of the stop on the carburetor

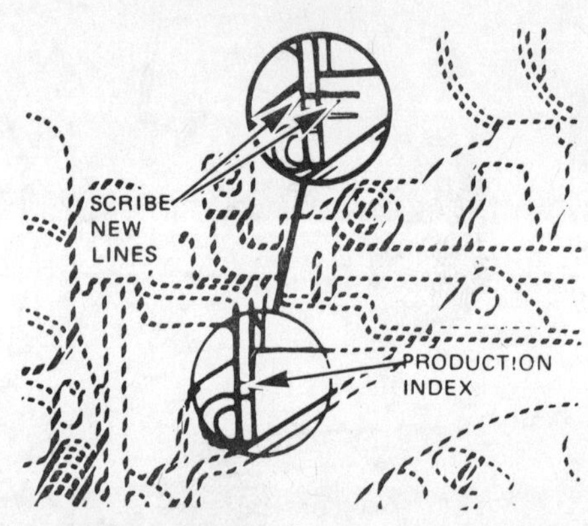

5.1 Be sure to scribe new lines on the accelerator pump cover and the carburetor top with the throttle held against the stop screw before beginning carburetor disassembly (4350 4V carburetor)

body. Ensure that the main jets, primary and secondary main well tubes and high speed bleeds are correctly installed in their respective positions.

Accelerator pump

27 When installing the return spring and pump diaphragm assembly, start the four cover screws, hold the pump lever partly open to align the gasket, then tighten the screws. **Note:** *If the plastic valve in the pump assembly was removed it must be replaced with a new valve.*

Automatic choke

28 When installing the diaphragm adjusting screw, initially adjust it so that the threads are flush with the inside of the cover. Install the fast idle rod with the end which has one tab in the fast idle adjustment lever, and the end which has two tabs in the primary throttle lever. Adjust the choke plate pull down as described in Section 8. Before installing the electric choke heater, make sure that the choke plate is either fully open or fully closed.

Upper body

29 When installing the enrichment valve vacuum diaphragm, depress the spring and install the screws and washers finger tight. Hold the stem so that the diaphragm is horizontal, then tighten the screws evenly.

5 Carburetor – disassembly and reassembly (4350 4V)

Refer to illustrations 5.1, 5.2, 5.13, 5.16, 5.17 and 5.19

1 Before starting disassembly, scribe a line on both the accelerator pump lever and the carburetor top cover with the throttle held against the stop screw (see illustration). Do not use the existing production index lines as reference.
2 Spread out a clean sheet of paper on the workbench and lay the components out in the order of removal (see illustration).

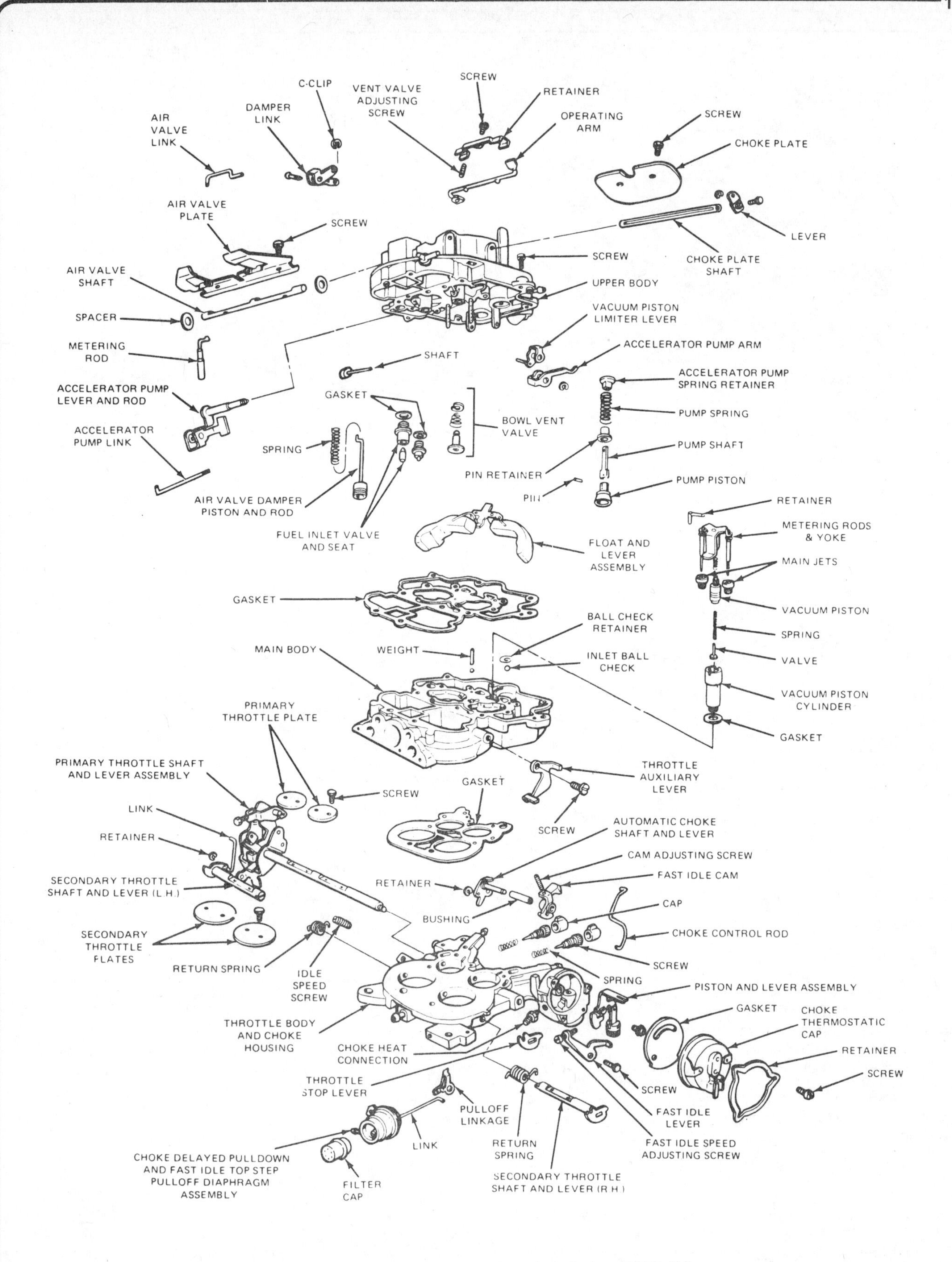

5.2 Carburetor components – exploded view (4350 4V)

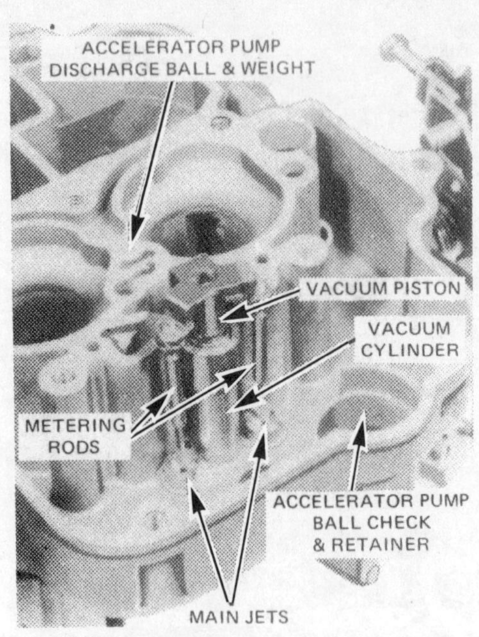

5.13 The fuel metering rods and main metering vacuum piston can be removed as an assembly (4350 4V carburetor)

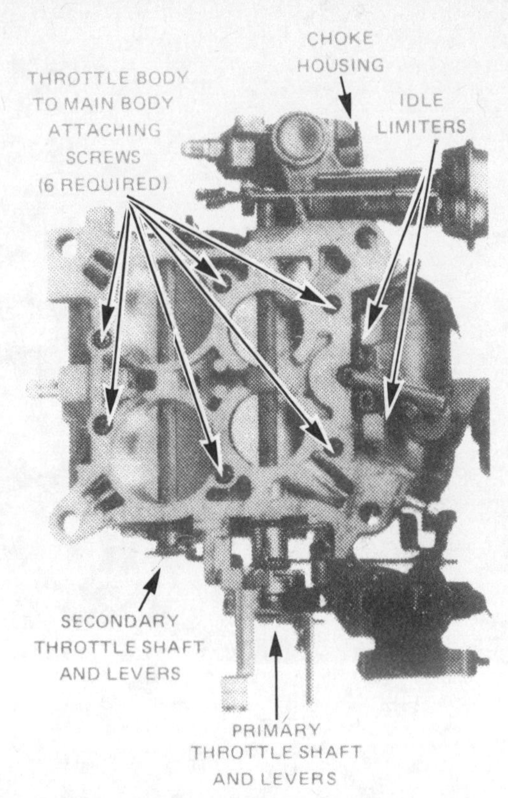

5.16 Throttle body-to-main body screw locations (4350 4V carburetor)

5.17 Removing the choke cover and thermostatic spring (4350 4V carburetor)

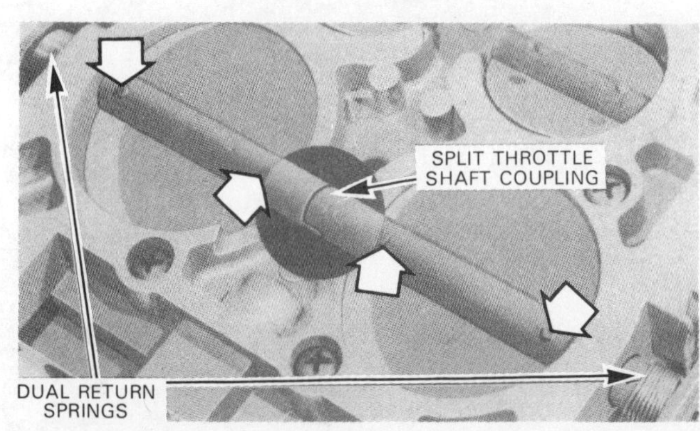

5.19 To remove the secondary throttle shaft, first file off the staked portions of the plate mounting screws (arrows) (4350 4V carburetor)

Disassembly

Upper body

3 Unscrew and remove the fuel filter retainer from the upper body. Recover the filter.
4 Disconnect the choke control rod at the upper end.
5 Loosen and remove the screws and washers retaining the upper body to the lower body. Lift away the upper body and the gasket.
6 Carefully remove the float pivot pin and lift out the float assembly, followed by the fuel inlet valve.
7 Unscrew the valve seat and remove the gasket.
8 Remove the pivot pin from the secondary air valve lever and the rod from the damper piston and air valve plate. Lift off the piston, rod and spring.
9 If required, disconnect the accelerator pump piston operating linkage and vacuum limiter lever. Slide the shaft out of the upper body.

Main body

Note: *The disassembly and reassembly of the carburetor main body requires a small screwdriver with a blade width not exceeding 3/32-inch.*

10 Turn the main body casting upside down and catch the accelerator pump discharge weight and ball valve as they drop out.
11 Turn the body right side up and depress the metering rod hanger. Measure and record the distance between the rod hanger and the vacuum cylinder for resetting during assembly.
12 Using needle nose pliers, remove the nylon vacuum piston stop from the channel in the main body casting.
13 Remove the main metering vacuum piston and fuel metering rods as a complete assembly (see illustration). **Caution:** *Do not disturb the setting of the piston metering rod adjustment screws.*
14 Using the small screwdriver, unscrew the vacuum piston cylinder and remove the cylinder, return spring and guide pin.
15 Remove the two main metering jets.

Throttle body

Caution: *Do not remove the idle mixture limiter caps or the mixture screws from the throttle body.*

16 Remove the screws retaining the throttle body to the main carburetor body and separate the two castings (see illustration).
17 Remove the choke housing cover screws and lift away the cover, gasket and thermostatic spring (see illustration). Remove the screw securing the choke piston and detach the piston and lever assembly.
18 Remove the retainers from the secondary throttle-to-primary throttle connecting link and remove the link.
19 If it is necessary to remove the two throttle shafts from the body, first file off the flared portion of the throttle plate screws, then remove

the screws. Slide the shafts out of the body, noting the secondary shaft is a two piece design (see illustration).
20 Disassembly is now complete and all parts should be cleaned with solvent (see Section 2). Remove any sediment in the float chamber and passages. Remove all traces of old gaskets using a sharp knife. When all parts are clean, reassembly can begin.

Reassembly

21 Reassembly of the carburetor is essentially the reverse of the disassembly procedure, but careful attention should be paid to the following points.

a) When installing new gaskets, check that all holes have been correctly punched and that no foreign material is stuck to the gaskets.
b) When installing the jets, make sure that the correct size wrench is used to avoid damage to the jet orifice.
c) When attaching the choke lever to the choke operating shaft, note that the end of the shaft is tapered and the retaining screw has a left-hand thread.
d) Note that the float pivot pin must be inserted from the accelerator pump side to ensure correct retention.

6 Carburetor – disassembly and reassembly (Holley 4180C 4V)

Refer to illustrations 6.2, 6.8, 6.14, 6.20, 6.25, 6.60 and 6.64

1 Remove the carburetor as described in Section 3 and place it on a carburetor repair stand. If a repair stand is not available, insert bolts about 2-1/4 inches long of the correct diameter through the carburetor retaining bolt holes. Attach nuts above and below the flange of the carburetor to keep the bolts in position. This will aid in working on the carburetor and prevent damage to the throttle plates.

Disassembly

2 Place a container under the primary fuel bowl and loosen the bowl retaining screws (see illustration). When the fuel is drained, remove the bowl, the gasket and the metering block and gasket.
3 Remove the pump transfer tube from the metering block or main housing and discard the O-rings.
4 Disconnect the fuel transfer tube and discard the O-ring.
5 On 1979 models, remove the balance tube, washer and O-ring. The O-ring can be discarded.
6 Remove the main jets from the metering block with a jet wrench.
7 Unscrew the power valve and remove it along with the gasket.
8 Unscrew the fuel level adjusting lock screw from the top of the primary fuel bowl and remove the gasket (see illustration). Loosen the adjusting nut until it can be removed with the gasket. Lift out the needle and seat assembly from the fuel bowl. The needle and seat assembly is a matched set and should not be disassembled.
9 Remove the float shaft retainer clip with needle nose pliers. Slide the float off of the shaft and remove the spring from the float.
10 Separate the baffle from the fuel bowl. Unscrew the fuel level sight plug from the side of the bowl.
11 Remove the fuel inlet fitting along with the filter, gasket, O-ring and filter screen.
12 Turn the bowl over and remove the accelerator pump cover, the diaphragm and the diaphragm return spring. The accelerator pump check ball should not be removed.
13 Remove the secondary fuel bowl from the carburetor.
14 Remove the metering block, body and gaskets with a clutch-type screwdriver (see illustration). Discard the gaskets.
15 Remove the balance tube, washer and O-ring on 1979 models.
16 Refer to Paragraphs 8 through 10, above to disassemble the secondary fuel bowl.
17 Remove the secondary diaphragm link retainer and the air cleaner mounting stud.
18 Turn the carburetor over and remove the screws securing the throttle body. Lift the throttle body and gasket out of position. Discard the gasket.
19 Disconnect the choke rod retainer from the housing shaft and lever assembly.
20 Remove the thermostat spring housing and gasket followed by the choke housing and gaskets, from the main body (see illustration).
21 Remove the nut from the choke housing shaft, followed by the star washer and spacer. Remove the shaft and the fast idle cam.
22 Remove the choke piston and the lever link assembly, then separate the choke rod and seal from the main body.
23 If necessary, remove the choke plate from the choke shaft and slide the choke shaft and lever out of the air horn. The screws securing the choke plate are staked to the choke shaft. It may be necessary to file off the flared portion of the screw to avoid damaging the threads in the choke shaft. Do not damage the venturi or the choke shaft while filing the screws.
24 Remove the diaphragm housing and gasket from the main body to gain access to the diaphragm housing cover. Remove the cover, followed by the spring and diaphragm and the vacuum check ball from the diaphragm housing.
25 Remove the screw from the accelerator pump discharge nozzle, then remove the nozzle and gaskets from the main body. Turn the main body over and catch the discharge needle as it falls out (see illustration).
26 The throttle body parts are assembled as a matched set and should not be disassembled.

Cleaning and inspection

27 Clean all parts with carburetor solvent. Do not immerse rubber or plastic parts as permanent damage will occur. Do not probe the jets, but blow through them with compressed air. Examine all fixed and moving parts for cracks, distortion, wear and other damage and replace parts as necessary. Discard all gaskets and the fuel inlet filter.
28 Check the secondary operating diaphragm and the accelerator pump diaphragm for cuts and tears and replace them as necessary. Be sure that all of the new gaskets are in good condition, that all of the holes are punched out properly and that no foreign material is stuck to the gasket.

Reassembly

29 Slide the accelerator pump discharge needle into position in the well and seat it lightly with a brass drift and small hammer.
30 Install the accelerator pump discharge nozzle and gaskets in the main body and secure them with the screw. Carefully stake the screw with a flat punch. Be careful that no chips fall into the carburetor.
31 Place the check ball in position in the vacuum port of the secondary diaphragm housing. Install the secondary diaphragm in the housing and place the spring in the cover. Secure the cover to the diaphragm housing. Before tightening the retaining screws completely, check that the vacuum opening is aligned and that the diaphragm is seated evenly in the housing. **Note:** *The cover can only be installed when the diaphragm is not connected to the main body.*
32 Position a new gasket on the main body and install the diaphragm housing. Tighten the housing retaining screws securely.
33 Slide the choke shaft into the air horn and attach the choke plate.
34 Position the rod seal on the choke rod, then slide the U-shaped end of the choke rod through the opening in the main body. Slip the rod end through the inner side of the choke lever bore.
35 With the rod end facing out, position the rod seal in the grooves on the lower side of the air cleaner mounting flange.
36 Place the piston assembly and the choke thermostat lever link in position in the choke housing.
37 Attach the fast idle cam assembly to the choke housing.
38 Install the choke housing shaft and the lever assembly.
39 With the lever and piston assembly in position on the choke housing shaft and lever assembly, install the spacer, the lockwasher and the securing nut.
40 Place the choke housing gasket in position on the main body while the main body is on its side. While the choke housing is being placed in position on the main body, insert the choke rod into the shaft lever of the choke housing. The projection on the choke rod must be positioned under the fast idle cam so that the cam will operate when the choke is closed. Install the choke housing retaining screws and lockwashers. Tighten the screws securely.
41 Install the choke rod cotter pin with needle nose pliers.
42 Position the thermostatic spring housing gasket on the choke housing. Attach the thermostatic spring to the spring lever. Install the housing, the clamp and the screws. The spring housing must be adjusted by aligning the mid-position mark on the choke housing with the index mark on the cover before the screws are tightened.
43 Turn the main body over and place the throttle body gasket in position. Position the throttle body so that the fuel inlet is on the same side as the accelerator pump operating lever.

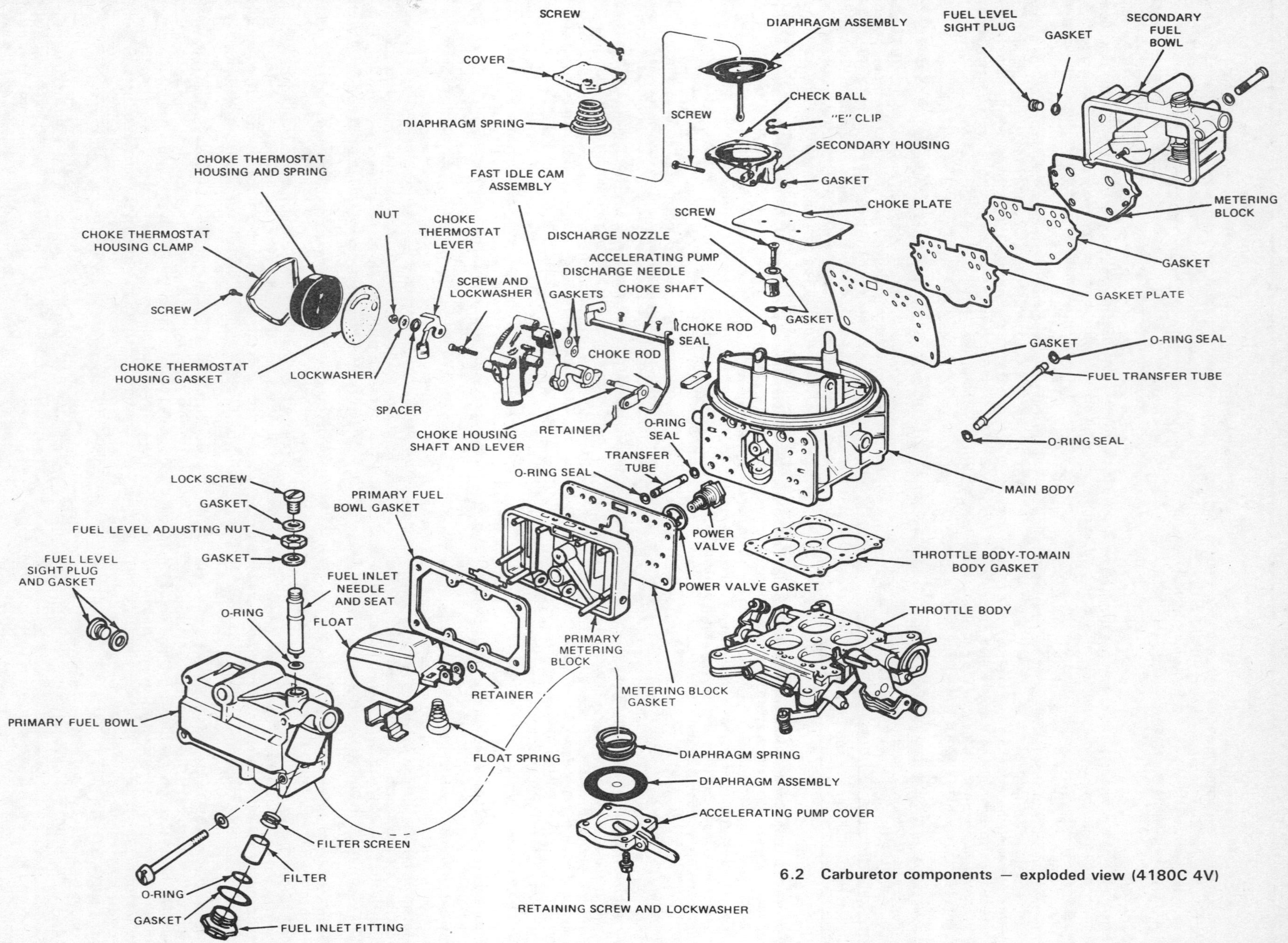

SCREW
COVER
DIAPHRAGM SPRING
DIAPHRAGM ASSEMBLY
CHECK BALL
SCREW
"E" CLIP
SECONDARY HOUSING
GASKET
FUEL LEVEL SIGHT PLUG
GASKET
SECONDARY FUEL BOWL
METERING BLOCK
CHOKE THERMOSTAT HOUSING AND SPRING
CHOKE THERMOSTAT HOUSING CLAMP
SCREW
CHOKE THERMOSTAT HOUSING GASKET
FAST IDLE CAM ASSEMBLY
NUT
CHOKE THERMOSTAT LEVER
SCREW AND LOCKWASHER
LOCKWASHER
SPACER
CHOKE HOUSING SHAFT AND LEVER
GASKETS
RETAINER
CHOKE PLATE
SCREW
DISCHARGE NOZZLE
ACCELERATING PUMP DISCHARGE NEEDLE
CHOKE SHAFT
GASKET
CHOKE ROD SEAL
CHOKE ROD
GASKET
GASKET PLATE
GASKET
O-RING SEAL
FUEL TRANSFER TUBE
O-RING SEAL
O-RING SEAL
TRANSFER TUBE
O-RING SEAL
MAIN BODY
POWER VALVE
THROTTLE BODY-TO-MAIN BODY GASKET
POWER VALVE GASKET
THROTTLE BODY
LOCK SCREW
GASKET
FUEL LEVEL ADJUSTING NUT
GASKET
FUEL LEVEL SIGHT PLUG AND GASKET
PRIMARY FUEL BOWL GASKET
FUEL INLET NEEDLE AND SEAT
O-RING
FLOAT
PRIMARY METERING BLOCK
RETAINER
METERING BLOCK GASKET
PRIMARY FUEL BOWL
FLOAT SPRING
DIAPHRAGM SPRING
DIAPHRAGM ASSEMBLY
ACCELERATING PUMP COVER
FILTER SCREEN
FILTER
O-RING
GASKET
FUEL INLET FITTING
RETAINING SCREW AND LOCKWASHER

6.2 Carburetor components — exploded view (4180C 4V)

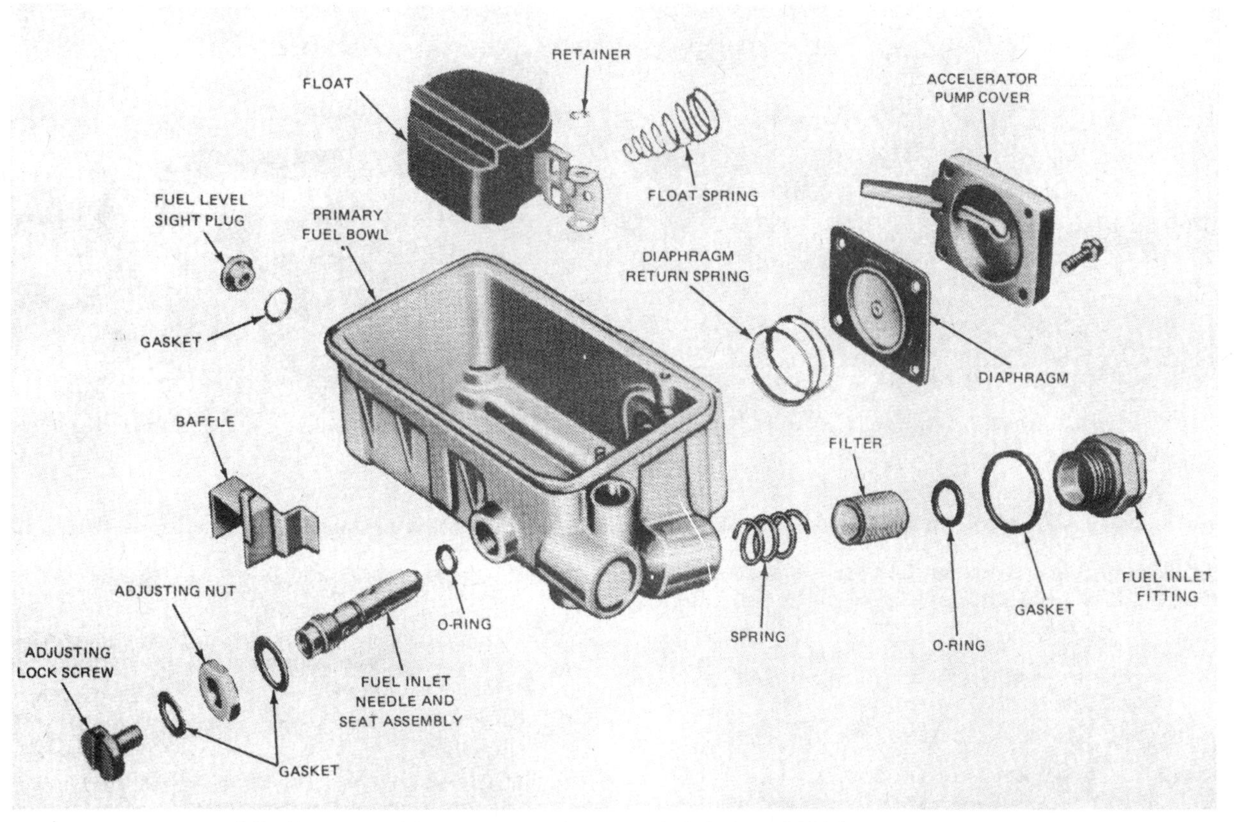

6.8 Primary fuel bowl components — exploded view (4180C 4V carburetor)

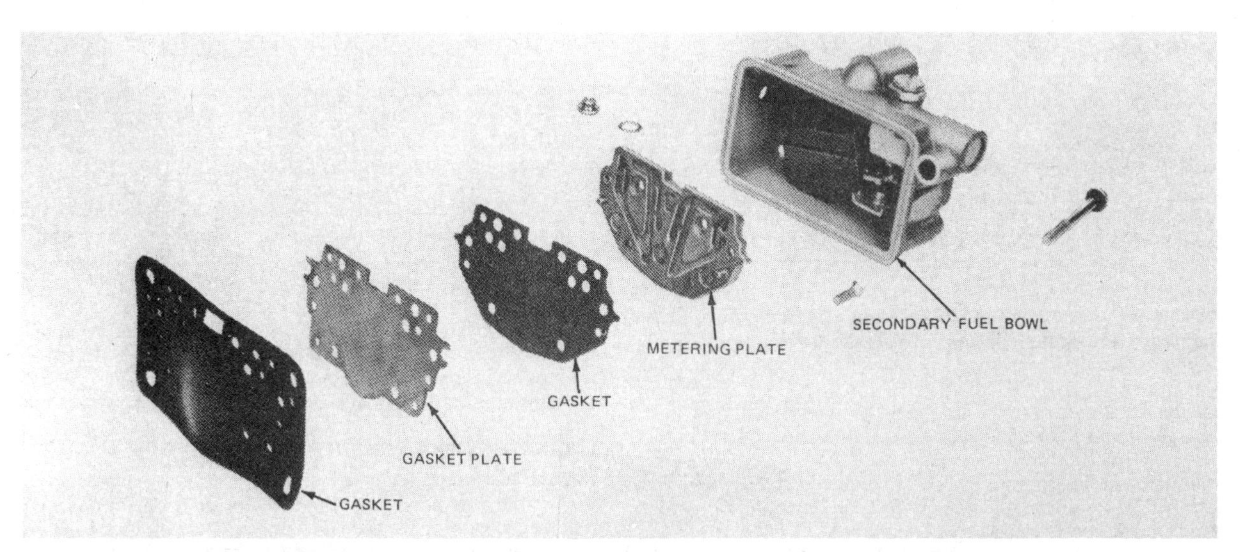

6.14 Secondary fuel bowl and metering block components — exploded view (4180C 4V carburetor)

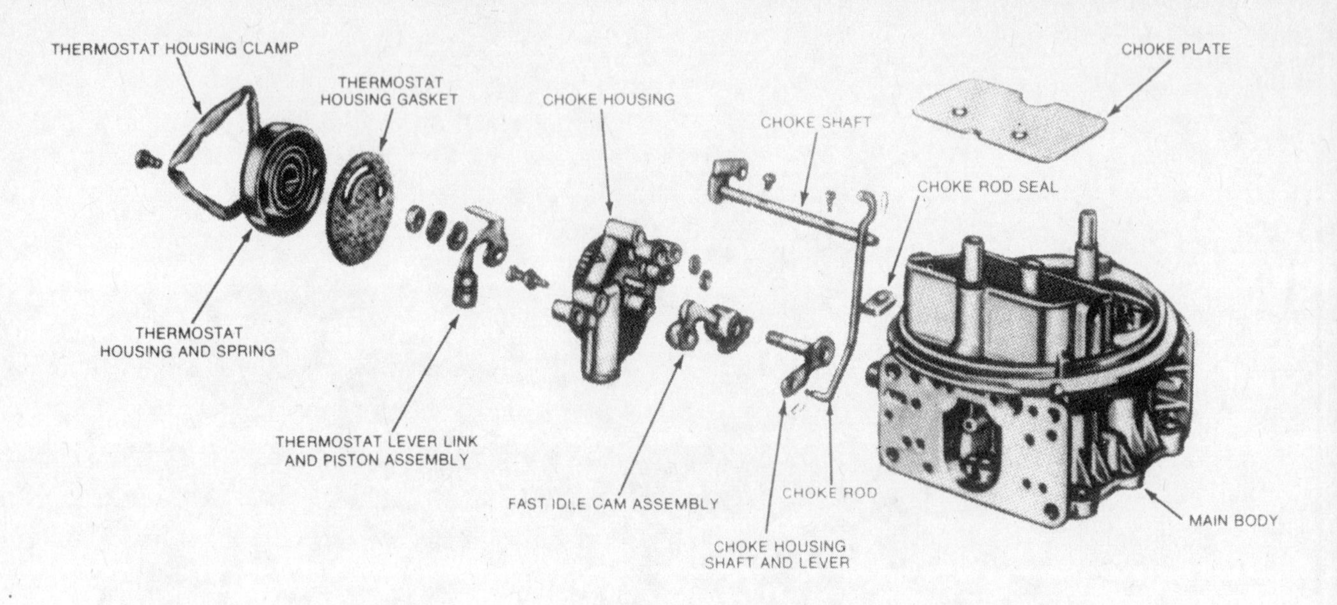

6.20 Choke components – exploded view (4180C 4V carburetor)

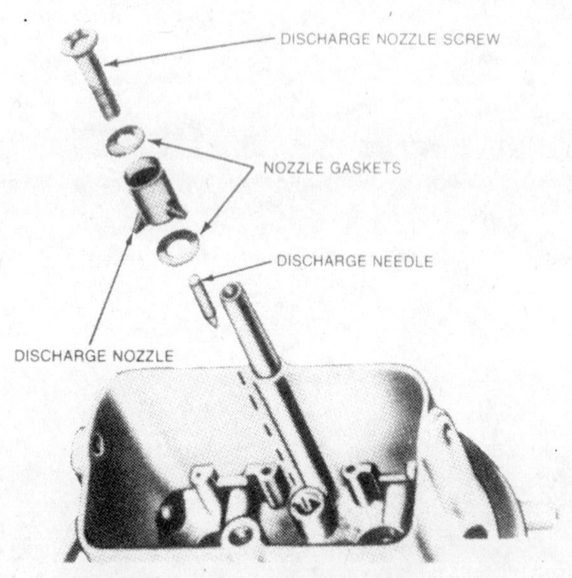

6.25 Accelerator pump discharge assembly components – exploded view (4180C 4V carburetor)

44 As the throttle body is placed in position, attach the secondary diaphragm rod to the operating lever. Secure the throttle body to the main body with the retaining screws and lockwashers and attach the retainer to the secondary diaphragm rod.
45 Install the accelerator pump diaphragm spring and diaphragm with the large end of the rivet against the operating lever in the pump chamber.
46 Install the cover but do not tighten the retaining screws until the diaphragm is centered. Compress the diaphragm with the operating lever, then tighten the cover screws securely.
47 Insert the filter, screen, new O-ring and new gasket, then thread the fuel inlet fitting into the fuel bowl.
48 Thread the fuel level sight plug into the side of the bowl with a new gasket.
49 Position the baffle plate on the ridges in the fuel bowl. Attach the spring to the float and attach the float to the float shaft. Be sure that the spring seats between the ridges on the boss at the bottom of the fuel bowl. Attach the float retainer with needle nose pliers.
50 Position a new O-ring, coated with petroleum jelly, on the fuel inlet needle and seat assembly. Place the needle and seat assembly in position through the top of the fuel bowl.
51 Thread the adjusting nut and a new gasket onto the needle and seat assembly. When the flat OD of the needle and seat assembly is aligned with the flat ID of the adjusting nut, install the adjusting lock screw and gasket.
52 Invert the fuel bowl and adjust the float level until the float is parallel to the bowl.
53 Thread the power valve into the metering block with a new gasket. Be sure that the power valve is the same as the one it is replacing. The number is stamped at the base of the valve.
54 Place a new gasket on the locating dowels on the back of the metering block, then attach the metering block to the main body. Insert the fuel bowl retaining screws through the bowl and position a new compression gasket on the bowl. Position the bowl over the metering block and tighten the retaining screws securely.
55 Install a new O-ring, coated with petroleum jelly, on the fuel transfer tube. The O-ring will fit against the flange on one end of the tube. This end of the tube can then be installed in the recess in the primary fuel bowl. Be sure that the O-rings are not pinched during installation.
56 Assemble the second fuel bowl and adjust the float until it is parallel with the bowl.
57 Place the metering plate gasket into position on the main body, over the balance tube, if so equipped. Attach the metering plate to the main body. If equipped with a balance tube, adjust the balance tube so that the end of the tube is one inch from the metering body.
58 Place a new O-ring, coated with petroleum jelly, against the flange of the fuel tube. Install the fuel bowl on the main body, inserting the fuel line into the recess in the bowl. Attach the retaining screws with new compression gaskets and tighten the screws securely.
59 Install the carburetor as described in Section 3. Adjust the carburetor as described below. Final adjustment of the carburetor must be performed by a Ford dealer or an emissions tune-up specialist.

Carburetor adjustments

Accelerator pump lever adjustment

60 Hold the primary throttle plate in the wide open position and insert a feeler gauge of the specified thickness between the adjustment screw head and the pump arm, when the pump arm is manually depressed (see illustration).
61 Loosen the locknut and turn the adjusting screw in to increase the clearance, or out to reduce the clearance. When the proper clearance is obtained, hold the adjustment screw with a wrench while tightening the locknut.

Float level adjustment (wet)

62 Run the engine until it reaches normal operating temperature and park the vehicle on level ground. Remove the air cleaner assembly.

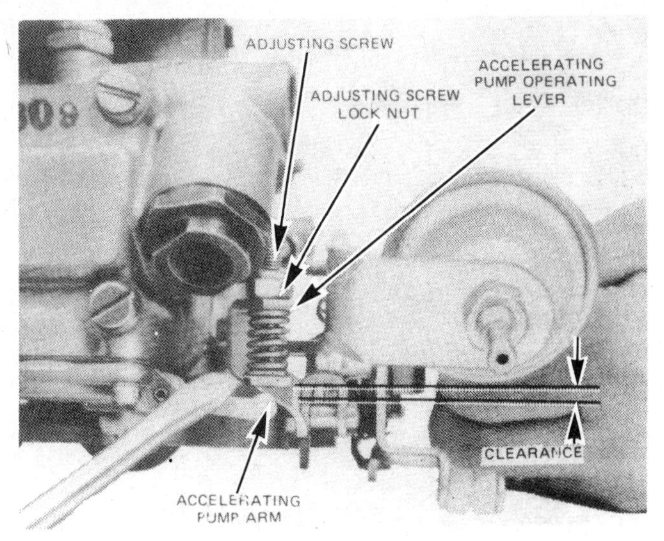

6.60 Accelerator pump lever clearance check and adjustment details (4180C 4V carburetor)

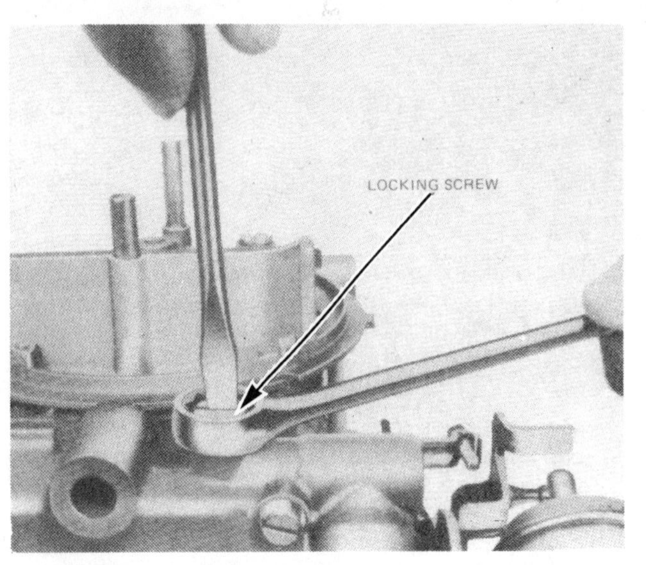

6.64 Wet float bowl fuel level adjustment details (4180C 4V carburetor)

63 Place a container under the fuel level sight plug and remove the sight plug. Only remove one sight plug and adjust one float level at a time. The proper float level exists when the fuel is right at the bottom of the plug opening.
64 If the level is too high loosen the lock screw at the top of the fuel bowl (see illustration). The lock screw should only be loosened enough to turn the adjusting nut. Lower the level by turning the nut clockwise, then raise the float to the desired level by turning the nut in the other direction.
65 Hold the adjusting screw in position while tightening the lock screw.
66 Install the sight plug and a new gasket.

7 Carburetor — disassembly and reassembly (2150 2V feedback)

Refer to illustrations 7.3a and 7.3b

1 Some late model carburetors have been equipped with electronic fuel metering devices (known as Feedback Duty Solenoids) to very accurately meter fuel. The solenoid is controlled by a computer that calculates the fuel/air ratio from information received by the sensors, ranging from engine load and RPM to present atmospheric pressure.
2 Due to the complexity of the EEC (Electronic Engine Control) system, the home mechanic is limited to visual inspection of the electrical connectors to be sure they are clean and secure. If a problem is suspected in the EEC system, the vehicle should be taken to a Ford dealer or a qualified service technician.
3 If a carburetor overhaul is to be done, the procedure in Section 4 should be followed with the addition of the following steps at the points indicated. Refer to the accompanying exploded view drawing of the 2150 2V feedback carburetor (see illustration).

a) *Step 17:* Remove the two retaining screws (rather than one) and lift out the booster venturi and metering rod assembly and gasket.
b) *Step 20:* On the throttle position sensor, mark the relationship of the two adjusting screw heads and the adjusting screw slots to simplify reinstallation (see illustration). Remove the two retaining screws and detach the sensor.
c) *Step 23:* Remove the three feedback solenoid mounting screws, the feedback solenoid and gasket from the main body.

8 Carburetor — external adjustments

Refer to illustrations 8.12a, 8.12b, 8.23a, 8.23b, 8.25a and 8.25b

Note: *All carburetors on late model US vehicles come equipped with adjustment limiters or limiter stops on the carburetor idle mixture screws. All adjustments to these mixture screws are to be made only within the range provided by the limiter devices. In addition, the following procedures are intended for general use only. The information given on the Emissions Control Information label located under the hood is specific for your engine and should be followed.*

The instructions included here should be regarded only as temporary adjustments. The vehicle should be taken to a dealer or repair facility equipped with the necessary instruments for adjusting the idle mixture as soon as possible after the vehicle is running. At the same time the idle mixture is set, the idle speed will also be reset, so the carburetor will operate within the range delineated on the Emissions Control Information label.

Note also that all necessary adjustments and/or inspection procedures discussed in Section 2 of this Chapter should be performed before carburetor adjustments are made.

Idle speed (preliminary)

1 Set the idle mixture screws to the full counterclockwise position allowed by the limiter caps.
2 Back off the idle speed adjusting screw until the throttle plates are seated in the throttle bore. Some vehicles are equipped with either a dashpot or a solenoid valve to hold the linkage open. Make sure these devices are not holding the throttle plates open when making this adjustment.
3 Turn the idle speed adjusting screw in until it initially contacts the throttle stop. Turn the screw an additional 1-1/2 turns to establish a preliminary idle speed adjustment.

Idle speed (engine running)

4 Set the parking brake and block the wheels to prevent movement. If equipped with an automatic transmission, have an assistant apply the brakes as a further safety precaution during the following procedures.
5 Start the engine and allow it to reach normal operating temperature.
6 Make sure that the ignition timing is set as described in Chapter 1.
7 On a vehicle with a manual shift transmission, the idle should be set with the transmission in Neutral. On vehicles with an automatic transmission, the idle adjustment is made with the transmission in Drive.
8 Make sure the choke plate is fully open.
9 Make sure the air conditioning is turned off.
10 Use a tachometer of known accuracy and connect it to the vehicle according to the manufacturer's instructions.
11 Adjust the engine curb idle rpm to the specification listed on the Emissions Control Information label under the hood. Make sure the air cleaner is installed during this adjustment.

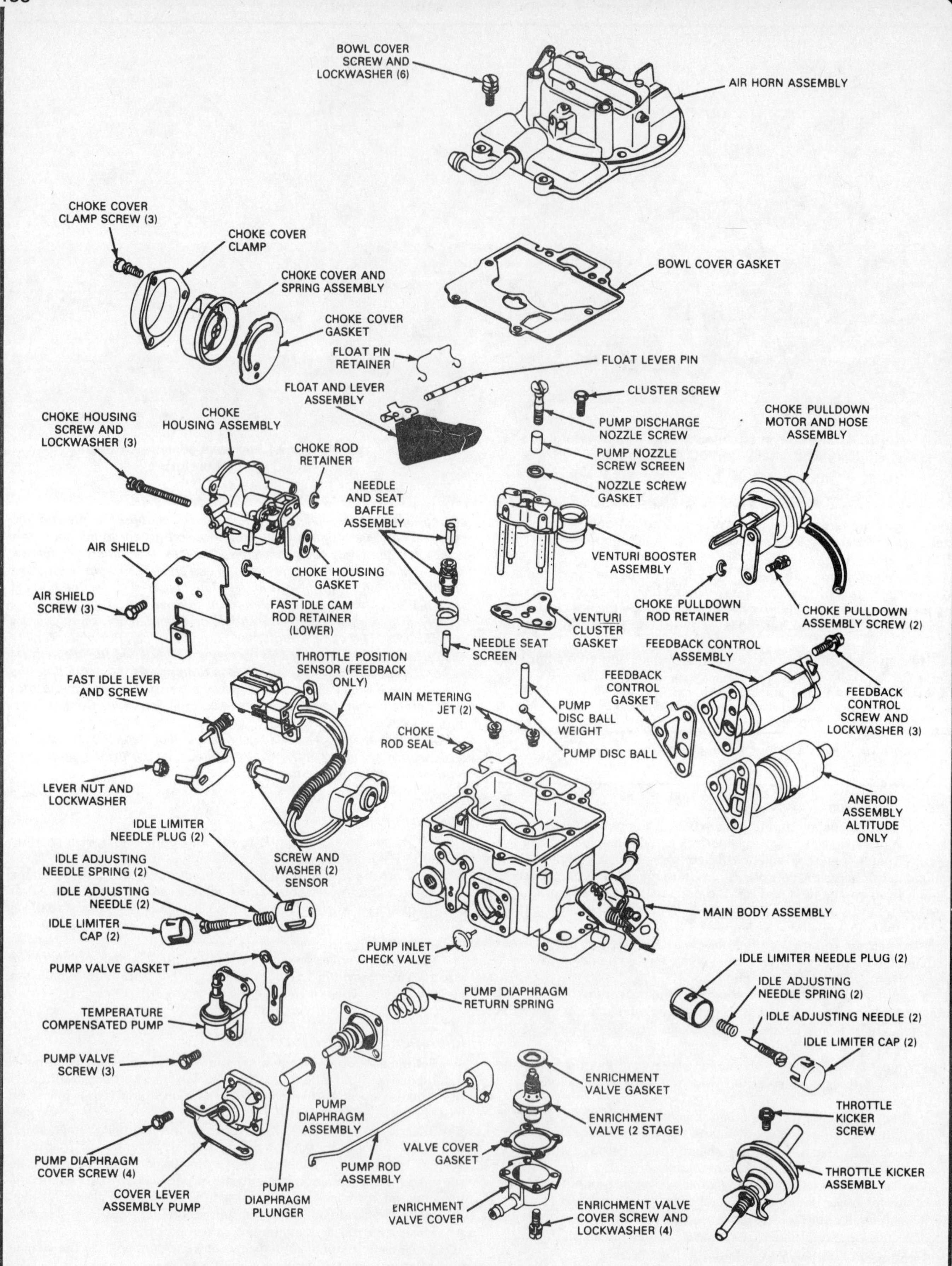

7.3a Carburetor components — exploded view (2150 2V carburetor with feedback solenoid)

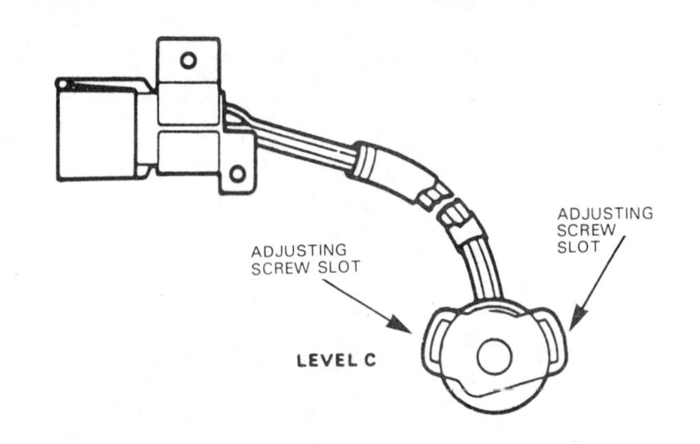

7.3b Before removing the throttle position sensor from the 2150 2V feedback carburetor, mark the screw heads and slots to ensure correct reinstallation

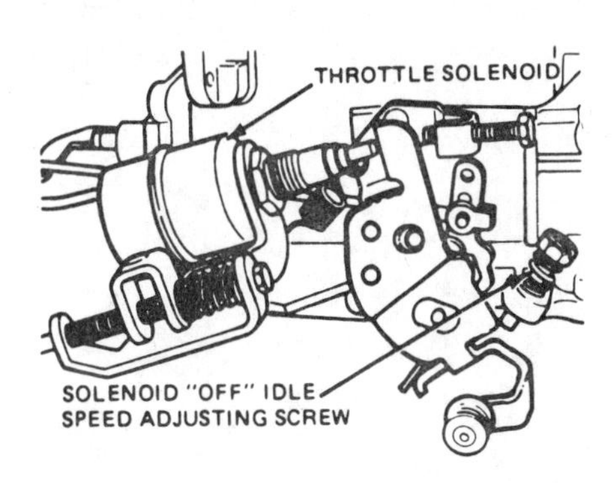

8.12a Throttle solenoid location (2150 2V carburetor)

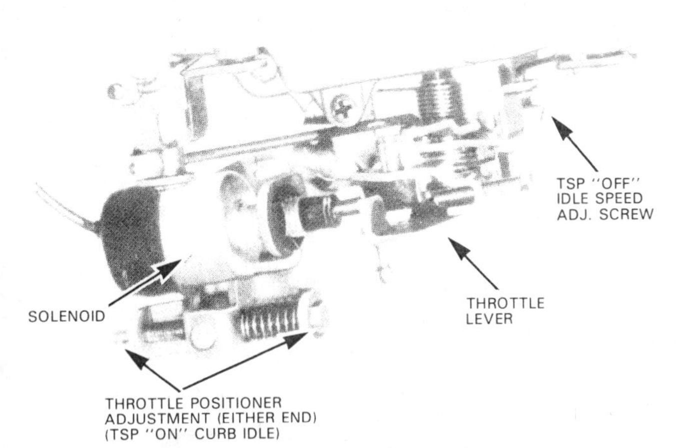

8.12b Throttle solenoid location (4350 4V carburetor)

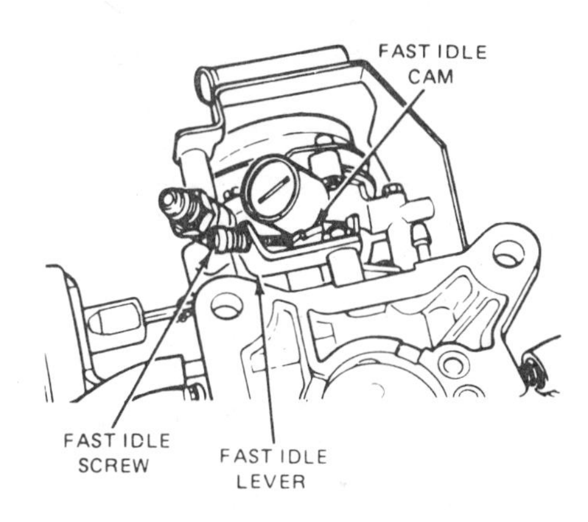

8.23a Fast idle speed adjusting screw location (2150 2V carburetor)

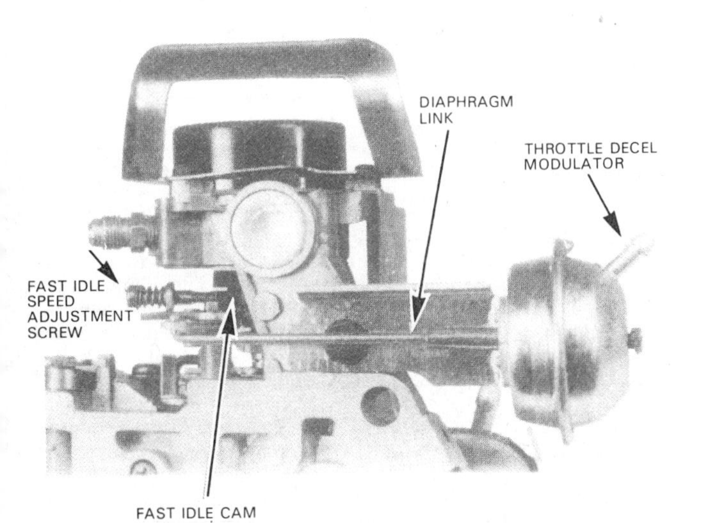

8.23b Fast idle speed adjusting screw location (4350 4V carburetor)

12 If so equipped, turn the solenoid assembly (see illustrations) to obtain the specified curb idle rpm with the solenoid activated.
13 Place the automatic transmission in Neutral.
14 Interrupt the power by disconnecting the solenoid lead wire at the connector.
15 Adjust the carburetor throttle stop screw to obtain 500 rpm in Neutral.
16 Connect the solenoid wire and open the throttle slightly by hand. The solenoid plunger should hold the throttle lever in the extended position and increase the engine speed.

Idle fuel/air mixture

17 With the engine running, carefully turn the idle mixture screws all the way in, but do not force them or they will be damaged.
18 Back the screws out very slowly until the smoothest idle is obtained. The idle mixture screws should be turned simultaneously in equal increments.
19 Take the vehicle to a Ford dealer or tune-up facility and have the mixture adjusted to conform to the applicable emission regulations and provide acceptable levels of driveability and performance.

Fast idle

20 The fast idle adjusting screw is provided to maintain engine idle rpm while the choke is operating and the engine has a limited air supply during the cold running cycle. As the choke plate moves through the range of travel from the closed to the open position, the fast idle cam rotates to allow decreasingly slower idle speeds until the normal operating temperature and correct curb idle rpm is reached.
21 Before adjusting the fast idle, make sure the curb idle speed is adjusted as previously discussed.
22 With the engine at normal operating temperature and the tachometer attached, manually rotate the fast idle cam until the fast idle adjusting screw rests on the specific step of the cam (see Emissions Control Information label for the proper step).
23 Turn the fast idle adjusting screw (see illustrations) in or out to obtain the specified fast idle speed.

Fast idle cam clearance

24 If the carburetor is in place on the engine, the air cleaner must be removed.

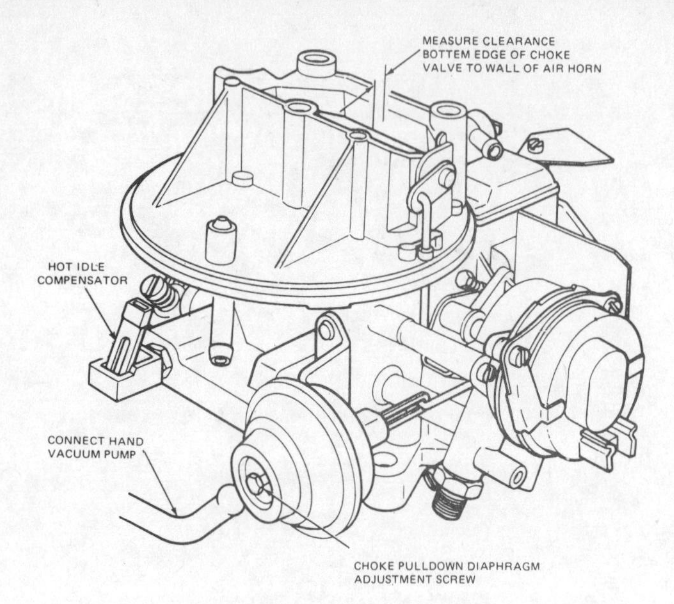

8.25a Fast idle cam/choke plate pulldown clearance check location (2150 2V carburetor)

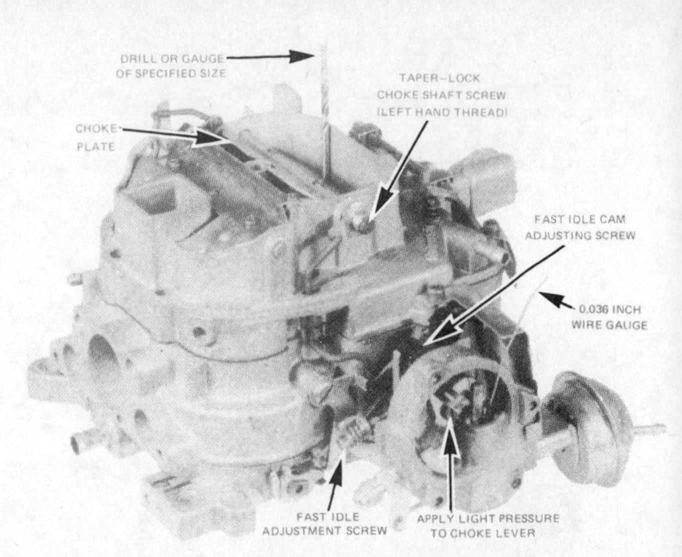

8.25b Fast idle cam/choke plate pulldown clearance check location (4350 4V carburetor)

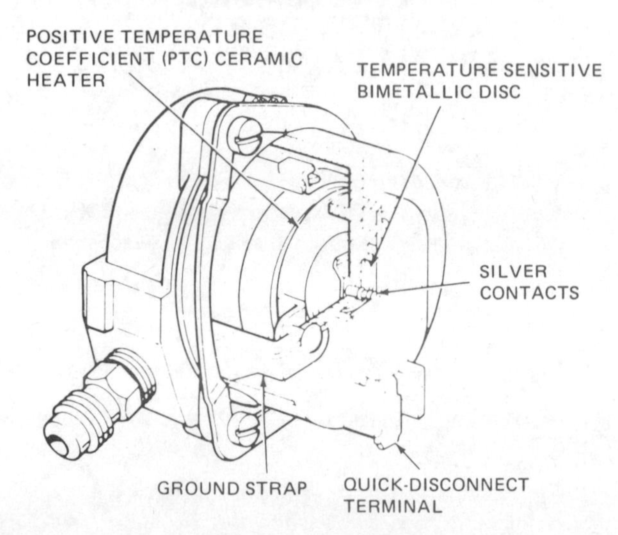

9.3 Typical electric choke components

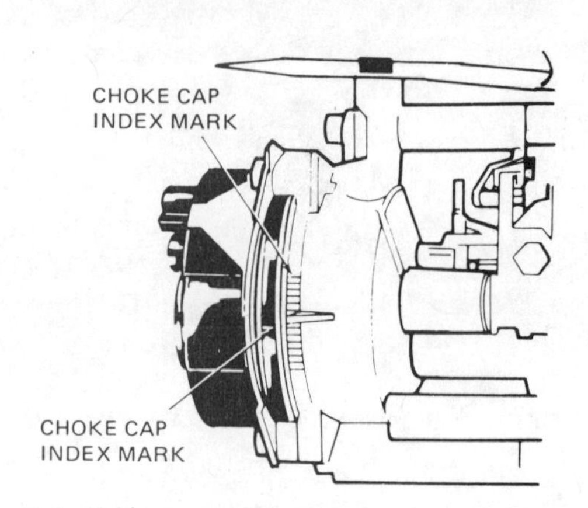

9.4 Typical choke housing and cap index marks

25 Insert the shank of a number 38 (0.1015-inch) or number 39 (0.0995-inch) drill bit between the lower edge of the choke plate and the air horn wall (see illustrations).

26 With the fast idle screw held on the bottom step of the fast idle cam, against the top step, the choke lever tang and the fast idle cam arm should just barely contact each other. Bend the choke lever tang up or down as necessary.

Choke vacuum pulldown

27 If the carburetor is in place on the engine, the air cleaner must be removed.

28 Remove the three screws and the ring retaining the choke thermostatic spring cover. Do not remove the screw retaining the water cover.

29 Pull the cover assembly away and remove the electric assist assembly.

30 Set the fast idle cam on the top step, then use a screwdriver to push the diaphragm stem back against the stop.

31 Insert the shank of a number 7 (0.201-inch) or number 8 (0.199-inch) drill bit between the lower edge of the choke plate and the air horn wall.

32 Adjust the choke plate-to-air horn wall clearance by turning the vacuum diaphragm adjustment screw as necessary with an Allen wrench.

9 Automatic choke – check and adjustment

Refer to illustrations 9.3 and 9.4

Note: *The choke operation checking procedure is outlined in Chapter 1.*

1 With the engine cold and not running, remove the air cleaner.

2 Rotate the throttle (or have an assistant depress the gas pedal) to the open position and see if the choke plate shuts tightly in the air horn opening. With the throttle held open, make sure that the choke plate can be moved freely and that it is not hanging up due to deposits of varnish. If the choke plate has excessive deposits of varnish it will have to be either cleaned with a commercial spray carburetor cleaner or the carburetor will have to be dismantled and overhauled or replaced (see Section 2). A spray cleaner will remove any surface varnish which may be causing sticky or erratic choke plate action. However, care must be used to prevent sediment from entering the throttle bore.

3 Start the engine. If equipped with an electric choke, use a voltmeter to make sure that voltage is present at the quick disconnect terminal (see illustration). Voltage should be constantly supplied to the temperature sensing switch as long as the engine is running. If no voltage is present, check the system circuit to locate the problem.

4 Some automatic chokes are equipped with a thermostatic spring housing which controls the choke action. To adjust this type of housing, loosen the three screws that attach the thermostatic spring housing to the choke housing. The spring housing can now be turned to vary the setting on the choke. Set the spring housing to the specified mark (see Emissions Control Information label in the engine compartment) and tighten the retaining screws (see illustration). Do not try to compensate for poor choke operation by varying the index setting from the

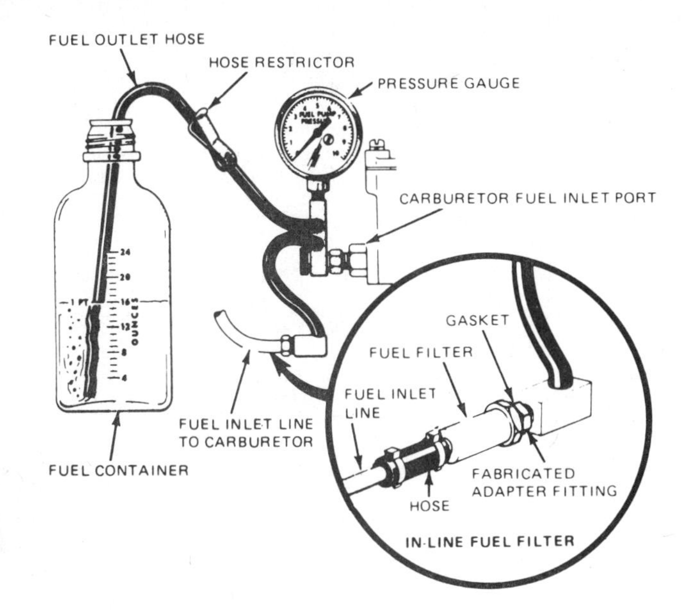

11.11 A fuel pressure gauge, some fuel hose, fittings and a calibrated container are required to check the fuel pump pressure and volume

specified spot. If the choke is not operating properly, the spring inside the housing may be worn or broken or other problems may exist in the choke system. If this situation exists, the spring housing should be replaced.

5 Allow the engine to cool completely (at least four hours – preferably overnight) and check for proper operation as described in Chapter 1.

10 Fuel lines – replacement

Warning: *Gasoline is extremely flammable, so extra precautions must be taken when working on any part of the fuel system. Do not smoke or allow open flames or bare light bulbs near the work area. Also, do not work in a garage if a natural gas appliance equipped with a pilot light is present.*

1 The fuel lines on these vehicles are generally made of metal with short lengths of rubber hose connecting critical flex points. The metal fuel lines are retained to the body and frame with clips and brackets. They generally will require no service. However, if they are allowed to come loose from their retaining brackets they can vibrate and eventually be worn through. If a fuel line must be replaced, leave it to a Ford dealer or repair shop as special flaring and crimping tools are required to build the lines. **Note:** *All EFI equipped vehicles and many late model vehicles are equipped with fuel lines that have special push connect fittings. Replacement of fuel lines with push connect fittings should be done by a Ford dealer service department (special tools and expertise are required to disconnect and reconnect the fittings). See Chapter 13 for more information.*

2 If a short section of fuel line is damaged, rubber fuel hose can be used to replace it if it is no longer than 12-inches. Cut a length of fuel quality rubber hose longer than the section to be replaced and use a tubing cutter to remove the damaged portion of the metal line. Install the rubber fuel line using two hose clamps at each end and check to make sure it is not leaking.

3 If new fuel lines are necessary, they must be cut, formed and flared out of fuel system tubing. If you have the equipment to do so, remove the old fuel line from the vehicle and duplicate the bends and length of the removed fuel line.

4 Install the new section of tubing and be careful to install new clamps and/or brackets where needed. Make sure the replacement tubing is of the same diameter, shape and quality as the original, that all flared ends conform to those on the original fuel line, that the fuel lines connected to fuel pumps or other fittings are of the double flare type and that all metal particles are removed from inside the tubing before installation.

5 Always check rubber hoses for any signs of leakage and deterioration.

6 If a rubber hose must be replaced, also replace the clamps.

7 If the vehicle is equipped with two tanks, it will also have a switching valve somewhere in the system. If a factory auxiliary tank is installed, the switching valve will be located next to the heater controls and will be electrically operated. **Warning:** *If this valve is being replaced, double-check to make sure that the cable is disconnected from the negative battery terminal before replacement, otherwise sparks could ignite gasoline present when removing the connection to the auxiliary fuel tank switching valve.*

11 Fuel pump – description and testing

Refer to illustration 11.11

Warning: *Gasoline is extremely flammable, so extra precautions must be taken when working on any part of the fuel system. Do not smoke or allow open flames or bare light bulbs near the work area. Also, do not work in a garage if a natural gas appliance equipped with a pilot light is present.*

1 Most vehicles are equipped with a single action mechanical fuel pump, while some may be equipped with an electric pump mounted in the tank. **Note:** *If a malfunction of an electric pump is suspected, take the vehicle to a Ford dealer to have it checked.*

2 The fuel pump on inline six cylinder engines is located on the lower left portion of the block midway between the front of the block and the distributor.

3 The fuel pump on V8 engines is mounted on the left side of the front cover.

4 All fuel pumps are permanently sealed and are not serviceable or rebuildable.

5 All fuel pumps are actuated mechanically by a rocker arm on the fuel pump, operating off an eccentric lobe on the nose of the camshaft.

Testing (preliminary)

6 Before inspecting the fuel pump, check all fuel hoses and lines and the fuel filter (Chapter 1).

7 Remove the air cleaner assembly.

8 Disconnect the fuel line at the carburetor inlet and unplug the high tension lead at the ignition coil to prevent the engine from starting.

9 Place a container at the end of the disconnected line and have an assistant crank the engine. A strong spurt of gasoline should shoot from the end of the line every second revolution.

Testing (pressure)

10 Disconnect the fuel line at the carburetor or at the fuel filter (if not already done).

11 Connect a fuel pressure gauge with a flexible hose between the fuel line and the carburetor (see illustration). Make sure the inside diameter of the hose is no smaller than the diameter of the fuel line.

12 Connect a T-fitting in the hose so the fuel line can be connected to the carburetor and the pressure gauge.

13 Make sure that the engine has been run and brought to normal operating temperature and that the idle is as specified on the Emission Control Information label.

14 Start and run the engine. Note the fuel pressure on the gauge. It should be within the Specifications listed at the front of this Chapter.

Testing (volume)

15 To check the fuel pump for volume, a T-fitting must be inserted in the fuel line with a flexible hose leading to a graduated fuel container. Graduated volume marks should be clearly visible on the container.

16 Install a hose restrictor, valve or other control device on the outlet line to allow the fuel to be shut off to the test container.

17 Start and run the engine with the fuel restrictor or valve shut.

18 Open the restrictor and allow the fuel to run into the container. At the end of the specified test time, close the restrictor and note the volume of fuel in the container. Compare the volume and time with those given in the Specifications at the front of this Chapter.

19 If the volume is below the specified amount, attach an auxiliary fuel supply to the inlet side of the fuel pump. A small gas can with a hose forced tightly into the cap can be used as an auxiliary fuel supply.

Repeat the test and check the volume. If the volume has changed or is now normal, the fuel lines and/or tank(s) are clogged. If the volume is still low, the fuel pump must be replaced with a new one.

12 Fuel pump (mechanical type only) – removal and installation

Warning: *Use extreme caution when working around the fuel pump, as the lines leading to and from the pump will be full of gasoline (possibly under pressure). Gasoline is extremely flammable, so extra precautions must be taken when working on any part of the fuel system. Do not smoke or allow open flames or bare light bulbs near the work area. Also, do not work in a garage if a natural gas appliance equipped with a pilot light is present.*

1 Disconnect the negative battery cable and then remove the inlet line at the fuel pump.
2 Plug the end of the line to prevent further leakage and possible contamination from dirt.
3 Remove the outlet line at the fuel pump and allow it to drain into a container.
4 Remove the two bolts and washers securing the fuel pump to the timing cover or engine block.
5 Remove the fuel pump and gasket. On some models a spacer plate may be included for heat insulation purposes.
6 Clean the mating surfaces of the fuel pump, timing cover or engine block and spacer (if so equipped). The mating surfaces must be perfectly smooth for a good gasket seal upon reinstallation.
7 Install a new gasket on the fuel pump mating surface using RTV-type sealant.
8 After cleaning the surfaces and attaching the gasket, apply sealant to the exposed side of the gasket and to the threads on the retaining bolts.
9 After installing the pump on the engine, make sure the rocker arm of the fuel pump is positioned correctly on the camshaft eccentric. It may be necessary to rotate the crankshaft until the eccentric is at its low position to facilitate fuel pump installation.
10 Holding the fuel pump tightly against the engine, install the retaining bolts and new lockwashers.
11 Tighten the retaining bolts to the specified torque.
12 Remove the plug from the inlet line and connect the inlet line to the fuel pump.
13 Connect the outlet line to the fuel pump.
14 Connect the cable to the negative battery terminal.
15 Start the engine and check for fuel and oil leaks.

13 Fuel tank – removal and installation

Refer to illustration 13.1

Warning: *Gasoline is extremely flammable, so extra precautions must be taken when working on any part of the fuel system. Do not smoke or allow open flames or bare light bulbs near the work area. Also, do not work in a garage if a natural gas appliance equipped with a pilot light is present.*

1 Some vehicles are equipped with a single tank, while others have a main and auxiliary tank, which allows for much greater fuel capacity (see illustration). **Note:** *Due to the need for special tools and expertise required to disconnect the fuel lines, the tank on all EFI equipped vehicles and certain other late model vehicles must be removed by a Ford dealer service department.*

Main tank

2 Disconnect the cable from the negative battery terminal.
3 Drain the fuel into a safety container by siphoning through the fuel hose at the fuel pump-to-fuel line connection.
4 Raise the rear of the vehicle and support it securely on jackstands. Block the front wheels to keep the vehicle from rolling.
5 Loosen the clamps and detach all hoses attached to the fuel tank, labeling the hoses to simplify installation.
6 Detach the wires from the fuel level sending unit on the tank.
7 Support the tank with a floor jack and section of plywood. Remove the nuts and bolts from the rear of the tank retaining straps and disengage the straps from the body brackets. Lower the tank enough to disconnect the hose from the vapor control valve (if equipped). Lower the tank the rest of the way and remove it from the vehicle.
8 If the tank is being replaced with a new one, remove the fuel level sending unit and vapor control valve by turning the retaining rings counterclockwise and pulling the units from the tank.
9 If the tank is to be reused, scrape away the old gasket material from the sending unit and vapor valve mounting surfaces on the tank. The sending unit will most likely require a new seal.
10 Installation is the reverse of the removal procedure. Do not over-tighten the strap nuts.

Auxiliary tank

11 Disconnect the cable from the negative battery terminal.
12 Siphon out enough fuel to make sure the tank is empty or nearly empty.
13 Support the tank and remove the restrictor brace from the front of the tank. Disengage the mounting strap ends attached to the frame member. Remove the other end of the straps by rotating them to free the L-shaped hook.
14 Lower the tank enough to detach the vapor valve hose, the filler hose and the outlet hose. Detach the sending unit wire lead.
15 Make sure that all hoses have been disconnected, then lower the tank completely and remove it from the vehicle.
16 Refer to Steps 8 through 10 above.

14 Fuel tank – cleaning and repair

1 If the fuel tank has a build-up of sediment or rust in the bottom it must be removed and cleaned.
2 When the tank is removed it should be flushed out with hot water and detergent or, preferably, sent to a radiator shop for chemical flushing. **Warning:** *Never attempt to weld, solder or make any type*

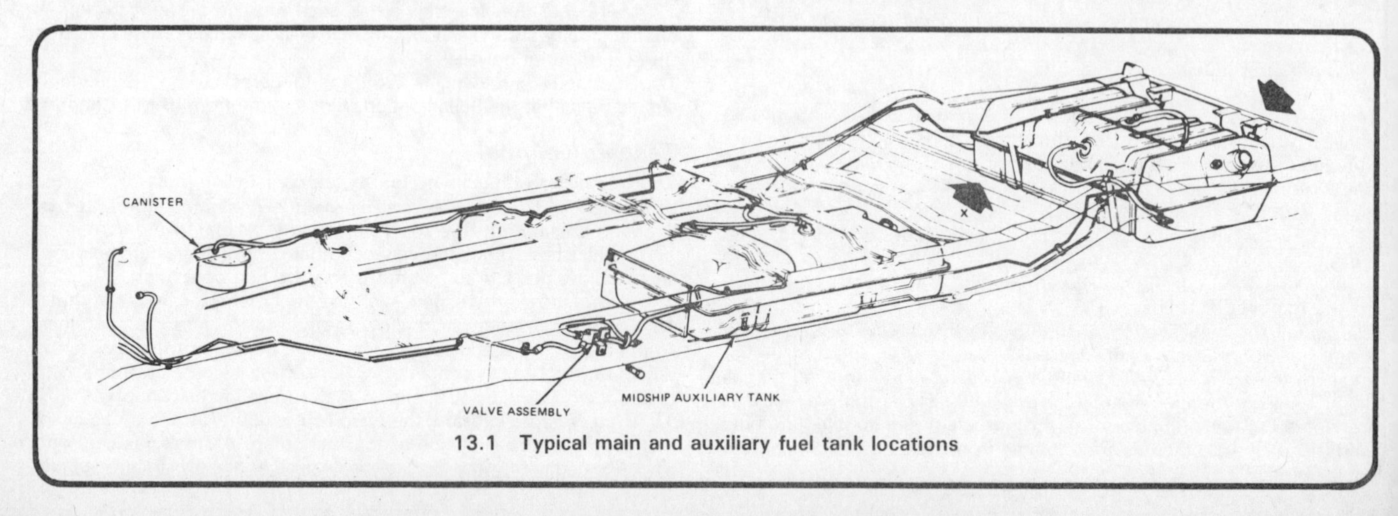

13.1 Typical main and auxiliary fuel tank locations

repairs on an empty fuel tank. Leave this work to an authorized repair shop.

3 The use of a chemical type sealer for on-vehicle repairs is advised only in case of emergency. The tank should be removed and sent to a shop for more permanent repairs as soon as possible.

4 Never store a gas tank in an enclosed area where gas fumes could build up and cause an explosion or fire.

15 Exhaust system — general information

1 The exhaust systems employed on the vehicles covered in this manual vary according to the engine, wheelbase, gross vehicle weight and emissions systems incorporated. Most vehicles are equipped with a catalytic converter as part of the emissions control system (refer to Chapter 6) and all vehicles have a single muffler and tailpipe.

2 Retention and alignment of the exhaust system is maintained through a series of metal and rubber clamps and metal brackets. Some systems, those in which excessive exhaust temperatures are created as a result of the installation of emissions control equipment, are equipped with heat shields.

3 Due to the high temperatures created in exhaust system components, any attempt to inspect or repair it should be done only after the entire system has cooled, a process which may take several hours to accomplish.

16 Exhaust system — component replacement

Refer to illustrations 16.2a, 16.2b and 16.5

1 Refer to the exhaust system check outlined in Chapter 1.

2 If the inspection reveals that the exhaust system, or portions of it, needs attention, first obtain the parts needed to repair the system. The components of the exhaust system can generally be split at their major divisions, such as the inlet pipe from the engine to the muffler or the muffler to the tailpipe. However, if corrosion is the cause for replacement it will probably be necessary to replace the entire exhaust system (see illustrations).

3 Raise the vehicle and support it securely on jackstands.

4 Make sure the exhaust system is cool.

5 Apply some rust penetrant to the bolts/nuts at the exhaust pipe flange (see illustration).

6 Remove the exhaust pipe flange nuts.

7 Remove the shields from the catalytic converter, if so equipped.

8 Remove the clamps retaining the muffler or converter to the pipe.

9 Remove the hanger supporting the muffler and/or catalytic converter from the vehicle.

10 Remove the clamps retaining the rear of the muffler to the tailpipe.

11 Remove the sections necessary for replacement. It may be necessary to allow the axle to hang free from the rear frame in order to get the curved section of the tailpipe over the rear axle housing. Be sure to support the frame securely before removing the support from

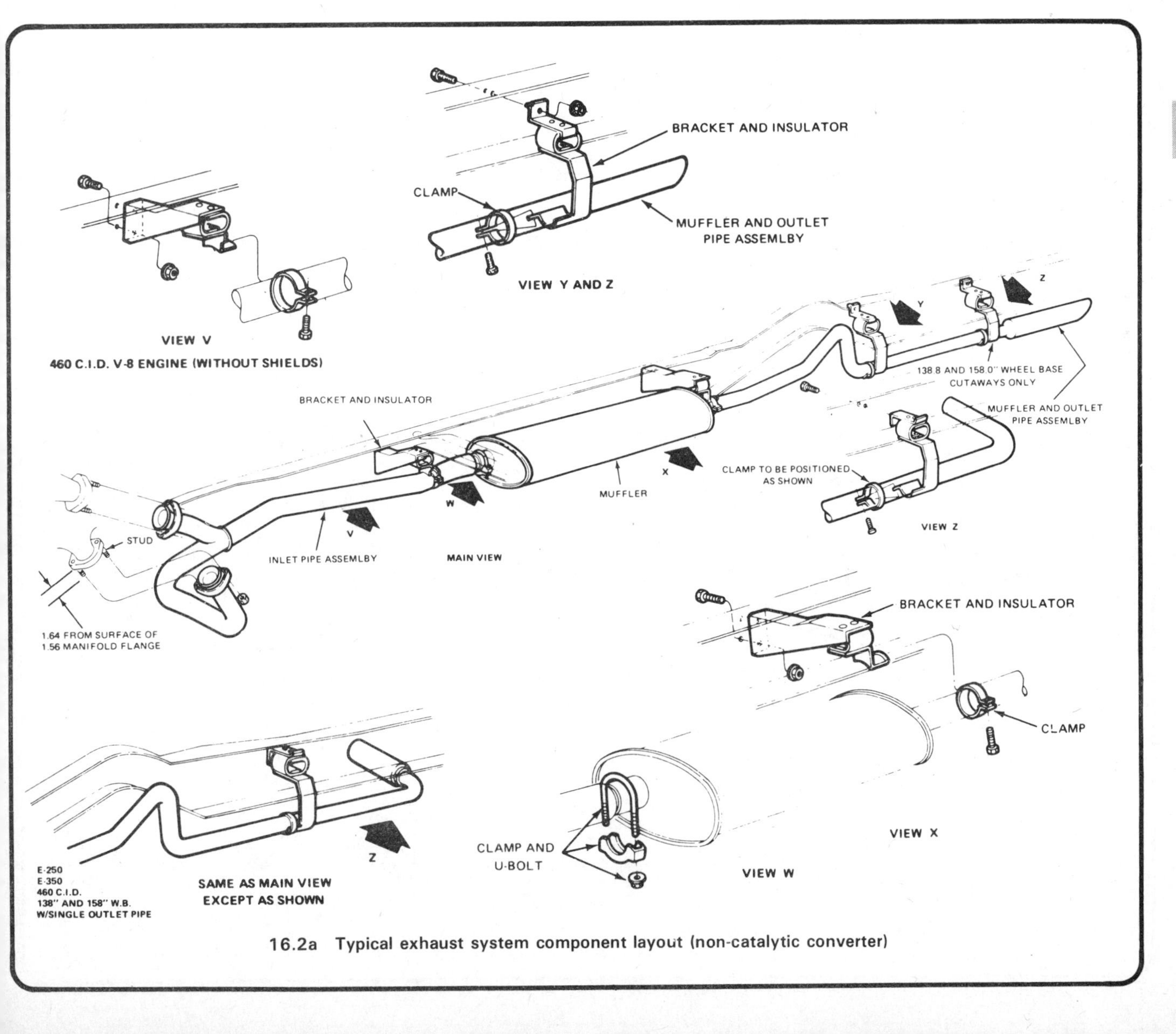

16.2a Typical exhaust system component layout (non-catalytic converter)

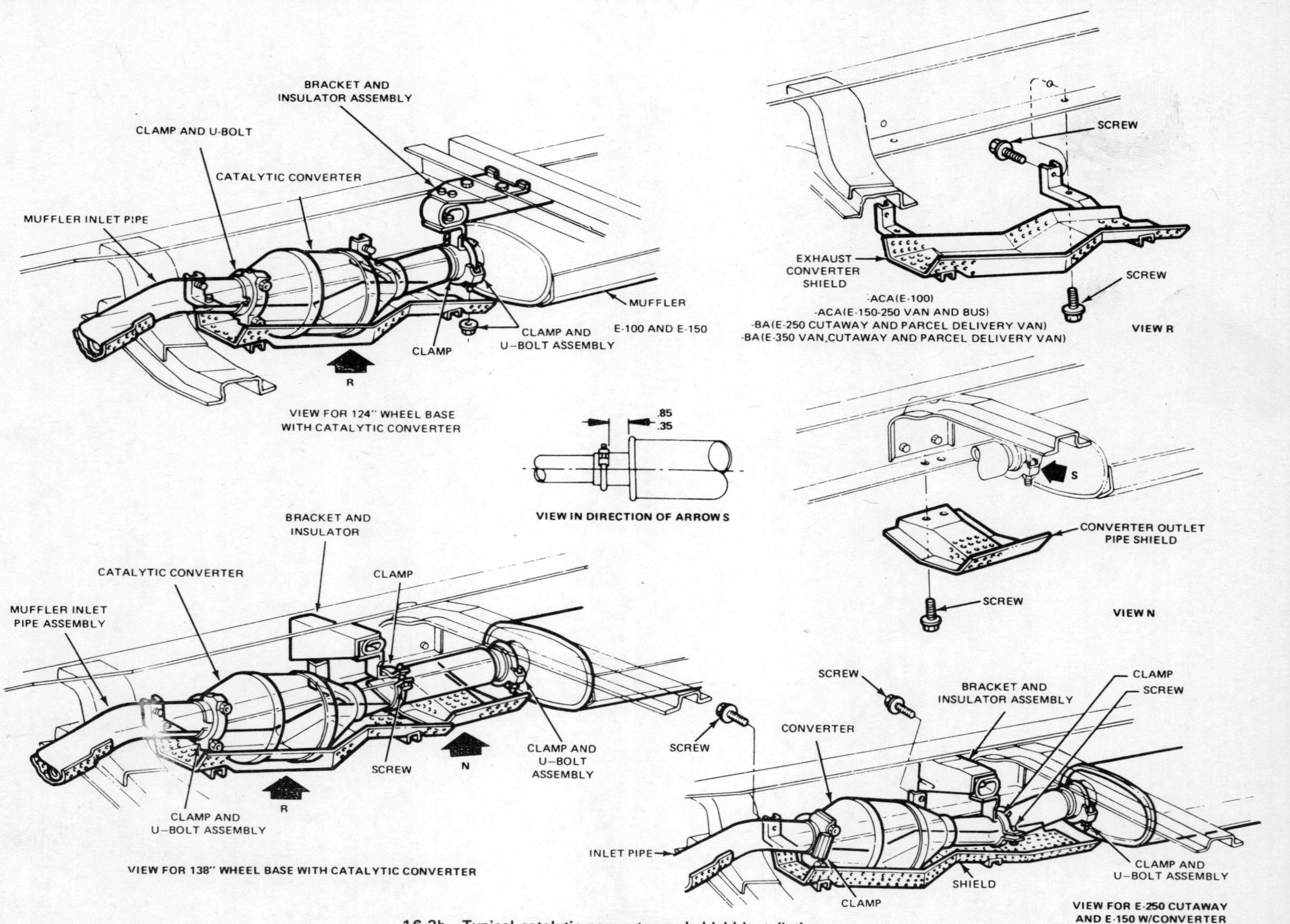

16.2b Typical catalytic converter and shield installation

the axle.

12 Installation is the reverse of removal. Always use new gaskets and retaining nuts whenever the system is being replaced. It is also a good idea to use new hangers and retaining brackets when replacing the exhaust system components.

13 Start the vehicle and check for exhaust leaks and rattles caused by misalignment.

17 Electronic Fuel Injection (EFI) — description and component removal and installation

Refer to illustrations 17.8, 17.11, 17.15, 17.20b, 17.20f, 17.22, 17.33, 17.34, 17.35, 17.38, 17.43 and 17.45

Description

The EFI system used on Ford vans is a multi-point, pulse time, mass air flow fuel injection system. Fuel is metered into the intake air stream in accordance with engine demand through eight injectors mounted on a tuned intake manifold.

An on-board computer (EEC-IV) accepts information from various engine sensors to determine the required fuel flow rate necessary to maintain a prescribed fuel/air mixture throughout the entire engine operational range. The computer then sends a command to the injectors to meter the required amount of fuel.

The EEC-IV computer also determines and compensates for the age of the vehicle and its unique characteristics. The system will automatically sense and compensate for changes in altitude and will also permit push starting of the vehicle (if a manual transmission is involved).

The fuel delivery subsystem consists of a low pressure in-tank fuel pump, a fuel filter/reservoir and a high pressure, chassis mounted electric fuel pump, which delivers fuel from the tank through a 20 micron filter to the fuel charging manifold assembly.

The fuel charging manifold incorporates electrically actuated fuel injectors directly above each of the eight intake ports. The injectors, when energized, spray a metered quantity of fuel into the intake air stream.

A constant fuel pressure drop is maintained across the injector nozzles by a pressure regulator. Excess fuel supplied by the pump, but not required by the engine, passes through the regulator and returns to the tank through a fuel return line.

One bank of four injectors is energized simultaneously, once every crankshaft revolution, followed by the second bank of injectors the next crankshaft revolution. The period of time that the injectors are energized is controlled by the computer. Air entering the engine is measured by a density meter. The computer receives input from various engine sensors and uses the information to compute the required fuel flow rate necessary to sustain the fuel/air mixture required for any given engine operating condition. The computer determines the needed injector pulse width and sends a command to each injector to meter the exact quantity of fuel.

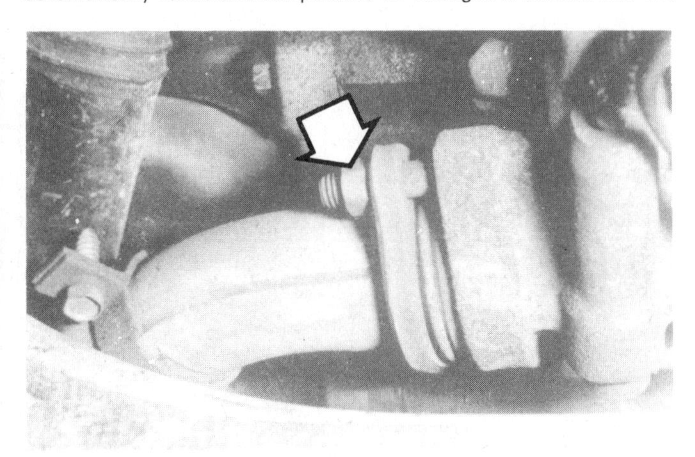

16.5 Apply penetrating oil to the manifold flange-to-pipe stud threads before attempting to loosen the nuts (arrow)

Component removal and installation

Warning: *Fuel supply lines on EFI equipped engines will remain pressurized for a period of time after the engine is shut off. The pressure must be relieved before servicing any component of the fuel system. Remember, gasoline is extremely flammable, so extra precautions must be taken when working on any part of the fuel system. Do not smoke or allow open flames or bare light bulbs near the work area. Also, do not work in a garage if a natural gas appliance equipped with a pilot light is present.*

Fuel system pressure relief

1 Locate and disconnect the wire from either the fuel pump relay, the inertia switch or the in-line high pressure fuel pump.

2 Crank the engine for approximately ten seconds. The engine may start and run, then stall. If it does, crank it for an additional five seconds after it stalls.

3 Reconnect the wire that was disconnected,

4 Disconnect the negative battery cable from the battery.

Upper intake manifold and throttle body

5 Open the hood and remove the engine cover. Cover the seats to protect the upholstery.

6 Relieve the fuel system pressure as described in Steps 1 through 4 above.

7 Remove the fuel filler cap.

8 Disconnect the wires from the air bypass valve, the throttle position sensor and the EGR position sensor on the valve (see illustration).

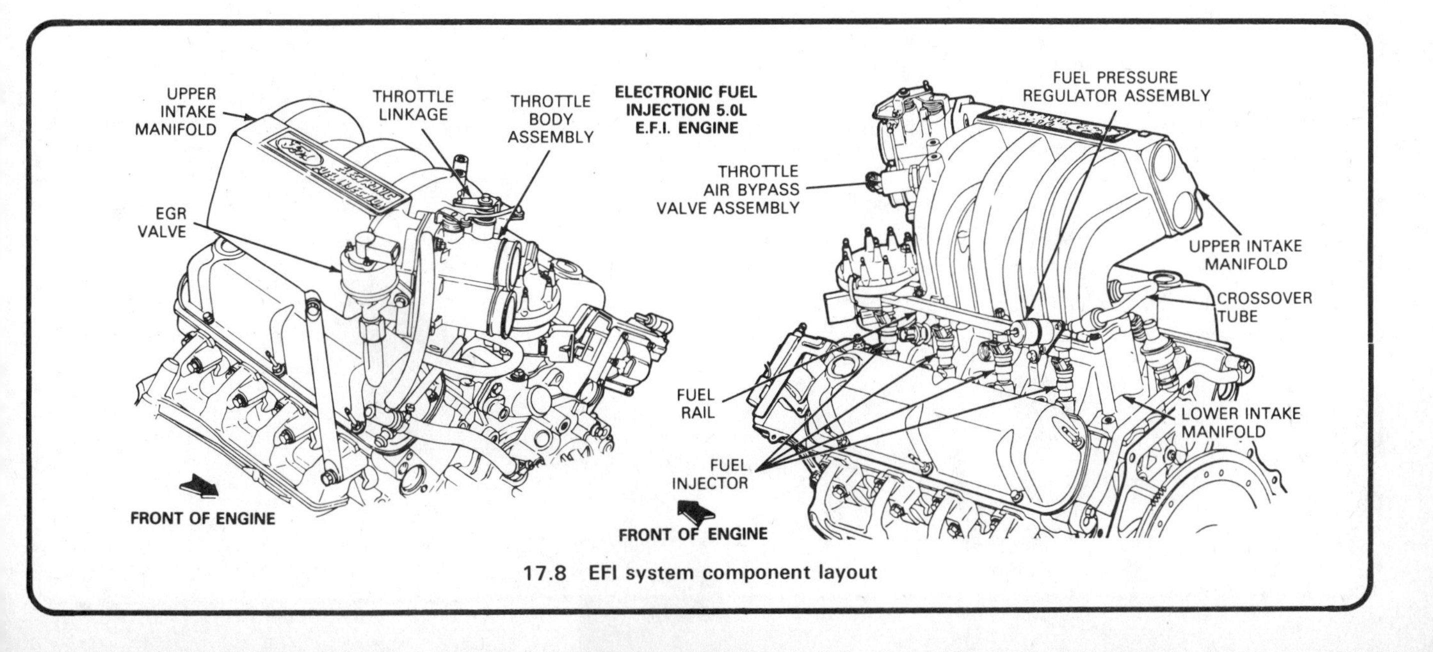

17.8 EFI system component layout

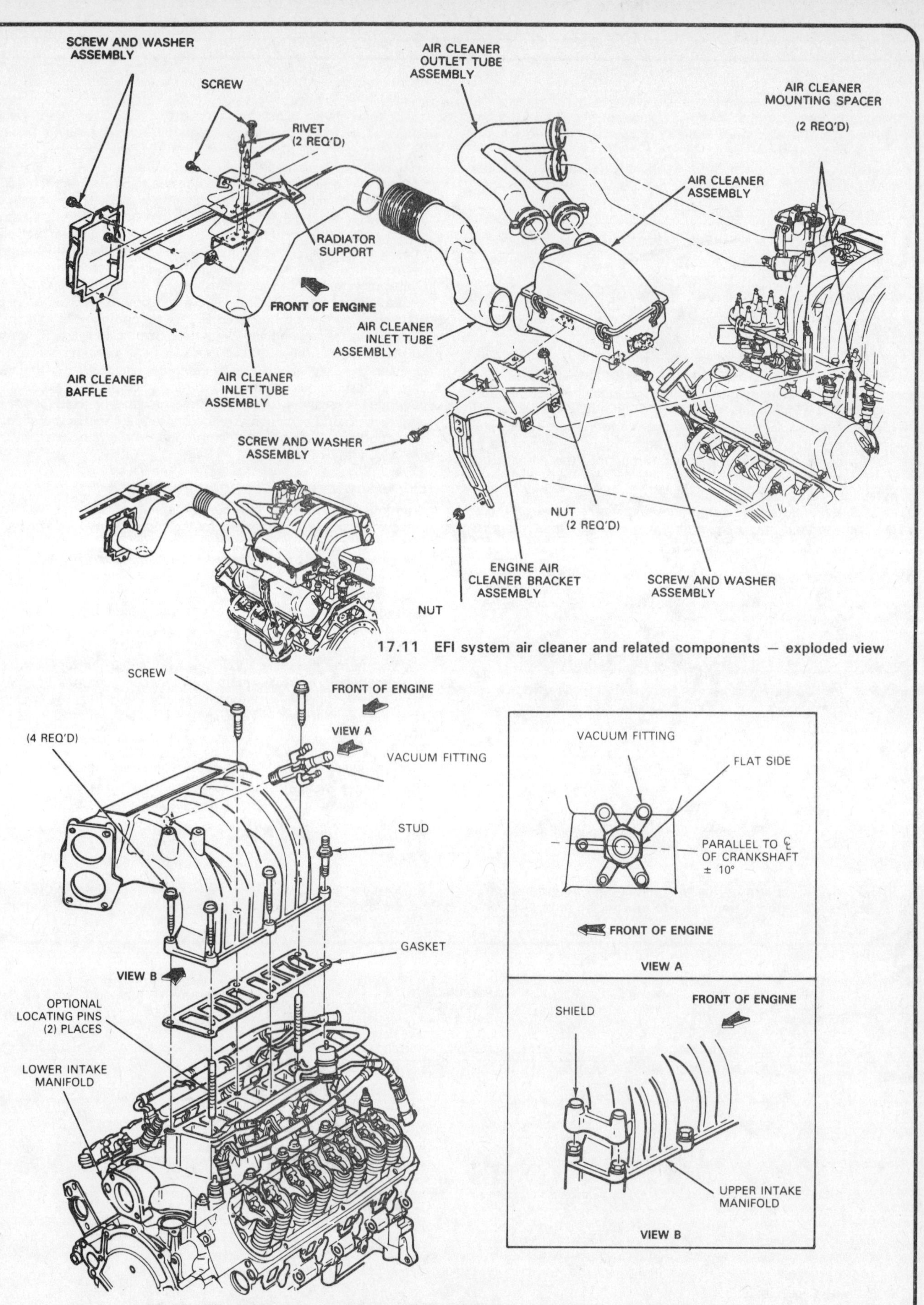

17.11 EFI system air cleaner and related components — exploded view

17.15 EFI engine upper intake manifold and related components — exploded view

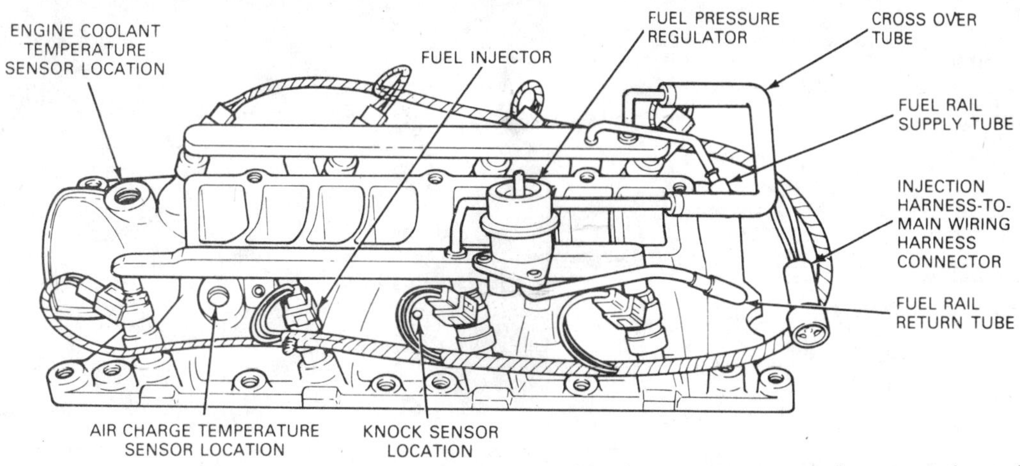

17.20b Be sure to remove or disconnect the sensor and injector wiring harnes connectors before attempting to lift off the EFI engine lower intake manifold

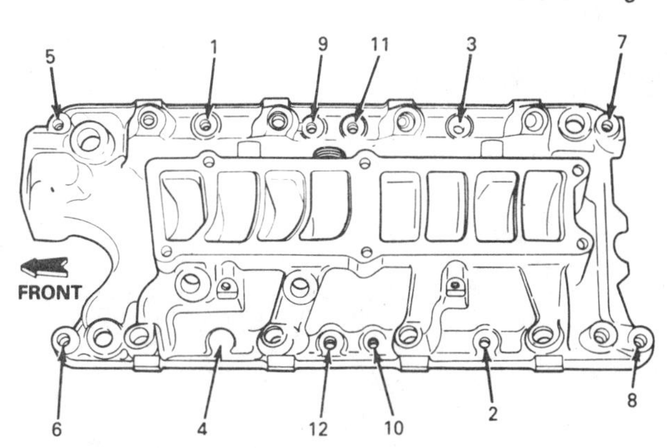

17.20f EFI engine lower intake manifold bolt tightening sequence

9 Detach the throttle linkage from the throttle ball and the AOD transmission linkage from the throttle body. Remove the two bolts and detach the cables and bracket (secure them out of the way).
10 Mark and detach the vacuum hoses from the vacuum tree, the EGR valve and the fuel pressure regulator.
11 Disconnect the PCV system hose from the fitting on the rear of the upper manifold. Detach the air cleaner-to-throttle body tubes from the throttle body (see illustration).
12 Remove the two canister purge lines from the fittings on the throttle body.
13 Detach the coolant lines from the throttle body.
14 Remove the flange nut and detach the tube from the EGR valve.
15 Remove the upper support bracket-to-upper manifold bolt and the six upper intake manifold retaining bolts (see illustration).
16 Carefully separate the upper manifold and throttle body, as an assembly, from the lower manifold.
17 Clean and check the mounting faces of the intake manifolds. Position the new gasket on the lower manifold mounting face.
18 Carefully lower the upper manifold into place (don't disturb the gasket in the process). Install the upper manifold retaining bolts and tighten them to the specified torque. Install and tighten the support bracket-to-upper manifold bolt.
19 The remaining steps are the reverse of removal. Make sure the vacuum lines and wires are routed and attached correctly.

Lower intake manifold

20 The procedure in Chapter 2, Part B, is essentially correct, but note the following points.

- a) Remove the upper intake manifold/throttle body assembly.
- b) Disconnect the wires from the engine coolant temperature sensor, the temperature sending unit, the air charge temperature sensor and the knock sensor (see illustration).
- c) Disconnect the injector wiring harness and remove the EGO ground wire from the intake manifold stud (the plated stud and ground wire must be installed in the exact same position they were removed from).

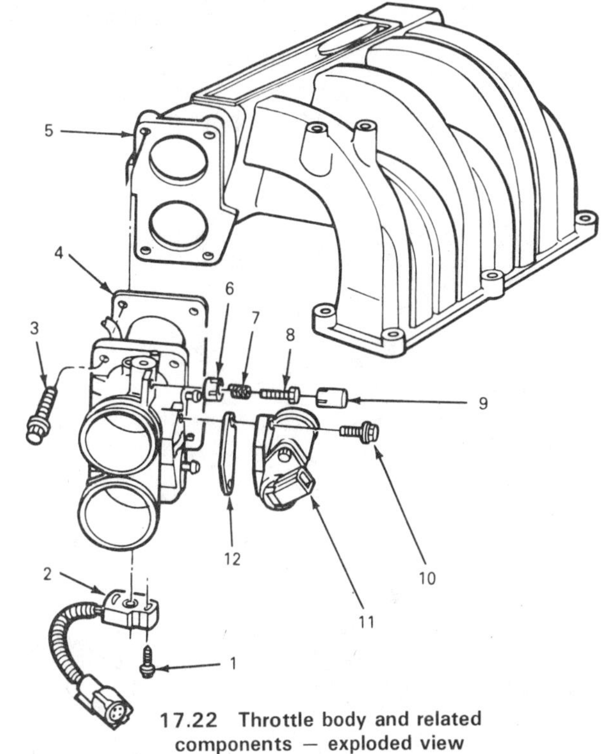

17.22 Throttle body and related components — exploded view

1 *Bolt/washer*
2 *Throttle position sensor*
3 *Bolt*
4 *Gasket*
5 *Upper intake manifold*
6 *Idle speed adjusting screw locking plug*
7 *Spring*
8 *Idle speed adjusting screw*
9 *Cap*
10 *Bolt*
11 *Air bypass valve*
12 *Gasket*

- d) Disconnect the fuel supply and return lines from the fuel rails (see *Fuel supply manifold assembly removal and installation* below).
- e) Remove the air cleaner bracket (two nuts on the intake manifold and one nut on the exhaust manifold).
- f) When installing the manifold, position the locating pins in opposite corners and make sure the gaskets interlock with the end seal tabs. Tighten the manifold bolts to the specified torque in the sequence shown (see illustration). Wait ten minutes, then tighten the bolts a second time to the same torque.

21 The remaining installation steps are the reverse of removal.

Throttle body

22 Open the hood and detach the air intake duct, then disconnect the throttle position sensor and air bypass valve wires (see illustration).

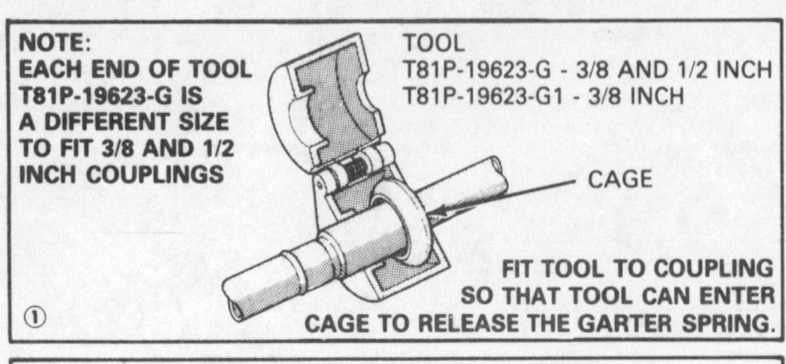

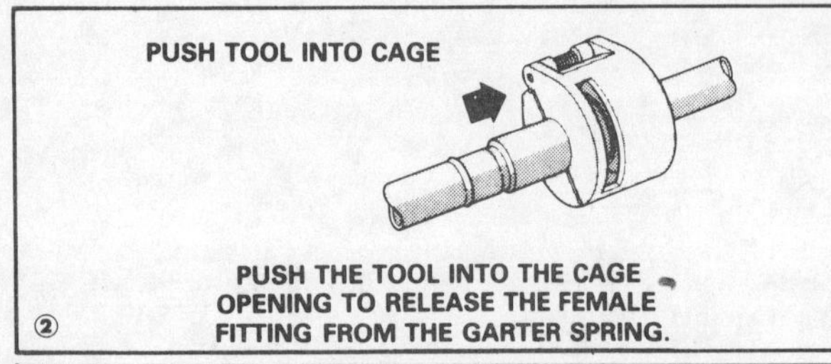

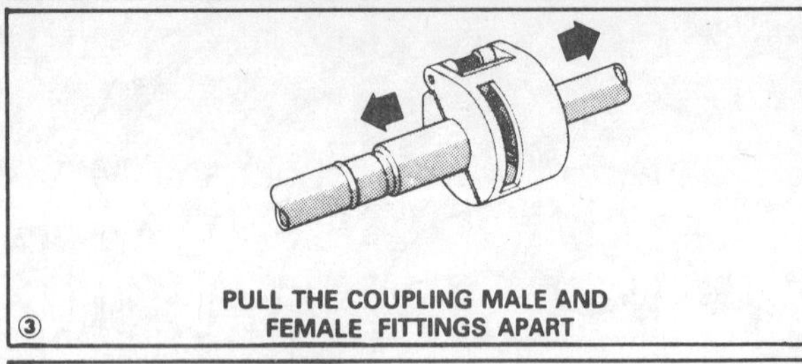

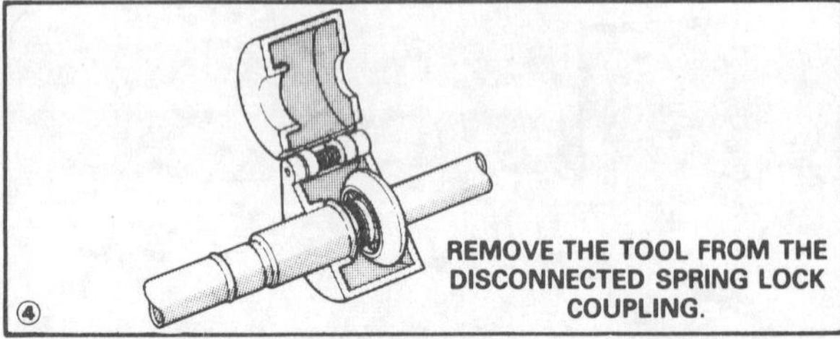

17.33 A special tool is required to disconnect the fuel lines on EFI engines

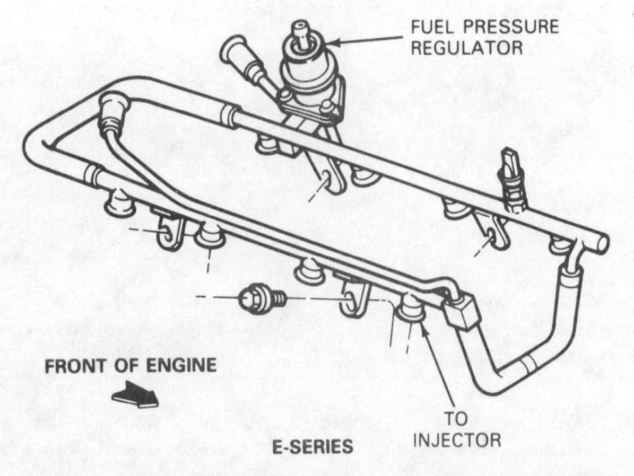

17.38 The fuel pressure regulator is held in place with three bolts

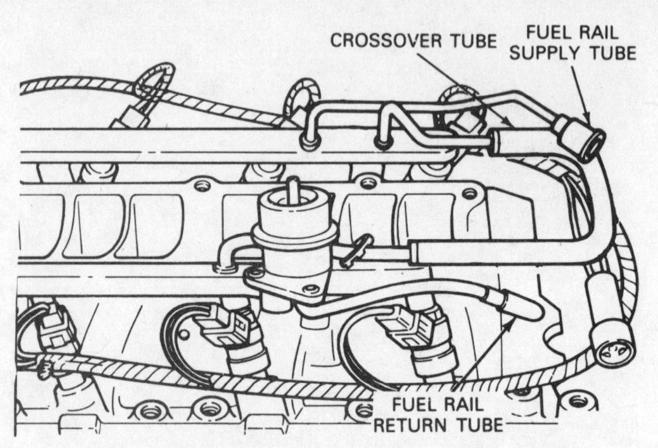

17.34 The fuel supply and return tubes must be disconnected before removing the manifold assembly

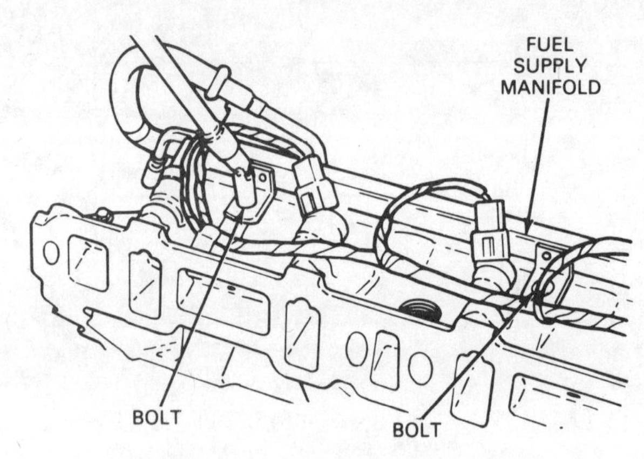

17.35 The fuel supply manifold is held in place with four bolts (two on each side)

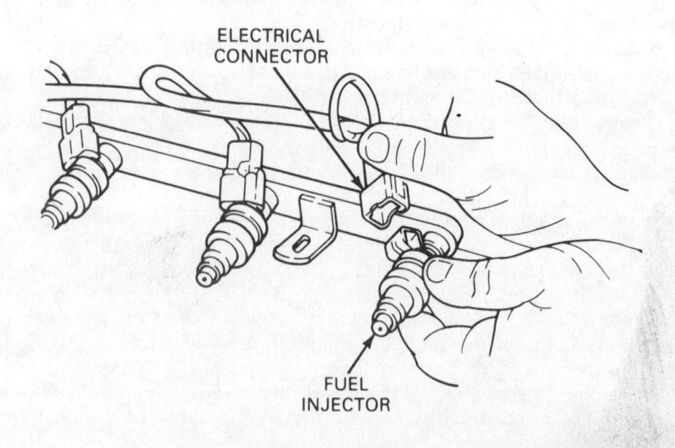

17.43 Carefully separate the wiring harness connectors before attempting to remove the injectors

23 Remove the four throttle body bolts and detach the throttle body from the upper intake manifold.
24 Remove and discard the gasket. Make sure the gasket surfaces of the manifold and throttle body are clean and smooth. If scraping is necessary, do not scratch or nick the manifold or throttle body and do not allow material to fall into the intake manifold.
25 Installation is the reverse of removal. Be sure to use a new gasket and tighten the bolts to the specified torque.

Throttle position sensor

26 Detach the sensor wiring harness, then scribe a reference mark across the edge of the sensor and the throttle body.
27 Remove the screws and detach the sensor.
28 Installation is the reverse of removal. Position the sensor with the wiring harness parallel to the throttle bores, then rotate it clockwise to align the reference marks before installing the screws. The wiring harness should point directly to the air bypass valve. The sensor should be adjusted by a Ford dealer service department.

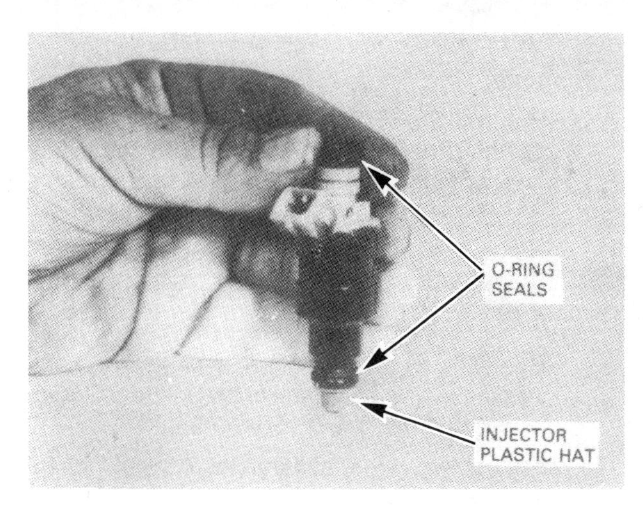

17.45 The O-rings and plastic 'hat' on each injector must be checked carefully for damage and deterioration

Air bypass valve

29 Disconnect the wiring harness, then remove the valve mounting bolts.

30 Detach the valve from the throttle body and remove the gasket.

31 Make sure the gasket surfaces of the valve and throttle body are clean and smooth. If scraping is necessary, do not scratch or nick the valve or throttle body and do not allow material to fall into the throttle body.

32 Installation is the reverse of removal. Be sure to use a new gasket.

Fuel supply manifold assembly

33 Remove the upper intake manifold assembly as described previously, then disconnect the crossover fuel hose from the fuel supply manifold. A special tool (number T81P-19623-G or G1), available from Ford dealers, must be used when disconnecting the hose (see illustration).

34 Disconnect the fuel supply and return lines at the fuel supply manifold (see illustration).

35 Remove the four (two per side) manifold retaining bolts (see illustration), and lift off the manifold.

36 Installation is the reverse of removal. Make sure the injector caps are clean before installing the manifold and that the injectors are seated properly. The fuel lines are connected by pushing the sections together carefully. Make sure they are locked.

Fuel pressure regulator

37 If the supply manifold is in place on the engine, relieve the fuel pressure as described above and remove the fuel tank cap.

38 Detach the vacuum line from the regulator and remove the three retaining screws from the regulator housing (see illustration).

39 Remove the regulator, gasket and O-ring. Discard the gasket and check the O-ring for cracks and deterioration.

40 Make sure the gasket surfaces are clean and smooth. If scraping is necessary, do not damage the regulator or fuel supply line surfaces.

41 Installation is the reverse of removal. Lubricate the O-ring with light oil, but do not use silicone grease as the injectors may clog. Turn the ignition switch On and Off several times without starting the engine to check for fuel leaks.

Fuel injectors

42 Remove the upper intake manifold and fuel supply manifold as described previously.

43 Carefully detach the wiring harness from individual injectors as required (see illustration).

44 Grasp the injector body and pull up while gently rocking the injector from side-to-side.

45 Check the injector O-rings (two each) for damage and deterioration (see illustration). Replace them with new ones if necessary.

46 Check the plastic "hat" covering the pintle for damage and deterioration. If it is missing, look for it in the intake manifold.

47 Installation is the reverse of removal. Lubricate the O-rings with light oil, but do not use silicone grease as the injector will clog.

Fuel pump — high pressure in-line

48 The high pressure, in-line fuel pump is attached to the left frame rail, just ahead of the fuel reservoir/filter described in Chapter 1.

49 Due to the need for special tools and expertise required to disconnect the fuel lines from the pump fittings, removal should be done by a Ford dealer service department.

Fuel pump — low pressure

50 The low pressure pump is mounted in the fuel tank and is attached to the fuel level sending unit.

51 If the pump must be replaced, the procedure must be done by a Ford dealer service department for the reasons outlined in Step 49 above.

Chapter 5 Engine electrical systems

Contents

Specifications

Ignition system

Distributor direction of rotation	
six-cylinder engines	Clockwise
V8 engines	Counterclockwise
Firing order	
six-cylinder engines	1-5-3-6-2-4
302 and 460 V8 engines	1-5-4-2-6-3-7-8
351 and 400 V8 engines	1-3-7-2-6-5-4-8
Number 1 cylinder	
six-cylinder engines	Front cylinder
V8 engines	Front cylinder on the right side (as seen from driver's seat)
Spark plug type and gap	See Emissions Control Information label in the engine compartment
Spark plug wire resistance	See Chapter 1
Ignition timing	See Chapter 1
Breaker point gap	See Chapter 1
Dwell angle	See Chapter 1

Charging system

Alternator brush length	
new	
thru 1980	1/2 in
1981 on	0.480 in
wear limit	
thru 1979	5/16 in
1980 on	1/4 in
Alternator drivebelt tension	See Chapter 1

Starting system

Starter brush length	
new	1/2 in
wear limit	1/4 in

Torque specifications

Distributor hold-down bolt	17 to 25 in-lbs
Distributor adapter-to-distributor base	18 to 23 in-lbs
Stator lower plate assembly-to-distributor base	15 in-lbs minimum
Diaphragm assembly-to-distributor base	15 in-lbs minimum
Spark plugs	See Chapter 1
Starter motor through bolts	55 to 75 in-lbs
Starter motor mounting bolts	15 to 20 Ft-lbs

1 General information

The engine electrical systems include the ignition, charging and starting components. They are considered separately from the rest of the electrical system (lights, etc.) because of their engine related functions.

Caution should be exercised when working on any of the systems. The components are easily damaged if checked, connected or stressed incorrectly. The alternator is driven by an engine drivebelt which could cause serious injury if your fingers or hands become entangled in it with the engine running. Both the starter and alternator are connected directly to the battery and could arc or even cause a fire if mishandled, overloaded or shorted out.

Never leave the ignition switch on for long periods of time with the engine off. Do not disconnect the battery cables while the engine is running. Be especially careful not to cross-connect battery cables from another source, such as another vehicle, when jump starting.

Additional safety related information on the engine electrical system can be found in *Safety first* near the front of this manual. It should be referred to before beginning any operation included in this Chapter.

2 Battery – emergency jump starting

Refer to the *Booster battery (jump) starting* procedure at the front of this Manual.

3 Battery cables – check and replacement

Refer to illustration 3.1

1 Periodically inspect the entire length of each battery cable (see illustration) for damage, cracked or burned insulation and corrosion. Poor battery cable connections can cause starting problems and decreased engine performance.

2 Check the cable-to-terminal connections at the ends of the cables for cracks, loose wire strands and corrosion. The presence of white, fluffy deposits under the insulation at the cable terminal connection is a sign the cable is corroded and should be replaced. Check the terminals for distortion, missing mounting bolts or nuts and corrosion.

3 If only the positive cable is to be replaced, be sure to disconnect the negative cable from the battery first. **Always disconnect the negative cable first and hook it up last.**

4 Disconnect and remove the cable from the vehicle. Make sure the replacement cable is the same length and diameter.

5 Clean the threads of the starter, solenoid or ground connection with a wire brush to remove rust and corrosion. Apply a light coat of petroleum jelly to the threads to ease installation and prevent future corrosion. Inspect the connections frequently to make sure they are clean and tight.

6 Attach the cable to the starter, solenoid or ground connection and tighten the mounting nut securely.

7 Before connecting the new cable to the battery, make sure it reaches the terminals without having to be stretched.

8 Connect the positive cable first, followed by the negative cable. Tighten the nuts and apply a thin coat of petroleum jelly to the terminal and cable connection.

4 Battery – removal and installation

1 The battery is located at the front of the engine compartment. It is held in place by a hold-down assembly which consists of a clamp which fits over the battery top and bolts which attach the ends of the clamp to the battery case. Many vehicles are equipped with an auxiliary battery as well, located in the opposite corner of the engine compartment.

2 Hydrogen gas is produced by the battery, so keep open flames and lighted cigarettes away from it at all times.

3 Always keep the battery in an upright position. Spilled electrolyte should be rinsed off immediately with large quantities of water. Always wear eye protection when working around a battery.

4 Always disconnect the negative (–) battery cable first, followed by the positive (+) cable.

5 After the cables are disconnected from the battery, remove the hold-down clamp bolts.

6 Carefully lift the battery out of the engine compartment.

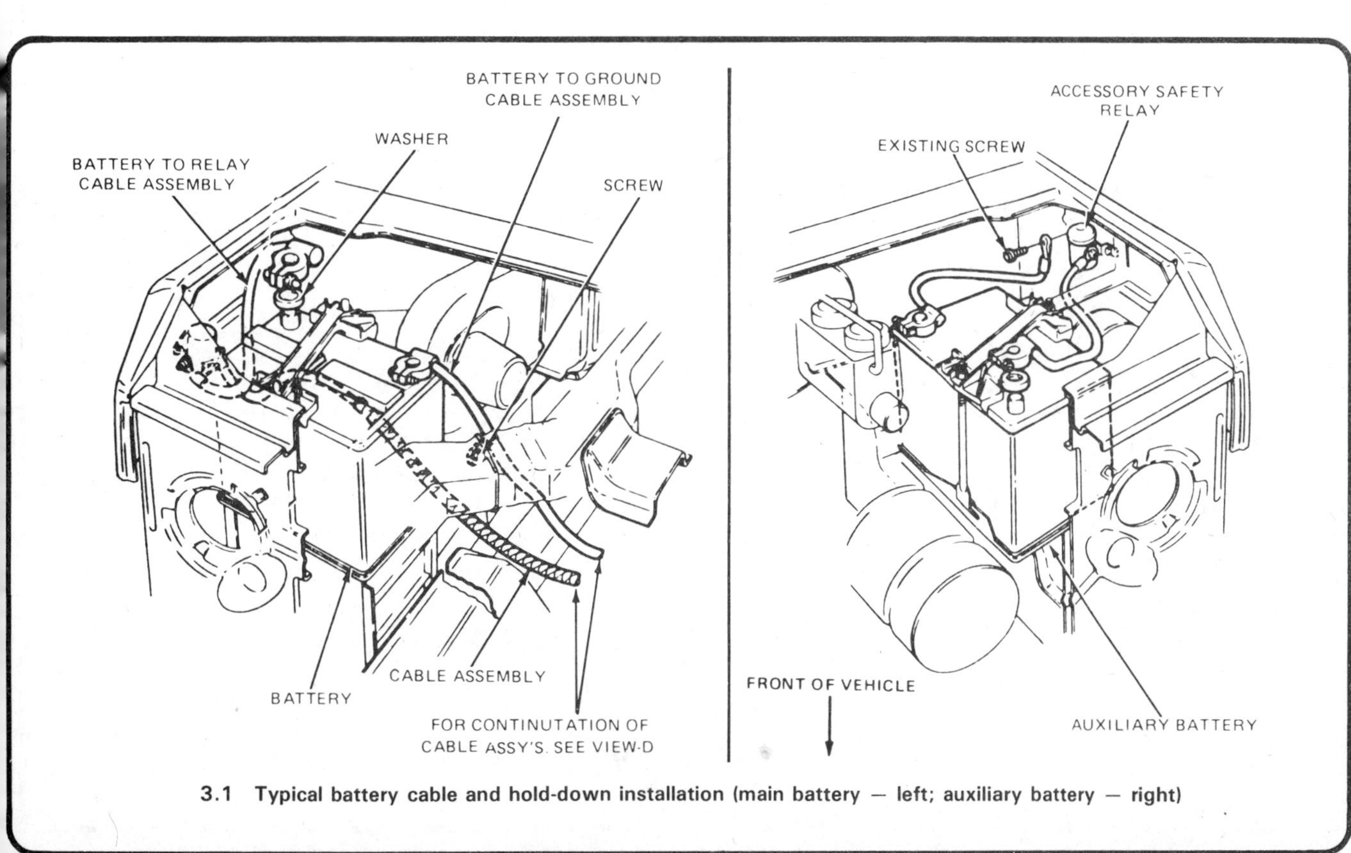

3.1 Typical battery cable and hold-down installation (main battery – left; auxiliary battery – right)

7 Installation is the reverse of removal. The clamp bolts should be tight, but do not overtighten them as damage to the battery case could occur. The battery posts and cable ends should be cleaned prior to connection (see Chapter 1).

5 Ignition system – general information

Refer to illustrations 5.1a, 5.1b, 5.1c, 5.1d and 5.5

The ignition system used on vehicles covered in this manual will be either a conventional breaker point type (early models only, through 1974), an electronic breakerless ignition system (1974 through 1977 models), a Duraspark II or III solid state system (1978 through 1986), or a TFI (Thick Film Integrated) solid-state system (1984 and later). The type of system used depends on engine type and model year (see illustrations).

Most of the systems mentioned incorporate centrifugal and vacuum advance mechanisms in the distributor body, while the spark advance function in the Duraspark III and TFI systems is dependent upon an Electronic Engine Control (EEC) system. The EEC computer controls the spark advance in response to various engine sensors. On all Duraspark III systems this includes a crankshaft position sensor, which replaces the stator assembly and armature normally located within the distributor body. On Duraspark III systems the distributor, therefore, serves only to distribute the high voltage generated by the ignition coil to the spark plugs. On the TFI ignition system, a Hall effect device replaces the stator assembly in the distributor.

The relationship of the distributor rotor to the cap is of special importance for proper high voltage distribution in the Duraspark III system. For this reason the distributor is secured to the engine and the distributor rotor, rather than the distributor body, is adjustable. Refer to Section 9 for this procedure.

In certain models the TFI ignition system features security hold-down bolts to secure the distributor. No distributor calibration is required and initial timing is not a normal adjustment.

The ignition system, especially if it is a solid state or electronic type, is very trouble free and requires only periodic maintenance as described in Chapter 1. Breaker point type systems require more frequent service and adjustment. When hooking up a tachometer/dwell meter to the coil terminal, make sure it is compatible with the type of ignition system on the vehicle (see illustration).

When working on an electronic ignition system, don't ground either of the ignition coil terminals, even momentarily. Also, most electronic ignition systems require that special silicone grease (Ford part number D7AZ-19A331-A) be applied to the inside of the spark plug wire boots. A small screwdriver can be used to apply the grease.

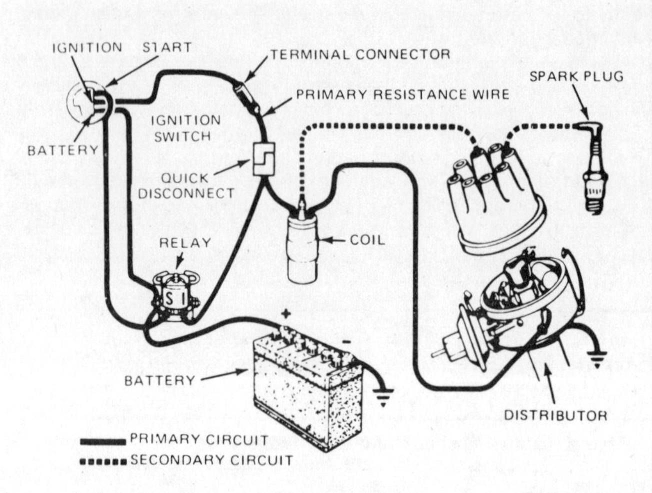

5.1a Contact breaker point ignition system diagram

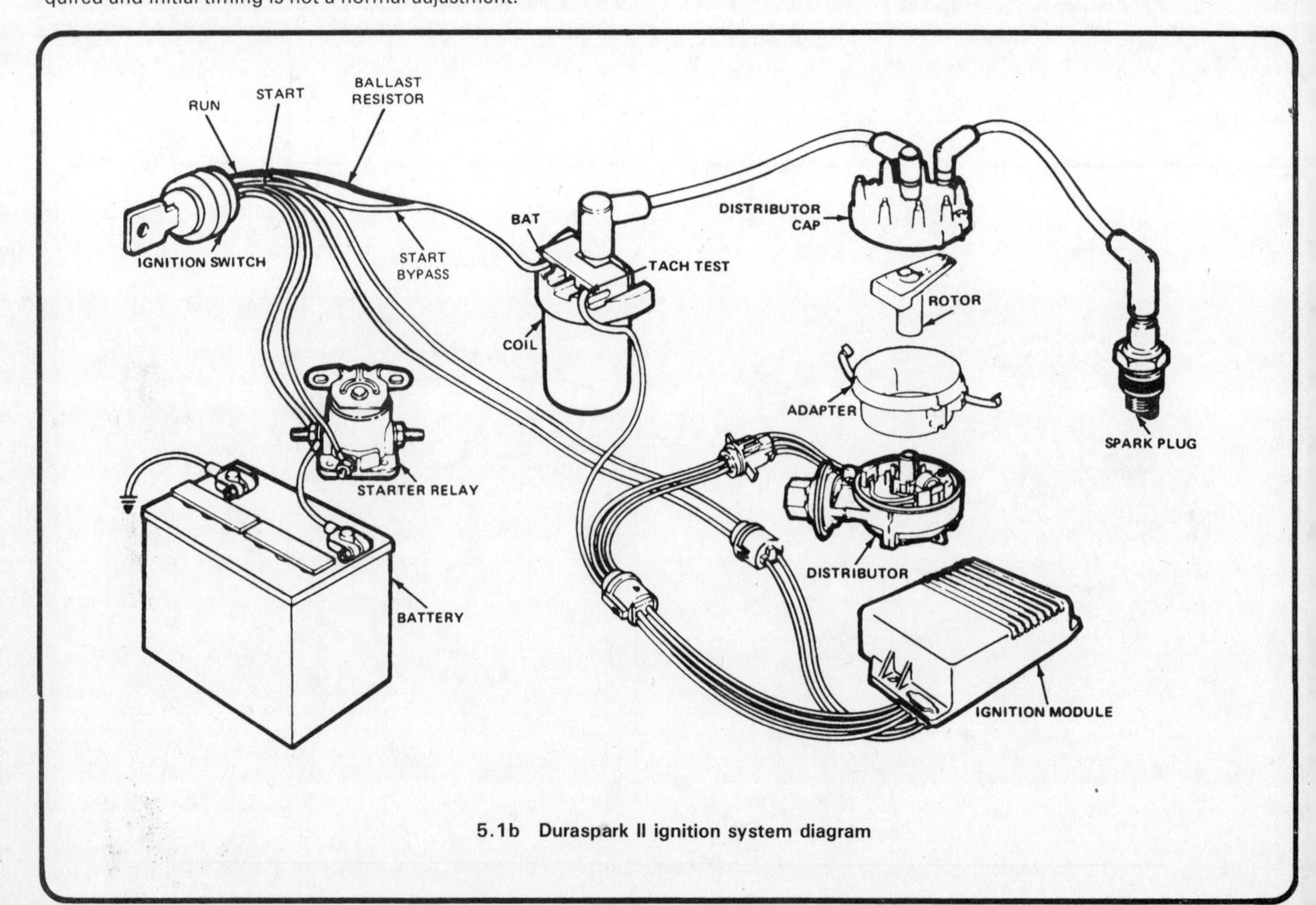

5.1b Duraspark II ignition system diagram

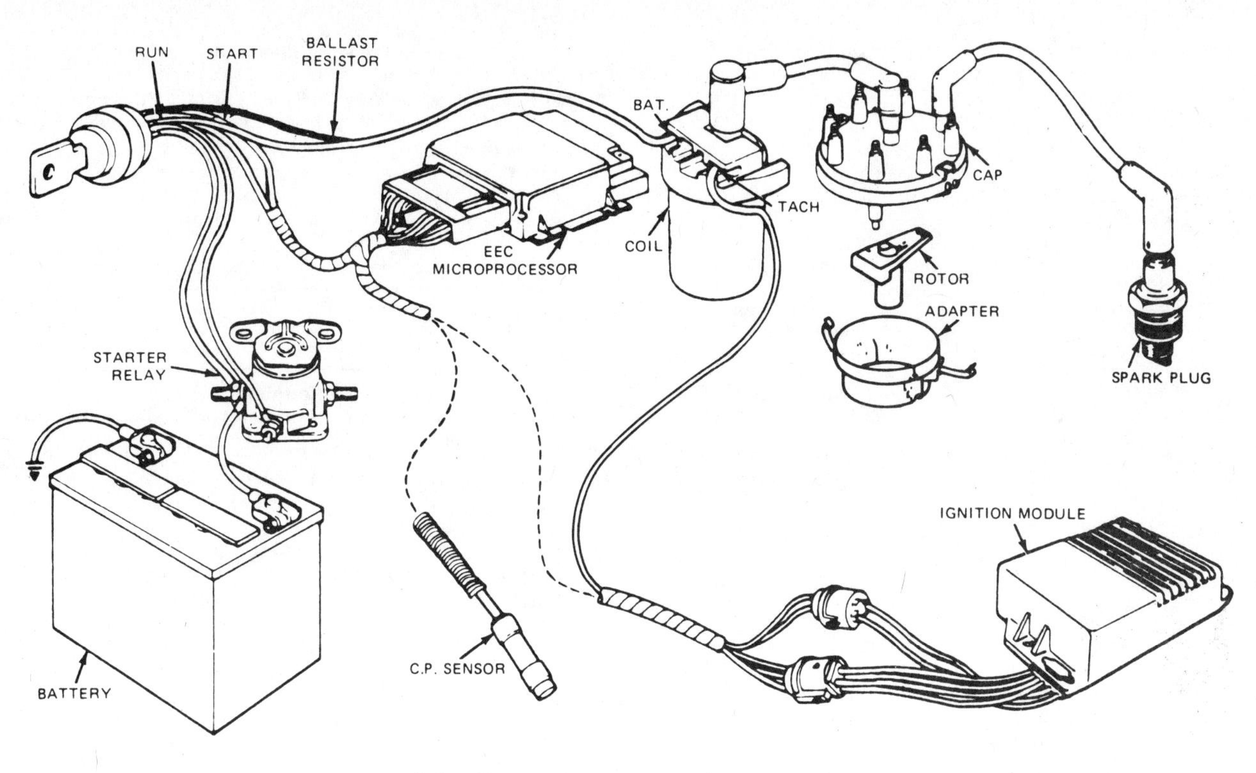

5.1c Duraspark III ignition system diagram

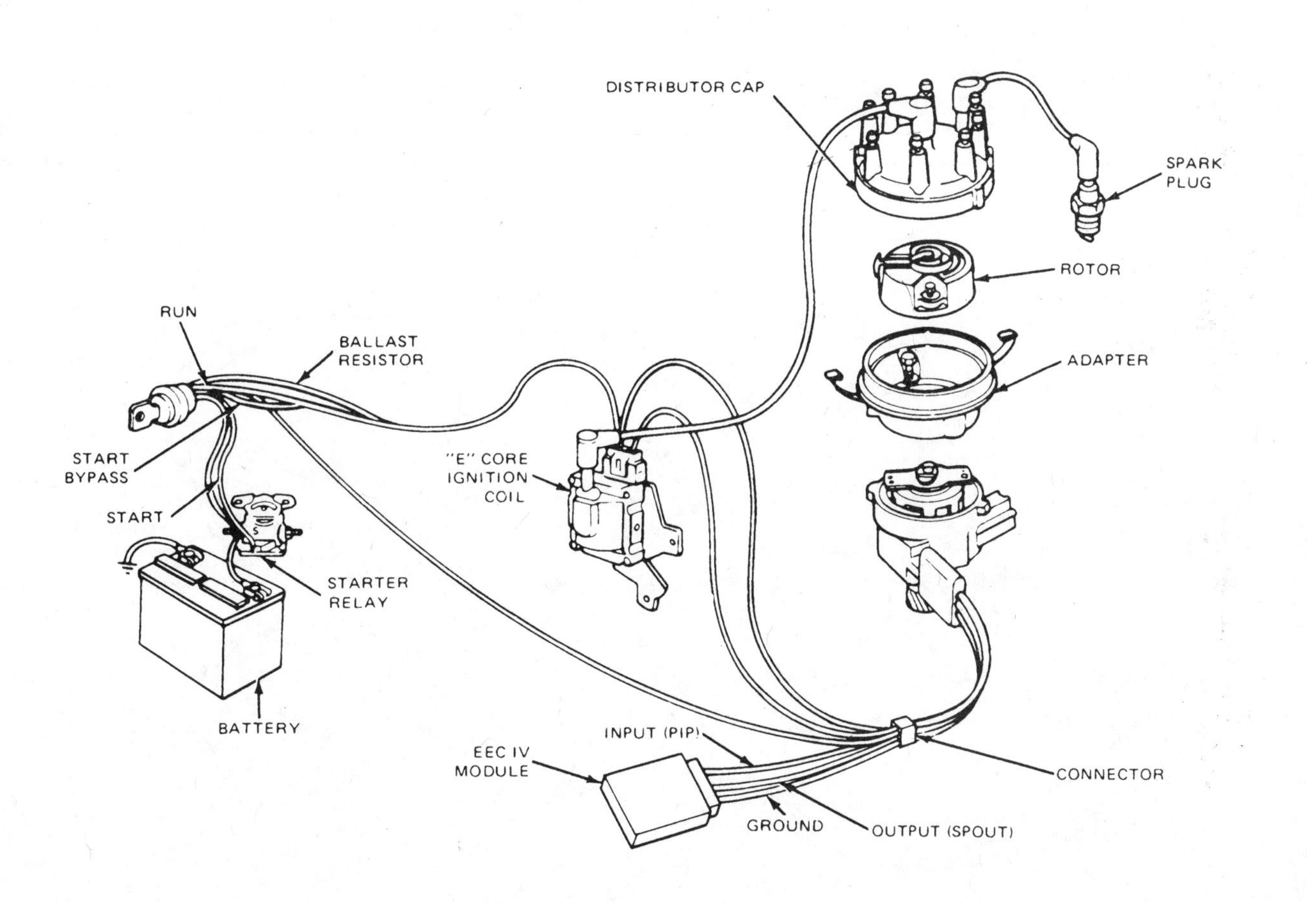

5.1d TFI ignition system diagram

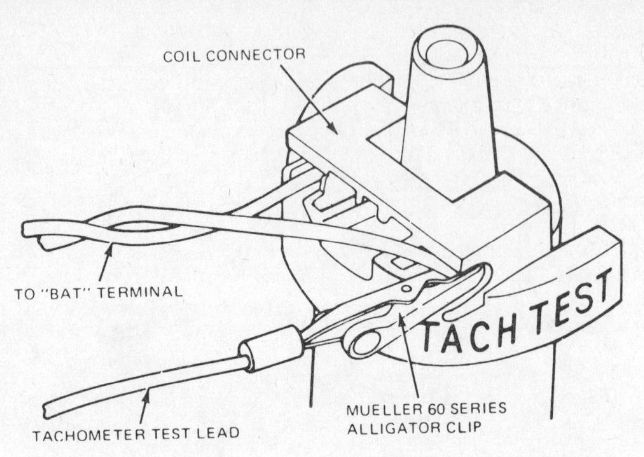

5.5 When attaching a tachometer, be sure to hook up the leads correctly, especially on electronic ignition systems

6 Ignition system inspection and testing – general information

Note: *Initial checking procedures for many ignition system components can be found in Chapter 1.*

Secondary ignition system problems (distributor cap, rotor, plug wires and spark plugs) and diagnosis are best handled with an automotive electronic oscilloscope. An experienced operator and an electronic oscilloscope can pinpoint such problems as worn spark plugs, high resistance spark plug wires, a damaged or cracked distributor cap and/or rotor, leakage between spark plug wires and other similar problems.

Primary system problems (breaker points, distributor internal components, electronic ignition module, etc.) can also be pinpointed with an oscilloscope. However, the main components sometimes require additional special test equipment and procedures. Testing of electronic ignition modules, for example, must be done by a Ford dealer service department.

A preliminary diagnosis of an ignition system can reveal such things as faulty or disconnected wires and current leakage from the coil and distributor (see Chapter 1).

With electronic ignition, the complexity of the components and testing procedures prevents in-field diagnosis of many problems by the home mechanic. If a preliminary overall visual check reveals no obvious problems such as disconnected, broken or damaged components, the vehicle will have to be taken to a professional mechanic to have the components diagnosed. Replacement of components diagnosed as defective may be made either by the service facility or the home mechanic, depending upon the ability of the home mechanic. Replacement parts are available from aftermarket manufacturers as well as Ford dealers.

7 Spark plugs – general information

Note: *Additional spark plug information can be found in Chapter 1.*

Properly functioning spark plugs are necessary if the engine is to perform well. At the intervals specified in Chapter 1, or your owner's manual, the spark plugs should be replaced with new ones.

It is important to replace spark plugs with new ones of the same heat range and type. A series of numbers and letters are included on the spark plug to help identify each variation.

The spark plug gap is very important. If it is too large or too small, the size of the spark and its efficiency will be seriously impaired. To set it, measure the gap with a wire-type feeler gauge, and then bend the outer plug electrode until the correct gap is achieved. The center electrode should never be bent as this may crack the insulator and cause plug failure.

The condition and appearance of the spark plugs will tell much about the condition and state of tune of the engine. If the insulator nose of the spark plug is clean and white with no deposits, it may be indicative of a weak mixture, or too hot a plug (a hot plug transfers heat away from the electrode slowly – a cold plug transfers it away quickly).

If the tip and insulator nose are covered with hard black deposits, it is indicative that the mixture is too rich. Should the plug be black and oily, it is likely the engine is fairly worn, as well as the mixture being too rich.

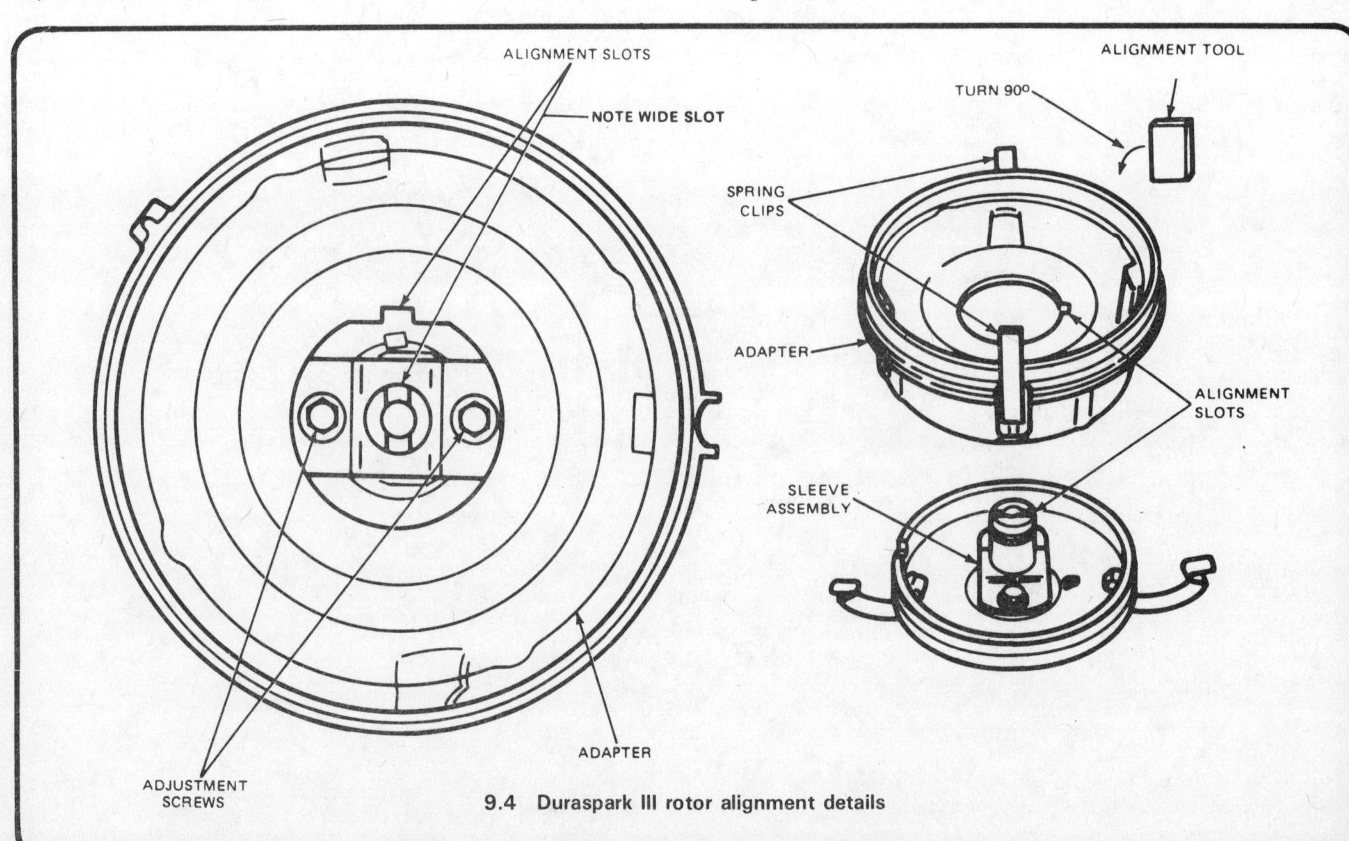

9.4 Duraspark III rotor alignment details

If the insulator nose is covered with light tan to greyish brown deposits, the mixture is correct and it is likely the engine is in good condition. See Chapter 1 for more information and color spark plug photos.

If there are long, brown tapered stains on the outside of the white portion of the plug, the plug must be replaced with a new one, as this indicates a faulty joint between the plug body and the insulator, allowing compression to leak away.

8 Ignition coil — check and replacement

Check

1 The ignition coil cannot be satisfactorily tested without special electronic diagnostic equipment. If a defect is suspected in the coil, have it checked by a dealer or repair shop specializing in electrical repairs. The coil can be replaced with a new one using the following procedure.

Replacement

2 Disconnect the negative battery cable from the battery.
3 Using coded strips of tape, mark each of the wires at the coil to help return the wires to their original positions during reinstallation.
4 Remove the coil-to-distributor high tension lead. Pull on the boot, not the wire.
5 Detach the wires at the coil. On electronic ignitions, the connections may be a push-lock connector type. Separate them from the coil by releasing the tab at the bottom of the connector.
6 Remove the bolts holding the coil bracket to the cylinder head or intake manifold.
7 Remove the coil from the coil bracket by loosening the clamp bolt.
8 Installation is the reverse of removal. Apply a thin layer of silicone grease to the inside of the coil-to-distributor high tension wire boot (see Section 5 for more information).

9 Duraspark III — rotor alignment

Refer to illustration 9.4

1 Disconnect the spring clips retaining the distributor cap to the adapter and position the cap and wires to one side.
2 Remove the rotor from the sleeve assembly.
3 Rotate the engine until the number 1 piston is on the compression stroke (see Chapter 2).
4 Slowly rotate the engine until a rotor alignment tool (Ford no. T79P-12200-A) can be inserted into the alignment slots in the sleeve assembly and adapter (see illustration).
5 Read the timing mark on the crankshaft damper indicated by the timing pointer.
6 If the timing mark reading is 0°, plus or minus 4°, alignment is acceptable.
7 If the timing mark reading is beyond the acceptable limit stated in Step 6, make sure the number 1 piston is on the compression stroke.
8 Slowly rotate the engine until the timing pointer aligns with the 0° timing mark on the crankshaft damper.
9 Loosen the two sleeve assembly adjustment screws and insert the rotor alignment tool into the alignment slots in the sleeve assembly and adapter.
10 Tighten the sleeve assembly adjustment screws and remove the alignment tool.
11 Attach the rotor to the sleeve assembly.
12 Install the distributor cap and ignition wires, making sure that the ignition wires are securely connected to the cap and to the spark plugs.

10 Stator assembly (Duraspark II only) — removal and installation

Refer to illustrations 10.5, 10.6 and 10.19

V8 engines

Removal

1 Remove the cable from the negative battery terminal.
2 Disconnect the spring clips retaining the distributor cap to the adapter and place the cap and wires aside.
3 Remove the rotor from the distributor shaft.
4 Unplug the distributor connector from the wiring harness.
5 Disconnect the spring clips retaining the distributor cap adapter to the distributor body and remove the adapter (see illustration).
6 Using a small gear puller or two screwdrivers, remove the armature from the sleeve and plate assembly (see illustration).

5

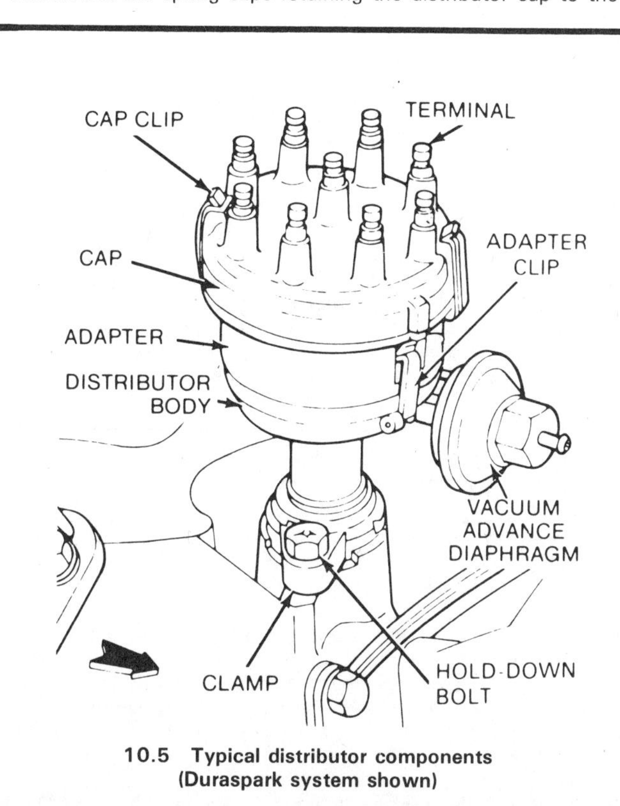

10.5 Typical distributor components (Duraspark system shown)

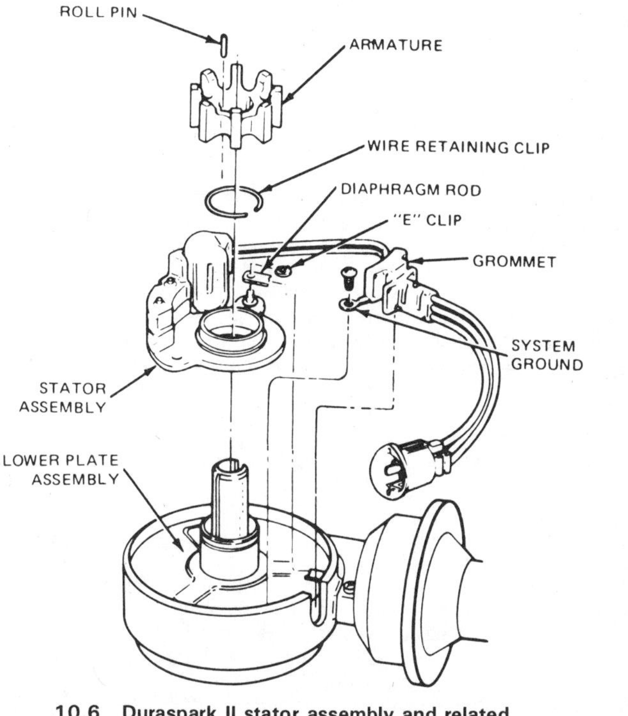

10.6 Duraspark II stator assembly and related parts — exploded view (V8 engine shown)

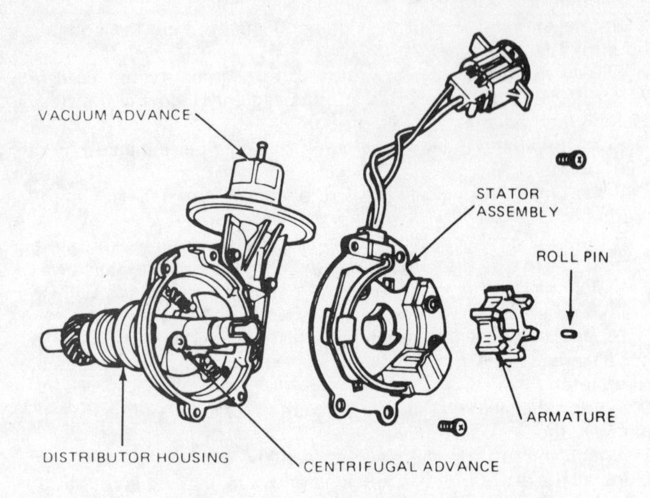

10.19 Duraspark II stator assembly and related parts – exploded view (six cylinder engine shown)

7 Remove the E-clip retaining the diaphragm rod to the stator assembly, then lift the diaphragm rod off the stator assembly pin.
8 Remove the screw retaining the ground strap at the stator assembly grommet.
9 Remove the wire retaining clip securing the stator assembly to the lower plate assembly.
10 Remove the grommet from the distributor base and lift the stator assembly off the lower plate assembly.

Installation

11 If the lower plate assembly is to be reused, clean the bushing to remove any accumulated dirt and grease.
12 Install the stator assembly by reversing the removal procedure. Note when installing the armature that there are two locating notches in it. Install the armature on the sleeve and plate assembly employing the unused notch and a new roll pin.
13 Check the initial timing (refer to Chapter 1).

Six cylinder engines

Removal

14 Disconnect the cable from the negative battery terminal.
15 Disconnect the spring clips retaining the distributor cap and place the cap and wires aside.
16 Remove the rotor from the distributor shaft.
17 Disconnect the distributor from the wiring harness.
18 Using a small gear puller or two screwdrivers as levers, remove the armature sleeve and plate assembly.
19 Remove the two screws retaining the lower plate assembly and stator assembly to the distributor base, noting that there are two different size screws employed (see illustration).
20 Remove the lower plate assembly and stator assembly from the distributor.
21 Remove the E-clip, flat washer and wave washer securing the stator assembly to the lower plate assembly, then separate the stator assembly from the lower plate assembly. Note the location of the wave washer.

Installation

22 Before installing the stator, remove any accumulated dirt or grease from parts that are to be reused.
23 Place the stator assembly on the lower plate assembly and install the wave washer (outer edges up) flat washer and E-clip.
24 Install the stator assembly/lower plate assembly on the distributor base, being sure to engage the pin on the stator assembly in the diaphragm rod.
25 Attach the lower plate assembly and stator assembly to the distributor base. Be sure to install the different size screws in their proper locations.
26 When installing the armature, note that there are two notches in it. Install the armature on the sleeve and plate assembly employing the unused notch and a new roll pin.
27 Reattach the distributor connector to the wiring harness.
28 Reinstall the rotor and distributor cap, making sure that the ignition wires are securely connected to the cap and spark plugs.
29 Connect the cable to the negative battery terminal.
30 Check the initial timing (refer to Chapter 1).

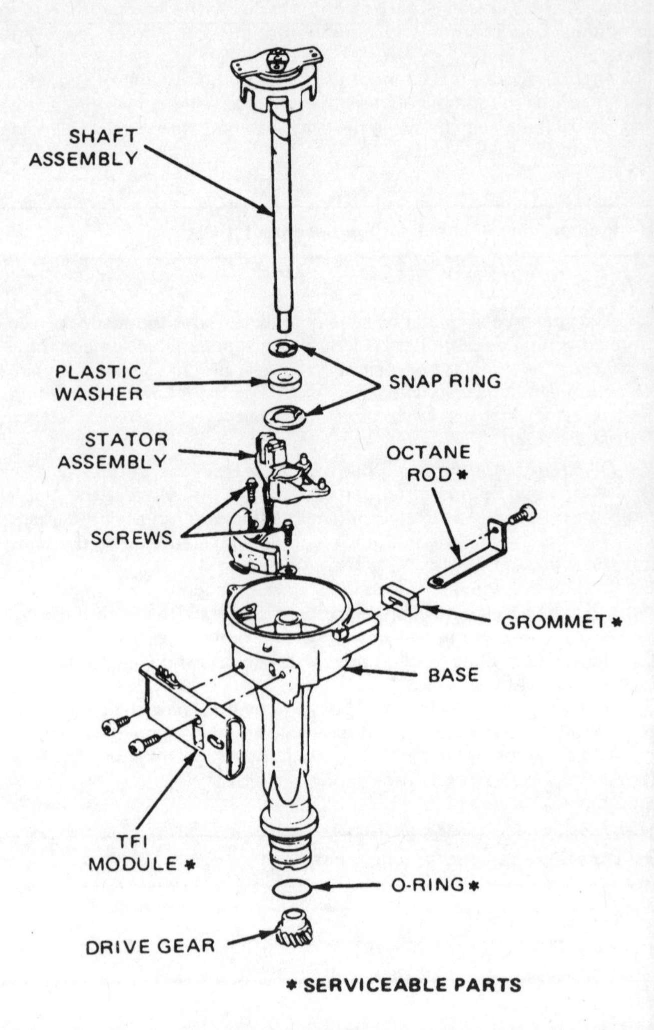

11.2 TFI system distributor components – exploded view (note TFI module location)

11 Ignition module (TFI system only) – removal and installation

Refer to illustration 11.2

1 Remove the distributor by referring to the appropriate Section.
2 With the distributor on the workbench, remove the two module mounting screws (see illustration).
3 Using a back-and-forth rocking motion, carefully disengage the module terminals from the connector in the base of the distributor.
4 With the terminals completely disengaged, slide the module down, pulling it gently away from the distributor. **Caution:** *Do not attempt to remove the module from the distributor until the connector pins are completely disengaged to avoid breaking the pins at the distributor/module connector.*
5 Installation is the reverse of removal.
6 Before installing the TFI module, coat the metal base with silicone dielectric grease approximately 1/32-inch thick.

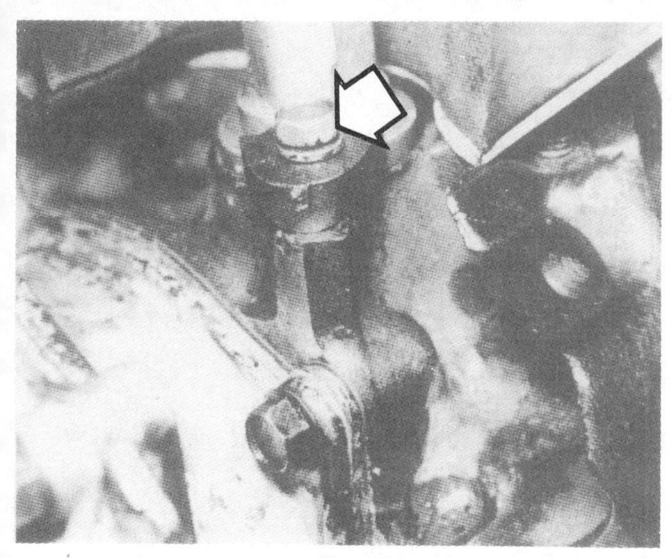

12.5 The distributor is attached to the block or manifold with a clamp and bolt (arrow)

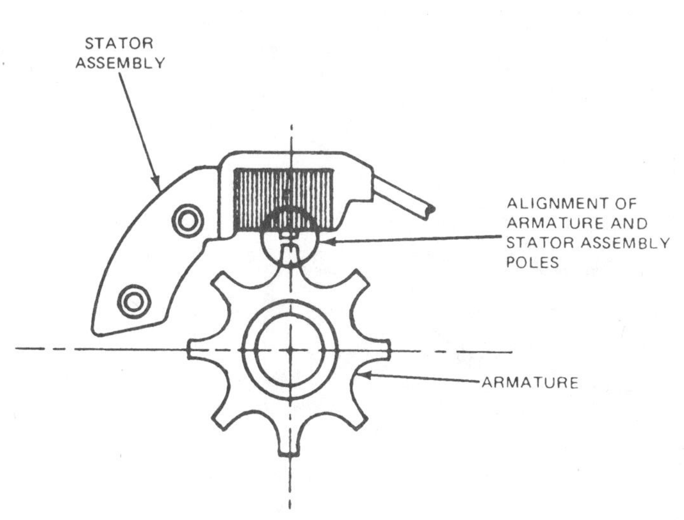

12.20 As the distributor is seated, the stator and armature poles must be aligned as shown (electronic ignition system)

12 Distributor – removal and installation

Refer to illustrations 12.5 and 12.20

Conventional breaker point distributor

Removal

1 Remove the air cleaner and duct assembly.
2 Release the two spring clips and remove the distributor cap and rotor.
3 Disconnect the primary wire and vacuum hose from the distributor body.
4 Rotate the crankshaft until the timing pointer is lined up with the timing mark on the crankshaft pulley. Check that the distributor rotor is lined up with the index mark on the top edge of the distributor body (that is, the rotor is pointed at the number one spark plug terminal in the distributor cap). **Note:** *If the rotor does not line up with the number one spark plug wire lead, rotate the crankshaft 360°.*
5 Remove the distributor hold-down bolt (see illustration) and lift out the distributor. **Note:** *The oil pump driveshaft may stick in the end of the distributor shaft and be withdrawn from the pump. If so, it must be reinstalled before the distributor is installed.*

Installation

6 If the oil pump driveshaft was withdrawn with the distributor, coat one end of it with grease and insert it into the end of the distributor driveshaft.
7 Insert the distributor into the engine block hole, making sure the pump driveshaft is correctly seated in the pump.
8 Notice that the rotor rotates as the gears mesh. The rotor must settle in exactly the same location it was in before the distributor was removed. To do this, lift out the distributor far enough to rotate the shaft one tooth at a time. When the rotor points in the desired direction with the distributor seated, install the distributor clamp plate, bolt and washer.
9 Reconnect the primary wire. Reconnect the lead to the center of the distributor cap and install the vacuum hose which runs from the intake manifold to the side of the vacuum advance unit.
10 Check the ignition timing as described in Chapter 1.

Duraspark II distributor

Removal

11 Disconnect the spring clips retaining the distributor cap to the adapter and position the cap and wires to one side.
12 Disconnect and plug the diaphragm assembly hose, if so equipped.
13 Unplug the distributor connector from the wiring harness.
14 Rotate the crankshaft to align the stator assembly pole and any armature pole.
15 Scribe a mark on the distributor body and engine block or intake manifold to indicate the position of the distributor in the engine and the position of the rotor in the distributor.
16 Remove the distributor hold-down bolt and clamp.
17 Pull the distributor out of the engine. **Note:** *The crankshaft should not be rotated while the distributor is out of the engine. However, if it is rotated, make sure you refer to the appropriate installation procedure below.*

Installation if crankshaft was rotated after distributor was removed

18 Rotate the crankshaft until the number 1 piston is at top dead center on the compression stroke (refer to Chapter 2 or Step 55 below).
19 Align the timing marks for the correct initial timing, determined by referring to the Emissions Control Information label in the engine compartment.
20 Install the distributor in the engine with the rotor pointing at the number 1 terminal in the cap and with the armature and stator assembly poles aligned (see illustration).
21 Make sure the oil pump intermediate shaft properly engages the distributor shaft.
22 If the distributor will not seat properly in the block or manifold, it may be necessary to crank the engine after the distributor gear is partially engaged in order to mesh the distributor shaft with the oil pump intermediate shaft and allow the distributor to seat.
23 If it is necessary to crank the engine, again rotate the crankshaft until the number 1 piston is at the top of the compression stroke and align the timing marks for correct initial timing.
24 Rotate the distributor in the engine to align the armature and stator assembly poles and verify that the rotor is pointing at the number 1 cap terminal.
25 Install the distributor hold-down clamp and bolt, but do not tighten it.

Installation if crankshaft was not rotated after distributor was removed and original distributor is being reinstalled

26 Install the distributor in the engine with the rotor and distributor aligned with the previously scribed marks. The armature and stator assembly poles should also align when the distributor is seated in the engine and properly installed.
27 If the distributor will not completely seat in the engine, crank the engine until the distributor shaft and oil pump intermediate shaft are meshed and the distributor is completely seated.
28 Install the distributor hold-down clamp and bolt, but do not tighten it.

Installation if crankshaft was not rotated after the distributor was removed and a new distributor is being installed

29 Install the distributor in the engine with the rotor aligned with the previously scribed mark on the block or manifold.

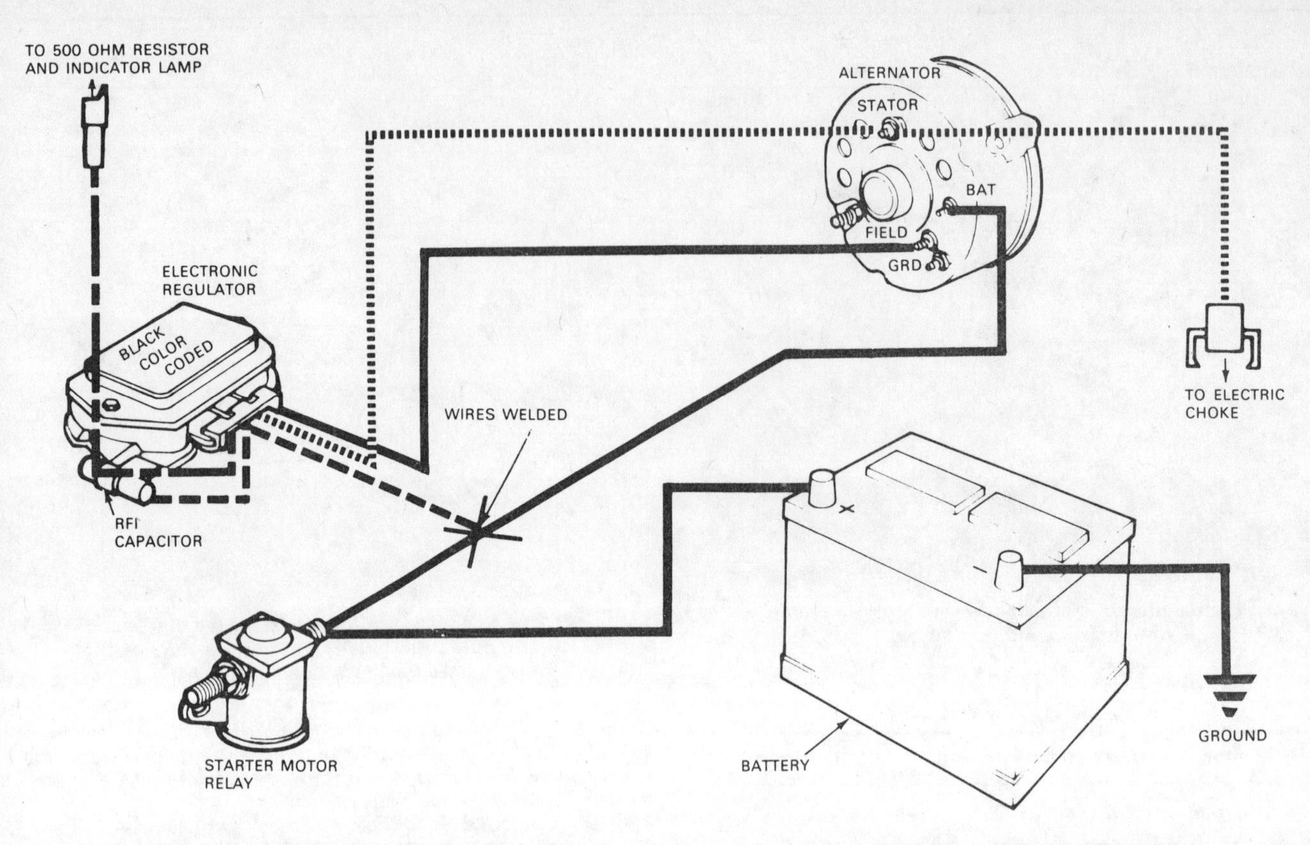

14.2a Typical charging system with electronic regulator and warning light indicator – 1979 model shown

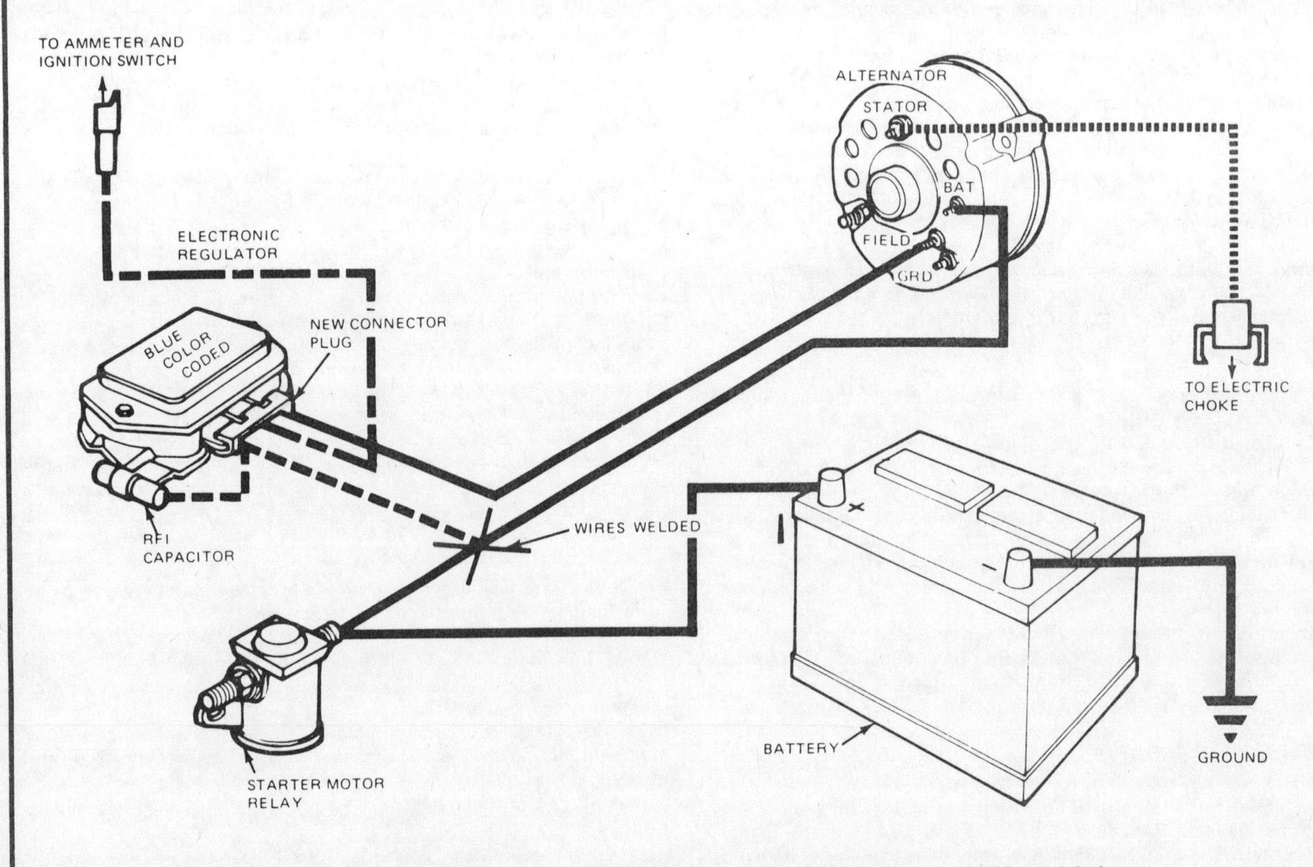

14.2b Typical charging system with electronic regulator and ammeter gauge – 1979 model shown

30 If necessary, crank the engine to seat the distributor.
31 Rotate the crankshaft until the timing marks for the correct initial timing (determined by referring to the Emissions Control Information label in the engine compartment) are aligned and the rotor is pointing at the number 1 cap terminal.
32 Rotate the distributor in the block to align the armature and stator assembly poles.
33 Install the distributor hold-down clamp and bolt, but do not tighten it completely at this time.

Installation (all circumstances)

34 If, in Steps 20, 24, 26 or 32 above, the armature and stator assembly poles cannot be aligned by rotating the distributor in the engine, pull the distributor out of the engine enough to disengage the distributor gear and rotate the distributor shaft to engage a different gear tooth, then reinstall the distributor and repeat the steps in the appropriate installation procedure as necessary.
35 Reattach the distributor connector to the wiring harness.
36 Install the distributor cap and ignition wires, making sure that the wires are securely connected to the distributor cap and spark plugs.
37 Set the initial timing according to the Emissions Control Information label located in the engine compartment.
38 Tighten the distributor hold-down bolt.
39 Recheck the initial timing and adjust as necessary (see Chapter 1).
40 Connect the vacuum hose, if so equipped.

Duraspark III distributor

Removal

41 Disconnect the spring clips retaining the distributor cap to the adapter and position the cap and spark plug wires to one side.
42 Remove the rotor from the distributor shaft.
43 Rotate the engine until the number 1 piston is at the top of the compression stroke (see Chapter 2 or Step 55 below) and the sleeve and adapter alignment slots are in line.
44 Remove the distributor hold-down clamp and bolt, then lift out the distributor. **Note:** *Do not rotate the engine with the distributor removed.*

Installation

45 Position the distributor in the engine so that the slot in the distributor base mounting flange is aligned with the hold-down bolt hole and the sleeve/adapter alignment slots are in line when the distributor is completely seated in the engine.
46 If the sleeve/adapter slots cannot be aligned, pull the distributor out of the engine enough to disengage the distributor gear and rotate the shaft to engage a different distributor gear tooth with the cam gear, then reinstall the distributor.
47 Install the distributor clamp and hold-down bolt and tighten the bolt.
48 Check the rotor alignment (refer to Section 9).
49 Install the distributor cap and wires, making sure that the wires are securely connected to the cap and spark plugs.

TFI distributor

Note: *Select models use a special distributor hold-down bolt. To remove the bolt you will need Ford tool number T82L-12270-A. A twelve point socket may also work.*

Removal

50 Unplug the distributor wiring harness connector.
51 Using paint or a scribe, mark on the distributor base the location of the number 1 plug wire tower to aid in reinstallation.
52 Remove the distributor cap and adapter from the distributor and position it and the ignition wires out of the way.
53 Remove the rotor.
54 Remove the distributor hold-down bolt and lift out the distributor.

Installation

55 Rotate the engine until the number 1 piston is at the top of the compression stroke. To do this, remove the number 1 spark plug (front cylinder on the right bank on V8 engines; front cylinder on six cylinder engines). With your thumb over the spark plug hole, slowly rotate the crankshaft until pressure is felt, indicating the number 1 piston is rising on the compression stroke.
56 Align the timing marks for the correct initial timing (see the Emissions Control Information label). Install the distributor in the engine.
57 Rotate the crankshaft until the rotor lines up with the mark on the base that you made earlier.
58 Continue rotating the crankshaft very slowly until the leading edge of the vane is centered in the vane switch stator assembly.
59 Rotate the distributor in the engine block to align the leading edge of the vane and the vane switch and verify that the rotor is pointing at the number 1 cap terminal. **Note:** *If the vane and vane switch stator cannot be aligned by rotating the distributor in the engine block, pull the distributor out of the block enough to disengage the gears and rotate the shaft to engage a different gear tooth.*
60 Install the distributor hold-down bolt and clamp. Do not tighten the bolt at this time.
61 Connect the distributor TFI and primary wiring harnesses.
62 Install the distributor rotor and tighten the screws.
63 Install the distributor cap adapter and tighten the screws.
64 Install the distributor cap and wires. Make sure the ignition wires are securely attached to the cap and spark plugs.
65 Set the initial timing, with a timing light, as described in Chapter 1.
66 Tighten the distributor hold-down bolt.
67 Check and readjust the timing if necessary.

13 Charging system — general information

The charging system is used to replenish the stored battery power that is consumed by running the starter, lights, accessories and ignition system. The alternator generates electrical power and is turned by a V-belt and pulley drive system that is driven by the crankshaft. It is generally located on the right side of the engine in varying positions depending on vehicle accessories.

The charging output is regulated by an external regulator on most models. Some later models are equipped with a regulator that is an integral part of the alternator. The regulator is connected to the alternator with a wiring harness utilizing quick release connectors. The system is protected from circuit overload by fusible links and the alternator is connected to the battery with heavy gauge wire. A charge indicator light or gauge is provided. Circuit diagrams for the charging system are provided at the end of Chapter 12

Adjustment of the drivebelt tension, along with battery terminal service, are the two primary maintenance items for the charging system. Details are provided in Chapter 1.

14 Charging system — check

Refer to illustrations 14.2a and 14.2b

1 As mentioned in the previous Section, the main components of the charging system are the alternator, voltage regulator and battery.
2 Little maintenance is required for this system to operate properly. A periodic check of the battery cables and connections (Chapter 1), alternator drivebelt tension (Chapter 1) and the various wiring and connectors is all that is necessary (see illustrations).
3 If an obvious charging system fault develops, the first checks should be those listed above. In many cases a fault is the result of a loose or corroded wire, loose drivebelt or other simple-to-fix problem. If, after all the visual checks have been performed, the system is still not charging the battery correctly, the system will have to be checked by a dealer service department or automotive electrical shop with the special diagnostic equipment required.
4 If the diagnostic tests indicate a faulty alternator, you can either replace it with a rebuilt unit as described in Section 15 or check and replace the brushes yourself as described in Section 16. Voltage regulators and batteries are not rebuildable and must be replaced with new parts.

15 Alternator — removal and installation

Refer to illustrations 15.2a and 15.2b

1 Disconnect the negative battery cable from the battery.
2 Carefully mark the wires and terminals at the rear or side of the

alternator and disconnect the wires. Most wires will have a retaining nut and washer holding them in place. However, some wires may have a plastic snap-fit connector with a retaining clip. If a terminal is covered by a slip-on plastic cover, be careful when pulling the cover back not to damage the terminal or connector (see illustrations).

3 Loosen the alternator adjustment arm bolt.

4 Loosen the alternator pivot bolt.

5 Pivot the alternator to allow the drivebelt to be removed from the pulleys.

6 Remove the adjustment arm bolt and pivot the arm out of the way.

7 Remove the pivot bolt and spacer and carefully lift the alternator up and out of the engine compartment. Be careful not to drop or jar the alternator as it can be damaged. **Note:** *If purchasing a new or rebuilt alternator, take the original one with you to the dealer or parts store so the two can be compared side-by-side.*

8 Installation is the reverse of removal. Be careful when connecting all terminals at the rear or side of the alternator. Make sure they are clean and tight and that all terminal ends are tight on the wires. If you find any loose terminal ends, make sure you install new ones, as any arcing or shorting at the wires or terminals can damage the alternator.

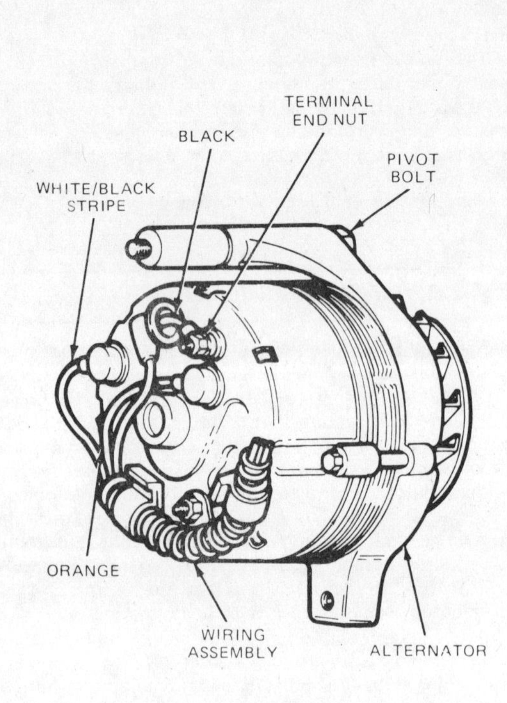

15.2a Typical rear terminal alternator wire harness connections

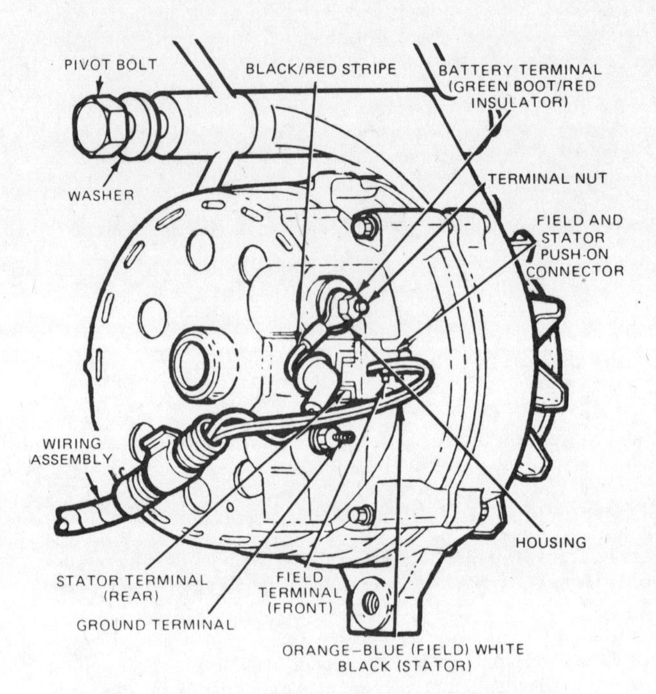

15.2b Typical side terminal alternator wire harness connections (with external regulator)

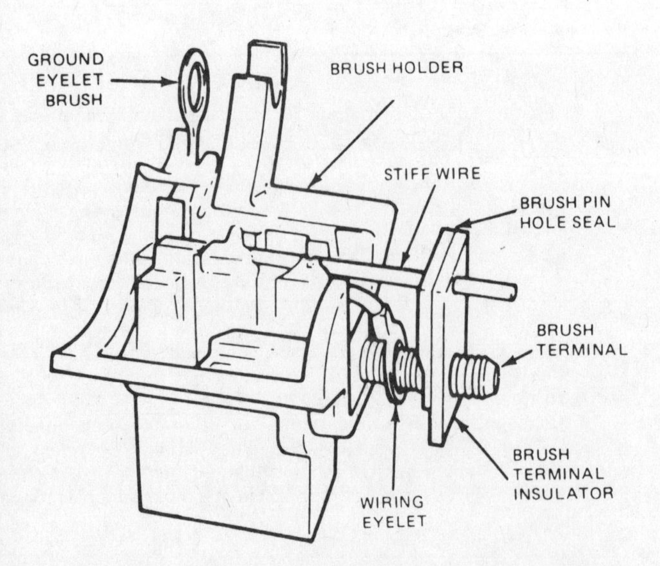

16.7 A stiff wire can be used to hold the alternator brushes in a retracted position while the brush holder assembly is reinstalled (rear terminal alternator shown)

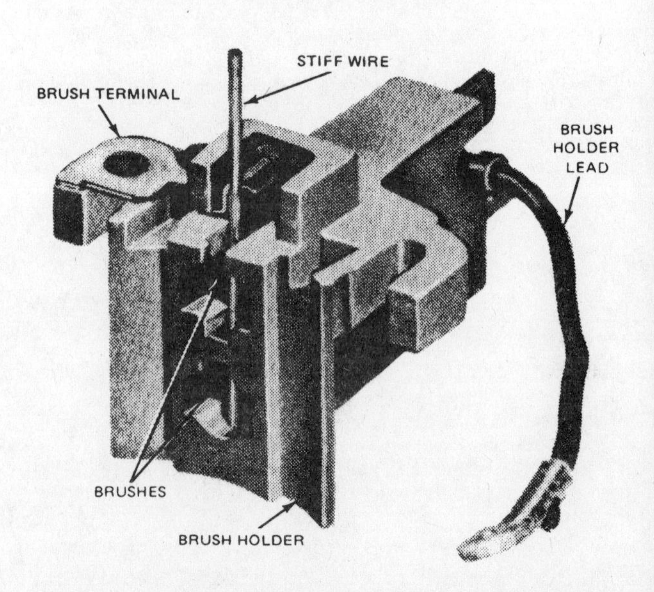

16.18 A stiff wire can be used to hold the alternator brushes in a retracted position while the brush holder assembly is reinstalled (side terminal alternator shown)

16 Alternator brushes — replacement

Refer to illustrations 16.7, 16.18, 16.24 and 16.25

Note: *Internal replacement parts for alternators may not be readily available in your area. Purchase the required parts before proceeding.*

Rear terminal alternator

1 Remove the alternator as described in Section 15.

2 Scribe a line across the length of the alternator housing to ensure correct reassembly.

3 Remove the housing through-bolts and the nuts and insulators on the rear housing. Make a careful note of all insulator locations.

4 Withdraw the rear housing section from the stator, rotor and front housing assemblv.

5 Remove the brushes and springs from the brush holder assembly, which is located inside the rear housing.

6 Check the length of the brushes and compare it to the Specifications at the beginning of this Chapter. Replace the brushes with new ones if necessary.

7 Install the springs and brushes in the holder assembly and retain them in place by inserting a piece of stiff wire though the rear housing and brush terminal insulator. Make sure enough wire protrudes through the rear housing so it can be withdrawn at a later stage (see illustration).

8 Attach the rear housing, rotor and front housing assembly to the stator, making sure the scribed marks are aligned.

9 Install the housing through-bolts and rear end insulators and nuts but do not tighten the nuts at this time.

10 Carefully extract the piece of wire from the rear housing and make sure that the brushes are seated on the slip ring. Tighten the through-bolts and rear housing nuts.

11 Install the alternator as described in Section 15.

Side terminal alternator

With external regulator

12 Remove the alternator as described in Section 15 and scribe a mark on both end housings and the stator to ensure correct reassembly.

13 Remove the through-bolts and separate the front housing and rotor from the rear housing and stator. Do not separate the rear housing and stator.

14 Use a soldering iron to unsolder and disengage the brush holder from the rear housing. Remove the brushes and springs from the brush holders.

15 Remove the two brush holder mounting screws and detach the brush holder from the rear housing.

16 Remove any sealing compound from the brush holder and rear housing.

17 Inspect the brushes for damage and measure the length. If they are worn out, replace them with new ones.

18 To reassemble, install the springs and brushes in the brush holders, inserting a piece of stiff wire to hold them in place (see illustration).

19 Place the brush holder in position in the rear housing.

20 Install the brush holder mounting screws and push the holder toward the shaft opening as you tighten the screws. Press the brush holder lead onto the rectifier lead and solder them in place. **Caution:** *The rectifier can be overheated and damaged if the soldering is not done quickly.*

21 Place the rotor and front housing in position in the stator and rear housing. After aligning the scribe marks, install the through-bolts.

22 Turn the fan and pulley to check for binding in the alternator.

23 Withdraw the wire which is retracting the brushes and seal the hole with waterproof cement. **Note:** *Do not use RTV-type sealant on the hole.*

With integral regulator

24 Remove the alternator as described in Section 15. The brush holder assembly is attached to the regulator, which is attached to the rear of the alternator with four Torx screws (see illustration).

25 After the screws have been removed and the regulator detached from the alternator, remove the *A* terminal insulator and the two regulator-to-brush holder Torx screws to separate the brush holder from the regulator (see illustration). Note how the brush leads are attached to the terminals.

26 Slide the brushes out of the holder and measure their length. If they are shorter than specified, replace them with new ones.

27 Reassembly is the reverse of disassembly. Be sure to insert the springs into the brush holder.

5

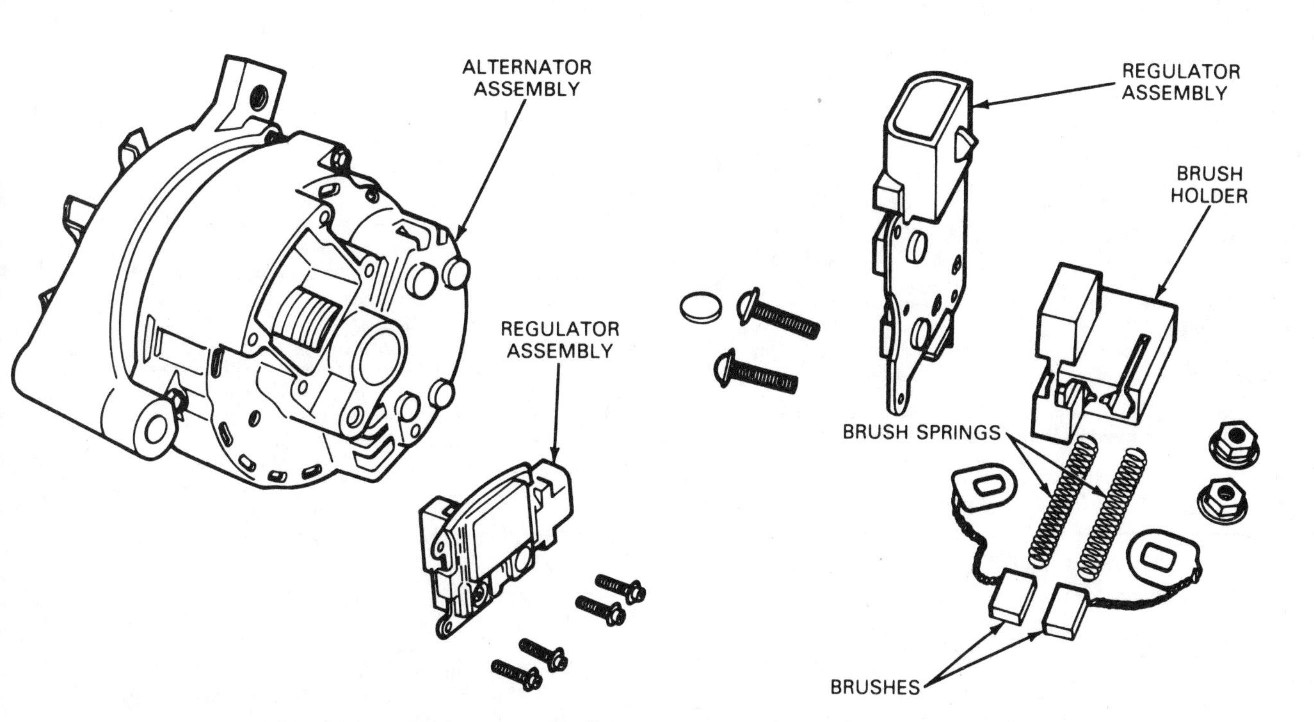

16.24 On vehicles with an integral regulator, remove the four Torx screws to detach the regulator from the alternator

16.25 The brush holder is attached to the regulator with two screws

17.2 The voltage regulator is typically mounted on the radiator support or fender well

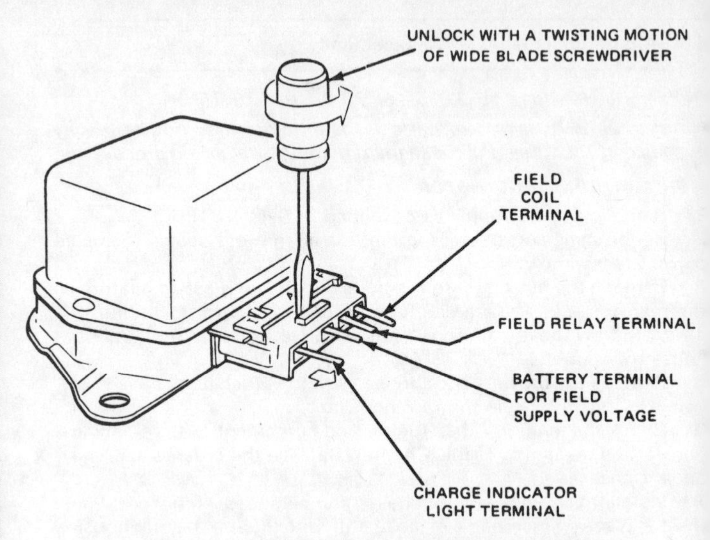

17.3 On some models, the regulator wiring harness can be detached by positioning a wide blade screwdriver between the tab on the connector and the edge of the receptacle – a twist of the screwdriver should push the connector sections apart

20.2 Loosen the nut and detach the cable from the starter before removing the starter mounting bolts

17 Voltage regulator – removal and installation

Refer to illustrations 17.2 and 17.3

External regulator

1 Remove the negative cable from the battery.
2 Locate the voltage regulator. It will usually be positioned on the radiator support or the fender well, near the front of the vehicle (see illustration).
3 Push the two tabs on either side of the quick release clip retaining the wiring loom to the regulator. Pull the quick release clip straight out from the side of the regulator. On some models, the regulator wiring harness connector can be detached with a screwdriver (see illustration).
4 Remove the two regulator mounting screws. Notice that one screw holds the ground wire terminal.
5 Remove the regulator.
6 Installation is the reverse of removal. Make sure you get the wiring clip positioned securely on the regulator terminals and that both tabs lock into place.

Integral regulator

7 The regulator is removed as part of the brush replacement procedure. Refer to Section 16.

18 Starting system – general information

The starting system consists of an electric starter motor with an integral positive engagement drive, the battery, a starter switch located inside the vehicle, a Neutral start switch (automatic transmission equipped vehicles only) a starter solenoid and wiring harnesses connecting the components.

When the ignition switch is turned to the Start position, the starter solenoid is energized through the starter control circuit. The solenoid then connects battery voltage to the starter motor.

Vehicles with an automatic transmission have a Neutral start switch in the starter control circuit which prevents operation of the starter if the gear selector lever is not in the N or P position.

When the starter is energized by the battery, current flows to the grounded field coil and operates the magnetic switch, which drives the starter drive plunger forward to engage the flywheel ring gear. When the drive plunger reaches a certain point in its travel, the field coil grounding contacts open and the starter motor contacts engage, allowing the starter to turn. A special holding coil is used to maintain the starter drive shoe in its fully seated position while the starter is turning the crankshaft. When the battery voltage is released from the starter, a retracting spring withdraws the starter drive pinion from the flywheel and the motor contact is broken.

19 Starter – in vehicle check

1 If the starter motor does not rotate when the starter switch is operated, make sure that the transmission selector lever is in N or P (automatic transmission vehicles only).
2 Check that the battery is fully charged and that all cables at the battery, starter and solenoid are tight and free of corrosion.
3 If the motor spins but the engine is not cranking, the starter drive is defective and the starter motor will have to be removed for drive replacement.

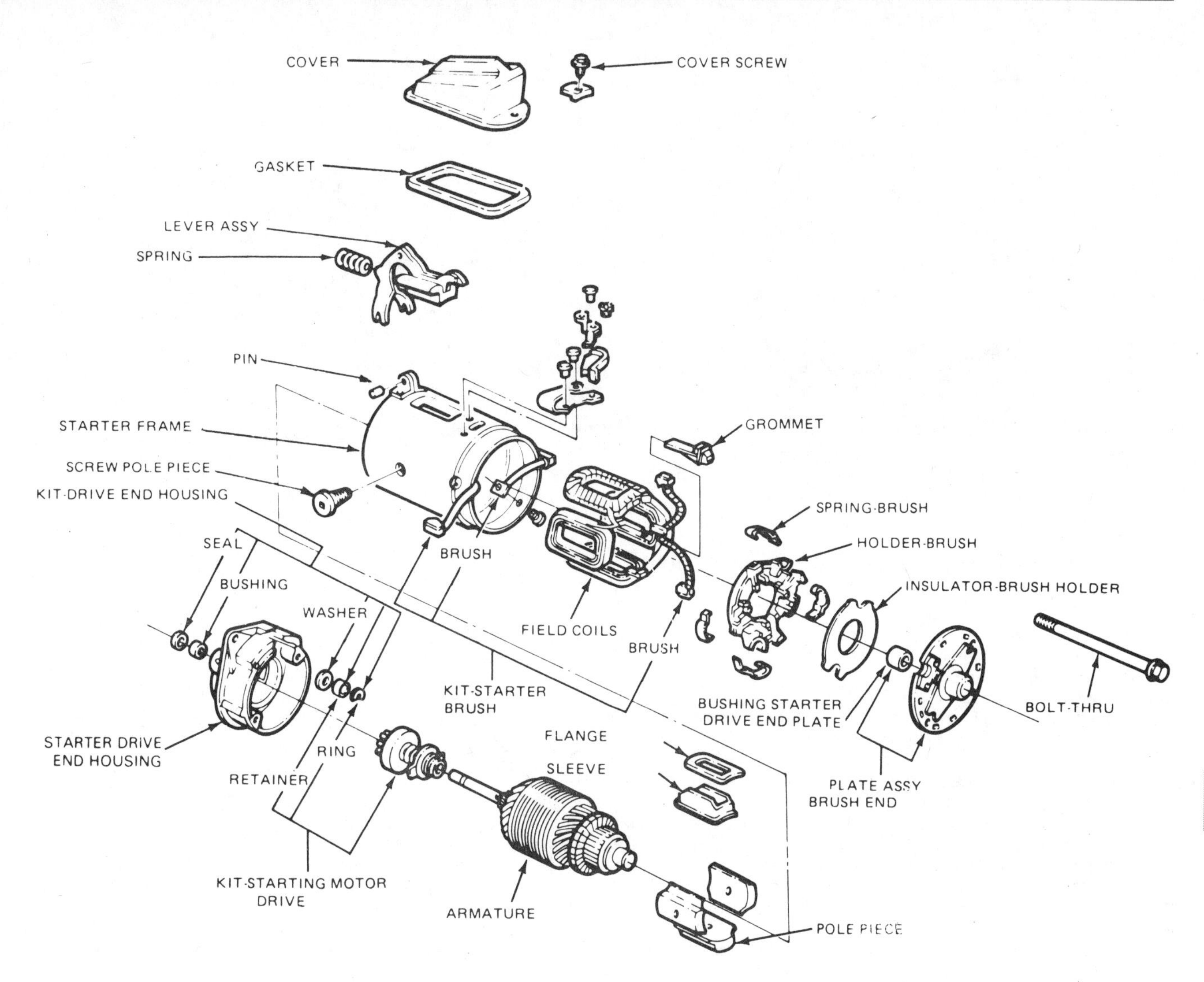

21.2 Typical starter motor components — exploded view

4 If the switch does not operate the starter at all but the drive can be heard engaging the flywheel with a loud "click", then the fault lies in the motor activating contacts within the motor itself. The motor will have to be removed and replaced or overhauled.
5 If the starter motor cranks the engine at an abnormally slow speed, make sure that the battery is fully charged and that all terminal connections are clean and tight. Verify that the engine oil viscosity is correct (not too thick) and the resistance is not due to a mechanical problem within the engine.
6 A voltmeter connected to the starter motor terminal of the solenoid and to ground will show the voltage being sent to the starter. If the voltage is adequate and the starter still turns slowly, the resistance is in the starter and the starter should be replaced or overhauled.
7 If a fault has been definitely traced to the starter, the original unit can be replaced by a rebuilt or new starter (Section 20) or the starter brushes can be checked and replaced (Section 21).

20 Starter — removal and installation

Refer to illustration 20.2

Note: *It may be helpful to raise the vehicle to gain access to the starter. If so, be sure to support it securely with jackstands.*

1 Disconnect the negative cable from the battery.
2 Disconnect the cable connecting the starter solenoid to the starter motor (see illustration).
3 Remove the retaining bolts securing the starter to the bellhousing.
4 Pull the starter out of the bellhousing and lower it from the vehicle.
5 Installation is the reverse of removal. When inserting the starter into the opening in the bellhousing, make sure it is situated squarely and that the mating surfaces are flush.

21 Starter brushes — replacement

Refer to illustrations 21.2 and 21.4

Note: *The starter must be removed from the vehicle before the brushes can be replaced. Before attempting to replace the brushes in the starter, make sure the problem you are having is related to the brushes. Often, loose connections, poor battery condition or wiring problems are a more likely cause of no start or poor starting conditions. Check on the availability of internal replacement parts before proceeding.*

1 Remove the starter from the vehicle (Section 20).
2 Remove the two through-bolts from the starter frame (see illustration).
3 Pull the brush endplate along with the brush springs and brushes from the holder.

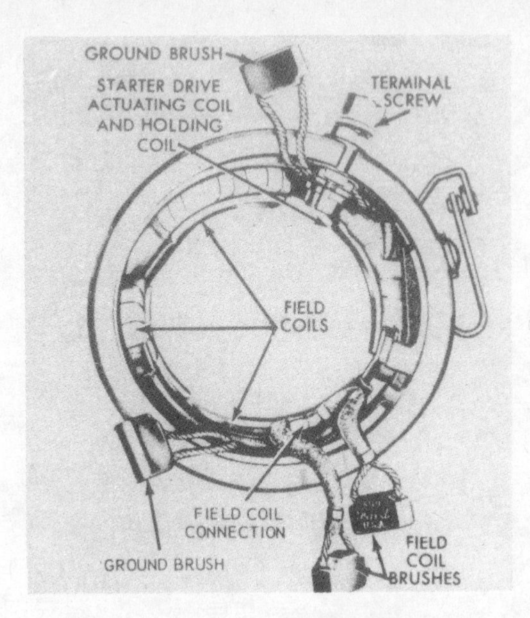

21.4 Starter motor ground and field coil brush locations

4 Remove the ground brush retaining screws from the frame. Remove the brushes from the frame (see illustration).
5 Cut the insulated brush leads from the field coils as close to the field connection point as possible.
6 Inspect the plastic brush holder for any signs of cracks or broken mounting pads. If these conditions exist, replace the plastic brush holder.
7 Place the new insulated field brush lead onto the field coil connection.
8 Crimp the clip provided with the brushes to hold the brush lead to the connection.
9 Using a low heat soldering gun (300 watts), solder the lead, the clip and the connection together using rosin core solder.
10 Attach the ground brush leads to the frame with the retaining screws.
11 Install the brush holder and insert the brushes into the holder.
12 Install the brush springs. Notice that the positive brush leads are positioned in their respective slots in the brush holder to prevent any chance of grounding the brushes.
13 Install the brush endplate in place. Make sure the endplate insulator is positioned correctly on the endplate.
14 Install the through-bolts in the starter frame and tighten them securely.
15 A battery can be used to check the starter by connecting jumper cables to the battery posts and the starter terminals.
16 Connect the ground lead to the starter body.

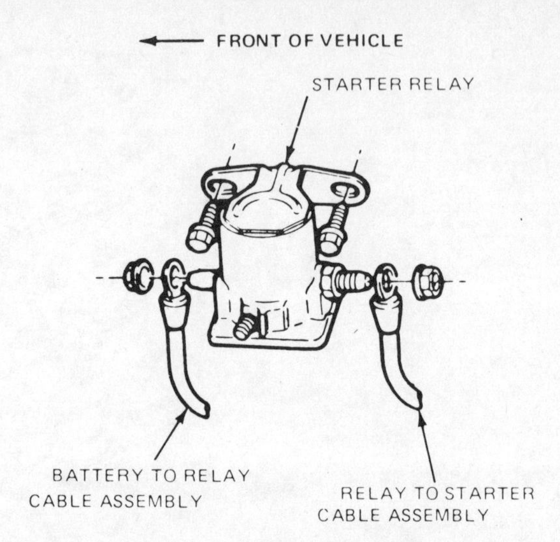

22.2 Typical starter solenoid (relay) mounting details

17 Secure the starter in a vise equipped with soft jaws.
18 Momentarily contact the starter connection with the positive cable from the battery.
19 The starter should spin and the solenoid drive should engage the gear in a forward position when this connection is made.
20 If the starter operates correctly, install the starter in the vehicle as described in the previous Section.

22 Starter solenoid (relay) — removal and installation

Refer to illustration 22.2

1 Disconnect the negative battery cable, followed by the positive cable, from the battery.
2 Disconnect the positive battery cable and the solenoid-to-starter cable from the terminals on the starter solenoid. Mark them to prevent mix-ups during installation (see illustration).
3 Disconnect the two starter solenoid control wires from the top posts on the solenoid. Make sure you mark or indicate the position of these wires as they can be cross-connected, which will damage the electrical system.
4 Remove the two starter solenoid mounting bolts and detach the starter solenoid.
5 Before installing the new or replacement solenoid, use a wire brush to carefully clean the mounting surface to ensure a good ground.
6 Install the starter solenoid and tighten the bolts. Use care when tightening the bolts as they are self-threading and can easily strip the mounting holes.
7 Reconnect all wires in their original positions.

Chapter 6 Emissions control systems

Refer to Chapter 13 for Specifications and information on 1987 and later models

Contents

1 General information

In order to meet US Federal anti-pollution laws, vehicles are equipped with a variety of emissions control systems, depending on the models and the states in which they are sold.

Since the emissions systems control so many engine functions, driveability and fuel consumption, as well as conformance to the law, can be affected if problems develop. Therefore, it is very important that the emissions system be kept operating at peak efficiency.

The information in this Chapter describes the subsystems within the overall emissions control system and the maintenance operations for these subsystems that are within the capability of the home mechanic. In addition, the Emissions Control Information label, located in the engine compartment, contains information required to properly maintain the emissions control system and keep the vehicle correctly tuned.

Due to the complexity of the subsystems, especially those under the control of electronic devices that require sophisticated electronic diagnosis, a Ford dealer service department should be contacted when you run into an emissions problem that cannot be readily diagnosed and repaired.

2 Positive Crankcase Ventilation (PCV) system

Refer to illustration 2.3

General description

1 The positive crankcase ventilation system is a closed recirculating system which is designed to prevent engine crankcase fumes from escaping into the atmosphere through the engine oil filler cap. The crankcase emission control system regulates blow-by vapors by circulating them back into the intake manifold, where they are mixed with the incoming fuel/air mixture.

2 The system consists of a replaceable PCV valve, a crankcase ventilation filter in the air cleaner and connecting hoses and gaskets.

3 The clean air source for the crankcase ventilation system is the air cleaner. Air passes through a filter located in the air cleaner to a hose connecting the air cleaner to the oil filler cap (see illustration). The oil filler cap is sealed at the opening to prevent the entrance of outside air. From the oil filler cap, the air flows into the rocker arm cover, down past the pushrods and into the crankcase. The air then circulates from the crankcase up into another section of the rocker arm cover. The air and crankcase gasses then enter a spring loaded regulator valve (PCV valve) which controls the amount of flow as operating conditions vary. Some engines have a fixed orifice PCV valve that meters a steady flow of gas regardless of the extent of engine blow-by gasses. In either case, the air and gas mixture is routed to the intake manifold through the crankcase vent hose tube and fittings. This process goes on continuously while the engine is running.

2.3 The PCV system inlet hose connects to the oil filler cap and the air cleaner (V8 engine shown — others similar)

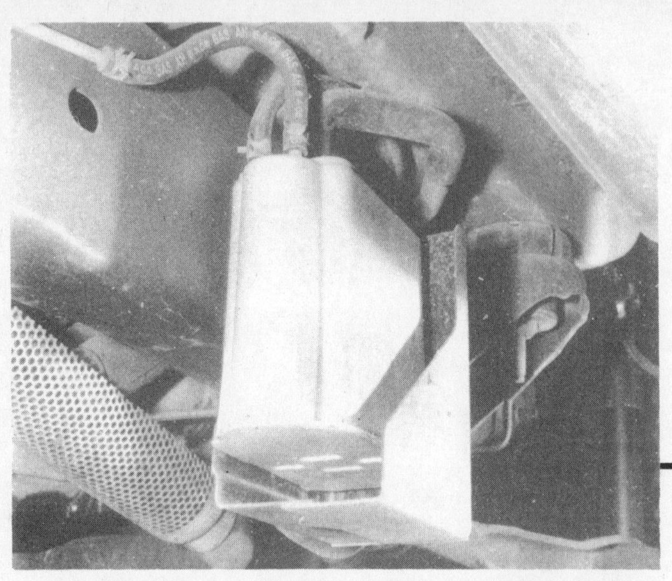

3.3 The ECS charcoal canister is bolted to the right frame rail in most instances and is protected by a metal shield

Checking

4 The checking procedure and additional illustrations for the PCV system are included in Chapter 1.

Component replacement

5 The replacement procedure for the PCV valve is included in Chapter 1.

3 Evaporative Control System (ECS)

Refer to illustrations 3.3 and 3.5

General description

1 This system is designed to limit fuel vapors released to the atmosphere by trapping and storing fuel that evaporates from the fuel tank

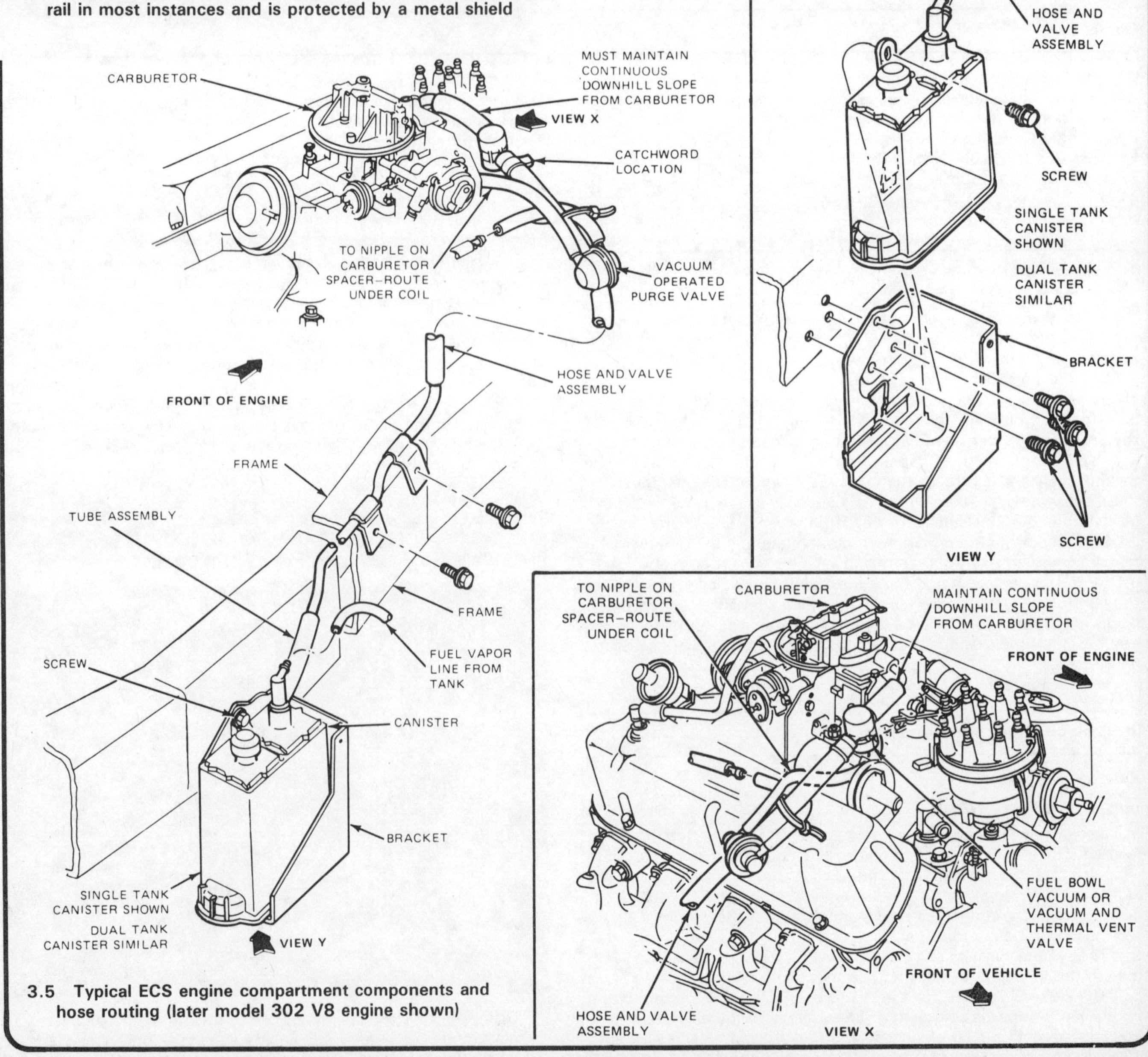

3.5 Typical ECS engine compartment components and hose routing (later model 302 V8 engine shown)

(and in some cases, from the carburetor) which would normally enter the atmosphere as hydrocarbon (HC) emissions.

2 The home mechanic can service parts of the system, including the charcoal filled canister, the lines to the fuel tank and carburetor and the fuel tank filler cap.

3 Fuel vapors are vented from the fuel tank (and in some cases also from the carburetor) for temporary storage in the canister, which is usually mounted on the right side frame rail in the engine compartment (see illustration). The canister outlet is connected to the carburetor air cleaner so the stored vapors will be drawn into the engine and burned. The fuel tank filler cap is a special design that vents air into the tank to replace the fuel being used, but does not vent fuel vapors to the outside air unless tank pressure builds up to more than approximately two psi above normal atmospheric pressure.

Checking

4 The checking procedures for the evaporative emissions control system are included in Chapter 1.

Component replacement

5 Replacement of the system canister and lines is accomplished by detaching the faulty component and replacing it with a new one. Be sure to mark all hoses and fittings prior to removal and use new clamps at all connections. Refer to the accompanying illustration of a typical ECS system.

4 Exhaust Gas Recirculation (EGR) system

Refer to illustrations 4.1, 4.3a 4.3b, 4.4a, 4.4b, 4.5, 4.7, 4.8 and 4.11

General description

1 The EGR system (see illustration) is designed to reintroduce small amounts of exhaust gas into the combustion cycle, reducing the generation of nitrous oxide (NOX) emissions. The amount of exhaust gas reintroduced, and the timing of the cycle, is controlled by various factors such as engine speed, altitude, engine vacuum, exhaust system back pressure, coolant temperature and throttle angle, depending on the engine calibration. All EGR valves are vacuum actuated and the vacuum diagram for your particular vehicle is shown on the Emissions Control Information label in the engine compartment.

2 For the vehicles covered by this manual there are three basic types of EGR valves: the ported valve, the integral back pressure transducer valve and the electronic sonic valve.

Ported valve

3 Two passages in the carburetor spacer block, connecting the exhaust system with the intake manifold, are blocked by the ported EGR valve, which is opened by vacuum and closed by spring pressure (see illustration). The valve may be either poppet or tapered stem design and may have a base entry or side entry, the function being the same (see illustration).

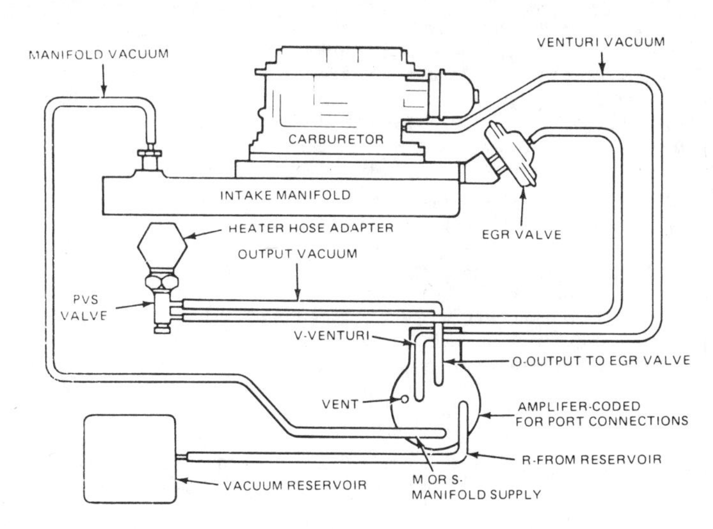

4.1 Typical EGR system diagram (early model shown)

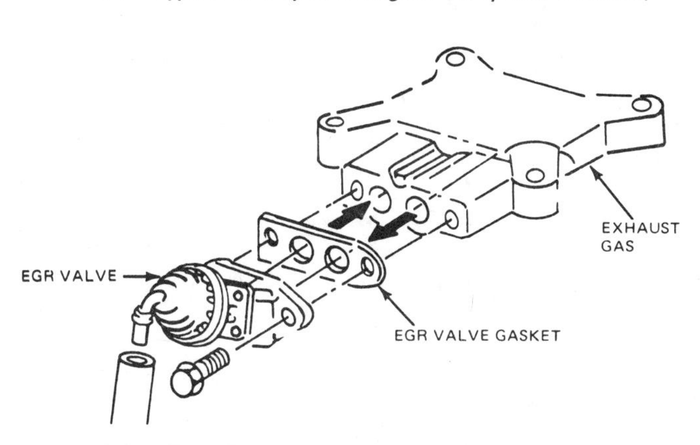

4.3a The EGR valve is normally bolted to the intake manifold near the carburetor

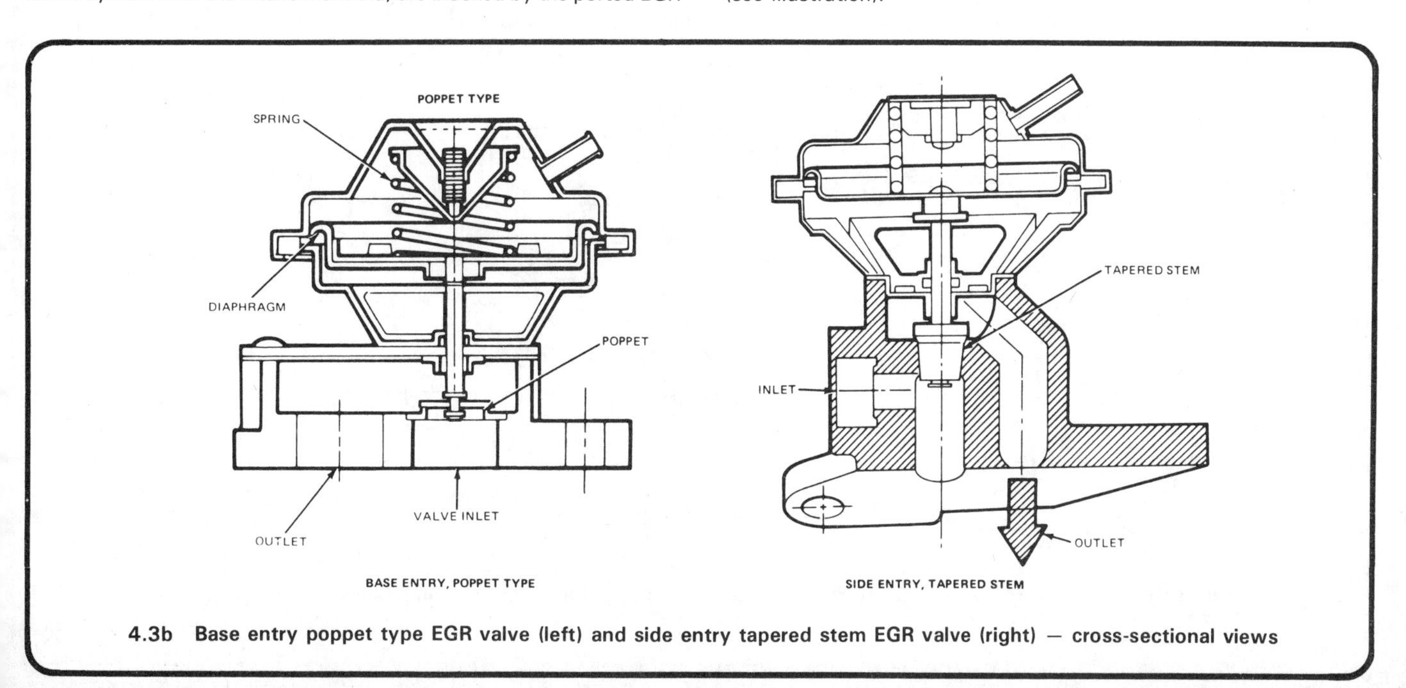

4.3b Base entry poppet type EGR valve (left) and side entry tapered stem EGR valve (right) — cross-sectional views

6

Integral back pressure transducer valve

4 This poppet-type or tapered (pintle) valve cannot be opened by vacuum until the bleed hole is closed by exhaust back pressure. Once the valve opens, it seeks a level dependent upon the exhaust back pressure flowing through the orifice and oscillates at that level. The higher the signal vacuum and exhaust back pressure, the more the valve opens. The valve body may be base entry or side entry (see illustrations).

Electronic sonic valve assembly

5 On vehicles equipped with an electronic engine control system, exhaust gas recirculation is controlled by the EEC through a system of engine sensors (see illustration). The EGR valve in this system resembles and is operated in the same manner as a conventional ported design valve. However, it uses a tapered pintle valve for more exact control of the flow rate, which is very closely proportional to the valve stem position. A sensor on top of the valve tells how far the EGR valve

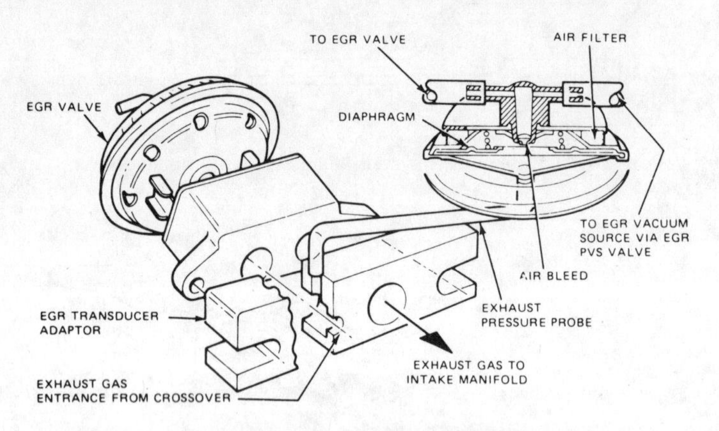

4.4a Some models are equipped with separate back pressure transducers as shown here, . . .

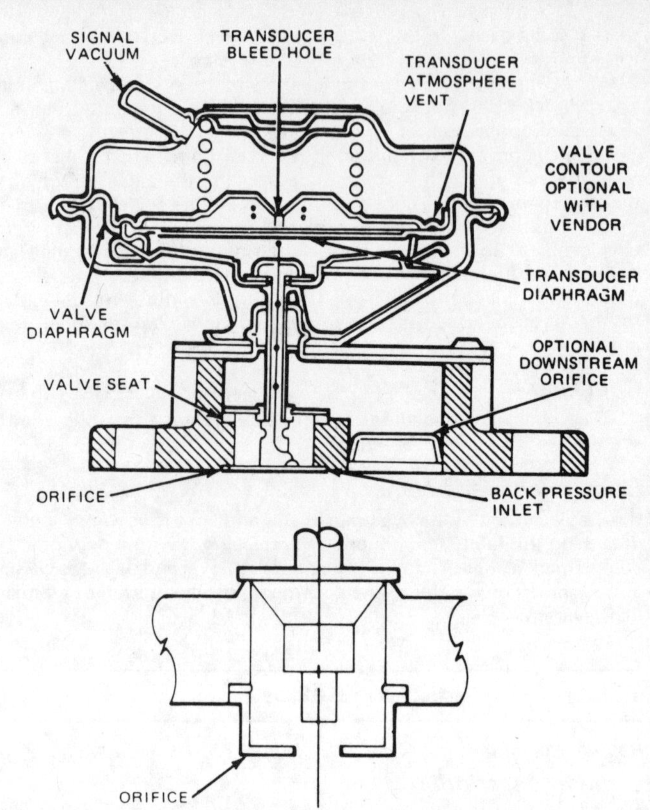

4.4b . . . while others have integral back pressure transducers as shown here

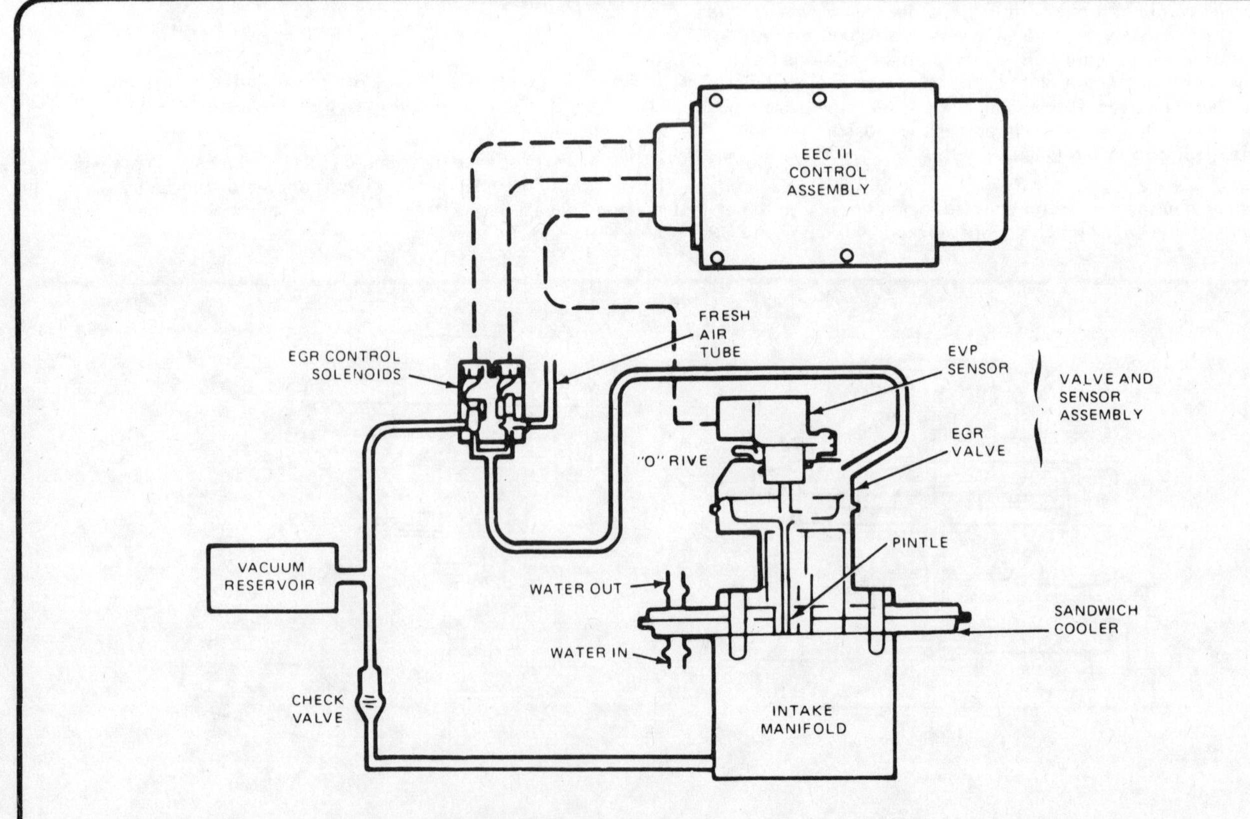

4.5 Diagram of a typical EGR system used with an EEC III system and incorporating an electronic sonic valve

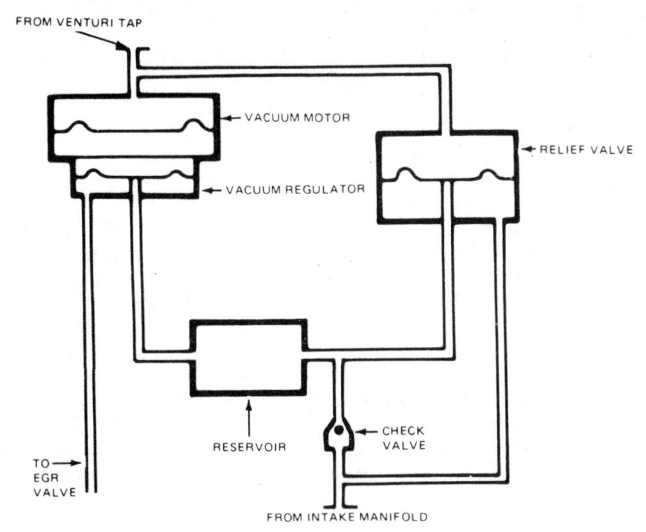

4.7 Vacuum diagram of a venturi vacuum amplifier and relief valve

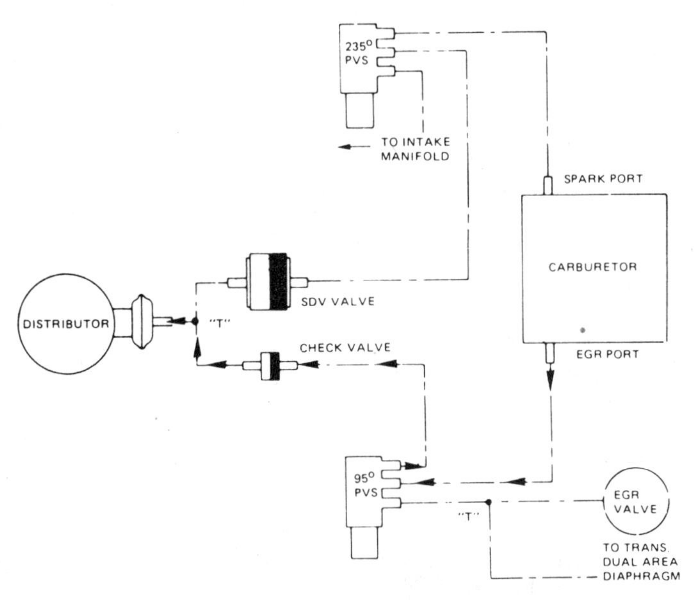

4.8 Vacuum routing with EGR/CSC system below 82°F

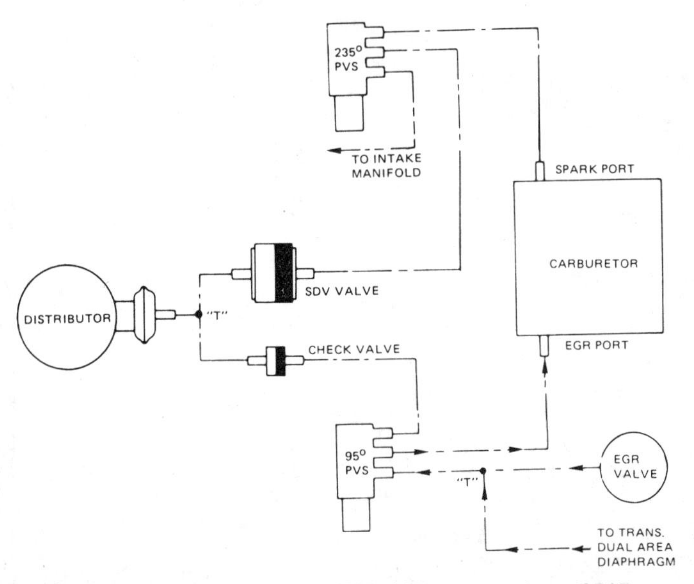

4.11 Vacuum routing with EGR/CSC system above 95°F

is open, sends an electrical signal to the EEC, which is also receiving several other signals such as temperature, rpm and throttle opening. The EEC then signals the EGR control solenoid to maintain or alter the flow as required by the engine operating conditions. Source vacuum from the manifold is either bled off or applied to the diaphragm depending on what the EEC commands. A cooler is used to reduce the temperature of the exhaust gasses, helping the gasses to flow better, reducing the tendency of the engine to detonate and making the valve more durable.

Operation

6 The EGR valve operates only in the part throttle mode. It remains closed in the cranking mode, the closed throttle mode and the wide open throttle mode.

7 On pre-1979 models a relief valve is used to modify the output EGR signal whenever venturi vacuum is equal to, or greater than, manifold vacuum. This allows the EGR valve to close at or near wide open throttle, when maximum engine power is required (see illustration).

8 The EGR/CSC (cold start cycle) regulates the distributor spark advance and EGR operation according to the engine coolant temperature by sequentially switching the vacuum signals. When the coolant temperature is below 82°F the EGR ported vacuum switch (PVS) admits carburetor EGR port vacuum (which occurs at approximately 2500 rpm) directly to the distributor advance diaphragm through the one-way check valve (see illustration). At the same time the PVS shuts off the carburetor vacuum to the EGR valve.

9 When the engine coolant is 95°F or above the EGR-PVS directs carburetor vacuum to the EGR valve.

10 At temperatures between 82 and 95°F the EGR-PVS may be closed, open or in the midposition.

11 A spark delay valve (SDV) is incorporated in the system to delay the carburetor vacuum to the distributor diaphragm unit for a predetermined time. During acceleration, little or no vacuum is admitted to the distributor diaphragm unit until acceleration is completed because of the time delay of the SDV and the rerouting of the EGR port vacuum at temperatures above 95°F (see illustration). The check valve blocks the vacuum signal from the SDV to the EGR-PVS so that carburetor vacuum will not be dissipated at temperatures above 95°F.

12 The 235°F PVS is not strictly part of the EGR system, but is connected to the distributor vacuum advance unit to prevent overheating while idling with a hot engine. At idle speeds no vacuum is generated at either of the carburetor ports and the engine timing is fully retarded. However, when the coolant temperature reaches 235°F the PVS is actuated to admit intake manifold vacuum to the distributor advance diaphragm. The engine timing is advanced, idling speed is increased and the engine temperature is lowered due to increased fan speed and coolant flow.

Checking

Ported valve and integral back pressure transducer valve

13 Make sure that all vacuum lines are properly routed, all connections are secure and that the hoses are not crimped, cracked or broken. If deteriorated hoses are encountered, replace them with new ones.

14 Visually inspect the valve for corrosion. If corrosion is encountered, clean the valve or replace it with a new one.

15 Using finger pressure, press in on the diaphragm on the bottom of the valve. If the valve sticks open or closed or does not operate smoothly, clean the valve or replace it with a new one.

Electronic sonic valve assembly

16 If a malfunction is suspected in the electronic sonic valve or the EVP sensor to which it is attached, have the system checked at a Ford dealer service department.

Cleaning

17 The EGR valve can be removed for cleaning, but if it is damaged, corroded or extremely dirty it is a good idea to install a new valve. If the valve is cleaned, make sure that the orifice in the body is clear but be careful not to enlarge it. If the valve can be dismantled, internal deposits can be removed with a rotary wire brush mounted in a drill motor. Deposits around the valve stem and discs can be removed by using a steel blade or shim approximately 0.028-inch thick in a sawing motion around the stem shoulder at both sides of the disc. Clean the cavity and passages in the main body and make sure the poppet valve moves freely.

Component replacement

18 When replacing any vacuum hoses, remove only one hose at a time and make sure that the replacement hose is of the same quality and size as the hose being replaced.
19 When replacing the EGR valve, label each hose as it is disconnected from the valve to ensure proper connection to the replacement valve.
20 The valve is easily removed from the intake manifold after disconnecting and labeling the attached hoses. Be sure to use a new gasket when installing the valve and checks for leaks when the job is complete.

5 Spark Control system

Note: *The information in this Section is applicable only to vehicles equipped with the Duraspark II ignition system. Because spark advance and retard functions on vehicles equipped with Duraspark III and TFI ignition systems are controlled by the EEC computer, checks and tests involving the spark control system on these vehicles must be performed by a Ford dealer service department.*

General description

1 The spark control system is designed to reduce hydrocarbon and oxides of nitrogen (NOX) emissions by advancing the ignition timing only when the engine is cold.
2 These systems are fairly complex and have many valves, relays, amplifiers and other components built into them. Each vehicle will have a system peculiar to the model year, geographic region and gross vehicle weight rating. A schematic diagram on the underside of the hood will detail the exact components and vacuum line routing of the particular system on your vehicle.
3 Depending on engine coolant temperature, altitude and the position of the throttle, vacuum is applied to either one or both of the diaphragms in the distributor vacuum unit and the ignition timing is changed to reduce emissions and improve cold engine driveability.

Checking

4 Visually check all vacuum hoses for cracks and hardening. Remove the distributor cap and rotor. Apply vacuum to the distributor advance port (and retard port, if so equipped) and see if the breaker or relay plate inside of the distributor moves. The plate should move opposite the direction of distributor rotation when vacuum is applied to the advance port and should move in the direction of rotation if vacuum is applied to the retard port (if so equipped).
5 Checking of the temperature relays, delay valves and other modifiers of the spark timing system is beyond the scope of the average home mechanic. Consult a Ford dealer service department if you suspect that you have other problems within the spark advance system.

Component replacement

6 When replacing any vacuum hoses, remove only one hose at a time and make sure that the replacement hose is of the same quality and size as the hose being replaced.
7 If it is determined that a malfunction in the spark control system is due to a faulty distributor, refer to Chapter 5 for the replacement procedure.

6 Thermactor system

Refer to illustrations 6.2a, 6.2b, 6.2c, 6.9, 6.20 and 6.26

General description

1 Thermactor systems are employed to reduce carbon monoxide and hydrocarbon emissions. Two types are found on Ford vans covered by this manual – a conventional thermactor air injection system and a managed air thermactor system.
2 The conventional thermactor air injection system consists of an air pump, an air bypass valve, a check valve, an air injection manifold and connecting hoses (see illustration). The managed air thermactor

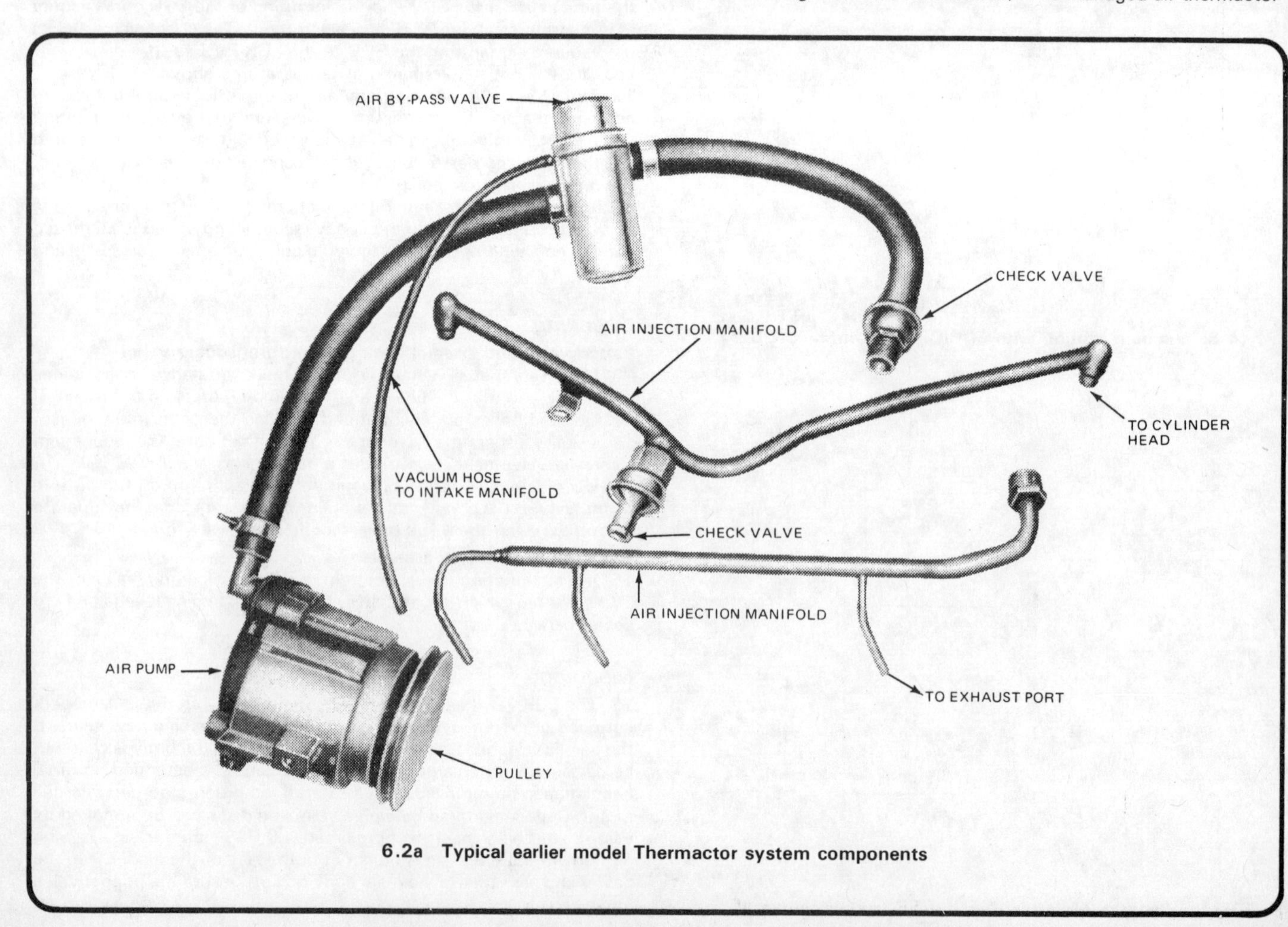

6.2a Typical earlier model Thermactor system components

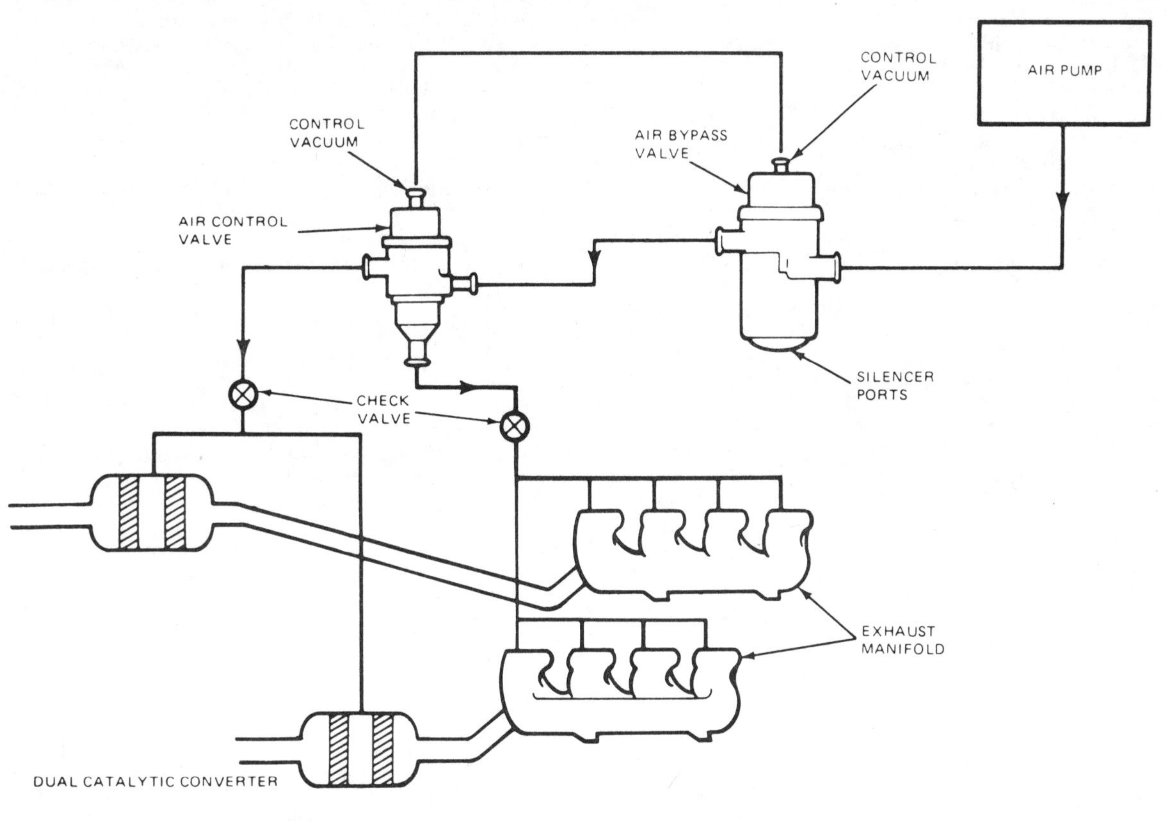

6.2b Diagram of a typical managed air Thermactor system with separate air control and air bypass valves

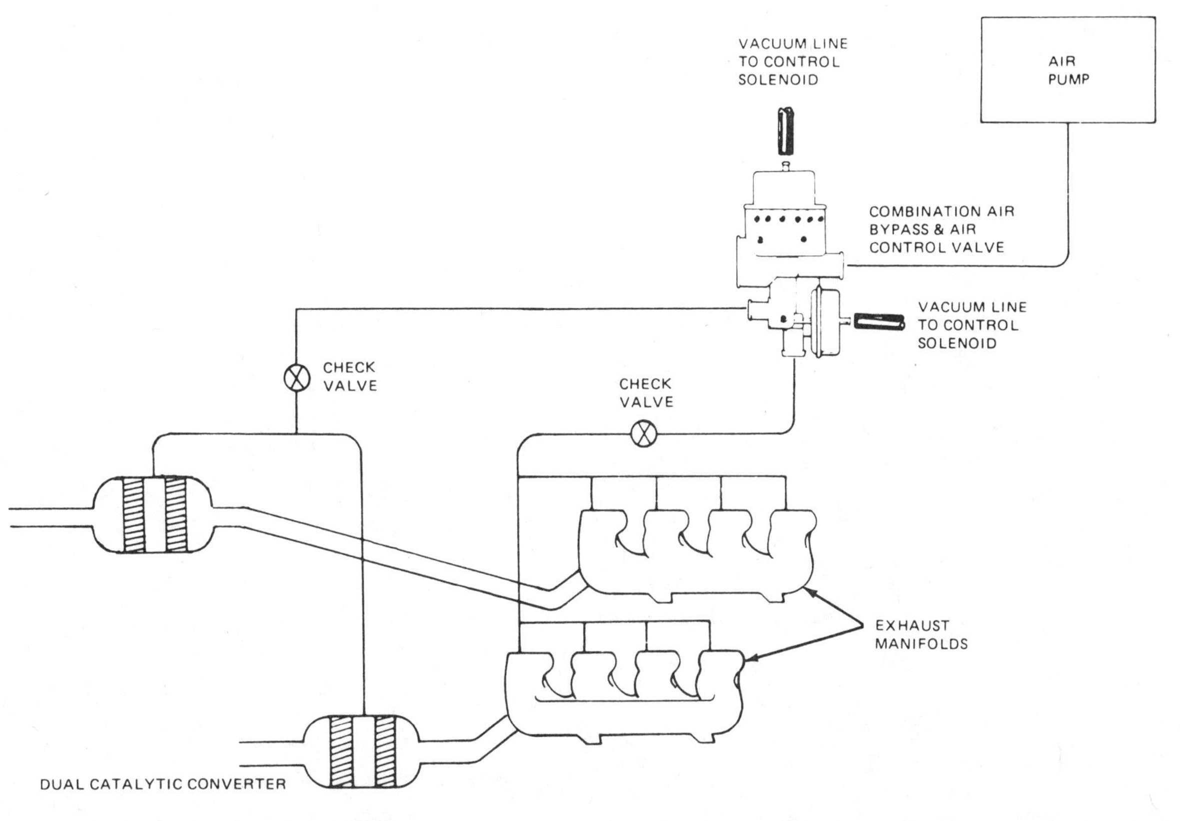

6.2c Diagram of a typical managed air Thermactor system with a combined air control/bypass valve

system adds a second check valve and an air control valve to the system. The air control valve may be incorporated with the air bypass valve into a single air/bypass control valve in some applications (see illustrations).

3 The thermactor air injection system functions by continuing combustion of unburned gasses after they leave the combustion chamber through an injection of fresh air into the hot exhaust system at the exhaust ports. At this point, the fresh air mixes with hot exhaust gas to promote further oxidation of both hydrocarbons and carbon monoxide, thereby reducing their concentration and converting some of them into harmless carbon dioxide and water. During some modes of operation, such as extended idle, the thermactor air is dumped into the atmosphere by the bypass valve to prevent overheating of the exhaust system.

4 The managed air thermactor system serves the same function as the thermactor air injection system, but, through the addition of the air control valve, diverts thermactor air either upstream to the exhaust manifold check valve, or downstream to the rear section check valve and on to the dual catalytic converter. The air is utilized in the converter to maintain feed gas oxygen content at a high level. This system comes in both electronic and nonelectronic controlled versions.

Checking (later models only)

Air supply pump

5 Check and adjust the drivebelt tension (refer to Chapter 1).

6 Disconnect the air supply hose at the air bypass valve inlet.

7 The pump is operating satisfactorily if air flow is felt at the pump outlet with the engine running at idle, increasing as the engine speed is increased.

8 If the air pump does not successfully pass the above check, replace it with a new or rebuilt unit.

Air bypass valve

9 With the engine running at idle, disconnect the hose from the outlet valve (see illustration).

10 Remove the vacuum line from the vacuum nipple and remove or bypass any restrictors or delay valves in the vacuum line.

11 Verify that vacuum is present in the vacuum line by putting your finger over the end.

12 Reconnect the vacuum line to the vacuum nipple.

13 With the engine running at 1500 rpm, air pump supply should be felt or heard at the air bypass valve outlet.

14 With the engine still running at 1500 rpm, disconnect the vacuum line. Air at the valve outlet should be decreased or shut off and air pump supply air should be felt or heard at the silencer ports.

15 Reconnect all disconnected hoses.

16 If the normally closed air bypass valve does not successfully pass the above tests, check the air pump (refer to Steps 5 through 7).

17 If the air pump is operating satisfactorily, replace the air bypass valve with a new one.

Air supply control valve

18 With the engine running at 1500 rpm, disconnect the hose at the air supply control valve inlet and verify the presence of air flow through the hose.

19 Reconnect the hose to the valve inlet.

20 Disconnect the hoses at the vacuum nipple and at outlets A and B (see illustration).

21 With the engine running at 1500 rpm, air flow should be felt at outlet B with little or no air flow through outlet A.

22 With the engine running at 1500 rpm, connect a line from any manifold vacuum fitting to the vacuum nipple.

23 Air flow should be present at outlet A with little or no air flow at outlet B.

24 Restore all connections.

25 If all conditions above are not met, replace the air control valve with a new one.

Combination air bypass/air control valve

26 Disconnect the hoses from outlets A and B (see illustration).

27 Disconnect the vacuum line at port D and plug the line.

28 With the engine running at 1500 rpm, verify that air flows from the bypass vents.

29 Unplug and reconnect the vacuum line to port D, then disconnect and plug the line attached to port S.

30 Verify that vacuum is present in the line to vacuum port D by momentarily disconnecting it.

31 Reconnect the vacuum line to port D.

32 With the engine running at 1500 rpm, verify that air is flowing out of outlet B with no airflow present at outlet A.

33 Attach a length of hose to port S.

34 With the engine running at 1500 rpm, apply vacuum by mouth to the hose and verify that air is flowing out of outlet A.

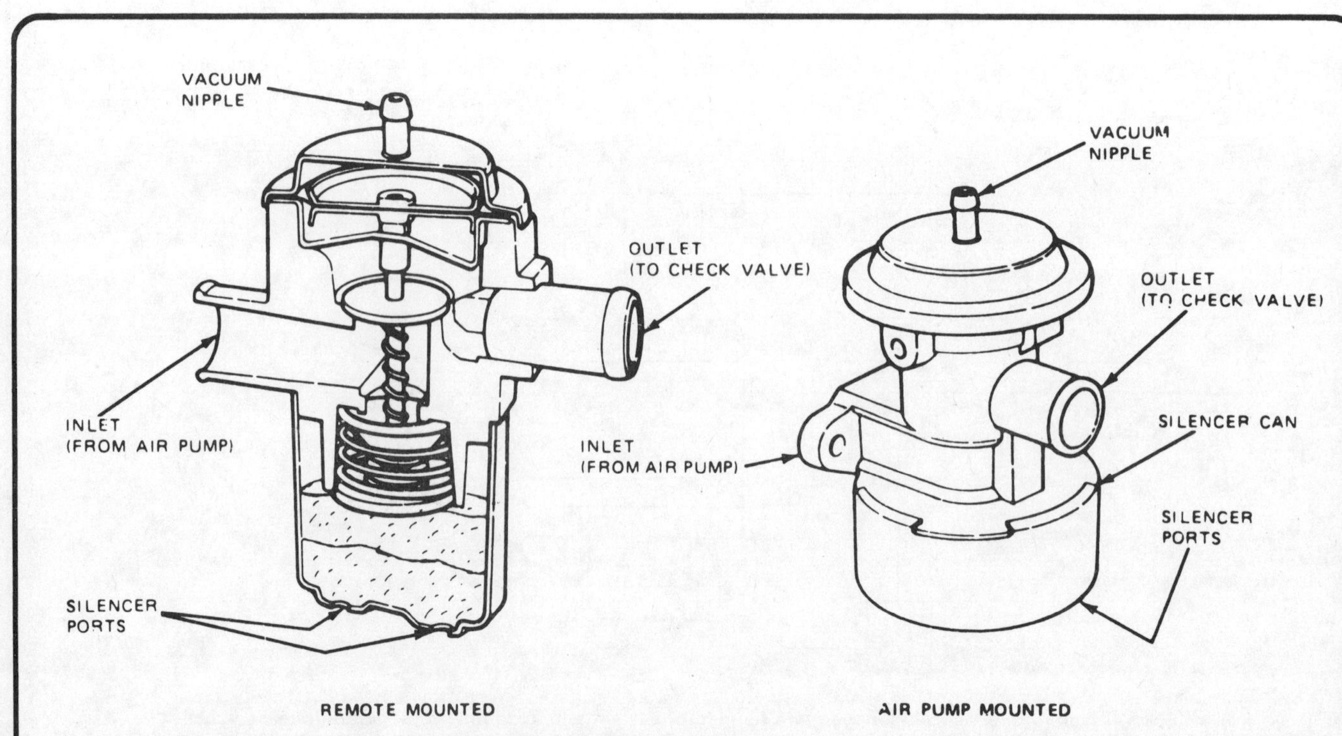

6.9 Two types of normally closed air bypass valves used in Thermactor systems — remote mounted (left) and air pump mounted (right)

35 Reconnect all hoses. Be sure to unplug the line to Port S before reconnecting it.
36 If all conditions above are not met, replace the combination valve with a new one.

Check valve

37 Disconnect the hoses from both ends of the check valve.
38 Blow through both ends of the check valve, verifying that air flows in one direction only.
39 If air flows in both directions, or not at all, replace the check valve with a new one.
40 When reconnecting the valve, make sure it is installed in the proper direction.

Component replacement

41 The air bypass valve, air supply control valve, check valve and combination air bypass/air control valves may be replaced by disconnecting the hoses leading to them (be sure to label the hoses as they are disconnected to facilitate reconnection), replacing the faulty component with a new one and reconnecting the hoses to the proper ports. Make sure that the hoses are in good condition. If they are not, replace them with new ones.
42 To replace the air supply pump, first loosen the appropriate engine drivebelt(s) (refer to Chapter 1), then remove the faulty pump from its mounting bracket, labeling all wires and hoses as they are removed to facilitate installation of the new unit.
43 After the new pump is installed, adjust the drivebelt(s) to the specified tension (refer to Chapter 1 Specifications).

7 Choke control system

General description

1 This system is installed to control emissions by precisely matching fuel supply to engine needs at all temperatures.
2 Choke system modifiers include a full electric choke or electrically assisted heating coil, and pulldown and pulloff assemblies. In addition, various control units which monitor engine temperature and choke position are utilized to provide appropriate choke action relative to the engine's ability to run on lean fuel mixtures as it warms up.
3 All of the systems combine to modify choke opening by lessening the time the choke plate is closed, thus lessening the amount of unburned hydrocarbons produced.

Checking

4 Information on the Emissions Control Information label located in the engine compartment and instructions in carburetor rebuild kits applicable to your particular carburetor can be used to service and rebuild these choke modifying devices. Also, refer to the Chapter 1 and the Chapter 4 carburetor and choke inspection and adjustment procedures.

Component replacement

5 Due to the variety of carburetors used on Ford vans covered in this manual, and the variety of choke control system components employed, it is not possible to include all the component replacement procedures. As is the case with general carburetor overhaul, the instructions accompanying the rebuild kit for your particular carburetor will explain these procedures.

8 Thermo-controlled air cleaner

Refer to illustrations 8.3a, 8.3b, 8.3c, 8.3d, 8.7, 8.8 and 8.12

General description

1 The air cleaner temperature control system is used to keep the air entering the carburetor at a warm and consistent temperature. The carburetor can then be calibrated much leaner for emissions reduction, improved warm-up and better driveability.
2 Two air flow circuits are used. They are controlled by various intake manifold vacuum and temperature sensing valves. A vacuum motor, which operates a heat duct valve (flapper door) in the air cleaner is

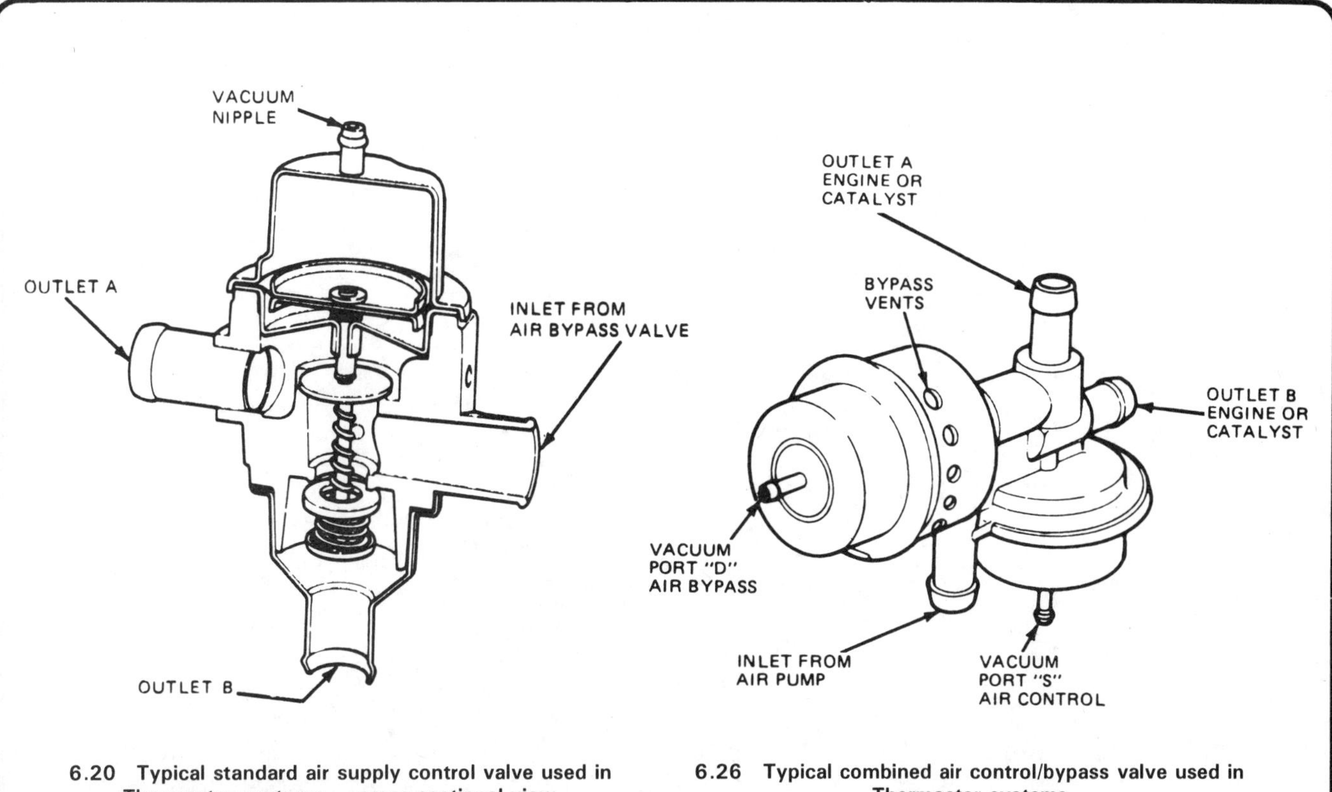

6.20 Typical standard air supply control valve used in Thermactor systems — cross-sectional view

6.26 Typical combined air control/bypass valve used in Thermactor systems

6

activated by these two previously mentioned circuits.

3 When the under hood temperature is cold, air is drawn through the shroud which fits over the exhaust manifold, up through the heat riser tube and into the air cleaner. This provides warm air for the carburetor, resulting in better driveability and faster warm-up. As the under hood temperature rises, the flapper door will be opened by the vacuum motor and the air that enters the air cleaner will be drawn through the cold air snorkel or duct. This provides a consistent intake air temperature (see illustrations).

4 An additional feature, incorporated on some models, is the Cold Temperature Actuated Vacuum (CTAV) system. It is designed to select either carburetor spark port vacuum or carburetor EGR port vacuum, as a function of ambient air temperature. The selected vacuum source is used to control the distributor vacuum advance diaphragm unit.

5 The system is composed of an ambient temperature switch, a three way solenoid valve, an external vacuum bleed and a latching relay.

6 The temperature switch activates the solenoid, which is open at temperatures below 49°F and is closed above 65°F. Within this

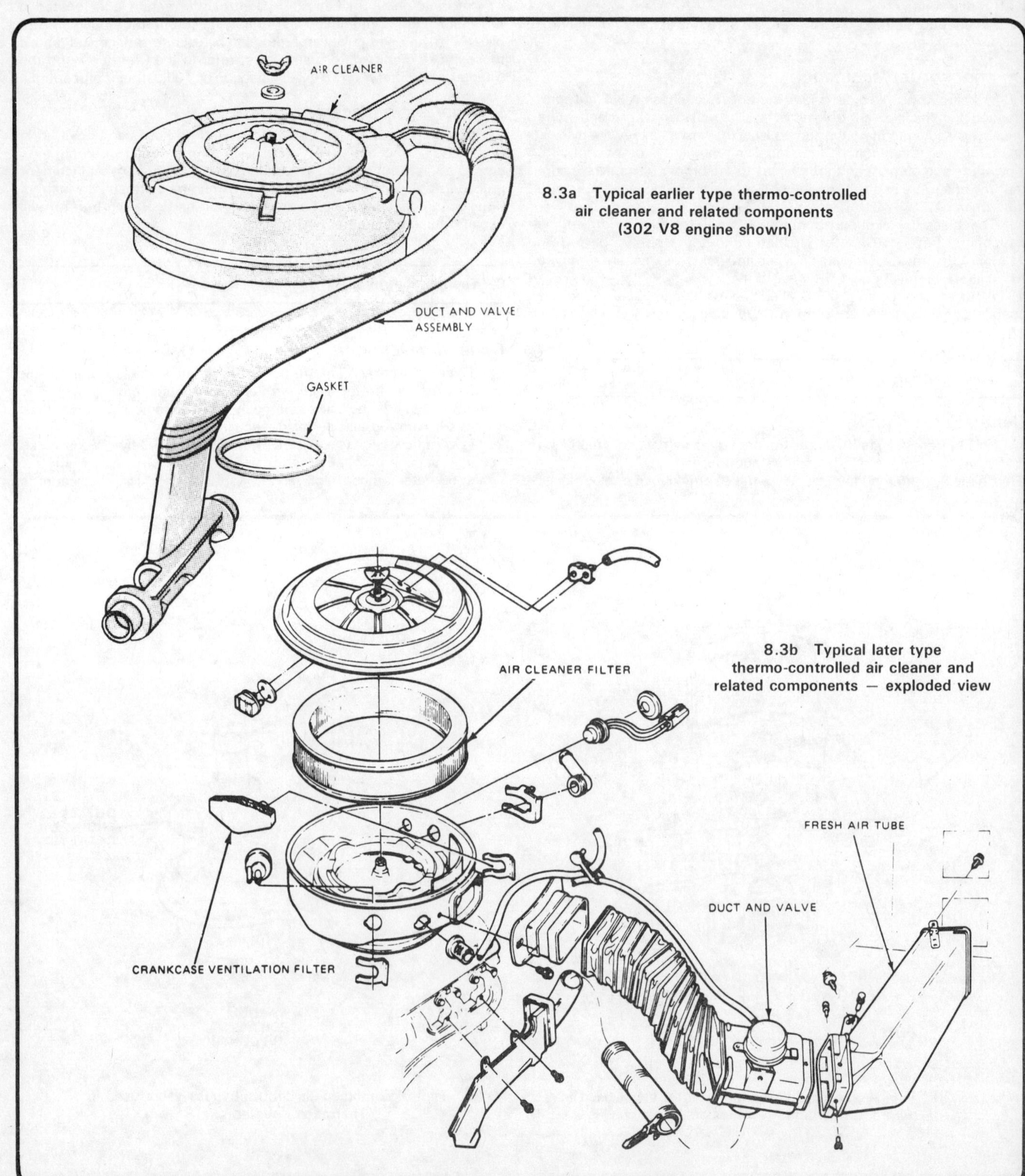

8.3a Typical earlier type thermo-controlled air cleaner and related components (302 V8 engine shown)

8.3b Typical later type thermo-controlled air cleaner and related components — exploded view

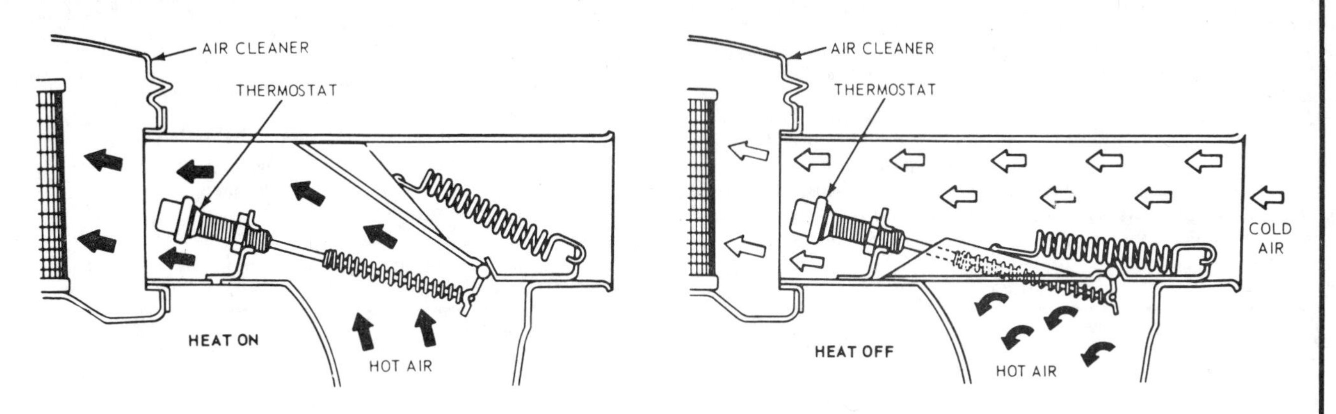

8.3c Operation of earlier type thermo-controlled air cleaner

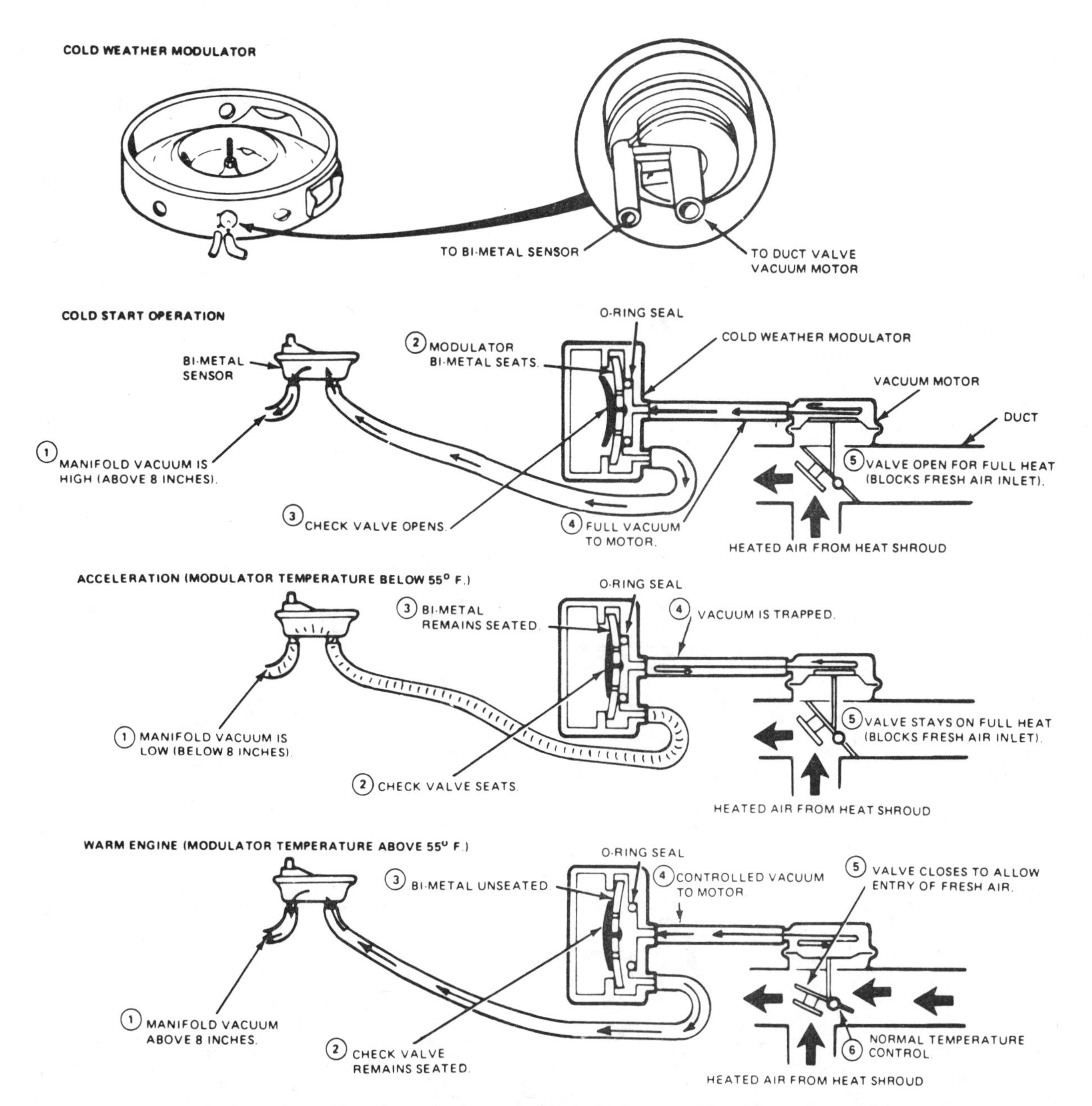

8.3d Operation of later type thermo-controlled air cleaner with cold weather modulator

6

temperature range the solenoid valve may be open or closed.

7 Below 49°F the system is inoperative and the distributor diaphragm receives carburetor spark port vacuum, while the EGR valve receives EGR port vacuum (see illustration).

8 When the temperature switch closes (above 65°F) the three way solenoid valve is energized from the ignition switch and the carburetor EGR port vacuum is delivered to the distributor advance diaphragm as well as to the EGR valve. The latching relay is also energized by the temperature switch closing, and will remain energized until the ignition switch is turned off, regardless of the temperature switch being open or closed (see illustration).

Checking

General

9 Refer to Chapter 1 for the thermo-controlled air cleaner check.

Flapper door

10 If the flapper door does not perform as indicated, see if it is rusted or stuck in an open or closed position by attempting to move it by hand. If it is rusted, it can usually be freed by cleaning and oiling it. Otherwise replace it with a new unit.

Vacuum motor

11 If the vacuum motor fails to open the flapper door, check carefully

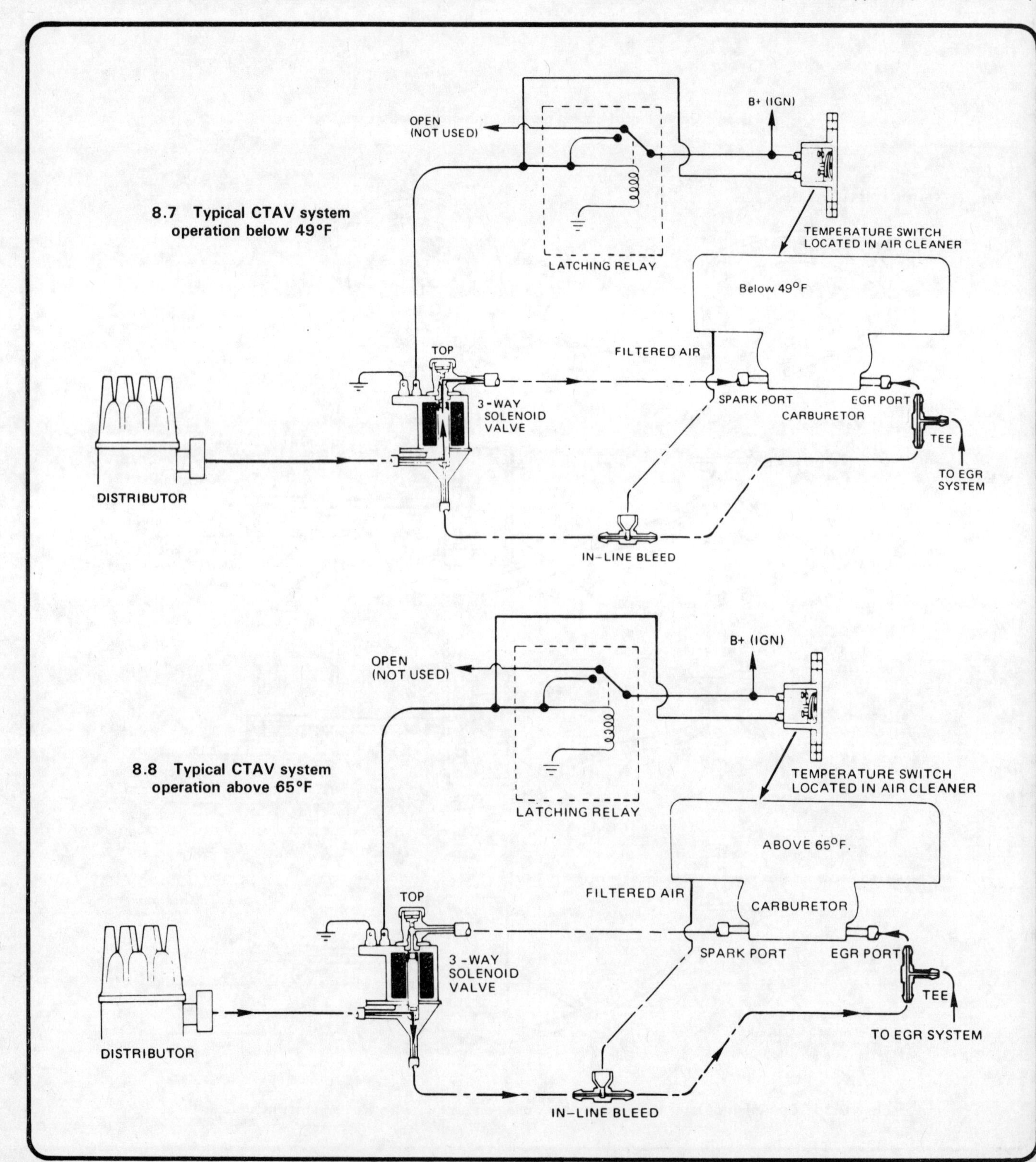

8.7 Typical CTAV system operation below 49°F

8.8 Typical CTAV system operation above 65°F

for a leak in the hose leading to it. If no leak is found, replace the vacuum motor with a new one.

CTAV system

12 Without special equipment the only possible check is electrical tests of the system circuitry (see illustration). Connect one terminal of a 12-volt low wattage bulb (such as an instrument panel bulb) to a good ground. Connect the other terminal at point B and remove the connector at point D. Turn on the ignition. If the light illuminates, replace the latching relay. If there is no light, reconnect at point D. There should now be a light. If not, check the temperature switch and the wiring back to the ignition switch. Provided that there is a light, disconnect at point D again. There should now be a light. If not, replace the latching relay. If it is possible to cool the temperature switch below 49 °F, check that the contacts are open at or below this temperature.

Component replacement

13 Replacement of the flapper door assembly and vacuum motor is accomplished by unbolting the faulty component, removing the vacuum lines leading to it (where appropriate) and installing the new component.

9 Decele tion throttle control system

Refer to illustrations 9.1, 9.3 and 9.4

General description

1 This system reduces the hydrocarbon and carbon monoxide contents of exhaust gasses by opening the throttle slightly during deceleration. On early models, the deceleration valve supplies a rich mixture to the intake manifold when decelerating with the throttle closed (see illustration).

2 The system consists of a throttle positioner, a vacuum solenoid valve, a vacuum sensing switch and electrical links to an electronic speed sensor/governor module, although not all components are in-

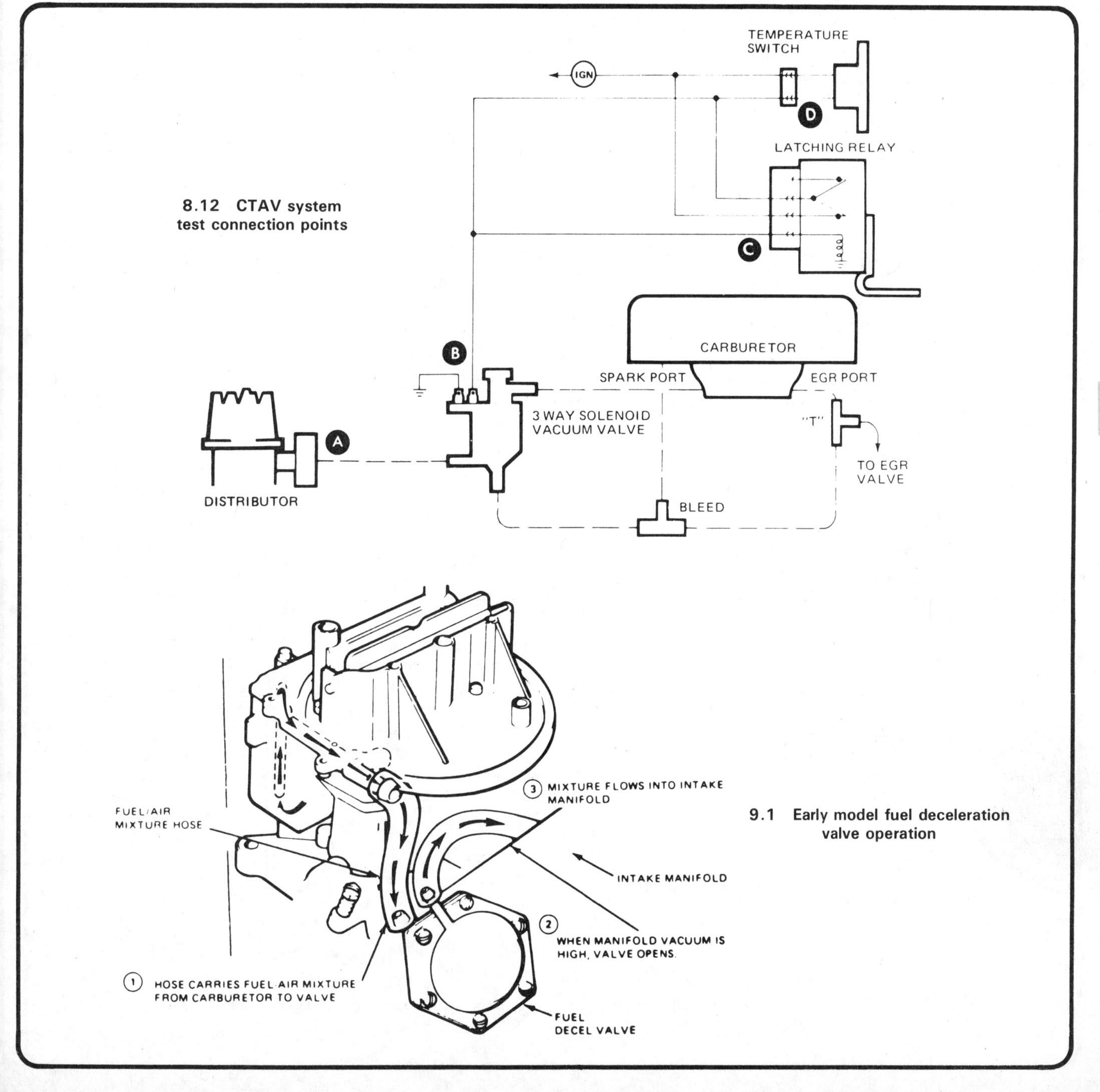

8.12 CTAV system test connection points

9.1 Early model fuel deceleration valve operation

cluded in all systems.

3 On those models without a vacuum sensing switch, when the engine speed is higher than a predetermined rpm a signal is sent to the solenoid, which allows manifold vacuum to activate the throttle positioner (see illustration).

4 On those models with a vacuum sensing switch, when the engine speed is higher that a predetermined rpm and manifold vacuum is at a certain value, the vacuum switch sends a signal to the module, then to the solenoid, allowing manifold vacuum to activate the throttle positioner (see illustration).

Checking

General

5 All of the following checks are to be made with the engine at normal operating temperature and with all accessories off unless otherwise noted.

6 With the engine at idle, accelerate to 2000 rpm or more, then let the engine return to idle while watching to see if the vacuum diaphragm plunger extends and retracts. If the plunger performs as indicated, the system is functioning normally. If not, check the throttle positioner.

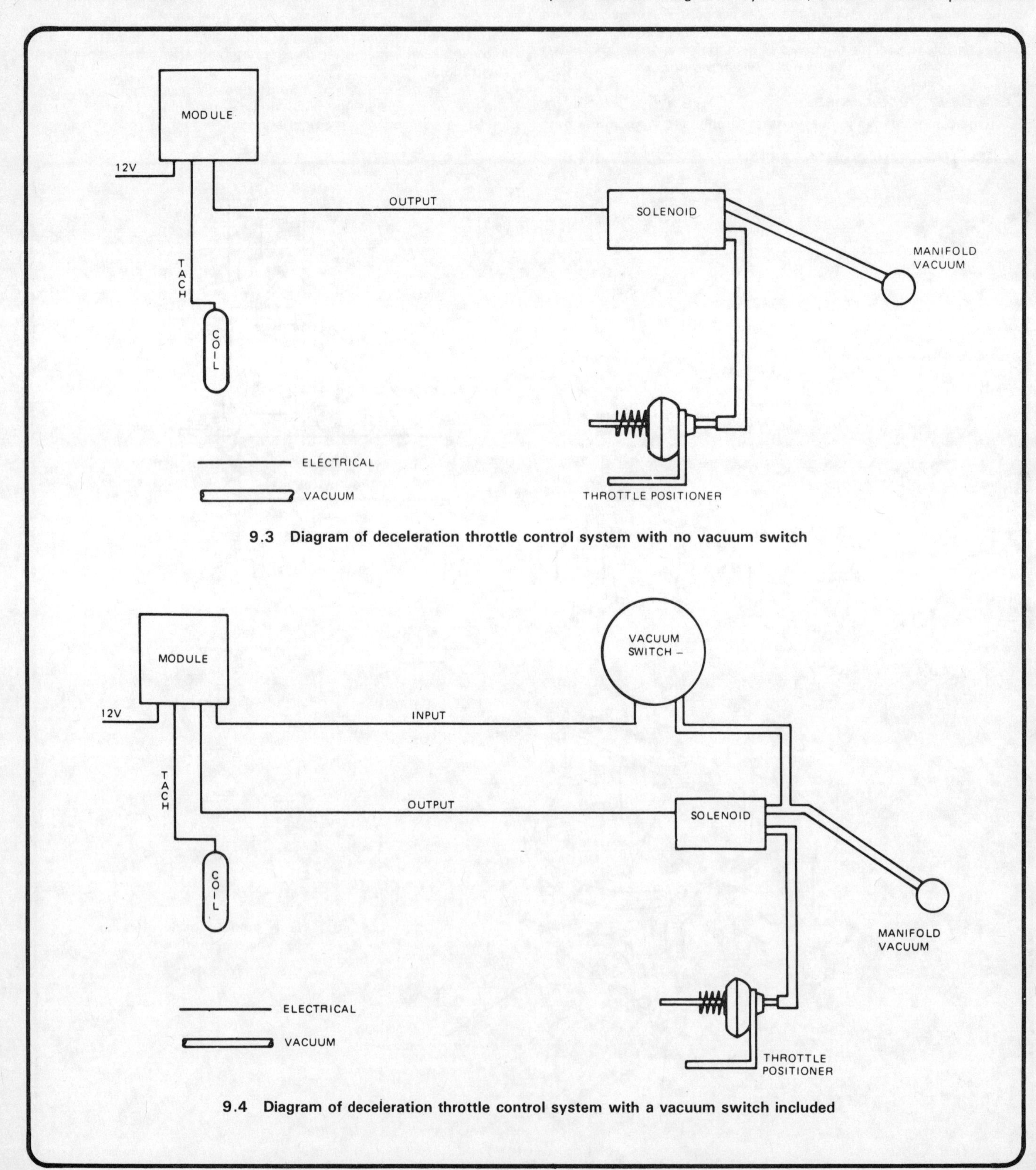

9.3 Diagram of deceleration throttle control system with no vacuum switch

9.4 Diagram of deceleration throttle control system with a vacuum switch included

Throttle positioner

7 Remove the hose from the throttle positioner and, using a hand operated vacuum pump, apply 19 in-Hg vacuum to the diaphragm and trap it. If the diaphragm does not respond or will not hold vacuum, replace the throttle positioner with a new one.

8 Remove the vacuum pump from the diaphragm and see if the diaphragm returns within five seconds. If not, replace the throttle positioner with a new one.

Vacuum solenoid valve

9 With the engine at idle, disconnect the vacuum supply hose from the solenoid valve and verify that the hose is supplying vacuum.

10 Disconnect the wires from the solenoid and apply battery voltage to both terminals, verifying that there is no increase in engine speed.

11 With battery voltage supplied to one terminal, ground the other terminal and verify that engine speed increases.

12 Remove the ground and verify that the engine returns to idle rpm.

13 If the solenoid fails to pass any of the above tests, replace it with a new one.

Vacuum sensing switch (if so equipped)

14 Disconnect the hose from the vacuum fitting on the switch.

15 Hook up a hand operated vacuum pump to the fitting.

16 Using an ohmmeter, verify that the switch is open (no continuity) while applying 19.4 in-Hg vacuum or less to the switch.

17 Increase the vacuum to 20.6 in-Hg or more and verify that the switch is closed (continuity).

18 If the switch fails to pass either of the above tests, replace it with a new one.

Electronic speed sensor/governor module (if so equipped)

19 If a fault is suspected in the module it must be checked by a Ford dealer service department.

Component replacement

20 Replacement of the throttle positioner, solenoid valve and sensing switch is accomplished by disconnecting the attached wires and/or hoses, removing the faulty component and replacing it with a new one.

10 Exhaust control system

General description

1 This system is employed to eliminate condensation of fuel on the cold surfaces of the intake tract during cold engine operation and to provide better evaporation and distribution of the air/fuel mixture. The result is better driveability, faster warm-up and a reduction in the release of hydrocarbons to the atmosphere.

2 The main component of this system is the exhaust heat control valve, which is controlled by either a thermostatic spring or a vacuum motor. It is mounted between the exhaust manifold and one branch of the exhaust pipe.

3 The valve operates by remaining closed when the engine is cold, routing hot exhaust gasses to the intake manifold through the heat riser tube, then slowly opening as the engine temperature increases.

Checking

4 Refer to Chapter 1 for the exhaust heat control valve checking procedure.

Component replacement

5 The replacement of the exhaust heat control valve is accomplished by unbolting the faulty valve from its location, cleaning the mating surfaces on the exhaust manifold and exhaust pipe and installing a new valve. Penetrating oil may be required on the nut/stud threads.

11 Catalytic converter

General description

1 The catalytic converter is designed to reduce hydrocarbon (HC) and carbon monoxide (CO) pollutants in the exhaust. The converter oxidizes these components and converts them to water and carbon dioxide.

2 The converter is located in the exhaust system and closely resembles a muffler. Some models have two converters, a light-off catalyst type, mounted just past the exhaust manifold pipe, and a conventional oxidation catalyst or three-way catalyst type mounted farther downstream.

3 **Note:** *If large amounts of unburned gasoline enter the converter it may overheat and cause a fire. Always observe the following precautions:*

Use only unleaded gasoline
Avoid prolonged idling
Do not prolong engine compression checks
Do not run the engine with a nearly empty fuel tank
Avoid coasting with the ignition turned Off

Checking

4 The catalytic converter requires little if any maintenance and servicing at regular intervals. However, the system should be inspected whenever the vehicle is raised on a hoist or if the exhaust system is checked or serviced.

5 Check all connections in the exhaust pipe assembly for looseness and damage. Check all the clamps for damage, cracks, and missing fasteners. Check the rubber hangers for cracks.

6 The converter itself should be checked for damage or dents (maximum 3/4-inch deep) which could affect performance. At the same time the converter is inspected, check the metal shield under it as well as the heat insulator above it for damage and loose fasteners.

Component replacement

7 Do not attempt to remove the catalytic converter until the complete exhaust system is cool. See Chapter 4 for exhaust system component illustrations. Raise the vehicle and support it securely on jackstands. Apply penetrating oil to the clamp threads and allow it to soak in.

8 Remove the nuts/bolts and the rubber hangers, then separate the converter from the exhaust pipes. Remove the old gaskets if they are stuck to the pipes.

9 Installation of the converter is the reverse of removal. Use new exhaust pipe gaskets and tighten the clamp fasteners securely. Replace the rubber hangers with new ones if the originals are deteriorated. Start the engine and check carefully for exhaust leaks.

Chapter 7 Part A Manual transmission

Contents

Specifications

Lubricant type	See Chapter 1
Countershaft gear-to-case end play	0.004 to 0.018 in
Reverse idler gear-to-case end play	0.004 to 0.018 in
Cluster gear-to-case end play (4-speed)	0.004 to 0.018 in

Torque specifications	**Ft-lbs**
3-speed	
Oil check/fill plug	25 to 30
Cover-to-case bolts	12 to 14
Transmission-to-bellhousing bolts	40 to 47
Bellhousing-to-engine bolts	22 to 27
Extension housing-to-case bolts	42 to 50
Front bearing retainer bolts	30 to 36
4-speed overdrive	
Oil check/fill plug	10 to 20
Input shaft bearing retainer-to-case bolts	19 to 25
Extension housing-to-case bolts	42 to 50
Cover-to-case bolts	20 to 25
Outer shift levers-to-cam and shaft	18 to 23
Detent bolt	10 to 15

1 General information

The manual transmission used in van models is either a 3-speed or a 4-speed overdrive. Transmission application depends on the year, engine and model of the vehicle in which it is installed. If you are in doubt as to which transmission is in your particular vehicle, check with your local Ford dealer or transmission repair shop.

All forward gears are synchronized to produce smooth, quiet shifts. All forward gears on the mainshaft and input shaft are in constant mesh with the corresponding gears on the countershaft gear cluster and are helically cut for quiet running.

The Reverse gear has straight cut gear teeth and drives First gear through a sliding idler gear.

The gears are engaged inside the case by sliding forks. The gears are selected by a steering column or floor mounted shift lever connected by rods and levers to the transmission. Where close tolerances are required during assembly of the transmission, selective thrust washers and snap-rings are used to eliminate excessive end play or backlash. This eliminates the need for matched assemblies.

Due to the complexity of transmissions and because of the special tools and expertise required to perform an overhaul, it should not be attempted by a novice mechanic. Depending on the expense involved in having a defective transmission overhauled, it may be a good idea to consider replacing it with a new or rebuilt unit. Your local dealer or transmission repair shop should be able to obtain one for you at reasonable cost. Regardless of how you decide to deal with a faulty transmission, costs can be reduced by removing and installing it yourself.

2 Shift linkage — adjustment

Refer to illustrations 2.1a, 2.1b, 2.1c and 2.5

1 The shift linkage consists of a lever and shift rods attached to the gear selector levers on the left side of the transmission. The 3-speed has a steering column mounted shift lever, while the 4-speed has a floor shift lever (see illustrations).

2 The most likely point of wear in the linkage is the bushings on the ends of the down tube levers or shift rods.

3 The bushings can be replaced by removing the spring clips and withdrawing the ends of the shift rods.

4 Replace the bushings and push the ends of the rods into place until they lock or snap in. Install the spring clips.

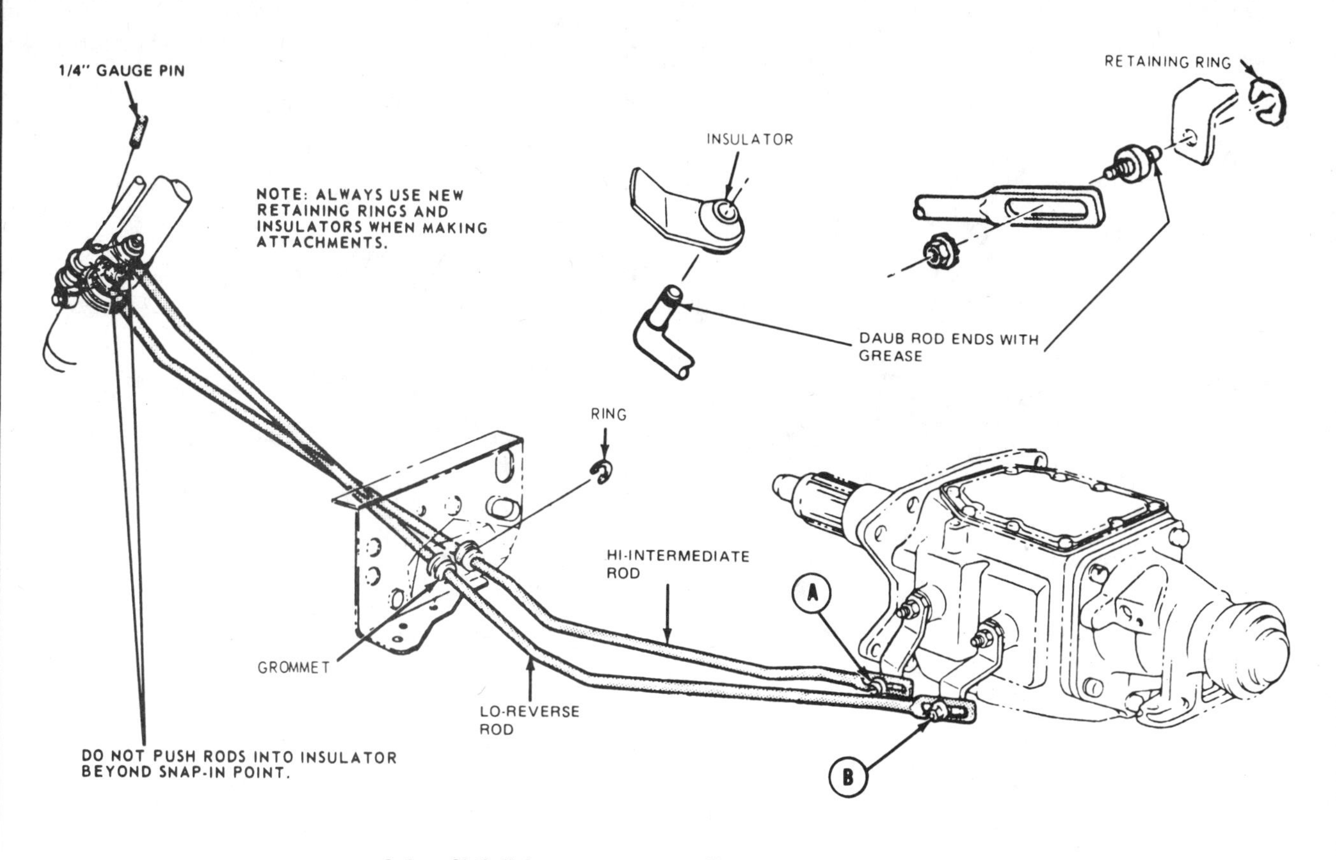

2.1a Shift linkage components (3-speed – column shift)

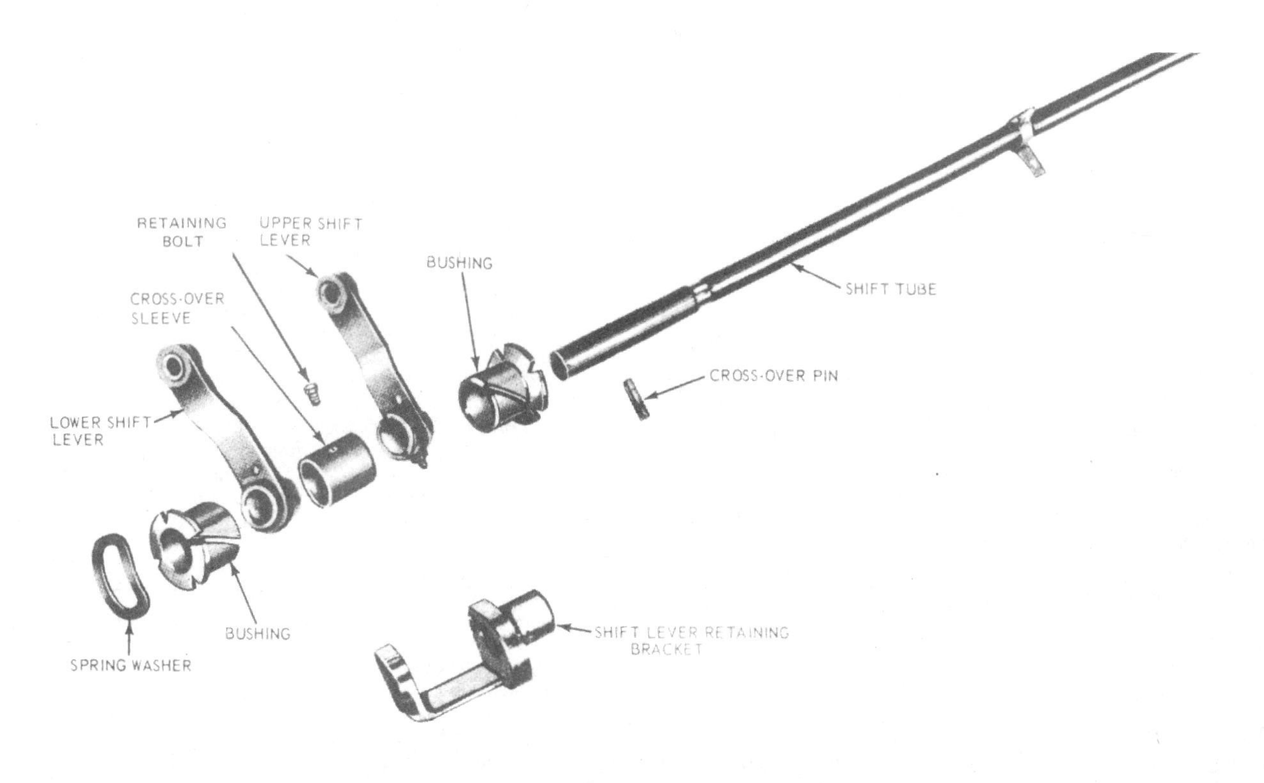

2.1b Shift linkage down tube and levers (3-speed – column shift)

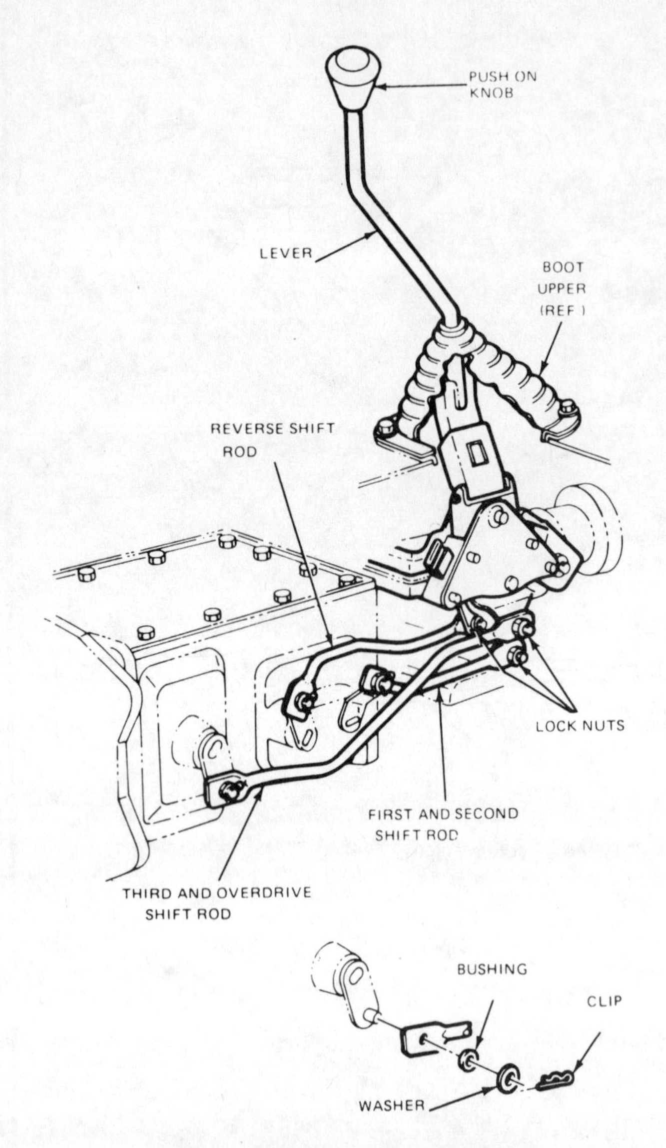

2.1c Shift linkage components (4-speed overdrive – floor shift)

Adjustment – 3-speed

5 To adjust the shift linkage, first position the shift lever in the Neutral position and install a 1/4-inch diameter rod through the column shift levers (see illustration).

6 Loosen nuts A and B and set the transmission shift levers in the Neutral position (see illustration 2.1a).

7 Tighten nuts A and B to 12 to 18 ft-lbs. Do not allow any movement between the stud and shift rod.

8 Remove the piece of rod from the column levers and check the linkage for correct operation. **Note:** *On some models, special grommets are used between the shift control rod ends and the down tube levers and a special tool is needed to remove and install them. Refer to Part B in this Chapter for a description of the removal and installation procedure and an illustration of the tool.*

Adjustment – 4-speed overdrive

9 Disconnect the three shift rods from the shifter assembly by removing the locknuts.

10 Make sure the levers are in the Neutral position and insert a 1/4-inch diameter steel pin through the alignment hole at the bottom of the shifter assembly.

11 Align the three transmission levers as follows: forward lever in the mid-position (Neutral); rear lever in the mid-position (Neutral); and the

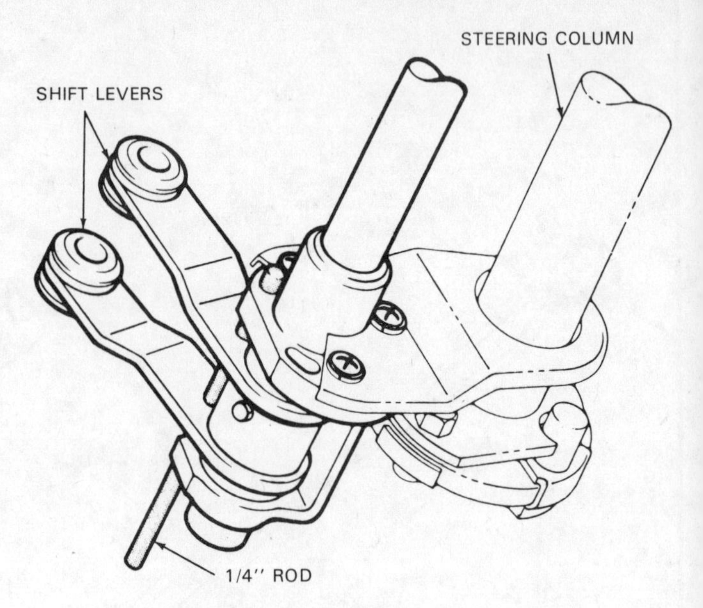

2.5 Insert the rod into the shift levers and bracket as shown here when adjusting the 3-speed column shift linkage

middle lever rotated counterclockwise to the Neutral position.

12 Raise the rear of the vehicle and support it on jackstands.

13 Turn the driveshaft by hand to make sure the transmission is in Neutral.

14 Move the middle lever to the Reverse position by turning it clockwise. This causes the interlock system to align the 1-2 and 3-4 shift rails in the precise Neutral position.

15 Reattach the 1-2 and 3-4 shift rods to the transmission levers and tighten the locknuts.

16 Rotate the middle lever counterclockwise back to the Neutral position and install the shift rod and locknut.

17 Remove the alignment pin from the shifter assembly.

18 Check the linkage for proper operation and lower the vehicle.

3 Transmission – removal and installation (3-speed)

Refer to illustration 3.9

1 If only the transmission is being removed from the vehicle, it can be taken out from below, leaving the engine in place. A lot of room is needed under the vehicle and a hoist should be used. However, provided that jacks and jackstands are available, the task can be accomplished without the use of special equipment.

2 Disconnect the negative cable from the battery.

3 Remove the lower extension housing-to-transmission bolt and drain the transmission oil into a container.

4 Disconnect the front of the driveshaft from the rear of the transmission (see Chapter 8) and support the shaft out of the way.

5 Disconnect the speedometer cable from the transmission extension housing.

6 Remove the spring clips and disconnect the two operating rods from the shift levers on the side of the transmission.

7 Disconnect the wire from the transmission back-up light switch (if equipped).

8 Place a jack under the transmission and secure the transmission to the jack with a chain or rope.

9 Raise the transmission slightly and remove the four bolts securing the rear support crossmember to the chassis side members. Remove the bolt securing the extension housing to the crossmember (see illustration).

10 Lower the transmission just enough to allow the four transmission-to-bellhousing bolts to be removed.

11 Wedge blocks of wood between the bottom of the engine and the crossmember to prevent the engine from dropping when the transmission is removed.

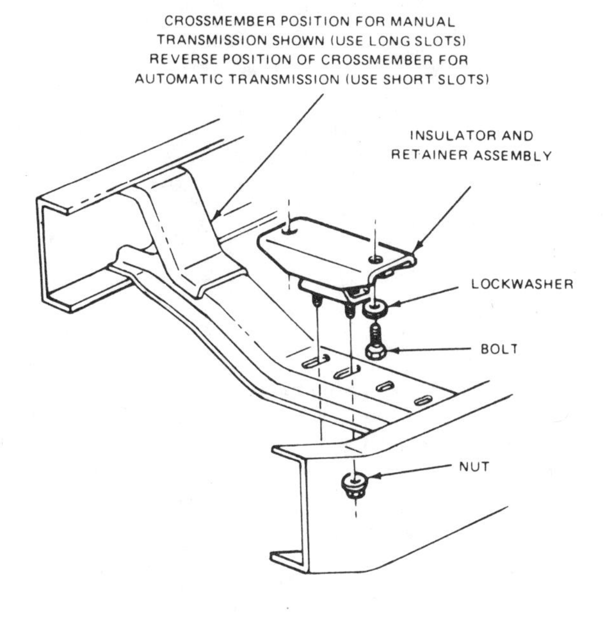

3.9 Transmission support and crossmember details

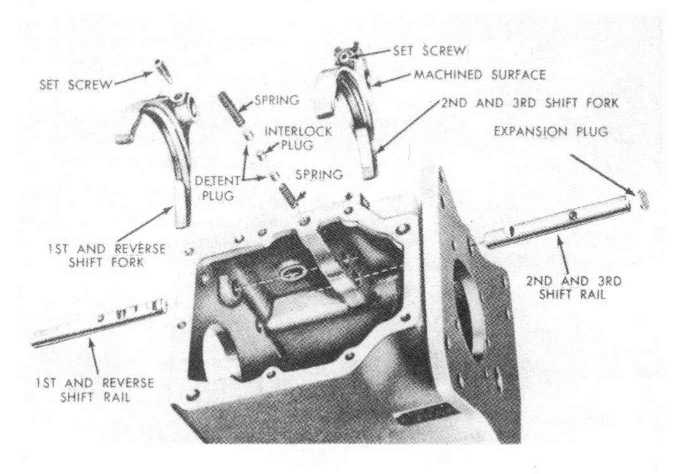

5.5 Shift forks and rails – exploded view

12 Carefully pull the transmission and jack to the rear until the input shaft clears the bellhousing, then lower the transmission to the floor.
13 When installing the transmission, make sure that the clutch release lever and bearing are correctly located in the bellhousing.
14 Apply light grease to the transmission input shaft splines and install the transmission by reversing the removal procedure. **Note:** *It may be necessary to rotate the engine crankshaft to align the clutch disc and input shaft splines. Do not force the transmission into place. If resistance is encountered, check the alignment of the clutch disc and input shaft.*

4 Transmission – removal and installation (4-speed overdrive)

1 Drain the transmission lubricant into a container (see Chapter 1).
2 Raise the vehicle and support it securely on jackstands.
3 Remove the speedometer cable from the extension housing and position it out of the way to avoid damage.
4 Mark the position of the driveshaft in relation to the transmission and the differential. Disconnect and remove the driveshaft.
5 Insert an appropriate tool into the rear of the extension housing to prevent lubricant leakage.
6 Support the transmission with a transmission jack. Make sure the transmission is securely clamped to the jack.
7 Separate the shift rods from the shift levers by removing the retaining clips and washers. Unscrew the bolts retaining the shift control to the extension housing and the nuts securing the shift control to the transmission case.
8 Raise the transmission enough to take the weight of the assembly off of the rear crossmember. Separate the crossmember from the frame rail side supports and the transmission and remove the crossmember.
9 Raise the rear of the engine with a jack (place a block of wood between the jack and the oil pan) until the weight is removed from the crossmember. Separate the crossmember from the side supports and the transmission and remove the crossmember.
10 Remove the transmission retaining bolts from the rear of the bellhousing.
11 Slowly withdraw the transmission from the bellhousing, making sure not to put unnecessary pressure on the input shaft. Once the input shaft has cleared the bellhousing, lower the transmission and remove it from beneath the vehicle. **Caution:** *Do not depress the clutch pedal while the transmission is removed from the vehicle.*
12 Apply a light coating of Ford C1AZ-19590-B lubricant or its equivalent to the clutch release lever fulcrum and fork. Apply this lubricant sparingly, as any excess will contaminate the clutch disc.
13 Secure the transmission to the transmission jack and raise it to the level of the bellhousing face. Make sure the clutch is aligned with an alignment tool if it was removed and replaced. Make sure the clutch throwout bearing and hub are properly positioned on the release lever fork.
14 Install guide studs in the retaining holes of the bellhousing. These studs can be purchased or made from bolts about two to three inches long. Cut the head off the bolts and grind off any burrs. A slot cut into the end of each bolt will enable you to thread them in and out with a screwdriver.
15 Raise the transmission and start the holes of the transmission face onto the guide studs. Slide the transmission forward on the studs until the input shaft engages the clutch splines. Continue to slide the transmission forward until the face of the case mates with the bellhousing face.
16 Remove the guide studs and install the retaining bolts.
17 The remaining steps are basically the reverse of removal.
18 When installing the shift control plate, note that the numbers 6 and 8 are stamped on the extension housing near the control plate bolt holes. These numbers indicate either a 6 or 8 cylinder engine application. Be sure the shift control plate bolts are placed in the right holes for proper positioning of the plate.
19 Fill the transmission to the proper level with the recommended lubricant (see Chapter 1).

5 Transmission – disassembly, overhaul and reassembly (3-speed)

Refer to illustrations 5.5, 5.6, 5.8, 5.12, 5.16, 5.25, 5.28, 5.35, 5.37, 5.38a, 5.38b, 5.43, 5.63 and 5.64

1 Place the transmission on a workbench and make sure that you have the following tools available in addition to common hand tools:

Snap-ring pliers (internal and external)
Brass, lead or copper hammer (at least 2 lbs)
A selection of steel and brass drifts
Small containers
A large vise mounted securely on a workbench
A selection of steel tubing or sections of pipe

2 Any attempt to disassemble the transmission without the tools and items mentioned is not impossible, but will certainly be very difficult and frustrating.
3 Read through the entire Section before starting work.

Transmission – disassembly

4 Remove the bolts and detach the cover plate and gasket from the top of the case. Drain out any oil remaining in the case.
5 Remove the spring and detent plug from the opening in the top left side of the case (see illustration).

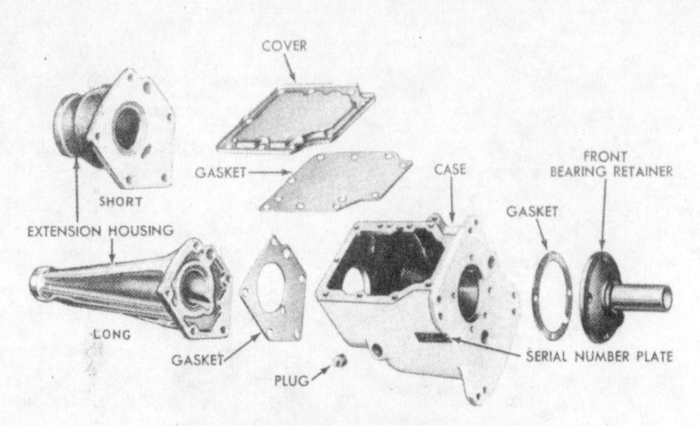

5.6 Transmission case and related components — exploded view

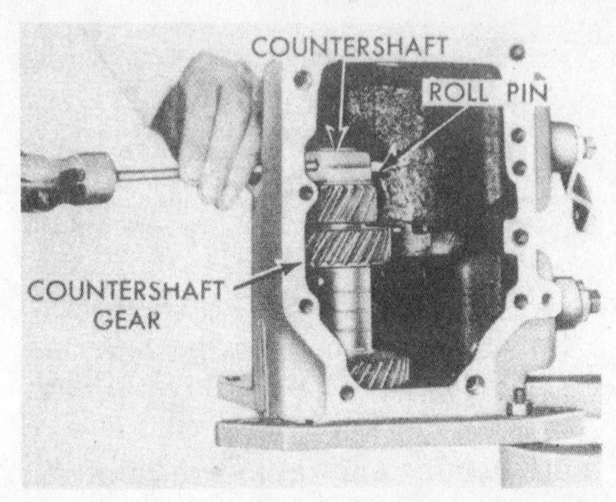

5.8 Driving the roll pin out of the countershaft

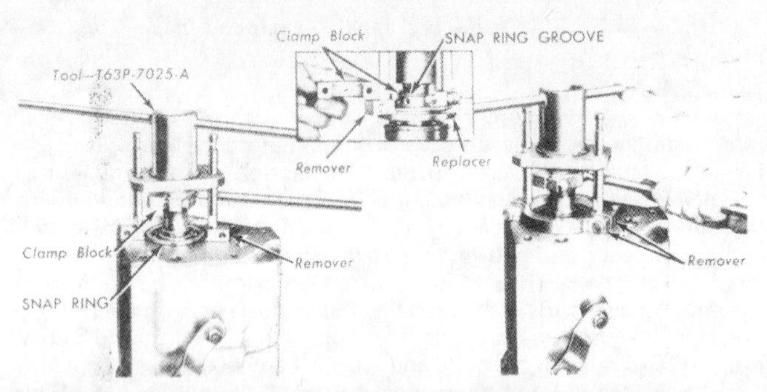

5.12 A special tool is required to remove and install the output shaft bearing

5.16 Turn the 2nd/3rd shift rail 90° to remove it

6 Remove the bolts and withdraw the extension housing from the rear of the case (see illustration). Remove the gasket.
7 Remove the bolts and detach the bearing retainer and gasket from the front of the case.
8 Remove the oil filler plug from the side of the case. Using a hammer and pin-punch inserted through the plug opening, drive out the roll pin that secures the countershaft to the case (see illustration).
9 Using a soft drift punch, carefully tap the countershaft out of the rear of the case while supporting the countershaft gear cluster with one hand.
10 When the shaft is removed, lower the countershaft gear cluster and thrust washers to the bottom of the case.
11 Remove the snap-ring securing the speedometer drive gear to the output (rear) shaft. Slide the gear off the end of the shaft and remove the gear locking ball from the shaft.
12 Remove the snap-ring securing the rear bearing to the output shaft and, using two screwdrivers placed between the outer snap-ring and case, carefully pry the bearing out of the case and slide it off the end of the output shaft. **Note:** *If difficulty is experienced when removing the bearing, a special tool (No. T63P-7025-A) is available from your Ford dealer (see illustration).*
13 Place the two shift levers on the side of the case in the neutral (center) position.
14 Remove the set screw securing the 1st/reverse shift fork to the shift rail. Slide the rail out the rear of the case.
15 Slide the 1st/reverse synchro-hub forward as far as possible, rotate the shift fork up and lift it out of the case.
16 Push the 2nd/3rd gear shift fork to the rear, to the 2nd gear position, to gain access the the retaining screw, then remove it. Rotate the shift rail 90° with pliers (see illustration).
17 Lift the interlock plug (see illustration 5.5) out of the case with a magnet.
18 Carefully tap the rear end of the 2nd/3rd gear shift rail to drive out the expansion plug from the front of the case. Withdraw the shift rail.
19 Remove the remaining detent plug and spring from the bore in the case.
20 Pull the input shaft and bearing from the front of the case.
21 Rotate the 2nd/3rd gear shift fork up and remove it from the case.

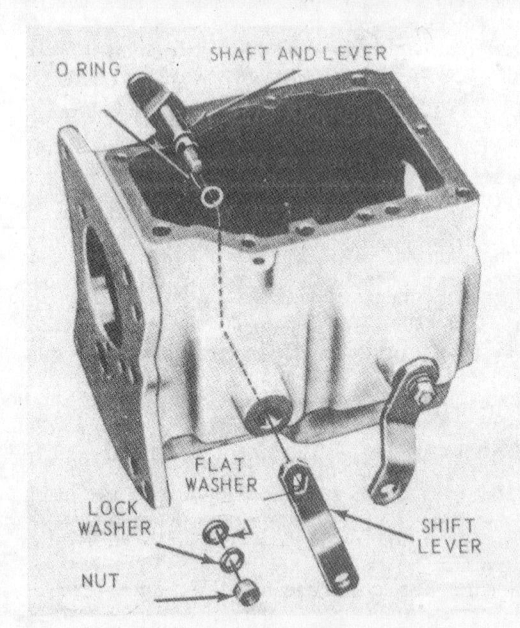

5.25 The shift levers are attached to the shafts with washers and nuts

22 Carefully lift the output shaft and gears out through the top of the case.
23 Slide the reverse gear shaft out of the case and lift out the reverse idler gear assembly and thrust washers.
24 Lift out the countershaft gear cluster and retrieve the thrust washers and any of the needle bearings that may have fallen out. Note that there are 25 needle bearings in each end of the shaft.
25 If required, the shift lever can be detached from the side of the case by removing the nuts and separating the levers and washers from the shafts. Slide the shafts out of the case and discard the O-rings (see illustration).

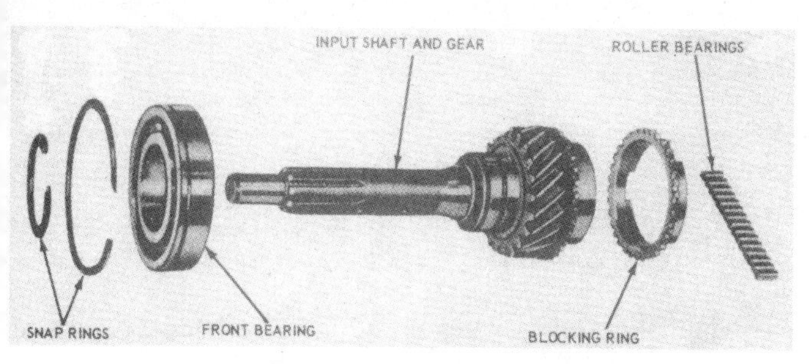

5.28 Input shaft and bearings — exploded view

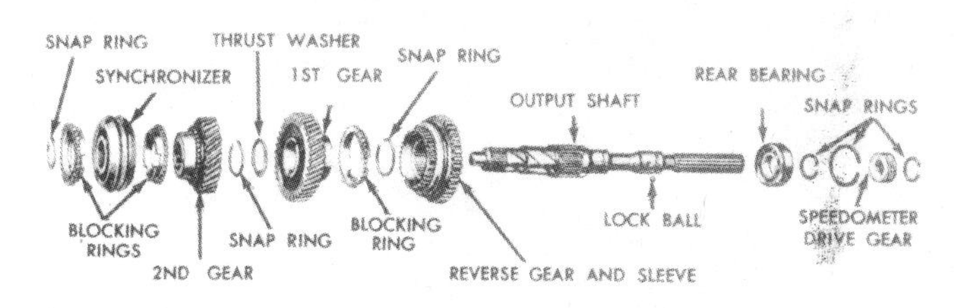

5.35 Output shaft components — exploded view

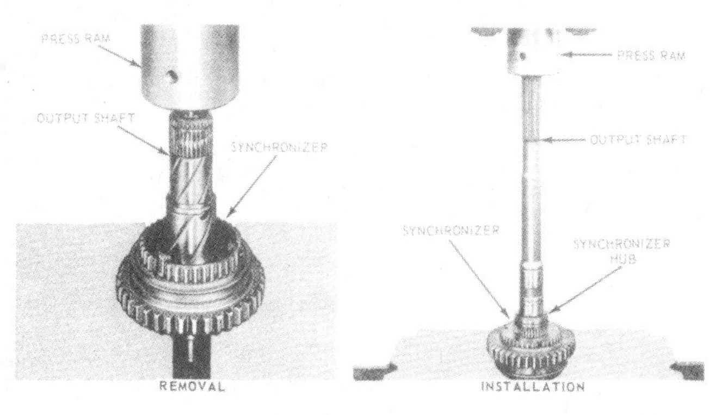

5.37 Removal and installation of the 1st/Reverse gear synchronizer hub requires a hydraulic press

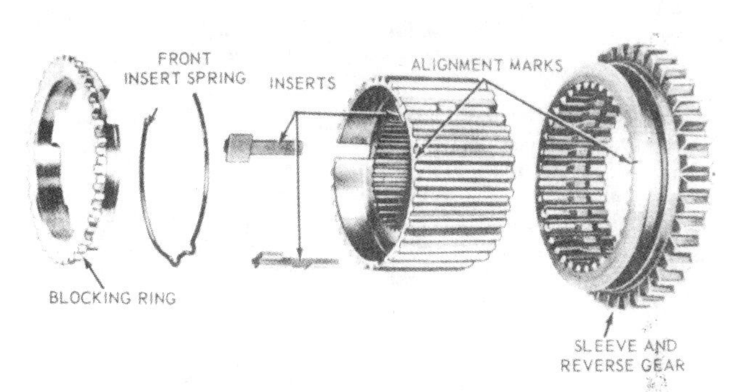

5.38a 1st/Reverse gear synchronizer components — exploded view

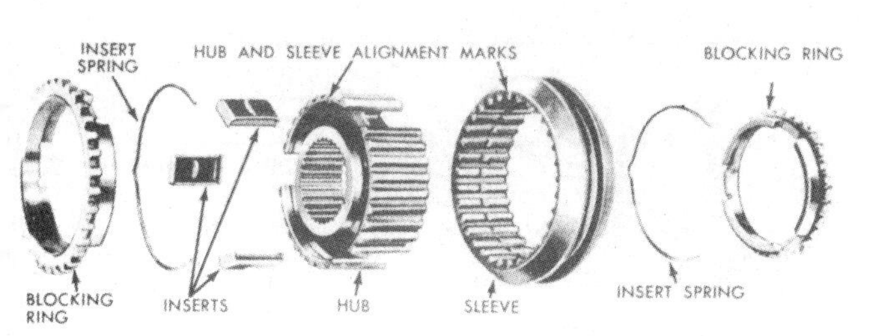

5.38b 2nd/3rd gear synchronizer components — exploded view

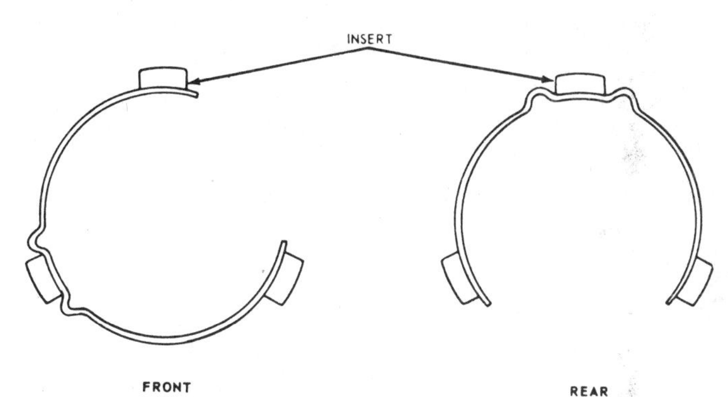

5.43 Correctly installed positions of the 1st/Reverse synchronizer insert springs

26 The transmission must now be thoroughly cleaned with solvent. If there are any metal chips and fragments in the bottom of the case it is obvious that several parts are badly worn. The components should be examined carefully for wear. The input and output shaft assemblies should be broken down further as described below.

Input shaft — disassembly and reassembly

27 The only reason for disassembling the input shaft is to install a new ball bearing or a new shaft.
28 Remove the small snap-ring which secures the bearing to the input shaft (see illustration).
29 Using a soft face hammer, gently tap the bearing forward and then remove it from the shaft.
30 When installing the new bearing, make sure that the groove cut in the outer edge faces away from the gear. If the bearing is installed wrong, it will not be possible to attach the large snap-ring which retains the bearing in the housing.
31 Using the jaws of a vise as a support behind the bearing, tap the bearing squarely into place by hitting the rear of the input shaft with a soft face hammer.
32 Install the snap-ring which holds the bearing to the input shaft.

Output shaft — disassembly and reassembly

33 The output shaft must be disassembled before some of the synchro-rings can be inspected. For disassembly, mount the plain section of the shaft in a vise.
34 As each component is removed from the shaft, make a careful note of its position and then place it on a clean sheet of paper in the order of removal.
35 Remove the snap-ring from the front of the output shaft and slide the blocker ring (if not already removed), the synchronizer hub and the 2nd gear off the shaft (see illustration).
36 Remove the next snap-ring and thrust washer and slide the 1st gear and blocking ring off the shaft.
37 Remove the final snap-ring and press the shaft out of the reverse gear and synchronizer sleeve assembly (see illustration). **Caution:** *Do not attempt to remove the 1st gear synchronizer hub from the shaft by hammering or prying it off, as damage will probably result.*
38 If it is necessary to disassemble the synchro-hubs and sleeves, refer to the accompanying illustrations before beginning disassembly. Also, etch alignment marks on each component.
39 Push the sleeves off the hubs and remove the inserts and insert springs, making careful note of how they are installed.
40 Do not mix the 1st/reverse synchro-hub components with those from the 2nd/3rd synchro-hub assembly.
41 To assemble a synchro-hub, first position the sleeve on the hub making sure the etched marks line up.
42 Install the three inserts and retain them in position with the springs. Make sure that the small end of the inserts faces the inside of the hub.
43 In the case of the 1st/reverse gear synchro assembly, make sure the inserts are retained by the springs (see illustration).

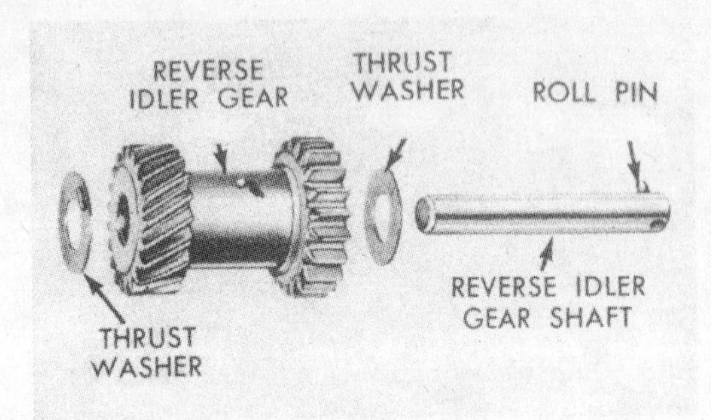

5.63 Reverse idler gear and shaft components — exploded view

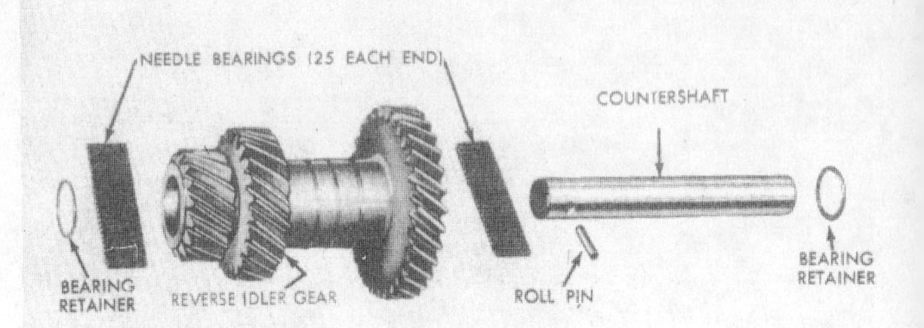

5.64 Countershaft assembly components — exploded view

44 To reassemble the gears and synchro-hubs on the output shaft, first lubricate the shaft and splines with clean transmission oil.
45 Carefully press the reverse gear and synchronizer assembly onto the shaft. Secure the gear in place with the snap-ring.
46 Place a blocking ring on the tapered surface of the 1st gear.
47 Slide the 1st gear onto the output shaft with the blocking ring facing toward the rear of the shaft. Rotate the gear as necessary to engage the three notches in the blocking ring with the synchronizer inserts.
48 Secure the 1st gear in position with the thrust washer and snap-ring.
49 Install a blocking ring onto the tapered surface of the 2nd gear. Slide the 2nd gear and ring onto the shaft and make sure that the tapered side of the gear faces toward the front of the shaft.
50 Slide the 2nd/3rd gear synchronizer onto the end of the shaft and secure it in place with the snap-ring.
51 The output shaft is now completely reassembled.

Inspection and overhaul

52 Carefully clean the transmission parts with solvent and dry them with compressed air. Check all the parts for wear, distortion, loose fit and damage to machined surfaces and threads.
53 Examine the gears for excessive wear and chipped teeth. Replace them as necessary.
54 Check the countershaft for signs of wear, especially where the countershaft gear cluster roller bearings ride. If a small ridge can be felt at either end of the shaft, replace it with a new one.
55 The three synchro blocking rings are bound to be badly worn and should be replaced with new ones. New rings will improve the smoothness and speed of shifts considerably.
56 The needle roller bearings located between the nose of the output shaft and the annulus in the rear of the input shaft are also liable to wear and should be routinely replaced.
57 Examine the condition of the two ball bearing assemblies, one on the input shaft and one on the output. Check them for noisy operation, looseness between the inner and outer races and general wear. Normally they should be replaced with new ones if the transmission is being rebuilt.
58 If either of the synchro-hubs are worn it will be necessary to buy a complete assembly, as the parts are not sold individually.
59 If the bushing in the extension housing is badly worn it should be replaced by a Ford dealer service department.
60 The oil seals in the extension housing, input shaft bearing retainer and selector levers should be replaced with new ones. Drive out the old seals with a drift or broad screwdriver. The seals come out quite easily.
61 With a piece of wood to spread the load evenly, carefully tap new seals into place. Make sure that they enter the housing squarely.
62 The only point on the output shaft that is likely to be worn is the nose where it enters the input shaft. However, examine it thoroughly for signs of scoring and distortion and replace it with a new one if damage is evident.

Transmission — reassembly

63 Insert the reverse gear idler shaft into the rear of the transmission case. Hold the spacer and reverse gear in position (helical gear teeth facing the front of the case) and slide the shaft into place (see illustration).
64 Push some grease into each end of the countershaft gear cluster and carefully insert the roller bearings (see illustration). Ideally, a dummy shaft of the same outside diameter as the countershaft, but shorter, should be placed inside the countershaft gear cluster to hold the bearings in place.
65 Stick the bearing retainer onto each end of the countershaft gear cluster with grease.
66 Carefully place the countergear assembly in the bottom of the case, but do not install the countershaft at this stage.
67 Install the output shaft and gear cluster into the case through the top access hole.
68 Install a new snap-ring in the groove around the rear bearing assembly and carefully drive the bearing along the output shaft until it enters the opening in the rear of the case. Note that the groove in the bearing must be positioned toward the rear of the case.
69 Install a new bearing retainer snap-ring on the output shaft.
70 Make sure that the roller bearings are correctly located in the end of the input shaft with grease.
71 Insert the input shaft, 3rd gear and synchro-ring assembly through the front of the case and make sure the end of the output shaft is correctly located in the roller bearing recess of the input shaft.
72 Install a new snap-ring in the front bearing groove and tap the bearing along the input shaft until it enters the opening in the case. Carefully tap the bearing into place until it seats.
73 Position the 2nd/3rd gear shift fork on the 2nd/3rd gear synchro-hub.
74 Refer back to illustration 5.5 and install the lower spring and detent plug in the case.
75 Push the 2nd/3rd gear synchro-hub as far as possible toward the rear of the case (2nd gear position).
76 Align the hole in the shift fork with the case and push the 2nd/3rd gear shift rail in from the front. It will be necessary to push the detent plug down to allow the shaft to enter the bore.
77 Push the shaft into the case until the detent plug engages in the forward notch (2nd gear position).
78 Secure the fork to the shaft with the set screw and then push the synchro unit forward to the neutral position.
79 Install the interlock plug in the bore in the side of the case so that it is resting on top of the shift rail.
80 Push the 1st reverse gear and synchro unit all the way forward to the 1st gear position.
81 Install the 1st/reverse shift fork in the synchro unit groove, rotate the fork into the correct position and slide the 1st/reverse shift rail in from the rear of the case.
82 Push the rail (shaft) in until the center notch (neutral) is in line with the detent bore.
83 Secure the fork to the shaft with the set screw.
84 Install a new shift rail expansion plug in the front of the case.
85 Place the case in a vertical position and insert a screwdriver through the oil filler hole to align the bore of the countershaft gear and the thrust washers with the bores in the case.
86 Insert the countershaft from the rear of the case and gently tap it through the gear cluster, driving out the dummy shaft.
87 If a dummy shaft was not used, insert the countershaft in the same way, but be careful not to dislodge any of the needle roller bearings or retainers.
88 Position the shaft so that the hole lines up with the hole in the case and when the shaft is in place, drive in a new roll pin.
89 Make sure that the countershaft gears rotate smoothly with no harshness that might indicate dislodged needle bearings.
90 Install a new front bearing retainer snap-ring on the input shaft.

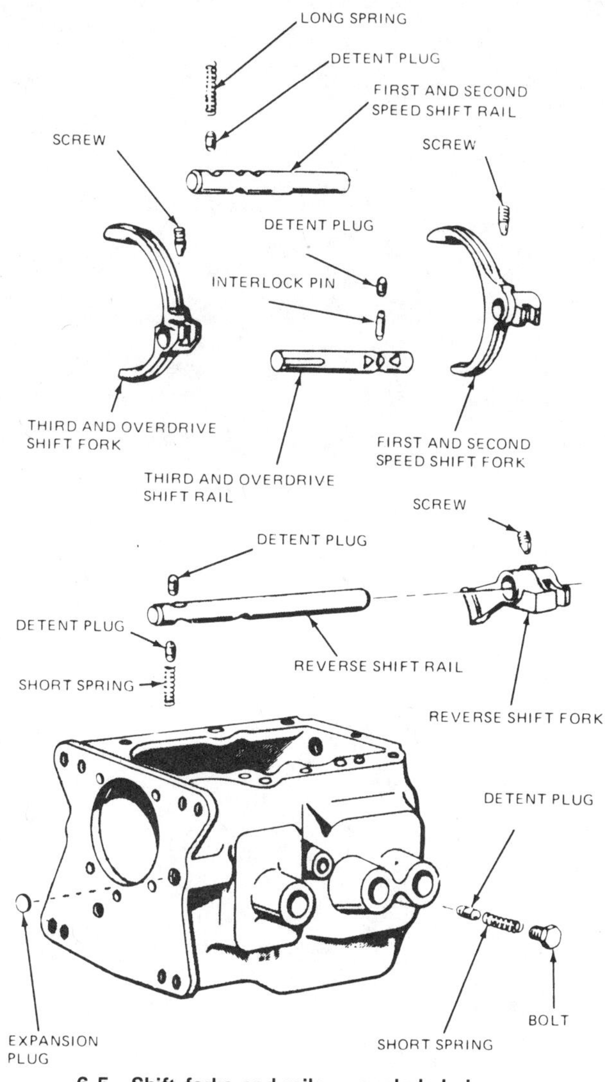

6.5 Shift forks and rails — exploded view

91 Install the input shaft retainer using a new gasket.
92 Apply RTV-type gasket sealer to the bearing retainer bolts and tighten them to the specified torque.
93 Insert the speedometer drive gear locking ball into the hole in the output shaft, hold the ball in position and slide the drive gear over it. Secure it with a snap-ring.
94 Place a new gasket over the rear of the transmission case and install the extension housing.
95 Apply RTV-type sealer to the extension housing retaining bolts and tighten them to the specified torque.
96 Rotate the input shaft by hand and make sure that all gears can be selected in order and that both shafts rotate smoothly.
97 Install the remaining detent plug and spring in the bore in the top of the case.
98 Coat a new cover gasket with sealant, install the top cover and tighten the bolts.
99 Temporarily install the oil filler plug.
100 After the transmission is reinstalled, refill it with the specified lubricant (Chapter 1).

6 Transmission — disassembly, overhaul and reassembly (4-speed overdrive)

Refer to illustrations 6.5, 6.16, 6.18, 6.22a, 6.22b, 6.25, 6.27, 6.36, 6.40, 6.42, 6.46, 6.47, 6.77, 6.81 and 6.82.

1 Place the transmission on a workbench or table and ensure the following tools are available in addition to normal hand tools.

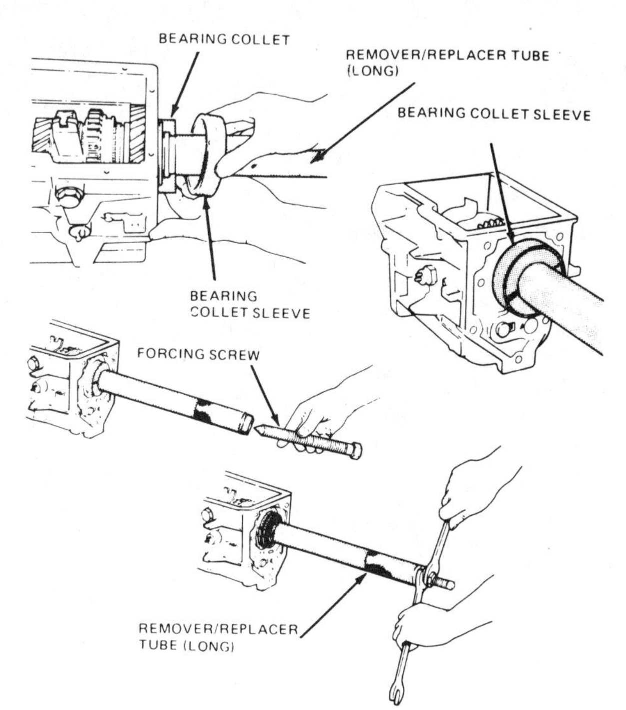

6.16 A special tool is required to remove the output shaft bearing

Snap-ring pliers (internal and external)
Brass, lead or copper hammer (at least 2 lbs)
A selection of steel and brass drifts
Small containers
A large vise mounted securely on a workbench
A selection of steel tubing or sections of pipe

2 Any attempt to overhaul the transmission without the items mentioned is not impossible, but it will certainly be difficult and inconvenient.
3 Read the complete Section before beginning the job.
4 Unscrew the bolts securing the cover to the case and lift the cover off along with the gasket.
5 Lift the detent plug retaining spring (long) from the case. Using a small magnet, remove the detent plug (see illustration).
6 Remove the bolts securing the extension housing to the rear of the transmission. Lift the housing away along with thes gasket.
7 Unscrew the bolts securing the input shaft bearing retainer. Remove the bearing retainer from the input shaft.
8 Use a wire hook to support the countershaft gear, then, working from the front of the case, push the countershaft out the rear of the case using a dummy countershaft such as an appropriate size wood dowel or steel rod. Lower the countershaft assembly, on the dummy shaft, to the bottom of the case with the wire hook. Remove the hook from the assembly.
9 Unscrew the set screw from the 1st and 2nd speed shift fork. Remove the 1st and 2nd speed shift rail from the rear of the case.
10 Remove the interlock detent from between the 1st/2nd and the 3rd/overdrive shift rails with a magnet.
11 Place the transmission in the overdrive position and remove the set screw from the 3rd and overdrive shift rails.
12 Remove the side detent bolt, the detent plug and the spring. Turn the 3rd and overdrive shift rail 1/4 turn (90°) clockwise and drive it out of the front of the case using a punch and hammer.
13 Use a magnet to remove the interlock pin from the top of the case.
14 Disconnect the speedometer drive gear from the output shaft by removing the retaining snap-ring. Slide the gear from the shaft, then remove the speedometer gear drive ball.
15 Remove the snap-rings securing the output shaft bearing to the shaft and the one from the outside of the output shaft bearing.
16 Using the appropriate special tools, remove the output shaft bearing from the output shaft (see illustration).
17 Remove the input shaft bearing retaining snap-ring and the snap-ring from the outside of the input shaft bearing, then remove the bearing retaining ring.

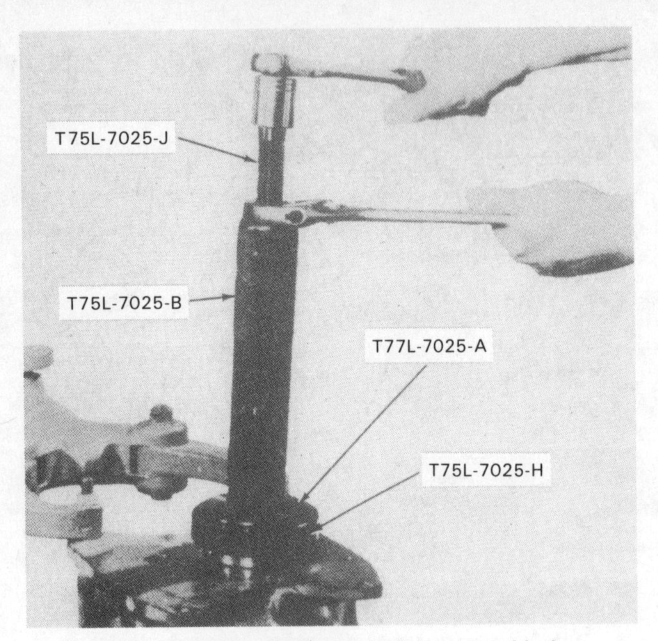

6.18 Removing the input shaft bearing with the special tool

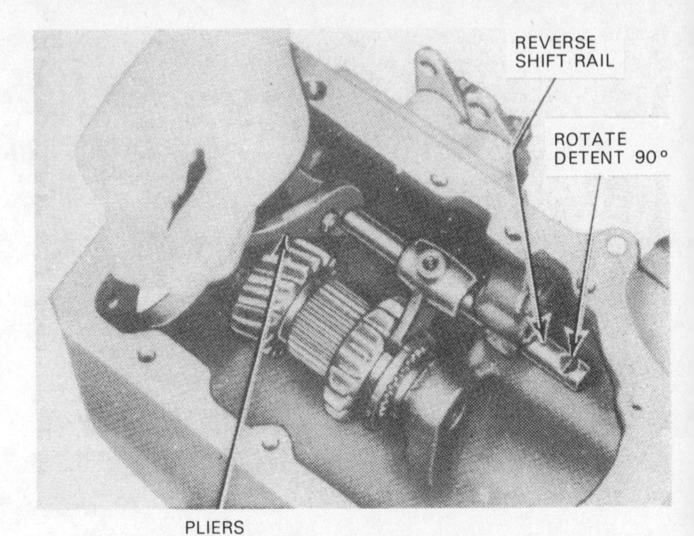

6.22a The reverse shift rail must be turned 90° with pliers before it can be removed

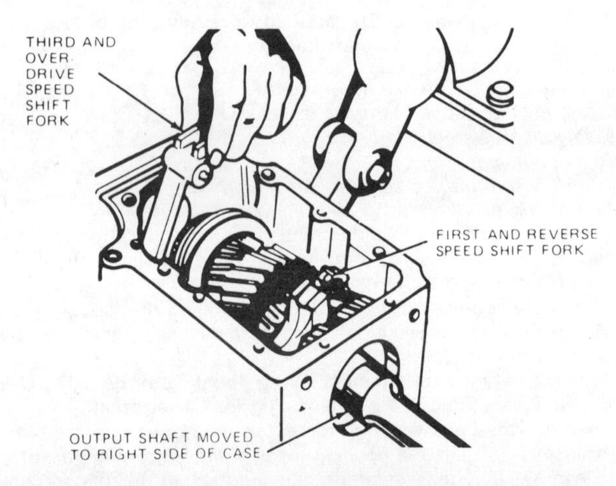

6.22b Remove the shift forks from the case after the output shaft is moved out of the way

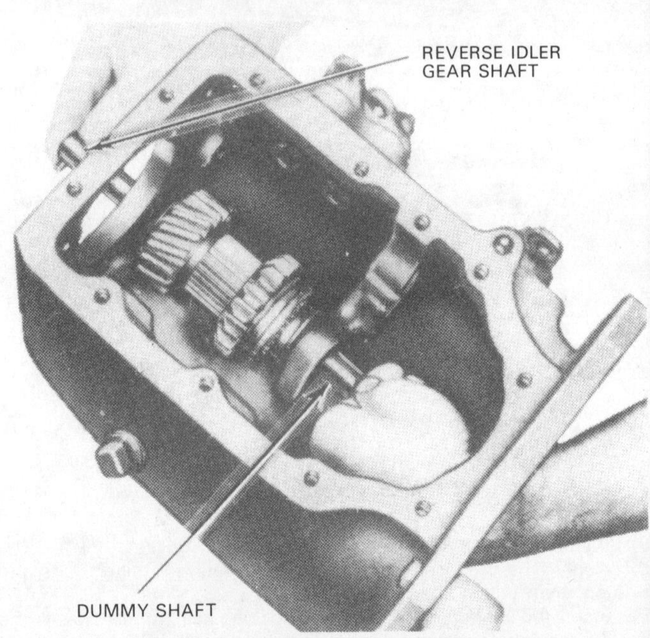

6.25 Insert a dummy shaft through the reverse idler gear to drive the shaft out of place

18 Using the appropriate special tools, separate the input bearing from the input shaft and the transmission case (see illustration).
19 Remove the blocking ring and the input shaft from the front of the transmission case.
20 To gain clearance for the shift forks, move the output shaft to the right side of the case. Turn the shift forks until they can be lifted from the case.
21 Carefully lift the output shaft assembly from the case while supporting the thrust washer and 1st gear so they will not slide off the shaft.
22 Unscrew the set screw in the reverse gear shift fork. Turn the shift rail 1/4 revolution (90°) and slide the shift rail out of the rear of the case (see illustration). Remove the reverse shift fork from the transmission case (see illustration).
23 Use a magnet to remove the reverse detent plug and spring from the case.
24 The countershaft assembly can now be removed from the case with all of the components attached to the dummy shaft. Be careful not to let any bearings or the dummy shaft drop from the countershaft gear.
25 Drive the reverse idler gear shaft out of the case by inserting a dummy shaft through the front of the case (see illustration).
26 Remove the reverse idler assembly from the case. Do not allow the bearings or the dummy shaft to drop from the reverse idler gear.
27 Slide the 3rd and overdrive synchronizer blocking ring and gear off of the output shaft after removing the snap-ring from the front of the shaft (see illustration).
28 Remove the next snap-ring and the 2nd gear thrust washer from the output shaft. Remove 2nd gear and the blocking ring from the output shaft.
29 Remove the final snap-ring.
30 Slide the thrust washer, 1st gear and the blocking ring from the rear of the shaft. The 1st and 2nd synchronizer hub is a slip fit on the output shaft.
31 Remove the retaining nut and washers from each shift lever and remove the three shift levers.
32 From inside the case, remove the three cams and shafts.
33 Remove the O-ring from each cam and throw the O-ring away. Coat the new O-rings with clean lubricant and install them on the cams.
34 Install each cam in its respective bore in the case.
35 Attach the three shift levers and secure them with a flat washer, lockwasher and nut.
36 Make alignment marks on the hub and the sleeve of each synchronizer before disassembly. Remove the synchronizer hub from each synchronizer sleeve (see illustration).

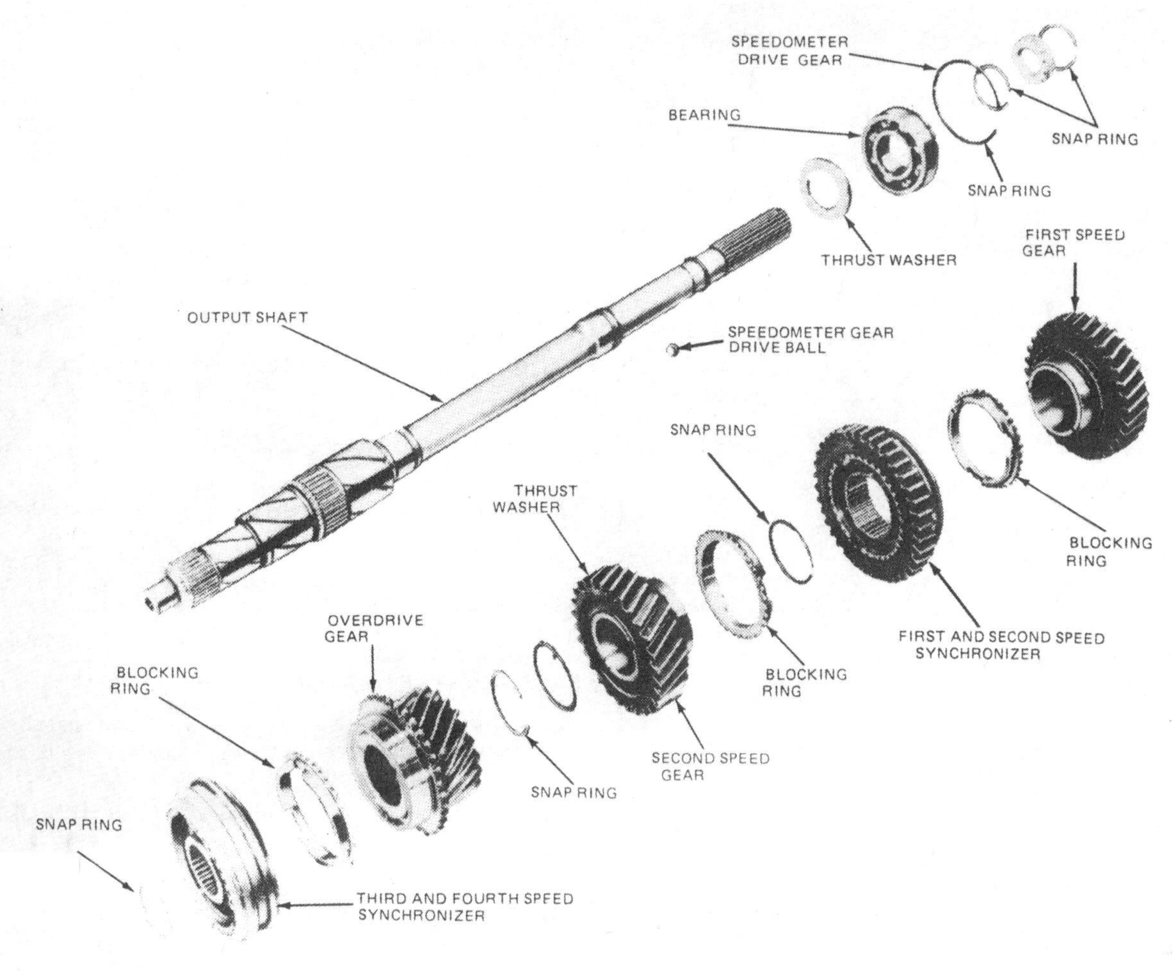

6.27 Output shaft components — exploded view

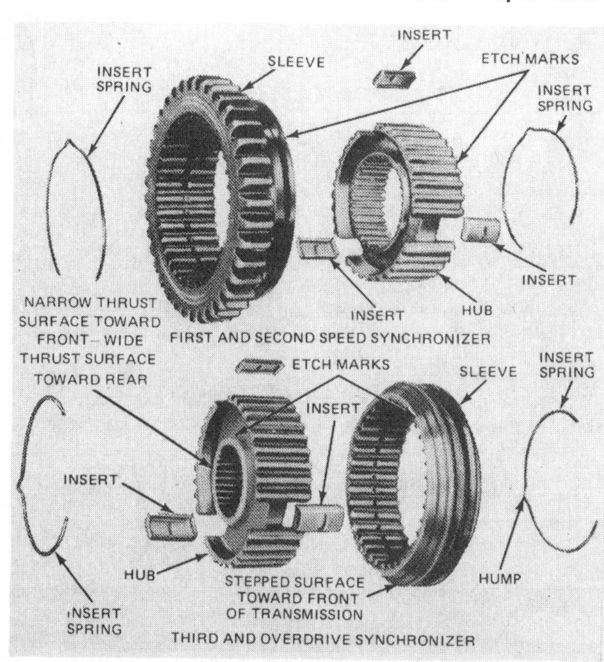

6.36 Synchronizer components — exploded view

37 Remove the inserts and insert springs from each hub. Be sure to keep the components to each synchronizer assembly separate from the components of the other assembly.

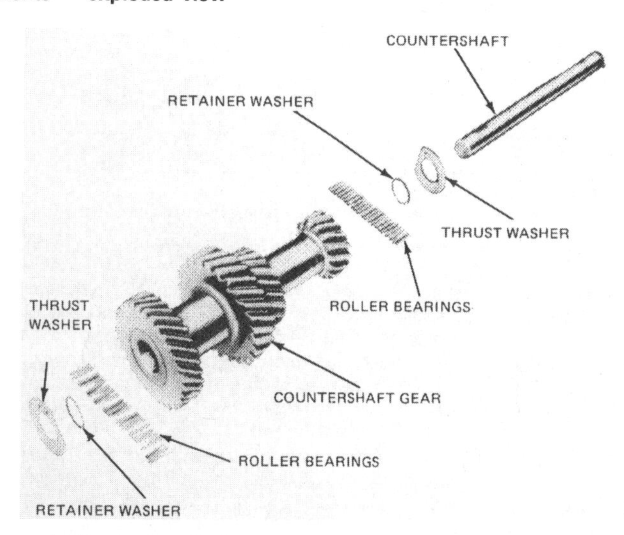

6.40 Countershaft gear components — exploded view

38 Place the hub in position in the sleeve with the alignment marks properly indexed.
39 Install the three inserts in the hub followed by the insert springs. The irregular surface (hump) must be seated in one of the inserts. Do not stagger the springs.
40 Separate the dummy shaft from the countershaft gear along with the bearing retainer washers and the 21 roller bearings from each end (see illustration).

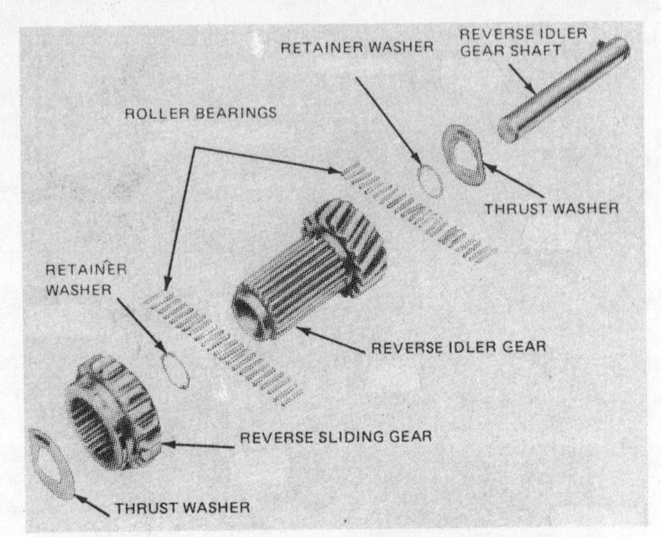

6.42 Reverse idler gear components — exploded view

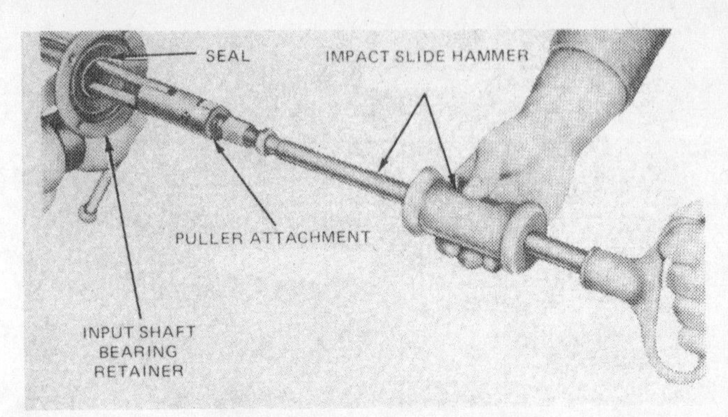

6.46 A special tool is required to remove the input shaft seal from the bearing retainer

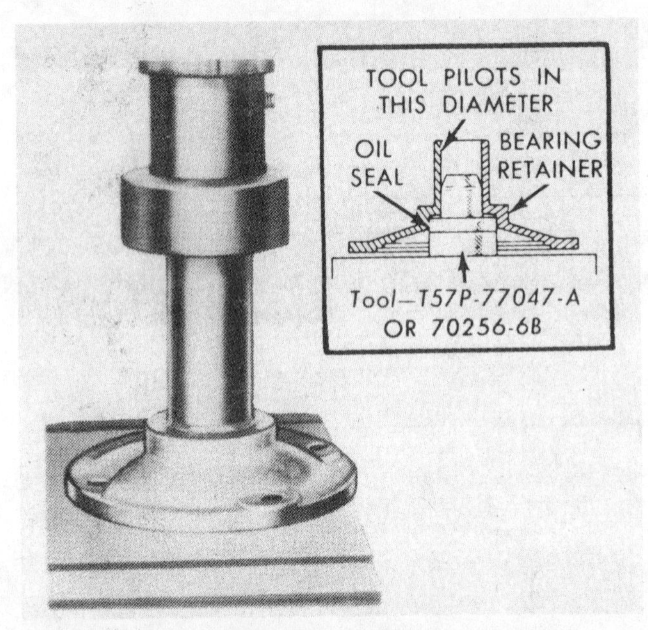

6.47 Pressing the new input shaft into position

41 Apply a coat of grease to the bore in each end of the countershaft gear. Insert the dummy shaft into the countershaft gear and place the 21 roller bearings into position on each end of the gear. Secure the bearings with a retaining washer on each side of the gear. Secure the bearings with a retaining washer on each side of the gear.
42 Remove the reverse sliding gear from the reverse idler gear (see illustration).
43 Remove the bearing retaining washers from each end of the reverse idler gear and carefully remove the dummy shaft. The roller bearings can now be removed from the gear. Be careful not to lose any of the roller bearings.
44 Apply a coat of grease to the bore in each end of the reverse idler gear and slide the dummy shaft into the gear. Install 22 roller bearings into each end of the gear and secure the bearings with a retaining washer on each end.
45 Slide the reverse idler sliding gear onto the reverse idler gear. Be sure that the shift fork groove faces forward.
46 Using the appropriate special tool, remove the input shaft seal from the bearing retainer (see illustration).
47 Coat the sealing surface of the new seal with grease or oil and drive the seal into the bearing retainer with the special seal installing tool (Ford T57P-77047-A) (see illustration).
48 Apply a thin coat of grease to the countershaft gear thrust surfaces in the case. Place a thrust washer in position at each end of the case.
49 Set the countershaft gear, dummy shaft and roller bearings into position in the case. Raise the case to a vertical position and align the thrust washer bores and the bore in the gear with the countershaft bores in the case. Insert the countershaft into the case, displacing the dummy shaft.
50 Lay the case down in a horizontal position and check the end play of the countershaft with a feeler gauge. If the end play is not within the specified limits, the thrust washers must be replaced with new ones. When the end play is correct, reinstall the dummy shaft, driving the countershaft out of position. Set the assembly in the bottom of the case.
51 Apply a thin coat of grease to the thrust surfaces for the reverse idler gear in the case. Place the two thrust washers in position in the case.
52 Place the reverse idler gear assembly in position in the case. The shift fork groove in the sliding gear must be facing the front of the case.
53 Align the bores of the thrust washers and the gear with the case bore and insert the reverse idler shaft, displacing the dummy shaft.
54 Measure the end play of the reverse idler shaft and compare it to the specified clearance. If the end play is not within the specified limits, the thrust washers must be replaced with new ones. If the end play is within the limits, leave the reverse idler shaft installed.
55 Position the detent spring and plug to the reverse gear shift rail in the case. Hold the reverse shift fork in place on the reverse idler sliding gear and install the shift rail from the rear of the case. Install the socket head set screw to secure the fork to the rail.
56 Attach the 1st and 2nd speed synchronizer to the front of the output shaft. Be sure that the shift fork groove faces the rear of the shaft. The 1st and reverse synchronizer hub is a slip fit on the output shaft. Position the synchronizer hub with the teeth end of the gear facing the rear of the shaft.
57 Place the blocking ring in position on the 2nd gear. Slip 2nd gear onto the front of the shaft. Be sure that the inserts in the synchronizer engage the notches in the blocker ring.
58 Install the thrust washer for 2nd gear and the snap-ring.
59 Place the overdrive gear onto the shaft. Be sure that the coned surface of the synchronizer is facing the front. Install a blocking ring on the overdrive gear.
60 Attach the 3rd and overdrive gear synchronizer to the shaft. Be sure that synchronizer inserts engage the notches in the blocking ring and the thrust surface faces the overdrive gear.
61 Attach the snap-ring to the front of the output shaft.
62 Place the blocking ring on 1st gear.
63 Install 1st gear on the rear of the output shaft, engaging the notches in the blocking ring with the synchronizer inserts.
64 Position the heavy thrust washer for the output shaft at the rear.
65 Hold the thrust washer and 1st gear securely on the output shaft and carefully lower the output shaft assembly into position in the case.
66 Place the 1st and 2nd shift fork and the 3rd and overdrive shift fork into position on their respective gears and turn them into place.
67 Install the detent plug and spring into the detent bore. Move the reverse shift rail into the Neutral position.

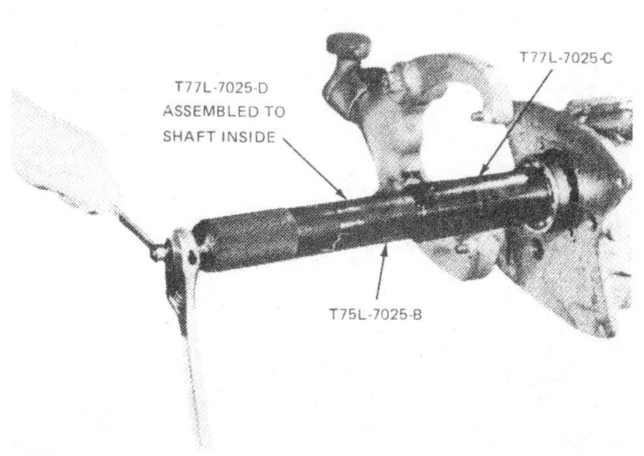

6.77 Use the special tool to install the input shaft bearing

68 Apply a thin coat of grease to the tapered 3rd and overdrive shift rail interlock pin and place the pin in position in the shift rail.
69 Align the 3rd and overdrive shift fork with the shift rail bores. Slide the shift rail into position. Be sure that the three detents are facing the outside of the case.
70 Set the front synchronizer in the overdrive position and install the set screw in the 3rd and overdrive shift fork. Slide the synchronizer into the Neutral position.
71 Install the detent plug, spring and bolt to the 3rd and overdrive shift rail in the left side of the case. Install the detent plug (tapered ends) in the detent bore in the case.
72 Position the 1st and 2nd-speed shift fork so it is aligned with the case bores and insert the shift rail into position. Install the set screw to retain the fork.
73 Apply a thin coat of grease to the input gear bore. Position the 15 roller bearings in the bore of the gear.
74 Attach the front blocking ring in the 3rd and overdrive synchronizer.
75 Attach a dummy bearing (special tool No. T77L-7025B) to the output shaft to support and align the shaft assembly in the case.
76 Carefully place the input shaft gear into the case so that the output shaft pilot engages the roller bearings in the input gear pocket.
77 Place the input shaft bearing on the input shaft along with the clamp, the sleeve and the replacer tube (special tool No. T77L-7025-D, T75L-7025-K and T77L-7025-C respectively). Press the bearing into the case and onto the shaft (see illustration).
78 Remove the special tools and install the snap-rings on the input shaft bearing and the input shaft.
79 Install a new gasket on the input shaft bearing retainer. Coat the threads of the retainer bolts with thread sealer and tighten the bolts to the specified torque.
80 Remove the dummy bearing from the output shaft and install the special tools required to press the bearing onto the output shaft. Be sure that the output shaft bearing is aligned with the case bore and that the countershaft is not interfering with the output shaft assembly.
81 Press the output shaft bearing into position on the output shaft and into the case (see illustration). Remove the special tools and install the proper snap-rings on the output shaft and bearing.
82 Place the transmission on its end, in a vertical position. Align the thrust washer bores and the countershaft gear bore with the case bores. Install the countershaft, driving the dummy shaft out (see illustration).
83 Attach the extension housing to the rear of the transmission case with a new gasket. Coat the threads of the mounting bolts with thread sealer and tighten the bolts to the specified torque.
84 Install the filler plug in the case.
85 Rotate the input shaft and simultaneously pour fresh lubricant over the entire gear train.
86 Test each shift fork for proper operation in all positions.

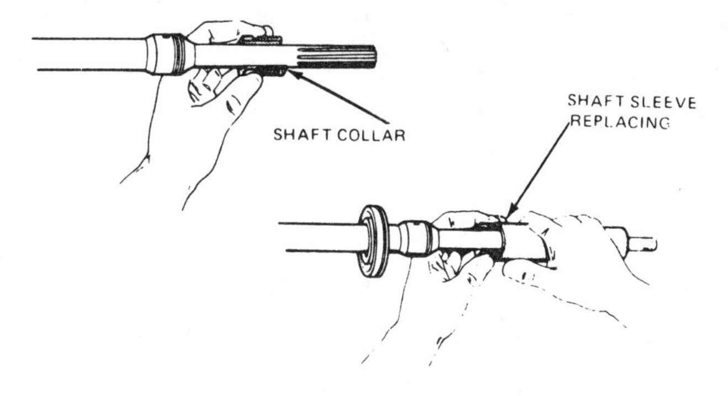

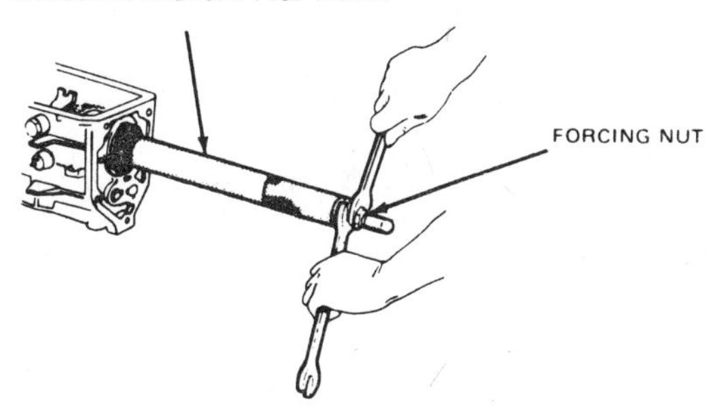

6.81 Installing the output shaft bearing with the factory special tool

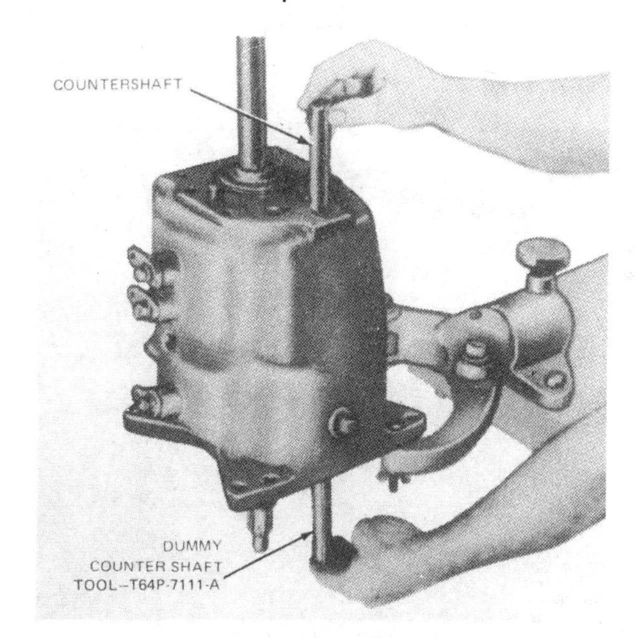

6.82 Align the bores of the countershaft gear assembly with the case bores and insert the countershaft, driving out the dummy shaft in the process

87 Place the remaining detent plug and the long spring into position and secure them by installing the cover. Be sure to use a new cover gasket and to coat the threads of the retaining bolts with thread sealer. Tighten the bolts to the specified torque.
88 Apply a coat of sealer to the 3rd and overdrive shift rail plug bore and install a new expansion plug.

Chapter 7 Part B Automatic transmission

Contents

Specifications

Transmission fluid type	See Chapter 1
Torque specifications	**Ft-lbs (unless otherwise noted)**
AOD transmission	
Filter-to-valve body	80 to 100 in-lbs
Oil pan bolts	12 to 16
Outer throttle lever-to-shaft	12 to 16
Neutral safety switch	7 to 10
Transmission-to-engine bolts	40 to 50
All other transmissions	
Torque converter-to-driveplate bolts	23 to 28
Converter housing-to-transmission case bolts	28 to 40
Oil pan-to-case bolts	12 to 16
Converter cover bolts	12 to 16
Transmission-to-engine bolts	23 to 33
Converter drain plug (if equipped)	20 to 30
Downshift lever-to-shaft nut	12 to 16
Filler tube-to-engine bolt	20 to 25
Filler tube-to-transmission bolt	32 to 42
Neutral switch-to-case bolts	55 to 75 in-lbs

1 General information

Ford vans are equipped with either a C4, C6 or AOD (automatic overdrive) transmission, depending on model, year and engine installed in the vehicle. Note that Ford normally specifies a different grade of transmission fluid than other manufacturers. It must be used when adding fluid or refilling the transmission (see Chapter 1).

Due to the complexity of the clutches and the hydraulic control system, as well as the special tools and expertise required to perform an automatic transmission overhaul, it should not be attempted by the novice home mechanic. The procedures in this Chapter are limited to general diagnosis, adjustment and transmission removal and installation.

If the transmission requires major repair work it should be left to a dealer service department or an automotive or transmission repair shop. You can, however, remove and install the transmission yourself and save the expense involved, even if the repair work is done by a professional.

Automatic transmission problems may be caused by four general conditions: poor engine performance, improper adjustments, hydraulic system malfunctions and mechanical malfunctions. Diagnosis of transmission troubles should always begin with a check of the easily repaired items: fluid level and condition (see Chapter 1) and linkage adjustment. Next, perform a road test to determine if the problem has been eliminated. If the problem persists after the preliminary checks and corrections, additional diagnosis should be done by a dealer service department or an automotive or transmission repair shop.

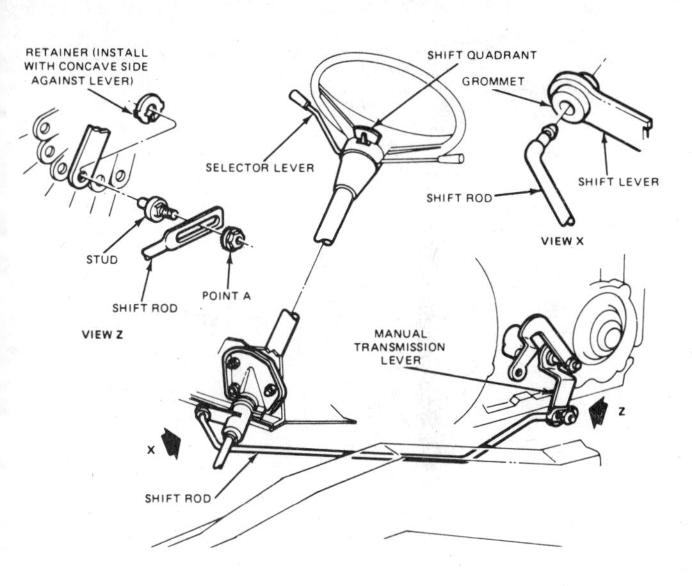

2.2a Shift linkage details (early C4/C6 transmissions)

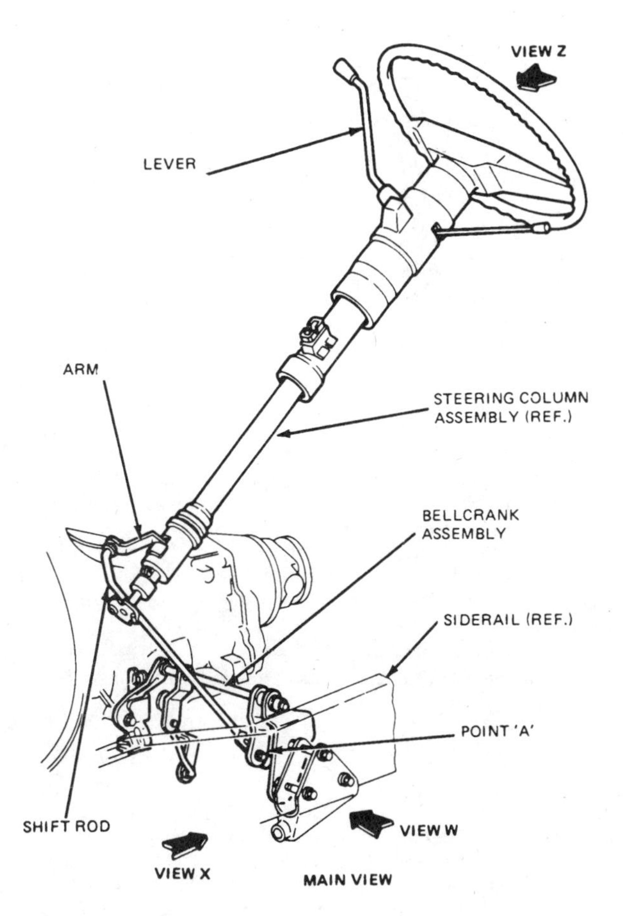

2.2c Shift linkage details (AOD transmission)

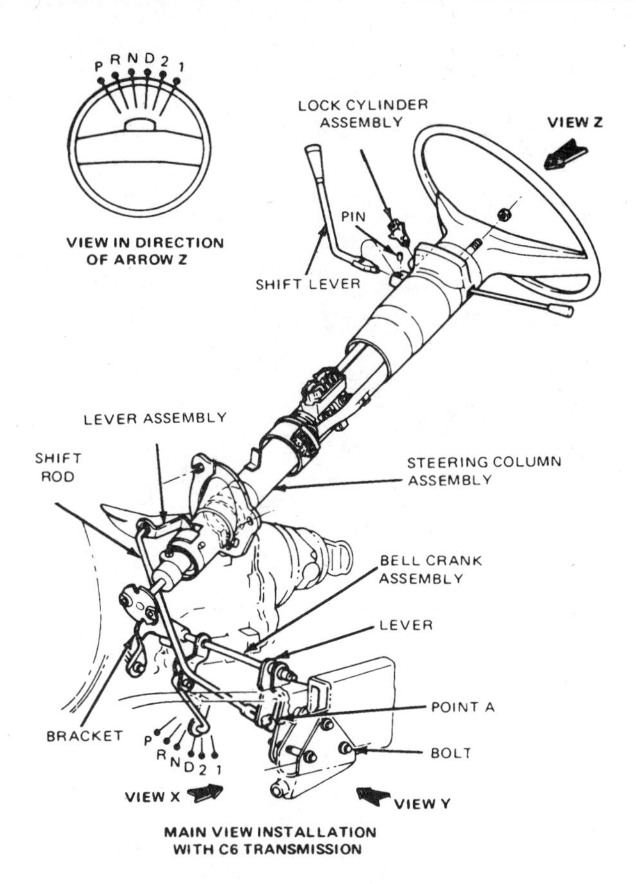

2.2b Shift linkage details (later C6 transmission)

2 Shift linkage – adjustment

Refer to illustrations 2.2a, 2.2b and 2.2c

1 With the engine stopped, place the transmission shift lever in the Drive position (Overdrive position on AOD transmission) and hold it against the stop by tying an eight pound weight to the end of the lever.
2 Loosen the shift rod retaining nut (point A) on the side of the transmission (see illustrations).
3 Place the selector lever on the side of the transmission in the Drive position by moving the lever to the rear as far as possible, then forward two clicks.
4 Hold the shift rod and lever stationary and tighten the nut to 12 to 18 ft-lbs. Do not allow any motion between the stud and the rod as the nut is tightened.
5 Remove the weight from the shift lever and check the operation of the transmission in all gear positions.

7B

3 Kickdown rod – adjustment

1 Disconnect the kickdown rod return spring and hold the throttle shaft lever in the wide open position.
2 Hold the kickdown rod against the throttle detent stop.
3 If necessary, turn the kickdown screw to provide a clearance of 0.060-inch between the screw tip and the throttle shaft lever tab.
4 Reconnect the kickdown rod return spring.

4 Throttle valve (TV) control rod linkage – adjustment (AOD transmission)

Refer to illustrations 4.4, 4.5, 4.6, 4.13, 4.15a and 4.15b

Adjustment at the carburetor

1 Adjust the idle speed to the specified rpm.
2 Make sure that the throttle lever is **not** at the fast idle position on the fast idle cam.
3 With the engine off, place the transmission in N (Neutral) and set the parking brake.

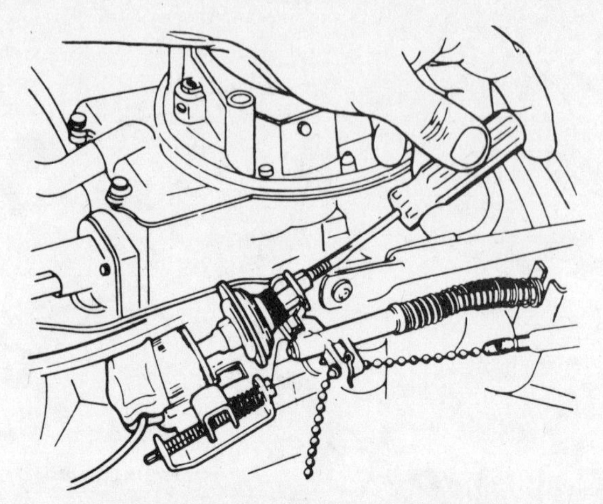

4.4 Unscrew the linkage lever adjusting screw until the screw end is flush with the lever face, ...

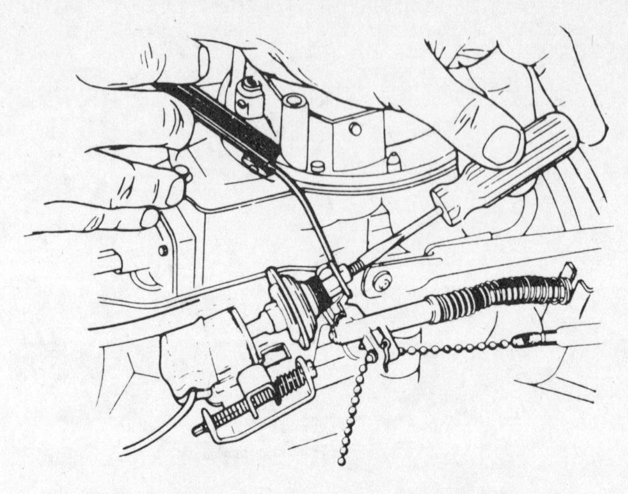

4.5 ... tighten the adjusting screw until there is a 0.005-inch gap between the screw and the throttle lever, then ...

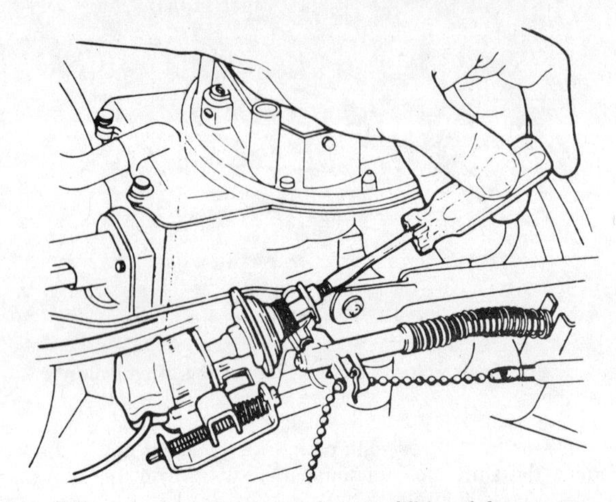

4.6 ... turn the screw in an additional four turns (two turns minimum)

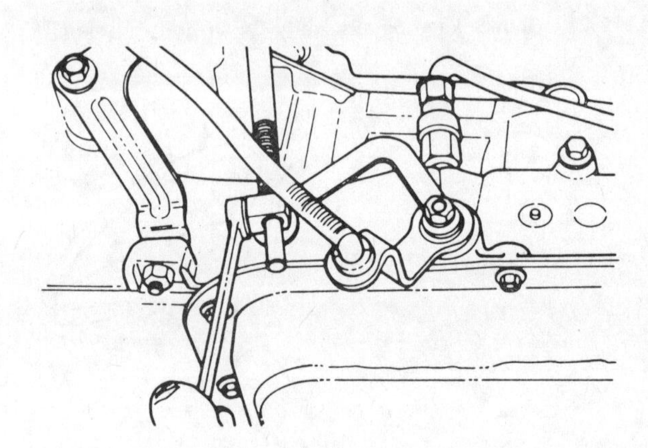

4.13 At the transmission end, loosen the bolt at the sliding trunnion block, ...

4 Turn the linkage lever adjusting screw counterclockwise until the screw end is flush with the lever face (see illustration).

5 Turn the screw in (clockwise) until there is 0.005-inch maximum clearance between the end of the screw and the throttle lever (see illustration). Push the linkage lever forward and release it before measuring the gap.

6 Turn the adjusting screw in (clockwise) an additional four turns. If screw travel is limited, a minimum of two is permissible (see illustration).

7 If the initial gap was not obtained or if the screw cannot be turned in at least two turns, the linkage will have to be adjusted at the transmission.

Adjustment at the transmission

8 Adjust the idle to the specified speed.

9 The vehicle should be raised and supported securely on jackstands to gain access to the linkage components at the TV control lever.

10 Place the transmission in N (Neutral) and set the parking brake. Turn the engine off. Be sure that the throttle lever is against its idle stop and not in the fast idle position on the fast idle cam.

11 Turn the linkage lever adjustment screw to approximately the midpoint.

12 Connect the TV control rods to the linkage lever at the carburetor if a new TV control rod assembly is being installed. **Warning:** *Be sure that the exhaust system is cool before continuing with the procedure.*

13 Loosen the bolt on the sliding trunnion block on the TV control rod assembly (see illustration). Clean the control rod, if necessary, so that the trunnion slides freely on it. Insert a pin into the lever grommet in the transmission.

14 Make sure that the linkage lever is firmly against the throttle lever at the carburetor by pushing up on the lower end of the control rod. Release the pressure on the control rod. The rod must stay up.

15 Push up on the TV control lever against its internal stop with approximately five pounds of pressure. Tighten the bolt on the trunnion block while pushing on the lever (see illustrations).

16 Lower the vehicle and check that the throttle lever is pressed against the idle stop. If the lever is not against the idle stop, repeat the adjustment procedure at the transmission (Steps 9 through 15).

5 Throttle valve (TV) cable — adjustment (AOD transmission with EFI system or six cylinder engine)

Refer to illustrations 5.1a, 5.1b, 5.1c, 5.1d, 5.5a, 5.5b, 5.6, 5.7 and 5.8

1 On six cylinder engines, the ISC plunger must be retracted before the TV cable is adjusted. Locate the Self Test connector and the Self Test Input connector near the passenger side fender well (see illustration). Connect a jumper wire between the STI connector and the ground wire in the Self Test connector (see illustrations). Turn the ignition switch to the Run position, but do not start the engine. The ISC plunger will retract (wait about 10 seconds to make sure it is fully retracted) (see illustration). Shut off the key, remove the jumper wire and proceed with the cable adjustment.

2 Set the parking brake and place the shift lever in the Neutral position.

3 Verify that the throttle lever is at the idle stop. If it isn't, check for binding and interference in the throttle linkage. **Caution:** *Do not attempt to adjust the idle stop.*

4 Make sure that the TV cable is not bent, under pressure or binding

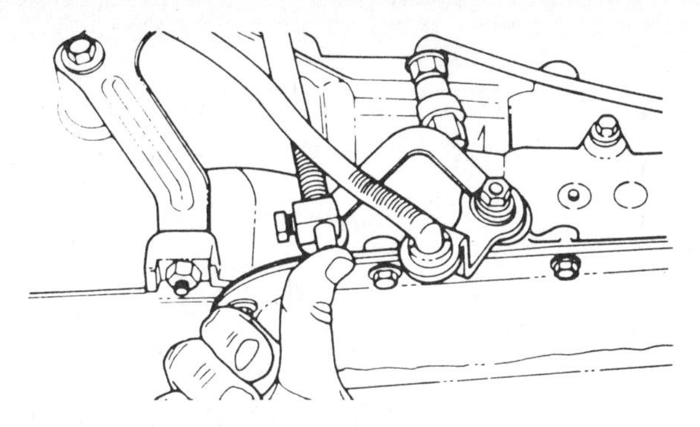

4.15a ... push up on the lower end of the control rod, then ...

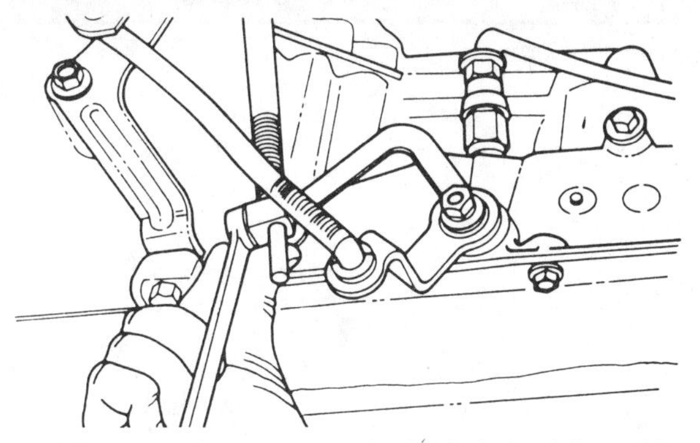

4.15b ... tighten the trunnion block bolt while pushing up on the TV control lever

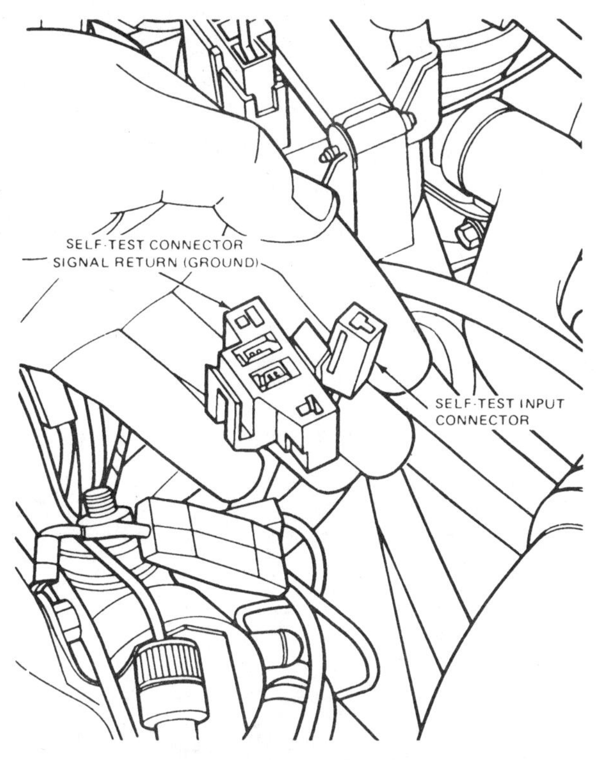

5.1a The Self Test and Self Test Input connectors are located near each other on the passenger side of the engine compartment

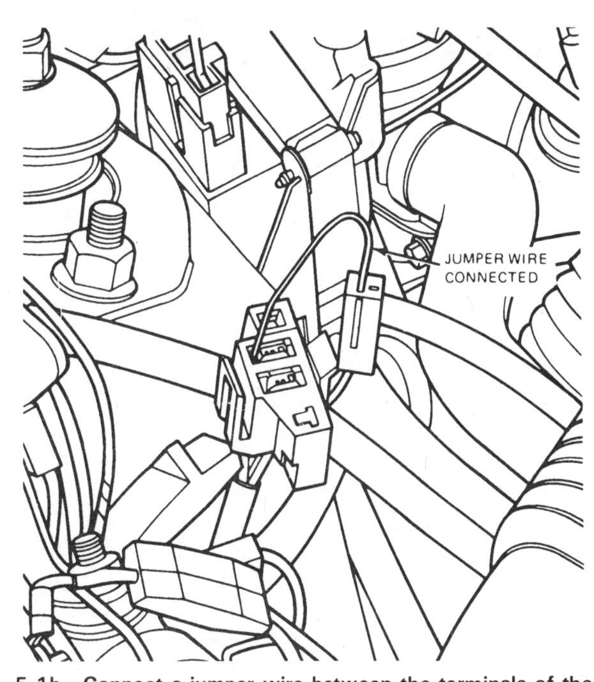

5.1b Connect a jumper wire between the terminals of the connectors as shown here

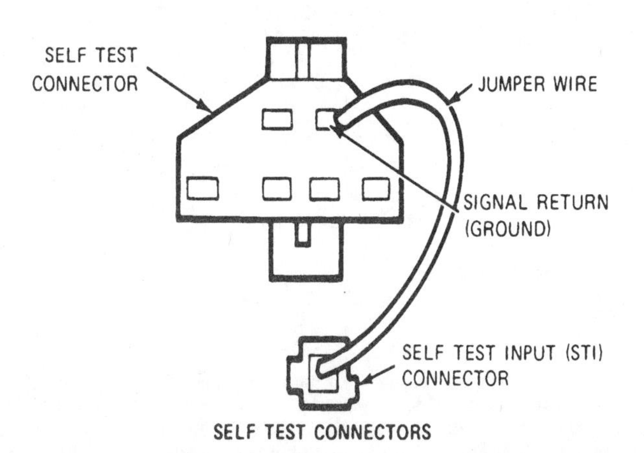

5.1c Make sure the jumper wire is attached to the ground terminal in the Self Test connector

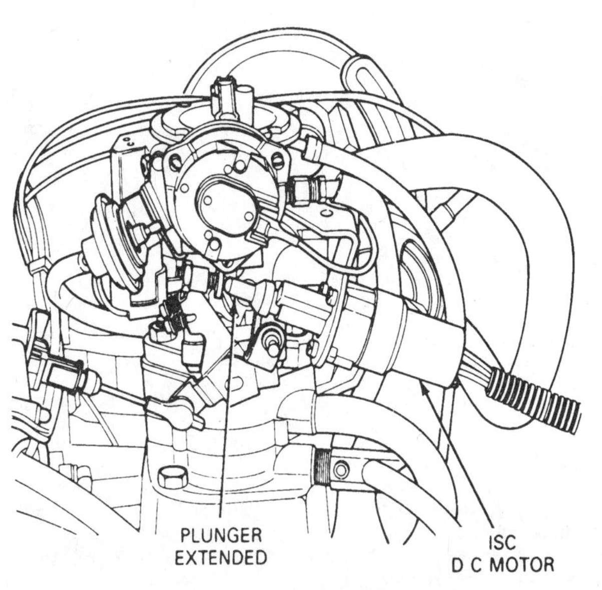

5.1d The ISC plunger (shown extended here) will retract automatically

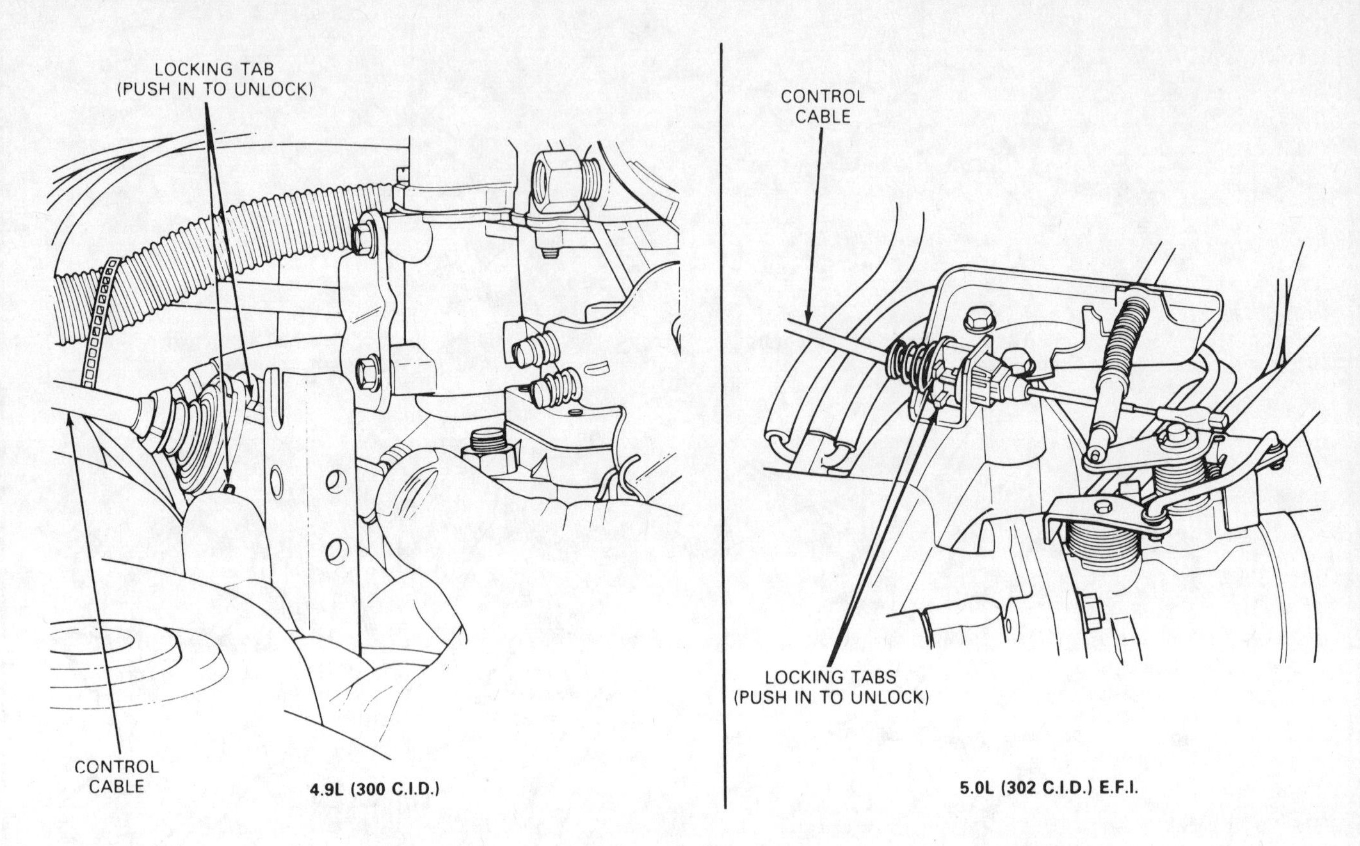

5.5a The locking tab can be released by pushing up on it . . .

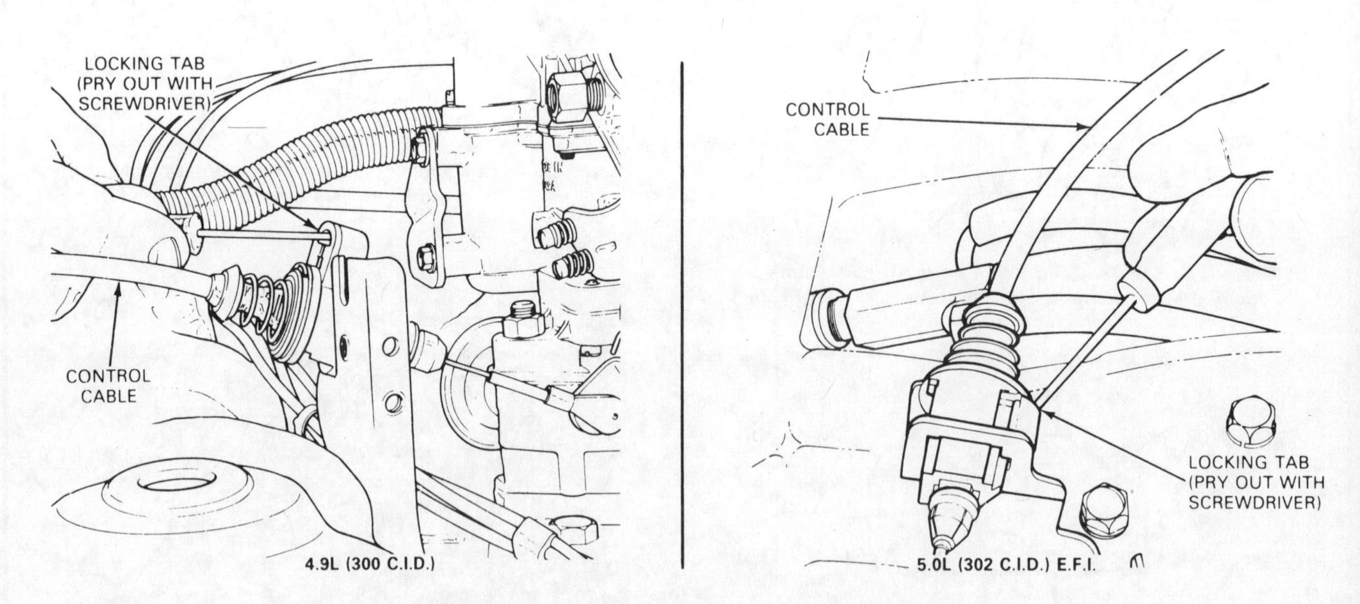

5.5b . . . and prying it out with a screwdriver

in any way. Look for damage at the cable and the rubber boot.

5 Unlock the tab at the carburetor or throttle body end of the cable by pushing up on it from below and then prying it up until the cable is free (see illustrations).

6 A spring (or springs) must be installed on the TV control lever to hold it in the idle position (as far to the rear as the lever will travel) with about ten pounds of force. Hook the spring to the transmission TV lever and the transmission case (see illustration).

7 On six cylinder engines, de-cam the carburetor (see illustration). The carburetor throttle lever must be in the anti-diesel idle position. Verify that the take-up spring (carburetor end of cable) tensions the cable properly and that the adjusting mechanism is not binding or sticking. If the spring is loose or bottomed out, check for bent cable brackets.

8 Push in on the locking tab until it is flush with the adjusting mechanism body (see illustration).

9 Remove the spring(s) installed in Step 5 above.

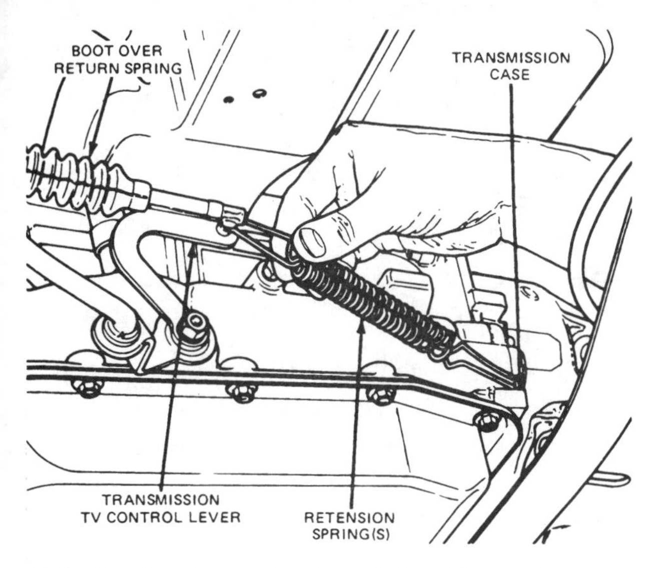

5.6 Attach springs to the transmission TV control lever to keep it in the idle position

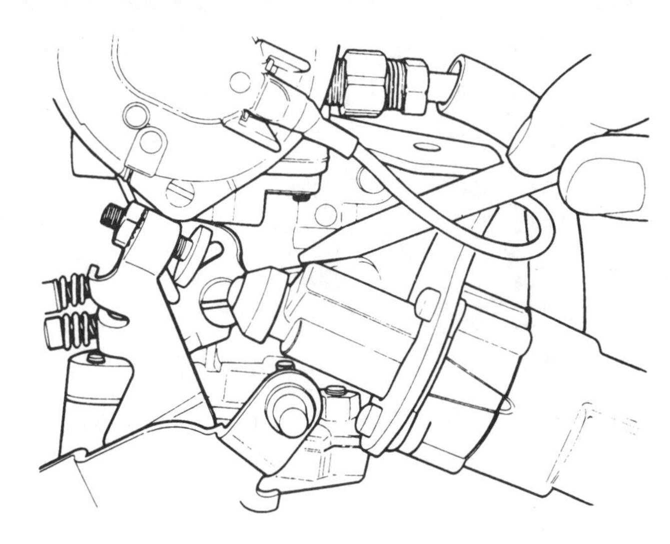

5.7 Make sure the fast idle cam is not affecting the carburetor throttle valve when adjusting the TV cable (six cylinder engine only)

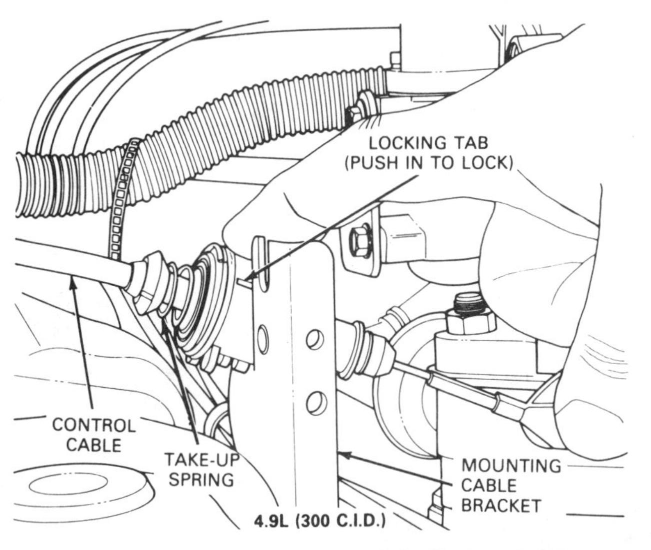

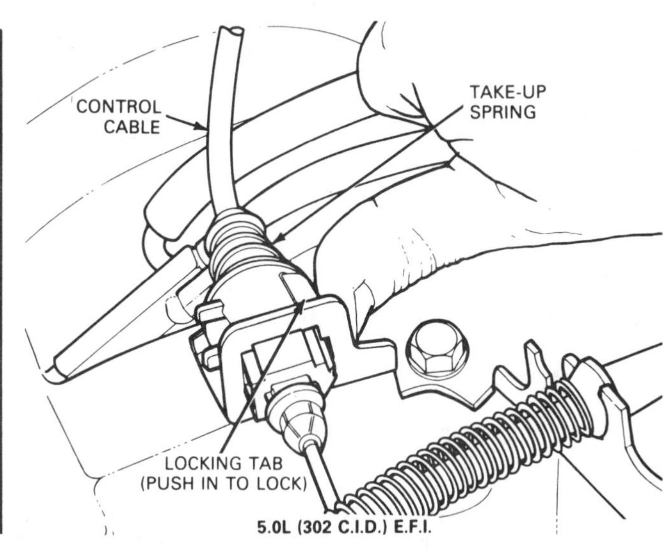

5.8 Push the locking tab in to retain the cable adjustment

7B

6 Neutral start switch – removal, installation and adjustment

Refer to illustration 6.8

1 Chock the front wheels, raise the rear of the vehicle and support it on jackstands.

2 Working under the vehicle, disconnect the kickdown linkage rod from the transmission kickdown lever.

3 Apply a little penetrating oil to the kickdown lever shaft and nut and allow it to soak for a few minutes.

4 Loosen and remove the transmission kickdown outer lever retaining nut and detach the lever.

5 Loosen and remove the two neutral start switch mounting bolts, then detach the switch.

6 Disconnect the multi-wire connector from the neutral switch.

7 To install the switch, position it on the transmission and install the bolts.

8 Place the selector lever in the Neutral position. Rotate the switch and insert a No. 43 drill bit into the gauge pin hole. It must be inserted a full 0.480-inch into the three holes of the switch (see illustration). Tighten the switch mounting bolts and remove the drill bit.

9 Reconnect the wire harness connector.

10 Make sure that the engine only starts with the selector lever in the N and P positions.

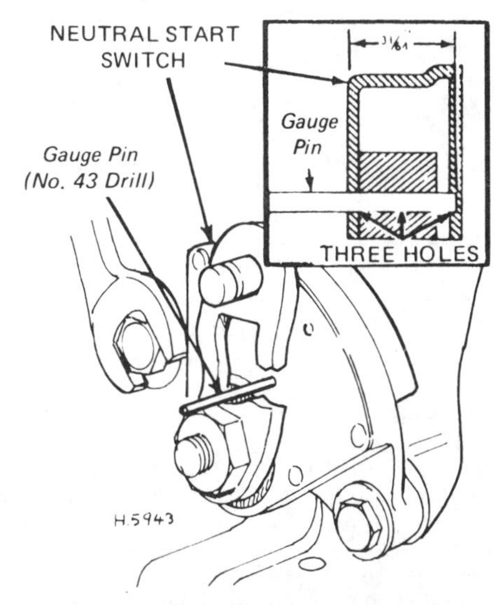

6.8 Neutral start switch adjustment details

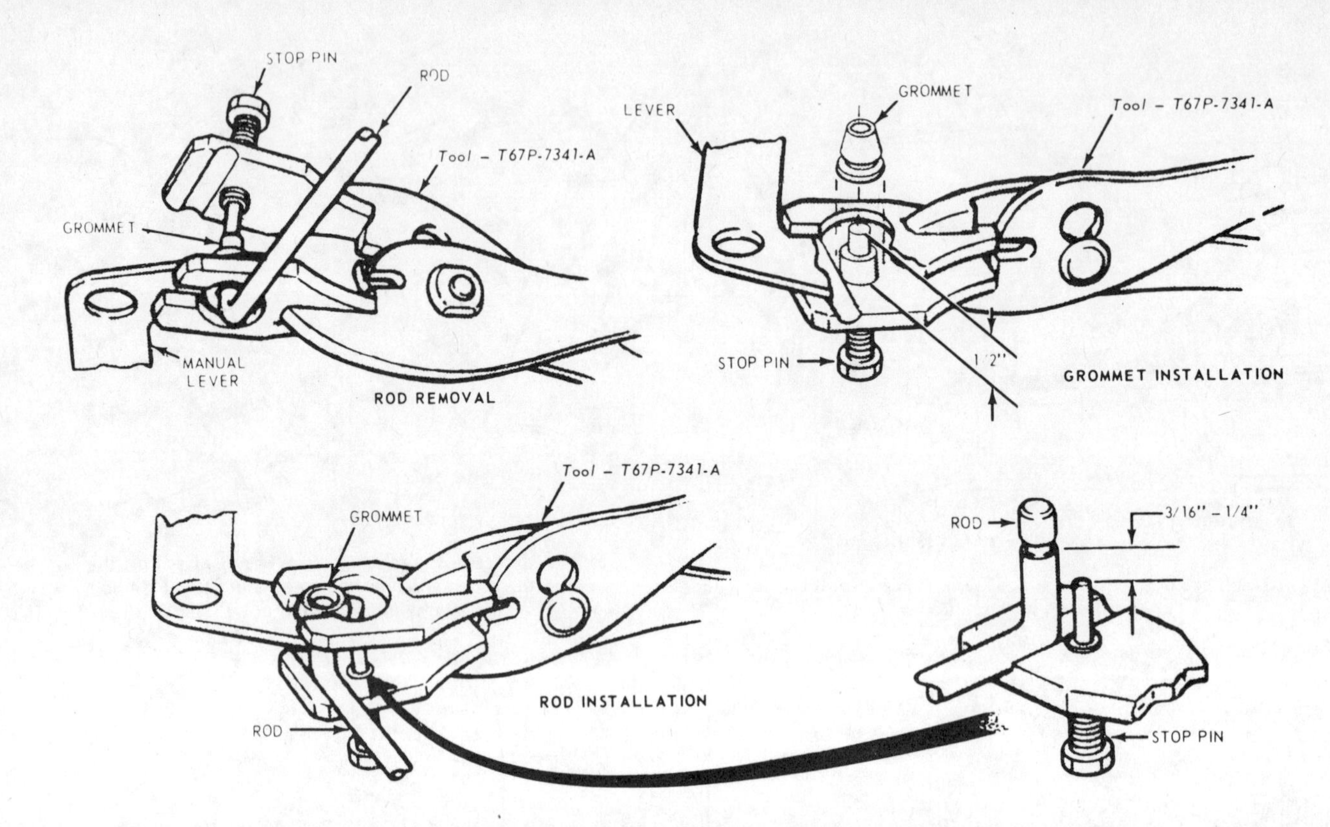

7.1 A special tool is required to replace the shift linkage grommets

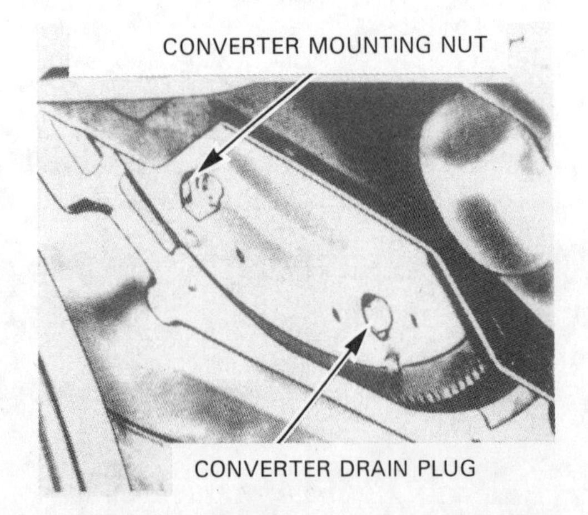

8.4 Converter mounting nut and drain plug locations

7 Shift linkage – removal and installation

Refer to illustration 7.1

1 The shift linkage rod is attached to the steering column lever through a special oil impregnated grommet. The removal and installation of this type of grommet requires Ford special tool No. T67P-7341-A (see illustration).

2 To remove the rod from the grommet, place the lower jaw of the tool between the lever and the rod. Position the stop pin against the end of the control rod and force the rod out of the grommet.

3 Remove the grommet from the lever by cutting off the largest shoulder with a knife.

4 Adjust the stop pin on the tool to 1/2-inch and coat the outside of a new grommet with lubricant.

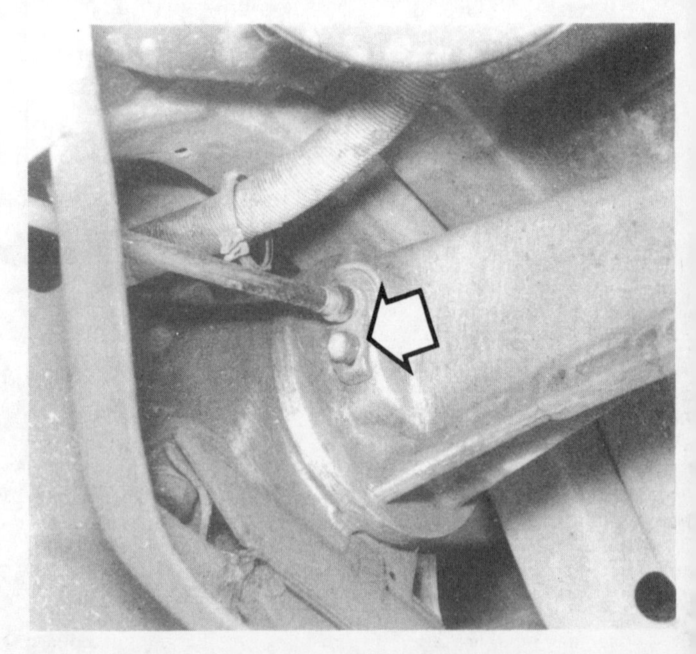

8.7 The speedometer drive cable is attached to the rear of the transmission with a clamp and bolt (arrow)

5 Position the new grommet on the stop pin and force it into the lever hole. Rotate the grommet to ensure that it is correctly seated.

6 Readjust the stop pin on the tool to the height shown in the illustration.

7 Position the rod on the tool and force the end of the rod into the grommet until the groove in the rod seats on the inner retaining lip of the grommet.

8 Adjust the shift linkage as described in Section 2.

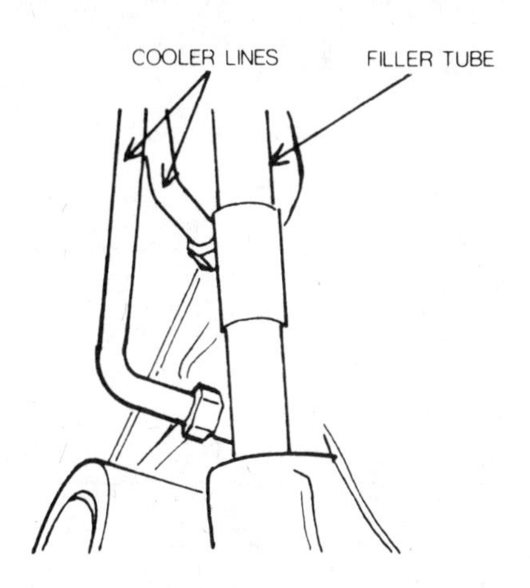

8.17 The transmission fluid filler tube and cooler lines must be disconnected before the transmission is removed

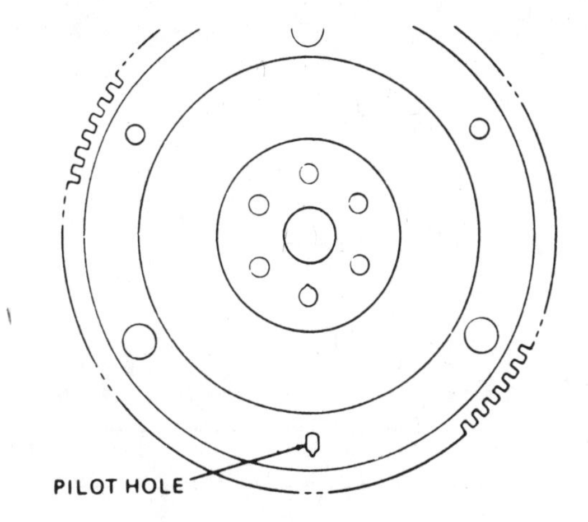

8.21 Make sure the driveplate pilot hole is in the position shown when attaching the torque converter

8 Transmission – removal and installation

Refer to illustrations 8.4, 8.7, 8.17 and 8.21

1 If possible, raise the vehicle on a hoist or place it over an inspection pit. As an alternative, raise the vehicle and support it securely on jackstands to provide the maximum possible amount of working room underneath. Disconnect the negative cable from the battery.

2 Place a large drain pan under the transmission oil pan, then, working from the rear, loosen the bolts and allow the fluid to drain out of the transmission. Remove all the bolts except the two front ones to drain as much fluid as possible, then temporarily install two bolts at the rear to hold it in place.

3 Remove the torque converter drain plug access cover and adapter plate bolts from the lower end of the converter housing.

4 Remove the driveplate-to-converter nuts. Turn the crankshaft as necessary to bring them into view (see illustration). **Caution:** *Do not rotate the engine backwards.*

5 Rotate the engine until the converter drain plug is accessible, then remove the plug, catching the fluid in the drain pan. Install and tighten the drain plug afterwards.

6 Remove the driveshaft (Chapter 8) and fasten a plastic bag over the end of the transmission to prevent the entry of dirt.

7 Detach the speedometer cable from the extension housing (see illustration).

8 Disconnect the shift rod at the transmission manual lever and the kickdown rod at the transmission kickdown lever. On AOD transmissions, detach the bellcrank bracket from the torque converter housing by unscrewing the two bolts.

9 Remove the starter motor mounting bolts and position the motor out of the way (support it to prevent strain on the wires).

10 Disconnect the neutral start switch wire connector.

11 Disconnect the vacuum lines from the vacuum unit.

12 Position a floor jack under the transmission and raise it so that it just takes up the transmission weight.

13 Remove the bolt and nut securing the rear mount to the crossmember.

14 Remove the four bolts securing the crossmember to the chassis. Raise the transmission slightly with the jack and remove the crossmember.

15 Disconnect the exhaust pipe flange(s) from the exhaust manifold(s).

16 Support the rear of the engine with a jack or large blocks.

17 Disconnect the oil cooler lines at the transmission and plug them to prevent dirt from entering (see illustration).

18 Remove the lower converter housing-to-engine bolts and the transmission filler tube.

19 Make sure that the transmission is securely strapped to the jack, then remove the two upper converter housing-to-engine bolts and the transmission filler tube.

20 Carefully move the transmission to the rear and down to separate it from the vehicle.

21 Installation of the transmission is essentially the reverse of the removal procedure, but the following points should be noted:

a) Rotate the converter to align the bolt drive lugs and drain plug with the holes in the driveplate.
b) Do not allow the transmission to take a nose-down attitude, as the converter will move forward and disengage from the pump gear.
c) When installing the three driveplate-to-converter bolts, position the driveplate so that the pilot hole is in the six o'clock position (see illustration). First install one bolt through the pilot hole and tighten it, followed by the two remaining bolts. Do not attempt to install it in any other way.
d) Adjust the kickdown rod and shift linkage (see Sections 2 and 3).
e) When the vehicle has been lowered to the ground, add transmission fluid as described in Chapter 1 until the level is correct.

Chapter 8 Clutch and driveline

Contents

Specifications

Clutch

Pedal height	See Chapter 1
Pedal free play	See Chapter 1

Rear axle

Type	
thru 1977	Semi-floating, removable differential carrier
1978 and later	Ford semi-floating, integral differential carrier (8.8 in ring gear)
	Ford semi-floating, removable differential carrier (9.9 in ring gear)
	Dana semi-floating and full-floating, integral differential carrier
Lubricant type	See Chapter 1
Axleshaft end play (1978 and later)	0.001 to 0.010 in

Torque specifications	**Ft-lbs**
Clutch and driveshaft	
Clutch housing (bellhousing)-to-engine bolts	40 to 50
Transmission-to-clutch housing bolts	37 to 42
Pressure plate-to-flywheel bolts	23 to 28
Driveshaft center bearing bolts	37 to 54
Rear axle (thru 1977)	
Carrier-to-housing nuts	25 to 40
Pinion retainer-to-carrier bolts	30 to 45
Axleshaft bearing retainer bolts	20 to 40
Pinion bearing preload (new bearing)	20 to 26
Rear axle (1978 and later — Dana)	
Pinion shaft nut	250 to 270
Pinion mate shaft lockpin	20 to 25
Cover-to-housing bolts	30 to 40
Axleshaft retaining bolts (full-floating axles)	40 to 50
Wheel bearing adjusting nut (full-floating axles)	120 to 140 (back off 1/8-to-1/4 turn on 1979 and 1980 models; 1/8-to-3/8 turn on 1981 and later models)
Rear axle (1978 and later — Ford 8.8 in integral carrier axle)	
Pinion mate shaft lock bolt (with Locktite)	15 to 30
Rear cover bolts	25 to 35

1 Clutch – general information

The clutch assembly consists of a pressure plate, a single clutch disc and a release (throwout) bearing, which pushes on the fingers of the pressure plate to release the clutch.

The pressure plate contains several springs which provide continuous surface pressure on the plate face for clutch engagement.

The clutch disc is splined and slides freely along the input shaft of the transmission. Friction lining material is riveted to the clutch disc. It has a spring cushioned hub to absorb driveline shocks and provide smooth engagement.

The clutch is operated by a series of levers and cranks known as the linkage. This linkage transfers motion at the clutch pedal into movement at the throwout bearing arm. When the clutch pedal is depressed, the throwout bearing pushes on the arms of the pressure plate assembly, which in turn pulls the pressure plate friction disc away from the lining material of the disc. Later models are equipped with a hydraulic clutch release system.

When the clutch pedal is released, the pressure plate springs force the pressure plate into contact with the lining on the clutch disc. Simultaneously, the clutch disc is pushed a fraction of an inch forward on the transmission splines by the pressure of the plate, which engages the disc with the flywheel. The clutch disc becomes firmly sandwiched between the pressure plate and the engine flywheel and engine power is transferred to the transmission.

2 Clutch – inspection (in vehicle)

1 Some vehicles are equipped with a removable inspection plate on the bottom of the clutch housing, accessible from under the vehicle. If your vehicle is so equipped, be sure to raise the vehicle and support it securely on jackstands before attempting to remove the plate.
2 Remove the bolts retaining the plate to the housing, then remove the plate.
3 Inspect the clutch assembly from the bottom of the housing. Look for broken, loose and worn parts. If no apparent defects are revealed compare the thickness of the clutch disc (sandwiched between the pressure plate and the flywheel) to a new disc. This comparison will give you some idea of the clutch disc life left and the need for replacement.
4 Reinstall the plate.
5 Remove the jackstands and lower the vehicle.

3 Clutch assembly – removal, inspection and installation

Refer to illustrations 3.1, 3.4, 3.13. 3.14a and 3.14b

Removal

1 Remove the transmission as described in Chapter 7. On vehicles with a hydraulic clutch release system, carefully detach the slave

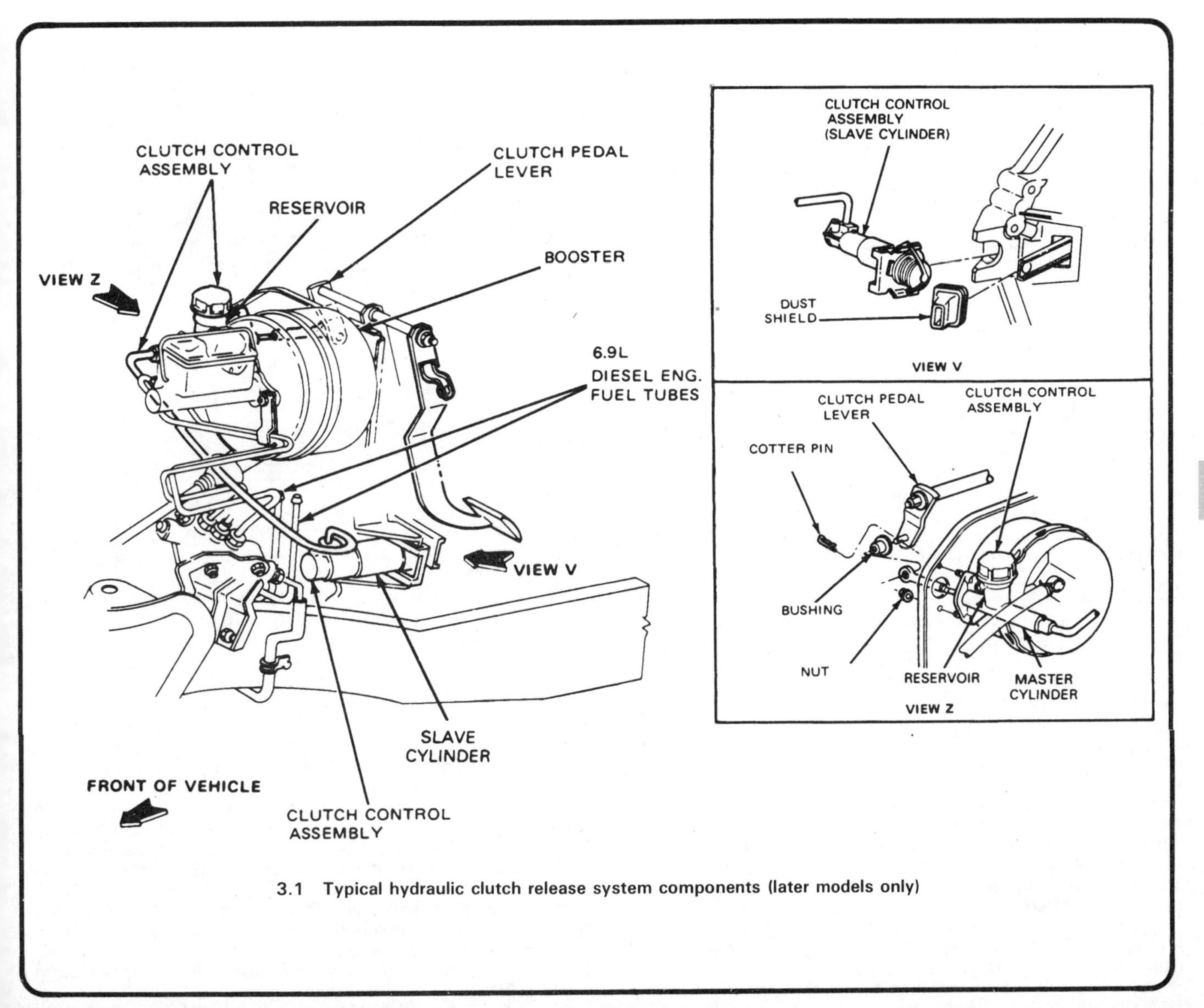

3.1 Typical hydraulic clutch release system components (later models only)

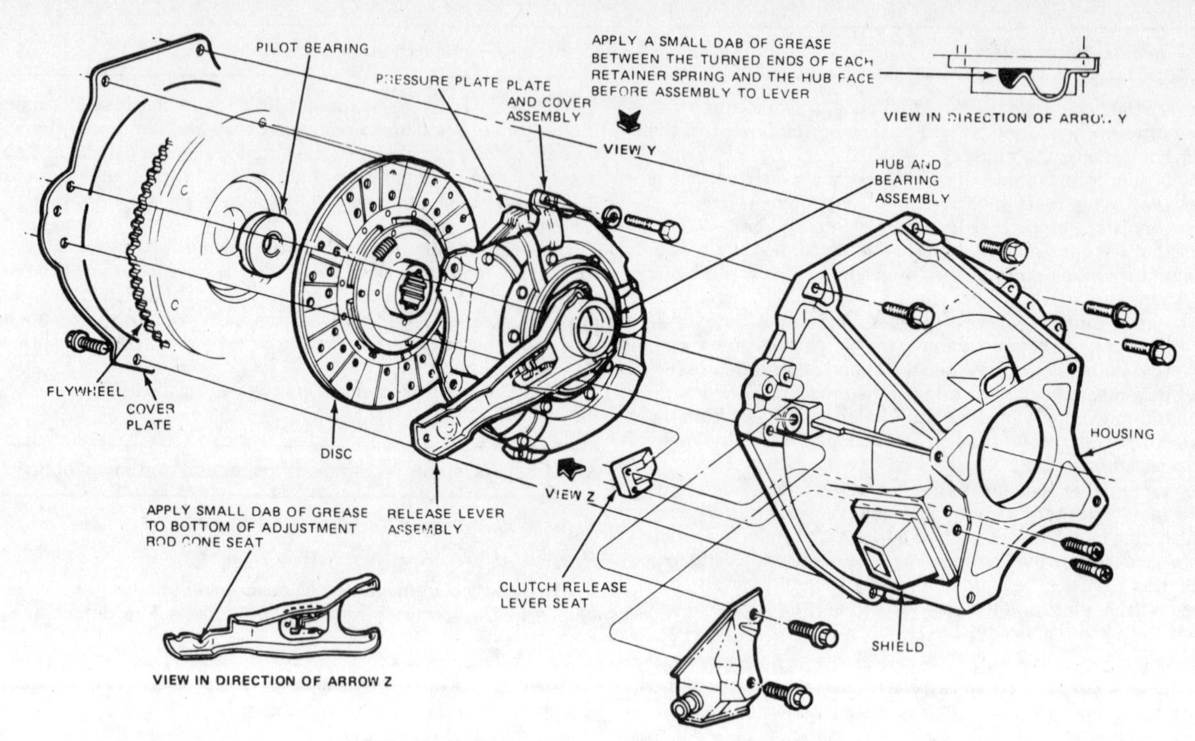

3.4 Typical clutch components — exploded view

cylinder from the bellhousing (see illustration).

2 Disconnect the clutch throwout bearing lever retracting spring and pushrod from the lever.

3 Remove the starter (Chapter 5).

4 If the bellhousing is not provided with a dust cover, the housing must be removed. Remove the bellhousing retaining bolts and remove the housing, then remove the release lever and the hub-to-release bearing assembly from the housing (see illustration).

5 If the bellhousing is provided with a dust cover, remove it from the housing, then remove the release lever and the hub and release bearing assembly from the bellhousing.

6 Scribe an indexing mark on the flywheel and pressure plate assembly if the pressure plate is to be reinstalled. This will ensure that the two components are installed in the same position relative to each other.

7 Working in a pattern around the circumference of the pressure plate assembly, loosen the retaining bolts a little at a time. This procedure is necessary to prevent warpage of the pressure plate.

8 Remove the pressure plate and clutch disc from the flywheel. These parts can be removed through the opening in the bottom of the clutch housing on models equipped with a dust cover.

Inspection

9 Closely check the clutch disc for glazed areas, cracks, warpage, the presence of oil or grease, broken hub springs and wear close to the rivets of the lining. If any of these conditions exist, replace the clutch disc with a new one. Always replace the disc if a new pressure plate is being installed.

10 Check the pressure plate for score marks, cracks, weak springs and heat marks (blue streaks on the friction surface). If any of these conditions exist, replace the pressure plate with a new one. If the clutch was chattering or rough during engagement while in operation, the pressure plate and clutch disc should be replaced.

11 Check the flywheel surface for score marks, cracks and heat marks. If these conditions exist, replace or resurface the flywheel (Section 4). Check the pilot bearing at this time.

12 Check the clutch release (throwout) bearing for roughness and excessive wear where the surface pushes on the clutch pressure plate fingers. If any of these conditions exist, replace the release bearing. It is usually a good idea to replace the release bearing any time the clutch is being serviced, as the cost is relatively small compared to the labor required to gain access to it.

3.13 Make sure the clutch disc is installed with the marked face against the flywheel

Installation

13 Place the clutch disc on the flywheel with the correct side facing the flywheel (see illustration). Make sure the disc is clean and free of grease, oil and other contaminants. Clean it off with an evaporative clutch and brake cleaner. Make sure the flywheel surface is clean and undamaged.

14 Position the pressure plate, matching the index marks made on the pressure plate and flywheel if the original pressure plate is being reused, over the disc and use an alignment tool to hold the clutch disc in correct alignment with the crankshaft axis (see illustrations).

15 Start the retaining bolts into the flywheel and tighten them finger tight. Slowly tighten the bolts, a little at a time, working around the circumference of the pressure plate. Tighten the bolts to the specified torque. Remove the clutch alignment tool.

16 The remainder of the installation procedure is the reverse of removal. Be sure to tighten all bolts securely.

3.14a Install the clutch disc/pressure plate assembly with the index marks aligned

3.14b A special tool or a transmission input shaft (shown here) can be used to center the clutch disc

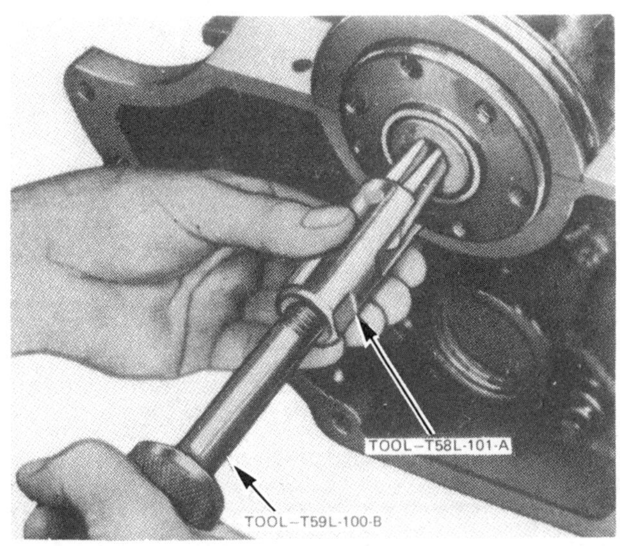

4.5 A special puller is required to remove the pilot bearing from the end of the crankshaft

4 Flywheel and pilot bearing – inspection and replacement

Refer to illustration 4.5

Inspection

1 Prior to inspecting the flywheel and pilot bearing, the transmission, clutch housing and clutch assembly must be removed.
2 Visually inspect the flywheel for score marks, cracks, warpage, and heat checking. If any of these conditions exist, the flywheel must be removed and replaced, or resurfaced at an automotive machine shop.
3 If the flywheel appears to be warped or has excessive runout, indicated by high spots on the friction surface, install a dial indicator on the rear of the engine block. Hold the crankshaft forward to take up any clearance in the crankshaft thrust bearing and slowly turn the engine through one revolution by hand. Observe the reading on the dial indicator. If the runout exceeds the Specifications (refer to Chapter 2), replace the flywheel.
4 Carefully insert your finger into the inner race of the pilot bearing located in the center of the crankshaft flange. Check for burrs and scoring. Rotate the bearing and feel for roughness and excessive play. If any of these conditions exist, replace the pilot bearing with a new one.

Replacement

5 Remove the pilot bearing with the special inside puller designed for this purpose (see illustration). You can purchase one from an auto parts store or they are often available from rental yards.
6 Install the bearing using the special tool.
7 If the flywheel is to be replaced, refer to Chapter 2.

5 Clutch linkage – removal and installation

Refer to illustration 5.3

1 The most likely source of wear in the clutch linkage is the equalizer shaft bushings.
2 To remove the shaft, jack up the front of the vehicle and support it on jackstands.
3 Disconnect the clutch release lever retracting spring (see illustration).

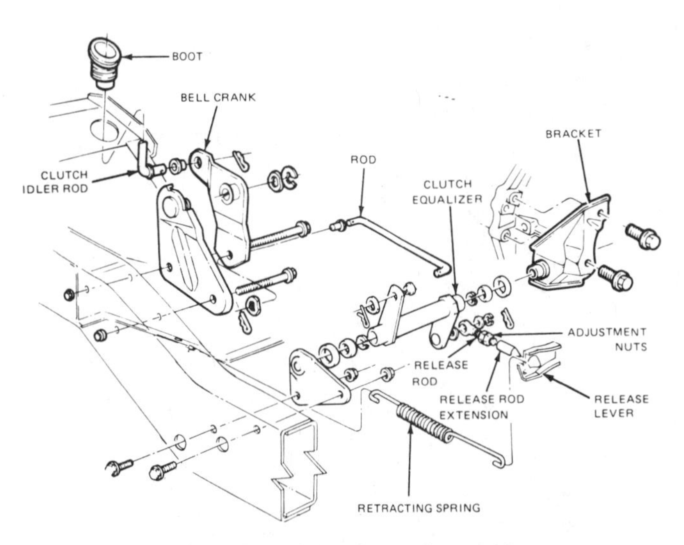

5.3 Typical clutch equalizer shaft components – exploded view

4 Remove the spring retainer and washer and detach the clutch release rod from the equalizer shaft.
5 Remove the spring clip and washer from the end of the clutch relay rod and detach the rod from the equalizer shaft.
6 Remove the nuts and bolts securing the equalizer shaft mounting bracket to the side of the chassis and slide the bracket and shaft assembly off the bellhousing mounting bracket.
7 Remove the bushings from inside the ends of the equalizer shaft and install new ones.
8 Reinstall the equalizer shaft and clutch linkage by reversing the removal procedure.
9 Adjust the clutch as described in Chapter 1.

6 Clutch pedal — removal and installation

Refer to illustrations 6.1 and 6.13

Mechanical clutch

1 From beneath the dash panel remove the spring clip and disconnect the operating rod from the clutch pedal lever (see illustration). Retrieve the bushing from the lever.
2 Remove the locknut securing the clutch lever to the clutch pedal shaft assembly and remove the lever.
3 Remove the shoulder bolt, bushing and locknut securing the master cylinder pushrod to the brake pedal.
4 Pull the clutch pedal and shaft out and remove them together with the brake pedal and bushings.
5 Examine the bushings and shaft for wear and replace them if necessary.
6 Grease the shaft and bushings and reinstall the clutch and brake pedal assembly using the reverse of the removal procedure.
7 Adjust the clutch and pedal height as described in Chapter 1.

Hydraulic clutch

8 Disconnect the clutch pedal return spring from the clutch pedal and bracket. Disconnect the barbed end of the clutch/starter interlock switch rod from the clutch pedal.
9 Remove the nut retaining the clutch pedal to the shaft and detach the clutch pedal. Remove the bushing from the shaft.
10 Install the bushing on the shaft. Position the clutch pedal on the shaft and tighten the nut.
11 Install the return spring. Make sure the spring engages the slots in the pedal and bracket.
12 Insert the barbed end of the clutch/starter interlock switch rod into the bracket on the clutch pedal. If required, adjust the clip on the interlock switch rod.
13 If the adjusting clip is out of position on the rod, remove both halves of the clip. Position both halves of the clip closer to the switch and snap the clips together on the rod. Depress the clutch pedal to the floor to adjust the switch (see illustration).

7 Driveshaft — general information

Refer to illustrations 7.1, 7.2a and 7.2b

The short wheelbase Econoline models are equipped with a one piece tubular driveshaft with a universal joint at each end. The front of the shaft has a slip yoke that is splined onto the transmission output shaft, while the rear of the shaft is attached to the rear axle pinion flange by two U-bolts (see illustration).

Long wheelbase models are equipped with a two-piece driveshaft made up of a front coupling shaft, a center support bearing and a rear driveshaft. If the vehicle is equipped with an automatic transmission, the front of the shaft is attached to the rear of the transmission output shaft by a splined slip yoke (see illustration). In the case of a manual transmission the shaft is secured by a flange and U-bolts (see illustration).

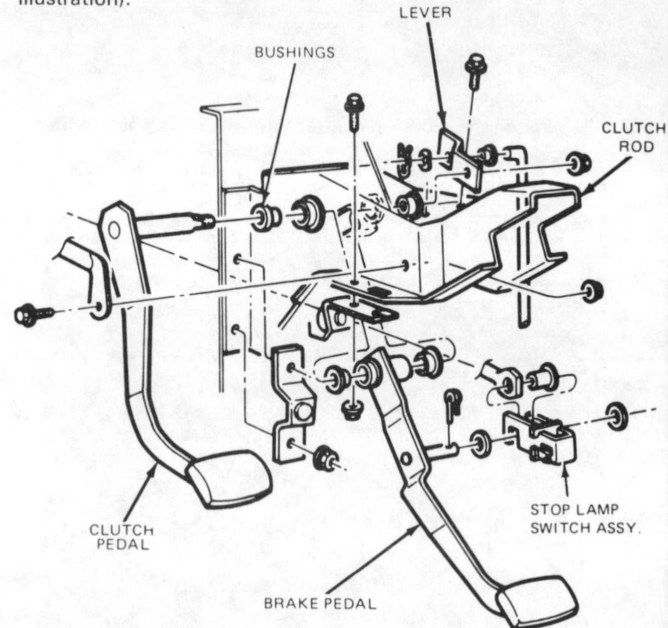

6.1 Typical clutch pedal components — exploded view

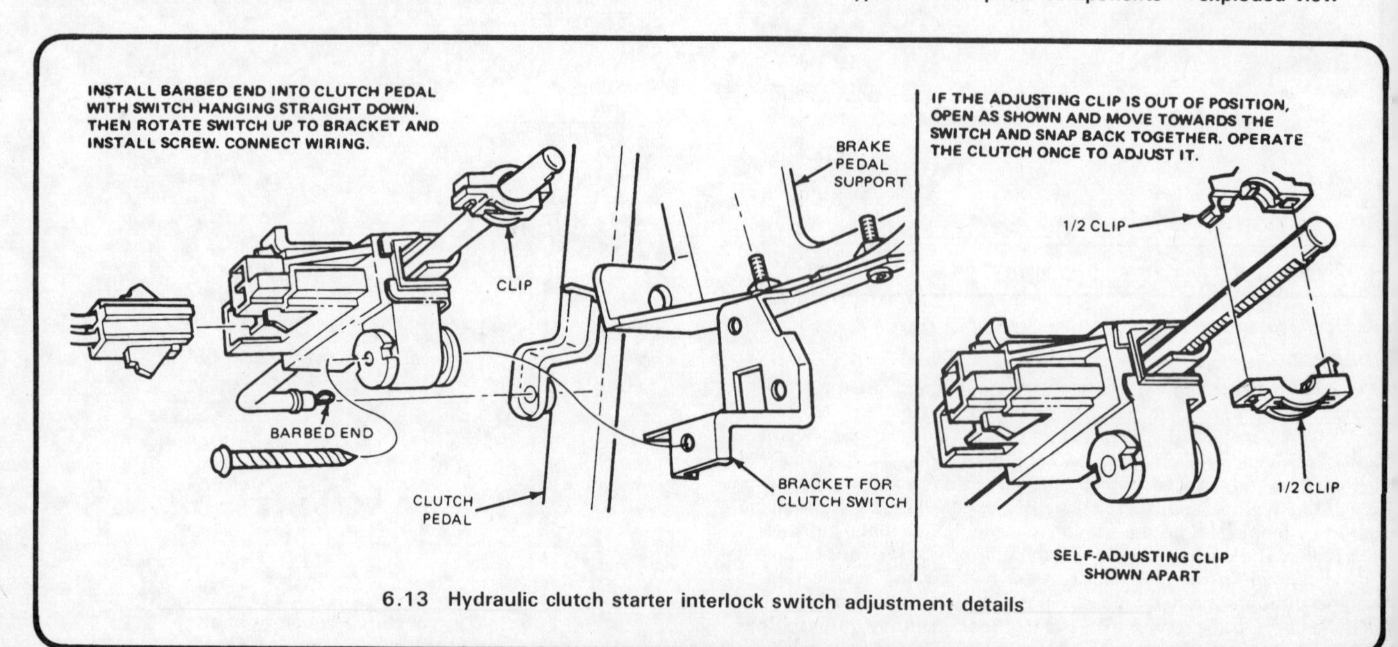

6.13 Hydraulic clutch starter interlock switch adjustment details

Some driveshafts have grease nipples on the universal joints and sliding yokes. They should be lubricated at periodic intervals. The center support bearing used on long wheelbase models is pre-lubricated and sealed for the life of the bearing.

The driveshaft is balanced during production and it is recommended that care be used when universal joints are replaced to help maintain this balance. It is sometimes better to have the universal joints replaced by a dealership or shop specializing in this type of work. If you replace the joints yourself, mark each yoke in relation to the one opposite it in order to maintain the balance. Do not drop the assembly during servicing operations.

8 Driveshaft – removal and installation

Refer to illustration 8.3

Note: *Where two piece driveshafts are involved, the rear shaft must be removed before the front shaft.*

1 Raise the vehicle and support is securely on jackstands.

2 Use chalk or a scribe to index the relationship of the driveshaft(s) to the mating flange. This ensures correct alignment when the driveshaft is reinstalled.

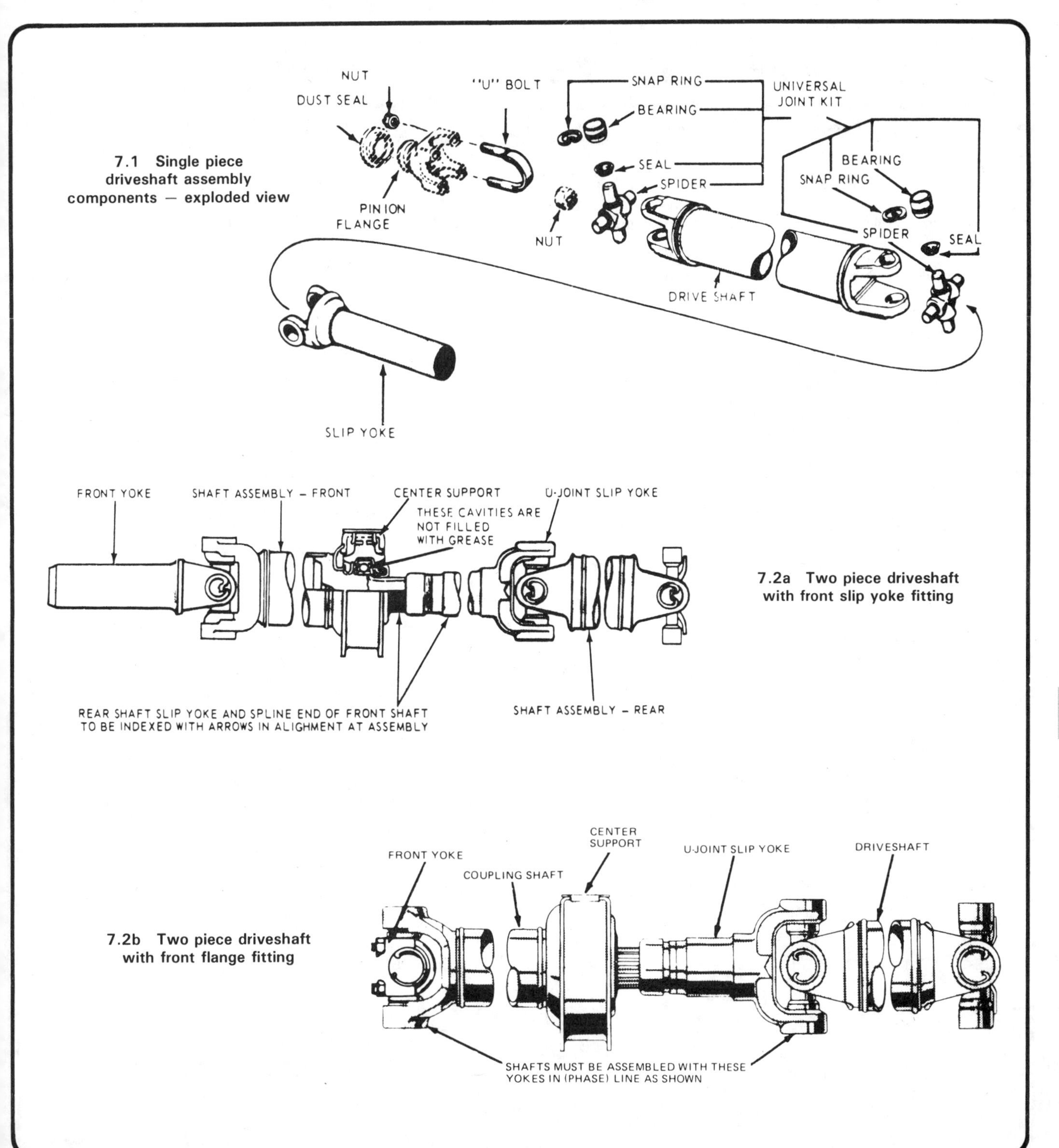

7.1 Single piece driveshaft assembly components – exploded view

7.2a Two piece driveshaft with front slip yoke fitting

7.2b Two piece driveshaft with front flange fitting

8

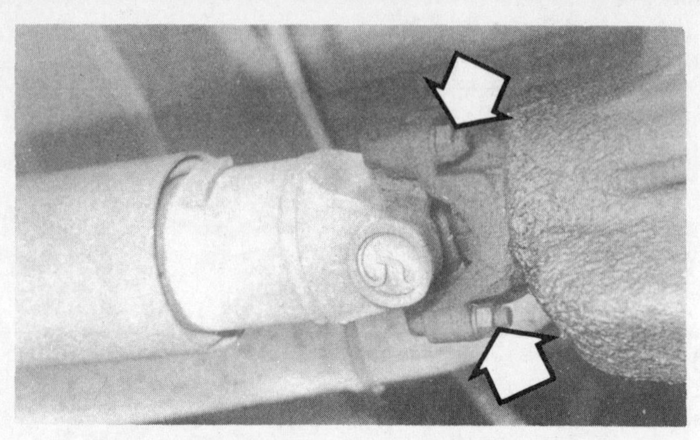

8.3 The U-bolt nuts (arrows) must be removed to separate the driveshaft from the pinion flange

3 Remove the nuts or bolts securing the universal joint clamps to the flange (see illustration). If the driveshaft has a spline on one end (either at the transmission or the center carrier bearing) be sure to place marks on the mating flange or shaft to retain proper alignment during reinstallation.
4 Remove the nuts or bolts retaining the straps or universal joint to the flange on the opposite end of the driveshaft (if so equipped).
5 Pry the universal joint away from its mating flange and remove the shaft from the flange. Be careful not to let the bearing caps fall off of the universal joint, which would cause contamination and loss of the needle bearings.
6 Repeat this process for the opposite end if it is equipped with a universal joint coupled to a flange.
7 If the opposite end is equipped with a sliding joint (spline), simply slide the yoke off the splined shaft.
8 If the shaft being removed is the front shaft of a two piece unit, the rear is released by unbolting the two bolts securing the center bearing assembly. Again, make sure both ends of the shaft have been marked for installation purposes.
9 Installation is the reverse of removal. If the shaft cannot be lined up because the differential or transmission has been rotated, put the transmission in Neutral or rotate one wheel to allow the original alignment to be achieved. Always tighten the retaining nuts or bolts to the correct torque and make sure the universal joint bearing caps are properly positioned in the flange seat.

9 Driveshaft center bearing — check and replacement

Refer to illustration 9.3

1 Remove the complete driveshaft and bearing assembly as described in the previous Section.
2 Check the bearing for wear or rough action by rotating it on the shaft. Examine the rubber support cushion for cracks and general deterioration. If any doubt exists as to the condition of the assembly, the best approach is to replace the bearing and rubber support.
3 Remove the bearing and rubber insulator (see illustration) from the front coupling shaft by either pressing it off or using a puller.
4 Press the new bearing and insulator onto the shaft until it is in contact with the flange. The new bearing is pre-packed with grease and no lubrication is required.
5 Install the support bearing and driveshafts on the vehicle as described in Section 9.

10 Universal joints — general information

Universal joints are mechanical couplings which connect two rotating components that meet each other at different angles.

These joints are composed of a yoke on each side connected by a cross piece called a trunnion or spider. Cups at each end of the trunnion contain needle bearings which provide smooth transfer of torque loads. Snap-rings, either inside or outside of the bearing cups, hold the assembly together.

11 Universal joints — check and lubrication

1 Refer to Chapter 1 for details on universal joint lubrication. Also see the routine maintenance schedule at the beginning of Chapter 1.
2 Wear in the needle bearings is characterized by vibration in the transmission, noise during acceleration, and in extreme cases of lack of lubrication, metallic squeaking and, ultimately, grating sounds as the bearings disintegrate.
3 It is easy to check if the bearings are worn with the driveshaft in position by trying to turn the shaft with one hand. Use your other hand to hold the rear axle flange when the rear universal joint is being checked and the front half coupling when the front universal joint is being checked. Any movement between the driveshaft and the couplings is indicative of considerable wear. Another method of checking for universal joint wear is to use a pry bar inserted into the gap between the universal joint and the driveshaft or flange. Leave the vehicle in gear and try to pry the joint both radially and axially. Any looseness should be apparent with this method. A final test for wear is to attempt to lift the shaft and note any movement between the yokes of the joints.
4 If any of the above conditions exist, replace the universal joints with new ones.

12 Universal joints — replacement

Refer to illustrations 12.2, 12.3a and 12.3b

1 Remove the driveshaft.
2 Extract the snap-rings from the ends of the bearing cups (see illustration).
3 Using sockets or pieces of pipe of suitable diameter, use a vise to press on the end of one cup to displace the opposite one into the

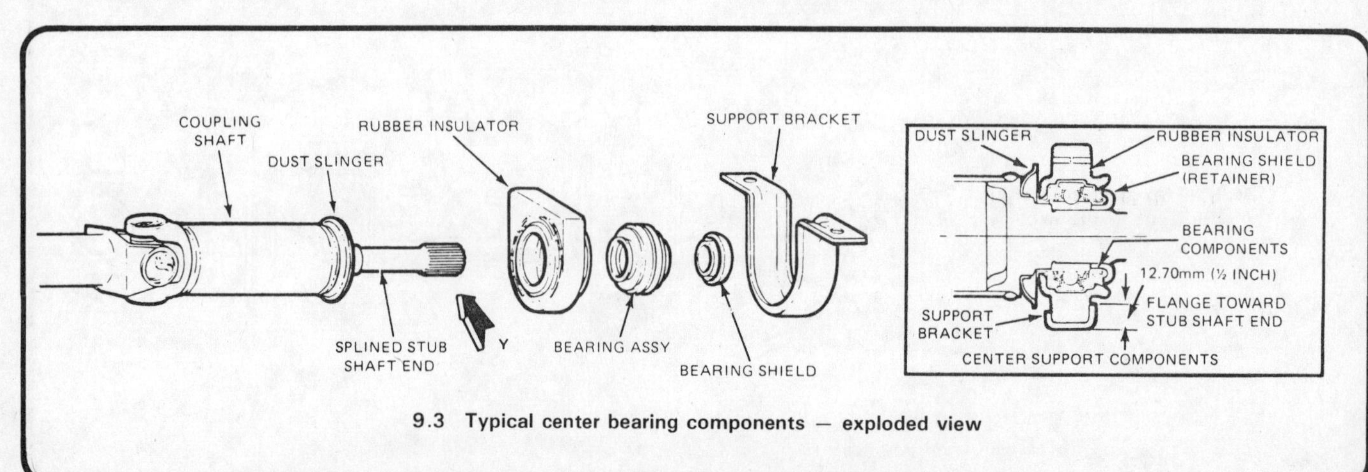

9.3 Typical center bearing components — exploded view

12.2 Removing the snap-ring from a bearing cup

12.3a Removing the bearing cups from the yoke with sockets and a large vise

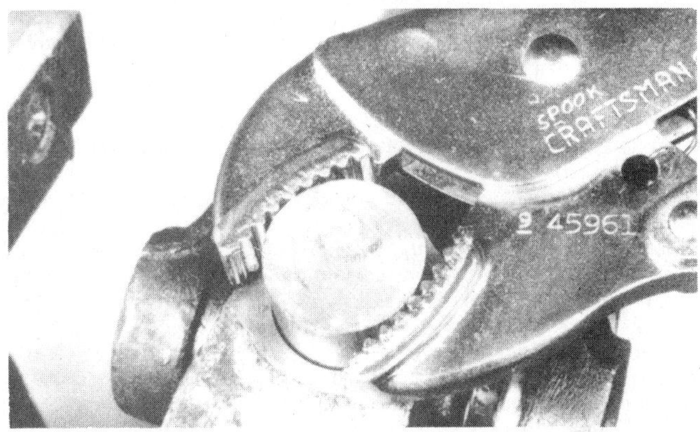

12.3b Pliers can be used to grip the cup to detach it from the yoke after it has been pushed out

larger socket or pipe. The bearing cup will not be completely ejected, so it should be gripped with pliers and twisted out of the yoke (see illustrations).

4 Remove the other bearing cup by pressing the spider in the opposite direction.

5 Clean the yoke and check it for damage and cracks.

6 Obtain the appropriate replacement U-joint.

7 Position the spider in the yoke, partially install the opposite cup, center the spider, then, using the vise, press both cups into position using sockets with a diameter slightly less than that of the bearing cups. Make sure that the needle bearings are not displaced and trapped during this operation.

8 Install the snap-rings.

13 Rear axle — general information

Refer to illustration 13.2

The rear axle assembly consists of a straight, hollow housing enclosing a differential assembly and axleshafts.

The axle assemblies employed on vehicles covered by this manual may be one of two designs: those with semi-floating axleshafts and those with full-floating axleshafts (see illustration). As a general rule,

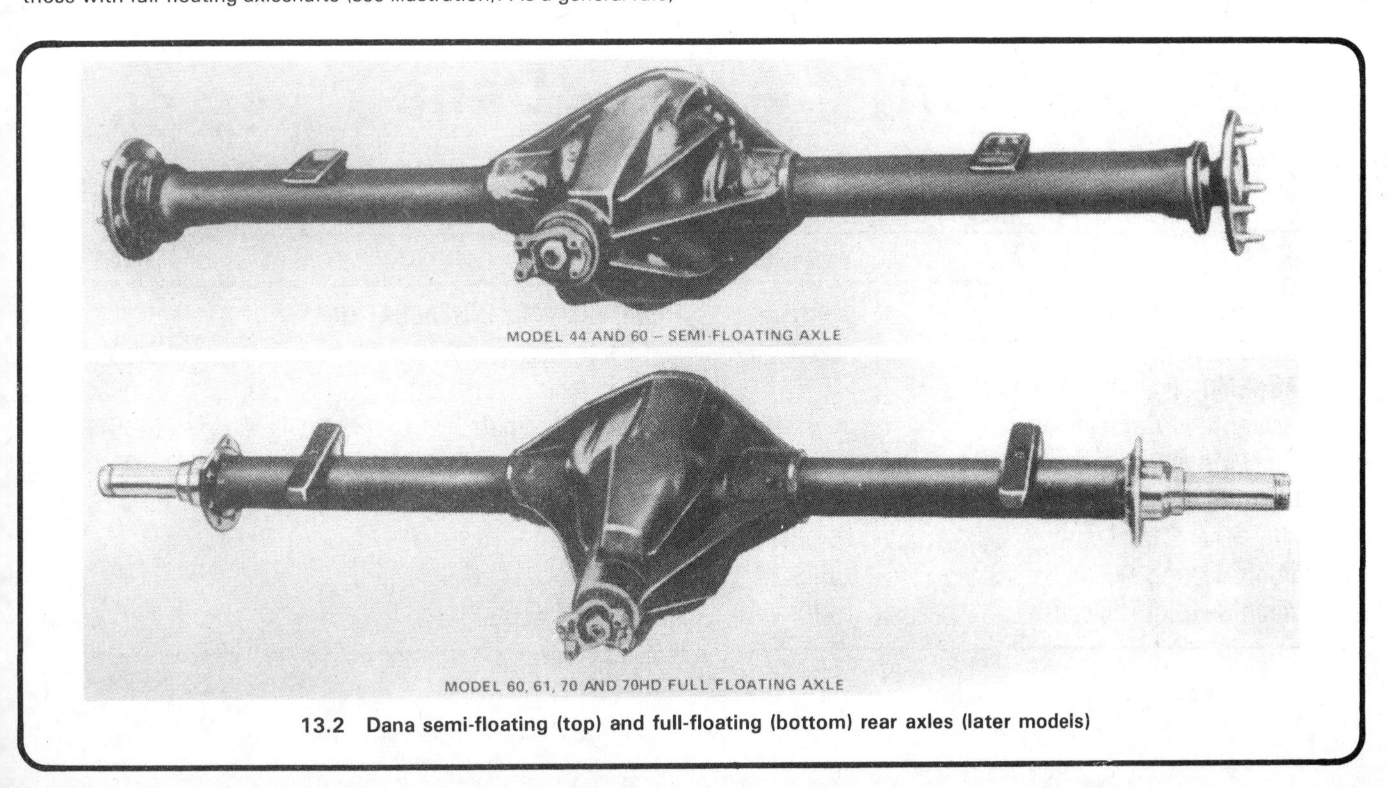

13.2 Dana semi-floating (top) and full-floating (bottom) rear axles (later models)

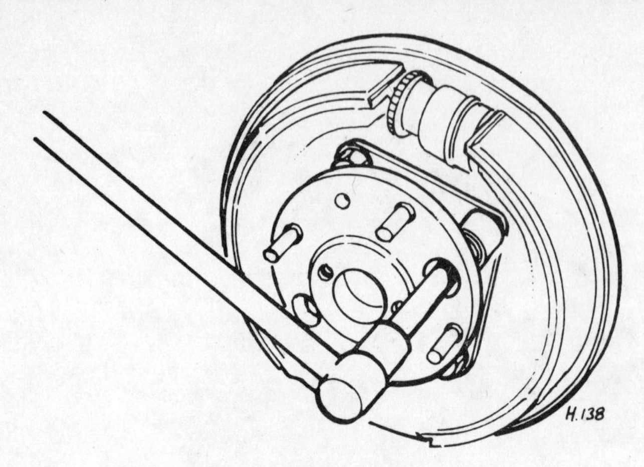

14.3 Removing the axleshaft retaining nuts

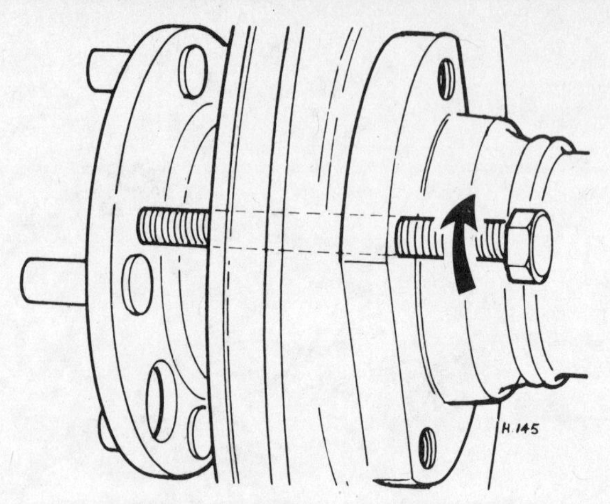

14.4 Long bolts threaded into the housing flange can be used to push a stubborn axleshaft out

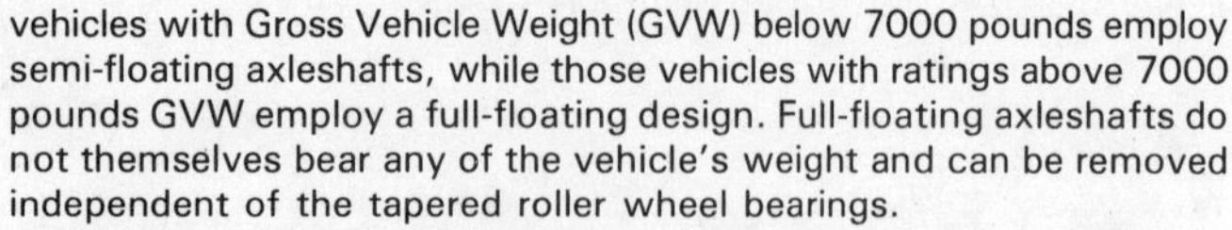

vehicles with Gross Vehicle Weight (GVW) below 7000 pounds employ semi-floating axleshafts, while those vehicles with ratings above 7000 pounds GVW employ a full-floating design. Full-floating axleshafts do not themselves bear any of the vehicle's weight and can be removed independent of the tapered roller wheel bearings.

Both types of rear end designs use hypoid gears with the ring gear centerline below the axleshaft centerline. The lighter duty type of rear axle has either a 9-inch ring gear differential with a drop out type of housing or an 8.8-inch ring gear differential with an integral carrier. The heavier duty rear ends are manufactured by Dana and have integral carrier housings.

Due to the need for special tools and equipment, it is recommended that operations on these models be limited to those described in this Chapter. Where repair or overhaul is required, remove the axle assembly and take it to a rebuilder or exchange it for a new or reconditioned unit.

Routine maintenance and minor repair procedures can be performed without removing the differential assembly from the axle housing or the rear axle assembly from the vehicle. The axleshafts, wheel hubs, wheel bearings, grease seals and wheel hub lugs can be serviced as described above.

14 Axleshaft – removal and installation (Ford removable carrier type)

Refer to illustrations 14.3 and 14.4

1 Place the vehicle on level ground, block the front wheels, loosen the rear wheel nuts on the side to be worked on (both sides if both axleshafts are being removed), then raise the rear of the vehicle and remove the wheels. Position jackstands underneath the vehicle.

2 Release the parking brake, then remove the brake drum.

3 Loosen and remove the four bolts retaining the axleshaft bearing housing to the axle housing. The bolts are accessible with a socket on an extension inserted through the holes in the axleshaft flange (see illustration).

4 It should be possible at this stage to remove the axleshaft by simply pulling on the flange, especially if tapered roller bearings are used. If this fails, thread long bolts into two of the axle housing flange holes from the back side, opposite each other, and push the axle out by tightening the bolts evenly (see illustration).

5 The axleshaft seals are made out of synthetic material and are easily

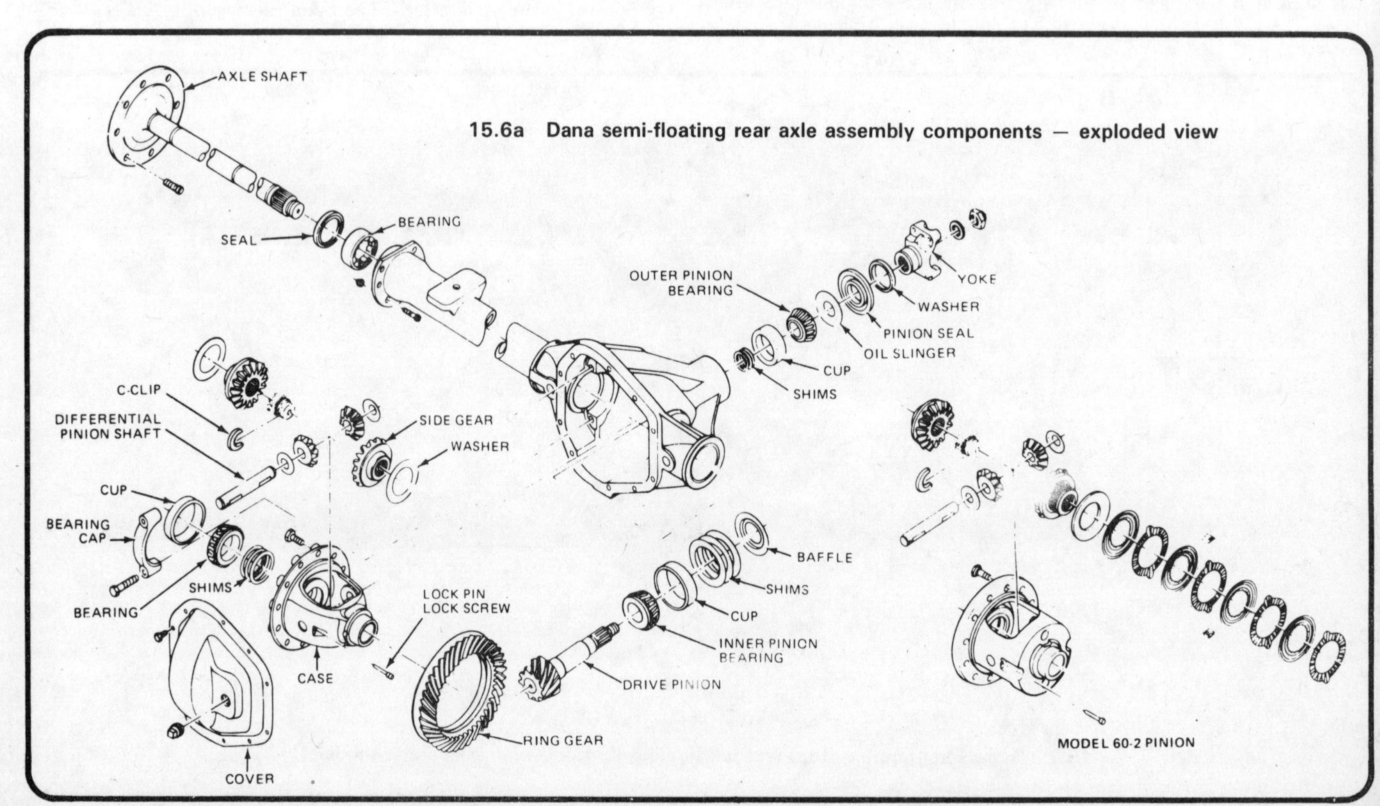

15.6a Dana semi-floating rear axle assembly components — exploded view

damaged. When withdrawing the shaft, be careful not to cut the seal inner lip with the axleshaft splines.

6 Installation is the reverse of removal, but again, care must be taken to avoid damaging the seal.

15 Axleshaft (semi-floating type) – removal and installation

Refer to illustrations 15.6a, 15.6b, 15.6c, 15.7 and 15.8

Note: *This procedure applies to Ford 8.8 inch and Dana equipped differentials, 1978 and later.*

1 Raise the rear of the vehicle and place it securely on jackstands.

2 Remove the wheel(s).

3 Release the parking brake and remove the brake drum(s).

4 Drain the rear axle lubricant into a container by removing the rear axle housing cover.

5 If still in place, discard the gasket.

6 Remove the differential pinion shaft lockpin or lockbolt (see illustrations) and discard it. **Note:** *It is possible for some Dana floating axles to be equipped with lockpins coated with Loctite (or equivalent) or with lockpins with torque-prevailing threads. The Loctite treated lockpins have a 5/32-inch hexagon socket head and the torque-prevailing lockpin has a 12-point drive head. If the axle is equipped with a Loctite treated lockpin, it must not be reused under any circumstances. If the lockpin is a torque-prevailing type, it may be reused up to four times (four removals and installations). When in doubt as to the number of times the torque-prevailing pin has been used, replace it with a new one.*

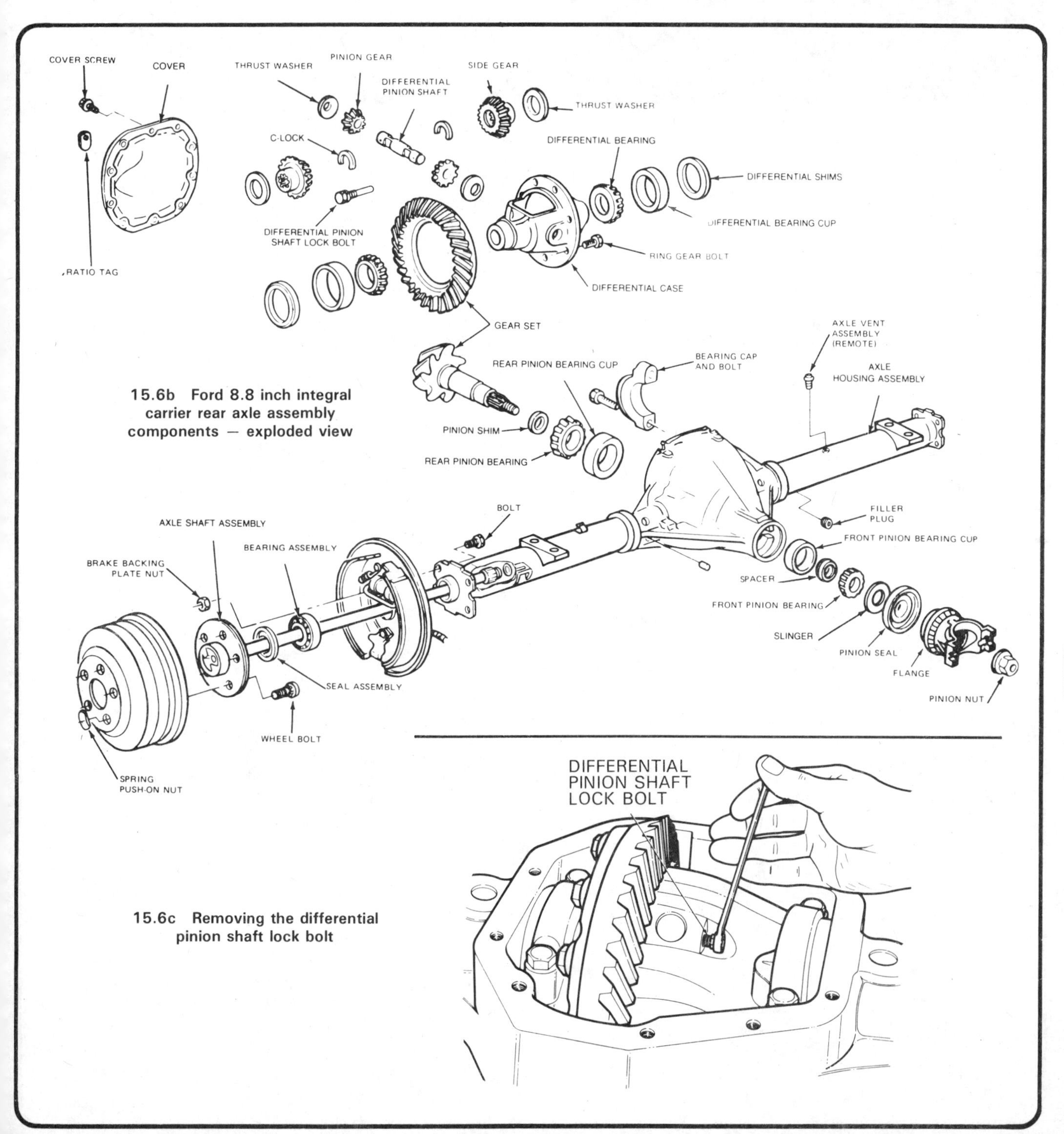

15.6b Ford 8.8 inch integral carrier rear axle assembly components – exploded view

15.6c Removing the differential pinion shaft lock bolt

8

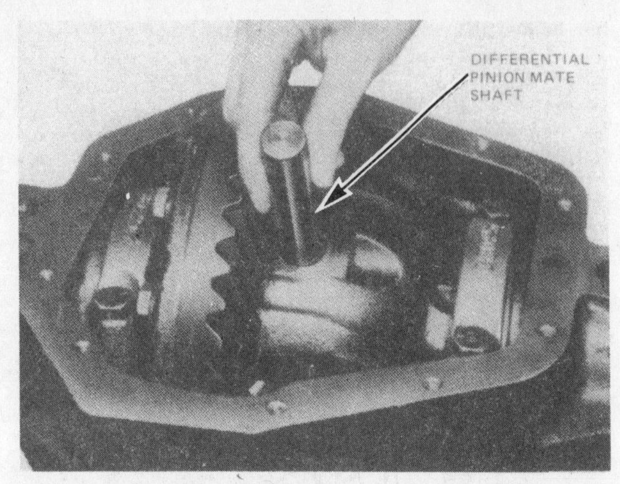

15.7 Withdraw the pinion mate shaft from the differential, . . .

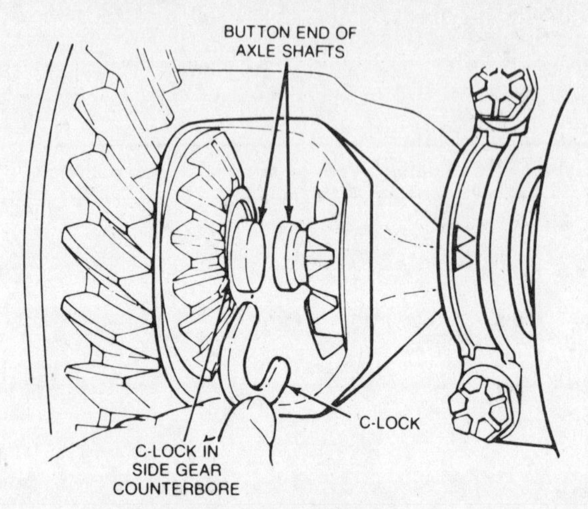

15.8 . . . then remove the C-lock from the end of the axleshaft

7 Lift out the differential pinion mate shaft (see illustration).

8 Push the flanged end of the axleshaft toward the center of the vehicle and remove the C-lock from the button end of the shaft (see illustration). **Note:** *Do not lose or damage the rubber O-ring which is in the axleshaft groove under the C-lock.*

9 Pull the axleshaft from the housing, taking care not to damage the oil seals.

10 Installation is basically the reverse of the removal procedure. Be very careful not to damage the axle seal when reinstalling the axleshaft (the splines on the end of the shaft are sharp). Tighten the pinion mate shaft lockpin to the specified torque.

11 Most axle housing covers are sealed with silicone rubber sealant rather than a gasket. Before applying this sealant (Ford No. D6AZ-19562-B or equivalent), make sure the machined surfaces on both cover and carrier are clean and free of oil. When cleaning the surfaces, cover the inside of the axle with a clean, lint-free cloth to prevent contamination. Apply a continuous bead of sealant to the carrier face, inside the cover bolt holes. Install the cover within 15 minutes of the application of the sealant and tighten the bolts in a crisscross pattern to the specified torque.

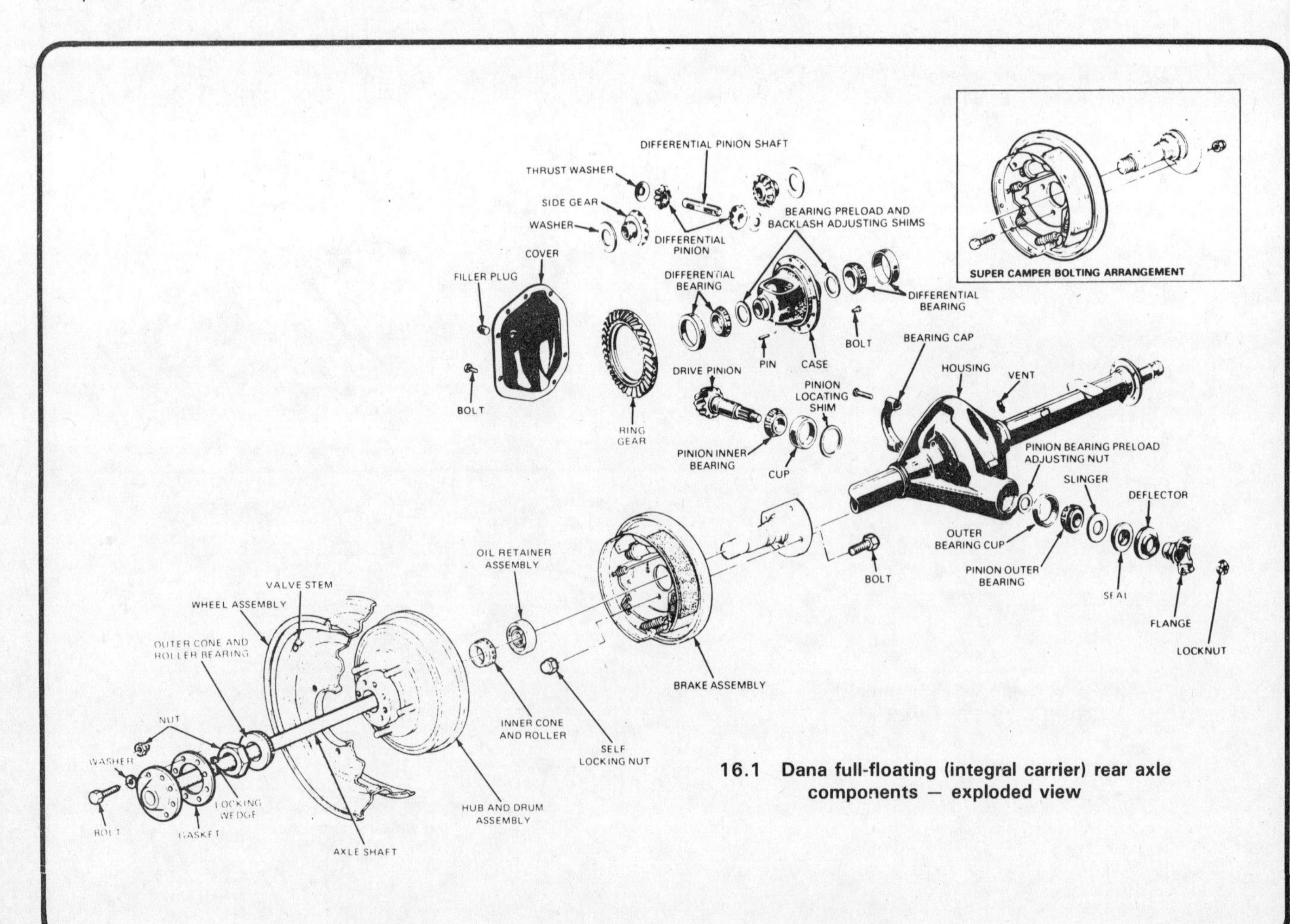

16.1 Dana full-floating (integral carrier) rear axle components — exploded view

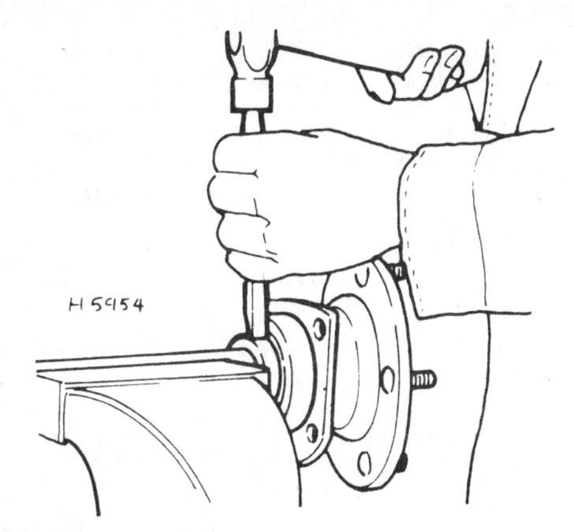

18.2 A chisel and hammer can be used to loosen the bearing retainer, but do not hit the axleshaft with the chisel!

16 Axleshaft (full-floating type) — removal and installation

Refer to illustration 16.1

Note: *This procedure applies to 1978 and later vehicles.*

1 Unscrew and remove the bolts which attach the axleshaft flange to the hub (see illustration). There is no need to remove the tire and wheel or jack up the vehicle.

2 Tap the flange with a soft face hammer to loosen the shaft and then grip the rib on the face of the flange with a pair of locking pliers. Twist the shaft slightly in both directions and then withdraw it from the axle tube.

3 Installation is the reverse of removal but hold the axleshaft level in order to engage the splines at the inner end with those in the differential side gear. Always use a new gasket on the flange and keep both the flange and hub mating surfaces free of grease and oil.

17 Axleshaft oil seal — replacement (ball bearing type)

1 Remove the axleshaft as described in Section 14.

2 The seal fits just inside the outer end of the axleshaft housing. Ideally, a slide hammer should be used to remove it. However, it can be removed using a hammer and chisel, but be very careful not to damage the axle housing. If it is damaged, oil will seep past the outside edge of the new seal.

3 Make a note of which way the seal is located in the housing before removing it. Usually the metal encased side of the seal faces out, toward the wheel.

4 Apply gasket sealer to the outside edge of the new seal and drive it evenly into place with a hammer and a block of wood or a section of pipe with the same outside diameter as the seal. Make sure it is seated squarely in the housing.

5 Install the axleshaft as described in Section 14.

18 Axleshaft bearing — replacement (ball bearing type)

Refer to illustration 18.2

1 Refer to Section 14 and remove the axleshaft.

2 Using a hammer and a sharp cold chisel, make several deep nicks in the bearing retainer ring (see illustration). This will release its grip on the shaft and allow it to be removed. If it is extremely tight, split it with a sharp chisel, but do not nick the axle.

3 Place the axleshaft upside-down in a vise, so that the bearing retainer is on the top of the jaws and the axleshaft flange is under them, and drive the axleshaft through the bearing with a soft face hammer. If this doesn't work a hydraulic press will be required. Note which way the bearing is installed.

4 Position the retainer plate and new bearing on the axleshaft. Make sure the bearing is installed facing the same way as the original.

5 Place the axleshaft vertically between the jaws of a vise with the flange facing up. The inner race must be resting on the top of the vise jaws. Using a soft face hammer, drive the axleshaft through the bearing until it is completely seated against the shaft shoulder.

6 The bearing retainer should be reinstalled in the same manner as the bearing. **Caution:** *Do not attempt to install the bearing and retainer at the same time.*

7 Pack the bearing with multi-purpose grease.

8 Install the axleshaft as described in Section 14.

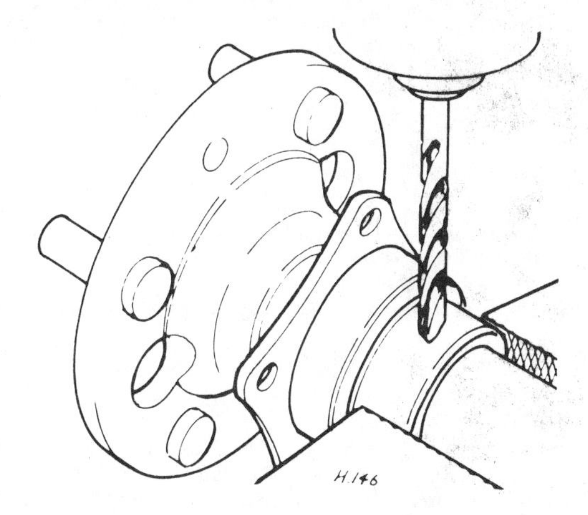

19.3 Drill a hole in the retainer, then use a chisel to split the retainer and remove it

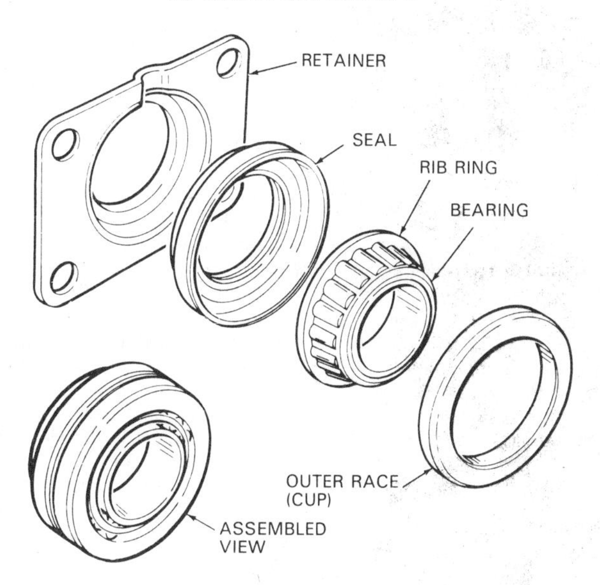

19.5 Tapered roller bearing type wheel bearing components — exploded view

19 Axleshaft bearing and oil seal — replacement (tapered roller bearing type)

Refer to illustrations 19.3 and 19.5

Note: *Due to the fact that a special press has to be used to remove and reinstall this type of bearing, it is recommended that the procedure be done by a dealer service department or an auto repair shop.*

1 Remove the axleshaft as described in Section 15.

2 Mount the axleshaft in a vise equipped with soft jaws.

3 Drill a hole in the bearing retainer (see illustration) and then remove it by splitting it with a cold chisel. Be careful not to damage the shaft during this procedure.

4 Using a large hydraulic press, remove the bearing and oil seal.

5 Install the bearing retainer plate, the new bearing (seal side facing the differential) and a new bearing retainer (see illustration) on the axleshaft.

6 Apply pressure to the retainer only, using a press, and seat the components against the shoulder of the axleshaft flange.
7 Install the axleshaft as described in Section 14.

20 Axleshaft bearing and oil seal — replacement (full-floating type)

Note: *The following brake drum/hub removal and installation procedure (Steps 3 through 5 and 14 through 16) applies only to 1975 and later models. For 1974 and earlier models, refer to Chapter 9 (Section 11) to remove and install the hub/drum assemblies and adjust the bearing preload.*

1 Raise the rear of the vehicle and place it securely on jackstands.
2 Remove the rear wheels, then remove the axleshafts as described in Section 16.
3 Remove the locking wedge from the adjusting nut keyway slot with a screwdriver (see illustration 16.1). **Caution:** *This must be done before the bearing adjusting nut is removed or even turned.*
4 Remove the wheel bearing adjusting nut.
5 Pull the brake drum and hub off the spindle. If the brake drum won't come off easily it may be necessary to retract the brake shoes slightly.
6 Remove the outer bearing assembly from inside the hub.
7 Use a brass drift to drive the inner bearing cone and inner seal out of the wheel hub.
8 Clean the inside of the wheel hub to remove all axle lubricant and grease. Clean the spindle.
9 Inspect the bearing assemblies for signs of wear, pitting, galling and other damage. Replace the bearings if any of these conditions exist. Inspect the bearing races for signs of erratic wear, galling and other damage. Drive out the bearing races with a brass drift if they need replacement. Install the new races with the special tool designed for this purpose. Never use a drift or punch for this operation as these races must be seated correctly and can be damaged easily.
10 Prior to installation, pack the inner and outer wheel bearing assemblies with the correct type of wheel bearing grease. If you do not have access to a bearing packer, pack each one carefully by hand and make sure the entire bearing is packed with grease.
11 Install the newly packed inner wheel bearing into the brake drum hub. Install a new hub inner seal with a drive tool (tubular drift, large socket, special tool) being careful not to damage the seal.
12 Wrap the spindle end and threaded area with tape to prevent damage to the inner wheel bearing seal during installation.
13 Carefully slide the hub and drum assembly over the spindle, being very careful to keep it straight so as not to contact the spindle with the seal, which would damage it. Remove the tape.
14 Install the outer wheel bearing. Tighten the wheel bearing adjusting nut by hand.
15 While rotating the hub/drum assembly, tighten the adjusting nut to 120 to 140 ft-lbs. Back off the nut enough to get 0.001 to 0.010-inch end play. This should require about 1/8-to-3/8 of a turn (1/8-to-1/4 turn on 1979 and 1980 models).
16 Position the locking wedge in the keyway and hammer it into position. **Note:** *It must not be bottomed against the shoulder of the adjusting nut when fully installed. The locking wedge and adjusting nut can be used over, providing the locking wedge cuts a new groove in the nylon retainer material within the specified nut loosening range. The wedge must not be pressed into a previously cut groove. If it is not possible to back the nut off within the specified limit, obtain the proper end play or align uncut nylon in which to press the locking wedge, discard the nut and wedge and replace them with new ones.*
17 Install the axleshaft with a new axle flange gasket, lockwashers and new axleshaft retaining bolts. Tighten the bolts to the specified torque.
18 Adjust the brakes if they were backed off to remove the drum.
19 Install the wheel, remove the jackstands and lower the vehicle.

21 Rear wheel bearing — replacement (semi-floating type)

Refer to illustration 21.2

1 Remove the axleshaft as described in Section 15.
2 When it is determined that the wheel bearing is to be replaced, the bearing cup, which normally remains in the axle housing when the axleshaft is removed, must also be replaced. Use a slide hammer puller to remove it from the housing (see illustration).
3 Before the wheel bearing and seal can be replaced, the inner retainer ring must first be removed. Never use heat to remove the ring, as it would damage the axleshaft.
4 Using a drill of 1/4-to-1/2 inch diameter, drill a hole in the outside diameter of the inner retainer approximately 3/4 of the way through the ring. Do not drill all the way through the retainer ring, as the drill would damage the axleshaft.
5 After drilling the hole in the retainer ring, position a chisel across the drill hole and strike it sharply to split the retainer ring.
6 Due to the need for a hydraulic press and various adapters to remove the wheel bearing, you must take the axleshaft(s) to an automotive machine shop or parts store with the equipment required for the work. On vehicles with tapered roller wheel bearings, the axleshaft seal should be replaced at the same time.
7 After the new bearing has been pressed onto the axleshaft, install the axle as described in Section 15.

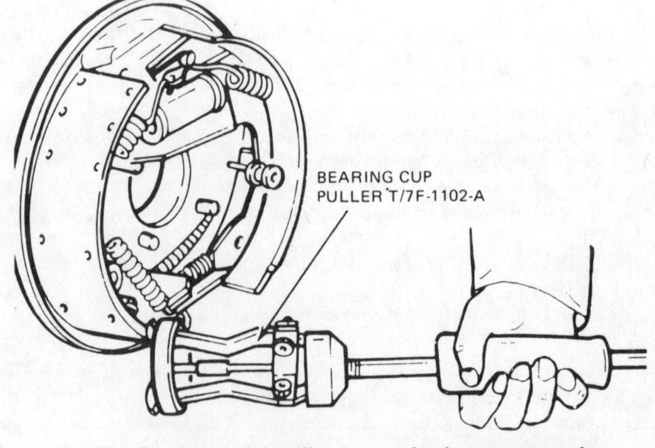

21.2 A special puller is required to remove the bearing cup from the axleshaft housing of a semi-floating type rear axle

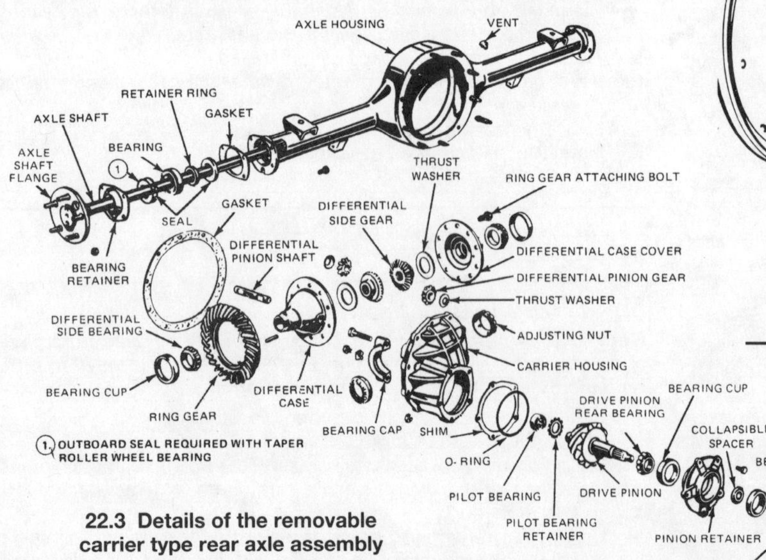

22.3 Details of the removable carrier type rear axle assembly

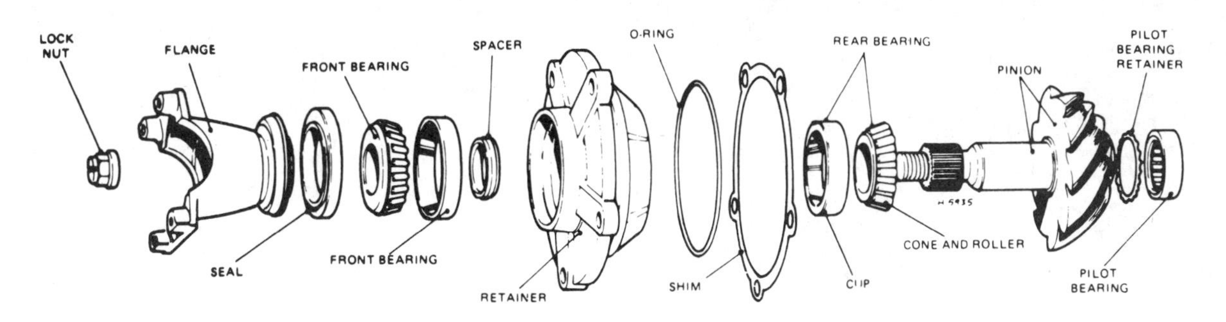

24.5 Pinion gear and related components (removable carrier type) – exploded view

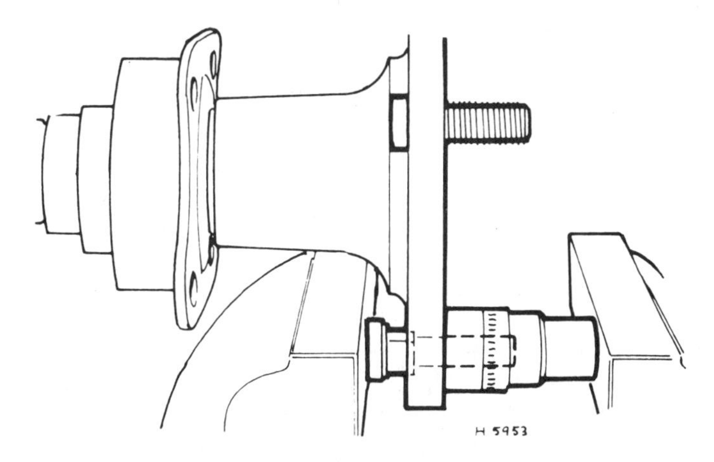

25.2 A large vise and socket can be used to press the wheel stud into place in the flange

22 Differential carrier - removal and installation

Refer to illustration 22.3

1 To remove the differential carrier assembly, block the front wheels, raise the rear of the vehicle and support it on jackstands. Drain the oil from the axle by removing the drain plug (if equipped) from the differential. Remove both wheels and brake drums and then partially withdraw both axleshafts as described earlier in this Chapter.

2 Disconnect the driveshaft from the differential and support it out of the way with a piece of wire or rope.

3 Remove the self-locking nuts holding the differential carrier assembly to the axle housing (see illustration). If an oil drain plug is not installed, pull the carrier forward slightly and allow the oil to drain into a pan. The carrier assembly can now be detached from the axle housing.

4 Before installation, carefully clean the mating surfaces of the carrier and the axle housing to remove all traces of the old gasket. Be sure to use a new gasket when the carrier is installed. The nuts retaining the differential carrier assembly to the axle housing should be tightened to the specified torque in a criss-cross pattern.

23 Rear axle assembly - removal and installation

1 Block the front wheels, raise the rear of the vehicle and support it on jackstands placed under the rear frame members.

2 Remove the wheels, brake drums and axleshafts as described earlier in this Chapter.

3 Disconnect the driveshaft from the differential and support it out of the way with a piece of wire or rope.

4 Disconnect the lower ends of the shock absorbers from the axle housing.

5 Remove the brake vent tube (if installed) from the brake line junction and retaining clamp (see Chapter 9).

6 Remove the brake lines from the clips that retain them to the axle, but do not disconnect any of the fittings.

7 Remove the brake linings and brake backing plates and support them with wire to avoid straining the rubber brake lines which are still attached (see Chapter 9).

8 Support the weight of the axle on a floor jack and remove the nuts from the spring U-bolts. Remove the bottom clamping plates.

9 Lower the axle assembly with the jack and withdraw it to the rear.

10 Installation is the reverse of removal. Tighten the U-bolt and shock absorber nuts to the specified torque (Chapter 10).

24 Pinion oil seal - replacement

Refer to illustration 24.5

1 Raise the rear of the vehicle and support it with jackstands. Block the front wheels to keep the vehicle from rolling.

2 Remove the rear wheels and brake drums.

3 Disconnect the driveshaft from the differential and support it out of the way with a piece of wire or rope.

4 Using a torque wrench (in-lbs) check and record the torque required to turn the pinion gear (except Dana axle).

5 Hold the drive pinion flange still with a suitable tool and remove the large locknut **(see illustration)**.

6 Remove the washer, drive flange and dust deflector, then pry out the oil seal. *Do not pry against the pinion shaft splines during this procedure.*

7 Tap in the new oil seal using a hammer and a piece of pipe as a drift. Do not strike the end of the pinion shaft.

8 Install the pinion drive flange and nut. If equipped with a Dana axle, tighten the pinon nut to 250 - 270 ft-lbs and proceed to step 9. If equipped with a Ford axle, tighten the pinion nut a little at a time, occasionally rotating the pinion and checking the turning torque with a wrench (in-lbs). Continue tightening and checking until the original recorded torque (step 4) is reached. Add 3 in-lbs to the turning torque to compensate for the new seal. If the pinion nut is overtightened, it cannot be loosened to correct the pinion shaft turning torque. This is because the pinion spacer has been compressed. A new spacer will have to be installed and the procedure repeated.

9 Install the brake drums, driveshaft and rear wheels and lower the vehicle.

25 Wheel stud - replacement

Refer to illustration 25.2

1 A wheel stud would normally require replacement because the threads have been damaged or the stud has been broken (usually caused by overtightening of the wheel nuts). To replace a wheel stud, first remove the axleshaft assembly. Using a large pin punch or drift and hammer, drive the old stud through the flange, towards the bearing.

2 To install a new stud, place it in the hole from the rear of the flange and position a socket over the hole opposite the stud. Use a bench vise to press the stud into the hole **(see illustration)**.

Chapter 9 Brakes

Contents

Specifications

General

Type of system	Pre-1974 — drum brakes on all four wheels 1974 and later — disc brakes on front wheels, drum brakes on rear wheels
Hydraulic system	Dual line, tandem master cylinder with power assistance
Brake fluid type	See Chapter 1

Front disc brakes

Type	
E100 and E150 models	Sliding caliper, single piston
E250 and E350 models	Floating caliper, twin pistons
Minimum rotor thickness	
sliding caliper type	1.12 in
floating caliper type	0.94 in
Minimum pad lining thickness	See Chapter 1

Drum brakes (front and rear)

Diameter	
E100 models	10 in
E200 models	11 in
E300 models	12 in

Drum brakes (rear only)

Diameter	
E150 models	11 in
E250 models	12 in
E350 models	12 in
Minimum brake shoe lining thickness	See Chapter 1

Master cylinder

Type	Tandem
Bore	0.938 in

Torque specifications	**Ft-lbs**
Parking brake assembly bolts	13 to 25
Master cylinder bolts	13 to 25
Wheel cylinder bolts	5 to 7
Pressure differential valve mounting nuts and bolts	7 to 11
Front backing plate nuts and bolts	9 to 14
Rear backing plate nuts and bolts	20 to 40
Brake booster (servo) mounting nuts and bolts	13 to 25
Parking brake bolts	20 to 25
Rear wheel bearing nuts (1974 and earlier full-floating axles)	
adjusting nut (inner)	50 to 80 (back off 3/8-turn)
locknut (outer)	90 to 110

1 General information

All Econoline models built prior to 1974 are equipped with drum type brakes on both front and rear wheels. The hydraulic system has a dual master cylinder and separate front and rear brake line circuits. A vacuum brake booster, which provides power assistance to the action of the brake pedal, is an option on all models.

Vans built after 1974 are equipped with disc brakes on the front wheels. E100 and E150 models have light duty, single piston calipers, while E250 and E350 models are equipped with heavy-duty, dual-piston calipers. Both the drum and disc brakes are self adjusting. The power assist unit is standard equipment on vehicles equipped with disc brakes. The dual master cylinder contains two hydraulic pistons (primary and secondary) fed by separate fluid reservoirs. A pressure differential valve and switch are an integral part of the master cylinder circuit.

In the event of failure of either the front or rear brake systems, the defective system is bypassed and the remaining system will continue to function, although obviously with considerably reduced braking efficiency. A warning light in the instrument panel will indicate when failure of one of the systems has occurred.

An independent parking brake system is provided, operated by a pedal on the drivers side of the cab. The parking brake operates the rear wheel brakes only through a system of cables.

2 Hydraulic system — bleeding

1 Removal of the air from the hydraulic fluid in the brake system is essential to the proper operation of the system. Before beginning the bleeding procedure, check the fluid level as described in Chapter 1.
2 Check all brake line fittings and connections for leaks and make sure the rubber hoses are in good condition.
3 If the condition of a caliper or wheel cylinder is in doubt, check for signs of fluid leakage around the seals.
4 If there is any possibility that the wrong fluid has been used in the system, drain all the fluid and flush the system with clean brake fluid. Replace all rubber brake parts, as they have been affected and could possibly fail under pressure.
5 Obtain a clean jar, a 12-inch length of clear tubing, which fits tightly over the bleeder valves, and a container of the recommended brake fluid.
6 The primary (front) and secondary (rear) hydraulic brake systems are individual systems and must be bled separately. Always bleed the longest line first.
7 To bleed the secondary system (rear), clean the area around the bleeder valve and start at the right-side wheel cylinder.
8 Remove the rubber cap over the end of the bleeder valve. Place the end of the tubing in the clean jar with enough fluid to keep the end of the tube submerged during the bleeding operation.
9 Open the bleeder valve approximately 1/4-turn with a wrench and have an assistant depress the brake pedal slowly through its full travel.
10 Close the bleeder valve and allow the pedal to return to the released position.
11 Continue the operation until no more air bubbles appear in the tubing. Give the brake pedal two more strokes to ensure that the line is completely free of air, then retighten the bleeder valve (make sure that the tube remains submerged until the valve is closed).
12 At regular intervals during the bleeding procedure, make sure that the reservoir is full, otherwise air will enter the system at this point. Do not reuse fluid bled from the system.
13 Repeat the whole procedure at the left-side rear brake.
14 To bleed the primary system (front), start with the front right-side and finish with the front left-side. The procedure is identical to the one for the rear brakes. **Note:** *Some models have a bleed valve incorporated in the master cylinder. Where this is the case, the master cylinder should be bled before the brake lines. The bleeding procedure is identical to that already described.*
15 Fill the master cylinder to within 1/4-inch of the top of the reservoirs, make sure that the diaphragm type gasket is correctly located in the cover, then install the cover.

3 Pressure differential valve — centering

1 After any repair or bleeding operation, it is possible that the dual brake warning light will come on due to the pressure differential valve remaining in an off center position.
2 To center the valve, first turn the ignition switch to the On or Accessory position.
3 Depress the brake pedal several times and the piston will center itself causing the warning light to go out.
4 Turn the ignition switch off.

4 Flexible hoses — inspection, removal and installation

Refer to illustration 4.3

1 Check the condition of the flexible hydraulic hoses leading to each of the front brakes and the one at the front of the rear axle. If they are swollen, damaged or abraded they must be replaced.
2 Clean the top of the brake master cylinder reservoir and remove the cover. Place a piece of polyethylene plastic over the top of the reservoir and install the cover. This is to stop hydraulic fluid from siphoning out during subsequent operations.

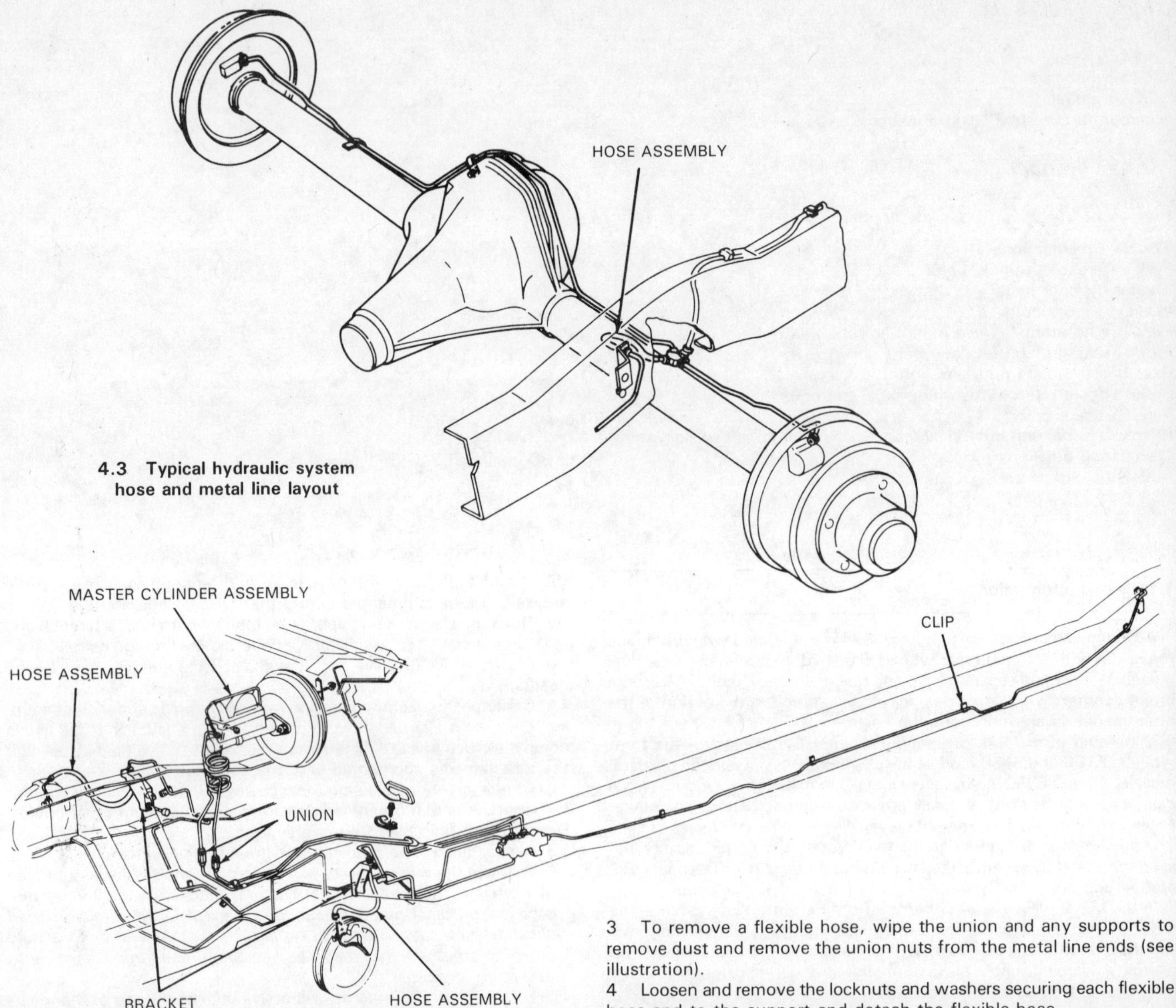

4.3 Typical hydraulic system hose and metal line layout

3 To remove a flexible hose, wipe the union and any supports to remove dust and remove the union nuts from the metal line ends (see illustration).

4 Loosen and remove the locknuts and washers securing each flexible hose end to the support and detach the flexible hose.

5 Installation is the reverse of removal. Be sure to bleed the brake hydraulic system as described in Section 2. If one hose has been removed it is only necessary to bleed either the front or rear brake hydraulic system.

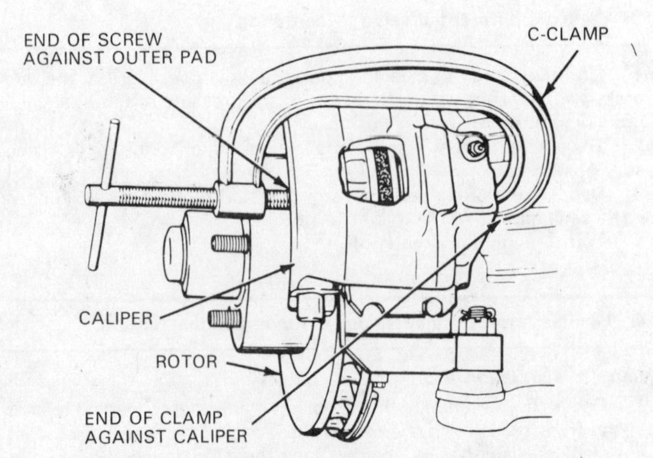

5.2 A large C-clamp can be used to compress the piston into the caliper bore to provide room for the new, thicker brake pads

5 Disc brake pads (sliding caliper) — replacement

Refer to illustrations 5.2, 5.3, 5.4, 5.5, 5.9, 5.11, 5.13, 5.15, 5.16, 5.17, 5.20a, 5.20b, 5.21, 5.22 and 5.28

Warning: *Disc brake pads must be replaced on both wheels at the same time — never replace the pads on only one wheel. Also, brake system dust contains asbestos, which is harmful to your health. Never blow it out with compressed air and do not inhale any of it. Do not, under any circumstances, use petroleum-based solvents to clean brake parts. Use brake cleaner or denatured alcohol only.*

1 Block the rear wheels, apply the parking brake, loosen the front wheel lug nuts, raise the front of the vehicle and support it on jackstands. Remove the wheels.

Single piston caliper

2 Before the caliper is removed, use a large C-clamp to push the piston into the bore. Position the screw end against the outer pad and the frame end against the caliper body, then slowly and carefully tighten the clamp screw. The piston should move into the bore and provide

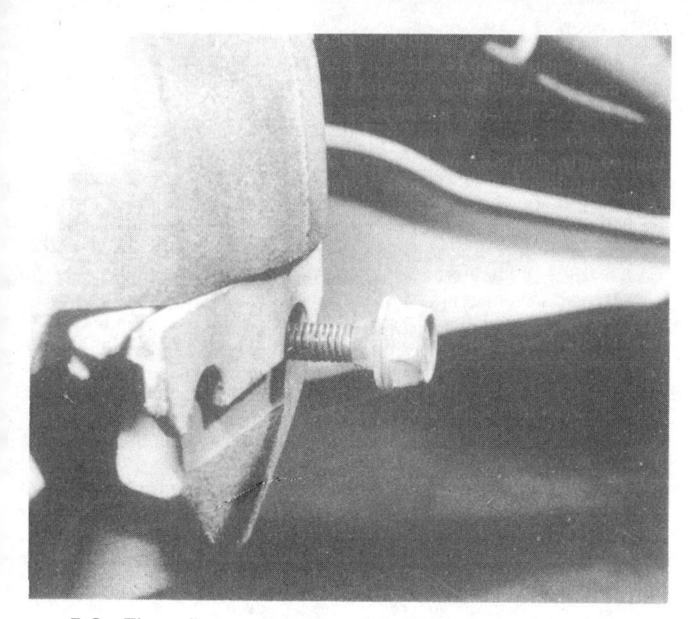

5.3 The caliper support key is held in place with a bolt that threads into the caliper bracket

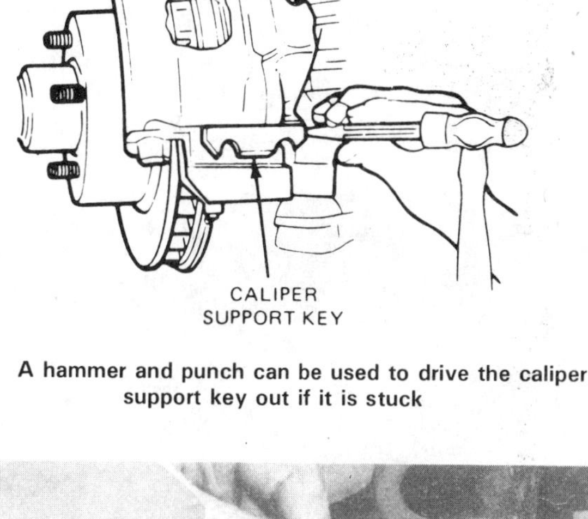

5.4 A hammer and punch can be used to drive the caliper support key out if it is stuck

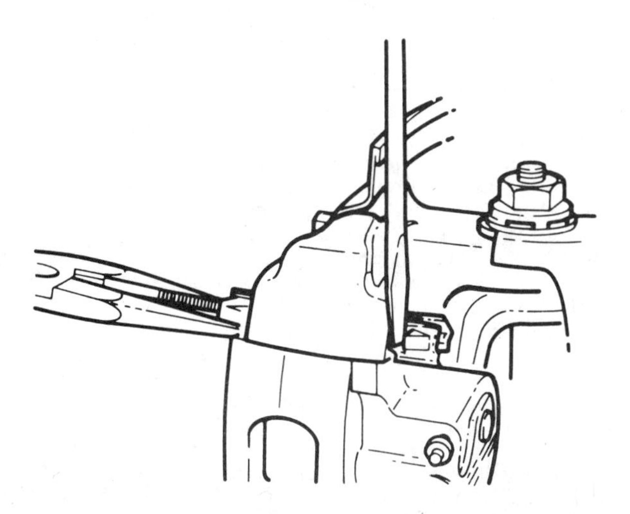

5.5 Remove the pin rail on 1986 models by directing the tabs into the grooves with a screwdriver and pulling on the rail with a pliers

5.9 The inner brake pad will remain in place in the caliper bracket and must be lifted out separately

room for the new, thicker pads (see illustration). If resistance is encountered the piston may be seized – do not apply undue force or the caliper may be damaged. Check the brake fluid level in the master cylinder as this is done. Fluid may have to be siphoned out to prevent overflow at the reservoir.

3 Most models are equipped with a key and spring that are held in place by a bolt, located at the bottom of the caliper. Remove the bolt to release the caliper support key (see illustration).

4 Using a hammer and punch, carefully tap the caliper key and spring out to remove them from the anchor plate. Do not damage the key (see illustration).

5 Starting in 1986, the caliper is held in place by two pin rails, one at the top and one at the bottom of the caliper. Use a wire brush to remove any dirt from the pin rail tabs (located at the ends of the pins). Using a punch and hammer, tap the upper pin rail in until it contacts the anchor plate. Insert a screwdriver blade into the slot behind each tab on the inner end of the rail. At the same time, compress the outer end of the pin rail with needle nose pliers and pull on it until the tabs slip into the grooves (see illustration). Pull the rail out with the pliers or tap it out with a hammer and punch. Repeat the procedure for the lower pin rail.

6 Push the caliper down and rotate the upper end up and out to detach the caliper from the anchor plate. The inner brake pad will remain in the anchor plate, while the outer pad will stay in the caliper.

7 Suspend the caliper assembly from a chassis member with a piece of wire. **Caution:** *Do not stretch or twist the rubber brake hose.*

8 If the pads are to be reused, mark them so they can be reinstalled in their original positions. They must not be interchanged.

9 The pad can now be removed from the anchor plate. The anti-rattle clips will probably come out when the pad is removed (see illustration). Note how the anti-rattle clips are installed – they must be repositioned in the exact same manner during installation.

10 Clean the caliper, anchor plate and disc (rotor) assembly and check for fluid leaks, wear and damage. The area on the anchor plate that mates with the caliper must be clean and smooth so the caliper slides freely without binding. Measure the thickness of the pad lining. If the lining has worn beyond the specified limits, the pads must be replaced with new ones. The service replacement outer pad is slightly different than the original. It has tabs on the flange at the lower edge of the pad and the distance between the upper tabs and lower flange is reduced to provide a slip-on interference fit.

11 Position a new anti-rattle clip on the lower end of the inner pad

(see illustration). Be sure the tab on the clip is positioned correctly and that the clip is seated completely. The loop-type spring on the clip must face away from the rotor. The installation directions that come with the new pads should contain detailed drawings showing the correct position of the anti-rattle clip(s).

12 If the old pads are being reused they must be installed in their original positions. Place the inner pad and anti-rattle clip in the anchor plate with the clip tab against the abutment. Compress the clip and slide the upper end of the pad into position.

13 Position the outer pad in the caliper and press the tabs into place. If necessary, use a C-clamp to seat the pad, but be careful not to damage the lining material (place small pieces of wood between the clamp and pad surface) (see illustration).

5.11 Don't overlook the anti-rattle clip attached to the pad

14 Detach the caliper from the wire.

15 On 1975 through 1985 models, position the caliper on the anchor plate by pivoting it around the upper mounting surface. Be careful not to tear the boot as it slips over the inner pad (see illustration).

16 Use a large screwdriver to hold the caliper against the anchor plate (see illustration).

17 Carefully insert the caliper key and spring (see illustration).

18 Remove the screwdriver and gently tap the caliper key into position.

19 Install the caliper key retaining bolt and tighten it to 12 to 20 ft-lbs.

20 On 1986 models, apply high-temperature disc brake grease (ESA-M1C72-A or equivalent) to the caliper and anchor plate grooves, then install the caliper and position the pin rails with the tabs adjacent to the grooves (see illustration). Tap the pins into place until the retention tabs on the inner end snap out of the groove and bear against the flanks of the anchor plate. Do not tap the pin rails in too far or the outer retention tabs may enter the grooves (the tabs on each end of the rail must be free to contact the anchor plate flanks) (see illustration).

Dual piston caliper

21 Remove the key retaining bolt. On most models, the key is at the top of the caliper, but on 1983 and later models it is at the bottom (see illustration).

22 Using a hammer and punch, drive the key out of the caliper and anchor plate. On calipers with the key and spring on top, drive the key in to remove it (see illustration). On calipers with the key and spring on the bottom, drive the key out (from the inside out). Starting in 1986, the caliper is held in place by two pin rails, one at the top and one at the bottom of the caliper. Use a wire brush to remove any dirt from the pin rail tabs (located at the ends of the pins). Using a punch and hammer, tap the upper pin rail in until it contacts the anchor plate. Insert a screwdriver blade into the slot behind each tab on the inner end of the rail. At the same time, compress the outer end of the pin rail with needle nose pliers and pull on it until the tabs slip into the grooves (see illustration 5.5). Pull the rail out with the pliers or tap it out with a hammer and punch. Repeat the procedure for the lower pin rail.

23 Rotate the caliper out and away from the rotor by pulling on the key end. Slide the opposite end of the caliper free from the support until the caliper clears the rotor. Suspend the caliper assembly with wire from a suspension member. **Caution:** *Be careful not to twist or stretch the flexible rubber brake hose.*

24 After making careful note of how it is installed, remove the large brake shoe anti-rattle spring.

25 Remove the inner and outer brake pads from the caliper.

26 Clean the caliper and piston faces to remove dust and accumulated debris. The area on the anchor plate that mates with the caliper must be clean and smooth so the caliper slides freely without binding.

27 To avoid brake fluid overflow when the caliper pistons are pressed into the bores, siphon or dip part of the brake fluid out of the large master cylinder reservoir (connected to the front brakes) and discard the removed fluid.

28 Use a C-clamp to compress the pistons into the bores (see illustration). **Caution:** *If both calipers have been removed, block the opposite side pistons to prevent them from popping out during this operation.*

29 Install the new pads and the anti-rattle spring.

30 Install the caliper assembly on the anchor plate and over the rotor.

31 On pre-1986 models, position the spring between the key and the caliper with the spring tangs overlapping the ends of the key. You may need a screwdriver or brake adjusting tool to support the caliper and start the key and spring into position.

32 Using a hammer and punch, tap the key and spring into place until the notch in the key lines up with the retaining bolt hole.

33 Install the bolt and tighten it to 12 to 20 ft-lbs.

34 On 1986 models, apply high temperature disc brake grease (ESA-M1C72-A or equivalent) to the caliper and anchor plate grooves, then install the caliper and position the pin rails with the tabs adjacent to the grooves (see illustration 5.20a). Tap the pins into place until the retention tabs on the inner end snap out of the groove and bear against the flanks of the anchor plate. Do not tap the pin rails in too far or

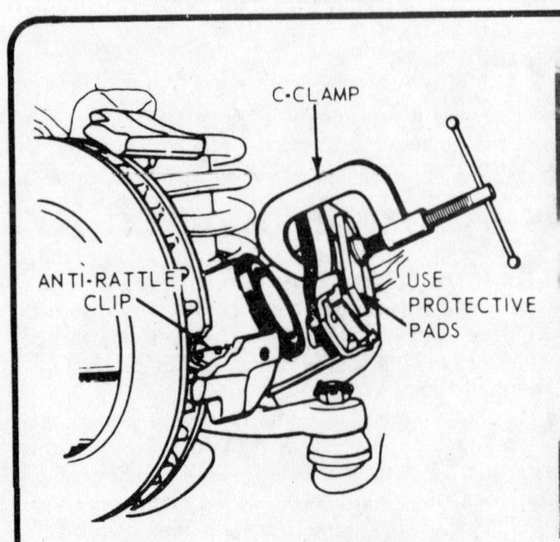

5.13 A C-clamp may be required to seat the outer pad in the caliper

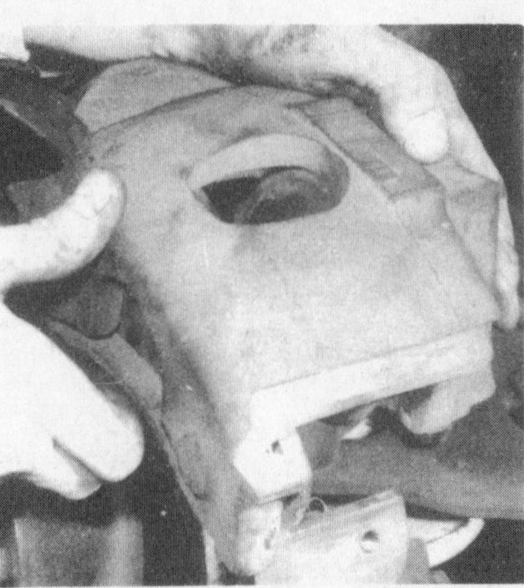

5.15 When installing the caliper, be careful not to strain the brake hose or dislodge the brake pads

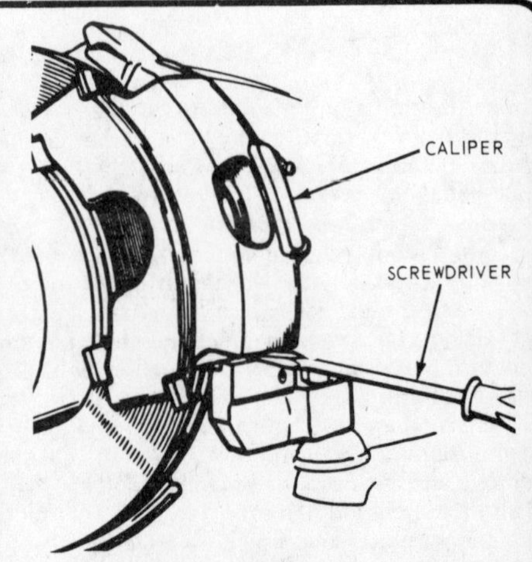

5.16 A large screwdriver can be used to hold the caliper in place as the support key and spring are installed

5.17 Be sure the key and spring are positioned with the spring tabs over the ends of the key

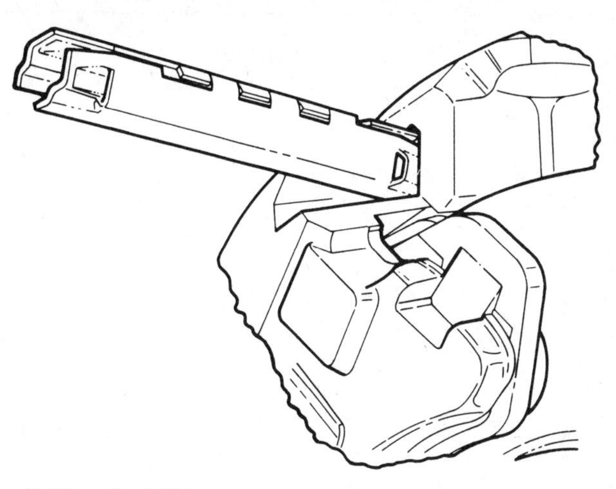

5.20a On 1986 models, the pin rail tabs must enter the anchor plate groove as shown here

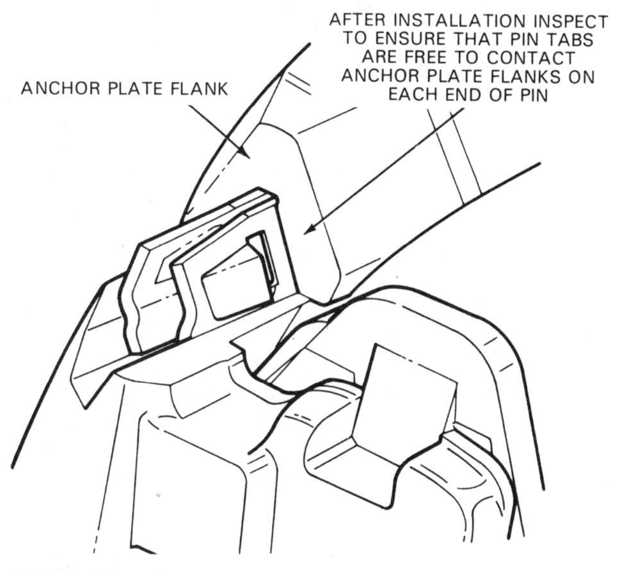

5.20b The tabs on each end of the pin rail must contact the anchor plate flanks and prevent the rail from sliding out of place

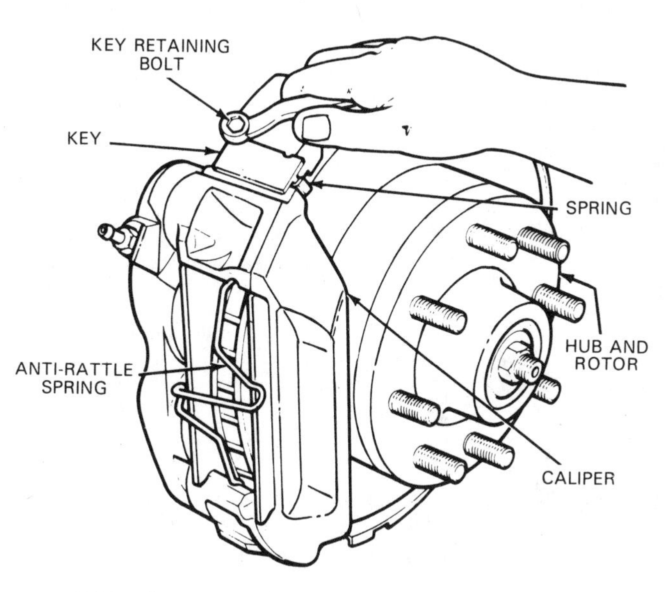

5.21 On most models the key is located at the top of the caliper

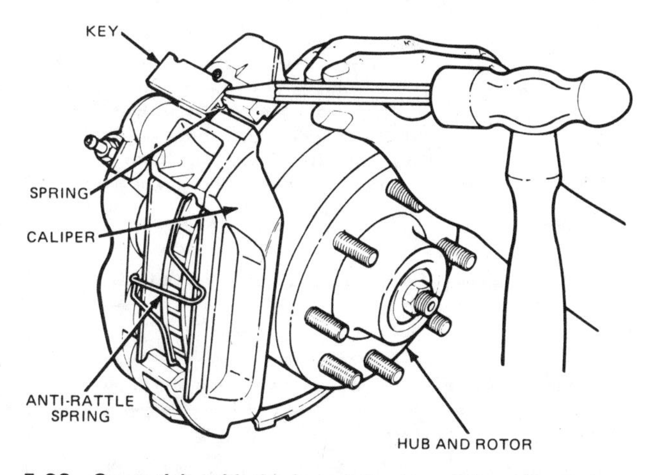

5.22 On models with the key at the top of the caliper, the key should be tapped from the outside in to remove it

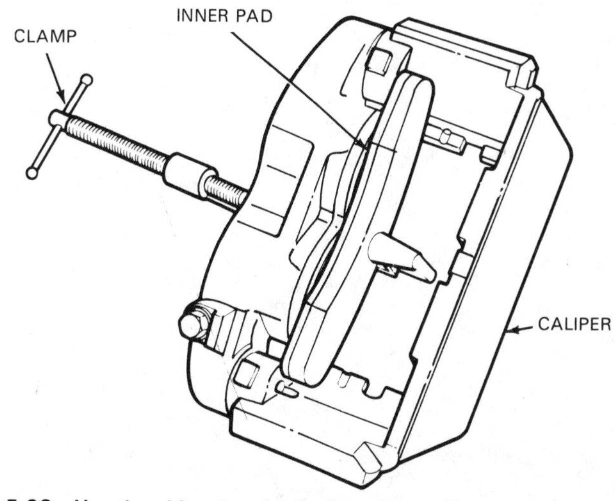

5.28 Use the old pad and a C-clamp to bottom the caliper pistons in the bores

9

the outer retention tabs may enter the grooves. The tabs on each end of the rail must be free to contact the anchor plate flanks (see illustration 5.20b).

All calipers

35 Repeat the procedure for the remaining caliper. Depress the brake pedal several times to seat the pads and center the calipers, then check the brake fluid level as described in Chapter 1.
36 Install the wheels, lower the vehicle and check the operation of the brakes before taking the vehicle in traffic. Try to avoid heavy brake application until after the brakes have been applied lightly several times to seat the pads.

6 Disc brake pads (floating caliper) — replacement

Refer to illustration 6.2

Warning: *Disc brake pads must be replaced on both wheels at the same time — never replace the pads on only one wheel. Also, brake system dust contains asbestos, which is harmful to your health. Never blow it out with compressed air and do not inhale any of it.*

1 Remove the front wheels as described in Step 1 of the previous Section.
2 Remove the nuts from the brake pad (shoe) mounting pins and withdraw the pins and anti-rattle coil springs (see illustration).
3 Using needle nose pliers, remove the two brake pads (shoes) from the caliper assembly.
4 Clean the inside of the caliper and remove all dust and dirt from the piston faces with brake cleaner or denatured alcohol. **Warning:** *Do not, under any circumstances, use petroleum-based solvents to clean brake system parts.*
5 Using a piece of wood as a lever, push the pistons into the caliper bores as far as possible. Be careful not to cock the pistons in the bores. Check the brake fluid level in the master cylinder as this is done. Fluid may have to be siphoned out to prevent overflow at the reservoir.
6 Position the new pads in the caliper assembly and insert the two mounting pins and anti-rattle springs. Make sure that the spring tangs engage in the holes in the pad plates.
7 Install the pin retaining nuts and tighten them to 17 to 23 ft-lbs.
8 Repeat the procedure for the remaining caliper, then install the wheels and lower the vehicle. Tighten the wheel lug nuts securely.
9 To seat the pads, pump the brake pedal several times. Check the brake fluid level as described in Chapter 1 and check the operation of the brakes before taking the vehicle in traffic. Try to avoid heavy brake application until after the brakes have been applied lightly several times to seat the pads.

7 Disc brake caliper (single piston) — overhaul

Refer to illustration 7.4

Note: *If brake fluid is leaking from the caliper, new seals will be required. If brake fluid is running down the side of the wheel, if a pool of fluid forms alongside one wheel or if the fluid level in the master cylinder drops excessively, seal failure is indicated. Be sure to purchase a brake caliper overhaul kit before beginning this procedure.*

1 Refer to Section 5 and remove the caliper, then detach the flexible rubber hose (see Section 4). Clean the exterior of the caliper with brake cleaner, denatured alcohol or clean brake fluid. **Warning:** *Do not, under any circumstances, use petroleum-based solvents to clean brake components.* Disassemble the caliper on a clean workbench.
2 Drain any remaining fluid from the caliper, then position a block of wood or a shop towel inside the caliper, in front of the piston. Apply compressed air to the hose port to ease the piston out of the bore. **Warning:** *Never place your fingers in front of the piston in an attempt to catch or protect it when applying compressed air — serious injury could result.*
3 If the piston has seized in the bore, carefully tap around the piston while applying air pressure. Remember, the piston may come out with considerable force.
4 Remove the rubber dust boot from the caliper assembly (see illustration).
5 Carefully remove the rubber piston seal from the cylinder bore with a wooden or plastic tool. Do not use a screwdriver or other metal tool,

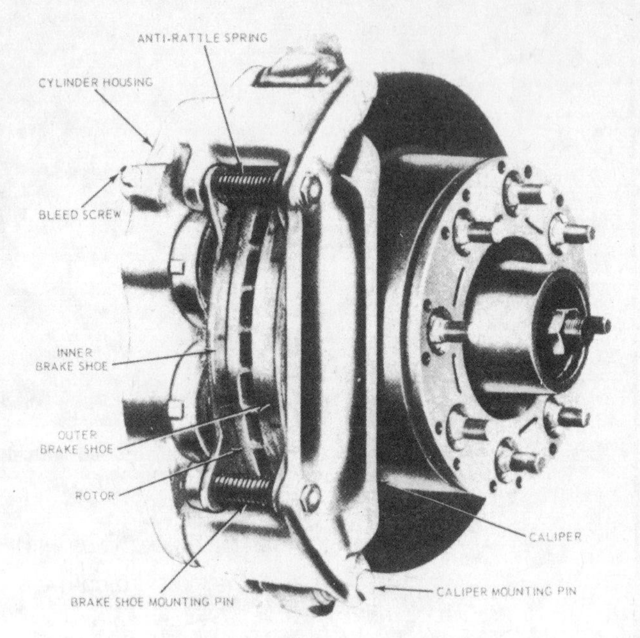

6.2 Floating caliper disc brake components

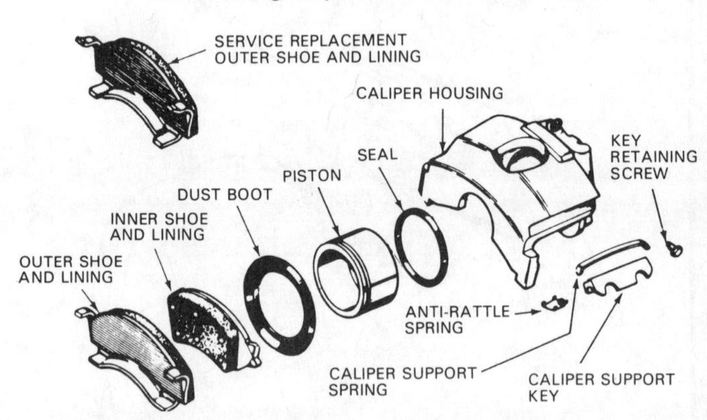

7.4 Single piston caliper components — exploded view

as it could damage the bore.
6 Clean all of the parts with brake cleaner and dry them with compressed air. During reassembly, new rubber seals must be used and they should be well lubricated with clean brake fluid before installation.
7 Inspect the piston and bore for signs of wear, score marks and other damage. If evident, a new caliper assembly will be required.
8 To reassemble the caliper, first place the new caliper piston seal in the groove in the caliper bore. The seal must not be twisted.
9 Install a new dust boot and make sure that the flange seats correctly in the outer groove of the caliper bore.
10 Carefully insert the piston into the bore. When it is about three-quarters of the way in, spread the dust boot over the piston and depress it into the bore the rest of the way.
11 Reassembly is now complete and the unit is ready for installation on the vehicle.
12 After installing the caliper and pads on the vehicle, bleed the brakes as described in Section 2.

8 Disc brake caliper (dual piston) — overhaul

Refer to illustration 8.2

Note: *If brake fluid is leaking from the caliper, new seals will be required. If brake fluid is running down the side of the wheel, if a pool of fluid forms alongside one wheel or if the fluid level in the master cylinder drops excessively, seal failure is indicated. Be sure to purchase a brake caliper overhaul kit before beginning this procedure.*

1 Refer to Section 5 or 6, remove the brake pads and disconnect

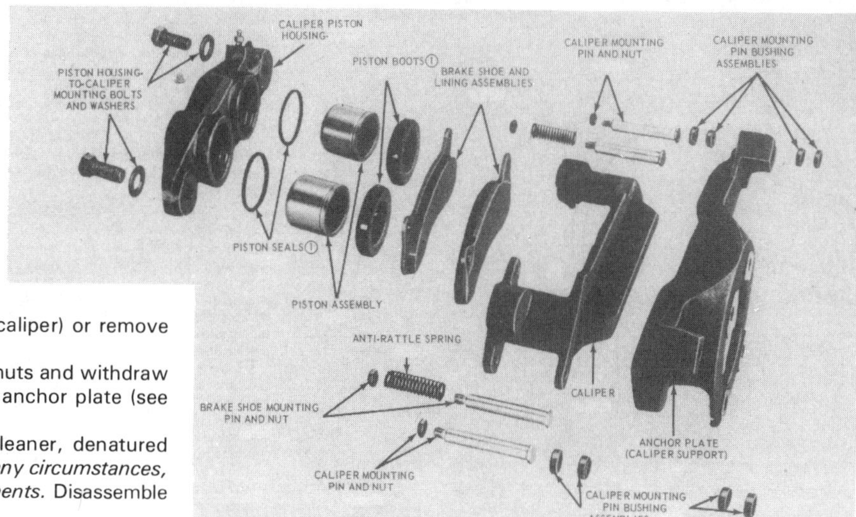

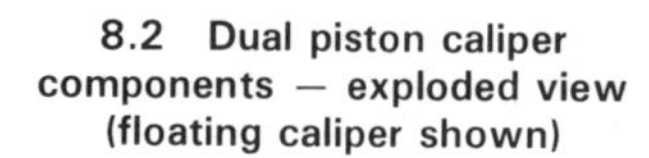
8.2 Dual piston caliper components — exploded view (floating caliper shown)

the flexible rubber hose from the caliper (floating caliper) or remove the caliper (sliding type).
2 On floating type calipers, remove the retaining nuts and withdraw the two pins securing the caliper assembly to the anchor plate (see illustration). Separate the caliper from the disc.
3 Clean the exterior of the caliper with brake cleaner, denatured alcohol or clean brake fluid. **Warning:** *Do not, under any circumstances, use petroleum-based solvents to clean brake components.* Disassemble the caliper on a clean workbench.
4 Drain any remaining fluid from the caliper, then position a block of wood or a shop towel inside the caliper, in front of the pistons. Apply compressed air to the hose port to ease the pistons out of the bores. **Warning:** *Never place your fingers in front of the pistons in an attempt to catch or protect them when applying compressed air — serious injury could result.*
5 If the pistons have seized in the bores, carefully tap around them while applying air pressure. Remember, the pistons may come out with considerable force.
6 Remove the wood block and pistons, then remove the bolts and separate the caliper from the piston housing.
7 Remove and discard the rubber piston boots.
8 Carefully remove the rubber piston seals from the bores with a wooden or plastic tool. Do not use a screwdriver, as it could damage the bores.
9 Clean all of the parts with brake cleaner and dry them with compressed air. During reassembly, new rubber seals must be used and they should be well lubricated with clean brake fluid before installation. Check the pistons and bores for wear and damage. If wear or damage is evident, a new piston housing is required.
10 Lubricate new piston seals with clean brake fluid and install them in the grooves in the piston housing bores.
11 Lubricate the housing bores with brake fluid and position the lips of the rubber boots in the bore grooves.
12 Insert the pistons through the rubber boots and push them into the cylinder bores, past the rubber seals. Be careful not to damage or dislodge the piston seals from the grooves in the bores.
13 Push both pistons all the way into the bores with a block of wood.
14 Rejoin the piston housing to the caliper and tighten the two retaining bolts to 155 to 185 ft-lbs.
15 Attach the caliper to the hub assembly and install the pins and retaining nuts (floating caliper).
16 Install the brake pads and attach the flexible brake hose to the caliper assembly.
17 Bleed the brakes as described in Section 2.

9 Front brake disc and hub — removal and installation

Refer to illustration 9.2

1 Refer to the appropriate Section and remove the caliper assembly. If the caliper does not require attention it is not necessary to disconnect the flexible rubber hose from the caliper. Suspend the caliper from the suspension arm with a piece of wire to avoid straining the flexible hose.
2 Carefully removed the grease cap from the hub (see illustration).

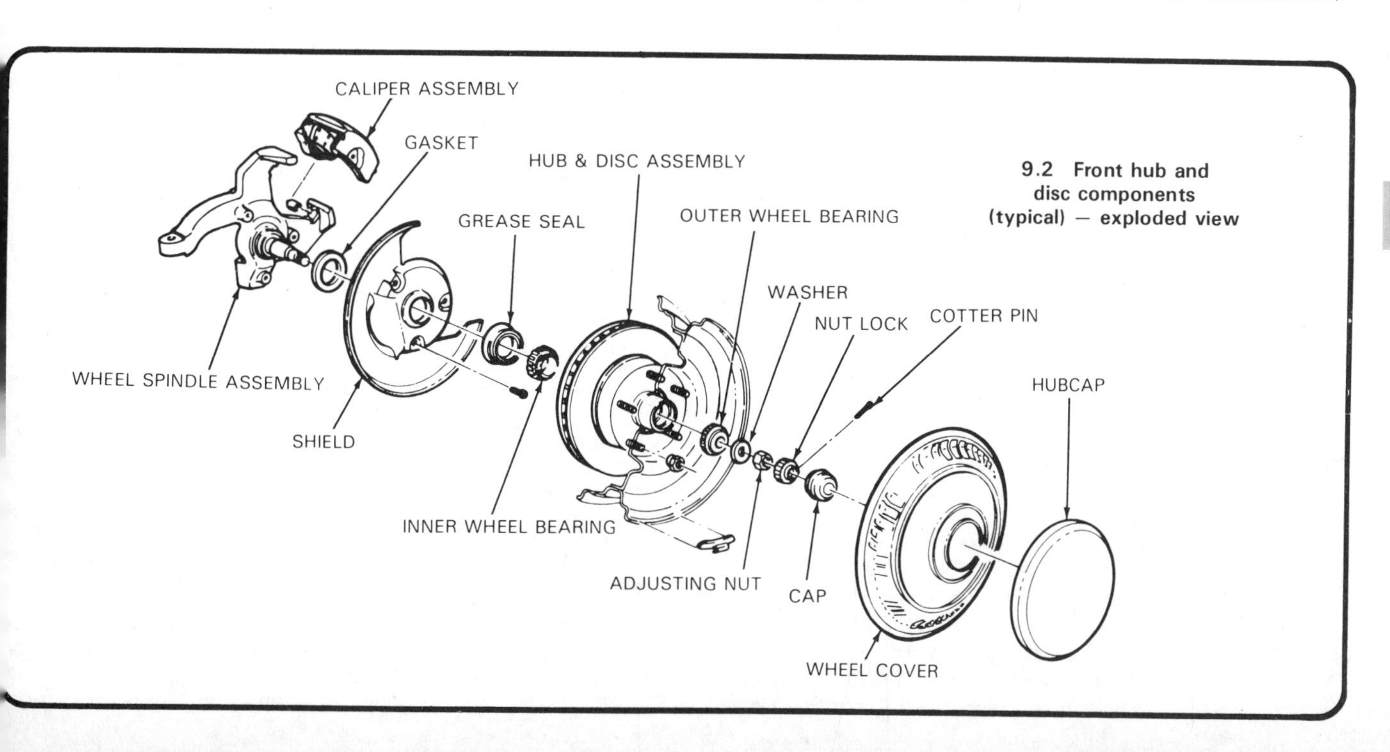

9.2 Front hub and disc components (typical) — exploded view

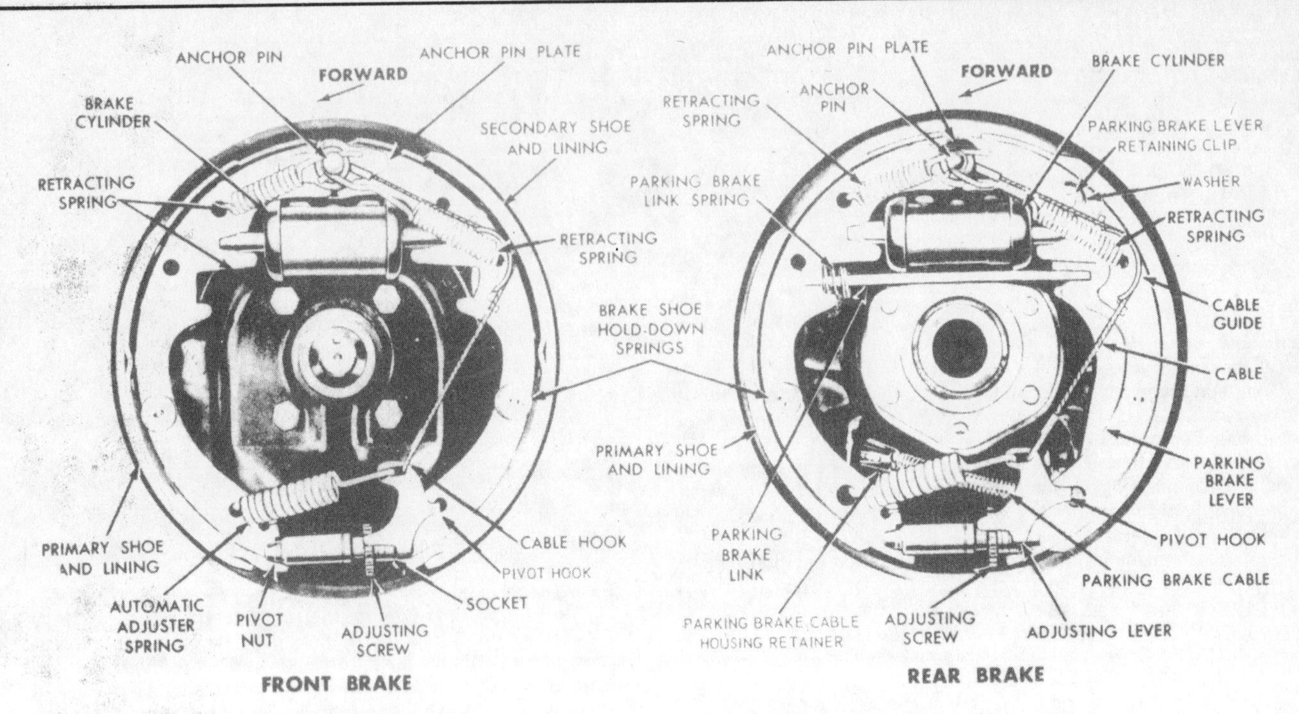

10.1 Front and rear drum brake components (E100 and E150 models)

3 Withdraw the cotter pin and nut lock from the wheel bearing adjusting nut.
4 Loosen and remove the wheel bearing adjusting nut from the spindle.
5 Grip the hub and disc assembly and pull it out far enough to loosen the washer and outer wheel bearing.
6 Push the hub and disc back onto the spindle and remove the washer and outer wheel bearing.
7 Grip the hub and disc assembly and carefully pull it off the wheel spindle.
8 Pry out the grease seal and remove the inner tapered roller bearing from the back of the hub assembly.
9 Clean out the hub and wash the bearings with solvent. Make sure that no grease or solvent is allowed to get on the brake disc.
10 Clean the disc with brake cleaner and inspect it for score marks and excessive wear. If damage is evident the disc may be resurfaced by an automotive machine shop, but the minimum thickness of the disc must not end up less than that specified.
11 To reassemble, pack the bearings with the recommended grease (see Chapter 1). Work the grease carefully into the bearing cage and rollers.
12 To reassemble the hub, install the inner bearing and then gently tap a new grease seal into the hub with a hammer and block of wood. Never reuse the old seal. The lip must face in, toward the hub.
13 Install the hub and disc assembly on the spindle, keeping it centered to prevent damage to the inner grease seal and the spindle threads.
14 Position the outer wheel bearing and flat washer over the spindle.
15 Thread the wheel bearing adjusting nut onto the spindle and tighten it finger tight so that the hub and disc will still rotate freely. Adjust the wheel bearings as described in the repacking procedure in Chapter 1.
16 Detach the caliper and anchor plate from the suspension arm and guide the assembly towards the disc. Be careful not to stretch or twist the rubber brake hose.
17 Install the caliper by referring to the appropriate Section.

10 Front and rear drum brake shoes (E100 and E150 models) – inspection, removal and installation

Refer to illustrations 10.1, 10.6, 10.8, 10.11, 10.15 and 10.31

Warning: *Brake dust contains asbestos, which is harmful to your health. Do not blow it out with compressed air and do not inhale any of it.*

1 The front and rear drum brakes used on E100 and E150 models are virtually identical in design. The main difference is that the rear brake incorporates the parking brake mechanism (see illustration).
2 Raise the front or rear of the vehicle, as required, and support it with jackstands. Remove the wheels.
3 If working on the front brakes, pry the dust cap from the center of the hub. Remove the cotter pin, nut lock and nut and pull off the hub and drum.
4 In the case of the rear brakes, remove the drum retainer (push-on nut) and pull off the hub and drum.
5 If the drum will not come off, remove the rubber cover from the brake backing plate and insert a narrow screwdriver through the slot. Disengage the adjusting lever from the adjusting screw.
6 While holding the adjusting lever away from the screw, back off the screw with a second screwdriver or a brake adjusting tool (see illustration). Be careful not to burr, chip or otherwise damage the notches in the adjusting screw.

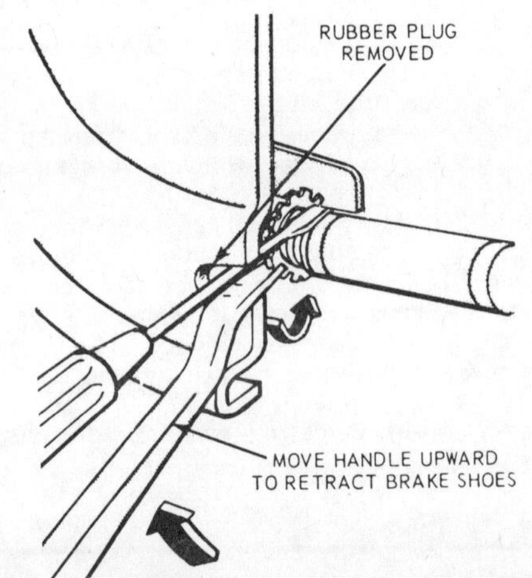

10.6 If the drum won't come off the brake shoes, retract them as shown here (E100 and E150 models)

7 The brake shoes should be replaced if they are worn beyond the specified limits or will be before the next routine check (see Chapter 1 for the lining thickness check).
8 To remove the brake shoes, detach and remove the secondary shoe retracting spring (see illustration).
9 Detach the primary shoe retracting spring and remove it.
10 Unhook the adjusting cable eye from the anchor pin.

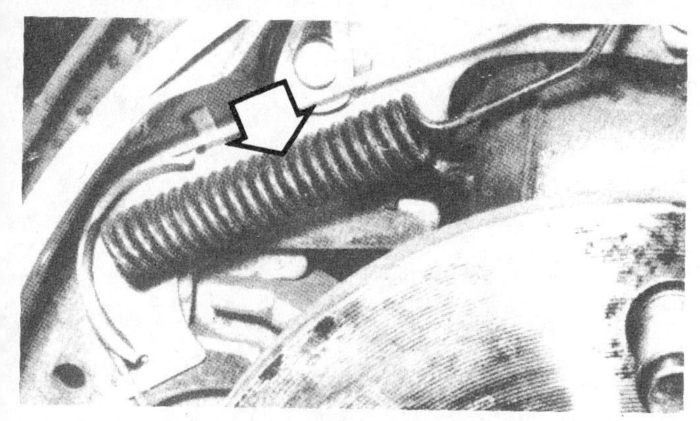

10.8 Remove the secondary shoe retracting spring from the post and shoe

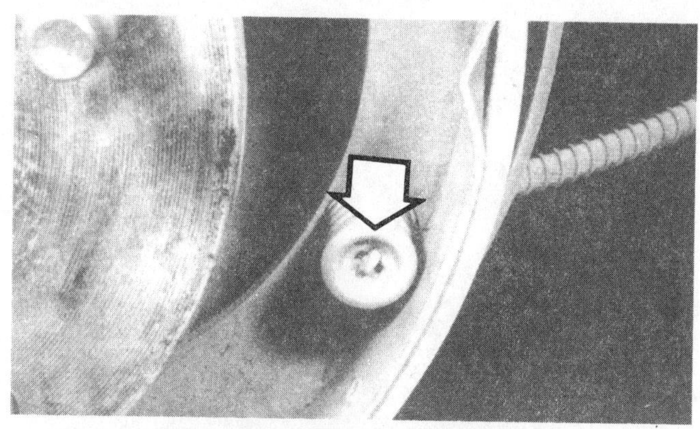

10.11 The shoe hold-down springs can be removed by depressing and turning them with a special tool

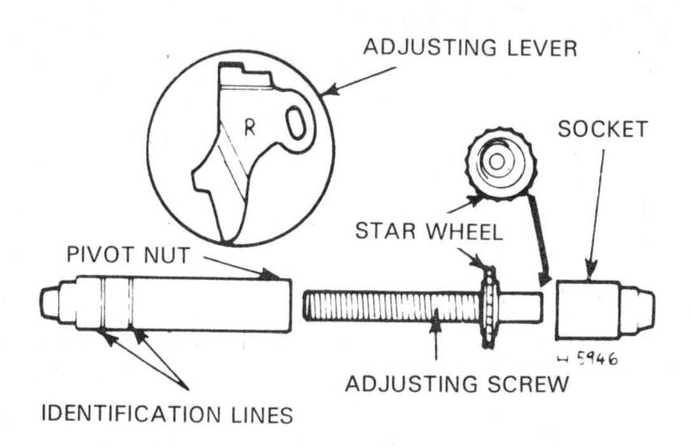

10.15 Automatic brake adjuster components — exploded view (note the identification marks for right side brakes)

11 Remove the shoe hold-down springs (see illustration), followed by the shoes, adjusting screw, pivot nut, socket and automatic adjuster parts.
12 Remove the parking brake link and spring. Disconnect the parking brake cable from the parking brake lever.
13 After the secondary shoe has been removed, the parking brake lever should be detached from the shoe.
14 Do not disassemble both brakes on one axle at the same time. This is because the brake shoe adjusting screw assemblies are not interchangeable and if mixed up could operate in reverse, increasing the drum-to-lining clearance every time the vehicle is backed up.
15 To prevent confusion, the socket end of the adjusting screw is stamped with an R or an L. The adjusting pivot nuts can be identified by the number of grooves machined around the body of the nut. Two grooves on the nut indicate a right-hand thread and one groove indicates a left-hand thread (see illustration).
16 If the brakes are to be left apart for a period of time, place a warning on the steering wheel, as accidental depression of the brake pedal will force the pistons out of the wheel cylinder.
17 Thoroughly clean all traces of dust from the shoes, backing plate and brake drums with brake cleaner. **Warning:** *Do not use compressed air to blow the dust out.*
18 Make sure that the pistons are free in the cylinder and that the rubber dust covers are undamaged and in position. Check for brake fluid leaks.
19 Prior to reassembly, apply a trace of white grease to the shoe support pads, brake shoe pivots and the adjusting screw star wheel face and threads.
20 To reassemble the brake, attach the parking brake lever to the secondary shoe and secure it with the spring washer and retaining clip.
21 Place the brake shoes on the backing plate and install the hold-down springs.
22 On the rear brakes only, install the parking brake link and spring. Back off the parking brake adjustment and connect the cable to the parking brake lever.
23 Position the shoe guide (anchor pin) plate on the anchor pin (if used).

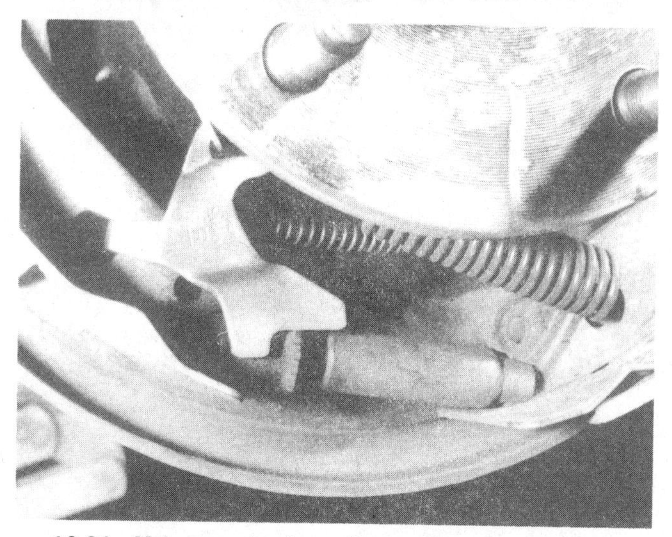

10.31 Make sure the shoe adjusting lever is installed as shown on the correct side

24 Place the cable eye over the anchor pin with the crimped side facing the backing plate.
25 Install the primary shoe retracting spring.
26 Position the cable guide in the secondary shoe web with the flanged hole in the hole in the secondary shoe web. Thread the cable around the cable guide groove. It is very important that the cable is positioned in the groove and not between the guide and the shoe web.
27 Install the secondary shoe retracting spring.
28 Make sure that the cable eye is not twisted or binding on the anchor pin. All parts must be flat on the anchor pin.
29 Apply some white grease to the threads and socket end of the adjusting screw. Turn the adjusting screw into the adjusting pivot nut as far as possible, then back it out 1/2-turn.
30 Place the adjusting socket on the screw and position the assembly between the shoe ends with the adjusting screw star wheel adjacent to the secondary shoe.
31 Insert the cable hook into the hole in the adjusting lever. The adjusting levers are stamped with and R or L to indicate which brake assembly they belong with (see illustration).
32 Position the hooked end of the adjuster spring completely into the large hole in the primary shoe web. The last coil of the spring must be at the edge of the hole.
33 Connect the looped end of the spring to the adjuster lever holes.
34 Pull the adjuster lever, cable and automatic adjuster spring down and towards the rear to engage the pivot hook in the large hole in the secondary shoe web.
35 After reassembly, check the action of the adjuster by pulling the section of the cable between the cable guide and the anchor pin towards the secondary shoe web far enough to lift the lever past a tooth on the adjusting screw star wheel.

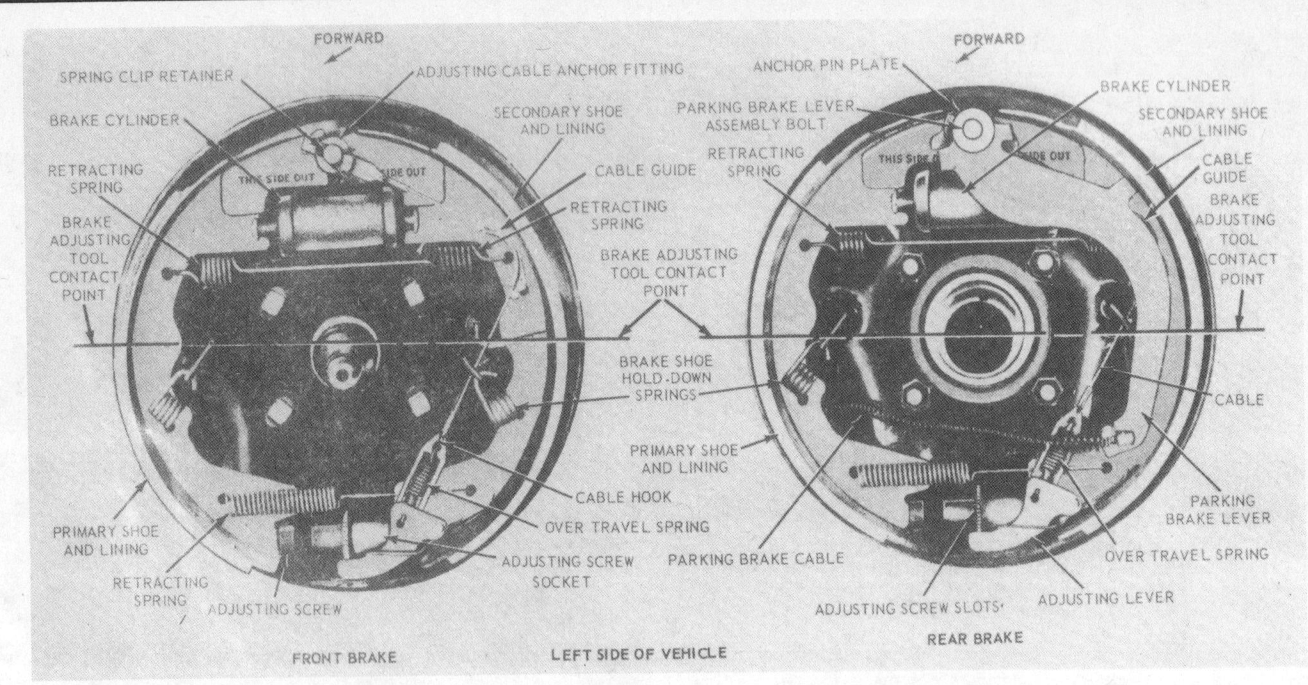

11.1 Front and rear drum brake components (E250 and E350 models)

11.3 The axleshaft must be removed on some models to detach the brake drum

36 The lever should snap into position behind the next tooth and releasing the cable should cause the adjuster spring to return the lever to its original position. This return motion of the lever will turn the adjusting screw slightly.

37 If pulling the cable does not produce the desired action, or if the lever action is sluggish instead of positive and sharp, check the position of the lever on the adjusting screw star wheel. With the brake unit in a vertical position (anchor pin at the top), the lever should contact the adjusting wheel 0.180 ± 0.030-inch above the centerline of the screw.

38 If the contact point is below this point, the lever will not lock on the teeth in the adjusting screw star wheel and the screw will not be turned as the lever is actuated by the cable.

39 Incorrect action should be checked as follows:

a) Inspect the cable and fittings. The cable should completely fill or extend slightly beyond the crimped section of the fittings. If it doesn't, the cable assembly should be replaced with a new one.
b) Check the cable length. The cable should measure 8.400-inch from the end of the cable anchor to the end of the cable hook.
c) Inspect the cable guide for damage. The cable groove should be parallel to the shoe web and the body of the guide should lie flat against the web. Replace the guide if it is damaged.
d) Inspect the pivot hook on the lever. The hook surfaces should be square to the body of the lever for correct pivot action. Replace the lever if the hook is damaged.
e) Make sure that the adjustment screw socket is correctly seated in the notch in the shoe web.

40 Repeat the entire procedure for the remaining brake assembly, then install the brake drums and wheels, lower the vehicle to the ground and check the operation of the brakes.

11 Front and rear drum brake shoes (E250 and E350 models) — inspection, removal and installation

Refer to illustrations 11.1, 11.3, 11.4, 11.5, 11.6, 11.7, 11.9 and 11.13

Note: *The rear brake drum removal and installation procedure described here applies only to 1974 and earlier models. Refer to Chapter 1 (Section 17) to remove and install the rear brake drums and adjust the bearing preload on 1975 and later vehicles with full-floating axles.*

1 Heavy duty drum brakes are used on the E250 and E350 models. Apart from the parking brake mechanism used on the rear brakes, both front and rear brakes are similar in design (see illustration).

2 If working on the front brakes, raise the vehicle and remove the brake drums as described in Paragraphs 2 and 3 of the previous Section.

3 To remove the rear brake drums, remove the eight bolts and withdraw the axleshaft and gasket (see illustration). Refer to Chapter 8 for more detailed information related to axleshaft removal.

4 Remove the large locknut and pry up the lockwasher tab (see illustration). A large socket and breaker bar will be needed to remove the nut.

5 Remove the lockwasher, adjusting nut and outer bearing and withdraw the hub and brake drum (see illustration).

6 If either the front or rear drums stick on the brake shoes, remove the rubber plug from the rear of the brake backing plate and back off the adjuster (see illustration).

7 On rear brakes, unscrew the nut from the rear of the backing plate, remove the pivot bolt and detach the parking brake lever (see illustration).

8 Disconnect the adjuster cable from the anchor pin and the adjuster lever. Unhook and remove the brake shoe retracting springs.

9 Unhook the brake shoe hold-down springs from the backing plate and detach the shoes and adjuster assembly (see illustration).

10 Refer to the procedure described in Paragraphs 14 through 19 of Section 10.

11 Begin reassembly by attaching the upper retracting pin to both shoes and then place them in position on the backing plate with the notches in the shoes engaged in the wheel cylinder piston slots.

12 Install the brake shoe hold-down springs.

13 Install the adjuster assembly. Make sure the adjuster screw is facing

11.4 A large screwdriver can be used to pry up the lockwasher tab so the adjusting nut can be removed

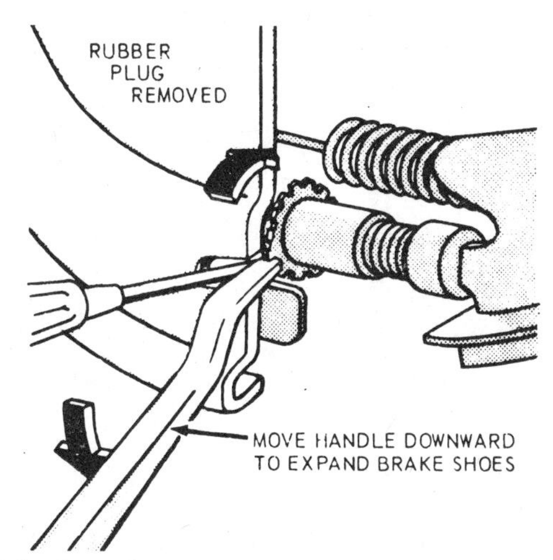

11.6 If the drum will not slide off, back off the shoe adjuster as shown here (E250 and E350 models)

11.9 Removing the brake shoe hold-down springs (E250 and E350 models)

11.5 After the adjusting nut and bearing are removed, the brake drum will slide off

11.7 The parking brake lever is secured in place by the pivot bolt at the top

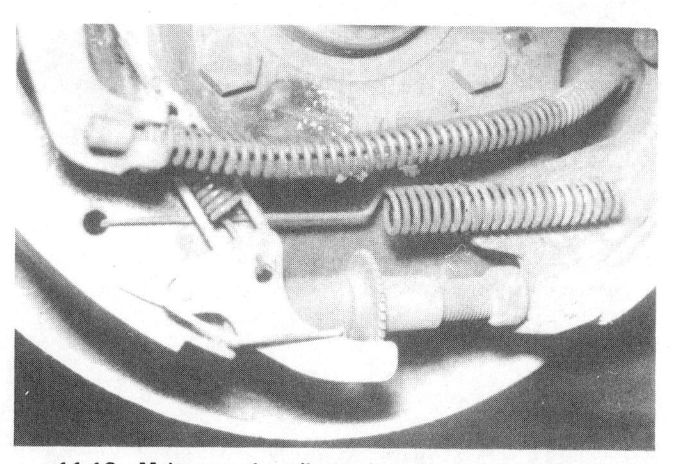

11.13 Make sure the adjuster lever contacts the wheel as shown here (E250 and E350 models)

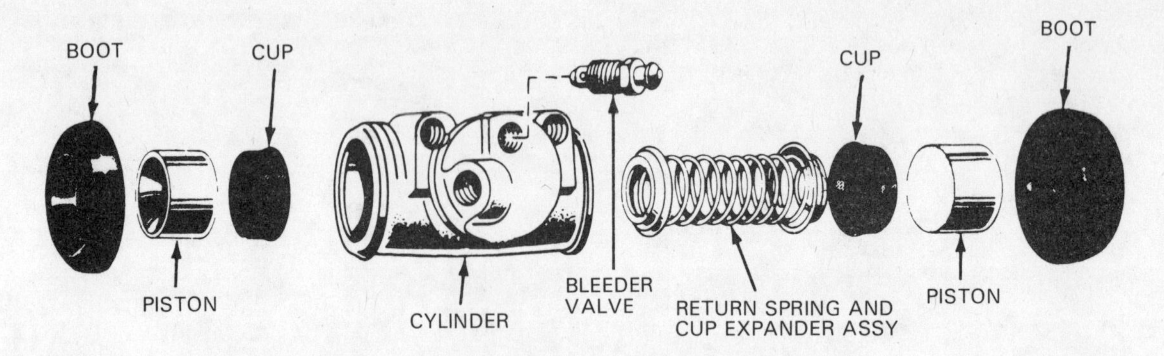

13.2 Wheel cylinder components — exploded view

toward the primary shoe (see illustration).

14 Install the bottom retracting spring, the adjuster lever spring and the adjuster lever.

15 Connect the adjuster cable to the adjuster lever. Make sure that the cable engages in the groove in the cable guide, then connect the upper end of the cable to the anchor pin.

16 On the rear brakes, connect the parking brake cable to the parking brake lever and attach the lever to the anchor pin. Thread on the nut at the rear of the backing plate to hold the parking brake lever in place.

17 Refer to Paragraphs 35 through 39 of Section 10 for the remaining steps in the brake shoe installation and adjustment procedure.

18 When reinstalling the rear hub/drum on 1974 and earlier models with full-floating axles, tighten the bearing adjusting nut (the inner nut) to 50 to 80 ft-lbs while turning the drum.

19 Back-off (loosen) the nut 3/8-turn, then coat a *new* lockwasher with axle lubricant and install it *smooth side out.*

20 Install the locknut and tighten it to 90 to 100 ft-lbs. The drum must turn freely with an end play of 0.001 to 0.0010-inch (the wheel may have to be installed to turn the drum easily). This is an important specification — check it with a dial indicator if one is available. The bearings must not be preloaded or damage will occur.

21 If the end play is acceptable and the wheel turns freely, bend two of the lockwasher tabs in, over flats of the adjusting nut, and two tabs out, over flats of the locknut, to lock the nuts in place.

22 Repeat the entire procedure for the remaining brake assembly, then install the wheels, lower the vehicle to the ground and check the operation of the brakes.

12 Drum brake wheel cylinder — removal and installation

1 The type of wheel cylinder and the method of removal is basically the same for all models.

2 Refer to Section 10 or 11 as appropriate and remove the brake shoes.

3 Unscrew the brake line fitting from the rear of the wheel cylinder. Do not pull the metal line from the cylinder, as it will bend, making installation difficult.

4 Loosen and remove the two bolts securing the wheel cylinder to the brake backing plate.

5 Detach the wheel cylinder and separate it from the brake line.

6 Plug the end of the brake line to prevent the loss of fluid.

7 Installation is the reverse of removal. Be sure to bleed the brakes as described in Section 2.

13 Drum brake wheel cylinder — inspection and overhaul

Refer to illustration 13.2

Note: *Purchase two wheel cylinder rebuild kits before beginning this procedure. Never overhaul only one wheel cylinder — always rebuild both of them at the same time.*

1 Remove the wheel cylinder as described in the previous Section.

2 To disassemble the wheel cylinder, first remove the rubber boot from each end and push out the two pistons, cup seals and return spring (see illustration). Discard the rubber parts and use the new ones from the rebuild kit when reassembling the wheel cylinder.

3 Inspect the pistons for scoring and scuff marks. If present, the pistons should be replaced with new ones.

4 Examine the inside of the cylinder bore for score marks and corrosion. If these conditions exist, the cylinder can be honed slightly to restore it, but replacement is recommended.

5 If the cylinder is in good condition, clean it with brake cleaner denatured alcohol or clean brake fluid. **Warning:** *Do not, under any circumstances, use petroleum-based solvents to clean brake parts.*

6 Remove the bleeder valve and make sure that the hole is clear.

7 Lubricate the new rubber cups with clean brake fluid and insert one into the bore, followed by one piston. Make sure the lip on the rubber cup faces in.

8 Place the return spring in the bore and push in until it contacts the first cup.

9 Install the remaining cup and piston in the cylinder bore.

10 Install the two rubber boots.

11 The wheel cylinder is now ready for installation.

14 Drum brake backing plate — removal and installation

Note: *For removal of the front brake backing plate, refer to Chapter 10*

1 Refer to Sections 10 and 12, and remove the brake shoes and wheel cylinder from the backing plate.

2 Disconnect the parking brake lever from the cable.

3 Refer to Chapter 8 and remove the axleshaft.

4 Disconnect the parking brake cable retainer from the backing plate.

5 The backing plate and gasket can now be detached from the end of the axle housing.

6 Installation is the reverse of removal. Be sure to bleed the brake system as described in Section 2. Don't forget to top up the rear axle oil level if necessary.

15 Drum brake shoes — adjustment

1 Automatic adjusters are included on drum brakes. They operate when the vehicle is backed up and stopped. If the vehicle is not backed up very often and the pedal movement has increased, then it will be necessary to adjust the brakes as follows.

2 Drive the vehicle in reverse and apply the brake pedal firmly. Now drive it forward and again apply the brake pedal firmly.

3 Repeat the cycle until desirable pedal movement is obtained. Should this not happen it will be necessary to remove the drums and inspect the adjuster mechanism as described in Section 10.

16 Brake master cylinder — removal and installation

1 For safety reasons, disconnect the negative cable from the battery.

2 Disconnect the wires from the brake light switch, located adjacent to the brake pedal.

3 Remove the nut and shoulder bolt securing the master cylinder

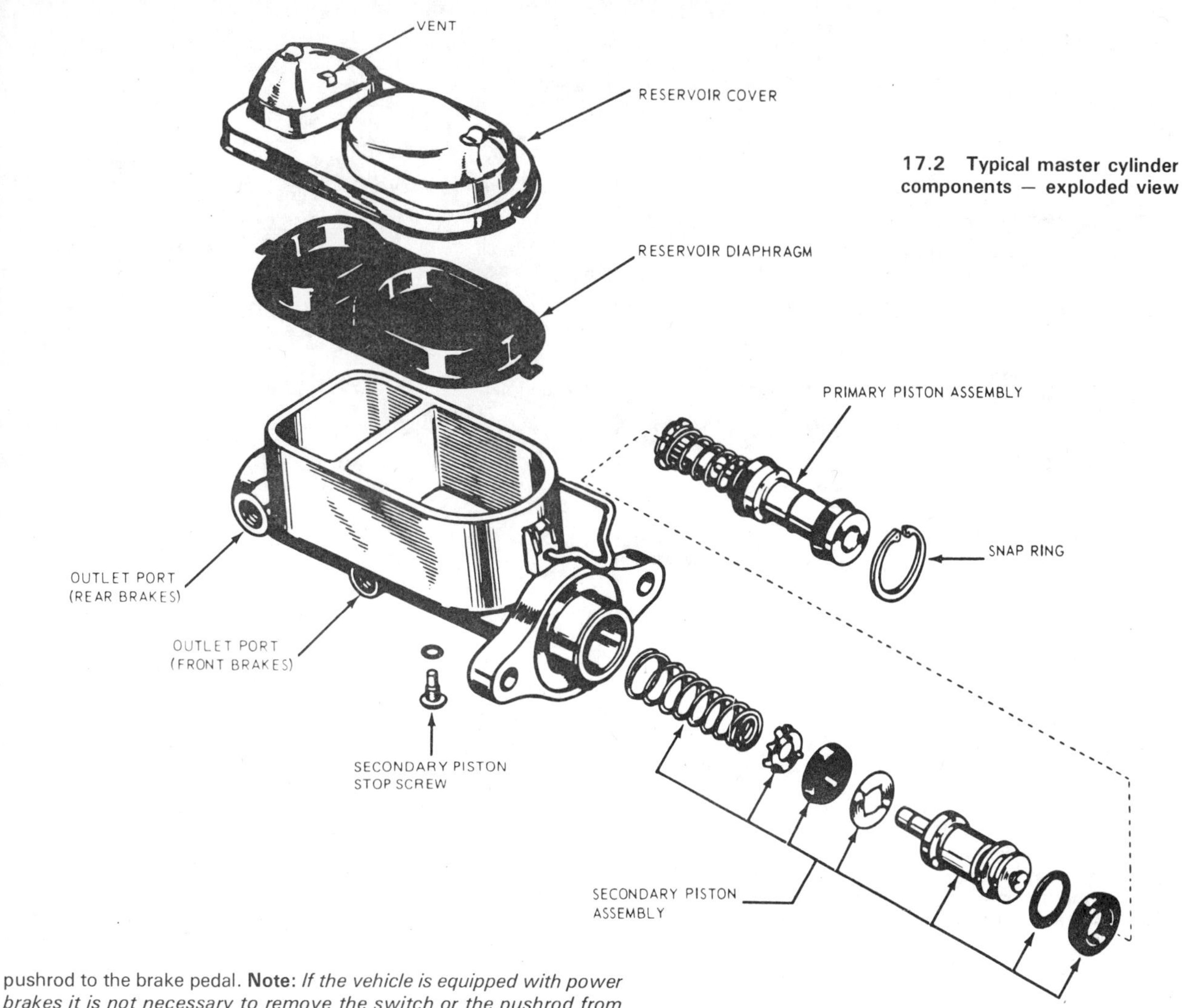

17.2 Typical master cylinder components — exploded view

pushrod to the brake pedal. **Note:** *If the vehicle is equipped with power brakes it is not necessary to remove the switch or the pushrod from the brake pedal.*

4 Unscrew the brake lines from the primary and secondary outlet ports of the master cylinder. Plug the ends of the lines to stop the entry of dirt. Take precautions to catch the brake fluid as the lines are detached from the master cylinder body.

5 Loosen and remove the two bolts securing the master cylinder to the dash or brake booster.

6 Pull the master cylinder forward and lift it up to remove it. Do not allow brake fluid to contact any painted areas, as it will ruin them.

7 Installation is the reverse of removal. Be sure to bleed the system as described in Section 2. If a new master cylinder is being installed, lubricate the seals before installation (they have a protective coating when originally assembled). Remove the plugs from the brake line ports, pour some clean brake fluid into the master cylinder and operate the pushrod several times to distribute it over all the internal components and surfaces.

17 Brake master cylinder — overhaul

Refer to illustrations 17.2 and 17.5

Note: *Purchase a master cylinder rebuild kit before beginning this procedure. The kit will include all of the replacement parts necessary for the overhaul. The rubber replacement parts, particularly the seals, are the key to fluid control in the master cylinder. As such, it's very important to install them correctly. Be very careful not to let them come in contact with petroleum-based solvents or lubricants.*

1 Clean the exterior of the master cylinder with brake cleaner and wipe it dry with a lint-free rag.

2 Remove the reservoir cover and diaphragm (see illustration) from the top of the reservoir and pour out any remaining brake fluid.

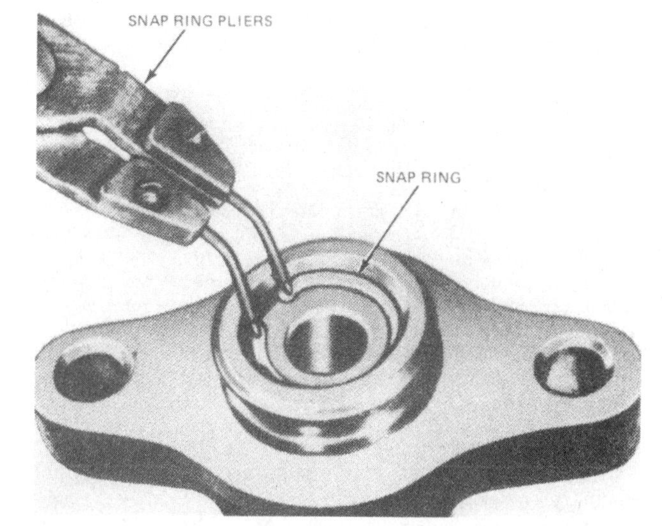

17.5 Use snap-ring pliers to remove the snap-ring that retains the master cylinder piston assembly in the bore

3 Loosen and remove the secondary piston stop screw from the bottom of the master cylinder body.

4 Loosen and remove the bleeder screw.

5 Depress the primary piston and remove the snap-ring from the groove at the rear of the master cylinder bore (see illustration).

6 **Do not** remove the screw retaining the primary return spring retainer, return spring, primary cup and protector on the primary piston.

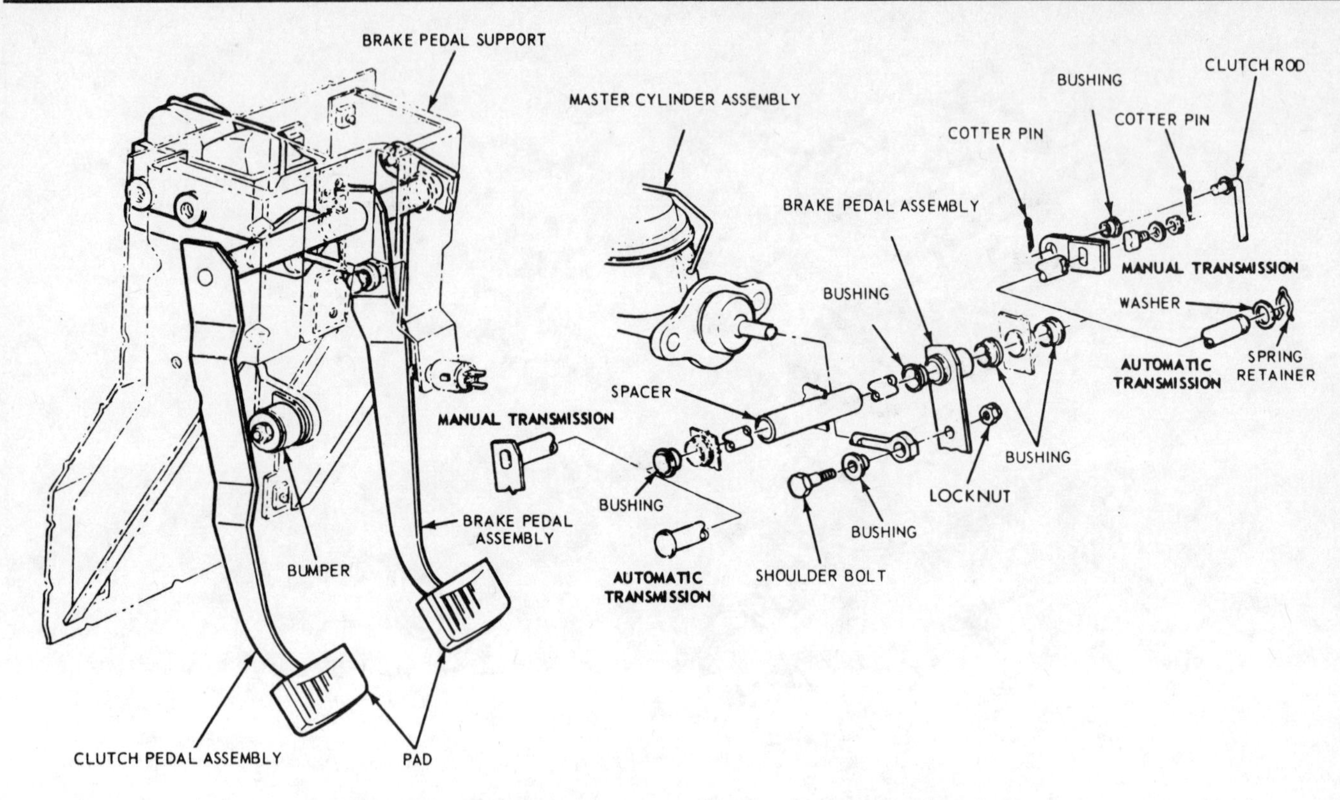

18.1 Brake pedal and related components – exploded view (typical)

This is factory set and must not be disturbed.
7 Remove the secondary piston assembly.
8 **Do not** remove the outlet line seats, outlet check valves and outlet check valve springs from the master cylinder body.
9 Examine the bore of the cylinder carefully for signs of scoring and scuffing. If it is in good condition, new seals can be installed. If, however, there is any doubt of the condition of the bore a new master cylinder must be installed.
10 If the rubber seals are swollen or very loose on the pistons, suspect oil contamination in the system. Oil will cause the rubber seals to swell. If one is swollen it is reasonable to assume that all seals in the brake system will need attention.
11 Thoroughly clean all parts with brake cleaner, denatured alcohol or brake fluid. **Warning:** *Do not, under any circumstances, use petroleum-based solvents to clean brake components.*
12 All components should be assembled wet after dipping them in new brake fluid.
13 Carefully insert the complete secondary piston and return spring assembly into the master cylinder bore. Ease the seals into the bore, taking care that they do not roll over. Push the assembly all the way in.
14 Insert the primary piston assembly into the master cylinder bore.
15 Depress the primary piston and install the snap-ring in the cylinder bore groove.
16 Install the pushrod boot and retainer on the pushrod and insert the assembly into the end of the primary piston. Make sure that the retainer is correctly seated and holding the pushrod securely.
17 Place the inner end of the pushrod boot in the master cylinder body retaining groove.
18 Install the secondary piston stop screw and O-ring in the bottom of the master cylinder body.
19 Position the diaphragm in the cover, make sure it is correctly seated and install the cover. Secure it in position with the spring retainer.

18 Brake pedal – removal and installation

Refer to illustration 18.1

1 On vehicles equipped with a manual transmission the brake pedal operates on the same shaft as the clutch pedal. Both pedals are removed as described in Chapter 8. Refer to the accompanying exploded view illustration of the brake pedal and related components.
2 On automatic transmission models, disconnect the master cylinder pushrod, then loosen the shoulder bolt nut and slide the shoulder bolt to the right until the brake pedal and pedal bushings can be removed.
3 Replace the bushings (if worn) and install the pedal and pushrod by reversing the removal procedure.

19 Pressure differential valve assembly – removal and installation

1 Disconnect the brake warning light connector from the warning light switch.
2 Disconnect the front and rear lines from the valve assembly. Plug the ends of the lines to prevent loss of brake fluid and entry of dirt.
3 Remove the bolts and nuts securing the valve assembly to the chassis.
4 Detach the valve assembly and bracket. Don't allow brake fluid to spill on painted surfaces as it will ruin them.
5 The valve assembly cannot be overhauled or repaired. If it is defective, replace it with a new one.
6 Installation is the reverse of removal. Be sure to bleed the system as described in Section 2.

20 Parking brake assembly – removal and installation

Refer to illustrations 20.1a and 20.1b

1 Loosen the adjusting locknut on the cable equalizer (see illustrations).
2 Working under the left front fender, remove the nuts securing the parking brake control assembly to the dash panel.
3 On E250 and E350 models only, remove the nut and bolt securing the parking brake bracket to the lower flange of the instrument panel.
4 Remove the cable ball end from the control assembly clevis and, if used, the cable retaining clip.
5 Remove the parking brake control assembly from the vehicle.
6 Installation is the reverse of removal. Check the cable adjustment as described in Section 21.

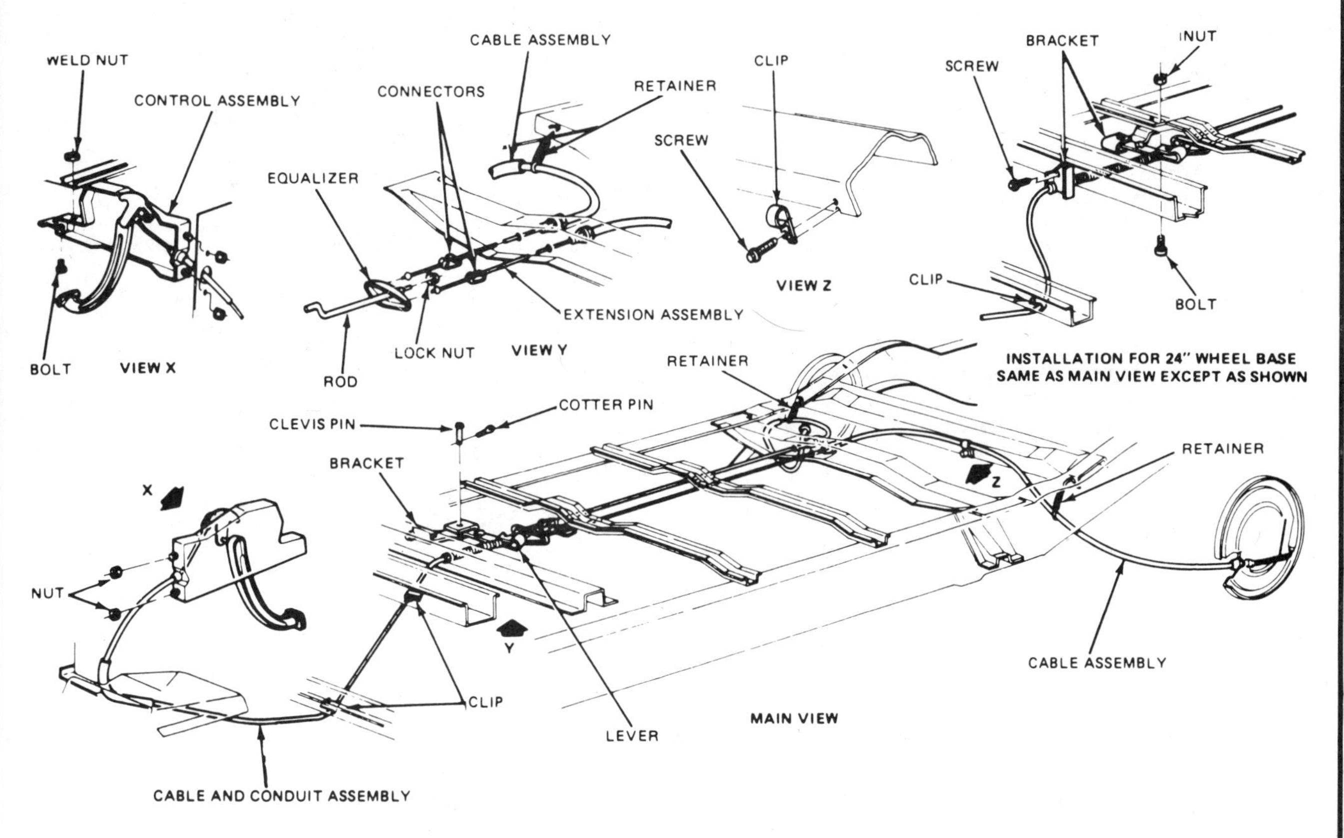

20.1a Parking brake system components – typical (E100 and E150 models)

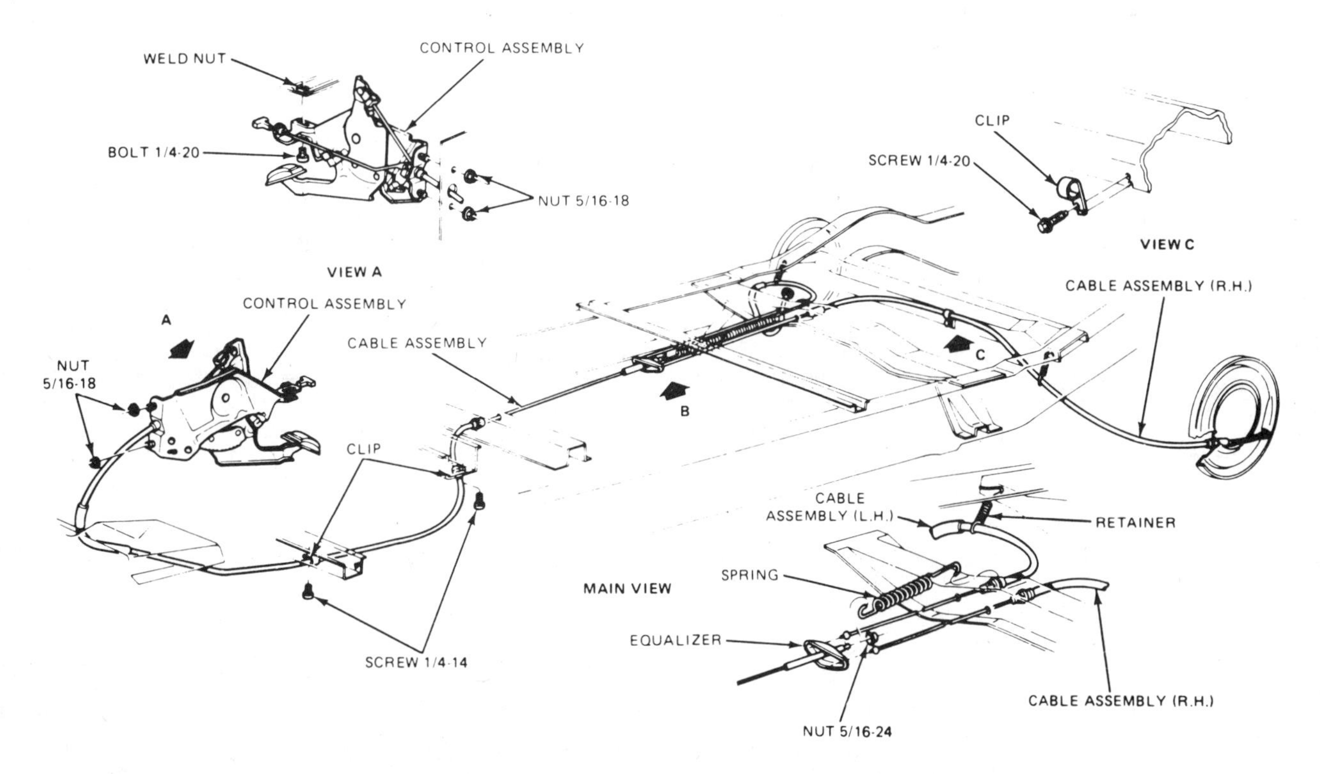

20.1b Parking brake system components – typical (E250 and E350 models)

21 Parking brake – adjustment

1 Refer to Section 15 and adjust the brakes.
2 Block the front wheels, raise the rear of the vehicle and support it on jackstands.
3 Completely release the parking brake and move the shift lever to the Neutral position.
4 Slowly tighten the adjustment nut on the cable equalizer until the rear brakes just begin to drag.
5 Back off the adjusting nut until the rear brakes are just released.
6 Lower the vehicle and check the parking brake lever free movement.

22 Parking brake cable – removal and installation

1 Block the front wheels, raise the rear of the vehicle and support it on jackstands. Remove the wheels.
2 Refer to Section 10 or 11 and remove the rear brake drums.
3 Release the parking brake and back off the adjusting nut.
4 Remove the cable from the equalizer.
5 Compress the retainer prongs and pull the cable to the rear, through the cable brackets, just enough to release the cable.
6 Remove the clips retaining each cable to the underside of the vehicle.
7 Remove the self adjuster springs and cable retainers from the brake backing plates.
8 Disconnect the ends of the cables from the parking brake levers on the secondary brake shoes.
9 Compress the cable retainer prongs and pull the cable ends from the backing plates.
10 Remove the nuts and bolts from the cable retainers on the chassis. Detach the cable from the retainers and remove it from under the vehicle.
11 Installation is the reverse of removal. Adjust the parking brake as described in Section 21.

23 Brake booster (servo) – general information

1 A power booster (servo) is included in the brake circuit to provide assistance to the driver when the brake pedal is depressed. This reduces the effort required by the driver to operate the brakes under all braking conditions.
2 The unit operates by vacuum obtained from the intake manifold and comprises basically a booster diaphragm and check valve. The braking effort is transmitted through another pushrod to the servo piston and its built-in control system. The servo piston does not fit tightly into the cylinder, but has a strong diaphragm to keep its edges in constant contact with the cylinder wall, assuring an air tight seal between the two parts. The forward chamber is held under the vacuum conditions created in the intake manifold of the engine. During periods when the brake pedal is not in use, the controls open a passage to the rear chamber, placing it under vacuum conditions as well. When the brake pedal is depressed the vacuum passage to the rear of the chamber is cut off and the chamber opened to atmospheric pressure. The consequent rush of air pushes the servo piston forward in the vacuum chamber and operates the main pushrod to the master cylinder.
3 The controls are designed so that assistance is given under all conditions. All air from the atmosphere entering the rear chamber is passed through a small air filter.
4 Under normal operating conditions the servo will give trouble free service for a very long time. If, however, a problem is suspected, such as increased foot pressure required to apply the brakes, it must be replaced with a new servo. No attempt should be made to repair the old servo as it is not a serviceable item.

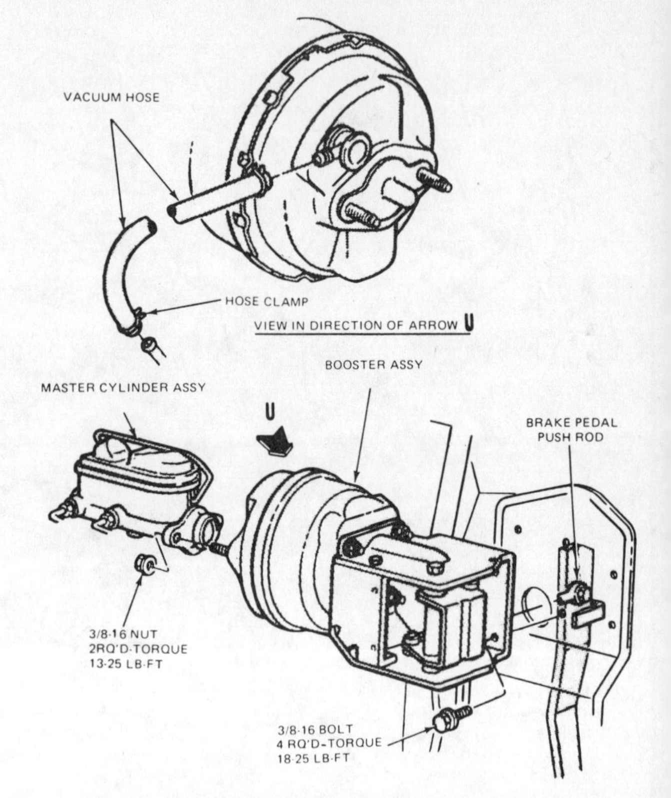

24.4 Typical power brake booster mounting details

24 Brake booster (servo) – removal and installation

Refer to illustration 24.4

1 Remove the brake light switch and actuating rod from the brake pedal as described in Section 18.
2 Working under the hood, remove the air cleaner from the carburetor and the vacuum hose from the servo.
3 Remove the master cylinder from the front of the servo and withdraw it just enough to allow removal of the servo. **Note:** *It is not necessary to disconnect the brake lines from the master cylinder, but great care must be taken to avoid bending them excessively.*
4 Remove the bolts securing the servo bracket to the engine compartment and lift the servo out of the vehicle (see illustration).
5 Installation is the reverse of removal. If it was necessary to disconnect the brake lines, the brake system must be bled as described in Section 2.

Chapter 10 Steering and suspension systems

Contents

Specifications

Front suspension

Type	Twin I-beam axles with coil springs and telescopic shock absorbers
Toe-in	0.125 in
Caster angle	5°
Camber angle	0.5°

Rear suspension

Type	Semi-elliptical leaf springs and telescopic shock absorbers

Steering (manual)

Type	Recirculating ball
Ratio	24:1
Number of turns lock-to-lock	6
Lubricant capacity	
thru 1977	0.87 lb
1978 and later	11 + 1 oz
Lubricant type	
thru 1977	C3AZ-19578-A
1978 and later	ESW-M1C87A or equivalent
Worm bearing preload (1978 and later)*	3 to 8 in-lbs
Total steering gear mesh load on center (1978 and later)	16 in-lbs maximum
Total center mesh load in excess of worm bearing preload (1978 and later)**	10 to 16 in-lbs
Lash adjuster end clearance (1978 and later)	0.000 to 0.002 in

* *Torque required to rotate the input shaft 1-1/2 turns past center (with the Pitman arm disconnected)*
** *Torque required to turn the worm assembly and input shaft past the center high point*

Steering (power assisted)

Type	Recirculating ball with integral piston and control valve
Pump type	Ford-Thompson with integral reservoir, belt driven from engine
Ratio	16:1
Number of turns lock-to-lock	4
Fluid capacity	2.44 US pints
Fluid type	See Chapter 1

Tires

Size and air pressure	Refer to the tire information decal located on the front of the left door pillar

Torque specifications	Ft-lbs
Front suspension	
Axle-to-pivot bracket	120 to 160
Radius arm-to-axle	180 to 220
Radius arm-to-rear bracket	75 to 125
Spindle bolt locking pin	40 to 55
Shock absorber lower mount	40 to 60
Shock absorber upper mount	15 to 20
Rear suspension	
U-bolt nuts (E100 – E200)	45 to 60
U-bolt nuts (E300)	150 to 200
Spring front hanger nuts	75 to 100
Spring rear hanger nuts	55 to 90
Shock absorber upper mount	40 to 60
Shock absorber lower mount	45 to 60
Steering (pre-1978 models)	
Steering gear-to-frame	55 to 75
Pitman arm-to-steering gear	170 to 230
Steering column retaining bolts	8 to 20
Balljoint nuts	50 to 75
Steering wheel nut	30 to 40
Steering (1978 and later models)	
Sector shaft cover bolts	30
Ball guide clamp screws	18 to 42 in-lbs
Preload adjuster locknut	85
Sector shaft-to-Pitman arm	170 to 230
Mesh load adjusting screw locknut	25
Drag link stud	50 to 75
Spindle connecting rod studs	50 to 75
Spindle connecting rod clamps	29 to 41
Power steering pump brackets	30 to 45

1 General information

The front suspension system used on the Econoline models covered by this manual is composed of twin I-beam axles. The wheel spindle end of each axle is located by a radius arm and supports the vehicle through a coil spring and telescopic shock absorber. The inner ends are attached to a pivot bracket at the opposite side of the vehicle.

The rear axle is supported on semi-elliptical leaf springs attached to the chassis by shackles and hanger assemblies. The springs are secured to the axle by U-bolts. Rear suspension damping is achieved by telescopic shock absorbers.

The steering gear is a recirculating ball type with power steering available as an option on all models.

The steering shaft rotates in two ball-type thrust bearings and at the steering wheel end in a bearing. The sector shaft moves in bushings and has a seal at the lower end. The upper end of the sector shaft engages a rack which is integral with the ball nut.

Steering shaft bearing adjustment is controlled by a large adjustment screw at the column end of the steering gear housing and an adjustment screw on the sector shaft end cover.

2 Front wheel spindles – removal and installation

Refer to illustration 2.6

1 Raise the front of the vehicle and support it on jackstands.
2 Support the weight of the axle on the side that is being worked on and remove the wheel.
3 Remove the brake drum (or disc) and hub as described in Chapter 1.

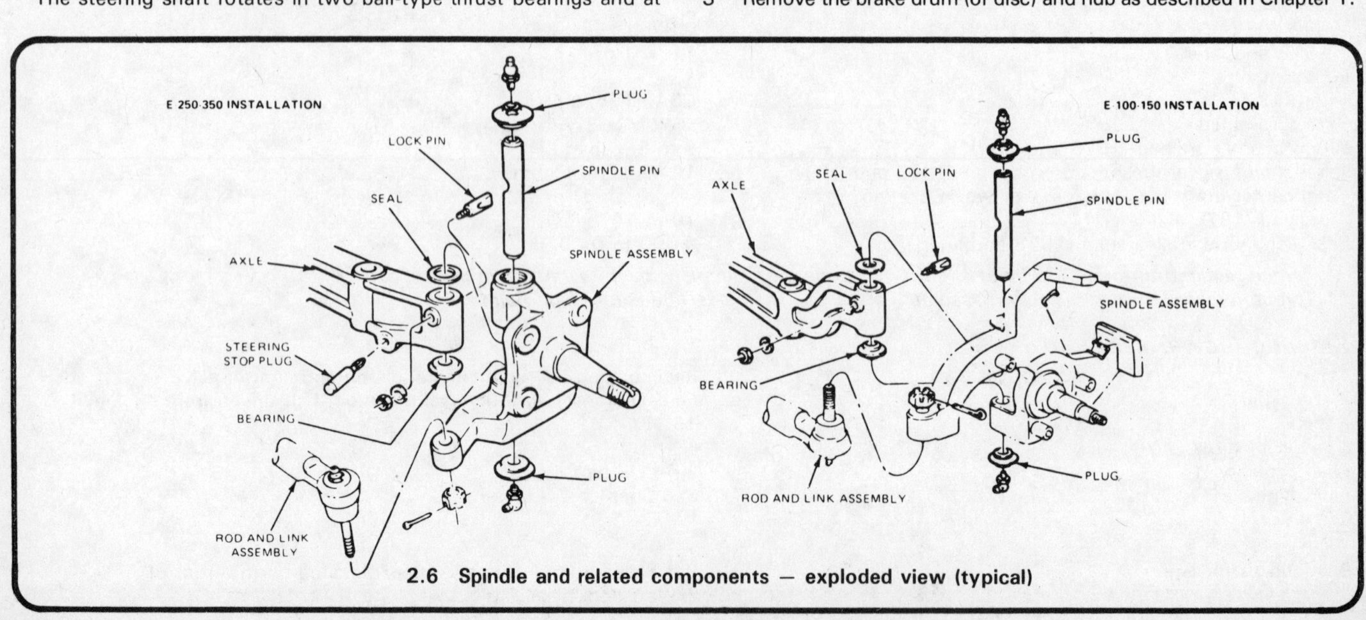

2.6 Spindle and related components – exploded view (typical)

4 On E100 and E150 models, remove the brake backing plate and steering arm from the spindle assembly.
5 On E350 models, disconnect the steering track rod balljoint from the steering arm with a large gear puller.
6 Remove the nut and washer from the spindle lock pin and drive out the pin (see illustration).
7 Remove the grease plugs from the top and bottom of the spindle pin, then drive it out from the top of the axle and remove the spindle assembly.
8 Remove the grease seal and bushings from the spindle assembly with a hammer and drift punch.
9 Check the spindle pin for wear and replace it with a new one if necessary.
10 Obtain new bushings, seals and a thrust bearing and install the bushings in the spindle assembly.
11 Place the spindle assembly in position on the axle. Pack the spindle thrust bearing with grease and insert it into the bottom of the spindle. Make sure that the lip side of the seal faces down, into the spindle.
12 Line up the notch in the spindle pin with the hole in the axle and tap the pin through the spindle and axle assembly until the notch is lined up with hole in the axle.
13 Carefully drive a new locking pin through the axle and secure it with the nut and washer. Tighten the nut and install the grease plugs at the top and bottom of the spindle pin.
14 Make sure that the spindle moves smoothly from lock-to-lock without any free play, then install the steering arm, backing plate, hub and brake drum/disc assembly by reversing the removal procedure.
15 Lubricate the spindle assembly with grease, install the wheel and have the toe-in checked by an alignment shop or dealer service department.

3 Front spring – removal and installation

Refer to illustrations 3.4 and 3.5

1 Raise the front of the vehicle and support it on jackstands.
2 Support the outer end of the axle with a floor jack and remove the wheel.
3 Disconnect the lower end of the shock absorber from the bracket.
4 Carefully peel back the front floor mat, remove the cover plate and remove the upper retainer bolts (see illustration).
5 Remove the nut securing the lower spring retainer to the seat (see illustration), lower the axle as far as possible and remove the spring retainers and spring.
6 Installation is the reverse of removal.

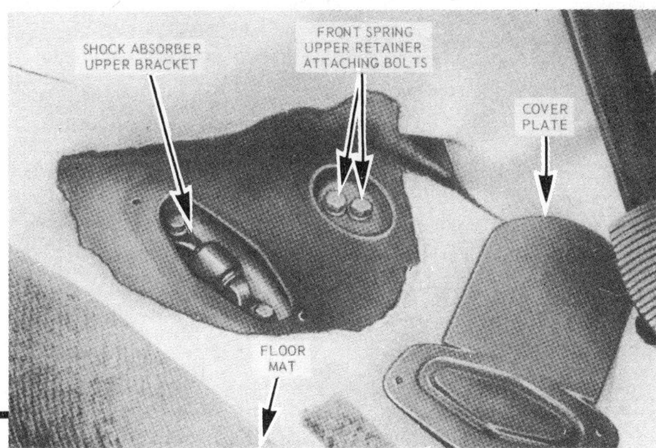

3.4 The front spring upper bolts and shock absorber upper mounting bolts are accessible through the holes in the floor pan inside the vehicle

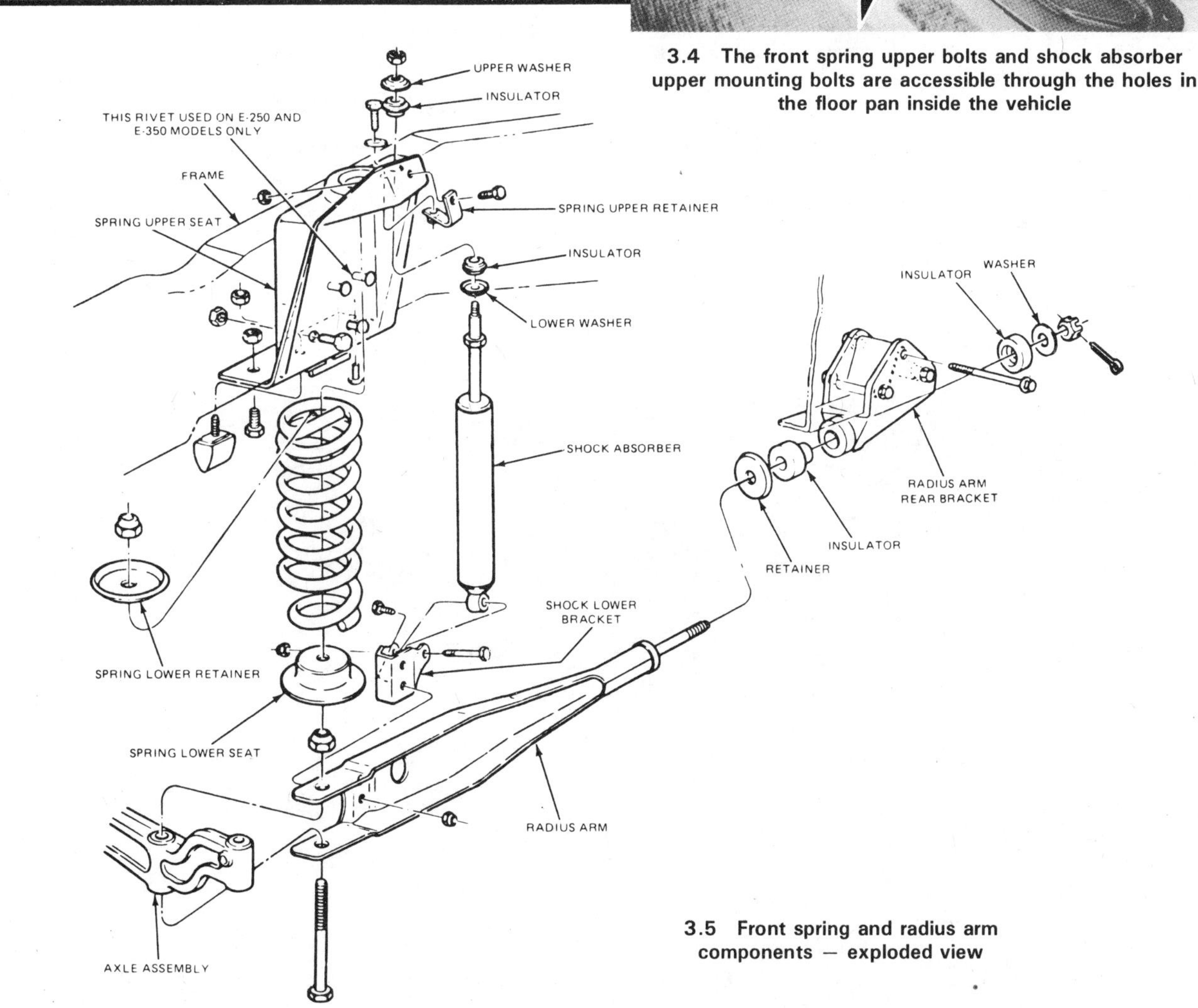

3.5 Front spring and radius arm components – exploded view

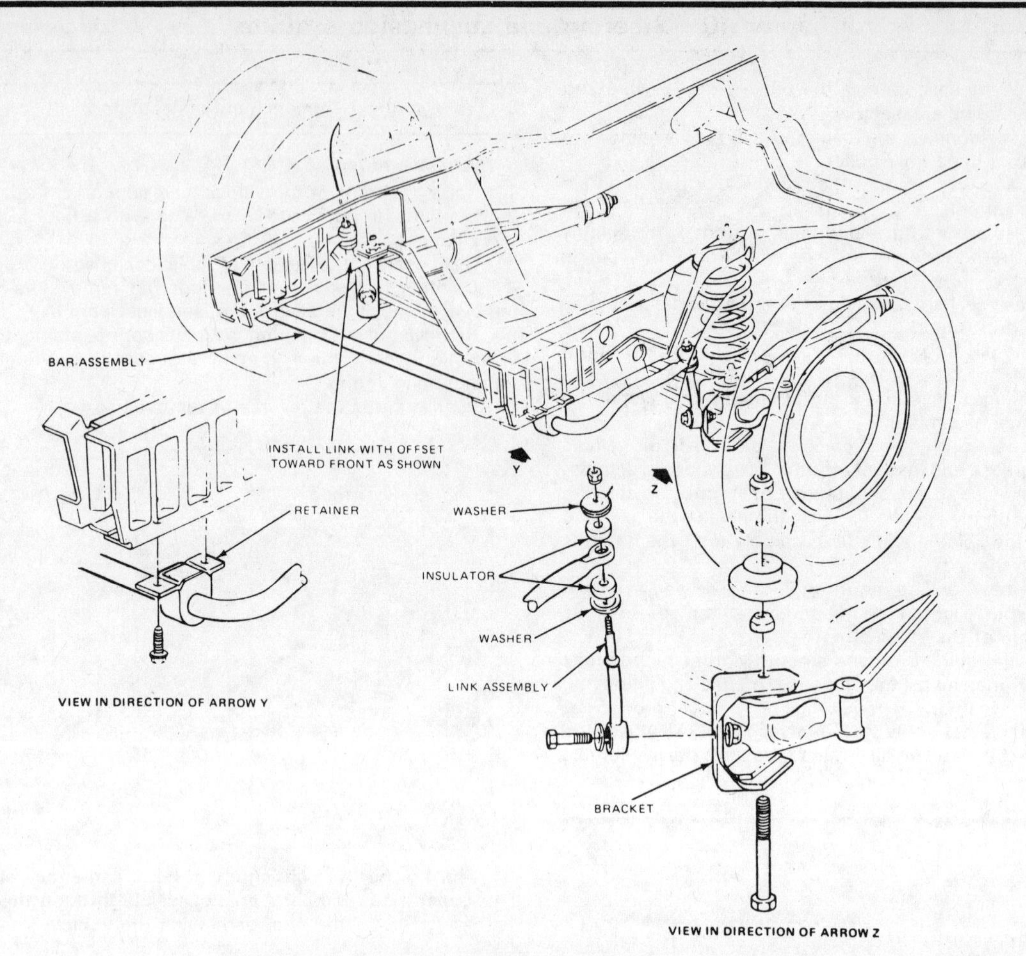

5.2 Stabilizer bar and related components — exploded view

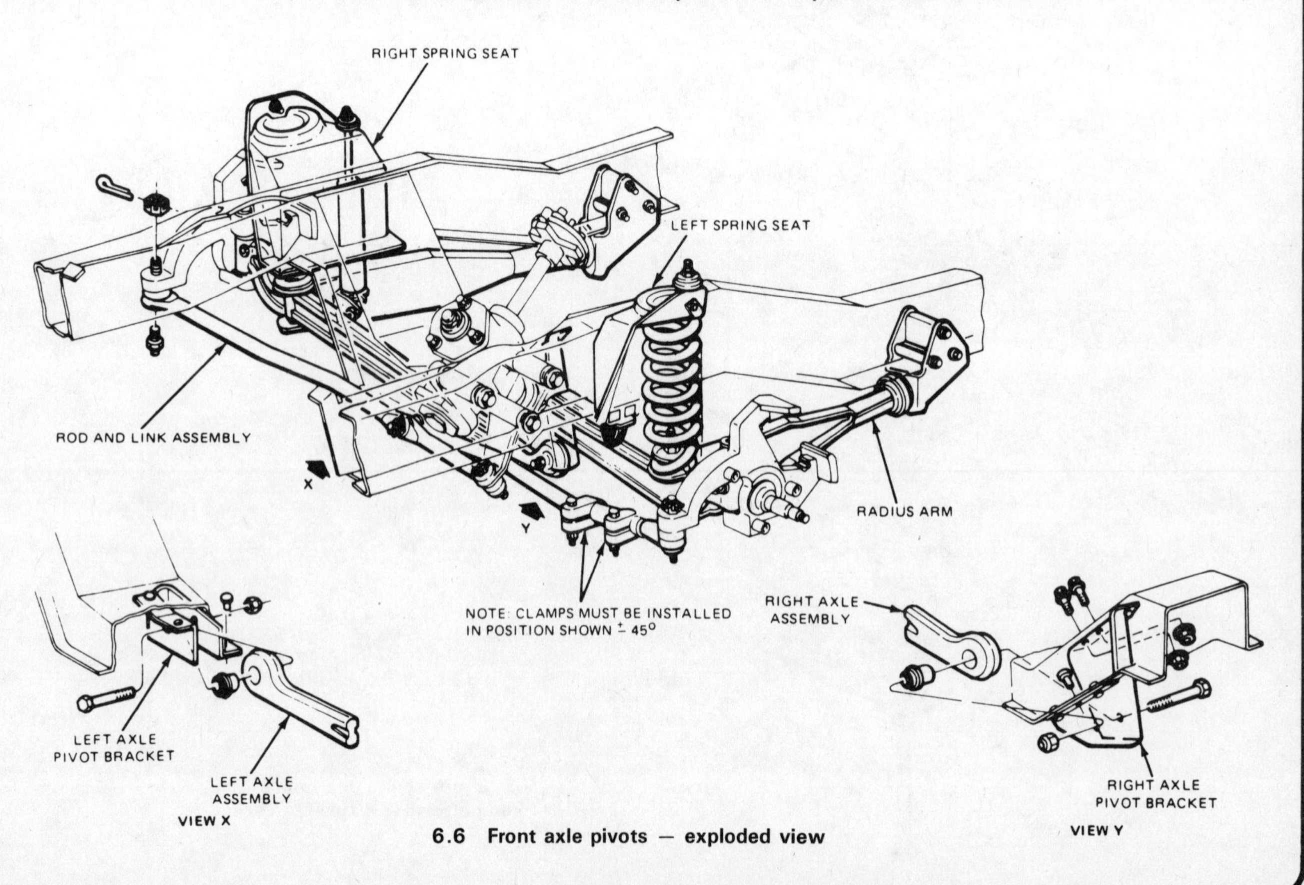

6.6 Front axle pivots — exploded view

4 Radius arm — removal and installation

1 Raise the front of the vehicle and support it on jackstands.
2 Position a jack under the outer end of the axle and remove the wheel.
3 Disconnect the lower end of the shock absorber from the support bracket.
4 Remove the front spring by referring to the appropriate Section.
5 Remove the lower spring seat from the radius arm and then remove the nut and bolt securing the front of the radius arm to the axle (see illustration 3.5).
6 Remove the nut, washer and rubber insulator from the rear of the radius arm.
7 Push the front of the radius arm away from the front axle and withdraw it from the rear bracket.
8 Remove the front retainer and rubber insulator from the radius arm and retrieve any shims that may be used.
9 Installation is the reverse of removal. Make sure that the rubber insulators are reinstalled in the correct order. Tighten the radius arm-to-axle bolt to the specified torque and secure it with the cotter pin. Tighten the rear nut to the specified torque.
10 After the radius arm has been installed, have the steering angles and toe-in checked by a dealer service department or an alignment shop.

5 Stabilizer bar — removal and installation

Refer to illustration 5.2

1 Raise the front of the vehicle and support it on jackstands.
2 Disconnect the left and right ends of the stabilizer bar from the link assemblies attached to the axle brackets (see illustration).
3 Remove the brackets securing the bar to the chassis and detach the bar and links from the vehicle.
4 Replace any worn bushings and install the bar by reversing the removal procedure. Make sure that the links are installed with the bend facing forward.

6 Front axle — removal and installation

Refer to illustration 6.6

1 Raise the vehicle and support it on jackstands.
2 Remove the front wheel spindle from the appropriate axle by following the procedure described in Section 2.
3 Remove the front spring as described in Section 3.
4 Remove the stabilizer bar (if used) as described in the previous Section.
5 Remove the spring lower seat from the radius arm and remove the nut and bolt securing the front of the radius arm to the axle.
6 Remove the nut and bolt securing the end of the axle to the pivot bracket and detach the axle from the vehicle (see illustration).
7 Examine the bushings and pivot bolt for wear and replace them if necessary. If the axle is bent, take it to a Ford dealer who will have the necessary equipment for straightening it.
8 Installation is the reverse of removal. Be sure to tighten the nuts to the specified torque.

7 Front shock absorber — removal and installation

1 Block the rear wheels, apply the parking brake, loosen the front wheel nuts, raise the front of the vehicle and support it on jackstands. Remove the wheels.
2 Remove the nut, washer and bushing from the top end of the shock absorber.
3 Remove the nut and bolt securing the bottom end of the shock absorber to the lower bracket.
4 Installation is the reverse of removal. Be sure to tighten the fasteners to the specified torque.

8 Rear suspension leaf spring — removal and installation

Refer to illustrations 8.2a and 8.2b

1 Block the front wheels, raise the rear of the vehicle and support it on jackstands. Support the weight of the rear axle on a jack.
2 Disconnect the lower end of the shock absorber from the axle bracket (see illustrations).
3 Remove the two U-bolts and the spring retainer plate and lower the axle.
4 Remove the upper and lower rear shackle bolts and nuts. Pull the rear shackle from the bracket and spring.
5 Remove the nut and bolt from the front spring bracket and remove the spring.
6 If the front bushing must be replaced, have it done by a Ford dealer service department as a special tool is required to draw out the old bushing and insert the new one. If you are inventive, the job can be done using a large bench vise and a selection of various size pipe sections.

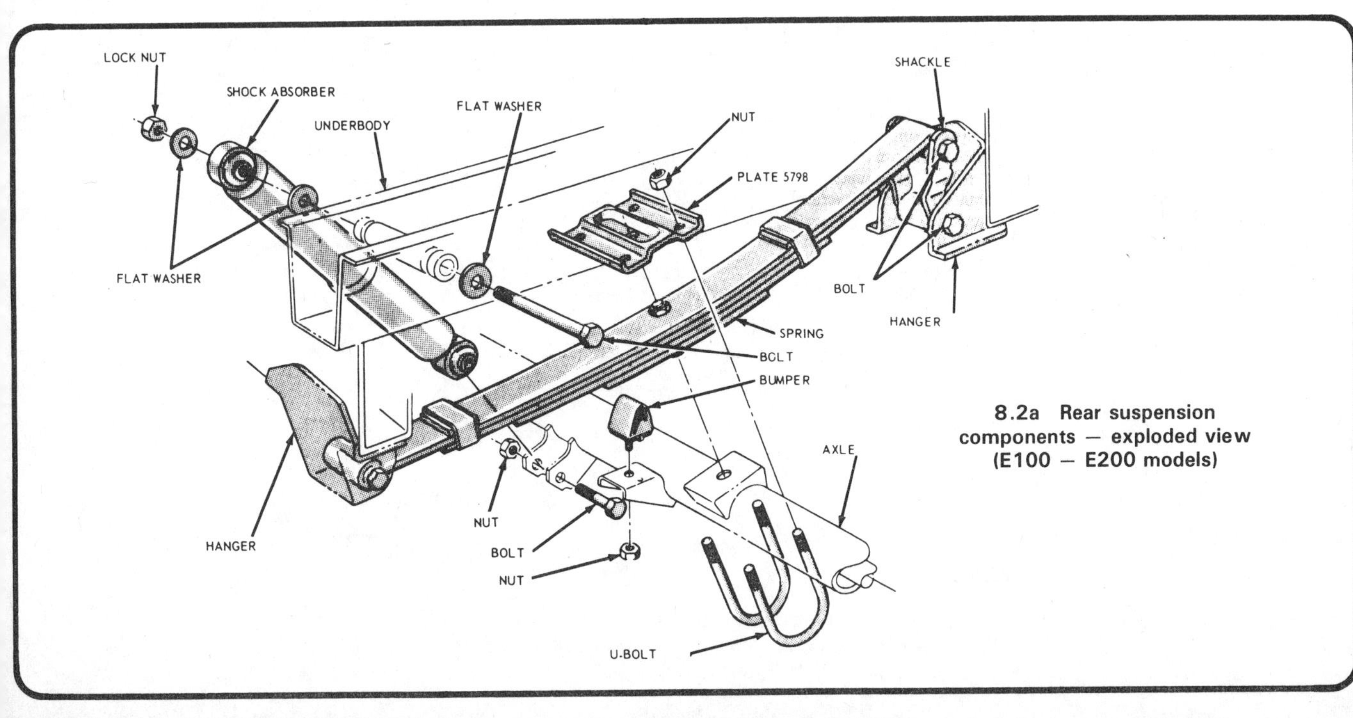

8.2a Rear suspension components — exploded view (E100 — E200 models)

7 The manufacturer recommends that when a rear spring is replaced, all old nuts and bolts should be discarded and new ones obtained. Always replace rear springs in pairs.
8 When installing the spring, attach the front end first, but do not tighten the nut completely at this stage.
9 Attach the rear spring shackle to the bracket and attach the spring to the shackle. Do not tighten the shackle pin nuts.
10 Raise the axle with a jack until the spring center bolt is located in the hole in the axle housing.
11 Install the spring retaining plate and U-bolts but do not tighten the nuts at this stage.
12 Attach the shock absorber to the axle bracket.
13 Raise the vehicle with the jack until the rear wheels are clear of the ground and then tighten the front and rear spring retaining nuts to the specified torque.
14 Install the wheels and lower the vehicle to the ground.

9 Rear shock absorber – removal and installation

1 Block the front wheels, raise the rear of the vehicle and support it on jackstands.
2 Remove the shock absorber lower mounting nut and detach the shock absorber from the mounting bracket.
3 Remove the nut from the upper mount stud and withdraw the shock absorber from the mount.
4 Installation is the reverse of removal. Be sure to tighten the fasteners to the specified torque.

10 Steering gear – removal and installation

Refer to illustration 10.3

1 Move the steering wheel to the straight ahead position and lock the steering with the ignition key.
2 Block the rear wheels, apply the parking brake, loosen the front wheel nuts, raise the front of the vehicle and support it on jackstands. Remove the wheels.
3 Remove the bolts from the lower half of the flex coupling (see illustration).
4 Using a large gear puller or Ford special tool No. 3290-C, disconnect the Pitman arm from the drag link.
5 Support the weight of the steering gear, remove the three mounting bolts and detach the steering gear from the vehicle.
6 Installation is the reverse of removal. Make sure that the input shaft is in the center position (approximately three turns from either stop) and the Pitman arm is located on the sector shaft so that it points down.

11 Steering gear – overhaul

Refer to illustrations 11.2, 11.8, 11.19, 11.23, 11.24 11.26, 11.28, 11.41 and 11.51

Thru 1977

1 Set the steering shaft to the center position and mark it in relation to the steering gear body.
2 Remove the large locknut and then unscrew the steering shaft adjuster nut. Lift out the upper bearing and cup (see illustration).
3 Remove the sector shaft adjuster screw locknut and loosen the adjuster screw approximately one turn.
4 Remove the screws securing the sector shaft cover and remove the cover by turning the adjuster screw clockwise. Retrieve the adjuster screw shim.
5 Withdraw the sector shaft from the housing.
6 Carefully withdraw the steering shaft and ball nut assembly and remove the lower bearing.
7 Check the condition of the steering shaft worm and ball nut. If the ball nut rotates roughly or the worm gear is pitted, replace them with new parts, as wear in one is certain to cause wear in the other.
8 Check the condition of the sector shaft bearing. If worn, drive it out of the housing with a hammer and drift punch or special tool No. T62F-3576A (see illustration).

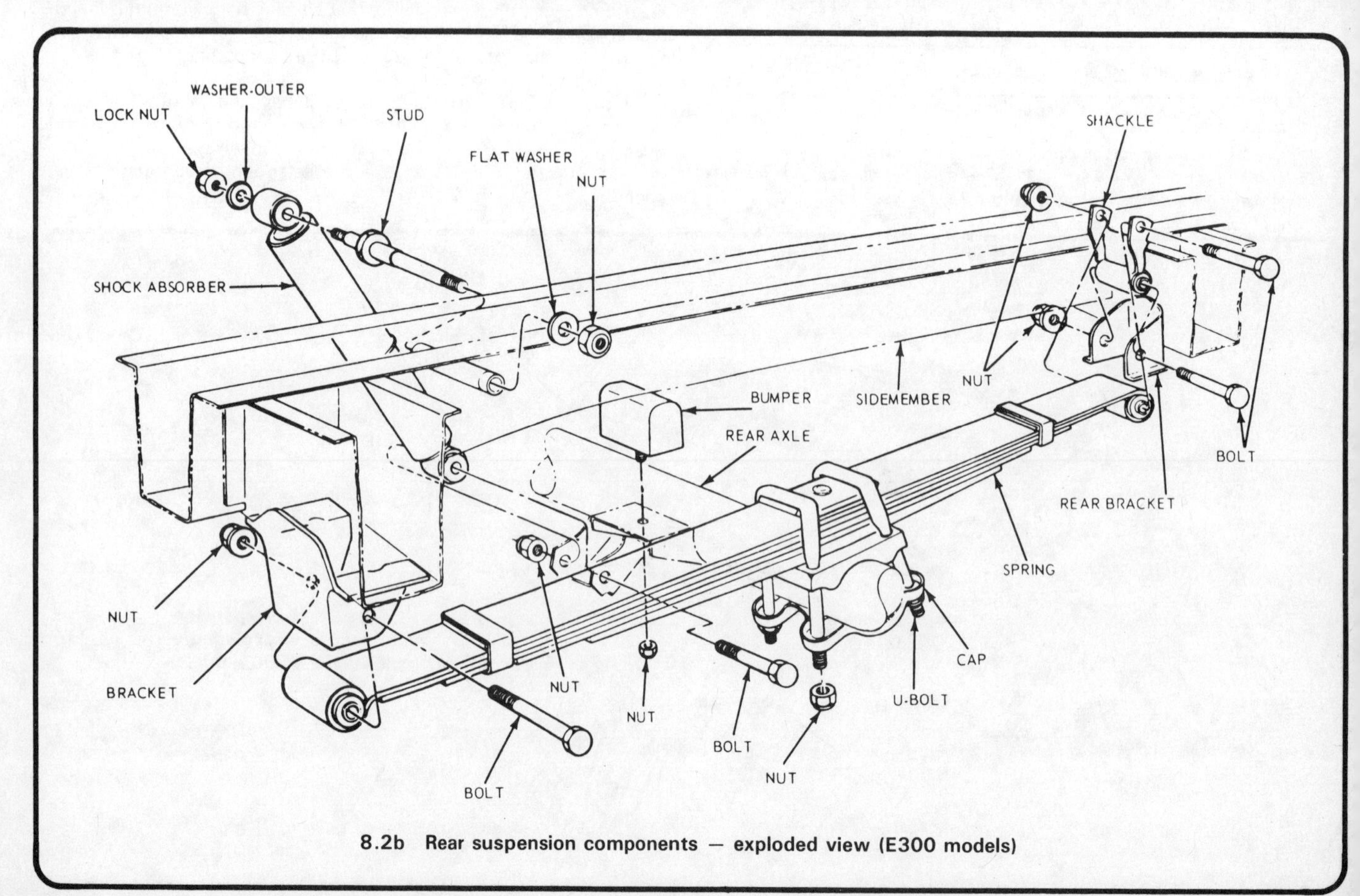

8.2b Rear suspension components – exploded view (E300 models)

9 Clean the housing assembly and replace the upper and lower steering shaft bearings and cups with new ones.
10 To reassemble the steering gear assembly, first install a new seal on the end of the sector shaft bearing. Apply some gear oil around the bearing.
11 Install the lower bearing in the steering gear housing.
12 Lubricate the worm gear and ball nut and insert them into the housing.
13 Place the upper bearing on the steering shaft.
14 Coat the threads of the adjuster nut with an oil resistant sealant and install it in the housing, but do not tighten it. Install the locknut with the flat side against the adjuster.
15 To adjust the worm bearing preload, attach an in-lbs torque wrench to the steering shaft and turn the adjuster nut in or out until the effort required to turn the steering shaft in either direction is 4-to-5 in-lbs.
16 Tighten the locknut and recheck the preload.
17 Rotate the steering shaft until the ball nut teeth are in position to mesh with the sector shaft. Lubricate the sector shaft and insert it into the housing.
18 Fill the housing with the specified amount of gear oil and install the cover and a new gasket. Make sure that the adjuster screw head is correctly engaged in the top of the sector shaft.
19 Attach an in-lbs torque wrench to the end of the sector shaft and turn the adjusting screw until an effort of 9-to-10 in-lbs is required to turn the shaft (see illustration). Tighten the locknut and recheck the setting.
20 The steering gear can now be reinstalled in the vehicle.

1978 and later

21 Remove the steering gear assembly from the vehicle as described in Chapter 11. Mount the assembly securely in a vise, clamping onto one of the mounting tabs. The workbench must be exceptionally clean and care must be taken to keep all of the components free of dirt.
22 With the worm shaft in the horizontal position, turn the worm shaft

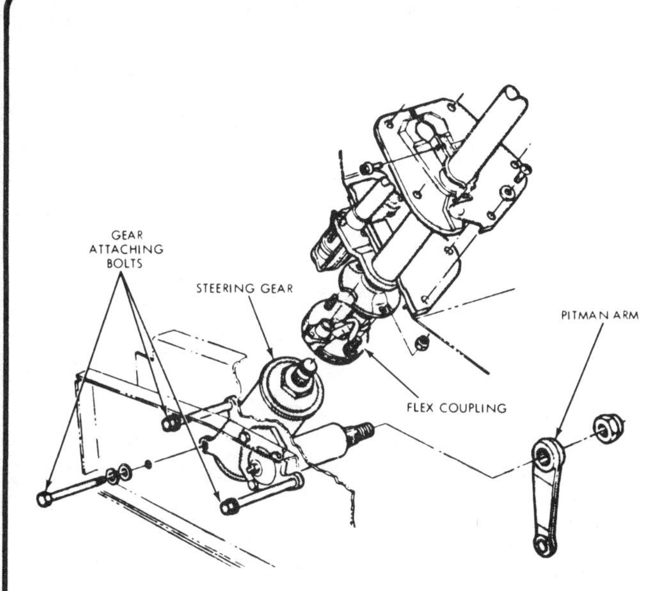

10.3 Steering gear mounting details

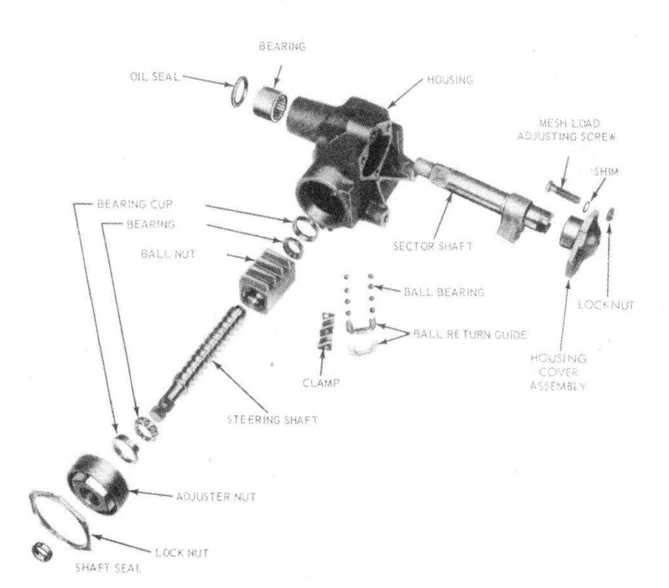

11.2 Steering gear components — exploded view (typical early models)

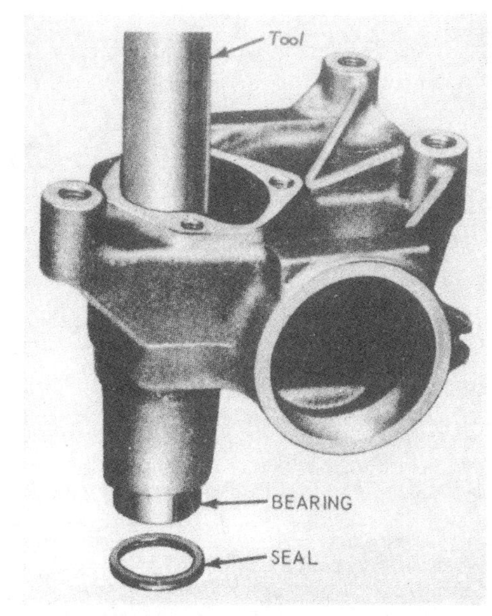

11.8 Driving out the sector shaft bearing

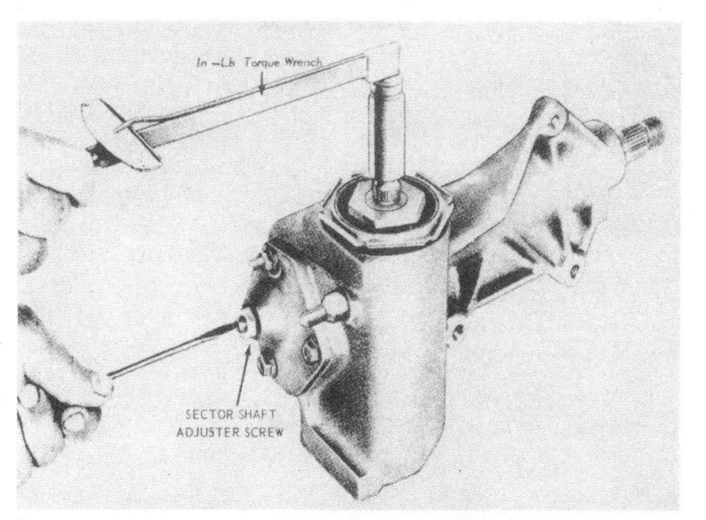

11.19 Adjusting the sector shaft preload

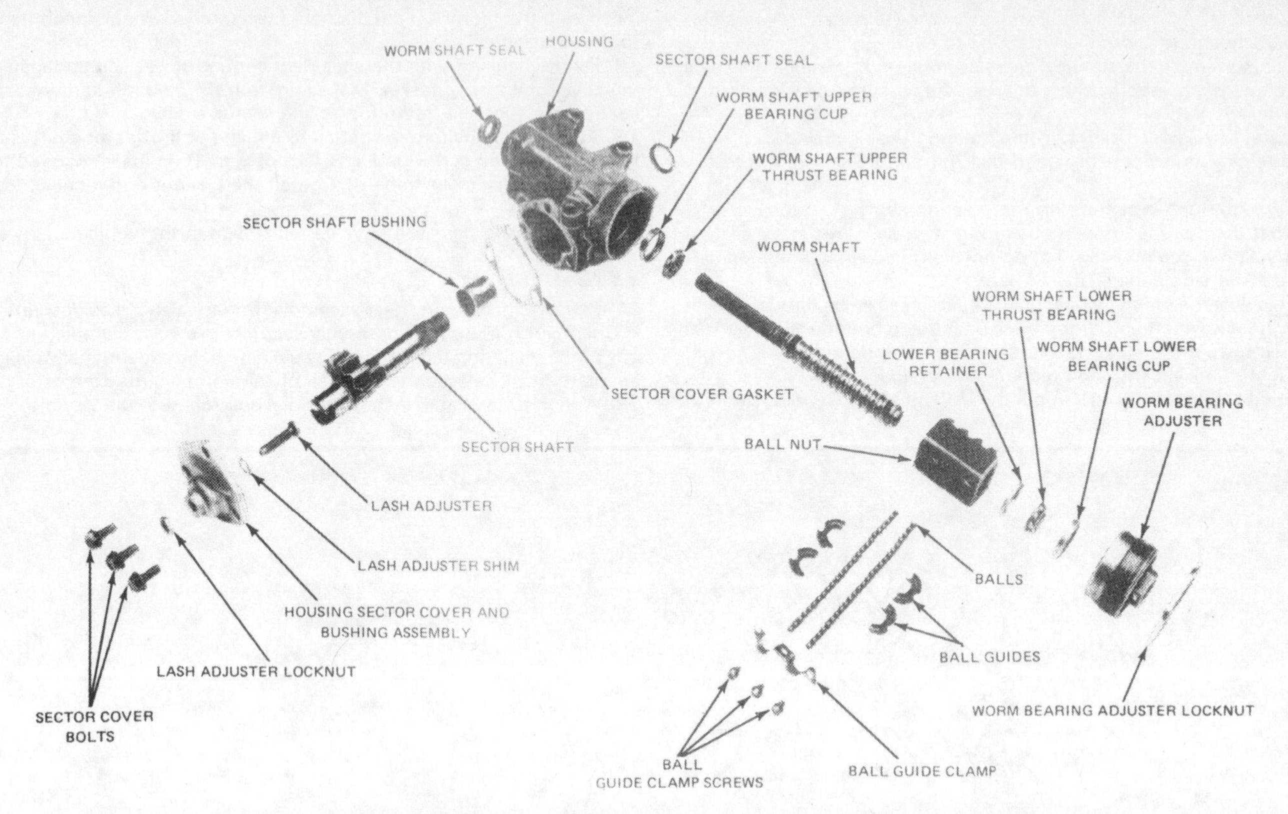

11.23 Steering gear components — exploded view (typical later models)

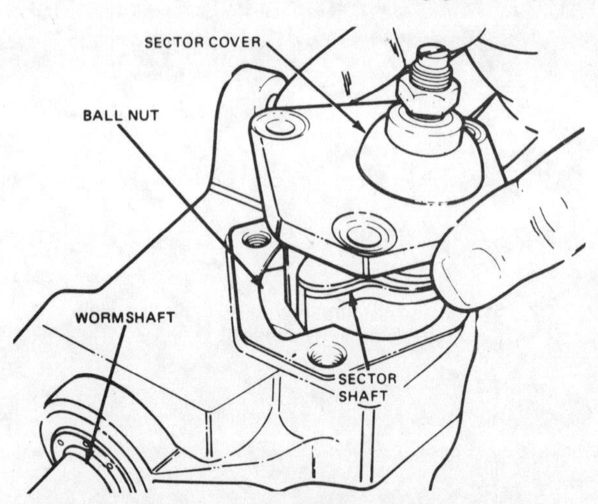

11.24 Lift the sector shaft assembly from the housing

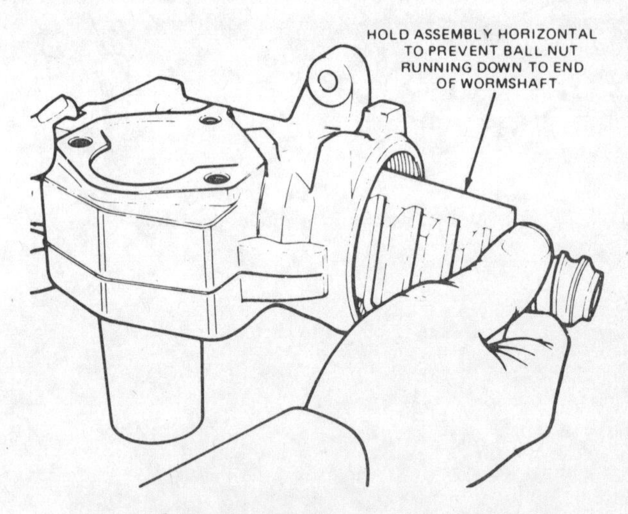

11.26 Carefully withdraw the wormshaft and ball nut from the housing

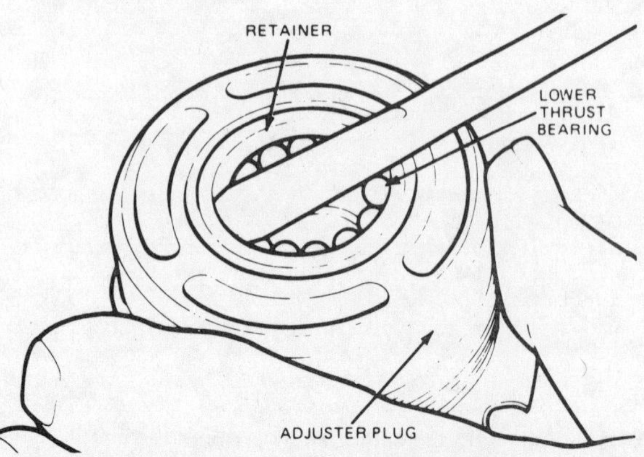

11.28 Prying the lower thrust bearing retainer from the adjuster plug with a large screwdriver

exactly halfway between stops, placing the gear on center.

23 Unscrew the sector cover mounting bolts from the housing (see illustration).

24 Using a soft face hammer, tap lightly on the end of the sector shaft, lifting the sector cover and sector shaft assembly from the housing (see illustration). It may be necessary to turn the worm shaft until the sector can pass through the opening in the housing.

25 Unscrew the locknut from the adjuster plug, then remove the adjuster plug assembly.

26 Carefully remove the worm shaft and ball assembly from the housing (see illustration). Do not allow the ball nut to run down to either end of the worm or the ball guides may be damaged.

27 From inside the gear housing, remove the worm shaft upper bearing.

28 Pry the worm shaft lower bearing retainer from the adjuster plug housing with a large screwdriver (see illustration). Remove the bearing.

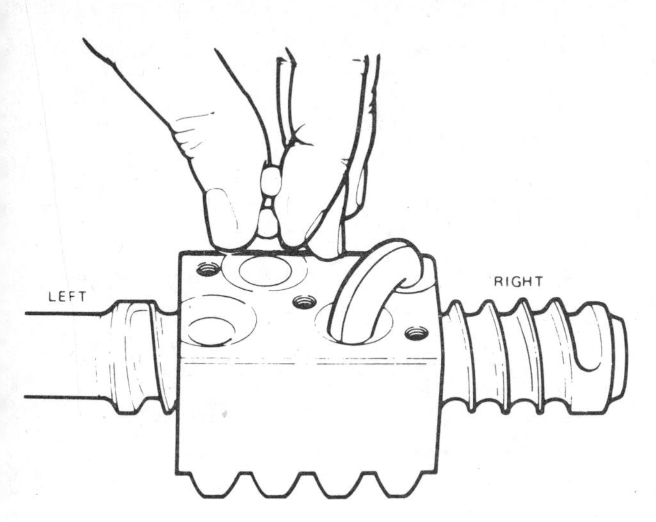

11.41 Install 25 balls in each ball circuit

29 Unscrew the locknut from the lash adjuster screw in the sector cover. Turn the lash adjuster screw clockwise to remove it from the sector cover.
30 Remove the adjuster screw and shim through the slot in the end of the sector shaft.
31 Pry the worm shaft and the sector shaft seals from position and discard them.
32 Thoroughly clean all of the steering assembly components in cleaning solvent and inspect them for wear. Check the bearings and bearing cups for signs of wear, scoring, pits, chipping or breakdown of the surface. If there is any doubt as to the condition of any components, they should be replaced with new parts.
33 Check the sector shaft for a good fit in the sector cover and the housing and inspect the worm shaft assembly to see if it is bent or damaged in any way.
34 Press new sector shaft and worm shaft seals into position using an appropriate size socket or special tool No. T75T-35527-A.
35 If the sector shaft bushing or the sector cover bushing needs replacing, it is best left to your Ford dealer to do because of the need for special tools. Removal of the worm shaft lower bearing cup requires a special puller and slide hammer the home mechanic is not likely to have. If the bearing cup needs replacing it should be taken to the dealer.
36 Carefully check the condition of the steering shaft worm and the ball nut. If the ball nut rotates harshly and there are signs of pitting on the worm gear, the best policy is to replace both the worm shaft and the ball nut with new components, as wear in one is certain to cause wear in the other.
37 Withdraw the ball guides from the ball nut after removing the screws and clamp securing them.
38 Hold the ball nut upside down over a clean receptacle and rotate the worm shaft back-and-forth until all the balls drop out of the ball nut. The ball nut can be removed from the worm after the balls are removed.
39 Inspect the worm, the ball nut and the balls and ball guides for signs of wear or damage. If in doubt about the condition of a component, replace it with a new unit.
40 With the shallow end of the ball nut teeth to the left (looking from the input shaft end) and the ball guide holes facing up, slip the ball nut over the worm. Look through the ball guide holes to align the grooves in the worm and the ball nut.
41 Separate the balls into two groups of 25. Hold the worm from rotating while installing the balls. Using the end of a suitable pin punch, press 20 balls into one of the guide holes. With the groove up, lay one half of the ball guide on the workbench and install the remaining five balls (see illustration).
42 Hold the other half of the ball guide in position on the ball guide with the balls in it. Plug the end with grease to hold the balls in place while installing the guides in the ball nut.
43 To complete one circuit of balls, push the guide into the guide holes of the ball nut. It may be necessary to tap the guide into position with the handle of a screwdriver if it does not push easily into place.

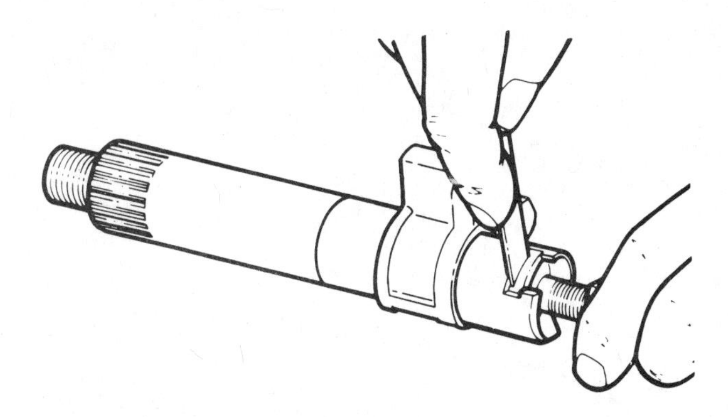

11.51 Measuring the lash adjuster end clearance with a feeler gauge

44 Assemble the other ball circuit as described in the previous steps. Assemble the ball guide clamp to the ball nut and tighten the screws to the specified torque.
45 Rotate the ball nut on the worm to be sure it does not bind. Do not rotate the ball nut to the end of the worm threads as damage to the ball guides may occur.
46 Lubricate the sector shaft bushings, the sector cover bushings and the worm shaft bearings with steering gear lubricant.
47 Mount the steering gear housing securely in a vise with the sector cover up and the worm shaft bore in the horizontal position. Be sure that the new sector shaft and worm shaft seals have been installed and that the worm shaft bearing cups and sector shaft bushings are installed in the housing.
48 Position the worm shaft upper bearing assembly over the worm shaft and install the worm shaft and ball nut assembly in the housing, feeding the ends of the shaft through the upper ball bearing cup and seal.
49 Install the worm shaft lower bearing assembly in the adjuster plug bearing cup and, with a suitable socket, press the stamped retainer into place.
50 Working from the lower end of the housing, install the adjuster plug and the locknut until almost all of the worm shaft end play is removed. At the same time, carefully guide the end of the worm shaft into the bearing.
51 Attach the lash adjuster and shim in the slotted end of the sector shaft and measure the end clearance (see illustration). If the end clearance is more that the specified limit, a different thickness shim must be installed.
52 Install 11 ounces of steering gear lubricant, then rotate the worm shaft until the ball nut is at the end of its travel. Install as much lubricant as possible into the housing without it coming out of the sector shaft opening. Rotate the worm shaft until the ball nut is at the other end of its travel and pack lubricant into the other opening.
53 Position the ball nut in the center of its travel by turning the worm shaft. Check that the sector shaft and the ball nut engage correctly, with the center tooth of the sector entering the center tooth space in the ball nut.
54 Install the sector shaft without the sector cover into the housing so the sector center tooth enters the center tooth space in the ball nut. Place some lubricant in the bushing hole in the sector cover and pack the remaining portion of lubricant into the housing. Place a new sector cover gasket in position on the housing.
55 Place the sector cover in position on the sector shaft. Insert a screwdriver through the cover and turn the lash adjuster screw counterclockwise until the screw bottoms. Turn the lash adjuster screw clockwise one-half turn.
56 While holding the cover away from the ball nut, install the sector cover bolts and tighten them to the specified torque.
57 Install a new locknut loosely on the adjuster screw and adjust the backlash of the sector shaft to the specified amount. Tighten the locknut to the specified torque.
58 Rotate the worm shaft from the gear center to both stops, counting the number of turns. Be sure there are three turns from center-to-stop and six turns from stop-to-stop.

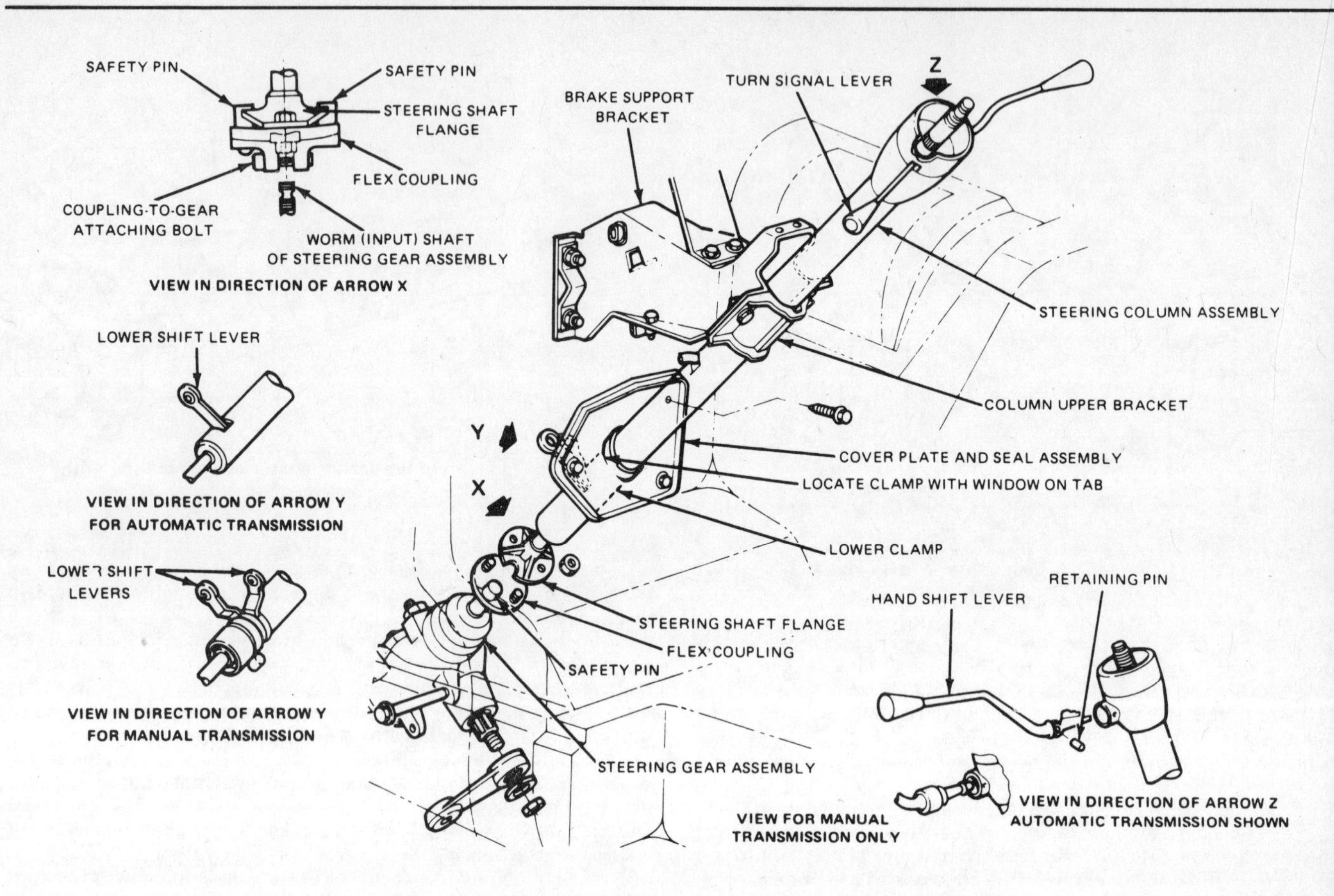

16.1 Typical steering column components — exploded view

12 Power steering system — general information

1 The power steering system available on Econoline vans has a Ford-Thompson type pump that is belt driven from the crankshaft to direct fluid to a servo assisted, recirculating ball type steering gear.
2 Servo assistance is obtained by a piston attached to the end of the worm shaft. The degree of assistance is controlled by a spool valve attached to the steering input shaft.
3 The power steering pump incorporates an integral fluid reservoir.
4 Due to the complexity of the components and the special tools required for overhaul and repair, the power steering pump and gearbox should not be disassembled. If problems are encountered, replace the units with new or factory rebuilt components.

13 Power steering system — bleeding

1 The power steering system will only require bleeding in the event that air has entered the system due to low fluid level, leaks and/or when lines are disconnected.
2 Open the hood and check the fluid level in the pump reservoir. Top up, if necessary, with the specified type of fluid.
3 If fluid is added, wait two minutes, then run the engine at approximately 1500 rpm. Slowly turn the steering wheel from lock-to-lock while checking and topping up the fluid level until the level remains steady and no more bubbles appear in the reservoir.
4 Clean and install the reservoir cap.

14 Power steering pump — removal and installation

1 Loosen the pump adjusting and mounting bolts.
2 Push the pump in toward the engine and remove the drivebelt.
3 Disconnect the lines from the pump and drain the fluid into a suitable container.
4 Plug or tape over the end of the lines to prevent the entry of dirt.
5 If necessary, remove the alternator drivebelt(s) as described in Chapter 5.
6 Remove the bolts attaching the pump to the engine bracket and detach the pump. **Note:** *On some engine installations it may be necessary to remove the pump and bracket as a unit.*
7 Installation is the reverse of removal. Top up the system with the specified fluid, adjust the drivebelt tension (see Chapter 1), then bleed the system as described in Section 13.

15 Power steering gear — removal and installation

The procedure for removing the power steering gear is similar to that described in Section 10, with the additional task of disconnecting the pump lines. Installation is the reverse of removal. Top up the system with the specified fluid, adjust the drivebelt tension (see Chapter 1), then bleed the system as described in Section 13.

16 Steering column — removal and installation

Refer to illustration 16.1

1 Set the wheels to the straight ahead position and mark the steering shaft flange and flexible coupling with a piece of chalk or paint to ensure correct alignment during reassembly (see illustration).
2 Remove the steering wheel as described in Chapter 12.
3 Remove the two screws retaining the steering column cover panel and remove the panel (later models only).
4 Disconnect the ignition switch wire connector and the back-up and Neutral start switch wires from the column.

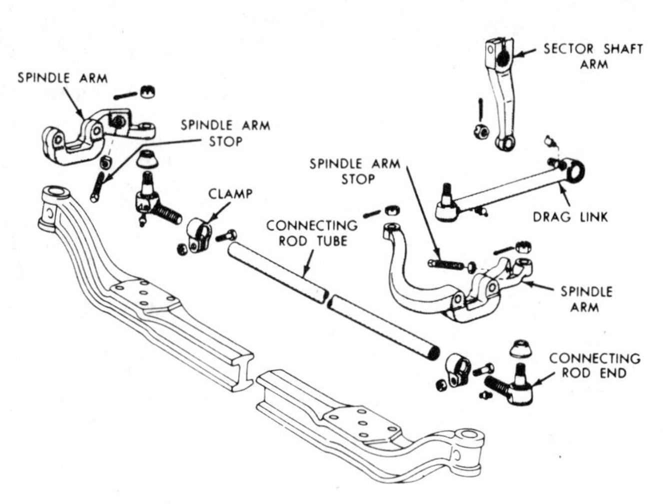

17.2 Steering linkage components — exploded view

5 Disconnect the shift rod from the column shift levers as described in Chapter 7.
6 Remove the two nuts securing the steering shaft flange to the flexible coupling.
7 Remove the three bolts retaining the steering column cover plate to the inside of the dash panel.
8 Remove the bolts securing the upper steering column bracket to the brake support bracket.
9 Withdraw the complete column assembly through the dash panel opening toward the inside of the vehicle.
10 If the steering column bearings require replacement, drive out the shift lever retaining pin and remove the lever.
11 Remove the turn signal lever and switch assembly as described in Chapter 12.
12 Remove the upper bearing snap-ring, push the steering shaft down and remove it along with the lower bearing.
13 Remove the upper shaft bearing from the top of the column.
14 Install new bearings in the column assembly and insert the steering shaft from the lower end of the column.
15 Push the shaft through the upper and lower bearings and install the upper bearing snap-ring.
16 Install the shift lever and turn signal switch, then install the steering column assembly in the vehicle by reversing the removal procedure.

17 Steering linkage — removal and installation

Refer to illustration 17.2

1 Raise the front of the vehicle and support it on jackstands.
2 Remove the cotter pins and nuts from the two connecting rod balljoints then disconnect them from the spindle arms with a puller (see illustration). Remove the connecting rod from beneath the vehicle.
3 Remove the cotter pins and nuts from the two drag link balljoints. Using a puller, disconnect the drag link from the sector shaft arm and spindle arm. Remove the drag link from the vehicle.
4 The drag link balljoints cannot be replaced. If worn, a new link must be obtained.
5 To replace the connecting rod balljoints, loosen the clamp bolts then unscrew the balljoints from the connecting rod, carefully counting the number of turns necessary to remove each balljoint. Note that this number probably will not be exactly the same for the left and right side balljoints.
6 Thread the new balljoints into the connecting rod the same number of turns as was required to remove the old balljoints. Tighten the clamp bolts and nuts.
7 Install the steering linkage in the vehicle by reversing the removal procedure.
8 Have the front wheel toe-in checked by a Ford dealer service department.

Chapter 11 Body

Contents

1 General information

The Econoline van body is an all steel, welded construction unit mounted on a separate frame. The frame consists of two main longitudinal members joined by crossmembers. This forms a ladder shaped chassis, which is extremely strong.

In the event of major body damage the chassis dimensions must be checked and restored by a body shop.

The Econoline is available with a side opening door and a variety of window layouts to suit all customer requirements. Rear passenger seats are available and, on later models, swivelling front seats are also available.

2 Body — maintenance

1 The condition of your vehicle's bodywork is of considerable importance, as it is on this that the resale value will mainly depend. It is much more difficult to repair neglected bodywork than mechanical components. The hidden portions of the body, such as the wheel arches, the underframe and the engine compartment, are equally important, although obviously not requiring such frequent attention as the immediately visible paint.

2 Once a year or every 12,000 miles it is a good idea to have the underside of the body steam cleaned. All traces of dirt and oil will be removed and the underside can then be inspected carefully for rust, damaged hydraulic lines, frayed electrical wiring and similar trouble areas. The front suspension should be greased on completion of this job.

3 At the same time, clean the engine and the engine compartment either using a steam cleaner or a water soluble cleaner.

4 The wheel arches should be given particular attention, as undercoating can easily come away here and stones and dirt thrown up from the wheels can soon cause the paint to chip and flake, and so allow rust to set in. If rust is found, clean down to the bare metal and apply an anti-rust paint.

5 Use a mild detergent and soft sponge to wash the exterior of the vehicle and rinse immediately with clear water. Owners who live in coastal regions and where salt or chemicals are used on the roads should wash the finish often to prevent damage to the finish. Do not wash the vehicle in direct sunlight or when the metal is warm. To remove road tar, insects or tree sap use a tar remover rather than a knife or sharp object, which could scratch the surface.

6 A coat of wax or polish may be your best protection against the elements. Use a good grade of pollsh or wax suitable for a high-quality synthetic finish. Do not use a wax or polish which contains large amounts of abrasives, as these will scratch the finish.

7 Bright metal parts can be protected with wax or a chrome preservative. During winter months or in coastal regions apply a heavier coating or, if necessary, use a non-corrosive compound such as petroleum jelly for protection. Do not use abrasive cleaners, strong detergents or materials such as steel wool on chrome or anodized aluminium parts, as these may damage the protective coating and cause discoloration or deterioration.

8 Interior surfaces can be wiped clean with a damp cloth or with cleaners specifically designed for car interior fabrics. Carefully read the manufacturer's instructions and test any commercial cleaners on an inconspicuous area first. The carpet should be vacuumed regularly and can be covered with mats.

9 Cleaning the mechanical parts of the vehicle serves two functions. First, it focuses your attention on parts which may be starting to fail, allowing you to fix or replace them before they cause problems. Second, it is much more pleasant to work on parts which are clean. You will still get dirty on major repair jobs, but it will be less extreme. Large areas should be brushed with detergent, allowed to soak for about 15 minutes and then carefully rinsed clean. Cover ignition and carburetor parts with plastic to prevent moisture from penetrating these critical components.

3 Upholstery and carpets — maintenance

1 Every three months remove the carpets or mats and clean the interior of the vehicle (more frequently if necessary). Vacuum the upholstery and carpets to remove loose dirt and dust.
2 If the upholstery is soiled, apply upholstery cleaner with a damp sponge and wipe it off with a clean, dry cloth.

4 Body repair — minor damage

Refer to the photo sequence.

Repair of minor scratches

If the scratch is superficial, and does not penetrate to the metal of the body, repair is simple. Lightly rub the area of the scratch with a fine rubbing compound to remove loose paint from the scratch and to clear the surrounding paint of wax buildup. Rinse the area with clean water.

Apply touch-up paint to the scratch, using a small brush. Continue to apply thin layers of paint until the surface of the paint in the scratch is level with the surrounding paint. Allow the new paint at least two weeks to harden, then blend it into the surrounding paint by rubbing with a very fine rubbing compound. Finally, apply a coat of wax to the scratch area.

Where the scratch has penetrated the paint and exposed the metal of the body, causing the metal to rust, a different repair technique is required. Remove any loose rust from the bottom of the scratch with a pocket knife, then apply rust inhibiting paint to prevent the formation of rust in the future. Using a rubber or nylon applicator, coat the scratched area with glaze type filler. If required, this filler can be mixed with thinner to provide a very thin paste, which is ideal for filling narrow scratches. Before the glaze filler in the scratch hardens, wrap a piece of smooth cotton cloth around the tip of a finger. Dip the cloth in thinner and then quickly wipe it along the surface of the scratch. This will ensure that the surface of the filler is slightly hollowed. The scratch can now be painted over as described earlier in this section.

Repair of dents

When denting of the vehicle's body has taken place, the first task is to pull the dent out until the affected area nearly attains its original shape. There is little point in trying to restore the original shape completely, as the metal in the damaged area will have stretched on impact and cannot be reshaped fully to its original contours. It is better to bring the level of the dent up to a point which is about 1/8-inch below the level of the surrounding metal. In cases where the dent is very shallow, it is not worth trying to pull it out at all.

If the underside of the dent is accessible, it can be hammered out gently from behind using a mallet with a wooden or plastic head. While doing this, hold a block of wood firmly against the metal to absorb the hammer blows and prevent a large area of the metal from being stretched.

If the dent is in a section of the body which has double layers, or some other factor making it inaccessible from behind, a different technique is in order. Drill several small holes through the metal inside the damaged area, particularly in the deeper sections. Screw long self-tapping screws into the holes just enough for them to get a good grip in the metal. Now the dent can be pulled out by pulling on the heads of the screws with locking pliers.

The next stage of repair is the removal of paint from the damaged area and from an inch or so of the surrounding metal. This is accomplished most easily by using a wire brush or sanding disk in a drill motor, although it can be done just as effectively by hand with sandpaper. To complete the preparation for filling, score the surface of the bare metal with a screwdriver or the tang of a file (or drill small holes in the affected area). This will provide a very good grip for the filler material. To complete the repair, see the Section on filling and painting.

Repair of rust holes or gashes

Remove all paint from the affected area and from an inch or so of the surrounding metal using a sanding disk or wire brush mounted in a drill motor. If these are not available, a few sheets of sandpaper will do the job as effectively. With the paint removed, you will be able to determine the severity of the corrosion and therefore decide whether to replace the whole panel, if possible, or repair the affected area. New body panels are not as expensive as most people think and it is often quicker to install a new panel than to attempt to repair large damaged areas.

Remove all trim pieces from the affected area (except those which will act as a guide to the original shape of the damaged body such as headlamp, shells etc.). Using metal snips or a hacksaw blade, remove all loose metal and any other metal that is badly affected by rust. Hammer the edges of the hole inwards to create a slight depression for the filler material.

Wire brush the affected area to remove the powdery rust from the surface of the metal. If the back of the rusted area is accessible, treat it with rust-inhibiting paint.

Before filling can be done it will be necessary to block the hole in some way. This can be accomplished with sheet metal riveted or screwed into place, or by stuffing the hole with wire mesh. Once the hole is blocked off, the affected area can be filled and painted.

Filling and painting

Many types of body fillers are available, but generally speaking, body repair kits which contain filler paste and a tube of resin hardener are best suited for this type of repair work. A wide, flexible plastic or nylon applicator will be necessary for imparting a smooth and contoured finish to the surface of the filler material.

Mix up a small amount of filler on a clean piece of wood or cardboard (use the hardener sparingly). Follow the instructions on the package, otherwise the filler will set incorrectly.

Using the applicator, apply the filler paste to the prepared area. Draw the applicator across the surface of the filler to achieve the desired contour and to level the filler surface. As soon as a contour that approximates the original one is achieved, stop working the paste. If you continue, the paste will begin to stick to the applicator. Continue to add thin layers of filler paste at 20 minute intervals until the level of the filler is just above the surrounding metal.

Once the filler has hardened the excess can be removed using a body file. From then on, progressively finer grades of sandpaper should be used, starting with a 180-grit paper and finishing with 600-grit wet-or-dry paper. Always wrap the sandpaper around a flat rubber or wooden block, otherwise the surface of the filler will not be completely flat. During the sanding of the filler surface, the wet-or-dry paper should be periodically rinsed in water. This will ensure that a very smooth finish is produced in the final stage.

At this point, the repair area should be surrounded by a ring of base metal, which in turn should be encircled by the finely feathered edge of the good paint. Rinse the repair area with clean water until all of the dust produced by the sand operation is gone.

Spray the entire area with a light coat of primer. This will reveal any imperfections in the surface of the filler. Repair these imperfections with fresh filler paste or glaze filler and once more smooth the surface with sandpaper. Repeat this spray-and-repair procedure until you are satisfied that the surface of the filler and the feathered edge of the paint are perfect. Rinse the areas with clean water and allow it to dry completely.

The repair area is now ready for painting. Paint spraying must be carried out in a warm, dry, windless and dustfree atmosphere. These conditions can be created if you have access to a large indoor working area, but if you are forced to work in the open, you will have to pick the day very carefully. If you are working indoors, dousing the floor in the work area with water will help settle dust which would otherwise be in the air. If the repair area is confined to one body panel mask off the surrounding panels. This will help to minimize the effects of a slight mismatch in paint color. Trim pieces such as chrome strips, door handles, etc., will also need to be masked off or removed. Use masking tape and several thicknesses of newspaper for the masking operations.

Before spraying, shake the paint can thoroughly, then spray a test area until the spray painting technique is mastered. Cover the repair area with a thick coat of primer. The thickness should be built up using several thin layers of primer rather than one thick one. After the primer has dried, use 600-grit wet-or-dry sandpaper, rub down the surface of the primer until it is very smooth. While doing this, the work area should be thoroughly rinsed with water and the wet-or-dry sandpaper periodically rinsed as well. Allow the primer to dry before spraying additional coats.

Spray on the top coat, again building up the thickness by using several thin layers of paint. Begin spraying in the center of the repair area and then, using a circular motion, work out until the whole repair area and about two inches of the surrounding original paint is covered. Remove all masking material 10 to 15 minutes after spraying on the final coat of paint. Allow the new paint at least two weeks to harden, then, using a very fine rubbing compound, blend the edges of the new paint into the existing paint. Finally, apply a coat of wax.

5 Body repair – major damage

1 Major damage must be repaired by an auto body/frame repair shop with the necessary welding and hydraulic straightening equipment.
2 If the damage has been serious, it is vital that the frame structure be checked for correct alignment, as the handling of the vehicle will be affected. Other problems, such as excessive tire wear and wear in the transmission and steering, may also occur.

6 Hinges and locks – maintenance

Every 3000 miles or three months, the door, hood, side and rear door hinges should be lubricated with a few drops of oil and the locks treated with dry graphite lubricant. The door striker plates should be given a thin coat of grease to reduce wear and ensure free movement.

7 Doors – finding and eliminating rattles

1 Make sure that the door is not loose at the hinges and that the latch is holding the door firmly in position. Also, the door must line up with the opening in the body.
2 If the hinges are loose or the door is out of alignment, the hinges must be reset.
3 If the latch is holding the door properly, it should close tightly when fully latched and the door should line up with the body. If it is out of alignment, adjust it. If the latch is loose, some part of the lock mechanism must be worn out.
4 Other rattles in the door could be caused by wear or looseness in the window mechanism, the glass channels and sill strips or the door buttons and interior latch release mechanism.

8 Windshield glass – removal and installation

Refer to illustration 8.11

1 If a shattered windshield is being replaced, the air vents should be covered before attempting removal. Adhesive sheeting can be used to stick to the outside of the glass to enable large areas of shattered glass to be removed.
2 Remove the windshield wiper arms and blades from the vehicle

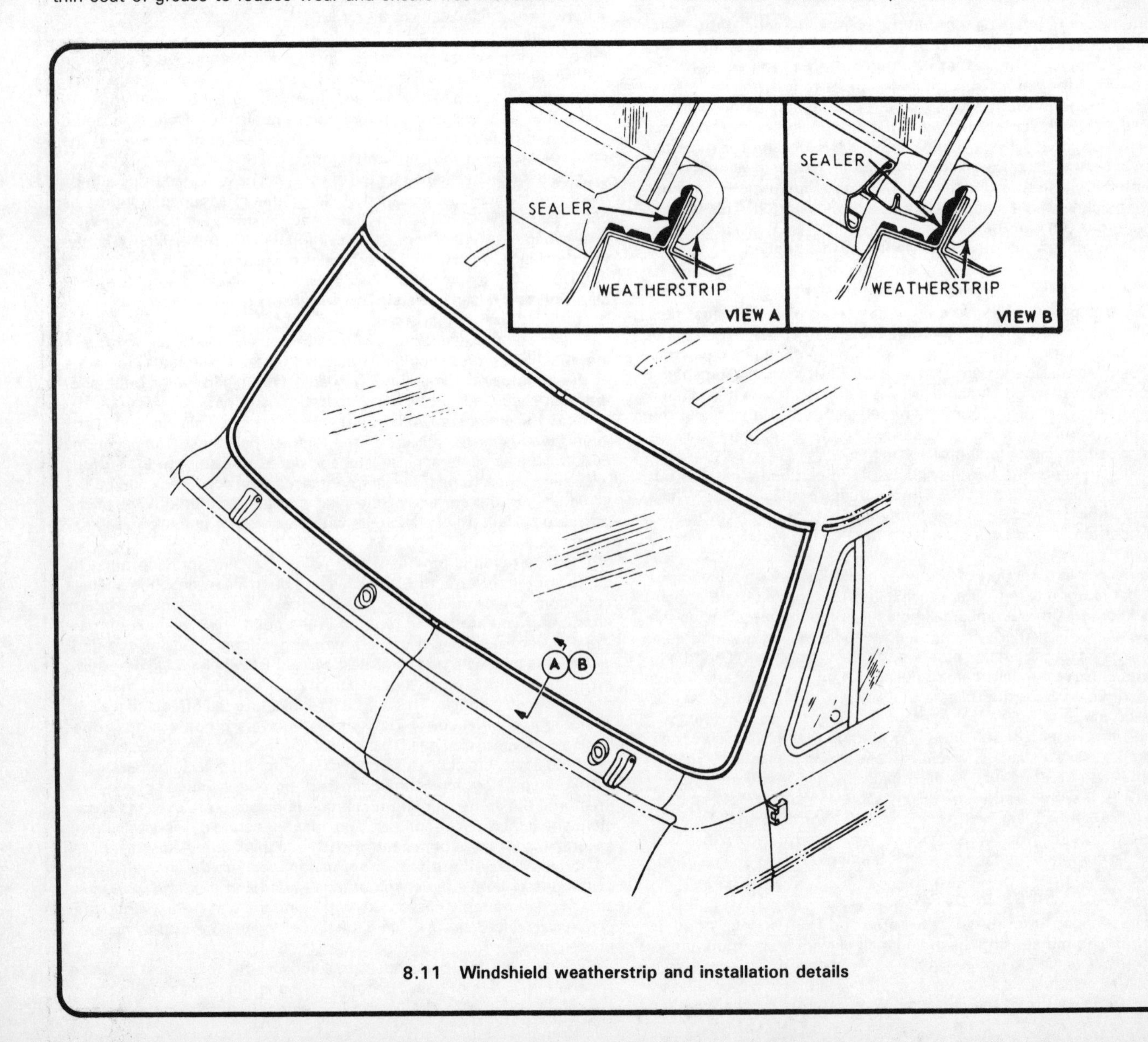

8.11 Windshield weatherstrip and installation details

(refer to Chapter 12 if necessary).

3 If the windshield is removed intact, an assistant will be required. Release the rubber surround from the body by running a small, blunt screwdriver around and under the rubber weatherstrip both inside and outside the vehicle. This operation will break the adhesion of the original sealant. Take care not to damage the paint or cut the rubber surround with the screwdriver.

4 Have your assistant push the inner lip of the rubber surround off the flange of the windshield body opening. Once the rubber surround starts to peel off the flange the windshield may be forced gently out with gentle hand pressure. The second person should support and remove the windshield, complete with rubber surround and metal beading as it comes out.

5 If you are removing a shattered windshield, remove all traces of sealing compound and broken glass from the weatherstrip and body flange.

6 Now is the time to remove all pieces of glass. Use a vacuum cleaner to extract as much as possible. Switch on the heater motor and adjust the controls to Defrost. Watch out for flying pieces of glass which might be blown out of the ducts.

7 Carefully inspect the rubber moulding for signs of splitting or deterioration.

8 To install the windshield, first position the weatherstrip on the windshield with the joint at the lower edge.

9 Insert a piece of thick cord into the channel of the weatherstrip with the two ends protruding at least 12-inches at the top center of the weatherstrip.

10 Apply sealer (type C5AZ-19554-A) in the windshield opening groove in the weatherstrip.

11 Position the windshield in the opening and, with an assistant to press the rubber surround hard against one end of the cord, move around the windshield, drawing the lip over the windshield flange of the body (see illustration). Keep the draw cord parallel to the windshield. Using the palms of your hands, press on the windshield from the outside to enable the lip to pass over the flange and to seat the windshield in the opening.

12 To ensure a good watertight joint, apply some sealer between the weatherstrip and the body and press the weatherstrip against the body all the way around the windshield.

13 Excess sealer can be removed with a solvent soaked cloth.

9 Door trim panel — removal and installation

Refer to illustrations 9.1a, 9.1b, 9.2 and 9.4

1 Remove the screws and detach the armrest from the inside of the door (see illustrations).

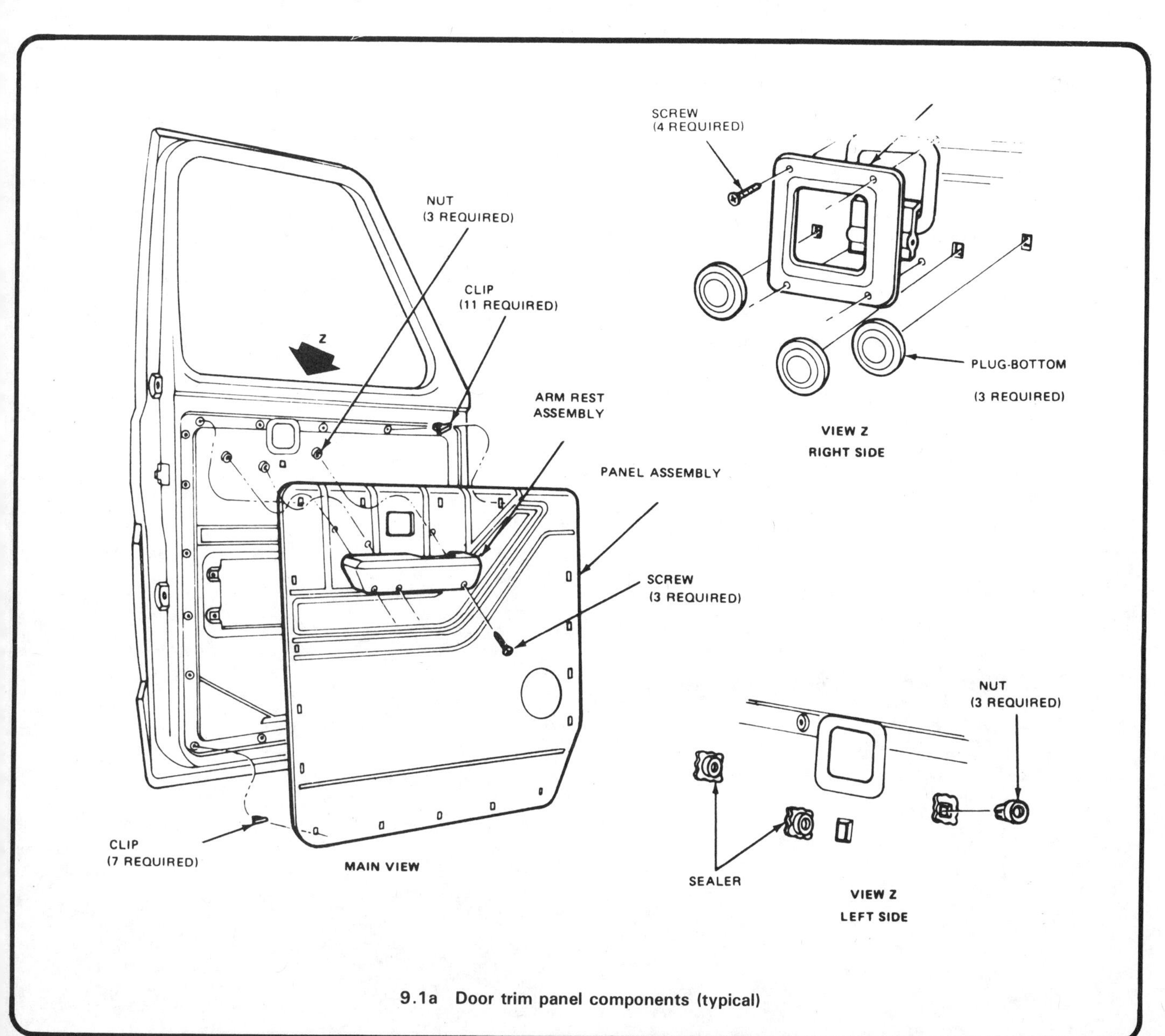

9.1a Door trim panel components (typical)

2 Remove the two nuts from the inside door handle and remove the handle and trim cup (see illustration).
3 If the vehicle is equipped with a radio, remove the four screws and detach the grille.
4 Remove the screw retaining the window regulator handle and remove the handle, washer and spacer (see illustration).
5 Carefully pry the clips out of the door and remove the trim panel.
6 Installation is the reverse of removal.

10 Front door latch – removal and installation

Refer to illustrations 10.2, 10.3 and 10.4

1 Remove the trim panel as described in the previous Section.
2 Remove the access panel from the door (see illustration).
3 Disconnect the control links from the latch assembly (see illustration).

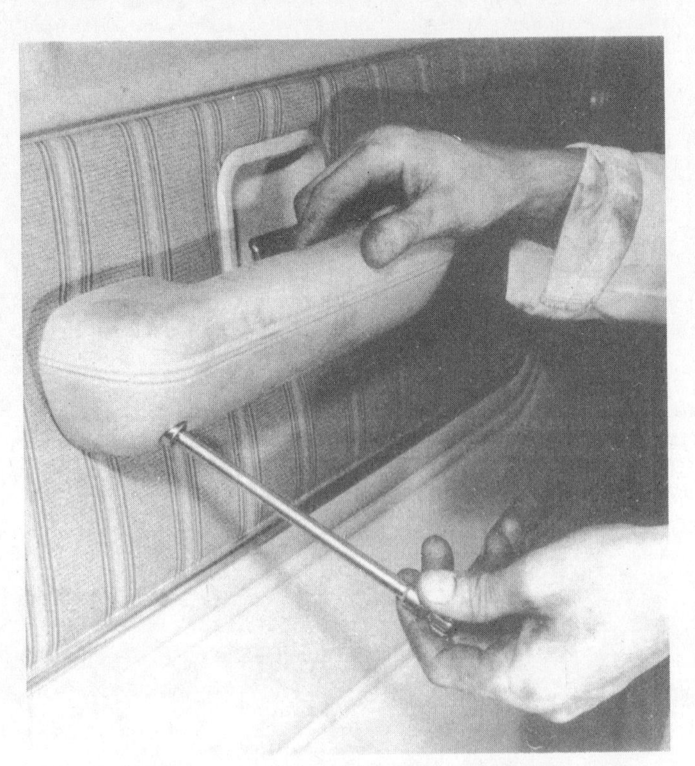

9.1b The armrest is attached to the door with two screws

9.2 Inside door handle and trim cup

9.4 Removing the window regulator handle

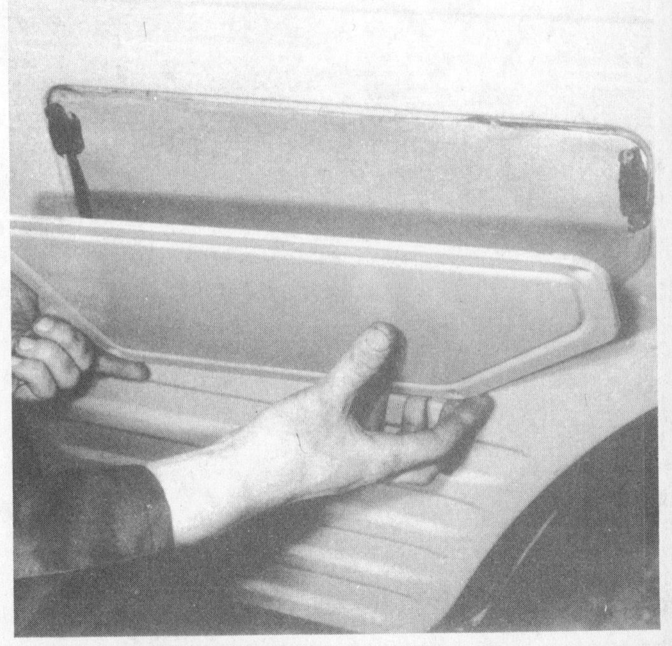

10.2 Removing the door access panel

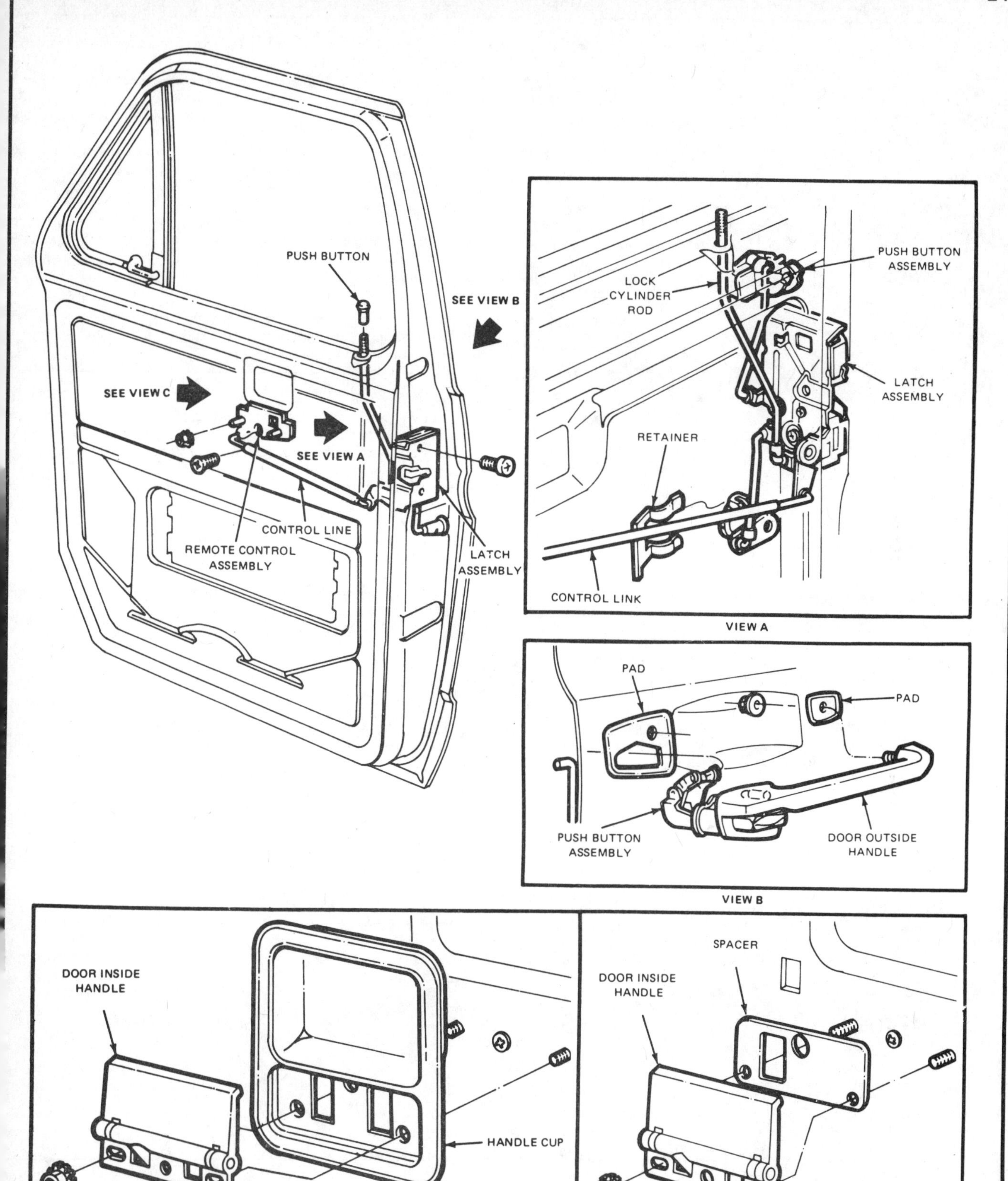

10.3 Front door latch assembly details

10.4 Withdrawing the door latch assembly through the access hole

4 Remove the three screws securing the latch assembly to the door and withdraw the latch (see illustration).
5 If required, the outside door handle can be detached by removing the two retaining nuts on the inside of the door.
6 The door lock cylinder can be removed by sliding out the retaining clip.
7 Installation is the reverse of removal. Be sure to adjust the stroke of the external push button in relation to the lock operating plate. Make sure that the plastic sheet beneath the interior trim panel is in good condition and in place.
8 Check the door closing operation. Adjust the position of the striker plate on the door pillar, if necessary, so that the door closes quietly and securely and the edge of the door is flush with the surrounding body surfaces when it is latched.

11 Window regulator mechanism and glass – removal and installation

Refer to illustration 11.2

1 Remove the door trim panel as described in Section 9 and remove the access panel from the door.
2 Remove the three screws securing the vent window assembly to the upper leading edge of the door (see illustration).
3 Remove the screw securing the front run retainer and division bar bracket to the door.
4 Slide the front run from the retainer and division bar.
5 Unsnap and remove the beltline weatherstrip from the door.

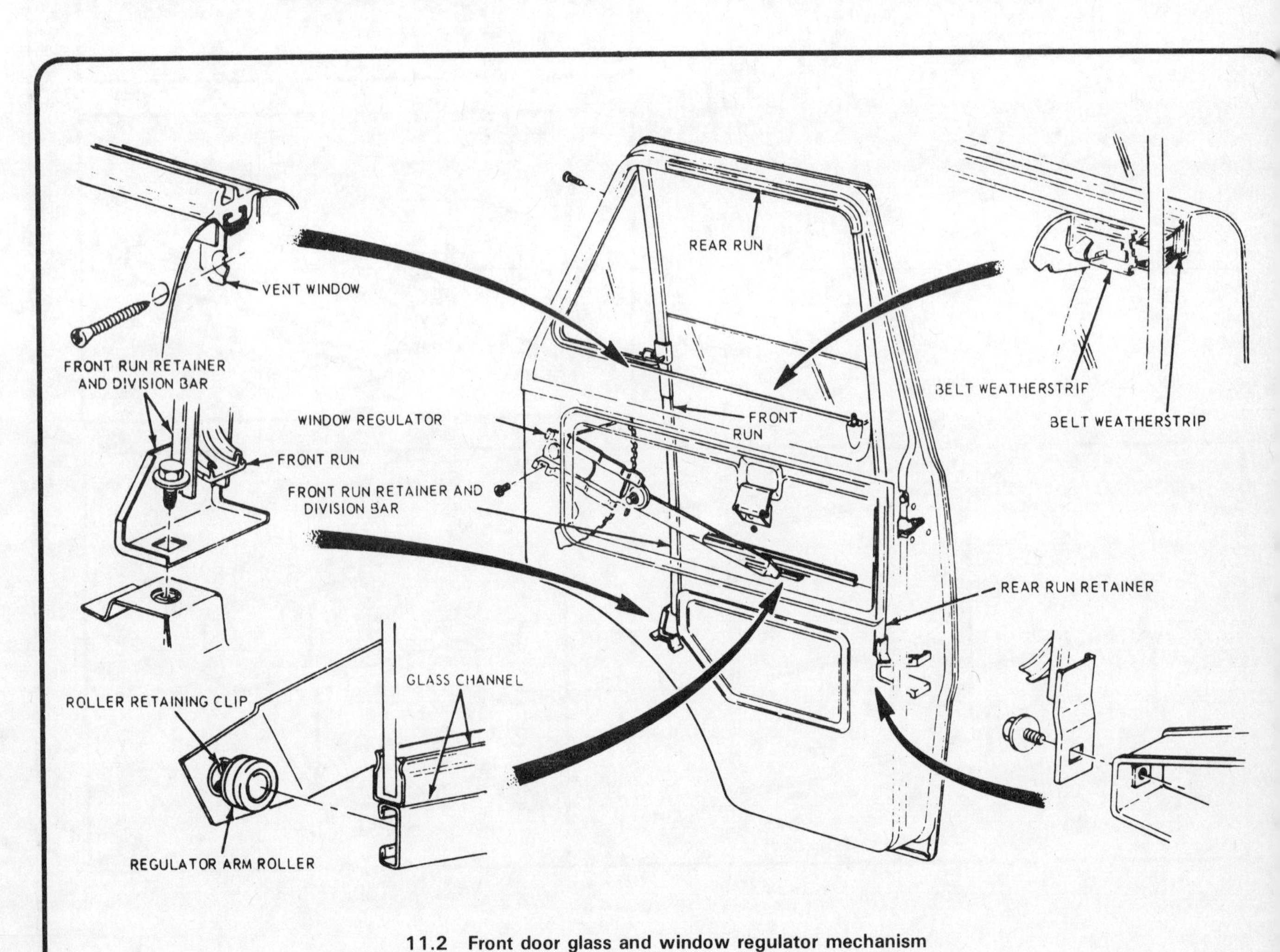

11.2 Front door glass and window regulator mechanism

6 Install the handle and lower the glass. Remove the two screws securing the rear run retainer to the door edge and lower the rear retainers to the bottom of the door.
7 Tilt the vent window and division bar assembly to the rear and separate the vent window from the front run retainer and division bar.
8 Remove the vent window and the front run retainer and division bar from the door.
9 Rotate the front edge of the glass down and remove the glass and channel assembly from the door, sliding the glass channel off the regulator arm roller.
10 Remove the regulator assembly retaining screws and withdraw the regulator through the opening.
11 Replace the regulator mechanism as an assembly if the spring is broken or other components are worn or damaged.
12 Installation is the reverse of removal. Lightly grease the channel at the bottom of the glass before engaging the operating arm rollers.
13 Wind the window up-and-down, to the fullest extent of its travel, and adjust the run retainer screws so that the window glass moves squarely and smoothly and yet the channels exert a slight grip to prevent the glass from dropping sharply when the regulator handle is first moved to wind the window down.

2 Side door latch – removal and installation

efer to illustration 12.2

Remove the door trim panel and the access panel.
Remove the two inside handle mounting nuts and remove the andle (see illustration).
Remove the three remote control bracket screws. Disconnect the emote control link and remove the remote control and bracket as an ssembly.
Unscrew and remove the door lock push button.
Disconnect the lock cylinder rod from the door latch.
Disconnect the door outside handle rod from the outside handle.
Remove the three latch mounting screws and remove the latch om the door.
Replace the latch assembly with a new one if it is defective.
Installation is the reverse of removal.

3 Rear door latch – removal and installation

efer to illustration 13.2

Remove the trim panel.
Remove the three screws securing the upper latch to the door (see ustration).
Disengage the lock rod clip, remove the latch rod from the upper tch and remove the upper latch from the door.
To remove the lower latch, remove the four screws and disengage e latch rod from the latch. Withdraw the latch from the door.
Installation is the reverse of removal.

4 Hinged doors – removal and installation

efer to illustrations 14.2a and 14.2b

The procedure for removing, installing and adjusting the front, side d rear doors is basically the same.
Remove the upper and lower hinge access hole cover plates (see strations).
Remove the door-to-lower hinge retaining bolts.
Have an assistant support the weight of the door and remove the or-to-upper hinge retaining bolts. Detach the door from the vehicle.
Install the door and the hinge bolts, but do not tighten the hinge lts completely at this stage.
The vertical position of the door can be adjusted by loosening the ge bolts and moving the door in the required direction. Fore-and- movement of the door is achieved by adding or removing shims

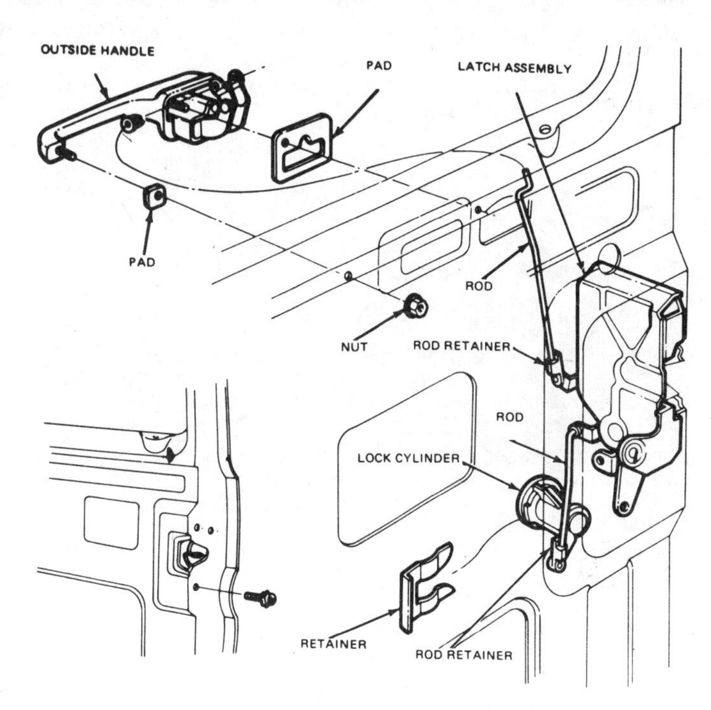

12.2 Side door latch assembly details

between the hinge and the door pillar.
7 When the door is correctly aligned in relation to the door opening, tighten all the hinge bolts.

15 Sliding door – adjustment

Refer to illustrations 15.1 and 15.3

1 To adjust the front edge of the door up-or-down, loosen the three lower guide mounting screws (see illustration). Rotate the guide at the lower mounting screw to obtain the required door position, then tighten the guide mounting screws.
2 In-or-out adjustment at the upper edge of the door is done by loosening the upper roller retaining nut and moving the roller in or out to obtain a flush fit with the body at the top edge of the door.
3 To adjust the lower front edge of the door either in-or-out, loosen the retaining screws on the guide assembly and move the guide assembly forward to move the door closer to the body and backward to move it away from the body at the pillar post (see illustration).

16 Sliding door latch – removal and installation

Refer to illustrations 16.2 and 16.5

1 Make a note of the inside door handle position and remove the mounting screw, handle and trim panel.
2 Remove the outside door handle retaining clip and withdraw the handle and shaft as a unit (see illustration).
3 Remove the screws securing the sleeve to the latch and withdraw the sleeve.
4 Remove the rear latch actuating rod assembly from the rear latch.
5 Remove the door lock rod from the lock cylinder and disengage it from the latch lever arm (see illustration).
6 Detach the stop actuating rod from the front latch.
7 Remove the three front latch retaining screws and disconnect the rear latch actuating rod from the latch.
8 Withdraw the latch assembly through the access hole in the door.
9 Installation is the reverse of removal.

These photos illustrate a method of repairing simple dents. They are intended to supplement *Body repair - minor damage* in this Chapter and should not be used as the sole instructions for body repair on these vehicles.

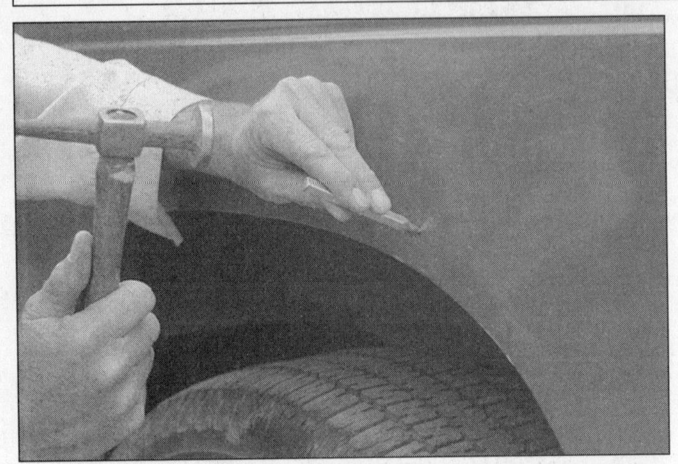

1 If you can't access the backside of the body panel to hammer out the dent, pull it out with a slide-hammer-type dent puller. In the deepest portion of the dent or along the crease line, drill or punch hole(s) at least one inch apart . . .

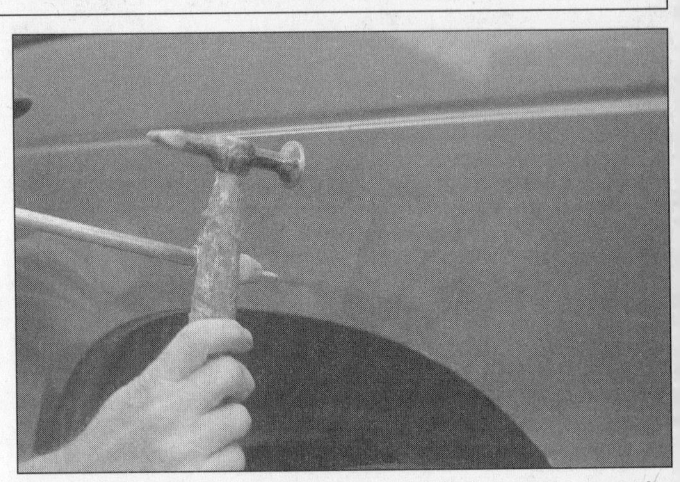

2 . . . then screw the slide-hammer into the hole and operate it. Tap with a hammer near the edge of the dent to help 'pop' the metal back to its original shape. When you're finished, the dent area should be close to its original contour and about 1/8-inch below the surface of the surrounding metal

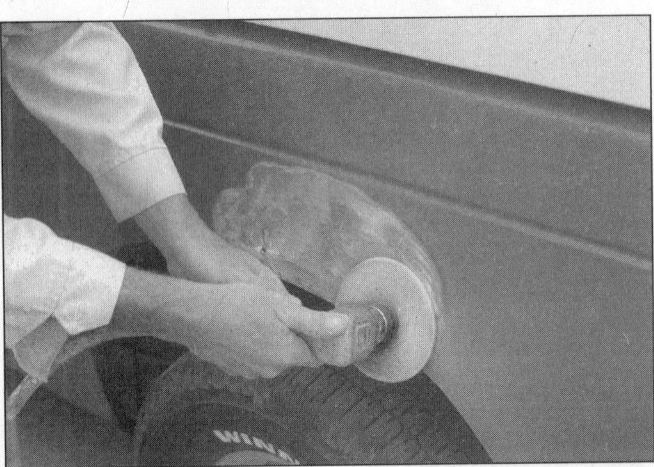

3 Using coarse-grit sandpaper, remove the paint down to the bare metal. Hand sanding works fine, but the disc sander shown here makes the job faster. Use finer (about 320-grit) sandpaper to feather-edge the paint at least one inch around the dent area

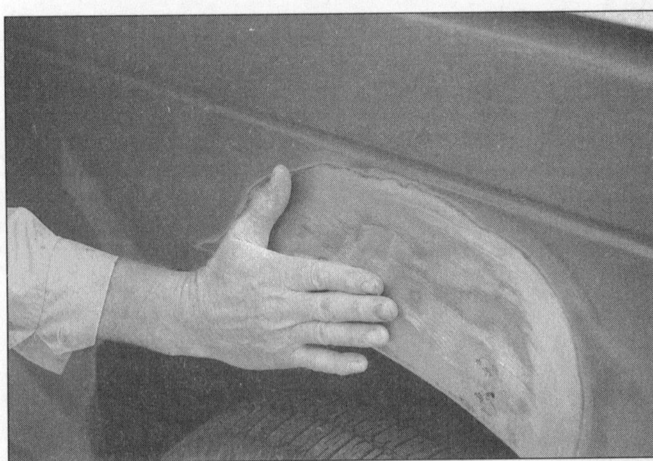

4 When the paint is removed, touch will probably be more helpful than sight for telling if the metal is straight. Hammer down the high spots or raise the low spots as necessary. Clean the repair area with wax/silicone remover

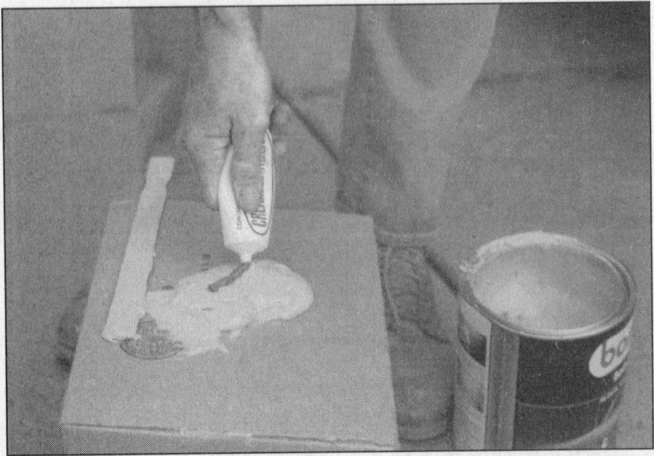

5 Following label instructions, mix up a batch of plastic filler and hardener. The ratio of filler to hardener is critical, and, if you mix it incorrectly, it will either not cure properly or cure too quickly (you won't have time to file and sand it into shape)

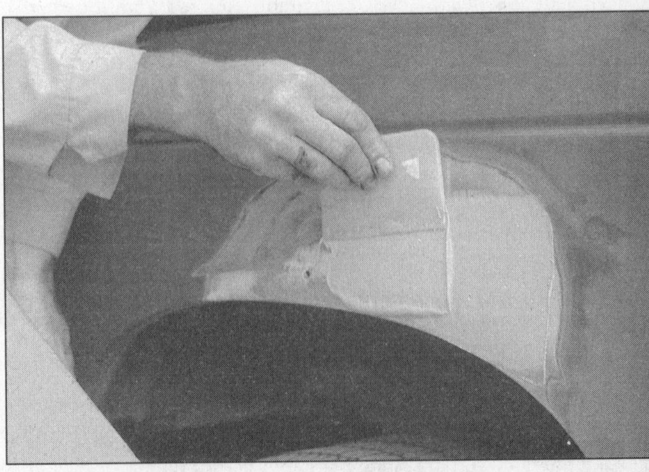

6 Working quickly so the filler doesn't harden, use a plastic applicator to press the body filler firmly into the metal, assuring i bonds completely. Work the filler until it matches the original contour and is slightly above the surrounding metal

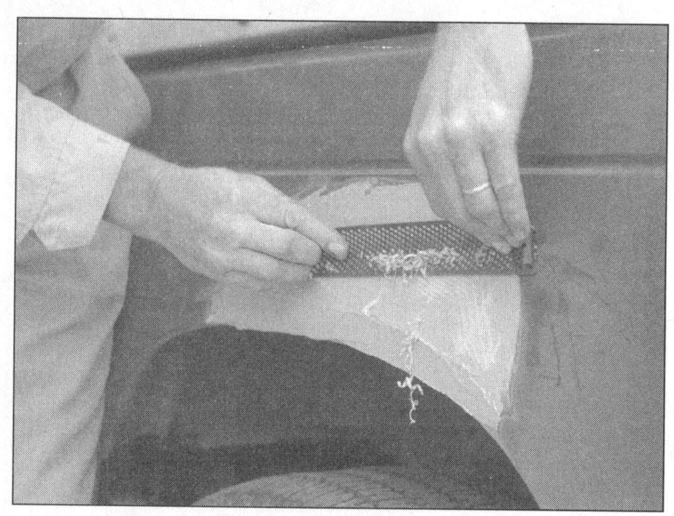

7 Let the filler harden until you can just dent it with your fingernail. Use a body file or Surform tool (shown here) to rough-shape the filler

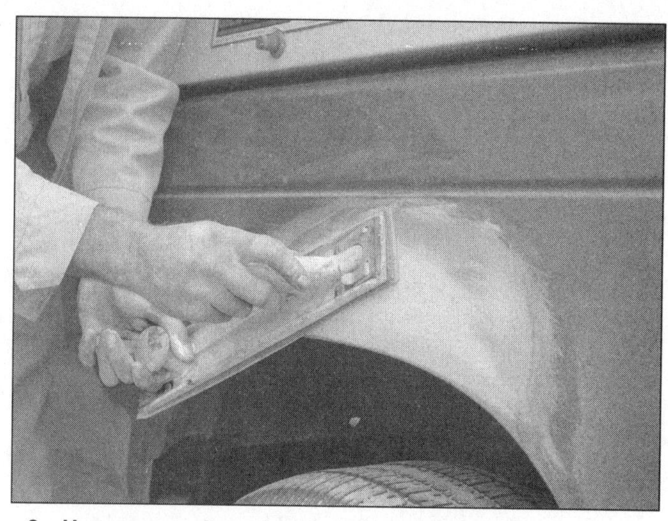

8 Use coarse-grit sandpaper and a sanding board or block to work the filler down until it's smooth and even. Work down to finer grits of sandpaper - always using a board or block - ending up with 360 or 400 grit

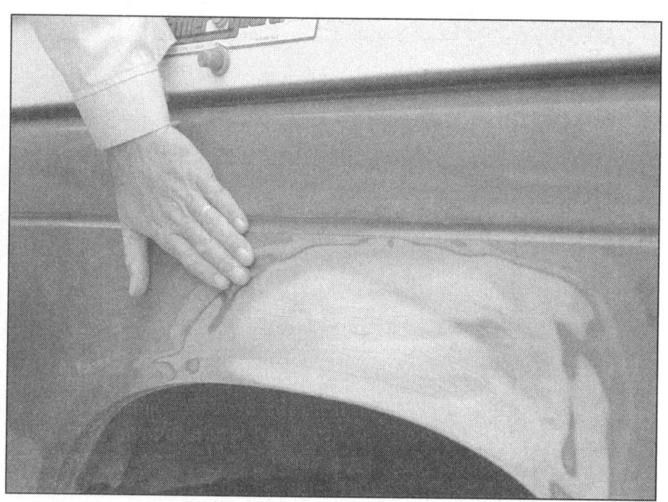

9 You shouldn't be able to feel any ridge at the transition from the filler to the bare metal or from the bare metal to the old paint. As soon as the repair is flat and uniform, remove the dust and mask off the adjacent panels or trim pieces

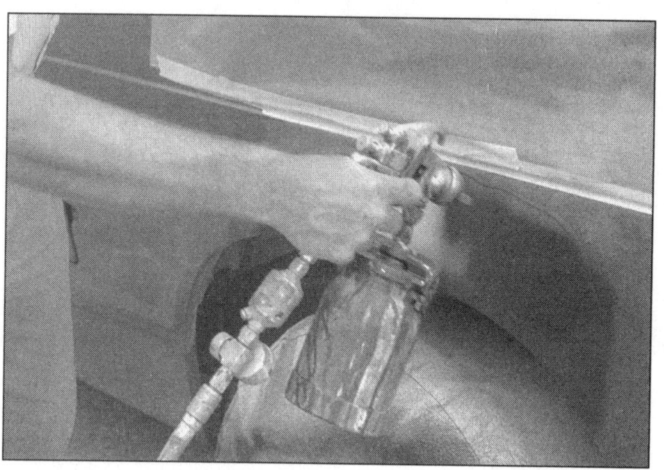

10 Apply several layers of primer to the area. Don't spray the primer on too heavy, so it sags or runs, and make sure each coat is dry before you spray on the next one. A professional-type spray gun is being used here, but aerosol spray primer is available inexpensively from auto parts stores

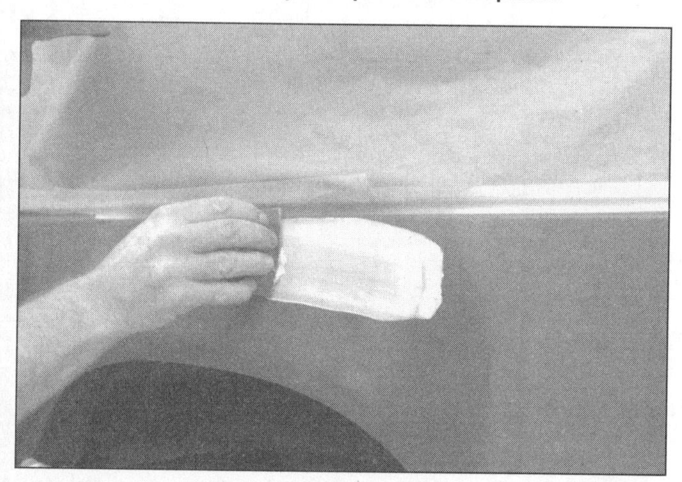

11 The primer will help reveal imperfections or scratches. Fill these with glazing compound. Follow the label instructions and sand it with 360 or 400-grit sandpaper until it's smooth. Repeat the glazing, sanding and respraying until the primer reveals a perfectly smooth surface

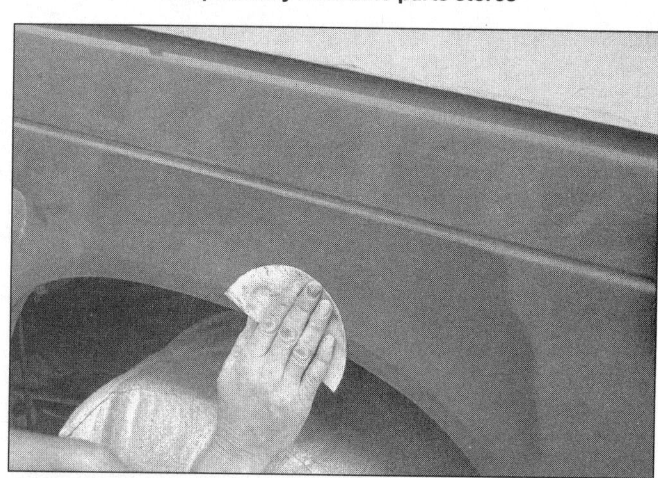

12 Finish sand the primer with very fine sandpaper (400 or 600-grit) to remove the primer overspray. Clean the area with water and allow it to dry. Use a tack rag to remove any dust, then apply the finish coat. Don't attempt to rub out or wax the repair area until the paint has dried completely (at least two weeks)

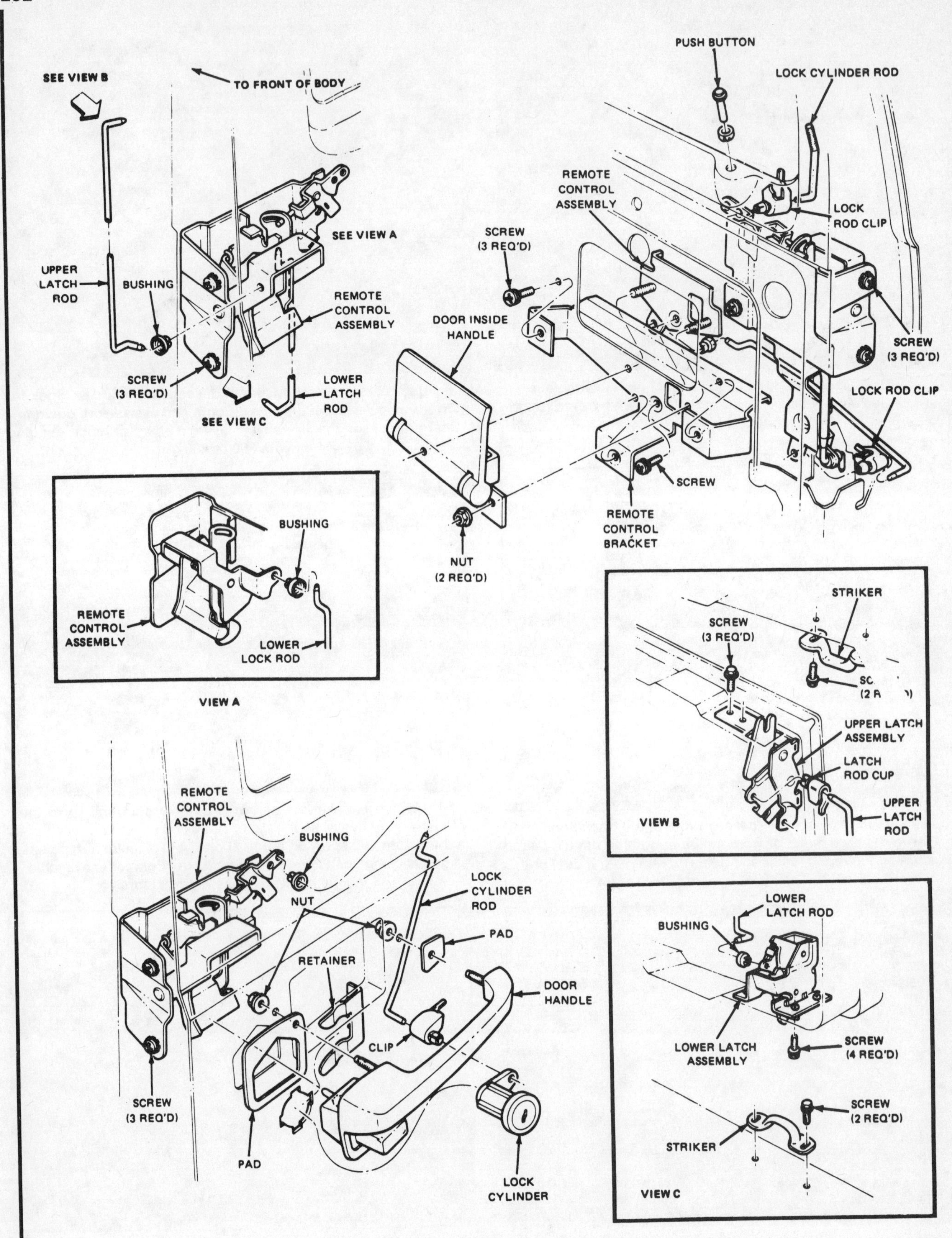

13.2 Rear door latch and related components

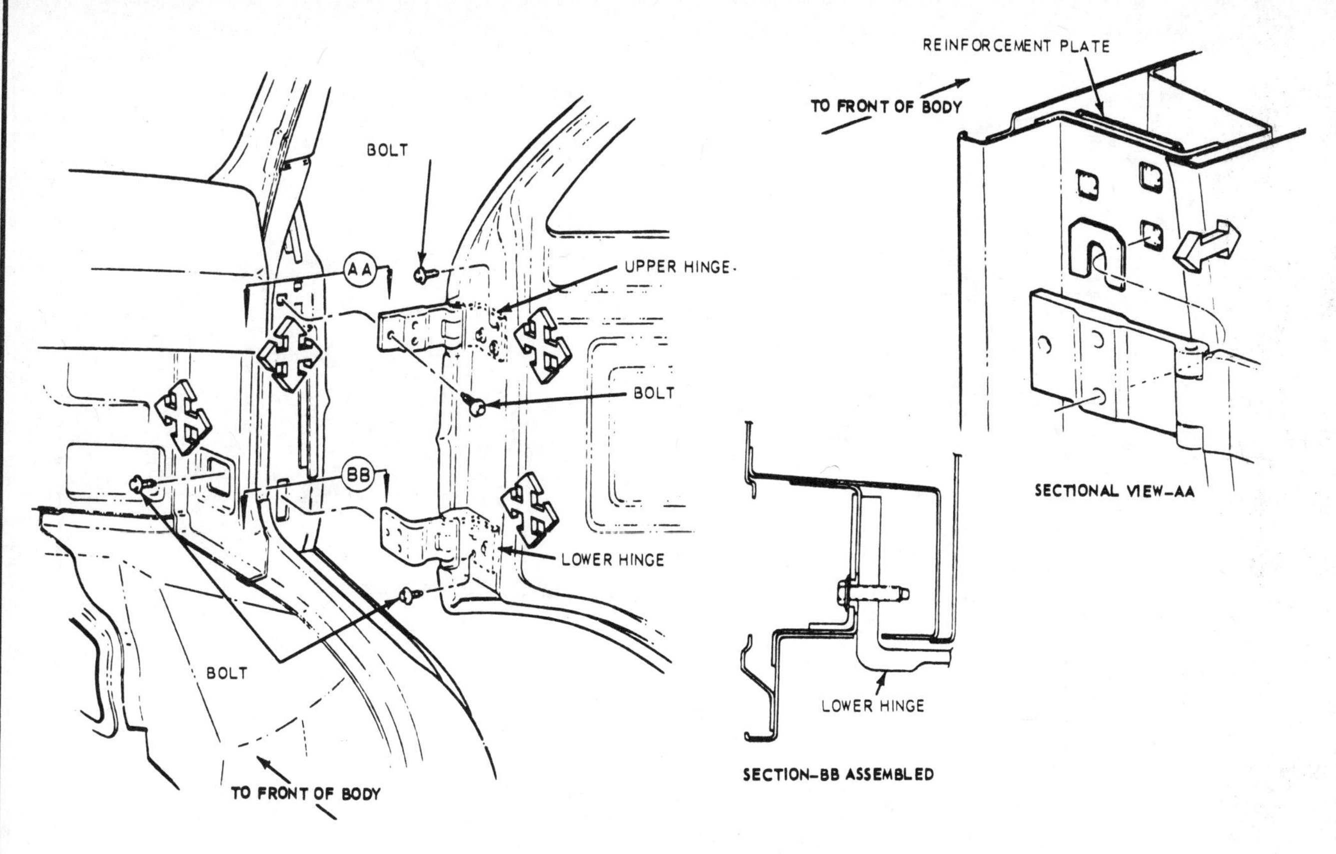

14.2a Front door hinge mounting and adjustment details

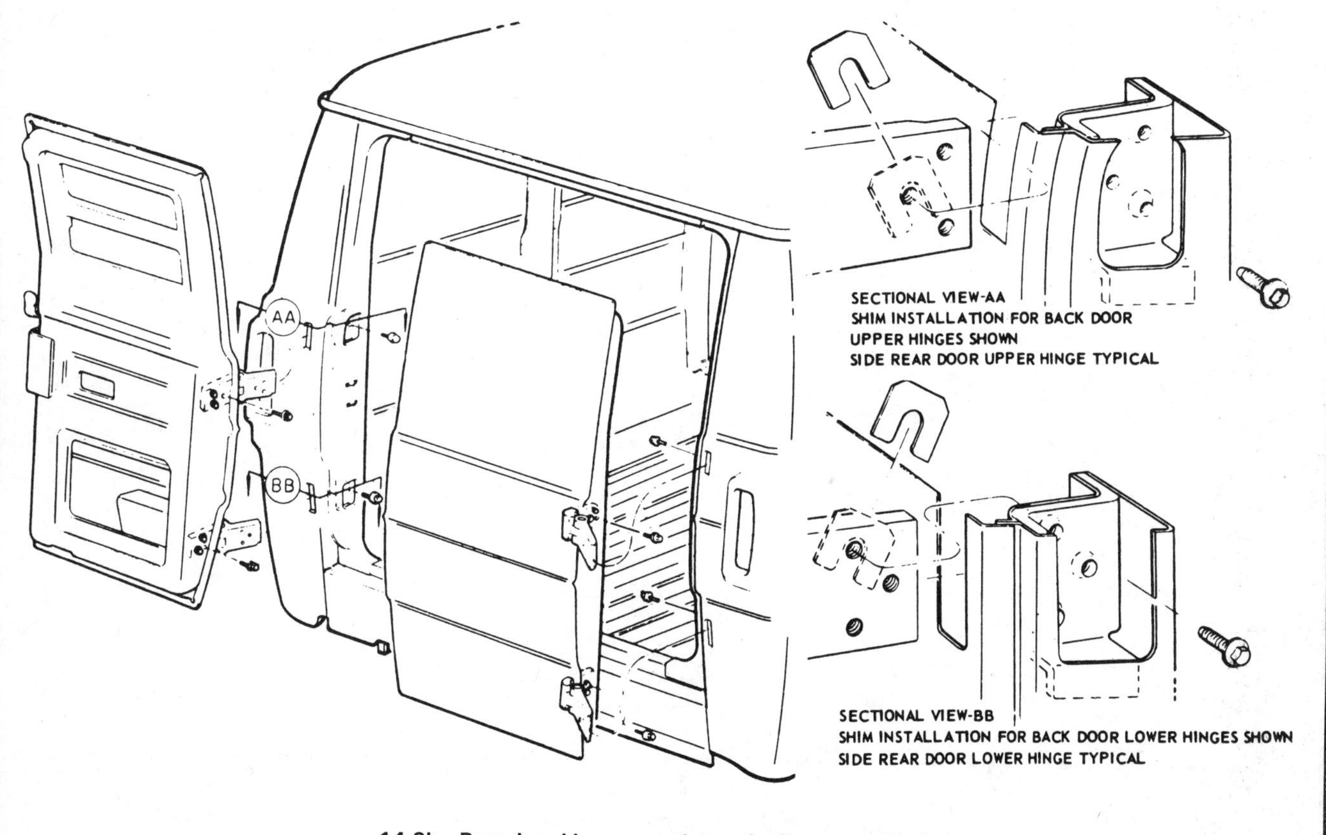

14.2b Rear door hinge mounting and adjustment details

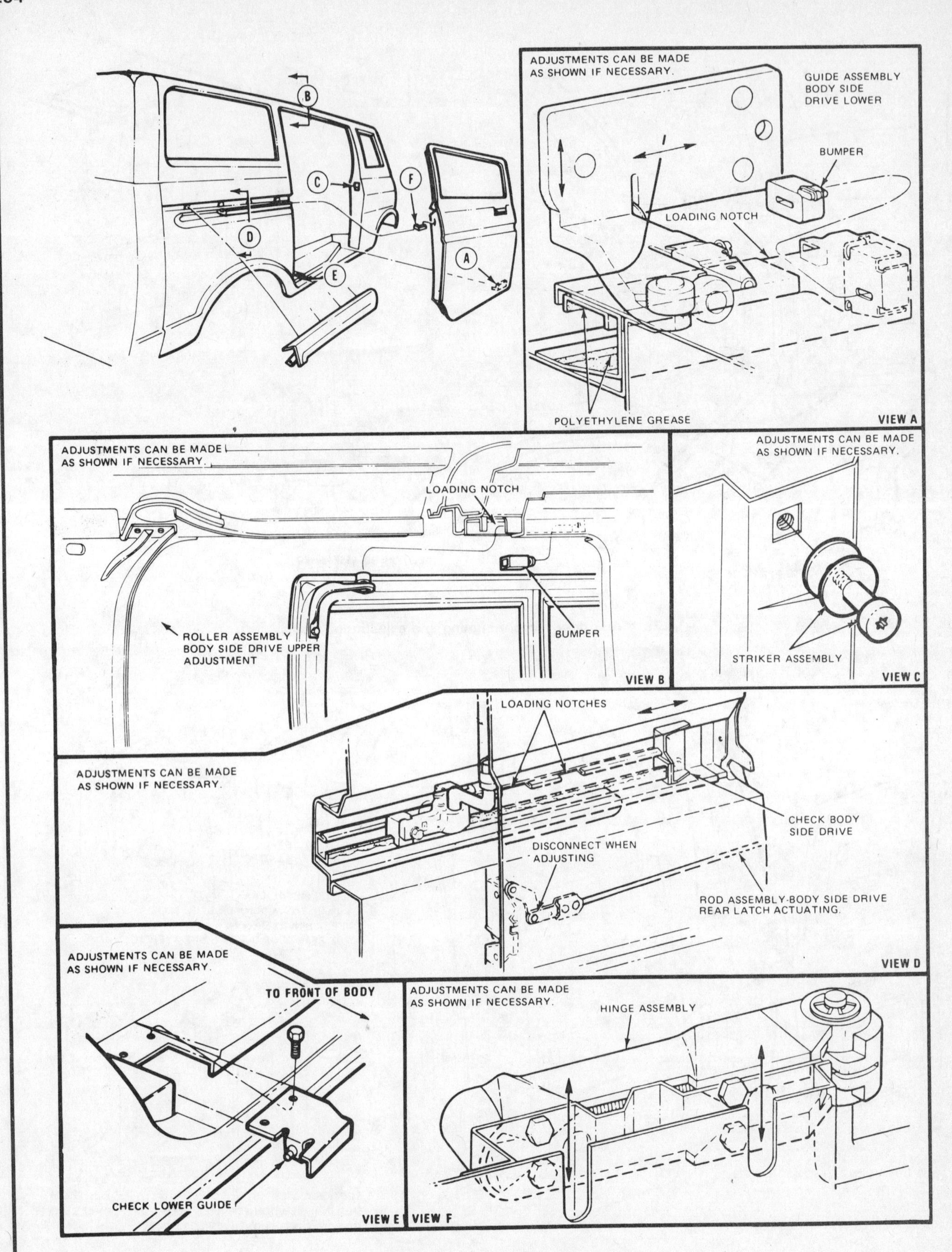

15.1 Sliding door adjustment details

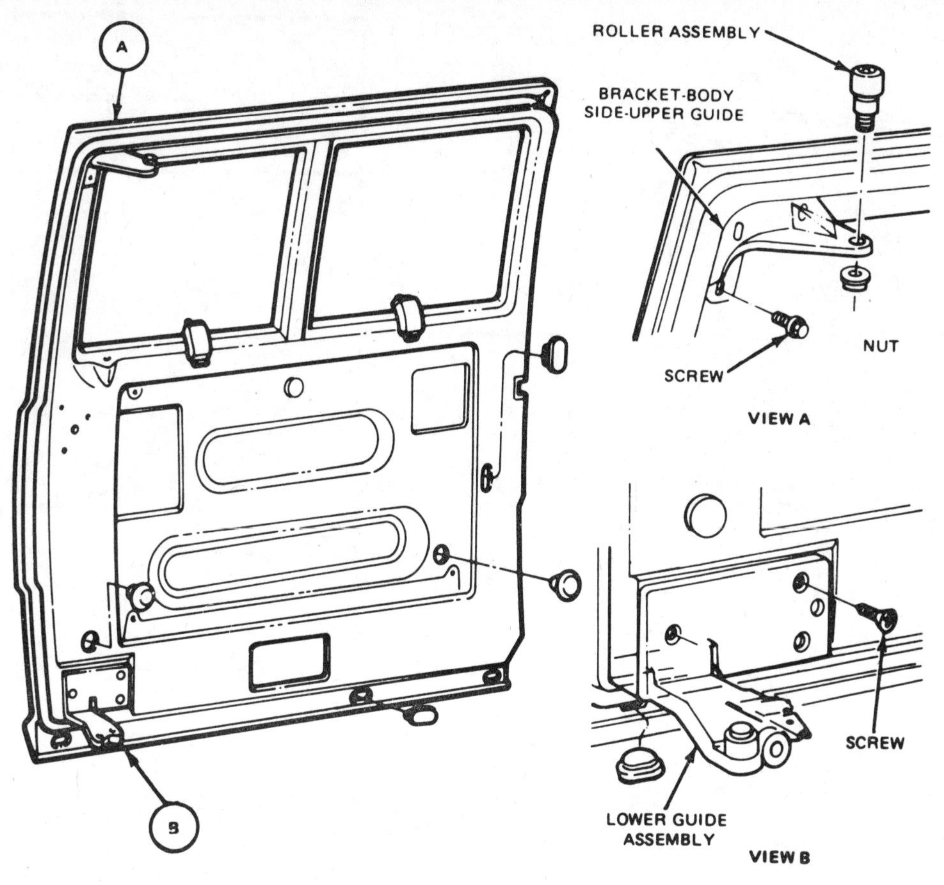

15.3 Sliding door guides and rollers

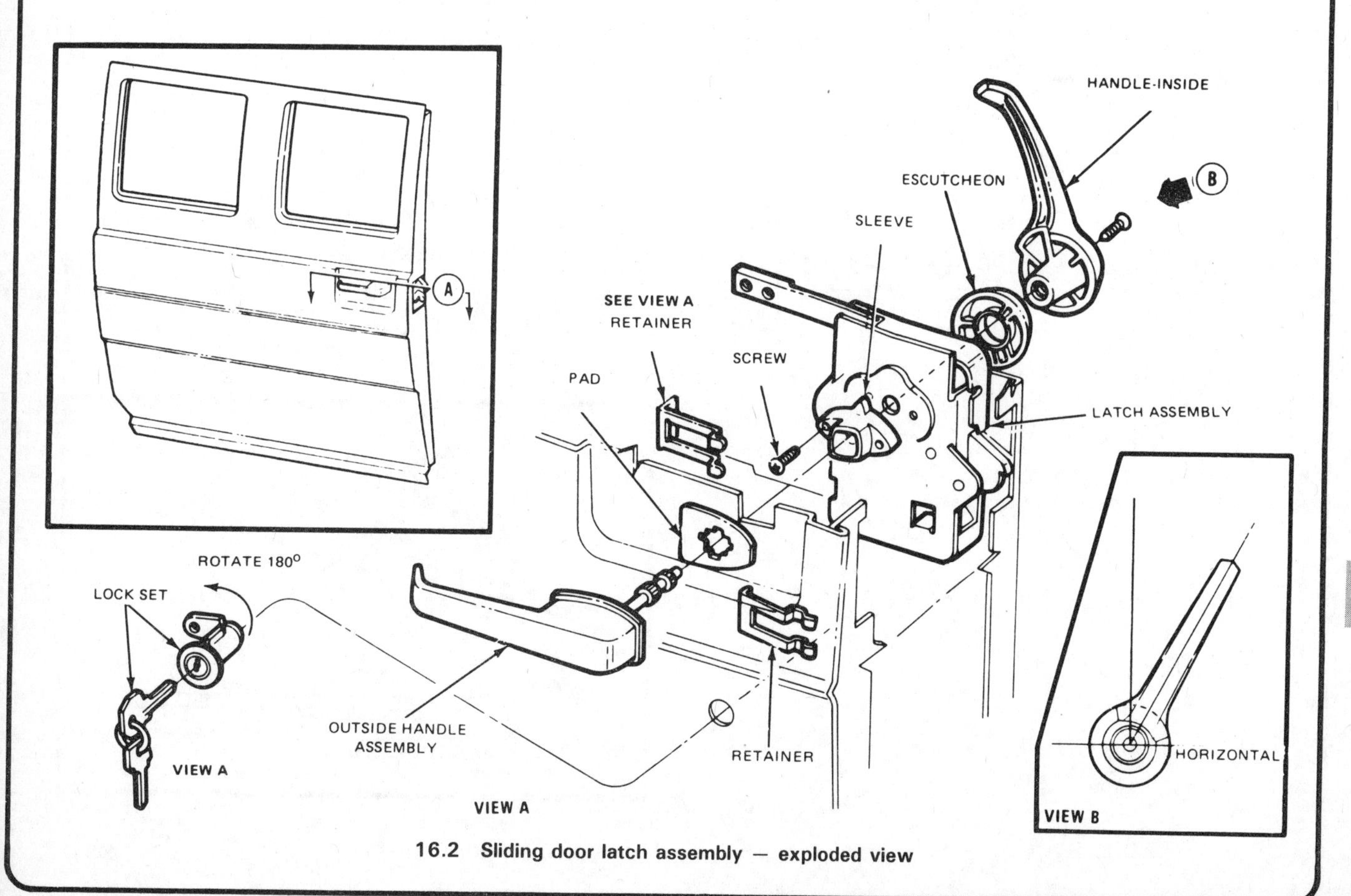

16.2 Sliding door latch assembly — exploded view

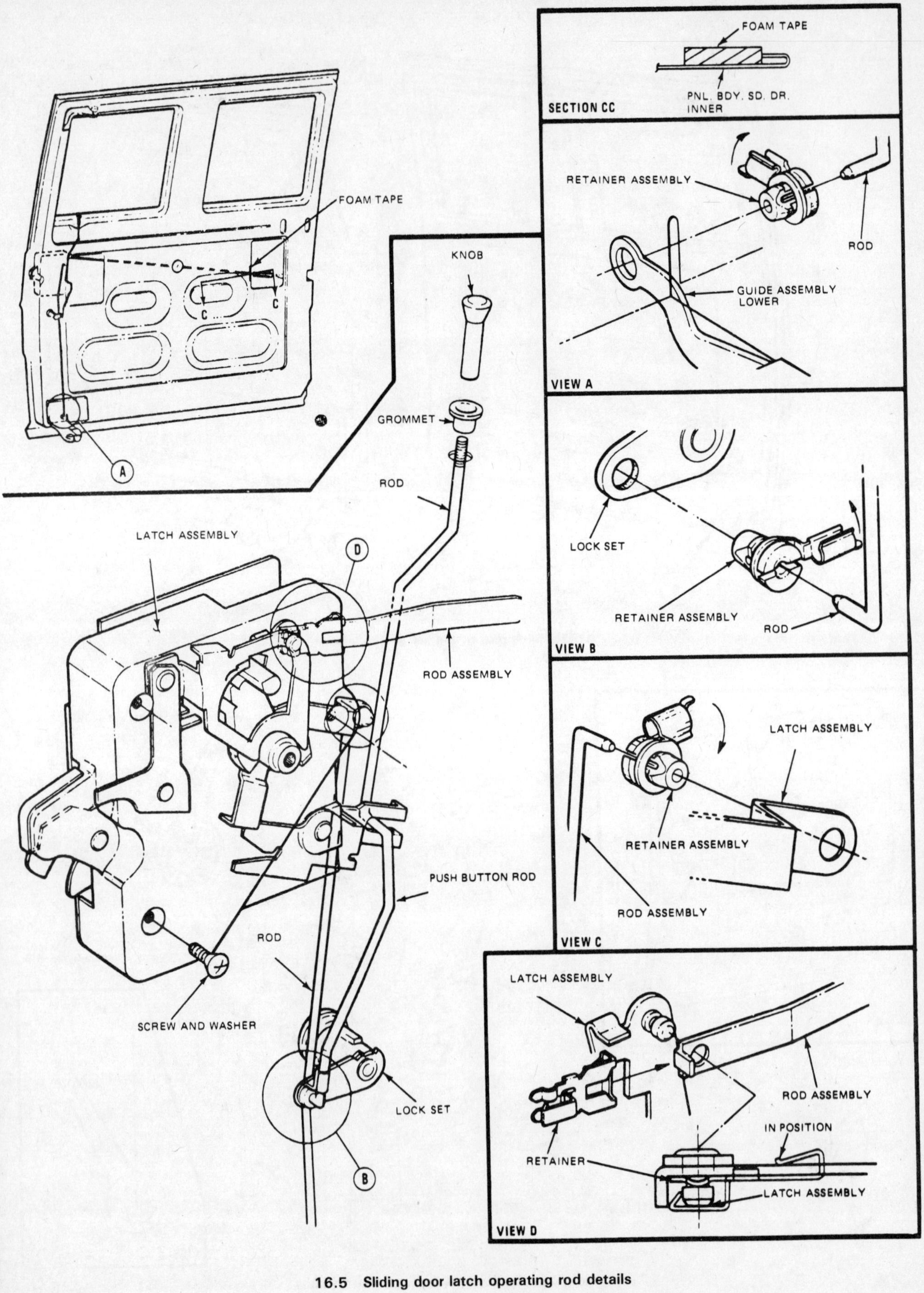

SECTION CC
FOAM TAPE
PNL. BDY. SD. DR. INNER

VIEW A
RETAINER ASSEMBLY
ROD
GUIDE ASSEMBLY LOWER

VIEW B
LOCK SET
RETAINER ASSEMBLY
ROD

VIEW C
LATCH ASSEMBLY
RETAINER ASSEMBLY
ROD ASSEMBLY

VIEW D
LATCH ASSEMBLY
ROD ASSEMBLY
IN POSITION
RETAINER
LATCH ASSEMBLY

FOAM TAPE
C
C
A
KNOB
GROMMET
ROD
LATCH ASSEMBLY
D
ROD ASSEMBLY
PUSH BUTTON ROD
ROD
SCREW AND WASHER
LOCK SET
B

16.5 Sliding door latch operating rod details

17 Sliding door – removal and installation

1 Remove the nuts on the inside of the vehicle and the two screws on the outside of the vehicle that retain the center track shield, then detach the shield.
2 Remove the lower guide check screws and detach the check.
3 Open the door and have an assistant support it.
4 Mark the location of the upper guide bracket on the door.
5 Remove the three retaining screws and detach the upper guide bracket.
6 Carefully separate the door from the vehicle.
7 Installation is the reverse of removal. Be sure to adjust the door by referring to the appropriate Section.

18 Hood – removal and installation

Refer to illustrations 18.3a and 18.3b

1 Raise the hood and prop it in the open position.
2 Place an old blanket below the rear of the hood to protect the paint.
3 Remove the screws retaining the hinges to the hood (see illustrations).
4 Carefully detach the hood from the vehicle.
5 Installation is the reverse of removal.

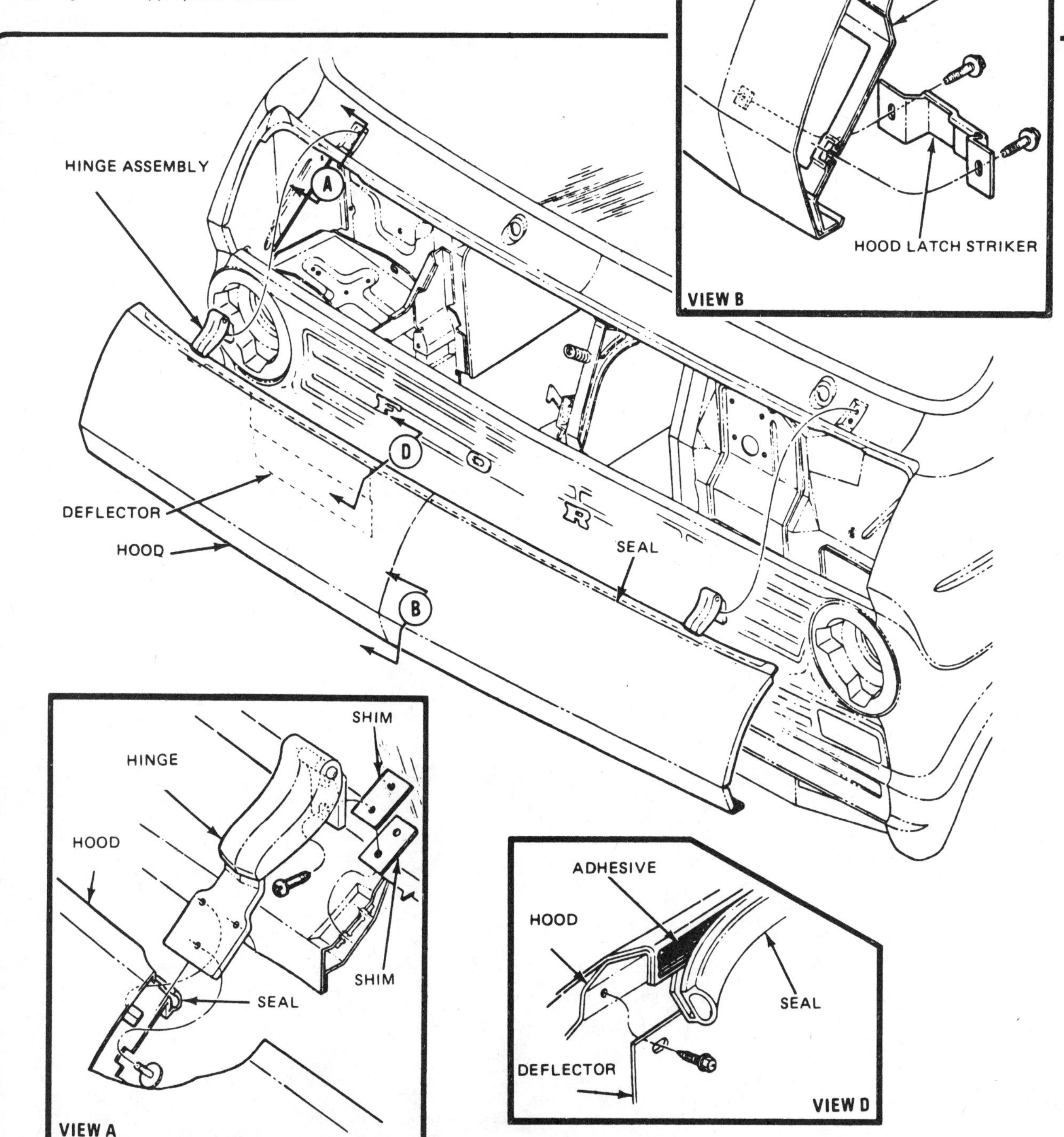

18.3a Hood and hinge assembly (early models)

6 To adjust the position of the hood after installation, loosen the hinge screws and move the hood to the required position. When the hood is adjusted correctly, tighten the hinge screws.

19 Radiator grille (thru 1973) – removal and installation

Refer to illustration 19.3

1 Raise the hood and support it in the open position.

2 Remove the headlight doors, headlights and ring assemblies (see Chapter 12).

3 Remove the five screws securing the bottom edge of the grille to the front body member (see illustration).

4 Remove the single screw attaching the hood latch support to the top edge of the grille.

5 Unplug the parking light wires at the connectors.

6 Remove the screws securing the grille to the upper right and left support brackets and detach the grille.

7 Installation is the reverse of removal.

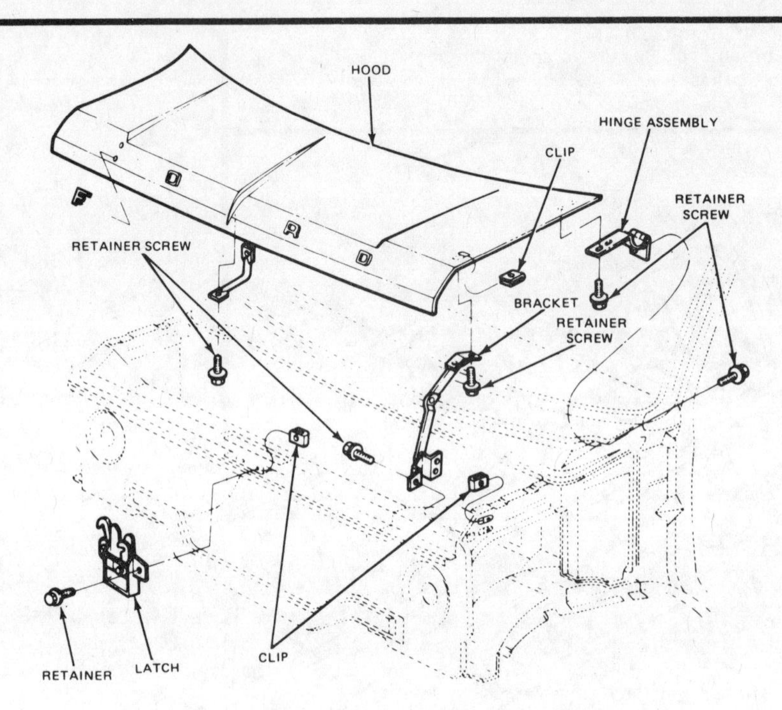

18.3b Hood and hinge assembly (later models)

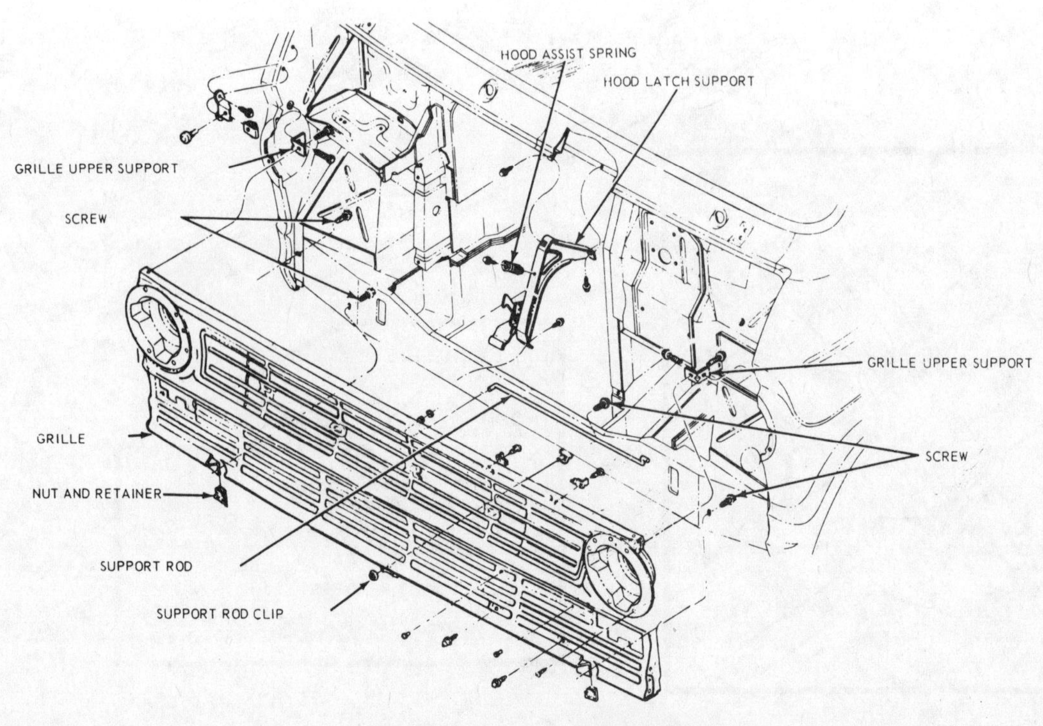

19.3 Radiator grille – early models

20 Radiator grille (1974 and later) – removal and installation

Refer to illustration 20.2

1 Raise the hood and support it in the open position.
2 Remove the two center screws securing the grille to the support bracket (see illustration).
3 Remove the six screws from the bottom of the grille.
4 Remove the nine screws located on the upper flange of the grille, which is attached to the radiator upper support and gussets.
5 Detach the grille.
6 Installation is the reverse of removal.

21 Front fenders (later models only) – removal and installation

Refer to illustration 21.4

1 The front fenders on later Econoline models are bolted in position and are fairly easy to remove, assuming that the mounting bolts are not badly corroded. If they are it will be necessary to cut them off with a bolt shearing tool or a hammer and cold chisel.
2 First remove the radiator grille as described in the previous Section.
3 Remove the front bumper as described in Section 22.
4 Remove the screws securing the front of the fender to the lower radiator grille panel (see illustration).

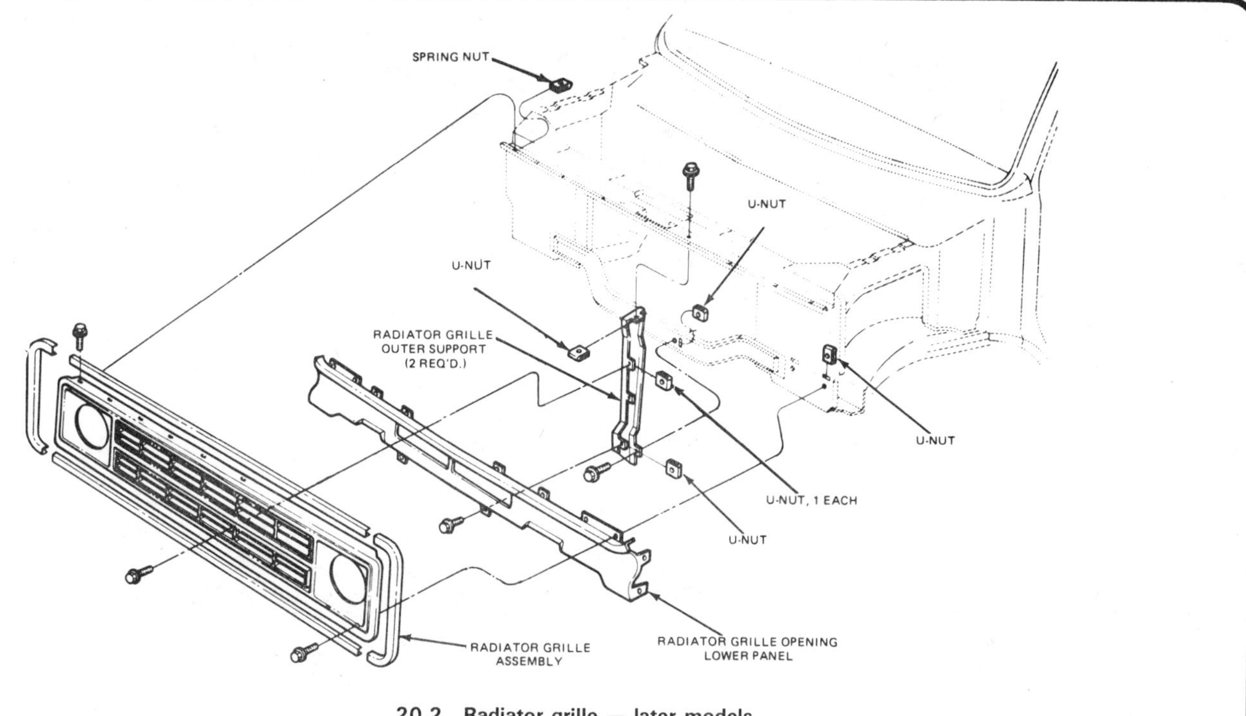

20.2 Radiator grille – later models

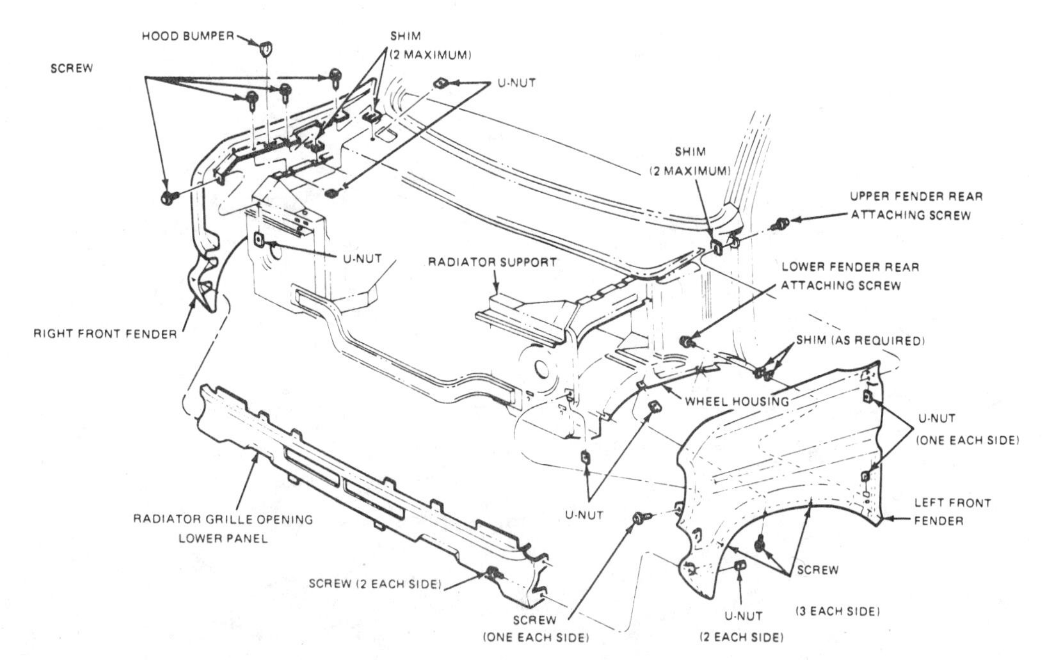

21.4 Bolt-on front fenders used on later models – exploded view

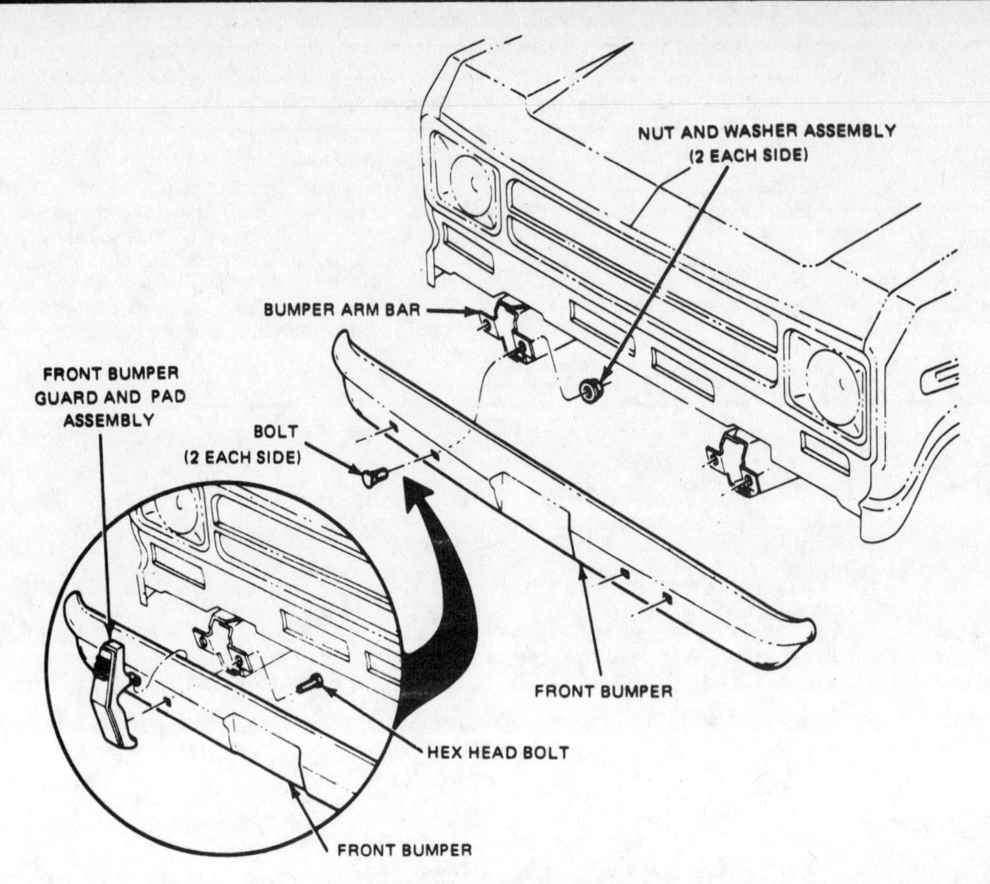

22.1 Front bumper assembly

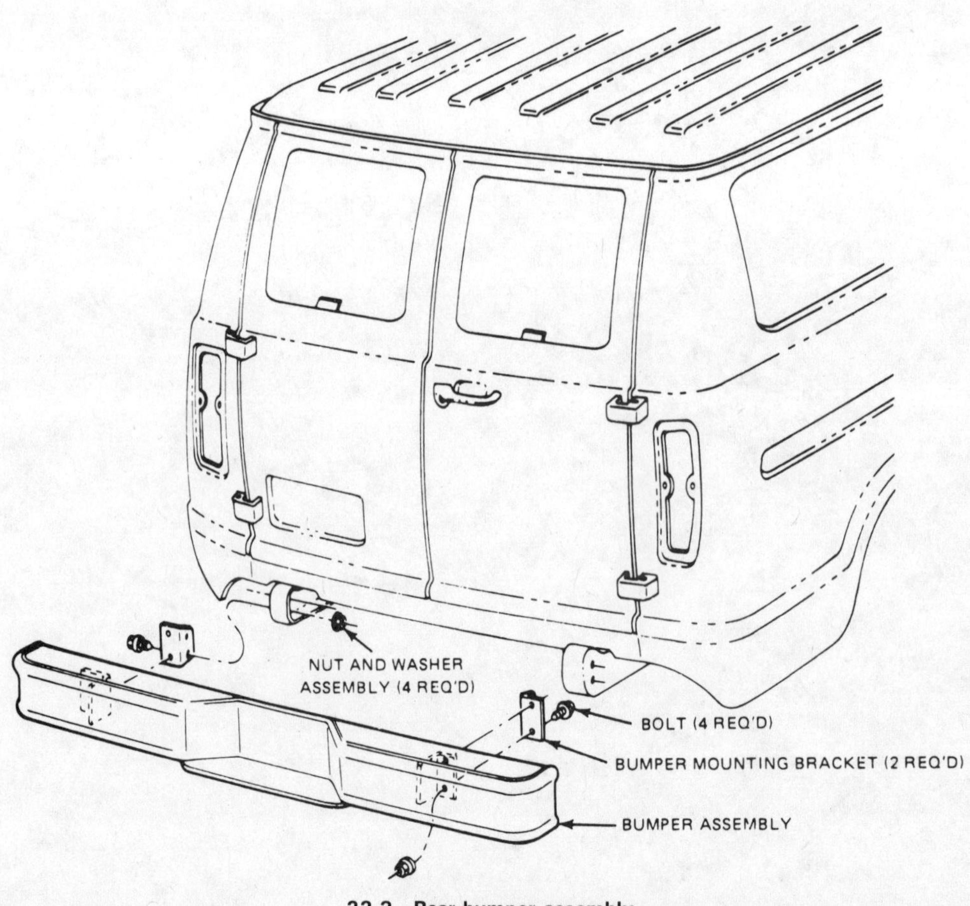

22.3 Rear bumper assembly

5 Remove the screws securing the front of the fender to the lower radiator grille panel.
6 Remove the three screws securing the lower edge of the fender to the wheel housing.
7 Remove the lower rear mounting screws and shims.
8 Open the front door and remove the upper rear mounting screw and shims.
9 Remove the two screws securing the front edge of the fender to the radiator support.
10 Remove the remaining three screws and shims securing the top of the fender to the body and detach the fender.
11 Installation is the reverse of removal.

22 Front and rear bumpers — removal and installation

Refer to illustrations 22.1 and 22.3

1 To detach the front bumper, remove the four nuts and bolts securing the bumper bar to the ends of the chassis members and remove the bumper (see illustration).
2 If the vehicle is equipped with bumper guards, they are retained by the inner bolt on each side of the bumper.
3 The rear bumper is detached by removing the four bolts securing the bumper bar to the mounting brackets (see illustration).
4 Installation is the reverse of removal.

Chapter 12 Chassis electrical system

Contents

Specifications

Fuses and circuit breakers

Circuit	Location	Circuit protection
Headlights	Integral with lighting switch	18 amp CB
Tail, marker, license, parking lights and horns	Integral with lighting switch	15 amp CB
Turn signal, back-up lights	Fuse panel	15 amp fuse
Windshield wiper system	Fuse pane	17.5 amp CB
Hazard warning and stop lights	Fuse pane	120 amp fuse
Courtesy lights	Fuse panel	15 amp fuse
Heater and/or A/C	Fuse panel	30 amp fuse
Instrument panel lights	Fuse panel	3 amp fuse
Auxiliary heater (same fuse as heater)	Fuse panel	30 amp fuse
Radio	Fuse panel	7.5 amp fuse
Air conditioner	Fuse panel	30 amp fuse and fuse link
Auxiliary air condiioner (same fuse as heater)	Fuse panel	30 amp fuse
School bus warning lights	Fuse panel	20 amp CB
Auxiliary fuel tank	Fuse panel	7.5 amp fuse
Speed control	Cartridge in-line fuse	5 amp fuse
Auxiliary battery	Fuse panel	20 amp fuse

Bulb specifications

Light description	Candlepower (C) or Wattage (W)	Trade number
Back-up light	32C	1156
Cargo light	12C	105
Clock	2C	1895
Dome light	12C	105
Amber fog light	35W	4415A
Amber fog light switch	1.5C	53X

Light description	Candlepower (C) or Wattage (W)	Trade number
Front parking light and turn signal	32C	1157
Handbrake signal	2C	1895
Headlight	50 and 40W	6012
Instrument panel high beam indicator, brake warning, oil and generator indicators, turn signal indicators and seat belt warning light	2C	1895
Parking brake release indicator	1C	257
License plate light	4C	1155 or 97
Radio pilot light	1.9C	1891
Rear tail/stop/turn signal	32C	1157
Stoplight (4.4 in diameter)	30W	4405

Torque specifications	Ft-lbs
Steering wheel nut	30 to 40

1 General information

The chassis electrical system is a 12-volt, negative ground system. Power for the lights and all electrical accessories is supplied by a lead/acid battery, which is charged by the alternator (see Chapter 5).

This Chapter covers repair and service procedures for the various lights and electrical components not directly associated with the engine. Information on the battery, alternator, voltage regulator and the starter motor is included in Chapter 5.

It should be emphasized that whenever work is done on the chassis electrical system the negative battery cable should be disconnected from the battery.

2 Electrical troubleshooting – general information

A typical electrical circuit consists of an electrical component, any switches, relays, motors, etc. relevant to that component and the wiring and connectors that connect the components to both the battery and the chassis. To aid in locating a problem in any electrical circuit, complete wiring diagams of each model are included at the end of this Chapter.

Before tackling any troublesome electrical circuit, first thoroughly study the appropriate diagrams to get a complete understanding of what makes up that individual circuit. Trouble spots, for instance, can often be narrowed down by noting if other components related to that circuit are operating properly or not. If several components or circuits fail at one time, chances are the fault lies in the fuse or ground connection, as several circuits often are routed through the same fuse and ground connections. This can be confirmed by referring to the fuse box and ground distribution diagrams in this Chapter.

Often, electrical problems stem from simple causes, such as loose or corroded connections, a blown fuse or melted fusible link. Prior to any electrical troubleshooting, always visually check the condition of the fuses, wires and connections of the problem circuit.

If testing instruments are going to be utilized, use the diagrams to plan ahead of time where you will make the necessary connections in order to accurately pinpoint the trouble spot.

The basic tools needed for electrical troubleshooting include a circuit tester or voltmeter (a 12 volt bulb with a set of test leads can also be used), a continuity tester (which includes a bulb, battery and set of test leads) and a jumper wire, preferably with a circuit breaker incorporated, which can be used to bypass electrical components.

Voltage checks should be performed if a circuit is not functioning properly. Connect one lead of a circuit tester to either the negative battery terminal or a known good ground. Connect the other lead to a connector in the circuit being tested, preferably nearest to the battery or fuse. If the bulb of the tester goes on, voltage is reaching that point, which means the part of the circuit between that connector and the battery is problem-free. Continue checking along the circuit in the same fashion. When you reach a point where no voltage is present, the problem lies between there and the last good test point. Most of the time the problem is due to a loose connecton. Keep in mind that some circuits only receive voltage when the ignition key is in the Accessory or Run position.

A method of finding shorts in a circuit is to remove the fuse and connect a test light or voltmeter in its place to the fuse terminals. There should be no load in the circuit. Move the wiring harness from side to side while watching the test light. If the bulb goes on, there is a short to ground somewhere in that area, probably where insulation has rubbed off a wire. The same test can be performed on other components of the circuit, including the switch.

A ground check should be done to see if a component is grounded properly. Disconnect the battery and connect one lead of a self-powered test light such as a continuity tester to a known good ground. Connect the other lead to the wire or ground connection being tested. If the bulb goes on, the ground is good. If the bulb does not go on, the ground is not good.

A continuity check is performed to see if a circuit, section of circuit or individual component is passing electricity properly. Disconnect the battery and connect one lead of a self-powered test light such as a continuity tester to one end of the circuit being tested and the other lead to the other end of the circuit. If the bulb goes on, there is continuity, which means the circuit is passing electricity properly. Switches can be checked in the same way.

Remember that all electrical circuits are composed basically of electricity running from the battery, through the wires, switches, relays, etc. to the electrical component (light bulb, motor, etc). From there it is run to the car body (ground) where it is passed back to the battery. Any electrical problem is basically an interruption in the flow of electricity from the battery to the component and back to the battery.

3 Fusible link – testing, removal and installation

Refer to illustration 3.1

1 A fusible link is used to protect the alternator from overload and is located in the wiring harness connecting the starter relay and *Bat* terminal on the alternator (see illustration).

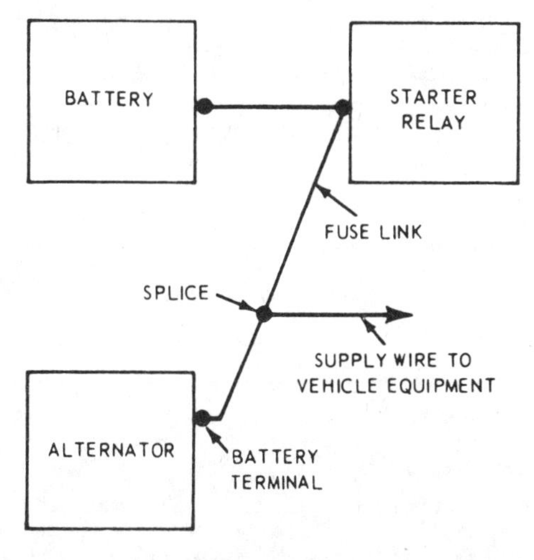

3.1 Fusible link location

12

4.1 Removing the headlight rim (early model)

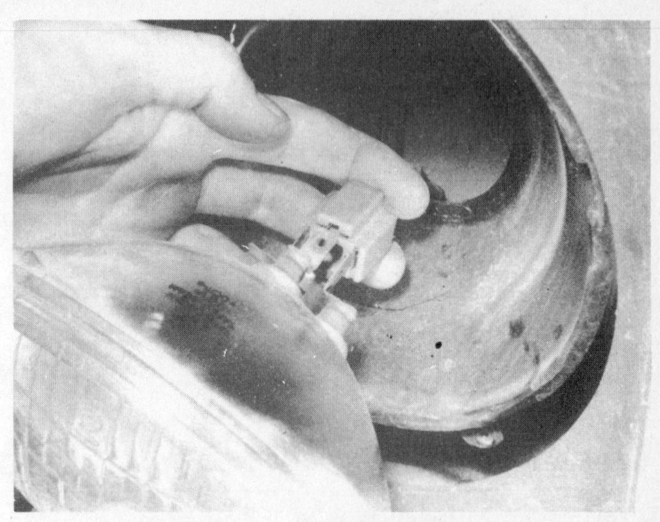

4.4 Disconnecting the wires from the sealed beam unit

2 If an alternator problem develops, first make sure that the battery is fully charged and the terminals are clean.
3 Using a voltmeter, check for voltage at the *Bat* terminal at the rear of the alternator. No reading indicates that the fusible link has probably burned out. Check the link visually (look for burned, swollen and melted insulation).
4 To install a new fusible link, disconnect the ground cable from the battery.
5 Disconnect the fusible link eyelet terminal from the battery terminal on the starter relay.
6 Cut the fusible link and the splices from the wires to which it is attached.
7 Splice and solder the new fusible link to the wire from which the old link was cut. Wrap the splice with electrical tape.
8 Reconnect the eyelet terminal to the battery terminal on the starter relay.
9 Reconnect the battery ground cable.

4 Headlight sealed beam – removal and installation

Refer to illustrations 4.1 and 4.4

Thru 1977

1 Remove the two screws securing the headlight rim to the fender, then detach the headlight rim (see illustration).
2 Remove the three screws securing the sealed beam unit retaining ring to the adjusting ring.
3 Rotate the retaining ring to disengage it from the adjusting screws.
4 Pull out the sealed beam and disconnect the wiring harness plug (see illustration).
5 The sealed beam is now free.
6 Installation is the reverse of removal. It is recommended that whenever the sealed beam unit is changed, the headlight alignment is checked. Further information will be found in Section 5.

1978 and later

7 Later model vehicles have a rectangular headlight instead of a round one. The procedure above is correct, but note that the retaining ring on later models is attached with four screws, all of which must be removed to detach the ring.

5 Headlight alignment

Refer to illustration 5.7

1 The headlights should be aligned by a shop equipped with special optical equipment, but the following procedure can be used to temporarily align them until they can be checked by a professional.
2 Position the vehicle on level ground 10 feet in front of a dark wall or board. The wall or board must be at a right angle to the centerline of the vehicle.

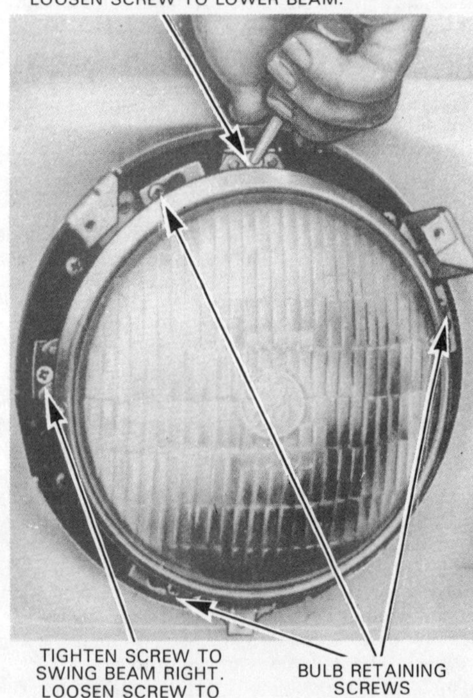

5.7 Headlight beam adjusting screw locations (early models)

3 Draw a vertical line on the board or wall in line with the centerline of the vehicle.
4 Rock the vehicle to settle the suspension components, then measure the height between the ground and the center of the headlights.
5 Draw a horizontal line across the board or wall at this measured height. On this horizontal line, make a vertical line on either side of the vertical centerline, equal to the distance between the center of each light and the center of the vehicle.
6 Remove the headlight rims and switch the headlights to high beam.
7 By careful adjusting of the horizontal and vertical adjusting screws on each light (see illustration), align the centers of each beam with the vertical lines which were previously marked on the horizontal line

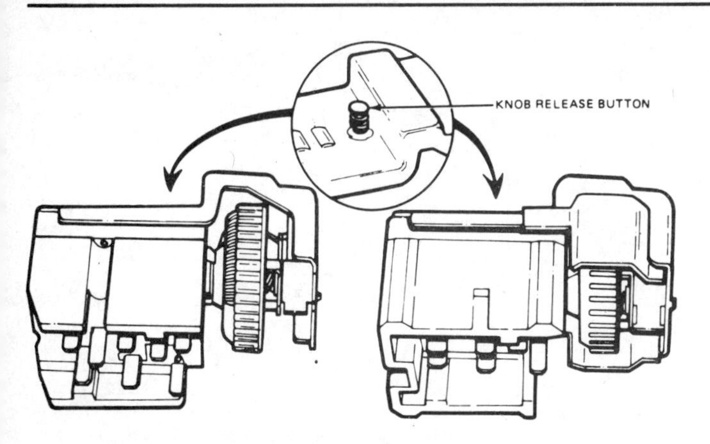

6.2 Light switch knob removal button location

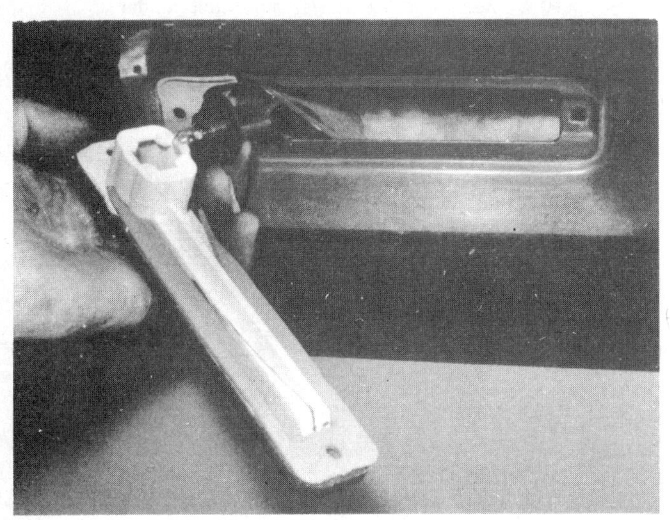

9.1 Removing a side marker lens and bulb holder

(the cross shaped marks should be at the center of each beam).
8 Rock the vehicle on its suspension again and check that the beams return to the correct positions. At the same time, check the operation of the dimmer switch, then install the headlight rims.

6 Headlight switch – removal and installation

Refer to illustration 6.2

1 Disconnect the negative cable from the battery.
2 Working through the access hole in the underside of the instrument panel, press the release button using a small screwdriver (see illustration) and remove the switch knob and shaft.
3 Remove the bezel nut and remove the switch body from behind the instrument panel. Disconnect the wiring harness connector.
4 Installation is the reverse of removal.

7 Headlight dimmer switch – removal and installation

1 Disconnect the negative cable from the battery.
2 Carefully pull back the carpet from around the dimmer switch, removing the scuff plate and cowl trim screws if necessary.
3 Remove the two screws securing the dimmer switch to the mounting bracket.
4 Move the switch forward and disconnect the wiring harness connector from the rear of the switch.
5 Installation is the reverse of removal.

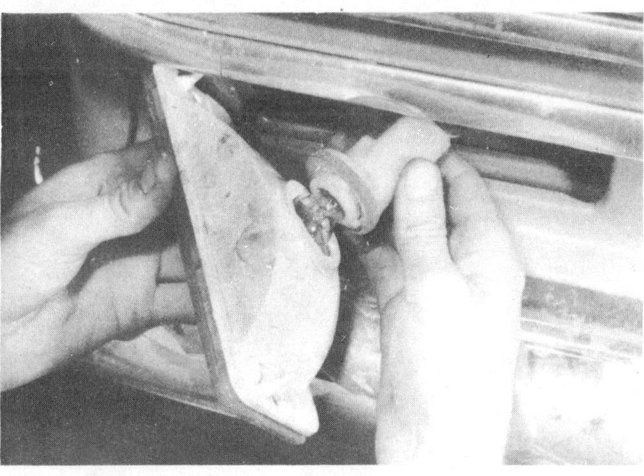

8.2 Removing the front parking/turn signal bulb

10.1 Removing the rear light cluster lens

8 Front parking and turn signal lights – bulb replacement

Refer to illustration 8.2

1 Remove the two mounting screws and detach the lens.
2 Remove the bulb from the holder by pushing it in and turning it counterclockwise (see illustration).
3 When installing the lens, do not overtighten the screws.

9 Front and rear side marker lights – bulb replacement

Refer to illustration 9.1

1 Remove the two mounting screws and withdraw the lens from the body just enough to disengage the bulb holder (see illustration).
2 Replace the bulb if necessary and position the bulb holder in the lens.
3 Attach the lens to the body. Do not overtighten the screws.

10 Rear light cluster – removal and installation

Refer to illustrations 10.1, 10.2 and 10.4

1 Remove the two mounting screws and pull out the lens unit (see illustration).

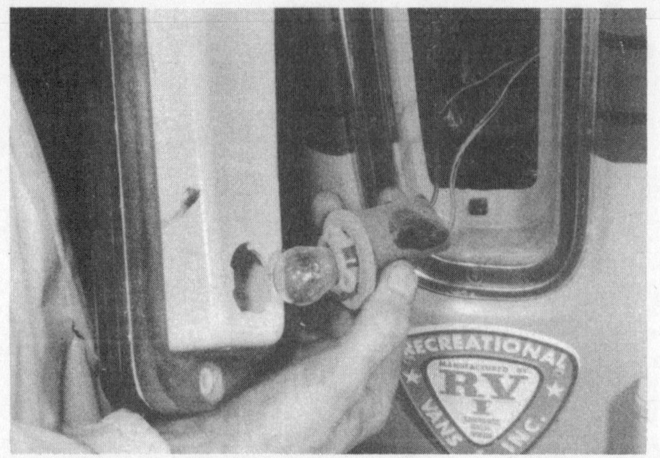

10.2 Disengaging the bulb holder from the rear light housing

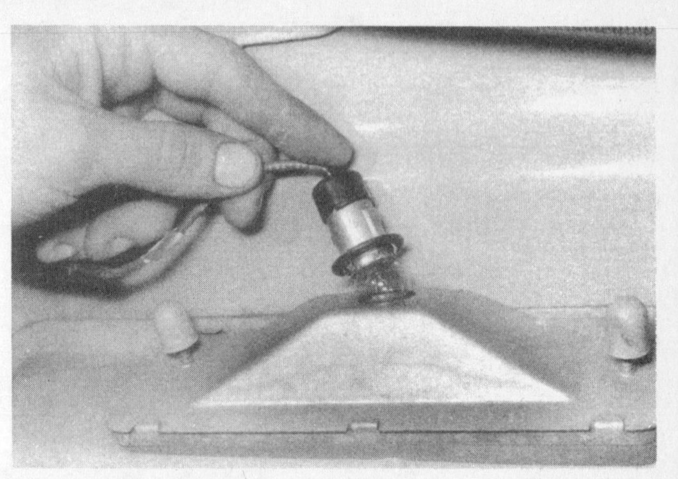

10.4 Removing the license plate bulb holder

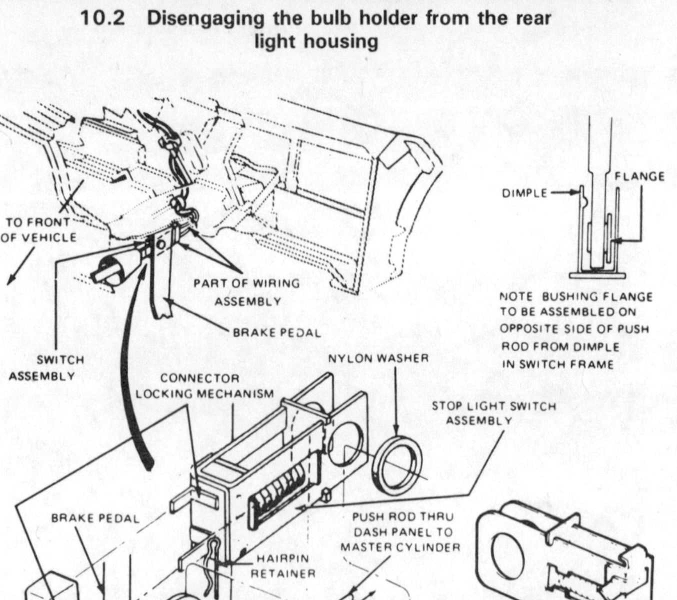

11.2 Stop light switch components – exploded view

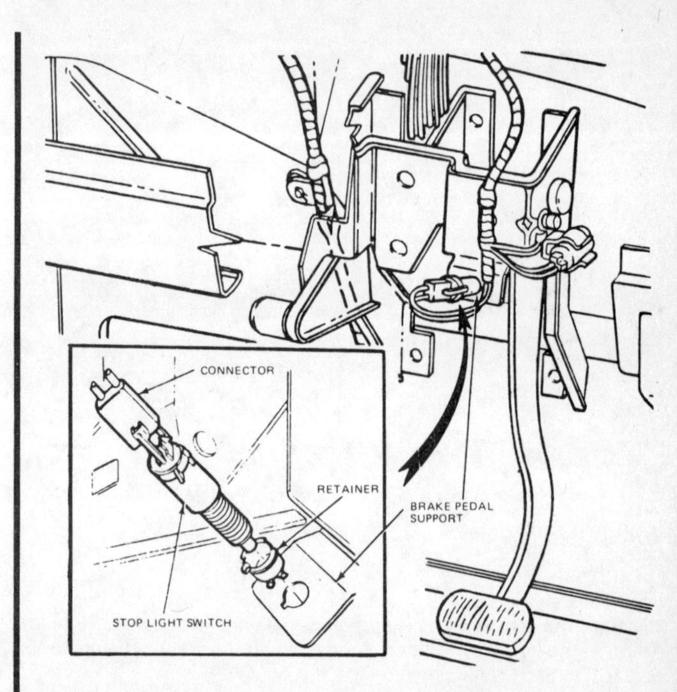

11.5 Stop light switch used on earlier vehicles

2 Twist the bulb holders to disengage them from the lens unit (see illustration).

3 Replace any burned out bulbs and install the bulb holders and lens unit.

4 The rear license plate light bulb is located inside the left rear door and can be replaced by simply pulling the bulb holder out of the light assembly (see illustration).

11 Stop light switch – removal and installation

Refer to illustrations 11.2 and 11.5

1 Disconnect the negative cable from the battery.

2 Disconnect the wires at the switch connector (see illustration).

3 Withdraw the hairpin clip and slide the stop light switch, the pushrod, nylon washers and bushings away from the brake pedal.

4 The switch can now be detached.

5 Installation is the reverse of removal. **Note:** *On earlier models the stop light switch is screwed into the brake pedal support bracket (see illustration).*

12 Turn signal switch – removal and installation

Refer to illustrations 12.4, 12.5, 12.6, 12.7 and 12.8

1 Disconnect the negative cable from the battery.

2 On earlier models, remove the horn button from the center of the steering wheel. On later models, remove the two screws from the back of the horn pad, detach the pad and disconnect the wires.

3 Mark the steering wheel hub and shaft so the steering wheel can be reinstalled in the same position on the shaft. Remove the steering wheel retaining nut and pull the wheel off the shaft with a puller. **Note:** *Do not hammer on the end of the shaft to remove the steering wheel.*

4 Remove the wiring harness cover from the steering column (see illustration).

5 Make a careful note of the color code and location of each wire before disconnecting the wires from the connector plug (see illustration).

6 Remove the turn signal switch lever by unscrewing it from the switch (see illustration).

7 Remove the screws securing the switch to the steering column (see illustration).

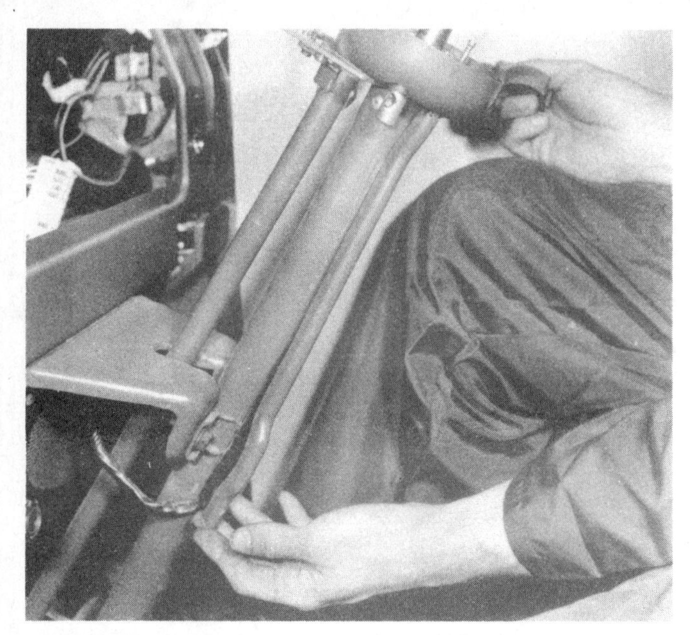

12.4 Removing the steering column wiring harness cover

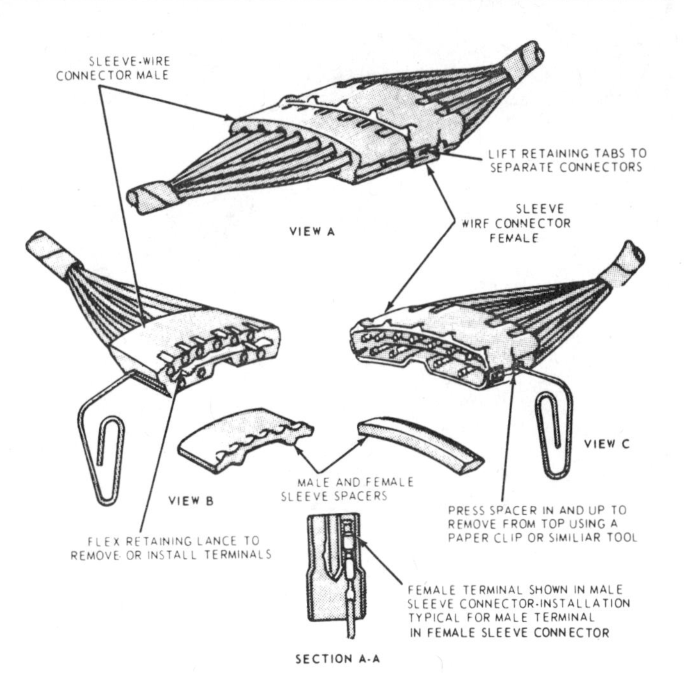

12.5 Steering column wiring harness connector details

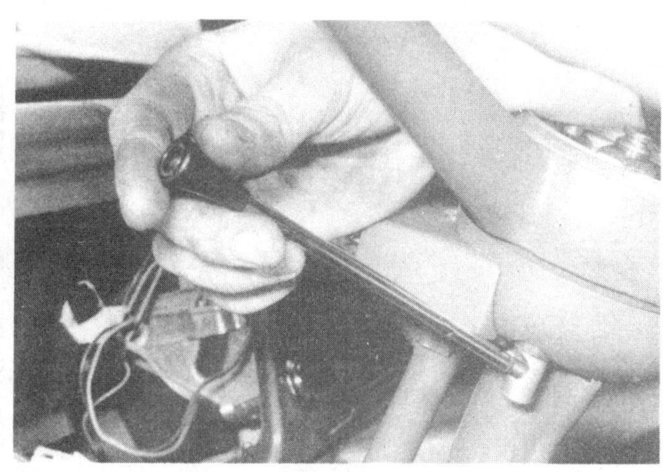

12.6 Unscrewing the turn signal lever

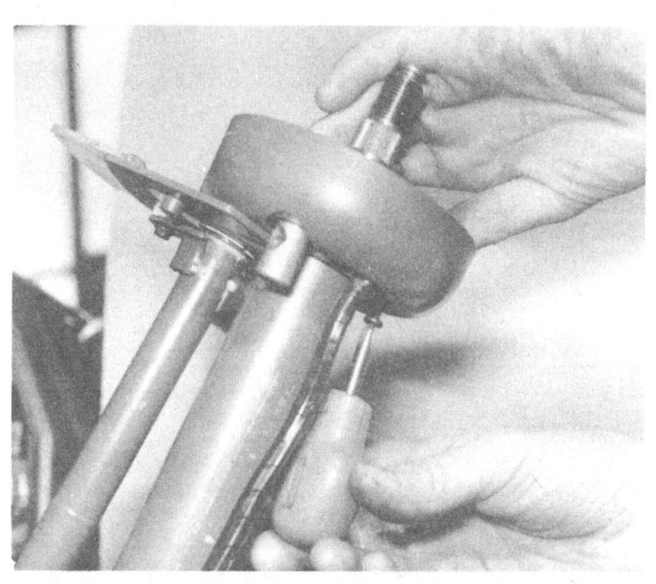

12.7 Removing the turn signal switch screws

8 Separate the turn signal and hazard warning switch from the steering column (see illustration).

9 The switch is not repairable and must be replaced with a new one if it is defective.

10 Installation is the reverse of removal. be sure to tighten the steering wheel nut to the specified torque.

13 Turn signal flasher unit - removal and installation

1 The turn signal flasher unit is located in various locations depending on model:

a) 1969 through 1974 models; the turn signal flasher unit is attached to the back of the instrument cluster in the upper left corner. The instrument cluster must be removed first (refer to section 16). Disconnect the wire connector from the terminals and remove it from the instrument cluster.

b) 1975 through 1990 models; the turn signal flasher unit is attached to the lower left instrument panel support. Disconnect the wire connector from the terminals and remove it from the mounting clip.

c) 1991 models; the turn signal flasher can be found on the front side of the fuse panel.

2 Make sure the new flasher unit has the same color code and part number as the original.

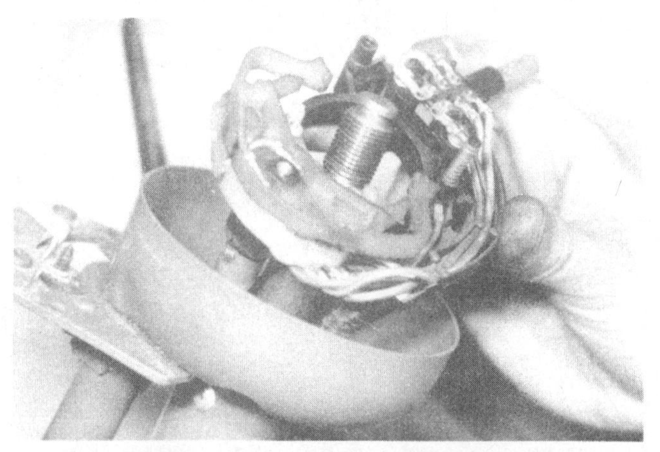

12.8 Withdrawing the turn signal switch assembly from the column

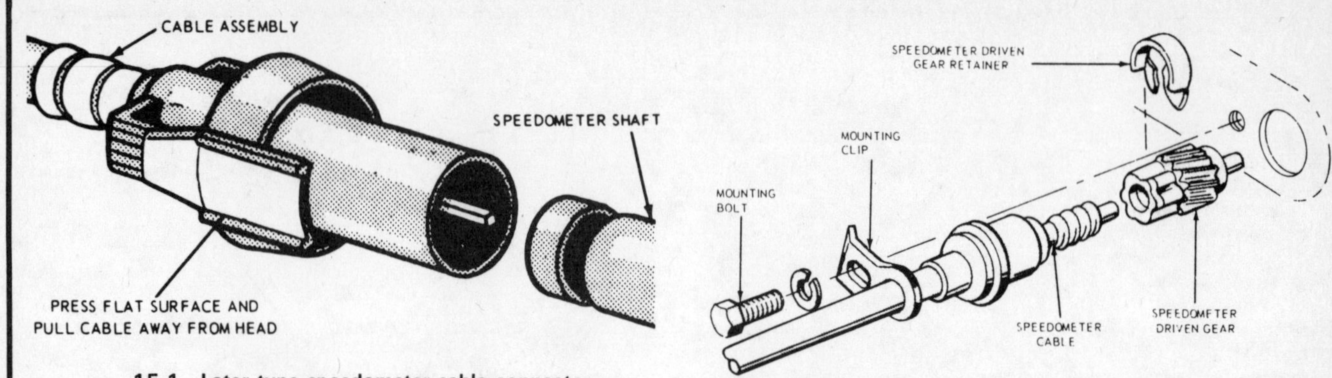

15.1 Later type speedometer cable connector

15.3 Speedometer cable-to-transmission mounting details

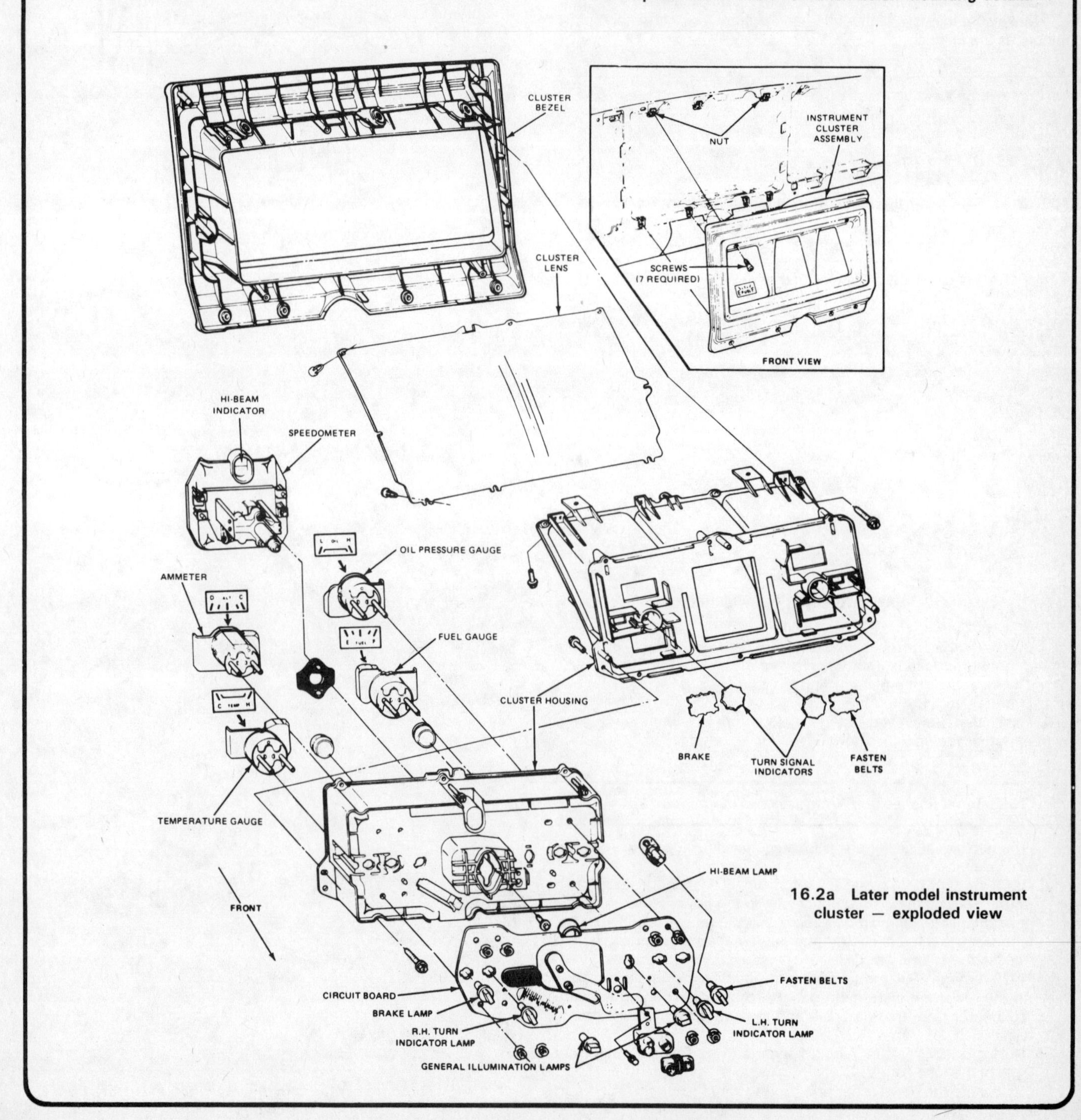

16.2a Later model instrument cluster – exploded view

14 Hazard warning light flasher unit - removal and installation

1 The hazard warning light flasher unit is located in various locations depending on model:

a) 1969 through 1974 models; the hazard warning light flasher is mounted to the right side of the brake pedal support. Disconnect the wire connector from the terminals and remove it from the mounting clip.
b) 1975 through 1990 models; the hazard warning light flasher is taped to the wiring harness under the left side of the instrument panel, disconnect the wire connector from the terminals and remove flasher unit.
c) 1991 models; the hazard warning light flasher is located on the back of the fuse panel, behind the turn signal flasher.

2 Make sure the new flasher unit has the same color code and part number as the original.

15 Speedometer cable — removal and installation

Refer to illustration 15.1 and 15.3

Inner cable

1 Working behind the instrument panel, disconnect the cable from the rear of the speedometer head (see illustration).
2 Carefully pull the inner cable out from the upper end of the outer cable.
3 If the inner cable is broken, raise the vehicle and, working underneath, remove the bolt securing the speedometer cable mounting clip to the transmission (see illustration).
4 Remove the speedometer cable shaft and driven gear from the transmission.
5 Remove the driven gear retainer and the driven gear and shaft from the cable.
6 Remove the lower part of the broken inner cable from the end of the outer cable.
7 Installation of the inner cable is the reverse of removal.
8 Lightly lubricate the inner cable and insert it into the outer cable. When the cable is nearly all the way in, turn it to ensure that the squared end engages with the speedometer driven gear.

Outer cable

9 Working behind the instrument panel, disconnect the cable from the rear of the speedometer head.
10 Push the outer cable and grommet through the opening in the dashboard panel.
11 Raise the vehicle and, working underneath, detach the cable from all of the retaining clips.
12 Disconnect the cable from the transmission as described above and withdraw it from under the vehicle.
13 Installation is the reverse of removal.

16 Instrument panel cluster — removal and installation

Refer to illustrations 16.2a, 16.2b, 16.5a, 16.5b, 16.5c, 16.7a, 16.7b, 16.9 and 16.10

1 Disconnect the negative cable from the battery.
2 Remove the screws securing the instrument cluster assembly to the dash panel (see illustrations).
3 Withdraw the panel just enough to enable the speedometer cable to be disconnected from the rear of the panel.
4 Disconnect the wiring harness multi-plug connector and detach the instrument panel cluster.
5 Unsnap the voltage regulator flasher unit and radio connector from the rear of the cover (see illustrations).

16.2b Removing the instrument cluster retaining screws

16.5a After the instrument cluster is pulled out of the dash, the flasher connector can be detached

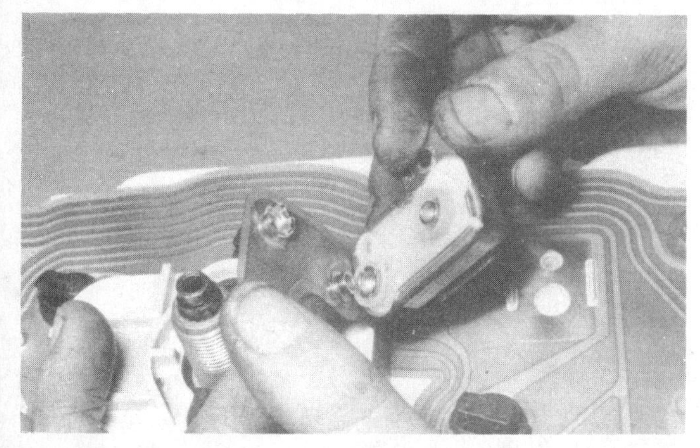

16.5b Disconnecting the voltage regulator from the rear of the instrument cluster

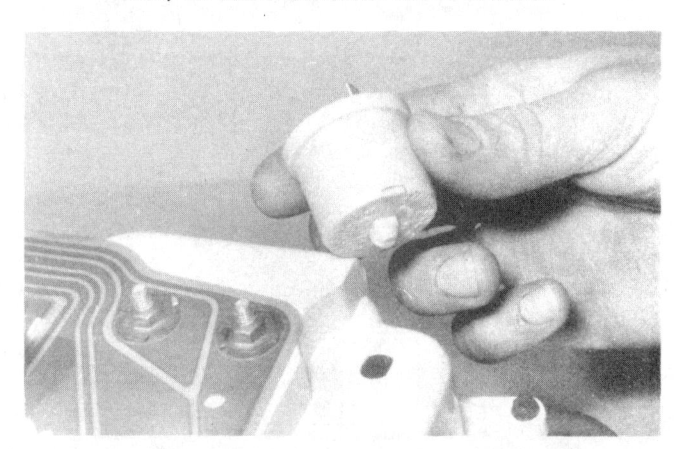

16.5c Removing the flasher unit from the rear of the instrument cluster

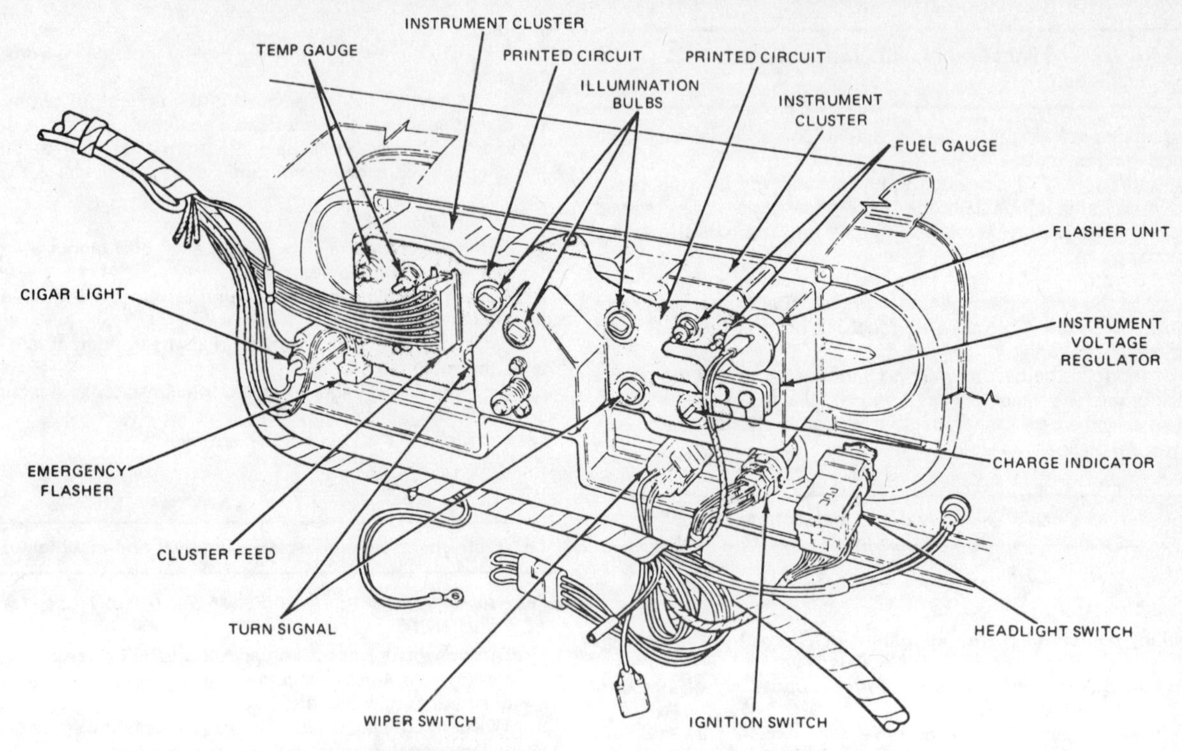

16.7a Instrument cluster components (early model)

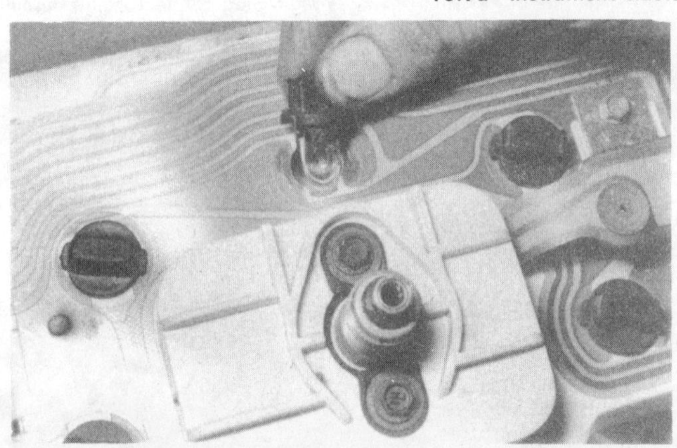

16.7b Removing an instrument cluster bulb

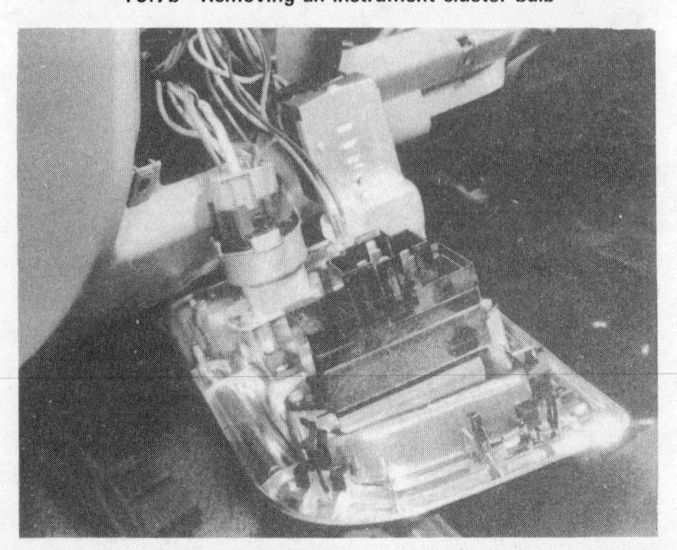

16.10 Removing the plug connectors from the rear of the switch panel (later models)

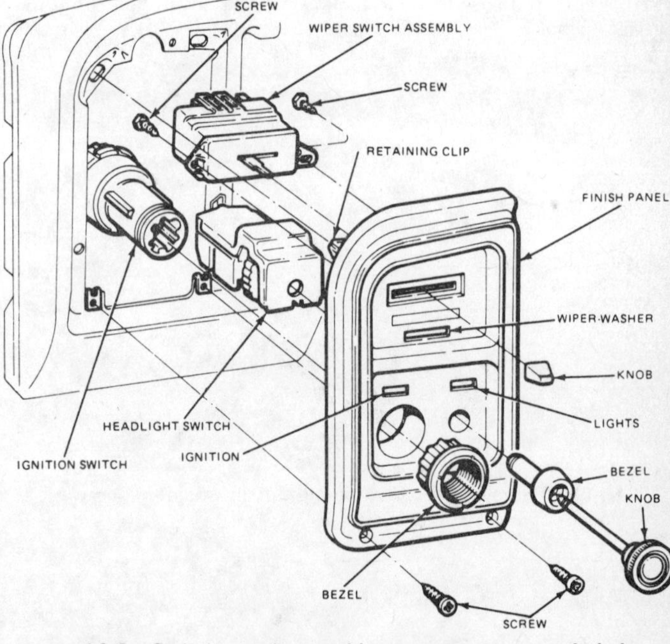

16.9 Switch panel assembly components – exploded view (later models)

6 Remove the nuts retaining the gauges to the circuit board and remove the gauges.

7 Remove all the indicator and light bulb holders from the rear of the housing by turning them counterclockwise (see illustrations).

8 Installation of the printed circuit and instrument cluster is the reverse of removal.

9 The switch panel, located on the left side of the instrument panel cluster, can be withdrawn after removing the mounting screws (see illustration).

10 To remove the panel completely, disconnect the wiring connectors from the rear of the panel switches (see illustration) and remove the panel from the vehicle.

11 Installation is the reverse of removal.

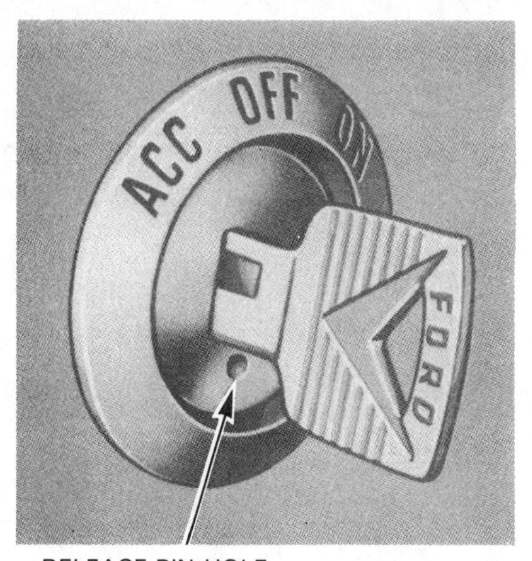

17.1 Ignition lock release hole (early models)

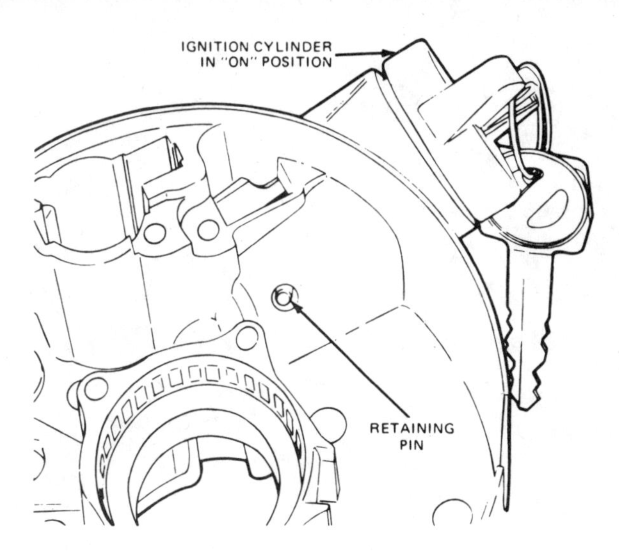

17.11a Location of the lock cylinder retaining pin on non-tilt steering column

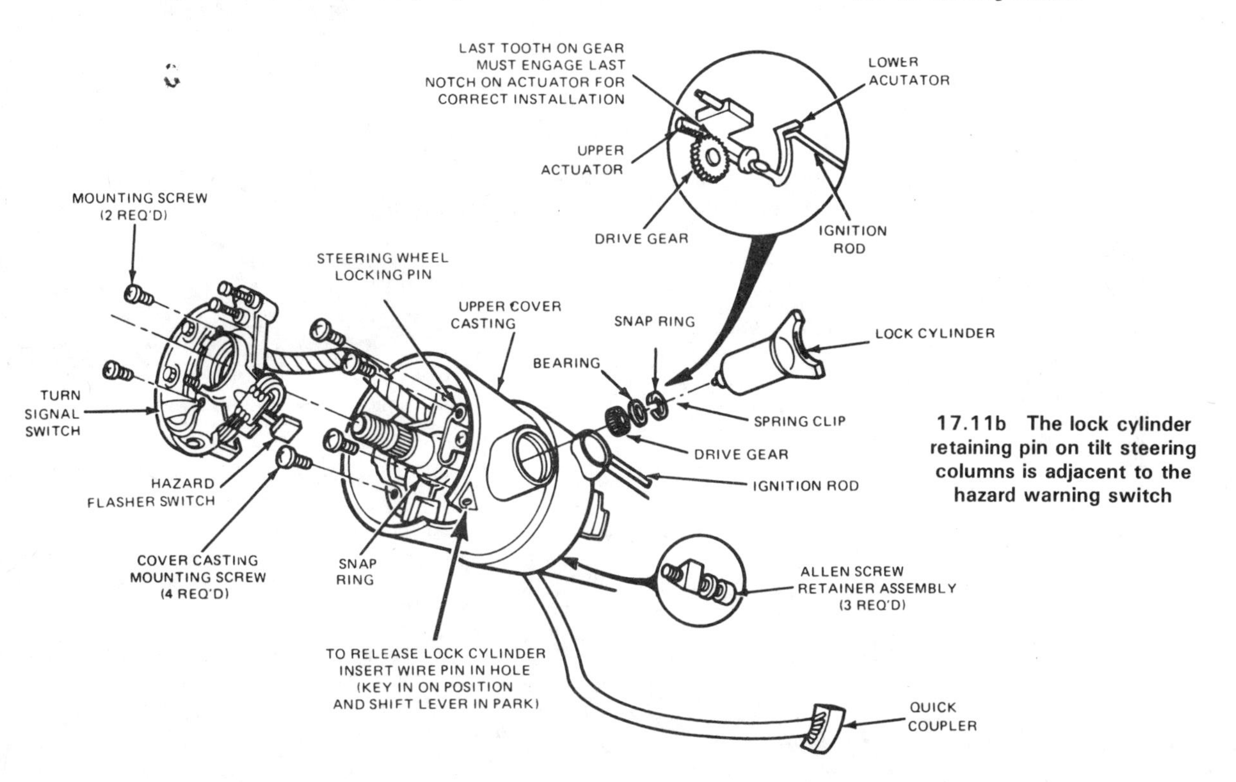

17.11b The lock cylinder retaining pin on tilt steering columns is adjacent to the hazard warning switch

17 Ignition switch – removal and installation

Refer to illustrations 17.1, 17.11a and 17.11b

Thru 1977

1 To remove the lock barrel of the ignition switch, insert the ignition key into the switch and insert a piece of wire into the hole in the front of the barrel (see illustration).

2 Push the wire in and turn the key counterclockwise, past the Accessory position, grip the key and withdraw the lock barrel from the switch assembly.

3 The lock barrel is not repairable and must be replaced with a new one if it is defective.

4 To install the lock barrel, insert the key and turn it to the Accessory position.

5 Push the barrel and key into the switch assembly until it is completely seated and then turn the key to the Lock position. Remove the wire and turn the key to check the operation of the lock and switch.

6 To remove the complete switch assembly from the instrument panel on earlier models, remove the bezel nut retaining the switch to the panel, withdraw the switch from the rear of the panel and disconnect the wiring harness from the rear of the switch.

7 On later models the switch can be removed after withdrawing the switch panel as described in the previous Section.

1978 and later

8 On later models the switch is attached to the steering column. Disconnect the negative cable from the battery.

9 Remove the steering wheel as described in Section 12.

10 Insert the key into the switch and turn it to the On position. On automatic transmission models, place the gear selector lever in Park.

11 Depress the lock cylinder retaining pin with a small punch or a 1/8-inch drill bit and pull out the lock cylinder. On non-tilt steering columns, the pin is near the base of the lock cylinder (see illustration). On tilt steering columns, the retaining pin is adjacent to the hazard warning switch (see illustration).

19.1 Fuse panel location (early model shown)

12 To install the switch, turn the cylinder to the On position and depress the retaining pin, then insert the cylinder into the housing in the flange casting.
13 Make sure that it is completely seated and aligned in the interlocking washer before turning the key to the Off position, which permits the retaining pin to extend into the cylinder casting hole.
14 Turn the key to all positions to make sure it works correctly.
15 The remaining installation steps are the reverse of removal. When the installation is complete, check to see that the engine starts in Park and Neutral and doesn't start in Drive or Reverse.

18 Seat belt/starter interlock system — general information

1 This system is installed on later models and is designed to prevent operation of the vehicle unless the front seat belts have been fastened.
2 If either of the front seats is occupied and the seat belts have not been fastened, then as the ignition key is turned to the II (ignition on) position, a warning light will flash and a buzzer will sound.
3 If the warning is ignored, the starter will not operate even when the key is turned to the Start position.
4 If the system fails to work, first check the fuse and then make sure the wires and connections are in good condition.

19 Fuses — general information

Refer to illustration 19.1

1 The fuse panel on early models is located on the right side of the steering column on a bracket attached to the pedal support (see illustration). On later models, the fuse box is on the left side of the steering column, under the instrument panel.
2 Most of the electrical circuits are protected by fuses. If a fuse blows, always try to find the cause and correct it before installing another fuse. **Caution:** *Never bypass a fuse with a piece of wire or other metal objects. Serious damage to the circuit could occur.*

20 Windshield wiper blades — removal and installation

Refer to illustration 20.5

The windshield wiper blades may be one of two types. With the bayonet type, the blade saddle slides over the end of the arm and is engaged by a locking stud. With the side saddle pin type, a pin on the arm indexes into the side of the blade saddle and engages a spring loaded clip in the saddle.

Bayonet type — Trico

1 To remove a Trico blade, press down on the arm to disengage the top stud.

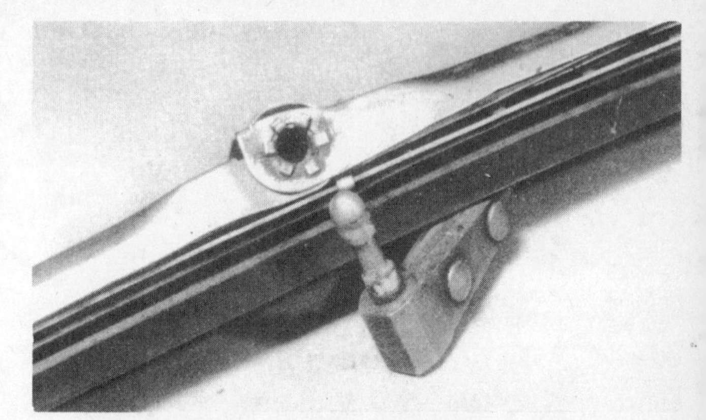

20.5 Detaching the wiper blade (later type shown)

21.3 Replacing the wiper arm (later type shown)

2 Depress the tab on the saddle to release the top stud and pull the blade from the arm.

Bayonet type — Anco

3 To remove an Anco blade, press in on the tab and pull the blade from the arm.

Saddle pin type — Trico

4 To remove a pin type Trico blade, insert a screwdriver into the spring release opening of the blade saddle, depress the spring clip and pull the blade from the arm.

Spring clip type

5 To remove this later type blade, pry the C-clip from the wiper arm pin and remove the blade (see illustration).

21 Windshield wiper arm — removal and installation

Refer to illustration 21.3

1 Before removing a wiper arm, turn the windshield wiper switch on and off to make sure the arms are in the normal parked position, parallel to the bottom of the windshield.
2 To remove the arm, swing it away from the windshield, depress the spring clips in the wiper arm boss and pull the arm off the spindle.
3 When installing the arm, position it in the parked position and push the boss onto the spindle (see illustration).

22 Windshield wiper mechanism — troubleshooting and repair

1 If the windshield wipers fail to operate, or work very slowly, check the terminals on the motor for loose connections and make sure that

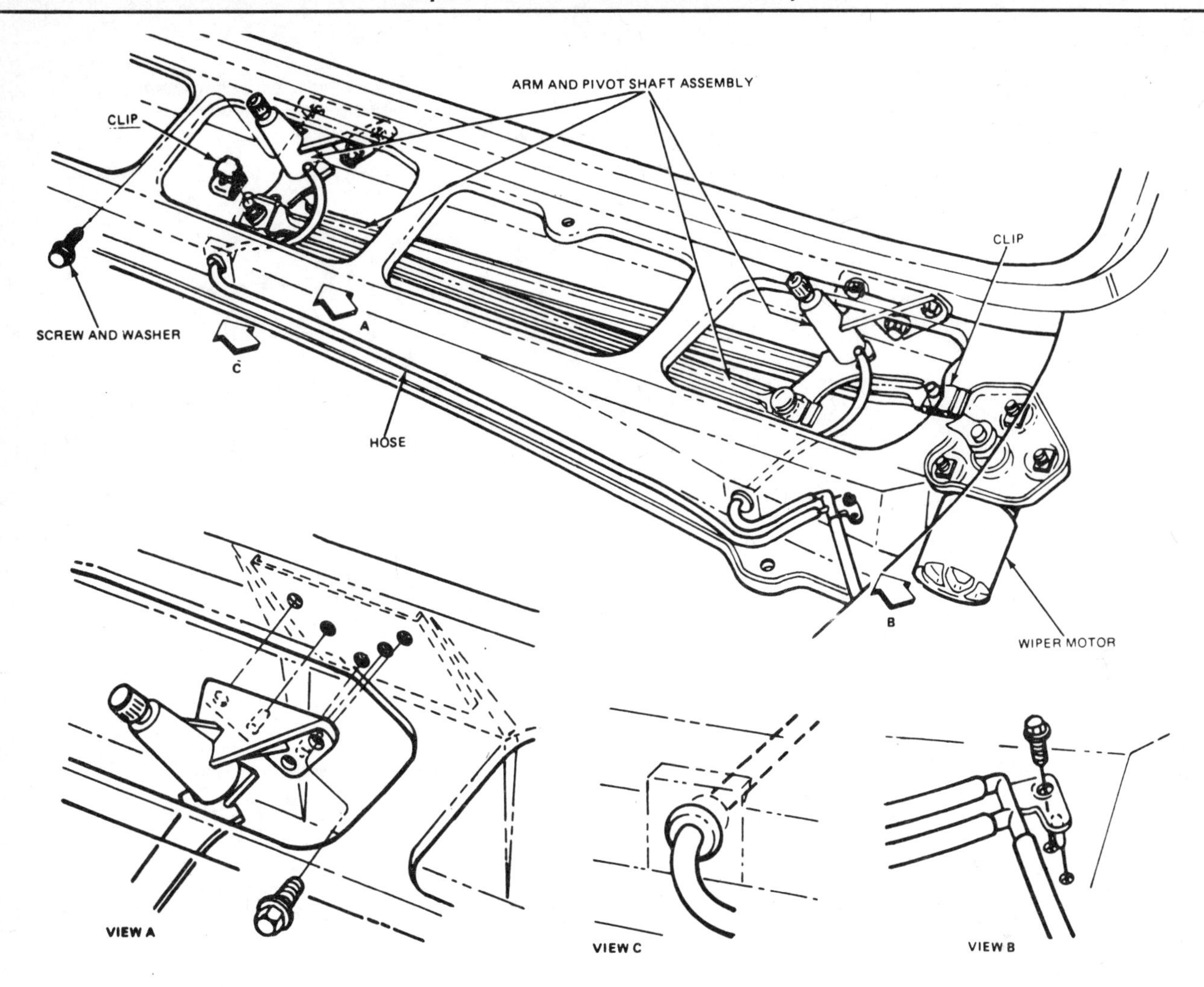

24.2 Windshield wiper motor and linkage

the insulation on the wires is not cracked or broken, causing a short circuit. If the wires are in good condition, check the current the motor is drawing by connecting an ammeter in series in the circuit and turning on the wiper switch. Current draw should be from 2.3 to 3.1 amps.
2 If no current is passing through the motor, make sure that the switch is operating correctly.
3 If the wiper motor draws excessive current, check the wiper blades for freedom of movement. If they are free, check the gearbox cover and gear assembly for damage.
4 If the motor draws very little current, make sure that the battery is fully charged. Check the brush gear and see if the brushes are bearing on the commutator. If not, check the brushes for freedom of movement and, if necessary, replace the tension springs. If the brushes are very worn they should be replaced.

23 Windshield wiper motor — removal and installation

1 Disconnect the negative cable from the battery.
2 Disconnect the wiper motor feed wires at the main harness connector.
3 Remove the clip securing the motor drive arm to the linkage arm and disconnect the motor from the linkage.
4 Remove the bracket mounting bolts and lift out the motor and bracket assembly. If the motor is to be dismantled, remove the bracket.
5 Installation is the reverse of removal. Note that the drive arm must be attached to the motor in the parked position — in the opposite direction to the ground strap.

24 Windshield wiper pivot shafts and link assembly — removal and installation

Refer to illustration 24.2

1 Remove the wiper arms and blades as described in Section 21.
2 Working from under the instrument panel inside the vehicle, remove the six screws retaining the pivot shaft assemblies (see illustration).
3 Remove the clip securing the linkage arm to the motor and withdraw the linkage and pivot shafts from the vehicle.
4 Installation is the reverse of removal. Lubricate all moving parts with light oil.
5 On later models, make sure that the windshield washer lines are correctly installed on the pivot shafts.

25 Windshield wiper switch — removal and installation

1 On earlier Econoline models, remove the small retaining screw on the side of the wiper control knob and pull off the knob.
2 Remove the bezel nut securing the switch to the instrument panel, withdraw the switch from the rear of the panel and disconnect the wires.
3 On later models, remove the switch panel from the main instrument panel as described in Section 16.
4 Withdraw the panel far enough to remove the two wiper switch retaining screws, pull off the control knob and remove the switch.
5 Installation of either type of switch is the reverse of removal.

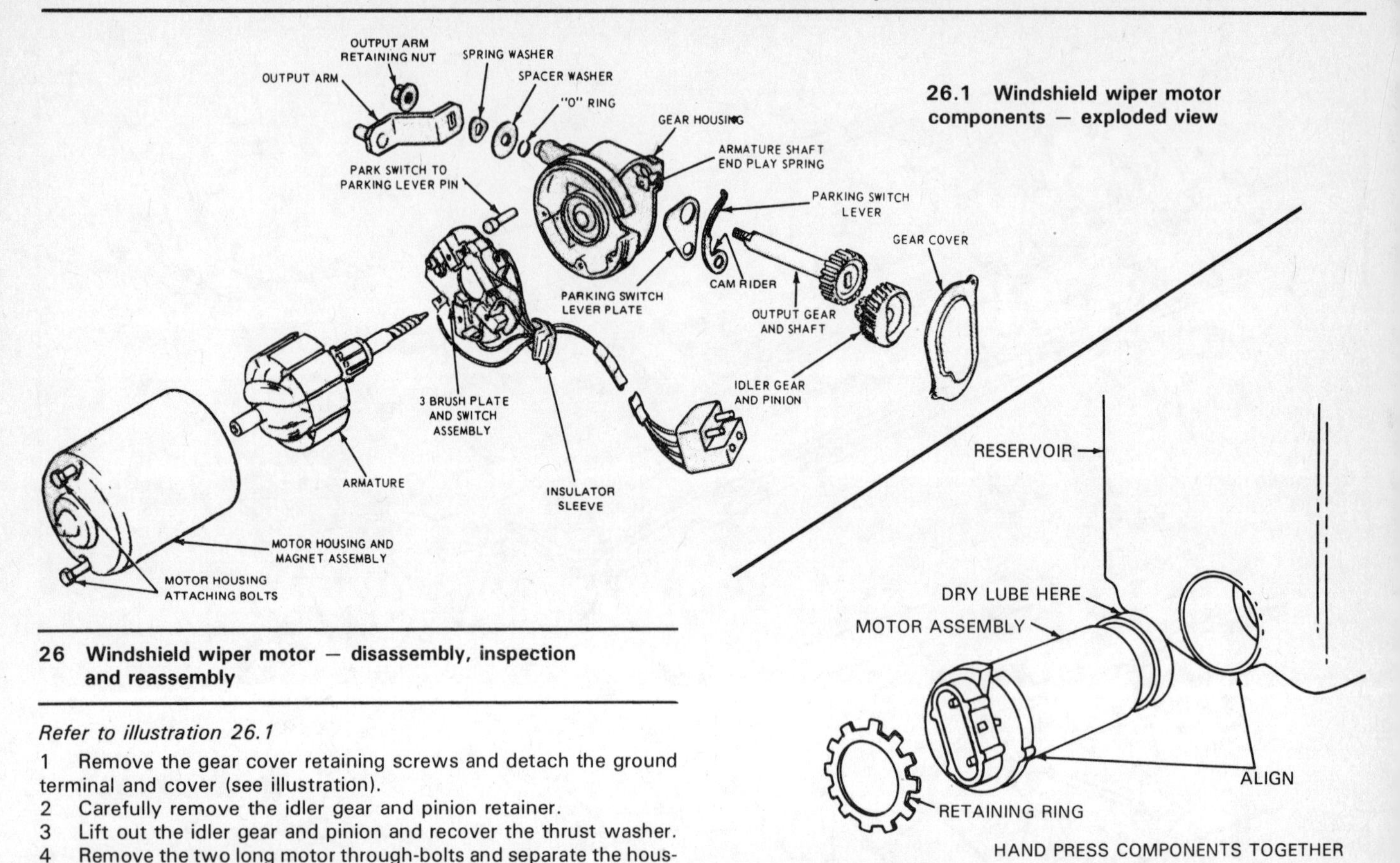

26.1 Windshield wiper motor components – exploded view

28.3 Windshield washer reservoir motor installation details

26 Windshield wiper motor – disassembly, inspection and reassembly

Refer to illustration 26.1

1 Remove the gear cover retaining screws and detach the ground terminal and cover (see illustration).
2 Carefully remove the idler gear and pinion retainer.
3 Lift out the idler gear and pinion and recover the thrust washer.
4 Remove the two long motor through-bolts and separate the housing, switch terminal insulator sleeve and armature.
5 Mark the position of the output arm relative to the shaft to ensure correct reassembly.
6 Remove the output arm retaining nut, output arm, spring washer, flat washer, output gearshaft assembly, thrust washer and parking switch lever and washer in that order.
7 Remove the brushes and brush springs.
8 Remove the brushplate and switch assembly and withdraw the switch contact-to-parking lever pin from the gear housing.
9 Thoroughly clean all parts and then inspect the gear housing for cracks, distortion and damage.
10 Carefully check all shafts, brushes and gears for signs of scoring and damage.
11 If the brushes are worn, replace them with new ones.
12 Any serious problem with the armature, such as a breakdown in insulation, means a new motor is required.
13 Reassembly is the reverse of disassembly.

27 Windshield washer nozzles – adjustment

To adjust the washer nozzles, carefully bend the nozzle in the required direction with a needle nose pliers. Do not squeeze the nozzle too hard or it may be crimped closed.

28 Windshield washer reservoir and pump – removal and installation

Refer to illustration 28.3

1 Remove the wiring connector plug and washer hose.
2 Remove the retaining screws and detach the washer and motor assembly from the left side fender apron.
3 To remove the pump motor from the reservoir, pry out the retaining ring and carefully pull the motor out of the reservoir recess (see illustration).
4 The motor and pump assembly cannot be repaired. If they are defective they must be replaced with new ones.
5 When installing the motor in the reservoir, make sure the projection on the motor body is lined up with the slot in the reservoir.
6 Press on the motor retaining ring and install the assembly by reversing the removal procedure.

29 Horn – troubleshooting and repair

1 If the horn sounds bad or fails completely, check the wire leading to the horn plug located on the body panel next to the horn itself. Also, make sure that the wire connection at the horn terminal is clean and secure.
2 The horn must be securely mounted and nothing should be contacting the horn body.
3 If the fault is not an external one, remove the horn cover and check the leads inside the horn. If they are sound, check the contacts. If they are burned or dirty, clean them with a fine file and douse them with contact cleaner.

30 Turn signal circuit – troubleshooting and repair

1 If the flasher unit fails to operate, or flashes very slowly or rapidly, check the turn signal circuit before assuming that the flasher is defective.
2 Examine the turn signal light bulbs, both front and rear, for broken filaments.
3 If the external flashers are working, but either of the internal flasher warning lights have ceased to function, check the filaments in the warning light bulbs.
4 If a flasher bulb is sound but does not work, check all the flasher circuit connections.
5 With the ignition switch on, make sure the correct voltage is reaching the flasher unit by connecting a voltmeter between the positive terminal and ground. If voltage is present at the unit, connect the two flasher unit terminals together and operate the turn signal switch. If one of the flasher warning lights comes on, the flasher unit must be replaced with a new one.

Component	Location
Air conditioner	
Blower motor	H–34
Blower motor & clutch control switch	E–34
Clutch solenoid	J–35
High blower relay	G–36
Blower motor resistor	G–34
Thermostat switch	E–35
Alternator	C–11
Alternator indicator	B–10
Regulator	G–11
Ambient sensor switch	D–14
Ammeter	D–13
Back-up lamps	K–27
Back–up lamp switch	F–27
Battery	C–2
Carburetor solenoid	J–17
Cargo lamp	J–54
Cigar lighter illumination lamp	E–42
Cigar lighter	E–53
Cluster illumination lamps	F–43
Constant voltage regulator	D–21
Distributor	J–2
Dome lamp	H–54
Door switches	
LH front	G–53 & C–57
RH front	G–53 & H–57
RH front (cargo)	F–57 & F–53
RH rear	53 & E–57
Dual brake warning lamp	H–18
Dual brake warning switch	J–18
ESC solenoid vacuum valve	K–13
Electronic dist. modulator	G–13
Emergency warning flasher	D–53
Emergency warning switch	C–51
Engine temperature switch	J–20
Fuel gauge	E–21
Fuel sender	J–21
Headlamps	
Dimmer switch	E–40
Hi-beam indicator lamp	F–40
Switch	C–41
LH	J–41
RH	J–42
Heater	
Auxiliary	J–31
Auxiliary switch	E–31
Blower motor	D–30
Blower motor resistor	E–30
Thermostat switch	D–32
Switch	H–30
Horn	
Standard	J–44
Switch	J–44
RPO	J–45
Relay	G–44
Ignition coil	F–3
Ignition switch	C–20
License lamp	J–51
Marker lamps	
LH front	J–47
RH front	J–48
LH rear	J–50
RH rear	J–53

Component	Location
Neutral start switch	C–16
Netural start & back-up lamp switch (auto trans)	F–28
Oil pressure	
Gauge	E–22
Sender (gauges)	J–22
Switch (lights)	J–23
Warning lamps	E–23
Park & turn signal lamps	
LH	J–46
RH	J–49
Radio receiver	F–32
Radio speaker	H–32
School bus warning lamps	
LH front	K–56
RH front	K–57
LH rear	K–59
RH rear	K–58
Speedo sensor	J–11
Starter motor relay	C–5
Starter motor	H–5
Stop lamp switch	D–48
Stop & turn signal lamps	
LH	J–51
RH	J–52
Suppression capacitor	G–9 & G–12
TRS solenoid vacuum valve	K–15
Temperature gauge	E–20
Transmission sensor switch	K–14
Turn signal flasher	S–53 & C–27
Turn signal indicator lamps	
LH	F–50
RH	F–50
Turn signal switch	C–50
Warning lamps	
Flasher	C–59
Relay	A–59
Switch	A–57
Windshield	
Washer motor pump	H–25
Wiper motor	J–24
Wiper switch	D–24
Washer switch	E–25

Wiring color key
Primary colors

Black	BK
Brown	BR
Red	R
Pink	PK
Orange	O
Yellow	Y
Green	G
Blue	B
Gray	GY
White	W
Violet	V

Wiring diagram key — pre-1974 models only

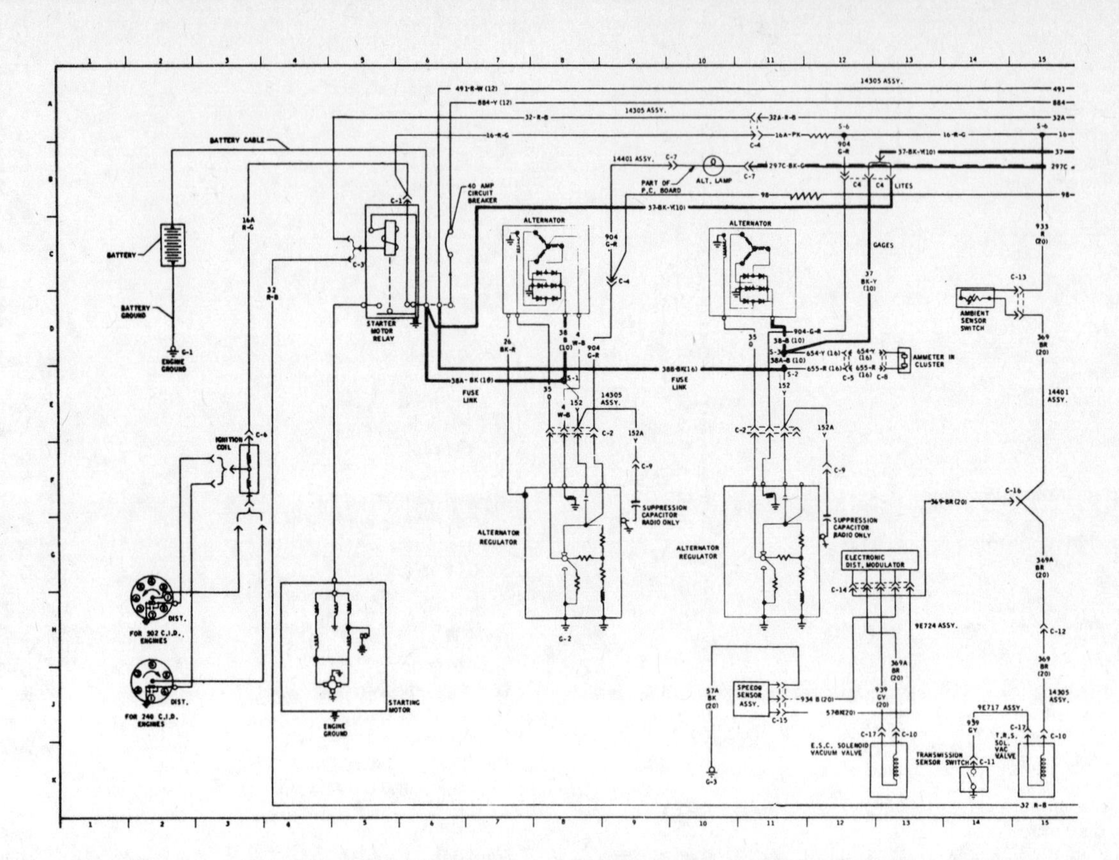

Wiring diagram – engine and related components (pre-1974 models)

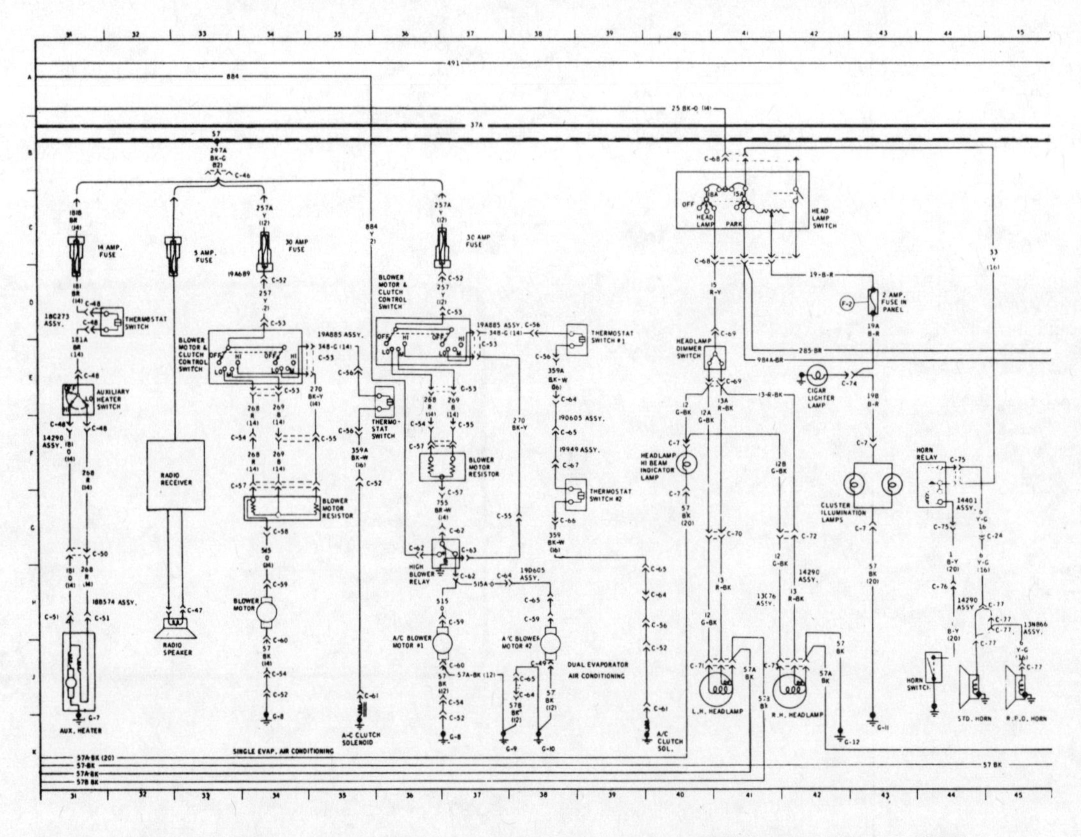

Wiring diagram – radio, heater, air conditioning system and lights (pre-1974 models)

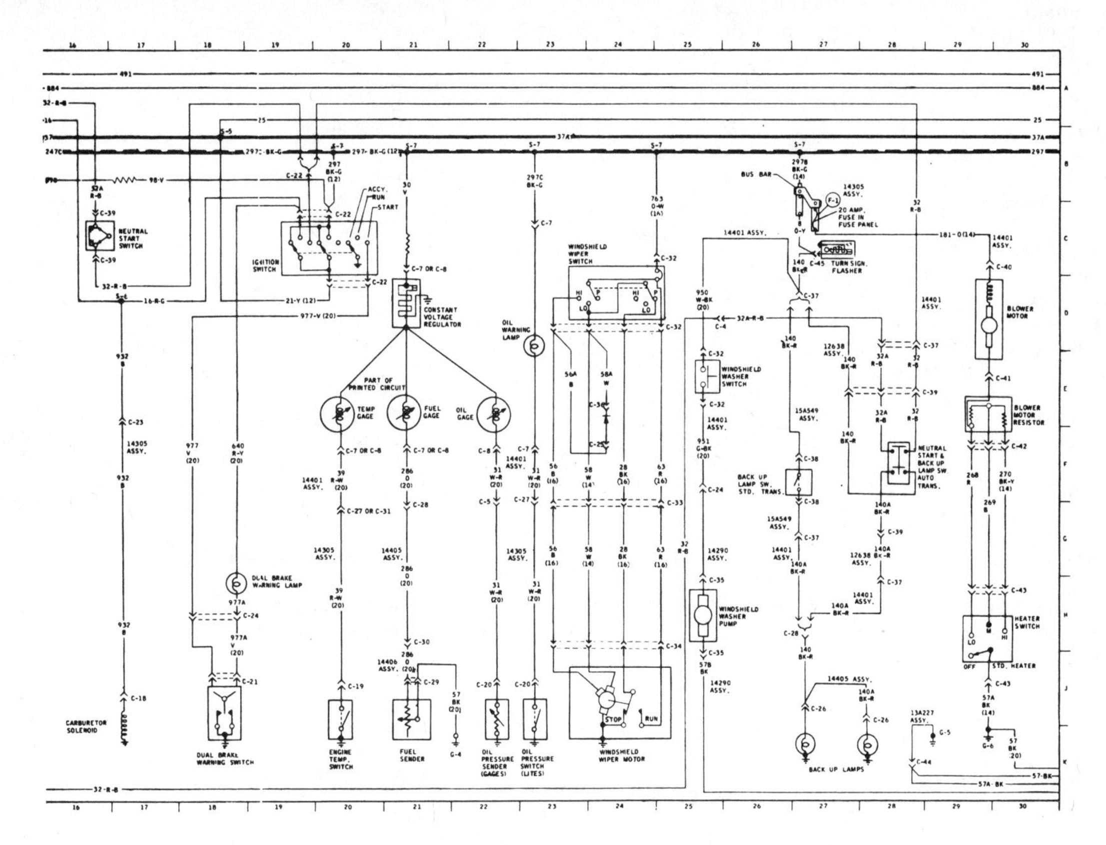

Wiring diagram — instrument panel and dashboard (pre-1974 models)

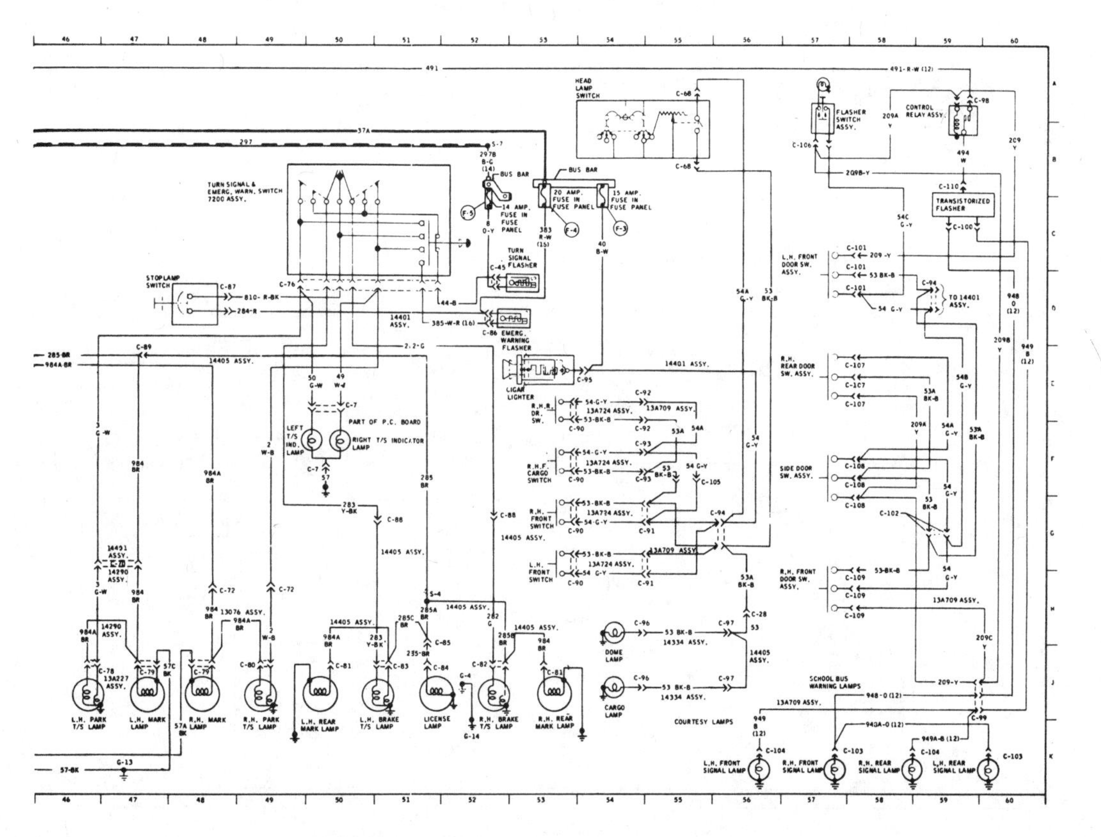

Wiring diagram — lights (pre-1974 models)

Component	Location
A/C clutch solenoid	F–51, E–53
Alternator	A–1, A–3
Regulator	F–2
Ammeter	B–8
Battery	E–5
Blower motor resistor	C–41, E–50, D–52, E–56
Cigar lighter	B–57
Constant voltage unit	C–46
Distributor (6 cylinder)	E–10
Distributor (8 cylinder)	F–10
Distributor breakerless	B, C–14, 15
Electric choke	D–4
Emergency warning flasher	F–21
Emission control solenoid	E–16
Engine temperature gauge	D–46
Engine temperature sender	F–46
Fuel gauge	D–47
Fuel gauge sender	F–47
Horn	E–39, F–39
Ignition coil	C–13
Illuminations	
Auxiliary fan	D–38
Cigar lighter	D–35
Cluster	E–36
Heater & A/C	D–36, D–38
Windshield wiper	D–37
Immersion heater	F–40
Indicators	
Alternator	B–19
Dual brake warning	D–18
Hi-beam	E–31
Oil warning	C–48
Turn signal	E–25, E–26

Component	Location
Lamps	
Back-up	D–23, F–23
Cargo	C–60
Dome	C–59
Headlamps	E–30, E–31
License	F–32
Marker	
LH front	E–28
RH front	E–29
LH rear	E–33
RH rear	E–35
Park and turn	
LH	E–27
RH	E–29
Roof	
LH front	F–63
RH	F–61
LH rear	F–64
RH	F–62
Stop and turn	
LH	F–33
RH	F–34
Modular assy breakerless ign.	F–14
Motors	
A/C Blower	F–49, E–52, F–56
Heater blower aux.	F–43
Heater blower	C–41
Starter	F–6
Windshield washer pump	F–45
Windshield wiper	E–44
Noise suppression capacitor	F–4
Oil pressure gauge	D–48
Oil pressure sender	F–47

Wiring diagram key – 1975 thru 1977 models only

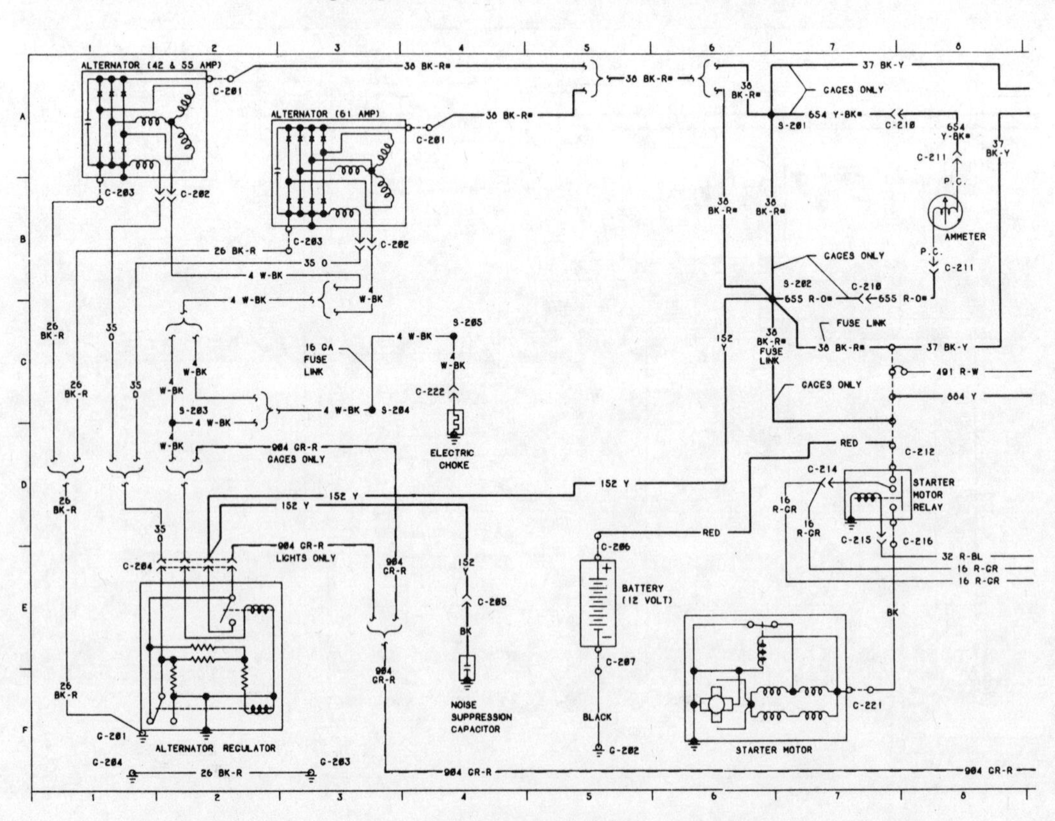

Wiring diagram – charging, ignition and starting systems (1975 thru 1977 models)

Component	Location
Radio	C–49
Radio suppression choke	C–46
Radio speaker	D–49
Relays	
A/C Hi-blower	C–54
Horn	B–39
Roof lamp	B–63
Starting motor	D–7
Switches	
A/C blower	C–51, B–53, B–56
Ambient sensor	C–43
Back-up lamp	D–22
Door jamb	
Front LH	E–59, B–60
Front RH	E–57, F–60
Side	D–57, E–60
Rear	C–57, C–60
Dual brake warning	F–18
Headlamp	B–33
Headlamp dimmer	B–30
Heater blower aux.	D–43
Heater blower	E–41
Horn	D–40
Ignition	D–19
Marker lamp flasher	A–60
Oil pressure	F–48
Start/interlock & back-up	C–21
Stoplamp	F–20
Temperature sensor	D–51, D–53
Thermostat	C–43
Turn and emergency signal	B–26
Windshield wiper/washer	B–44
Transistorized flasher	C–64
Turn signal flasher	E–26
Vacuum valve solenoid	D–17

Wiring color key primary colors	
Black	BK
Brown	BR
Tan	T
Red	R
Pink	PK
Orange	O
Yellow	Y
Dark green	DG
Light green	LG
Dark blue	DB
Light blue	LB
Purple	P
Gray	GY
White	W
Hash	(H)
Dot	(D)

Wiring diagram key — 1975 thru 1977 models only (continued)

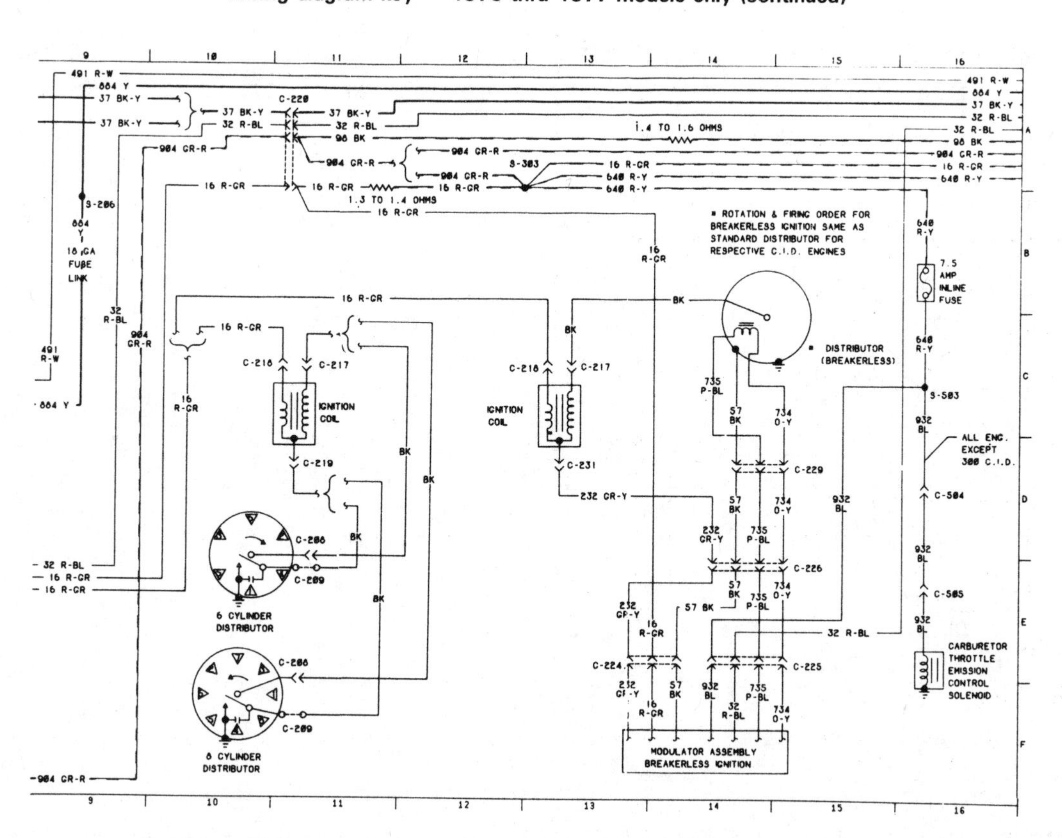

Wiring diagram — breakerless ignition system (1975 thru 1977 models)

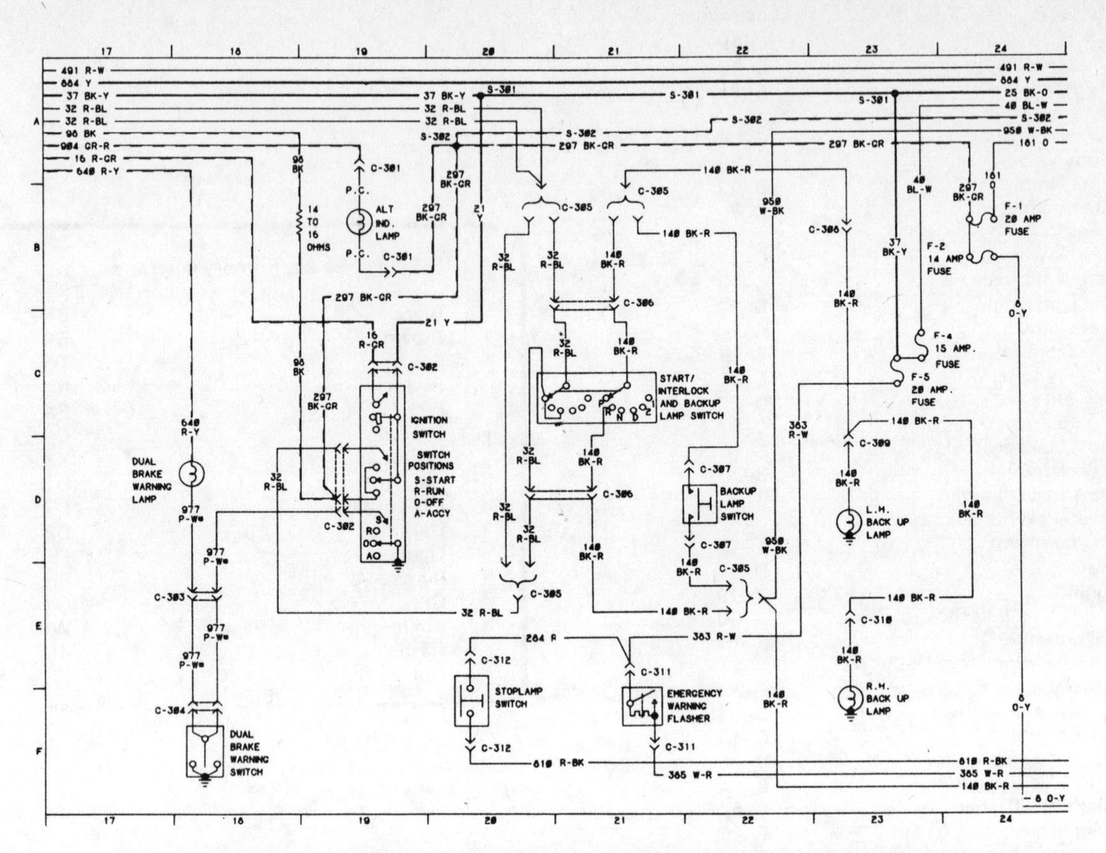

Wiring diagram – ignition switch, switches and lights (1975 thru 1977 models)

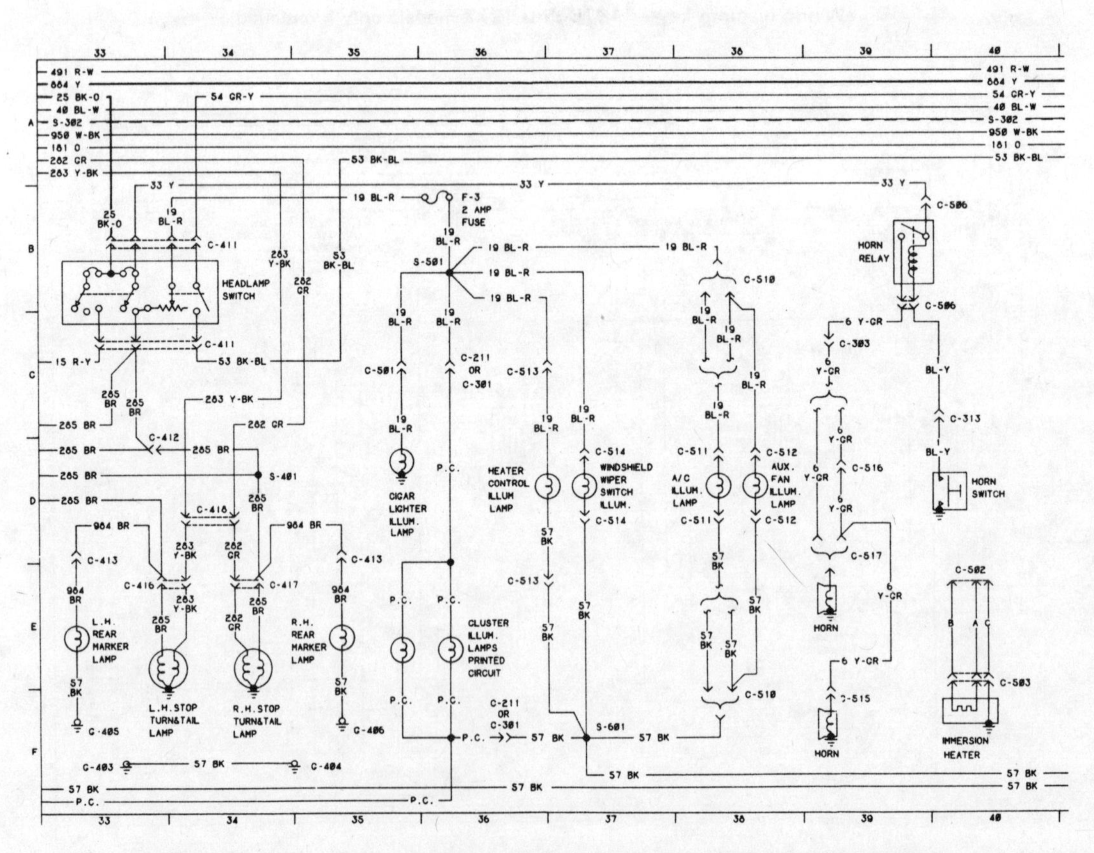

Wiring diagram – instrument panel switches and lights (1975 thru 1977 models)

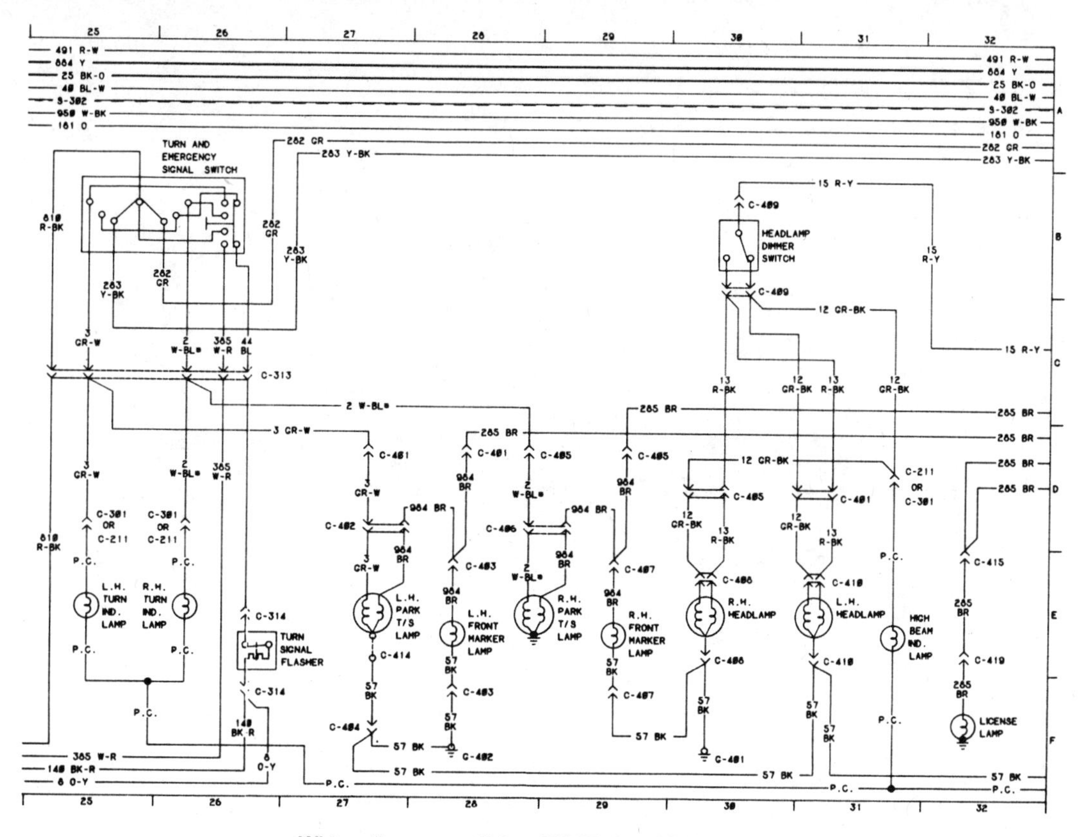

Wiring diagram — lights (1975 thru 1977 models)

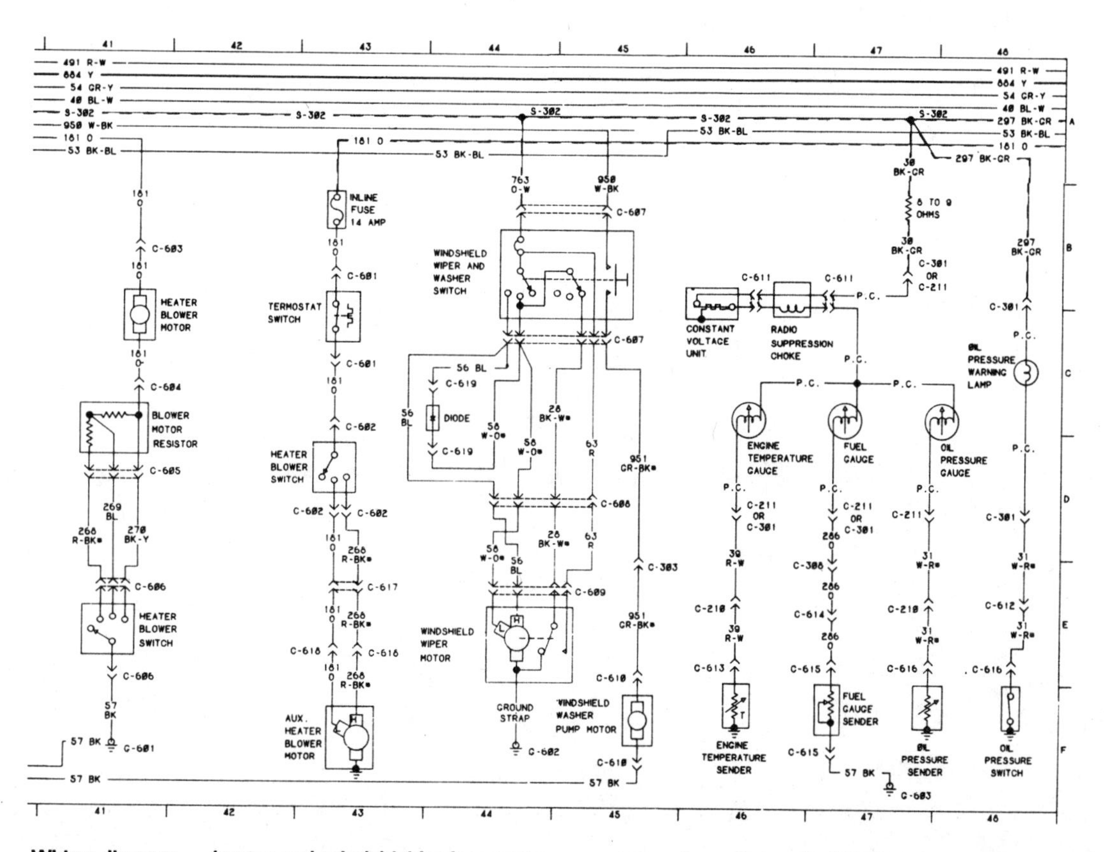

Wiring diagram — heater and windshield wiper system, gauges and sending units (1975 thru 1977 models)

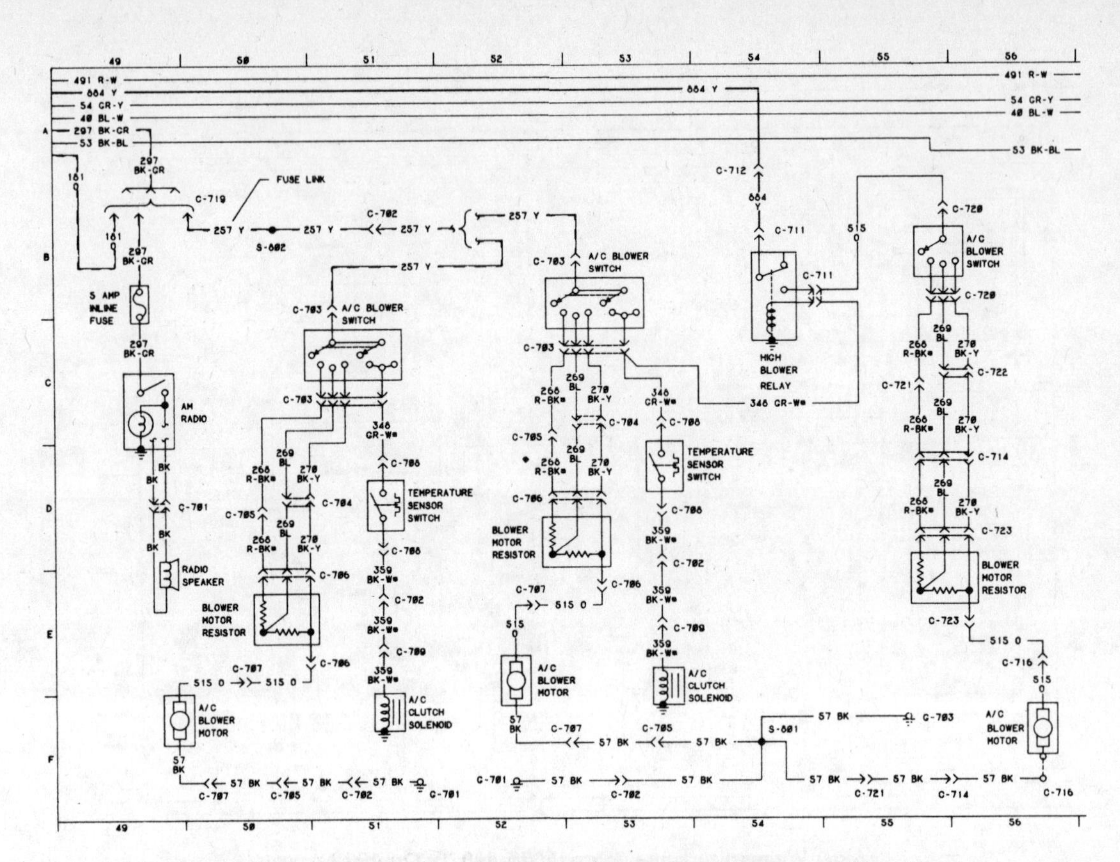

Wiring diagram — radio and air conditioning system (1975 thru 1977 models)

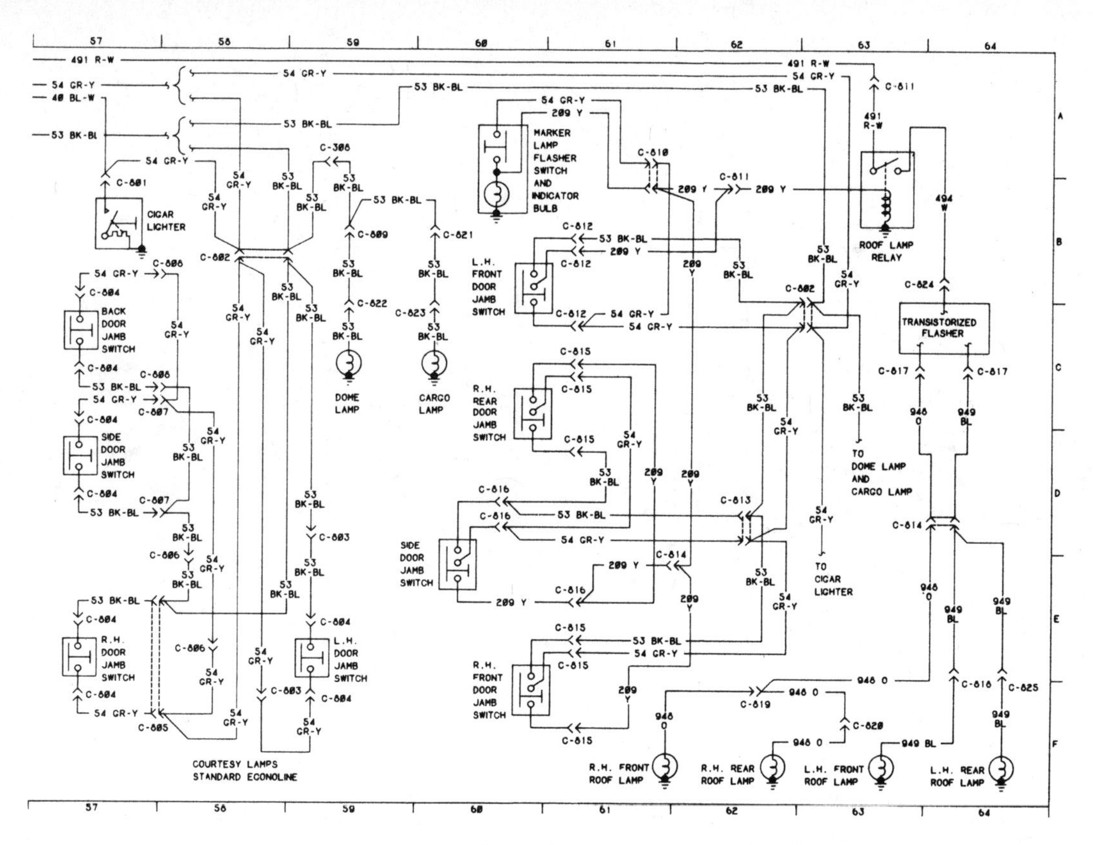

Wiring diagram — lights and switches (1975 thru 1977 models)

Wiring diagram color code

Color	Code
Black	BK
Blue	BL
Brown	BR
Gray	GY
Green	GR
Orange	O
Purple	P
Red	R
White	W
Yellow	Y

Stripe is understood and has no color code

Color code key for wiring diagrams on following pages

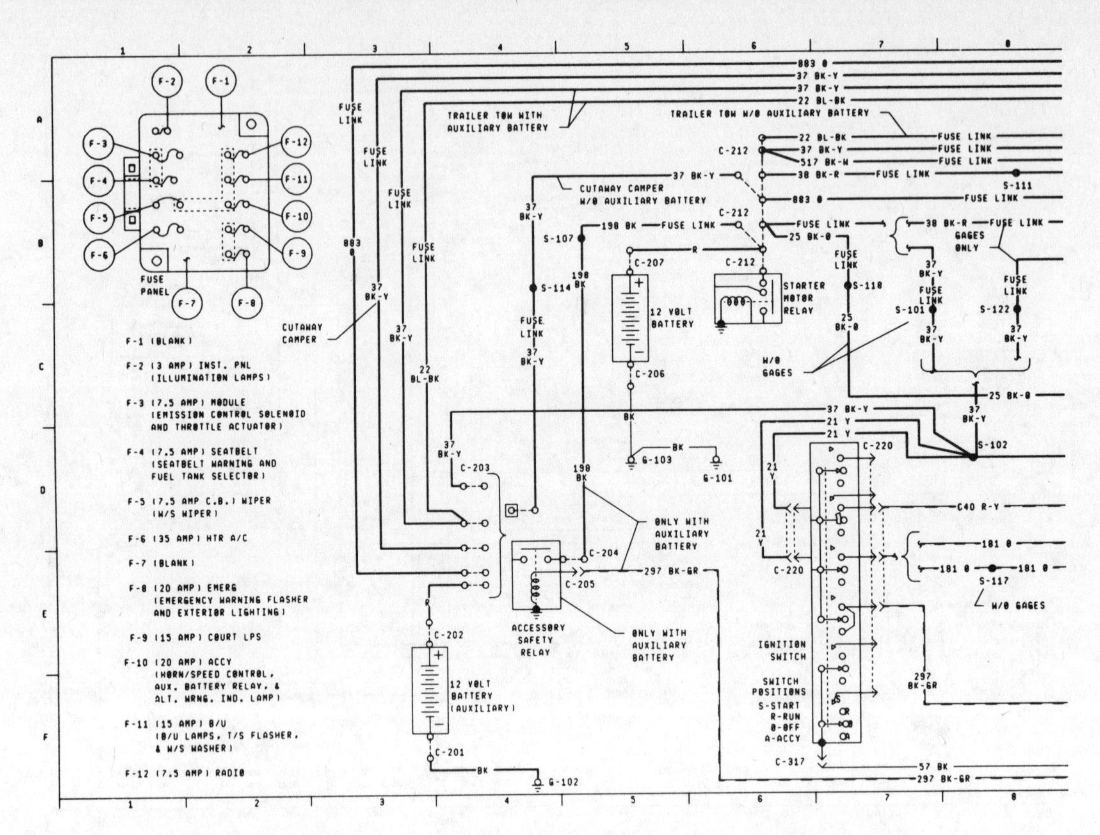

Wiring diagram — power distribution (1978 and later — typical) 1 of 2

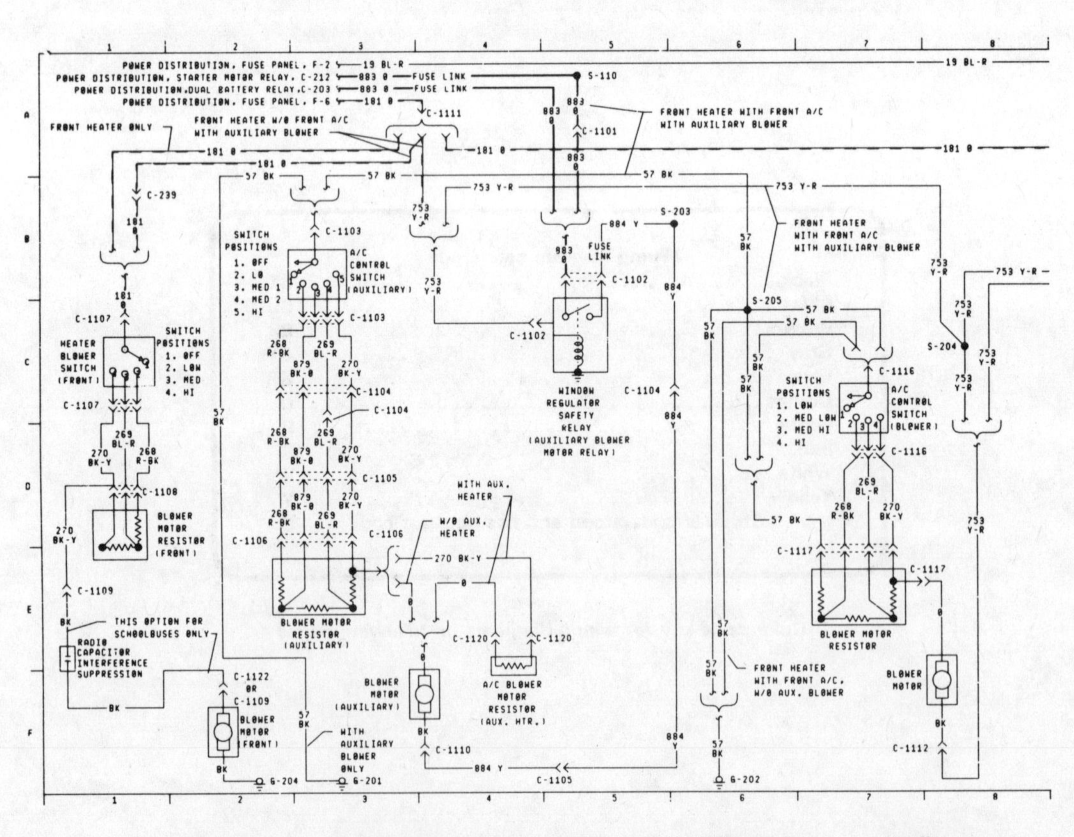

Wiring diagram — heater and air conditioning system (1978 and later — typical)

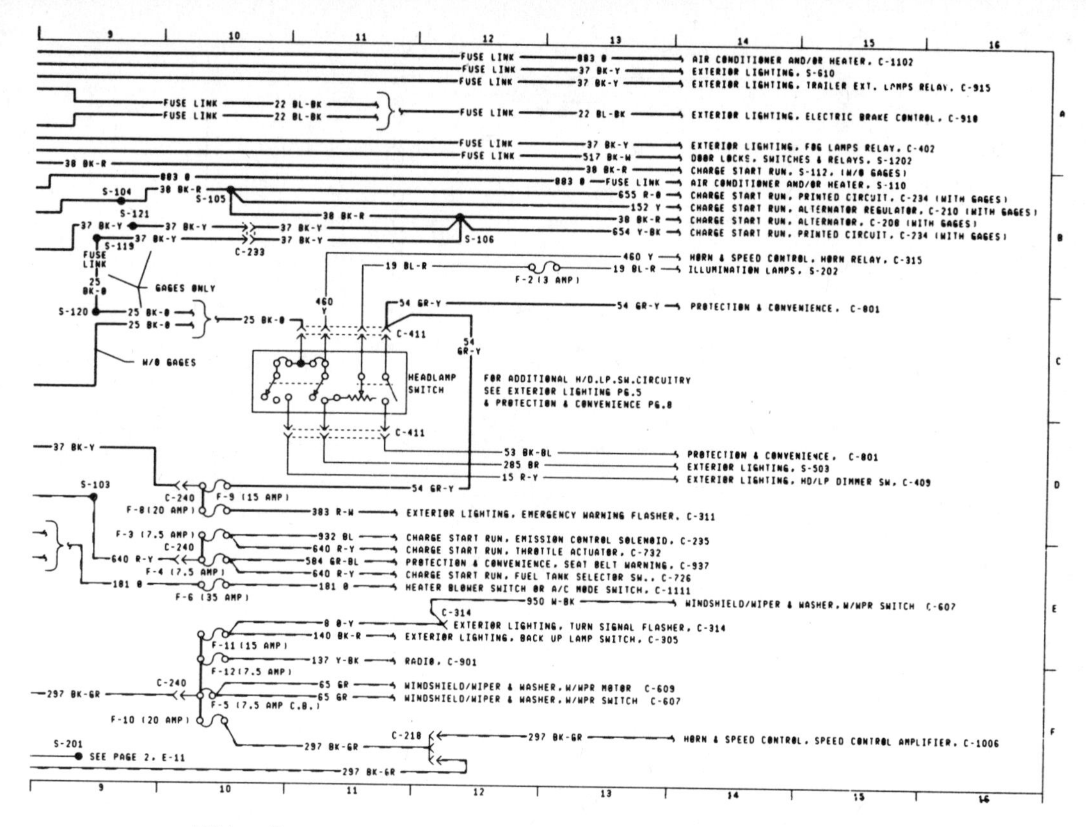

Wiring diagram — power distribution (1978 and later — typical) 2 of 2

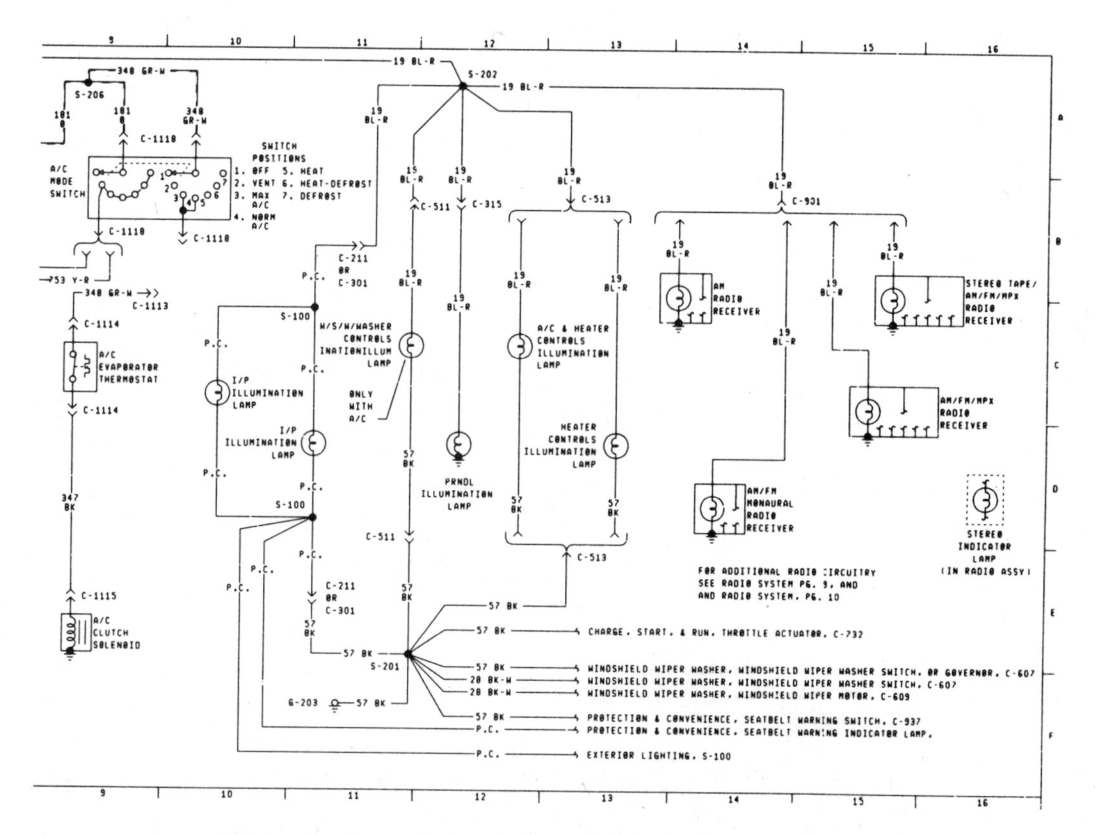

Wiring diagram — dash panel lights (1978 and later — typical)

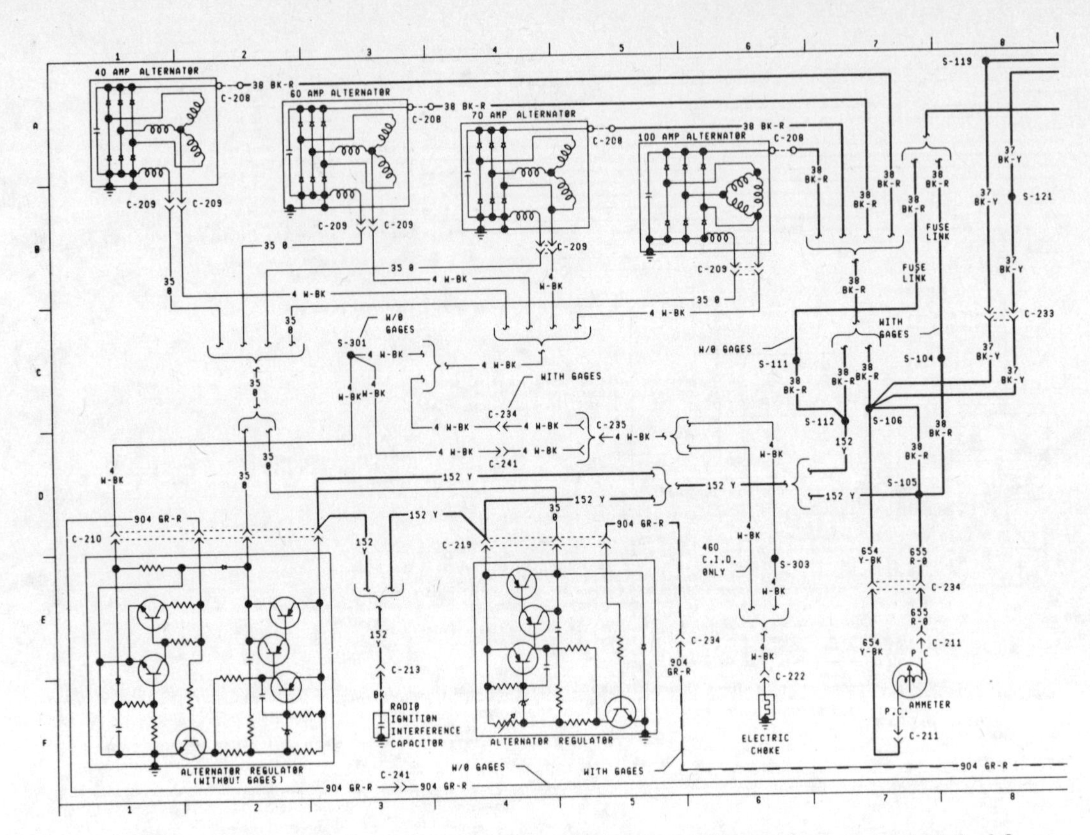

Wiring diagram – charging, starting and ignition systems (1978 and later – typical) 1 of 6

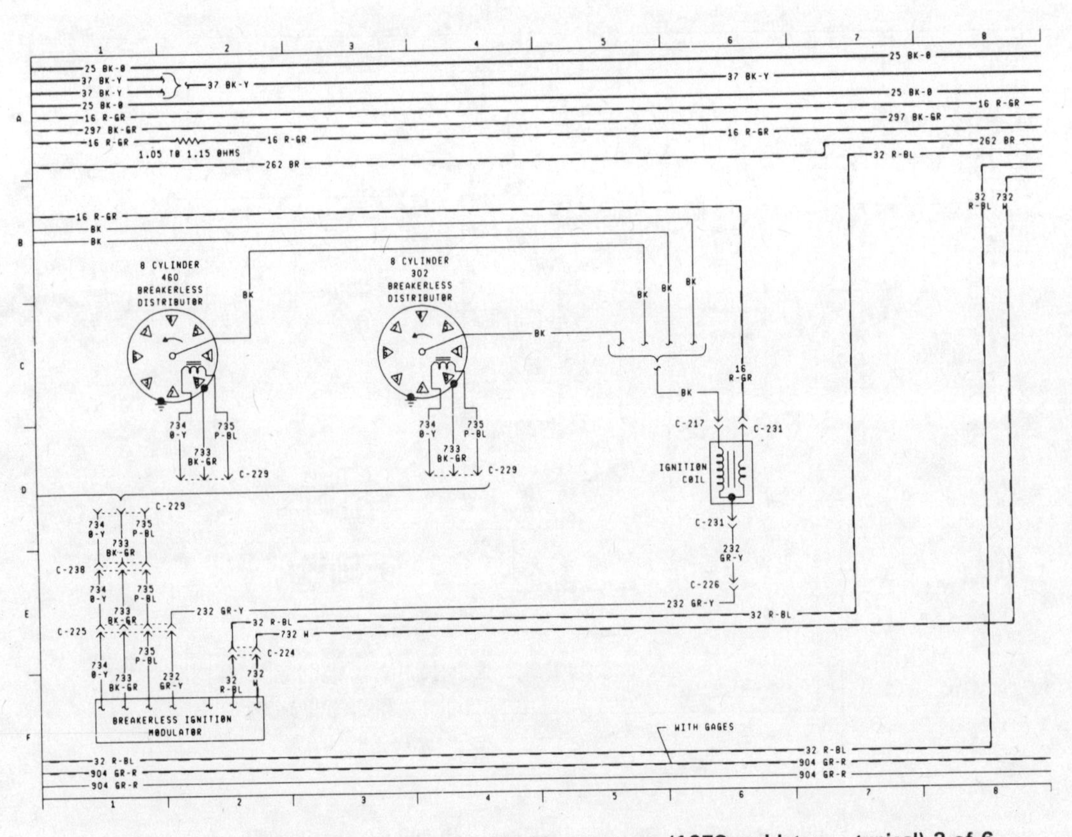

Wiring diagram – charging, starting and ignition systems (1978 and later – typical) 3 of 6

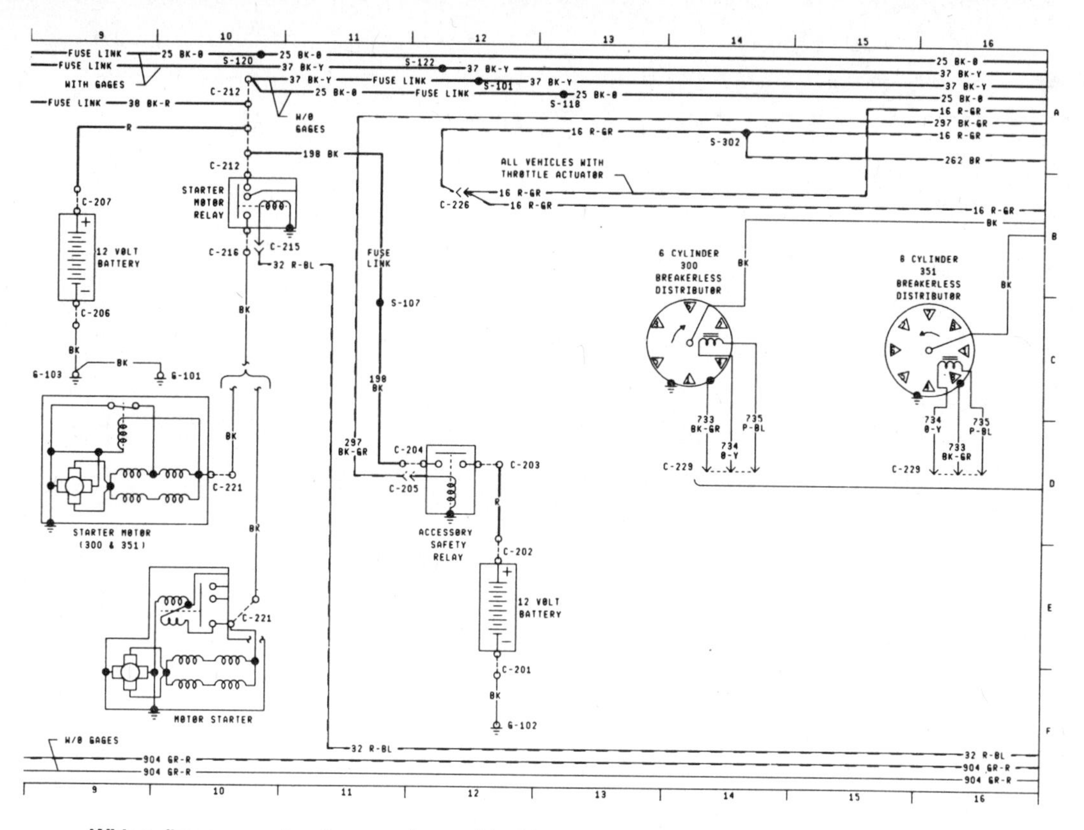

Wiring diagram — charging, starting and ignition systems (1978 and later — typical) 2 of 6

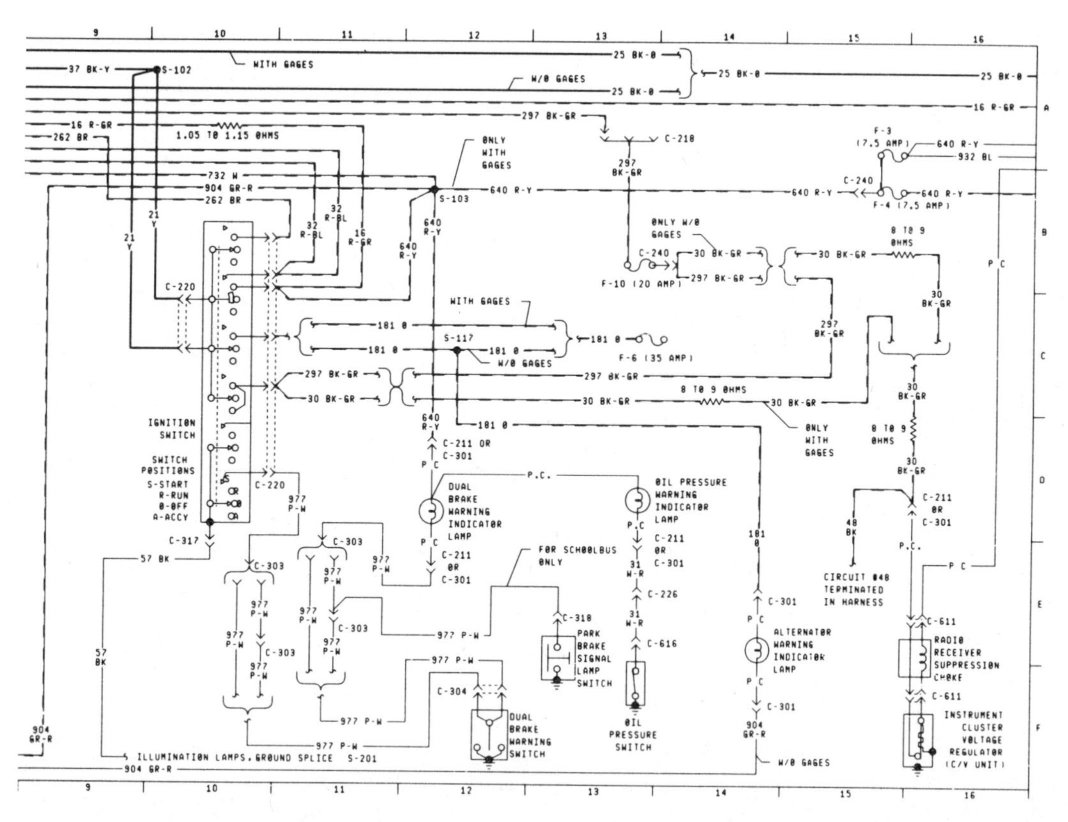

Wiring diagram — charging, starting and ignition systems (1978 and later — typical) 4 of 6

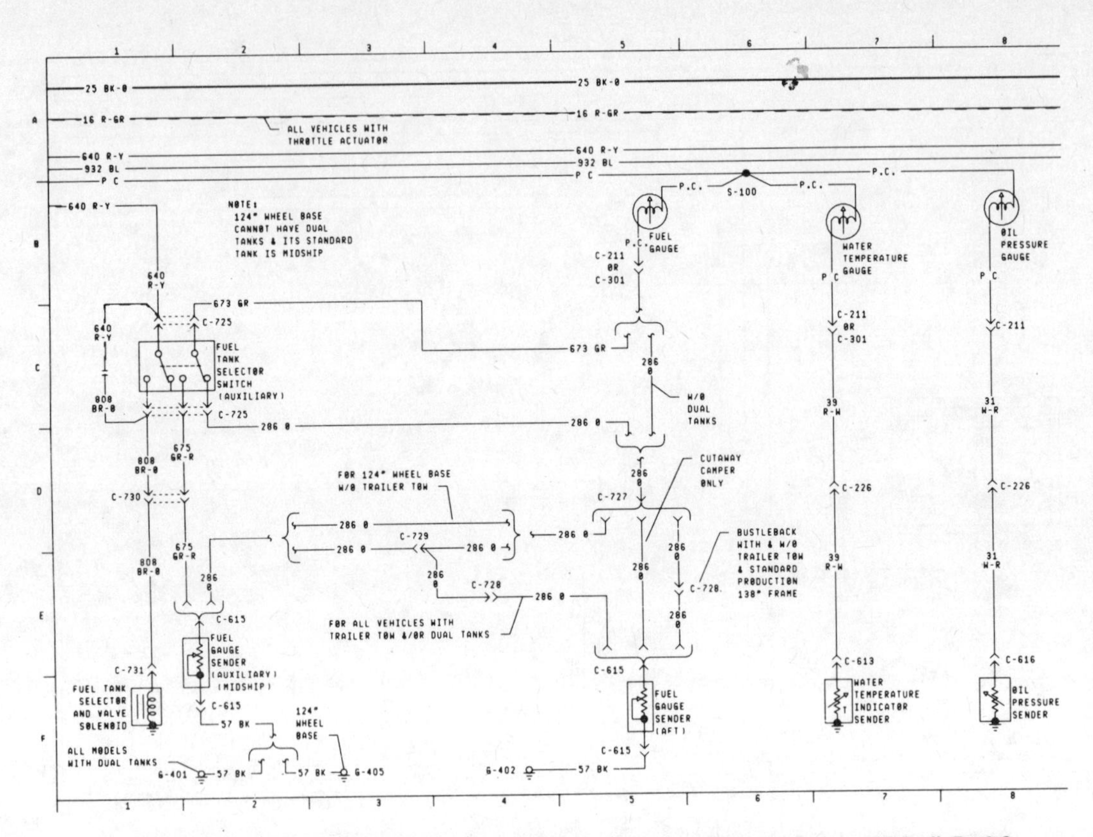

Wiring diagram – charging, starting and ignition systems (1978 and later – typical) 5 of 6

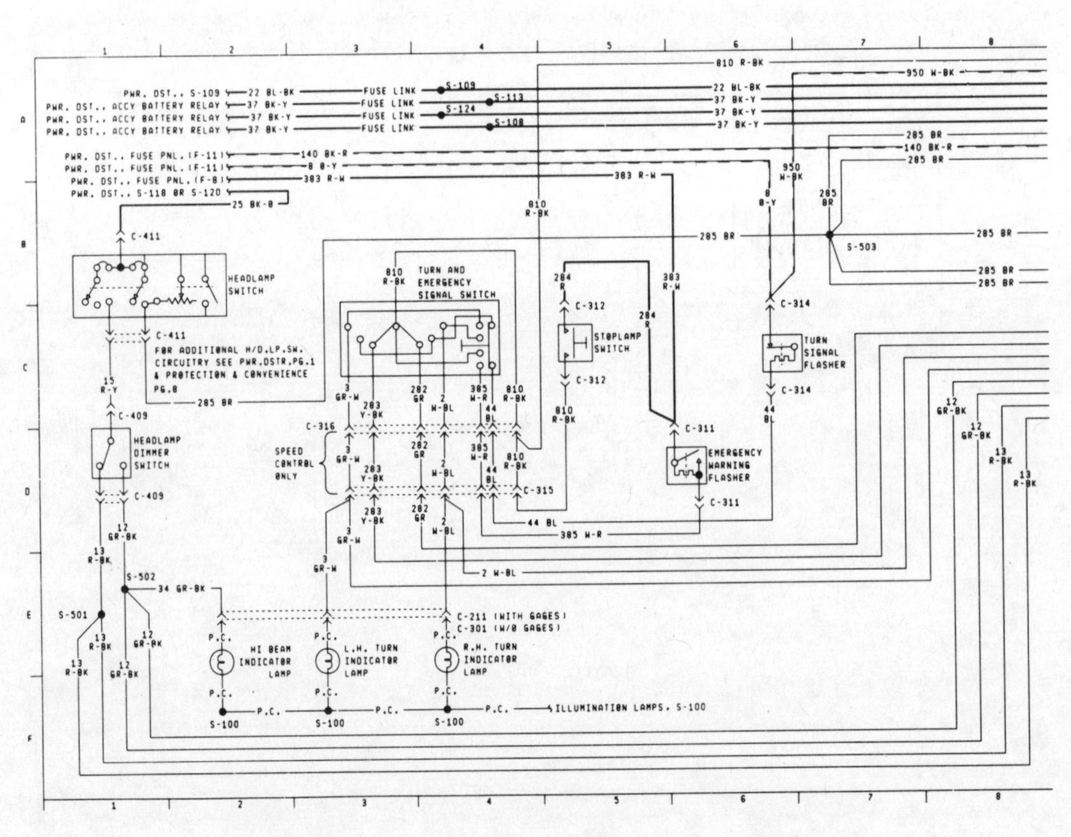

Wiring diagram – exterior lights (1978 and later – typical) 1 of 4

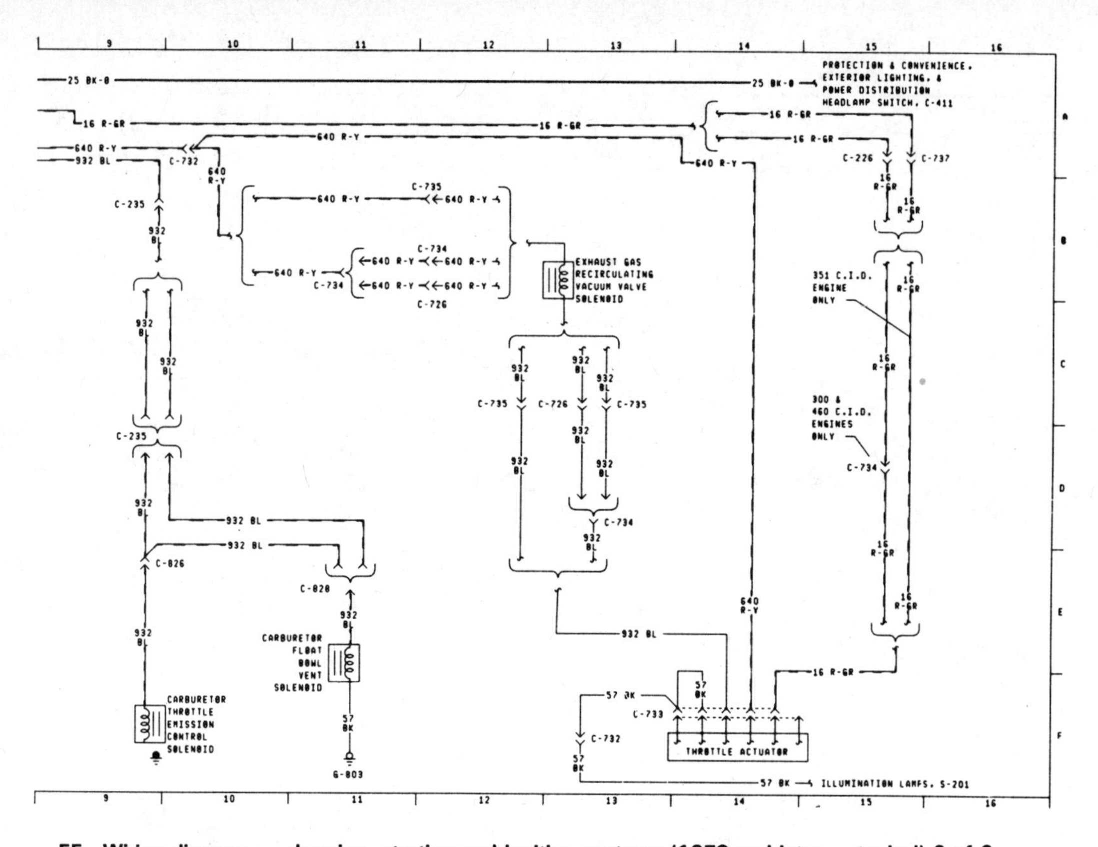

FF Wiring diagram — charging, starting and ignition systems (1978 and later — typical) 6 of 6

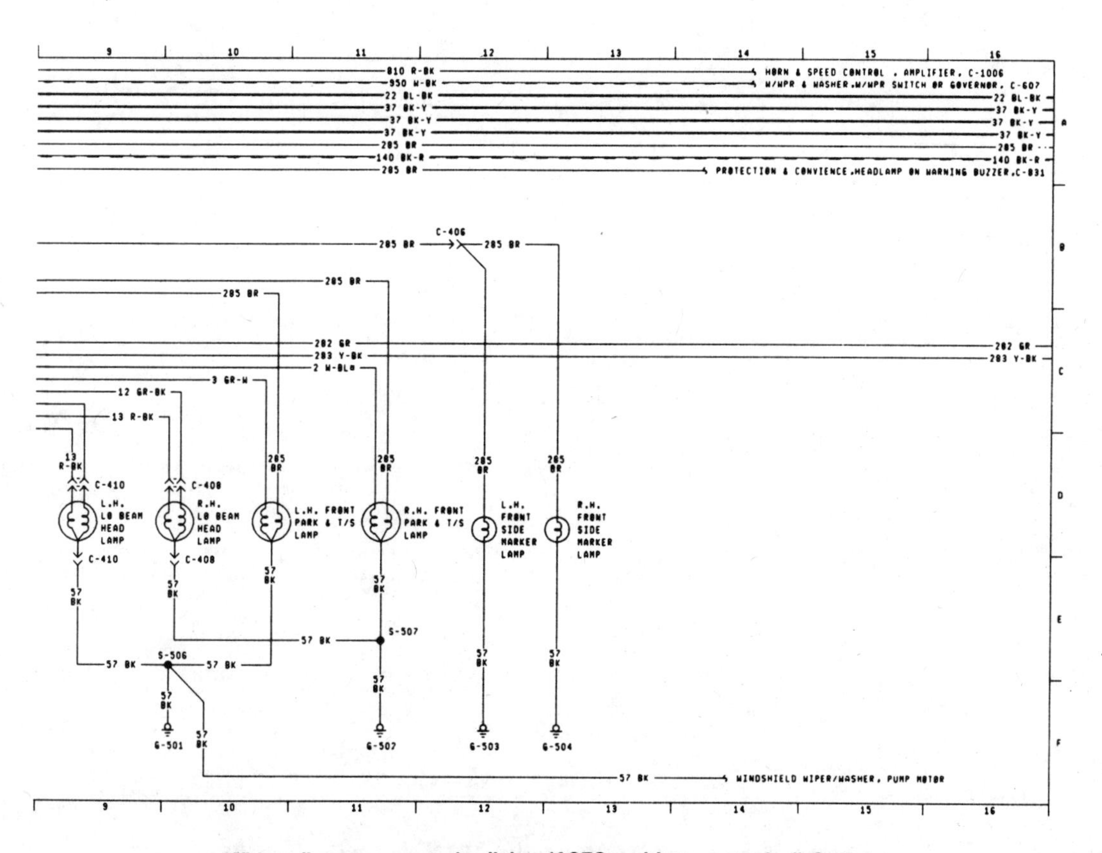

Wiring diagram — exterior lights (1978 and later — typical) 2 of 4

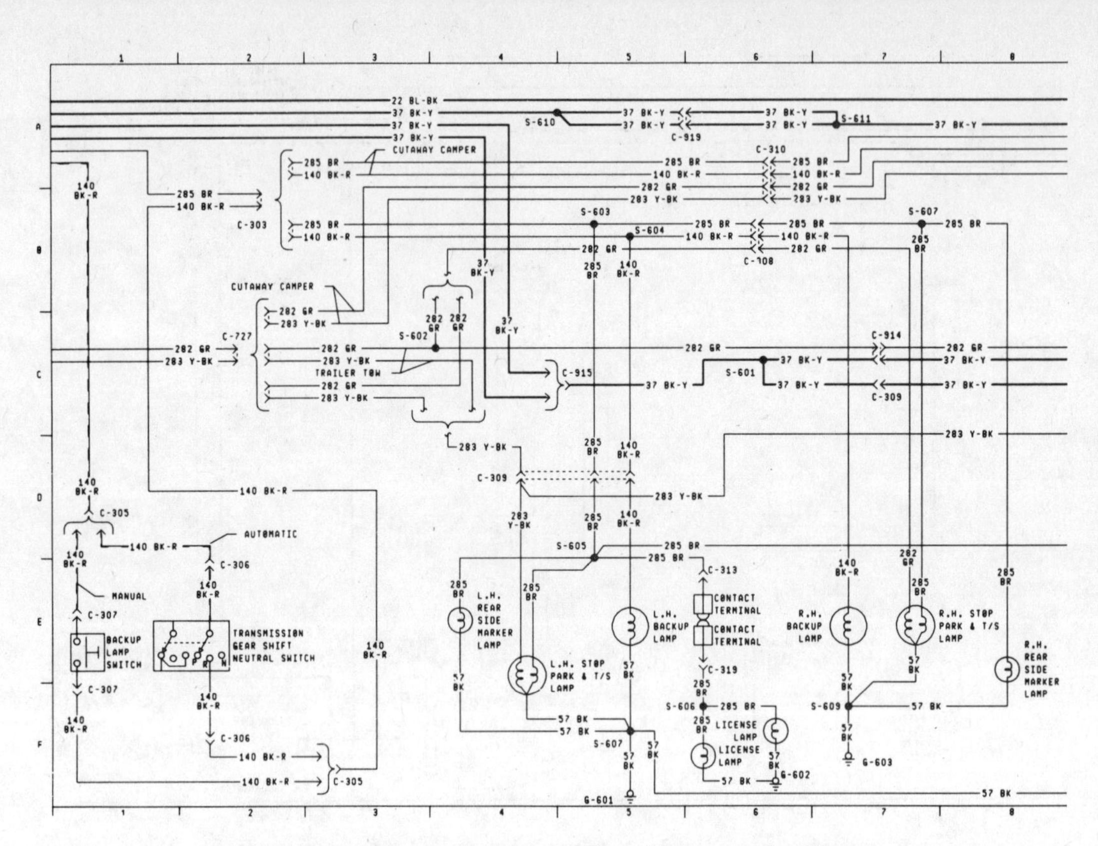

Wiring diagram — exterior lights (1978 and later — typical) 3 of 4

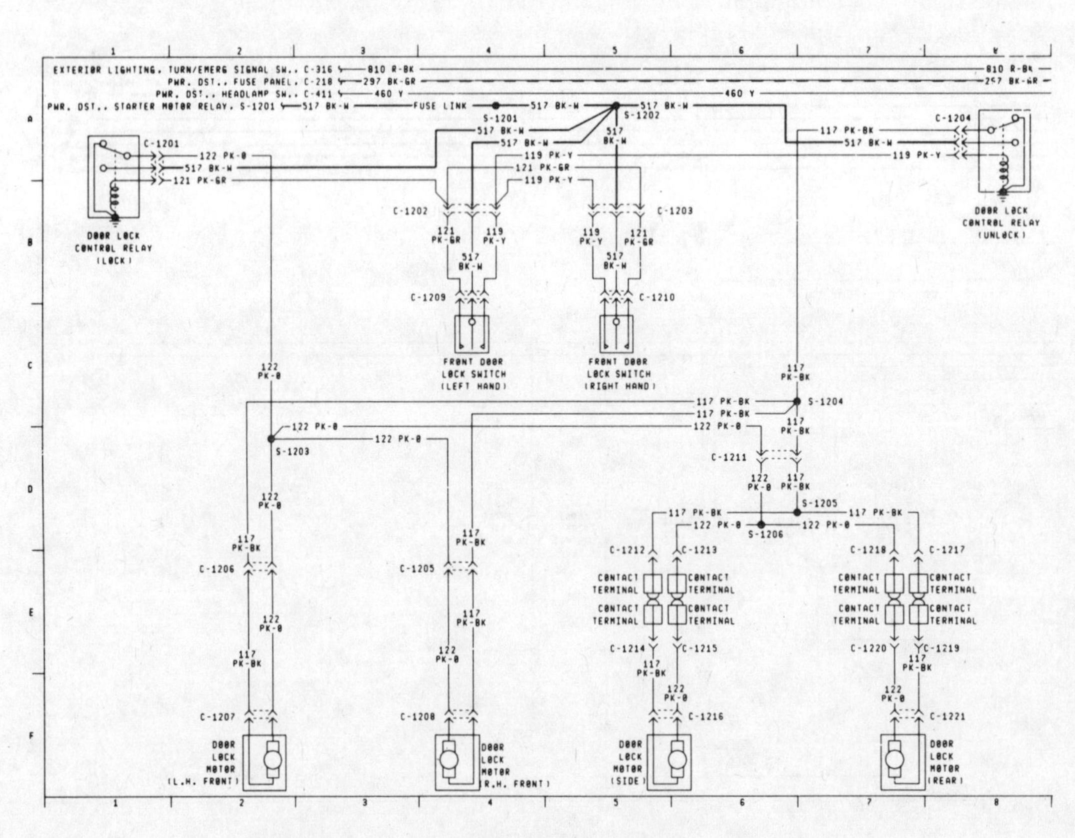

Wiring diagram — power door locks (1978 and later — typical)

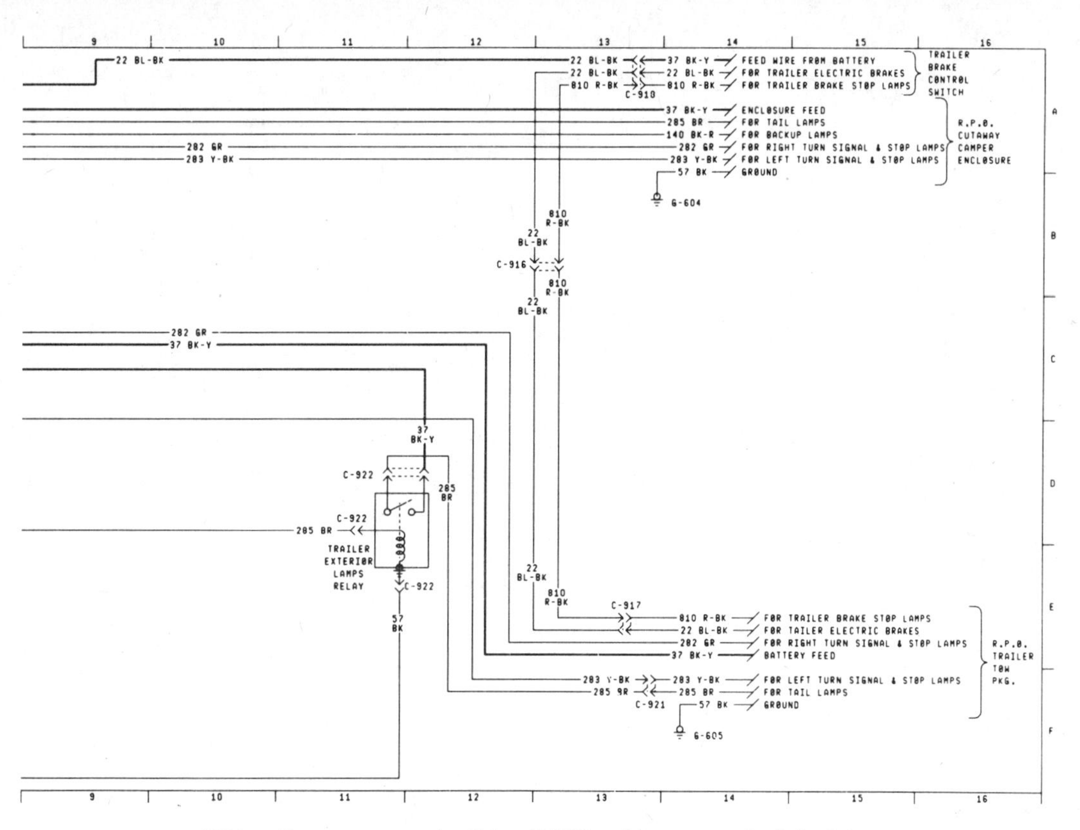

Wiring diagram — exterior lights (1978 and later — typical) 4 of 4

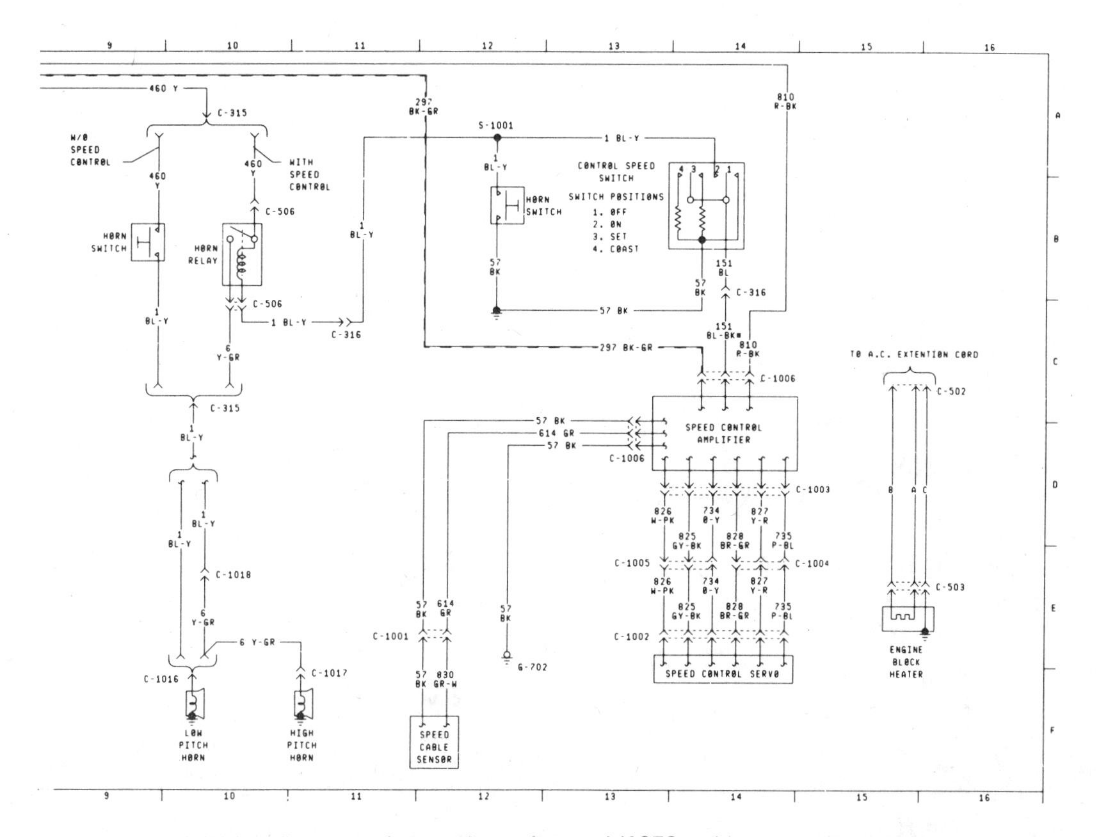

Wiring diagram — horn and speed control (1978 and later — typical)

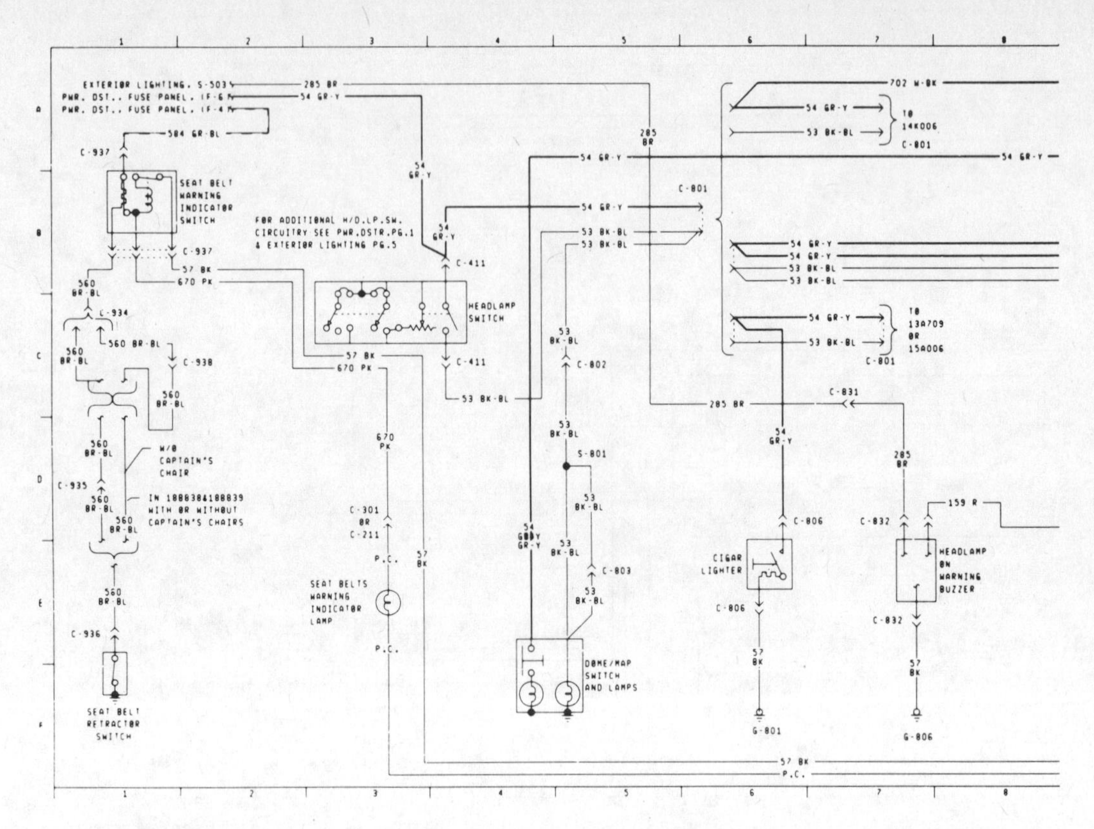

Wiring diagram – protection and convenience devices (1978 and later – typical) 1 of 2

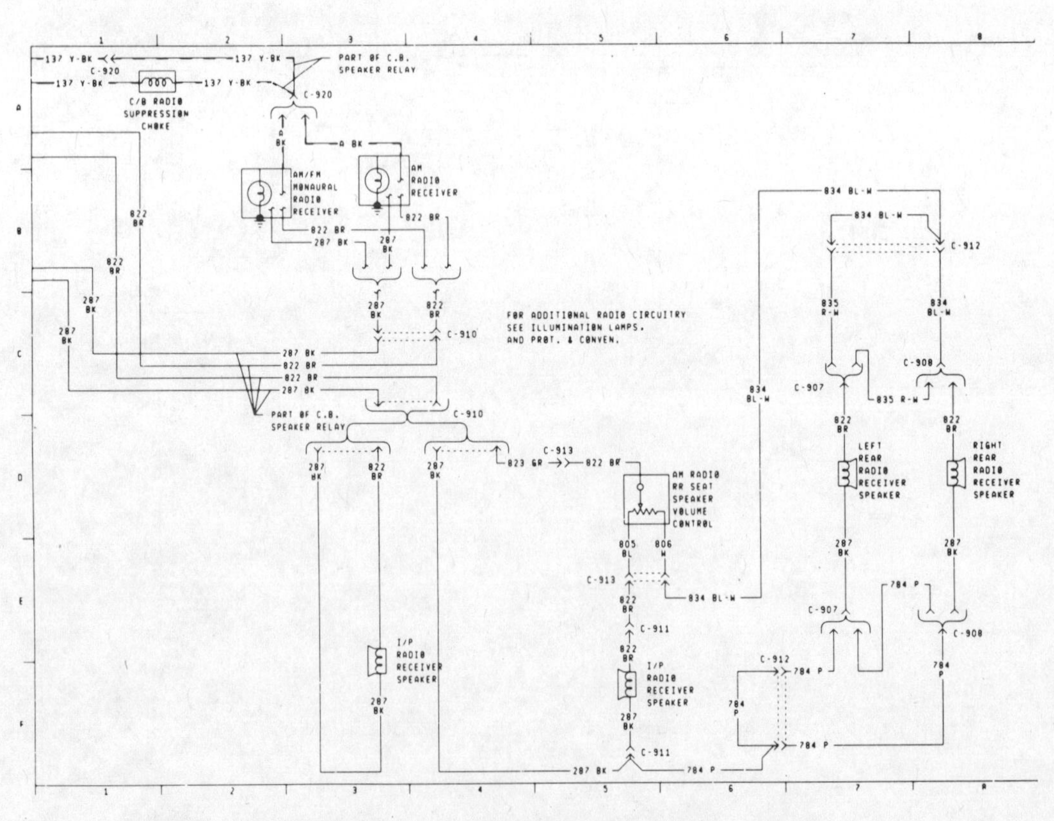

Wiring diagram – standard radio (1978 and later – typical)

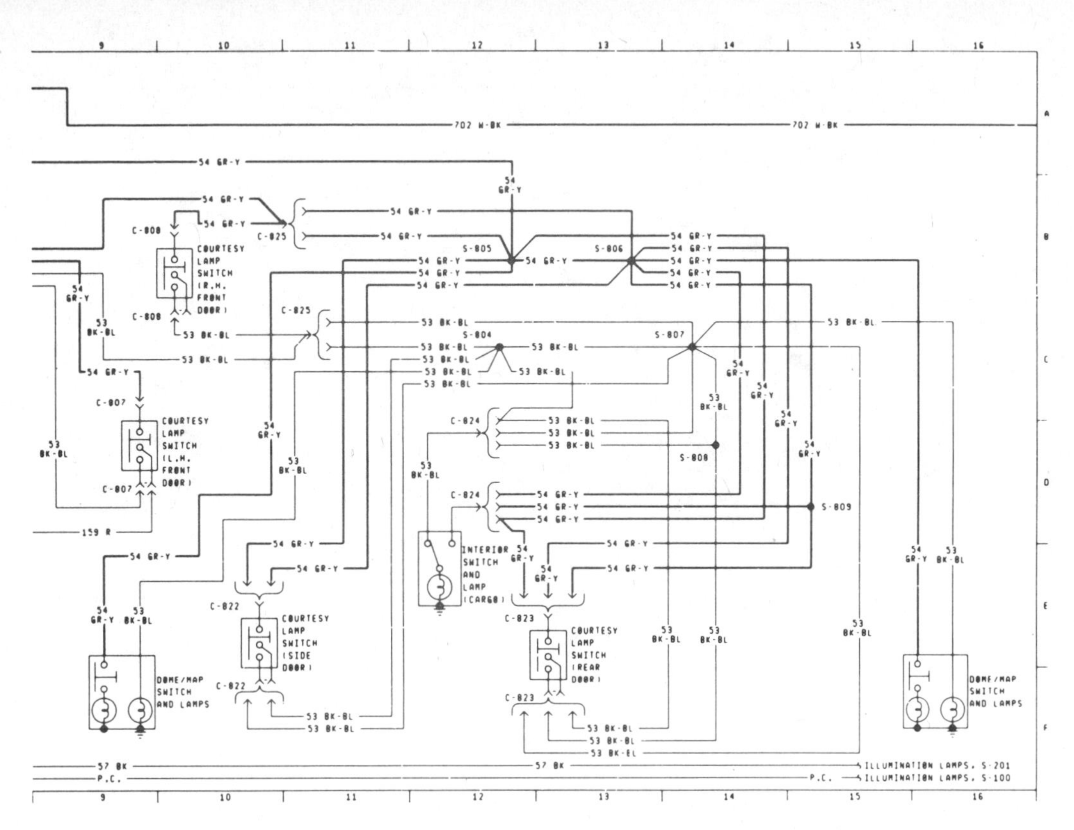

Wiring diagram — protection and convenience devices (1978 and later — typical) 2 of 2

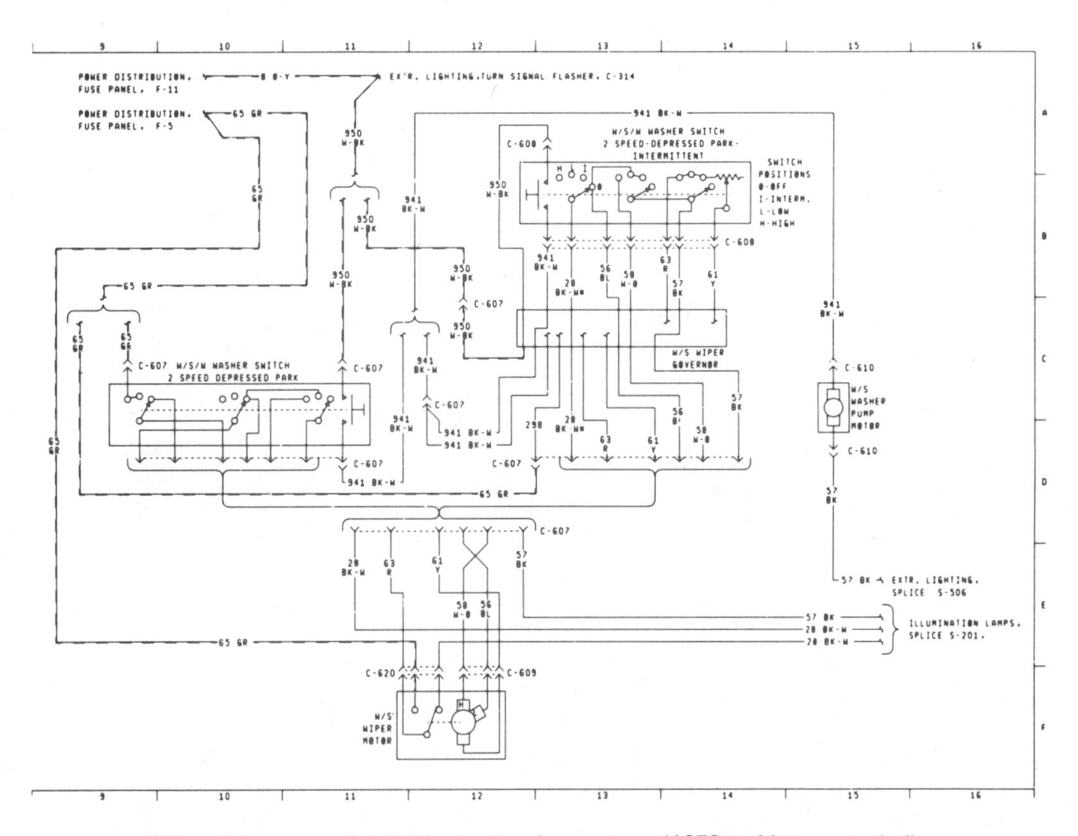

Wiring diagram — windshield wiper/washer systems (1978 and later — typical)

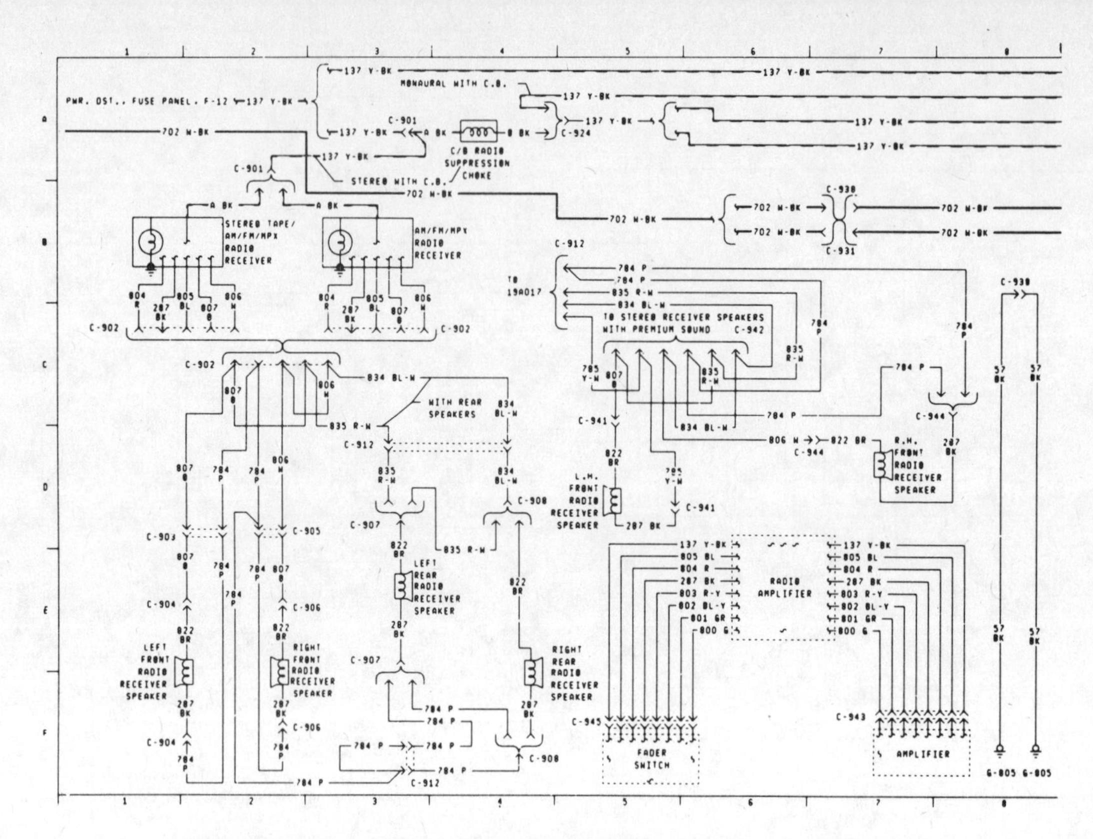

Wiring diagram — optional stereo system (1978 and later — typical) 1 of 2

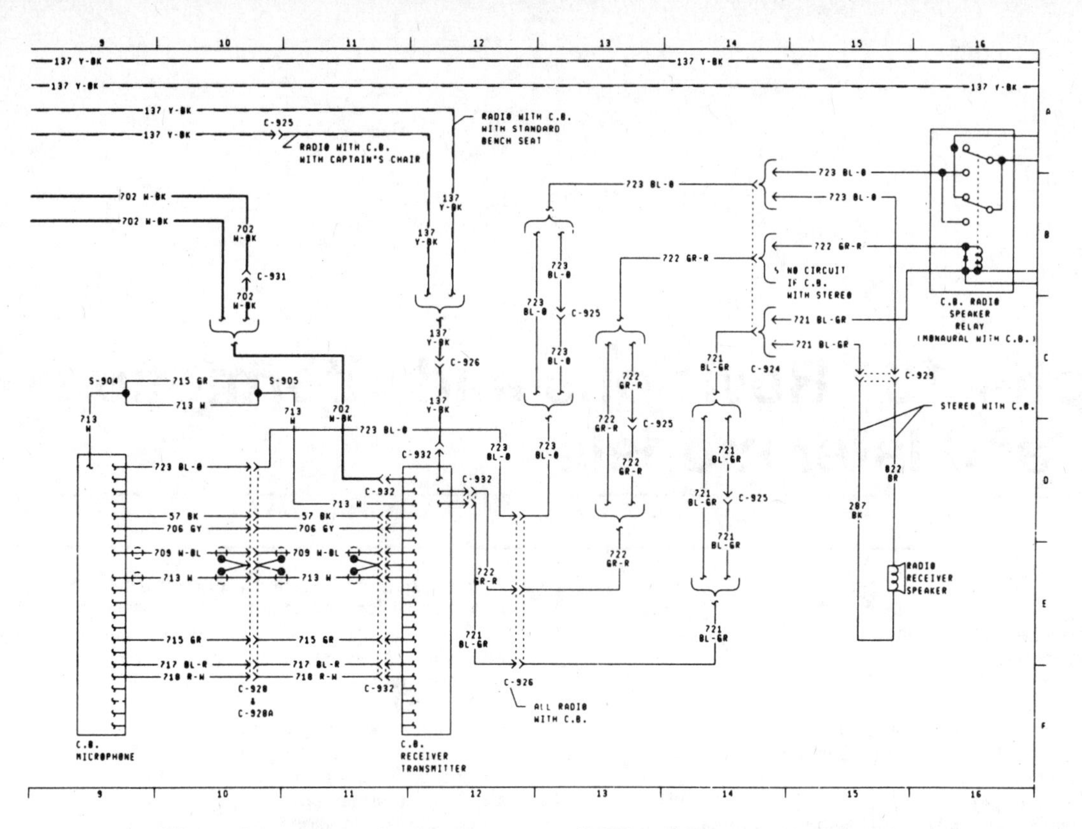

Wiring diagram — optional stereo system (1978 and later — typical) 2 of 2

Chapter 13 Revisions and information on 1987 and later models

Contents

1 Introduction

This Supplement contains specifications and service procedure changes that apply to Ford full-size vans manufactured in 1987 and later. Also included is information related to previous models that was not available at the time of original publication of this manual.

Where no differences (or very minor differences) exist between 1986 models and later models, no information is given. In those instances, the original material included in Chapters 1 through 12 should be used.

Before beginning a service or repair procedure, check this Supplement for new specifications and procedure changes. Make note of the supplementary information and be sure to include it while following the original procedures in Chapters 1 through 12.

2 Specifications

Note: *The following specifications are revisions of or supplementary to those listed at the beginning of each Chapter of this manual. The original specifications apply unless alternative information is included here.*

Tune-up and routine maintenance

Recommended lubricants and fluids

Note: *Listed here are manufacturer recommendations at the time this manual was written. Manufacturers occasionally upgrade their fluid and lubricant specifications, so check with your local auto parts store for current recommendations.*

Engine oil type (1986 on)	SAE grade SF
Automatic transmission fluid type (1988)	**Mercon ATF** (Ford part no. XT-2-QDX)
Manual transmission lubricant type (1988)	**Mercon ATF** (Ford part no. XT-2-QDX)

Manual transmission capacity

Mazda M5OD 5-speed	7.6 pints
S5BZF 5-speed	3.5 quarts

Torque specifications

	Ft-lbs
Wheel lug nuts(1987 on)*	
1/2 in nut	100
9/16 in nut (single rear wheels)	140

** Retorque after 500 miles each time lugs nuts are loosened*

Engine

Torque specifications

	Ft-lbs *(unless otherwise indicated)*
Six cylinder inline engine (1987 on)	
Lower intake manifold bolts	22 to 32
Oil pan-to-block bolts	15 to 18
302/351 cu in (5.0/5.8L) V8 engines	
Lower intake manifold bolts	23 to 25
Lower intake manifold stud nut	8 to 10
460 cu in (7.5L) V8 z	
Camshaft thrust plate bolts	70 to 105 in-lbs
Exhaust manifold bolts	22 to 30
Intake manifold bolts	
1st step	8 to 12
2nd step	12 to 22
3rd step	22 to 35
Oil filter insert-to-engine block/adapter	45 to 150
Oil filter adapter-to-engine block	40 to 65
Rocker arm cover bolts	6 to 9
Water outlet housing bolts	12 to 18

Fuel, exhaust and emission control systems

Fuel system pressure (EFI)

4.9 L	45 to 60 psi
All others	30 to 45 psi

Torque specifications

	Ft-lbs *(unless otherwise indicated)*
Six-cylinder inline engine (1987 on)	
Air bypass valve-to-throttle body	71 to 102 in-lbs
Cooling manifold	35 to 50
EGR tube	25 to 35
EGR valve-to-upper intake manifold	13 to 19
Fuel injector manifold-to-fuel charging assembly	12 to 15
Fuel pressure regulator-to-injector manifold	27 to 40 in-lbs
Throttle body-to-upper intake manifold	12 to 18
Throttle position sensor-to-throttle body	14 to 16 in-lbs
Upper intake manifold-to-lower intake manifold	12 to 18
302/351 cu in (5.0/5.8L) V8 engines (1987 on)	
Air bypass valve-to-throttle body	71 to 102 in-lbs
EGR tube	25 to 35
EGR valve-to-upper intake manifold	13 to 19
Fuel injector manifold-to-fuel charging assembly	12 to 15
Fuel pressure regulator-to-injector manifold	27 to 40 in-lbs
Throttle position sensor-to-throttle body	14 to 16 in-lbs
Throttle body-to-upper intake manifold	12 to 18
Upper intake manifold-to-lower intake manifold	15 to 22
460 cu in (7.5L) V8 engine (1988 on)	
Air bypass valve-to-lower manifold	70 to 100 in-lbs
Air supply tube clamps	12 to 20 in-lbs
EGR tube	25 to 35
EGR valve-to-upper intake manifold	70 to 100 in-lbs
Fuel Pressure regulator-to-injector manifold	27 to 40 in-lbs
Fuel injector manifold-to-fuel charging assembly	70 to 105 in-lbs
Throttle body-to-upper intake manifold	70 to 100 in-lbs
Throttle position sensor-to-throttle body	11 to 16 in-lbs
Upper intake manifold-to-lower intake manifold	12 to 18
Thermactor pump-to-pump bracket	30 to 40
Thermactor pump pulley-to-pump hub	100 to 130 in-lbs
Thermactor pump-to-pump bracket	30 to 40

3 Tune-up and routine maintenance

Routine maintenance intervals

Every 5000 miles (8000 Km)

Check the manual transmission lubricant level (should be done every time the engine oil is changed)

Every 60,000 miles (96,000 Km)

Replace the EGR valve (all 1987 and later models with EFI)
Replace the heated exhaust gas oxygen sensor (all models with EFI)
Clean the idle speed control bypass valve (all models with EFI)

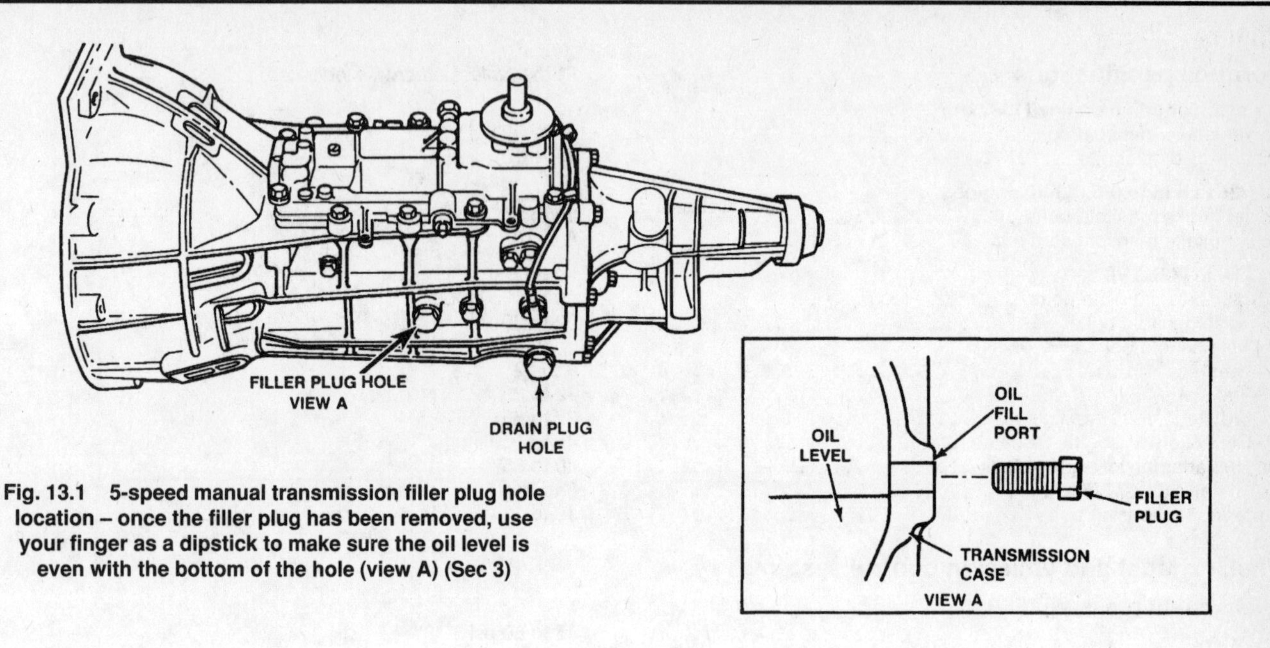

Fig. 13.1 5-speed manual transmission filler plug hole location – once the filler plug has been removed, use your finger as a dipstick to make sure the oil level is even with the bottom of the hole (view A) (Sec 3)

Manual transmission lubricant level check

1 Manual transmissions don't have a dipstick. The oil level is checked by removing a plug from the side of the case (Fig. 13.1). Locate the plug and use a rag to clean the plug and the area around it. If the vehicle is raised to gain access to the plug, be sure to support it safely on jackstands – DO NOT crawl under the vehicle when it's supported only by a jack!

2 With the engine and drivetrain components cold, remove the plug. If lubricant immediately starts leaking out, thread the plug back into the case – the level is correct. If it doesn't, completely remove the plug and reach inside the hole with your little finger. The level should be even with, or very near, the bottom of the plug hole.

3 If the transmission needs more lubricant, use a syringe or small pump to add it through the hole.

4 Thread the plug back into the case and tighten it securely. Drive the vehicle then check for leaks around the plug.

EGR valve replacement

5 The EGR valve is mounted on the upper intake manifold (see Fig. 13.2). Follow the procedure in Section 4 of Chapter 6, but note that the EGR valve used on EFI models has a wire harness that must be disconnected.

Heated exhaust gas oxygen sensor replacement

6 Locate the oxygen sensor between the exhaust manifold and the catalytic converter.

7 Allow the exhaust system to cool completely before removing the oxygen sensor.

8 Follow the sensor wire to the connector, then unplug it – be careful, it's made of plastic and will break if mishandled.

9 Unscrew the sensor. It may be necessary to use a crowfoot wrench on some vehicles.

10 Installation is the reverse of removal.

Idle speed control bypass valve cleaning

11 Locate the ISC bypass valve on the upper intake manifold or throttle body (see Fig. 13.2). Unplug the wire harness connector (be careful, don't break the plastic body of the connector).

12 Remove the bolts and detach the ISC bypass valve from the throttle body or upper intake manifold.

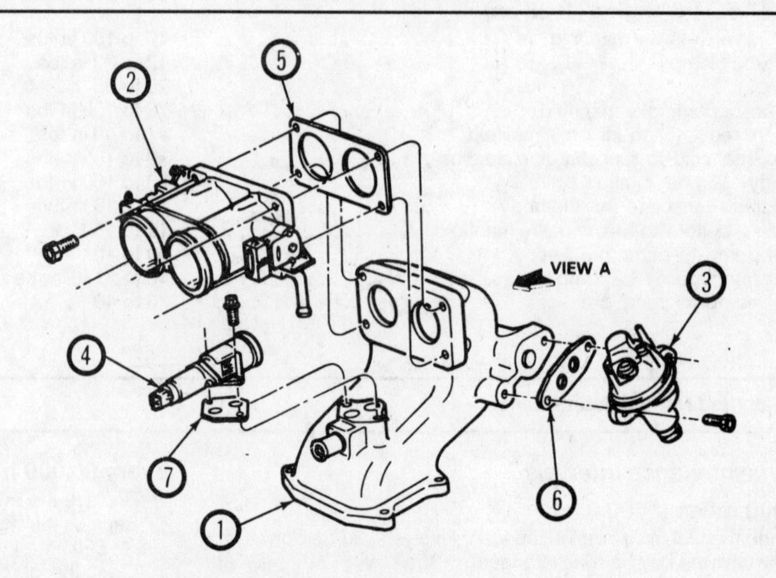

Fig. 13.2 The EGR valve (number 3 in this illustration) is attached to the upper intake manifold on EFI engines (460 cu in/7.5L V8 shown) (Sec 3)

1 Upper intake manifold
2 Throttle body
3 EGR valve
4 ISC bypass valve
5 Throttle body gasket
6 EGR valve gasket
7 Air bypass valve gasket

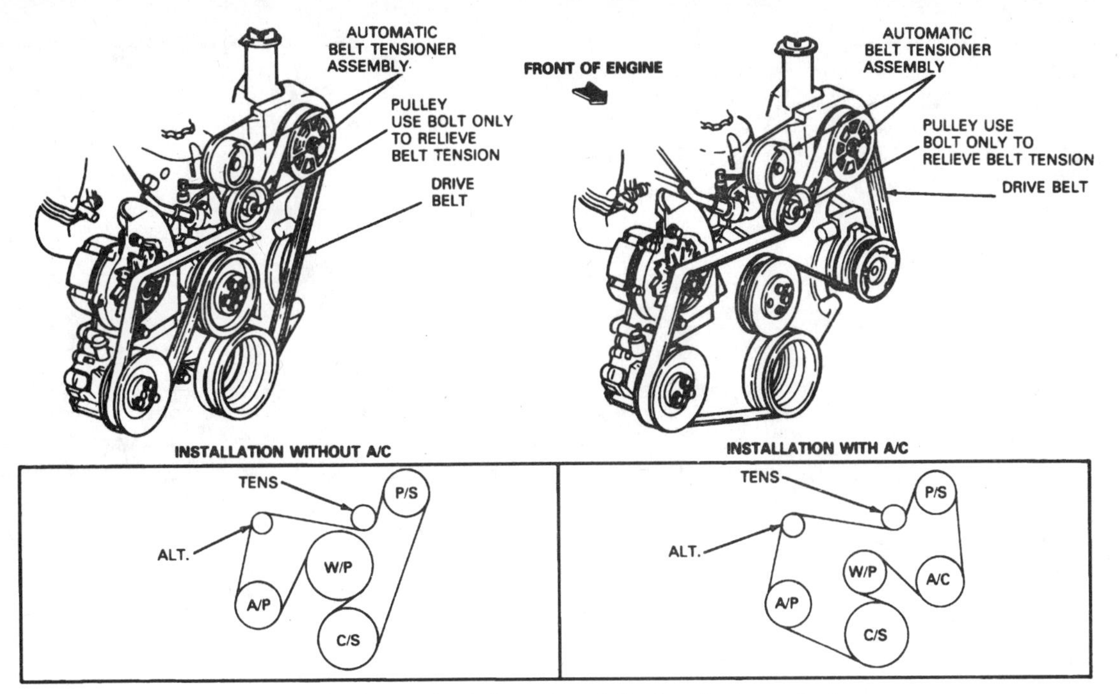

13.3 Serpentine belt routing (4.9L six-cylinder engine)

13 Remove the electrical solenoid from the bypass valve.

14 Soak the mechanical portion of the valve in Carburetor and combustion chamber cleaner (Ford part no. D9AZ-19579-B) or equivalent for 2 or 3 minutes maximum. **Warning:** *Do not exceed the 3 minutes soak time and don't use choke cleaner, as the internal O-ring may begin to deteriorate.*

15 While soaking, shake the valve and push the rod in and out. Dry it with compressed air.

16 Clean the gasket surfaces. Be careful not to drop gasket material into the throttle body or manifold.

17 Reassemble the bypass valve and install it on the throttle body or upper intake manifold. Use a new gasket and don't overtighten the bolts.

Drivebelt check and replacement – serpentine type

18 Many later models are equipped with a single serpentine drivebelt. This drivebelt is located at the front of the engine. The belt drives the water pump, alternator, power steering pump, air conditioner compressor and thermactor air pump. The condition and tension of the drivebelt is critical to the operation of the engine and accessories. Excessive tension causes bearing wear, while insufficient tension produces slippage, noise, component vibration and belt failure. Because of the belt composition and the high stress to which it subjected, the drivebelt will stretch and continue to deteriorate as it gets older. As a result, it must be periodically checked. The serpentine belt has an automatic tensioner and requires no adjustment for the life of the belt.

Check

19 The "serpentine" type belt is a single V-ribbed belt and is used to drive all accessories and is so called because of the winding path it follows between the various drive, accessory and idler pulleys (Figs. 13.3, 13.4 and 13.5).

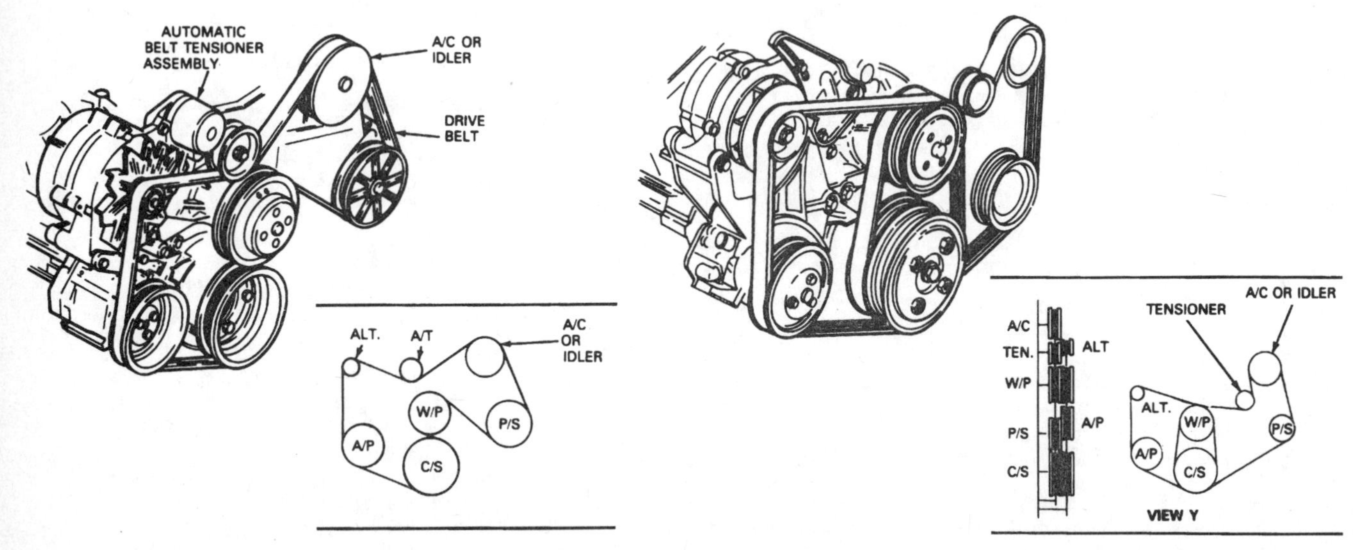

13.4 Serpentine belt routing (5.0L and 5.8L V8 engines)

13.5 Serpentine belt routing (7.5L V8 engine)

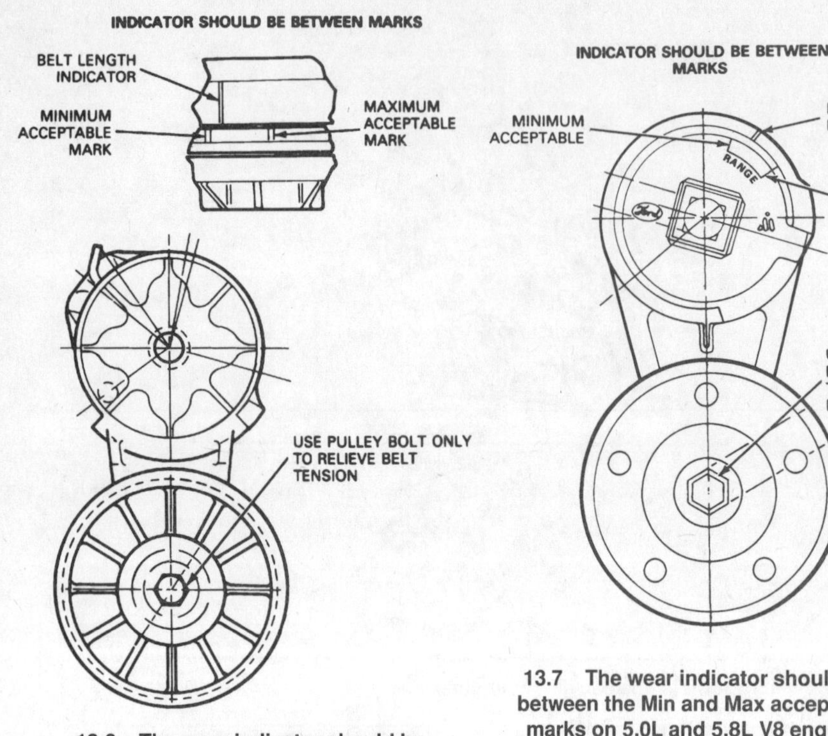

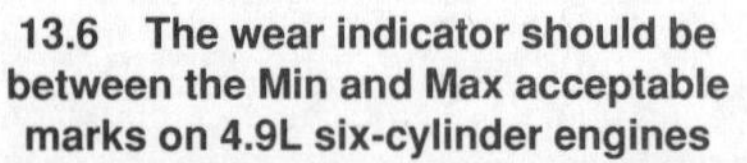

13.6 The wear indicator should be between the Min and Max acceptable marks on 4.9L six-cylinder engines

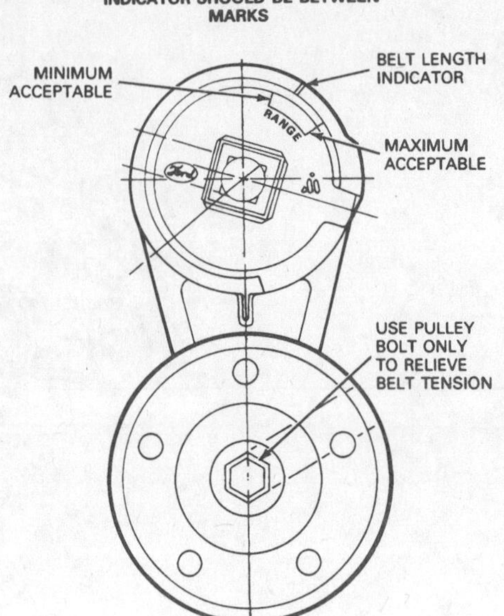

13.7 The wear indicator should be between the Min and Max acceptable marks on 5.0L and 5.8L V8 engines

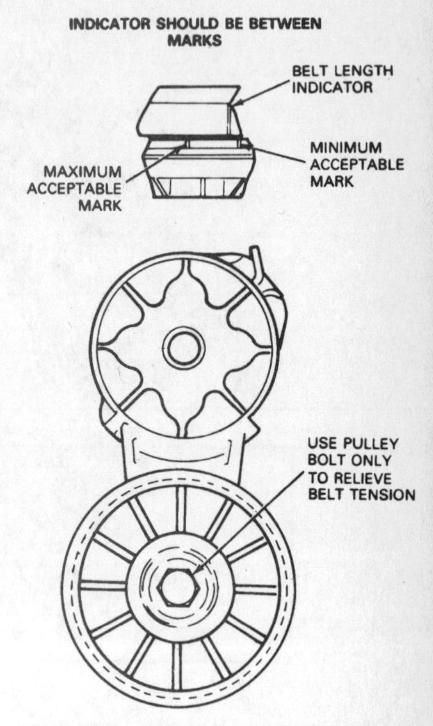

13.8 The wear indicator should be between the Min and Max acceptable marks on the 7.5L V8 engine

20 With the engine off, open the hood and locate the drive belt at the front of the engine. With a flashlight, check the belt for separation of the rubber plies from each side of the core, a severed core, separation of the ribs from the rubber, cracks, torn or worn ribs and cracks in the inner ridges of the ribs. Also check for fraying and glazing, which gives the belt a shinny appearance. Cracks in the rib side of the V-ribbed belts are acceptable, as are small chunks missing from the ribs. Both sides of the belt should be inspected, which means you'll have to twist it the check the underside. Use your fingers to feel a belt where you can't see. If any of the above conditions are evident, replace the belt as described in the following Steps.

21 To check the tension of the serpentine belt, look at the wear indicator (Figs. 13.6, 13.7 and 13.8). It should be between the Min and Max acceptable marks. If it is not between the two marks, replace the belt (see Step 5).

Replacement

22 Install a 5/8-inch or 16 mm box-end wrench on the tensioner pulley bolt and lift the tensioner arm and pulley away from the belt.

23 Remove the old belt and release the tensioner slowly. Caution: Do not allow the tensioner to snap back after the belt is removed because this may damage the tensioner.

24 Install a new belt over each pulley, making sure that all six belt ribs are correctly seated in each pulley groove (Fig. 13.9).

25 Move the tensioner arm back away and place the belt under the pulley. Slowly release the tensioner back onto the belt.

4 Six cylinder inline engine

Intake manifold – removal and installation (EFI only)

1 Make sure the ignition is off, then disconnect the negative battery cable from the battery.

2 Remove the fuel filler cap to relieve fuel tank pressure.

3 Disconnect the accelerator cable and remove the cable bracket.

4 Remove the upper intake manifold and throttle body.

5 Move the vacuum harness away from the lower intake manifold.

6 Release the fuel pressure from the fuel system (see Chapter 4).

7 Remove the injector cooling manifold from the lifting eye attachment.

8 Remove the sixteen bolts that attach both the lower intake and exhaust manifolds to the cylinder head. Do not remove the bolts that attach only the exhaust manifolds.

9 Detach the lower intake manifold assembly from the cylinder head.

10 Remove all traces of old gasket material with a scraper, then clean the mating surfaces with lacquer thinner or acetone. Check the mating surfaces of the lower intake manifold and the cylinder head for nicks and other damage that could prevent proper sealing.

11 Clean and oil the manifold bolt threads.

12 Position the lower intake manifold and a new gasket on the cylinder head and install the bolts. Tighten the bolts to the specified torque in the sequence shown (see Fig. 13.10).

13 The remaining installation steps are the reverse of removal.

5 460 cu in (7.5L) V8 engine

Intake manifolds – removal and installation (EFI only)

Upper manifold

The procedure in Chapter 4 (Section 17) is essentially correct, but note that the 460 engine doesn't have a support bracket and only four bolts, rather than six, are used to attach the upper manifold to the lower manifold.

Lower manifold

1 The procedure in Chapter 2, Part B, is essentially correct, but note the following points:

2 Remove the fuel filler cap to relieve fuel tank pressure.

3 Remove the upper intake manifold/throttle body assembly (see Chapter 4). **Warning:** *The fuel pressure must be relieved before disconnecting any lines! See Chapter 4 for the procedure.*

4 Disconnect the wiring harness from the main wiring harness.

5 Disconnect the fuel lines at the fuel rail and remove the fuel rail (see Chapter 4).

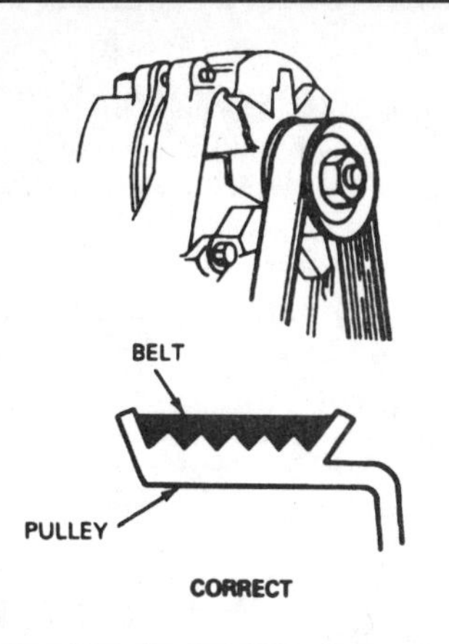

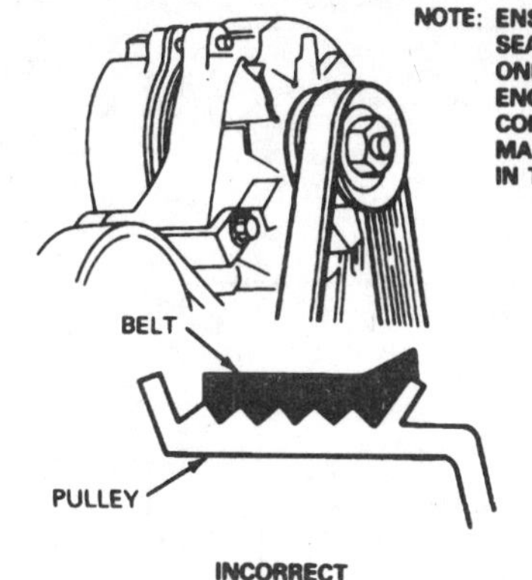

13.9 V-ribbed serpentine drivebelts should be centered in the pulleys, not offset

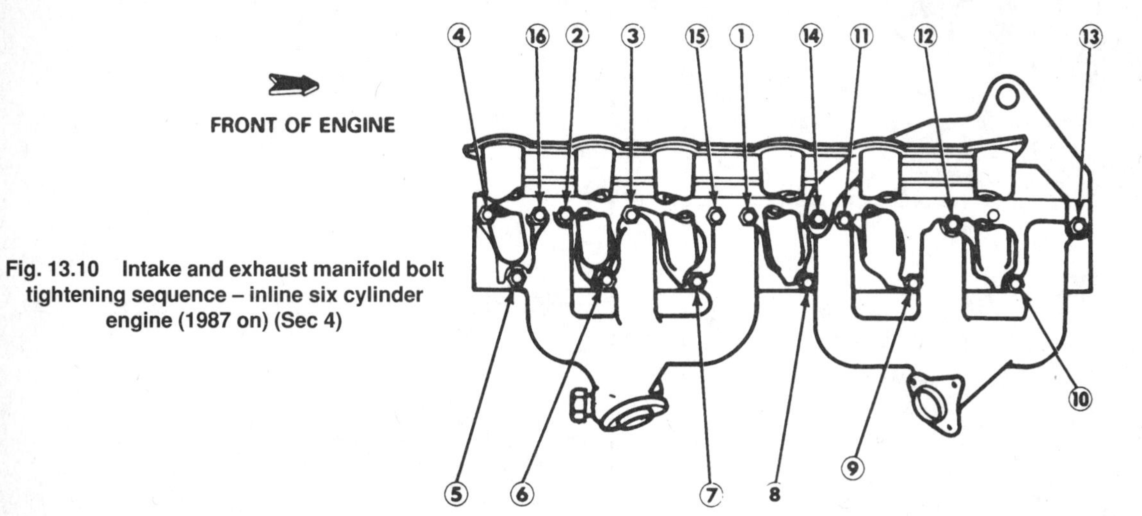

Fig. 13.10 Intake and exhaust manifold bolt tightening sequence – inline six cylinder engine (1987 on) (Sec 4)

6 Remove the intake manifold bolts. Note the locations of the stud bolts and different length bolts.
7 Remove the intake manifold. If necessary, pry the manifold away from the cylinder heads, but be careful not to damage the gasket sealing surface.
8 After the gasket sealing surfaces have been cleaned, inspect the manifold for cracks, damaged gasket surfaces and anything that would make it unfit for installation.
9 Install the intake manifold bolts and nuts and tighten them to the specified torque in the recommended sequence.
10 The remaining steps are the reverse of removal.

Cylinder heads – removal and installation

The procedure in Chapter 2 is correct, but be sure to place the two long cylinder head bolts in the two lower rear bolt holes in the lefthand cylinder head. Also, place a long cylinder head bolt in the lower rear bolt hole in the right-hand cylinder head. Use rubber bands to hold the bolts in position, above the head-to-block mating surface, until the cylinder heads are installed.

Rocker arm covers – removal and installation

Follow the procedure in Chapter 2, but note that the bolts on later models are in the center of the cover, rather than at the edges. A special tightening sequence is recommended by the manufacturer (Fig. 13.11).

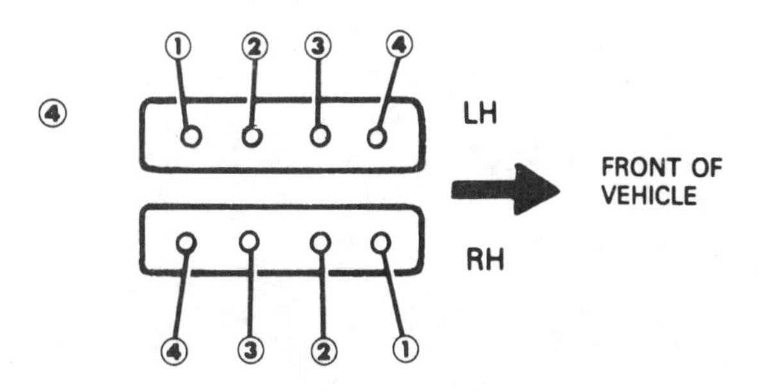

Fig. 13.11 Tightening sequence for 460 cu in (7.5L) V8 engine rocker arm cover bolts (1988) (Sec 5)

6 Cooling, heating and air conditioning systems

Quick disconnect heater hoses – general information

Some later models are equipped with quick disconnect heater hoses, which require two special tools to disconnect. When reconnecting the hoses, always use new O-rings (Fig. 13.12).

Quick disconnect air conditioning lines – general information

Some later models are equipped with quick disconnect a/c lines that require special tools for disassembly (Fig. 13.13). Warning: The air conditioning system is under high pressure. Do not loosen any hose fittings or remove any components until after the system has been discharged by a dealer service department, automotive air conditioning shop or service station.

7 Fuel and exhaust systems

Plastic fuel lines – general information

The plastic fuel lines used on later models can be damaged by torches, welding sparks, grinding and other operations which involve heat and high temperatures. Plastic fuel lines must not be repaired with hoses and hose clamps. Push connect fittings cannot be repaired except to replace the retaining clips. If the plastic lines, push connect fittings or steel tube ends

TO CONNECT COUPLING

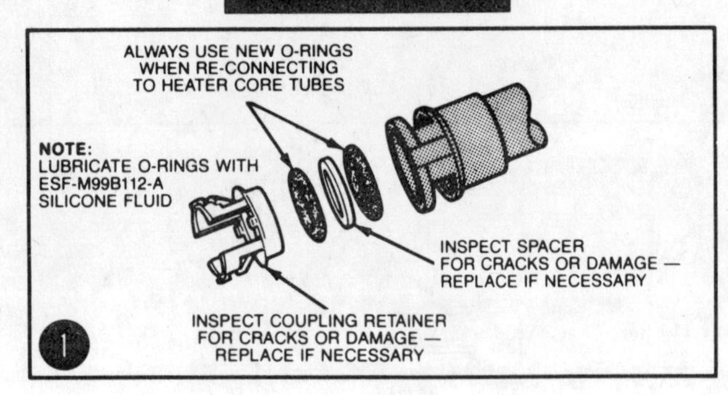

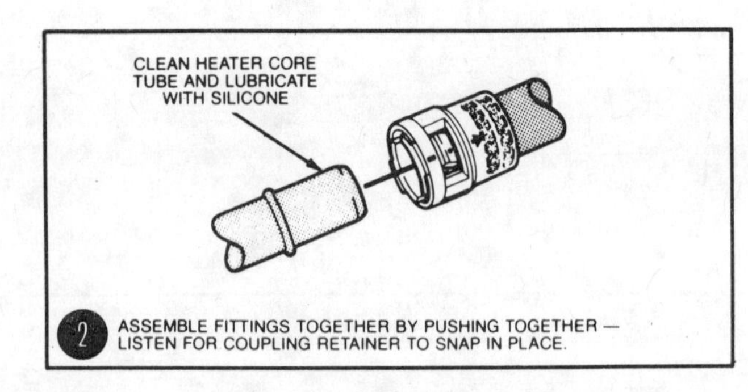

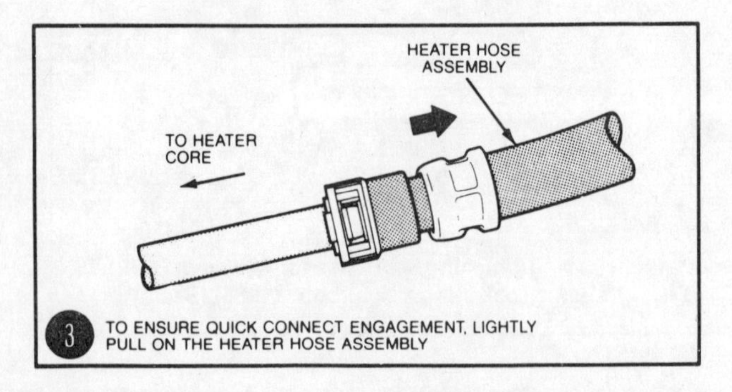

TO DISCONNECT COUPLING

CAUTION — ENGINE SHOULD BE OFF BEFORE DISCONNECTING COUPLING

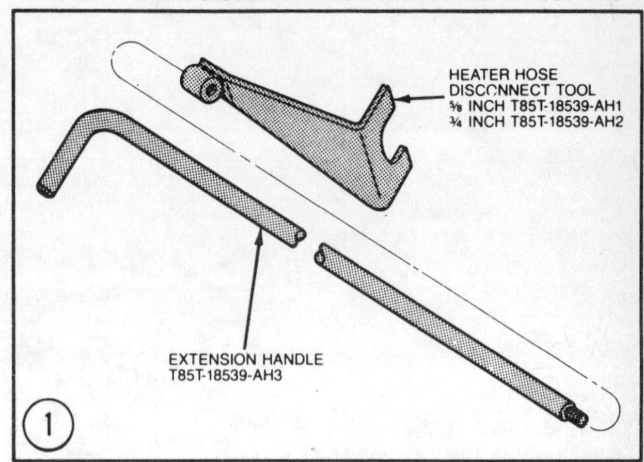

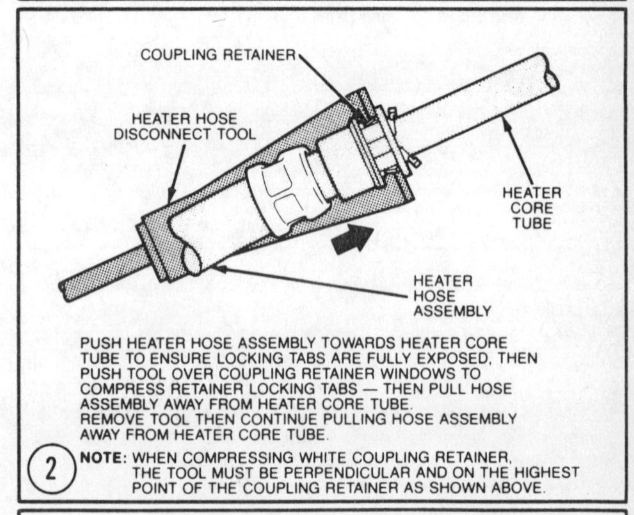

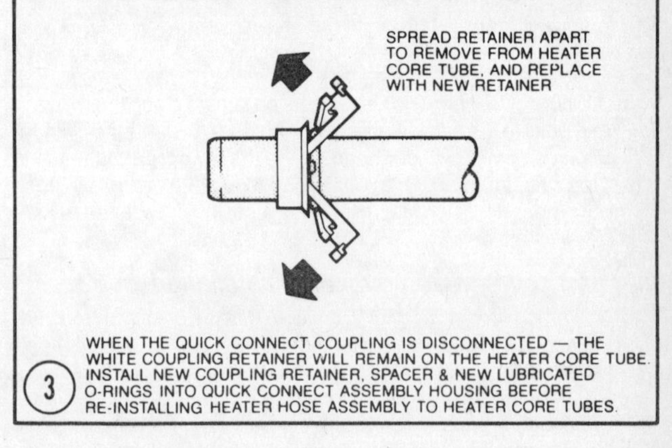

Fig. 13.12 Quick disconnect heater hose details (two special tools are needed to disconnect the hoses – the tools can be purchased from a Ford dealer or parts store); always use new O-rings when reconnecting the hoses (Sec 6)

*ALSO SUPPLIED IN KIT E35Y-19D690-A WITH GARTER SPRINGS
† ALSO SUPPLIED IN KIT E1ZZ-19B596-A

O-RINGS — 3/8" — 389157*†
1/2" — 389158*†
5/8" — 389623*
3/4" — 390209-S

FEMALE FITTING
GARTER SPRING
MALE FITTING
CAGE

SPRING LOCK COUPLING DISCONNECTED

TO CONNECT COUPLING

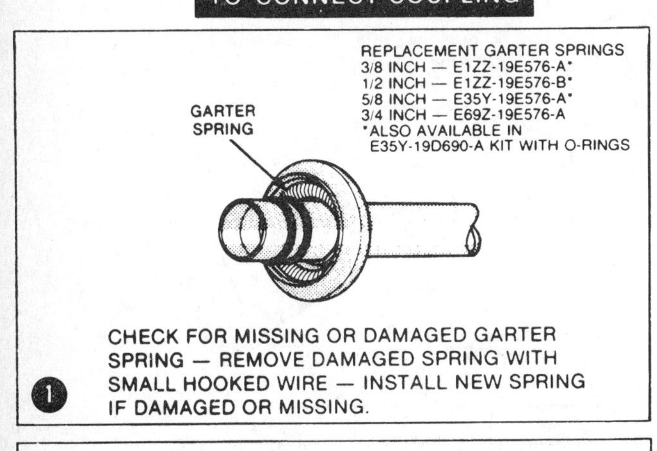

1 CHECK FOR MISSING OR DAMAGED GARTER SPRING — REMOVE DAMAGED SPRING WITH SMALL HOOKED WIRE — INSTALL NEW SPRING IF DAMAGED OR MISSING.

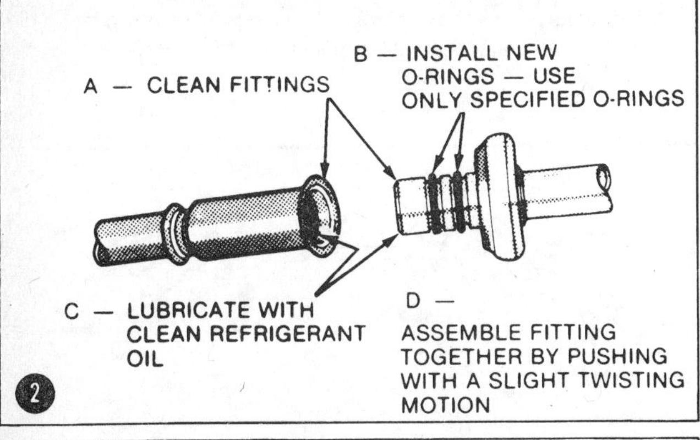

2

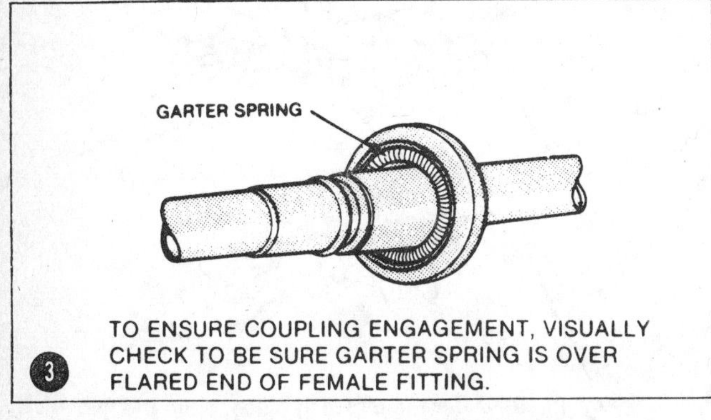

3 TO ENSURE COUPLING ENGAGEMENT, VISUALLY CHECK TO BE SURE GARTER SPRING IS OVER FLARED END OF FEMALE FITTING.

TO DISCONNECT COUPLING

CAUTION — DISCHARGE SYSTEM BEFORE DISCONNECTING COUPLING

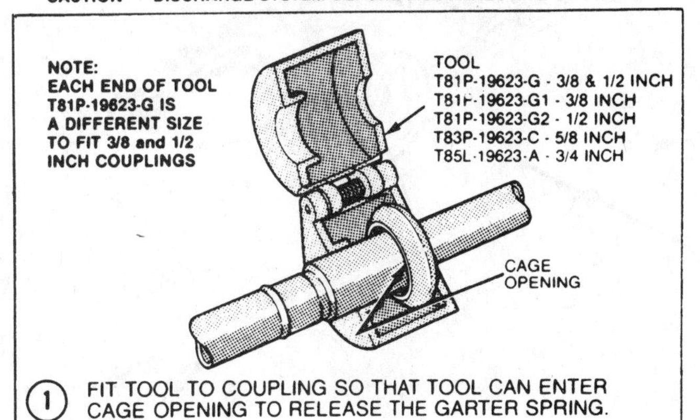

1 FIT TOOL TO COUPLING SO THAT TOOL CAN ENTER CAGE OPENING TO RELEASE THE GARTER SPRING.

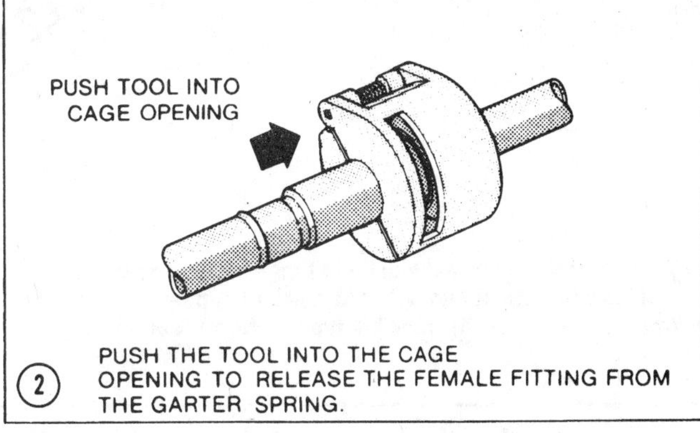

2 PUSH THE TOOL INTO THE CAGE OPENING TO RELEASE THE FEMALE FITTING FROM THE GARTER SPRING.

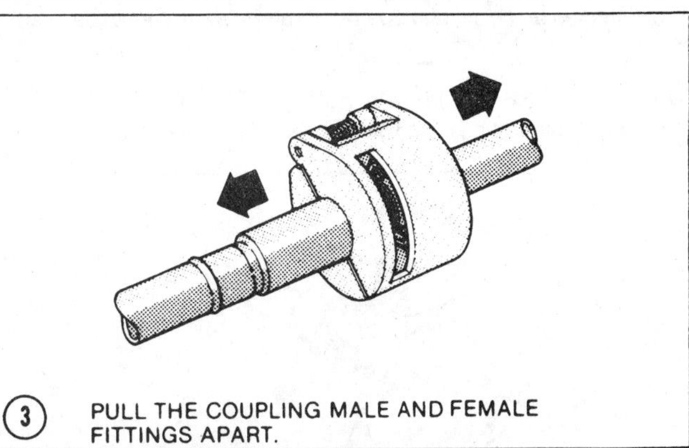

3 PULL THE COUPLING MALE AND FEMALE FITTINGS APART.

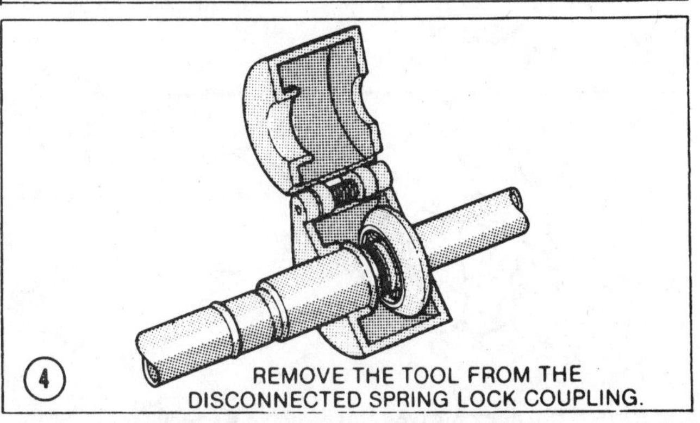

4 REMOVE THE TOOL FROM THE DISCONNECTED SPRING LOCK COUPLING.

Fig. 13.13 Quick disconnect air conditioning system line details (make sure the A/C system has been discharged before disconnecting the lines – special tools are needed and can be purchased from a Ford dealer or parts store) (Sec 6)

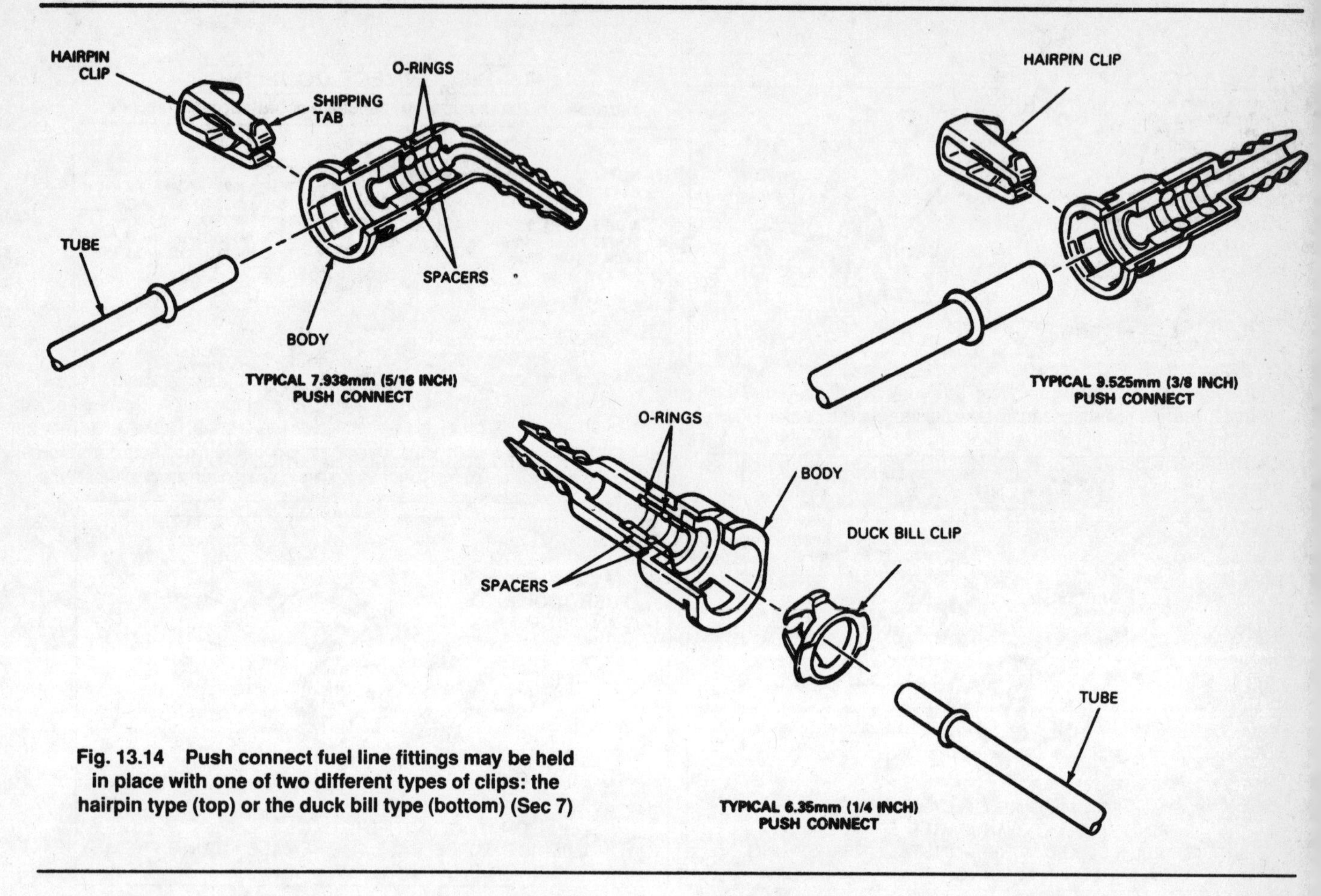

Fig. 13.14 Push connect fuel line fittings may be held in place with one of two different types of clips: the hairpin type (top) or the duck bill type (bottom) (Sec 7)

become damaged and leak, only approved service parts should be used for repairs.

Push connect fuel line fittings – general information

The push connect fittings used on later model fuel lines are designed with two different retaining clips. The fittings used to connect 3/8-inch and 5/16-inch lines have a "hairpin" clip. The fittings used to connect 1/4-inch diameter lines have a "duck bill" clip and require a special tool for disassembly. Use new clips each time the lines are disconnected (see Figs. 13.14, 13.15 and 13.16).

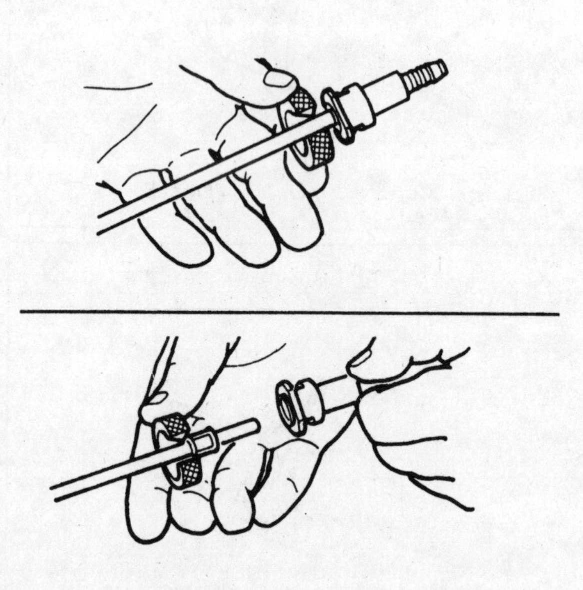

Fig. 13.15 The duck bill type clip requires a special tool for removal (Sec 7)

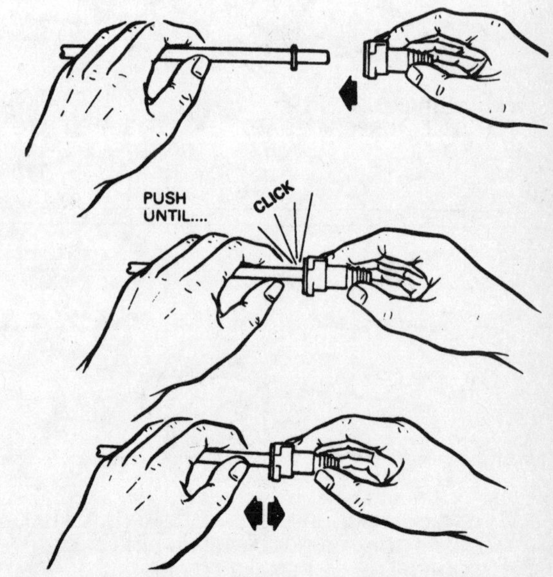

Fig. 13.16 After the clip is in place, push the two sections of the fuel line together until you hear the clip snap into place, then try to pull them apart (Sec 7)

8 Emissions control systems

Check engine warning light

1 On all EFI equipped vehicles, the check engine warning light is used to indicate malfunctions in the electronic engine control system. If the system is functioning properly, the indicator light will come on when the ignition key is turned to the On position prior to engine cranking and go out when the engine starts.

2 If the indicator light fails to come on when the ignition key is turned on, or comes on and remains on while driving, contact a dealer for service as soon as possible.

3 If the indicator light comes on and later goes out while driving, it is an indication that a temporary condition has corrected itself. Under such circumstances, it is not necessary to take the vehicle to a dealer. However, if the frequency of the intermittent problem becomes troublesome, a Ford dealer service department or repair shop will identify and correct the cause.

Emission maintenance warning light

4 Some models are equipped with an emission maintenance warning light. When the ignition key is initially placed in the On position, the light will come on for 2 to 5 seconds to indicate proper functioning of the warning system. The purpose of the emission light is to alert the driver that the 60,000 mile emission system maintenance is due (see Section 3).

9 Manual transmission

General information

1 Later models may be equipped with one of two new 5-speed transmissions – the Mazda M50D or the S5BZF.

2 The Mazda M5OD, introduced in 1988, is a top shift, fully synchronized, manual transmission equipped with an overdrive fifth gear ratio. The transmission main case, top cover and extension housing are constructed of luminum alloy.

3 The S5BZF 5-speed transmission features an aluminum case with an integral clutch housing.

5-speed transmission – overhaul

4 Overhauling a transmission is a difficult job for the do-it-yourselfer. It involves the disassembly and reassembly of many small parts. Numerous clearances must be precisely measured and, if necessary, changed with select fit spacers and snap-rings. As a result, if transmission problems arise, it can be removed and installed by a competent do-it-yourselfer, but overhaul should be left to a transmission repair shop. Rebuilt units may be available – check with your dealer parts department and auto parts stores. At any rate, the time and money involved in an overhaul is almost sure to exceed the cost of a rebuilt transmission.

5 Nevertheless, it's not impossible for an inexperienced mechanic to rebuild a transmission if the special tools are available and the job is done in a deliberate step-by-step manner so nothing is overlooked.

6 The tools necessary for an overhaul include internal and external snap-ring pliers, a bearing puller, a slide hammer, a set of pin punches, a dial indicator and possibly a hydraulic press. In addition, a large, sturdy workbench and a vise or stand will be required.

7 During disassembly of the transmission, make careful notes of how each piece comes off, where it fits in relation to other pieces and what holds it in place.

8 Before taking the transmission apart for repair, it will help if you have some idea what area is malfunctioning. Certain problems can be closely tied to specific areas, which can make component examination and replacement easier.

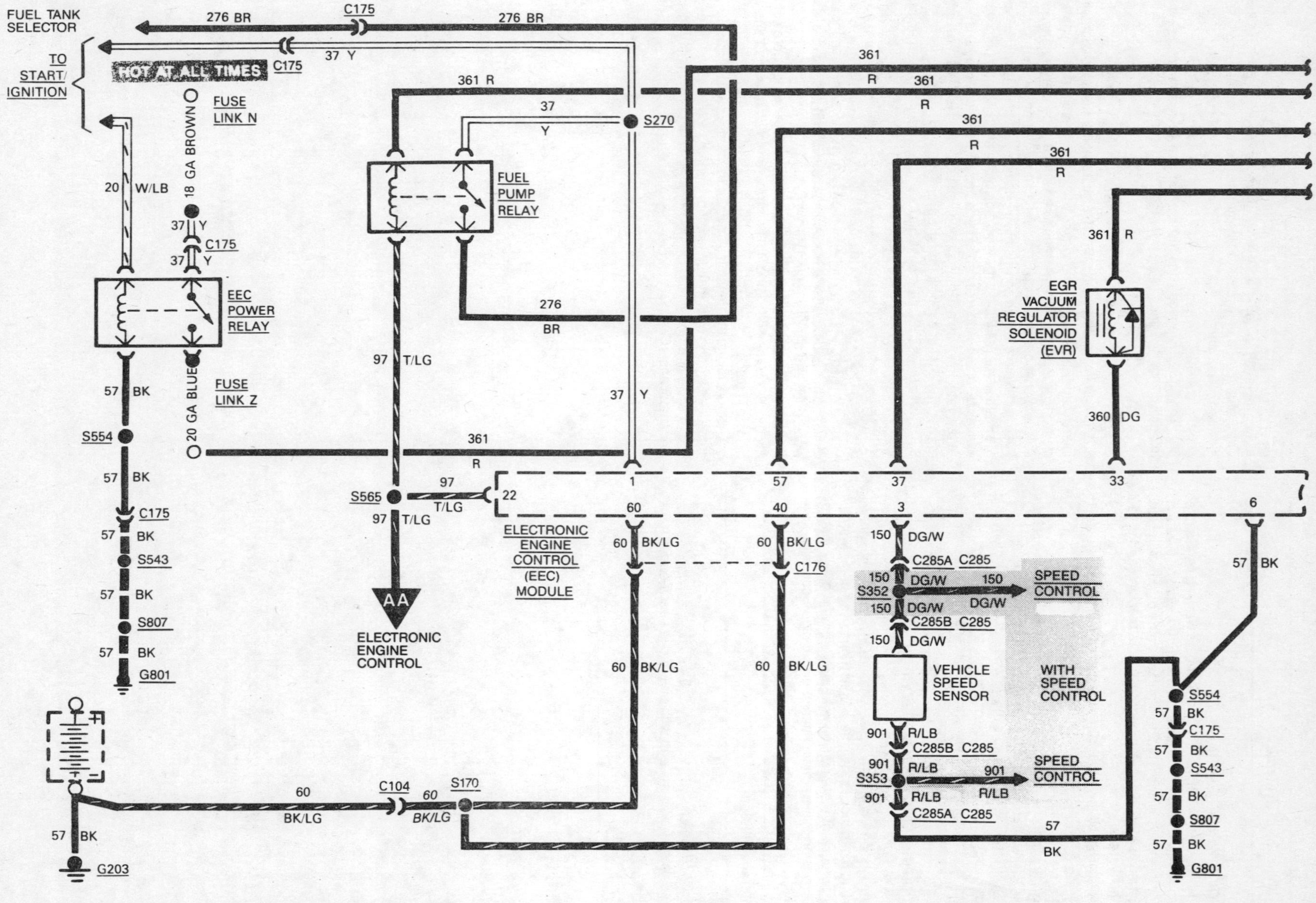

FUEL TANK SELECTOR
TO START/ IGNITION
HOT AT ALL TIMES
276 BR
C175
276 BR
37 Y
C175
361 R
361 R
361 R
361 R
361 R
FUSE LINK N
18 GA BROWN
20 W/LB
37 Y
C175
37 Y
EEC POWER RELAY
20 GA BLUE
FUSE LINK Z
57 BK
S554
57 BK
C175
57 BK
S543
57 BK
S807
57 BK
G801
FUEL PUMP RELAY
37 Y
S270
276 BR
97 T/LG
37 Y
361 R
S565
97 T/LG
97 T/LG
AA
ELECTRONIC ENGINE CONTROL
22
1
57
37
33
60
40
3
6
ELECTRONIC ENGINE CONTROL (EEC) MODULE
EGR VACUUM REGULATOR SOLENOID (EVR)
361 R
360 DG
60 BK/LG
60 BK/LG
C176
60 BK/LG
60 BK/LG
150 DG/W
C285A C285
150 DG/W
S352
150 DG/W
SPEED CONTROL
150 DG/W
C285B C285
150 DG/W
VEHICLE SPEED SENSOR
WITH SPEED CONTROL
901 R/LB
C285B C285
901 R/LB
S353
901 R/LB
SPEED CONTROL
901 R/LB
C285A C285
57 BK
57 BK
S554
57 BK
C175
57 BK
S543
57 BK
S807
57 BK
G801
60 BK/LG
C104
60 BK/LG
S170
57 BK
G203

Electronic engine control wiring diagram — 4.9L engine — 1987 on (1 of 4)

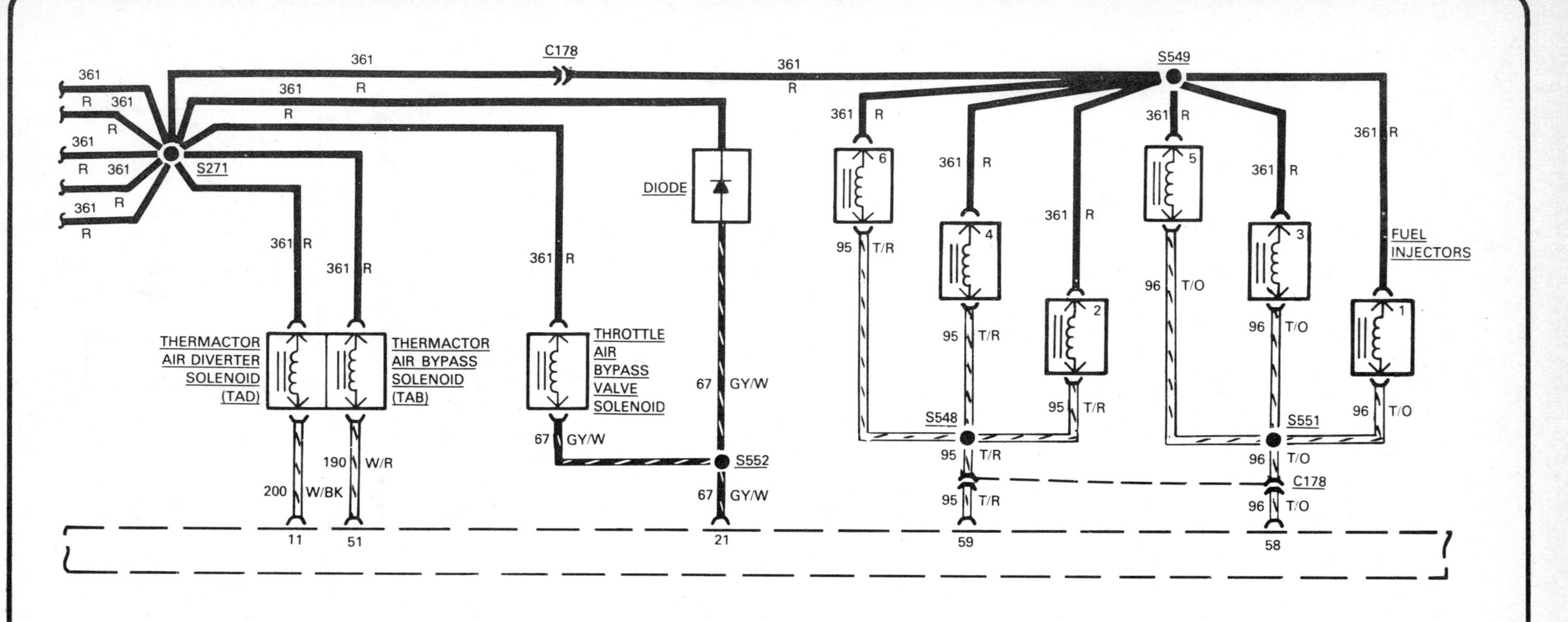

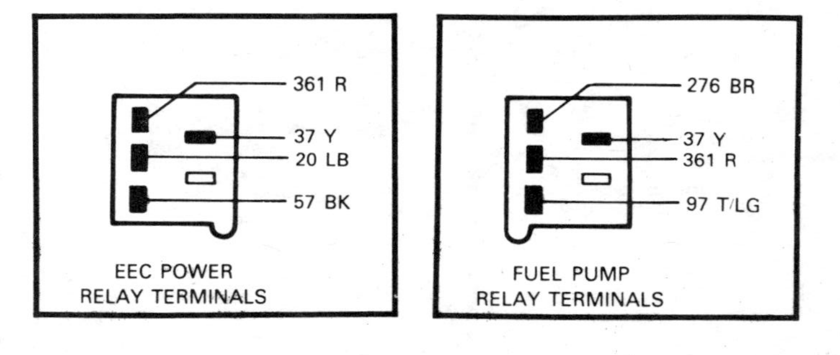

Electronic engine control wiring diagram — 4.9L engine — 1987 on (2 of 4)

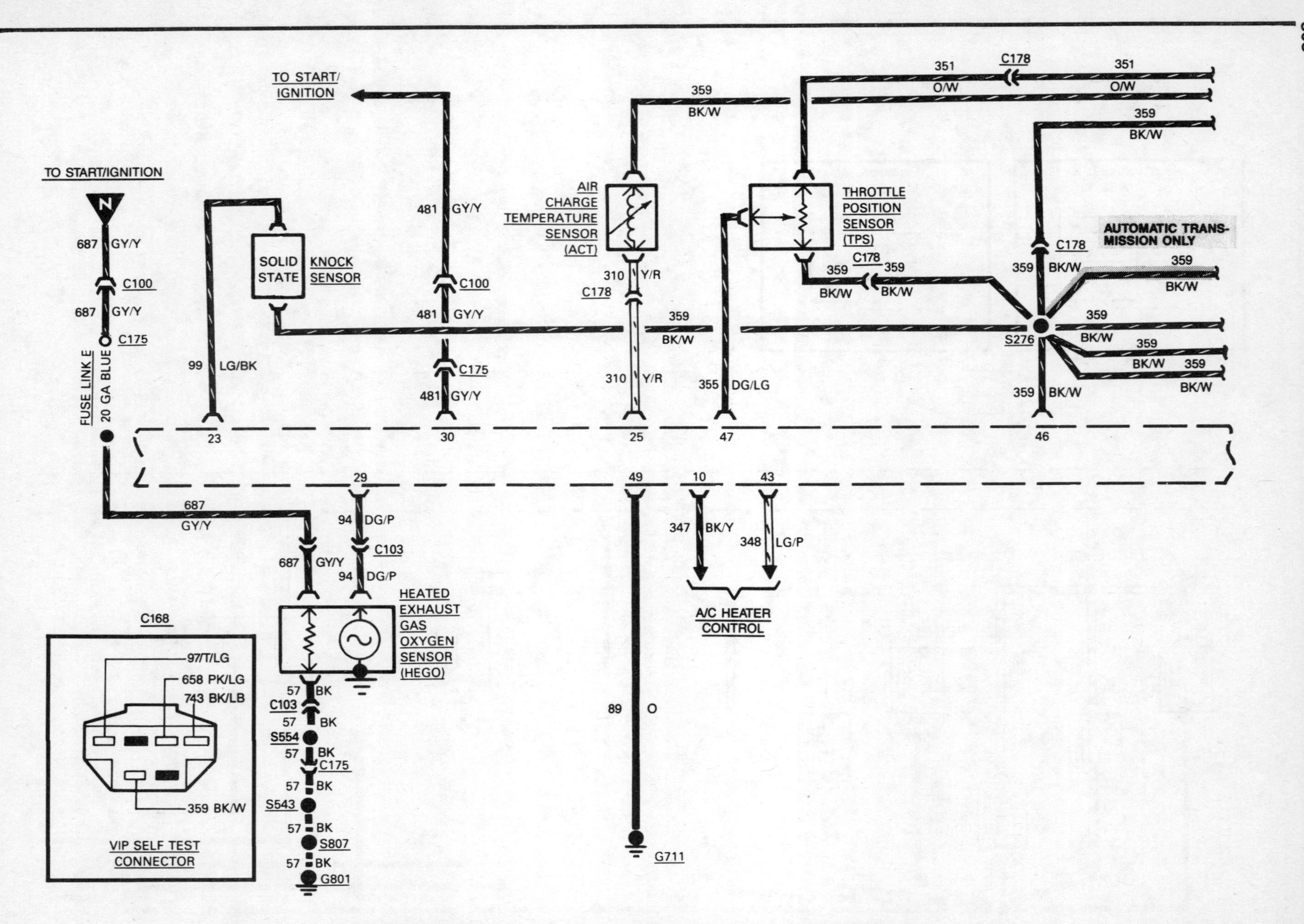

Electronic engine control wiring diagram — 4.9L engine — 1987 on (3 of 4)

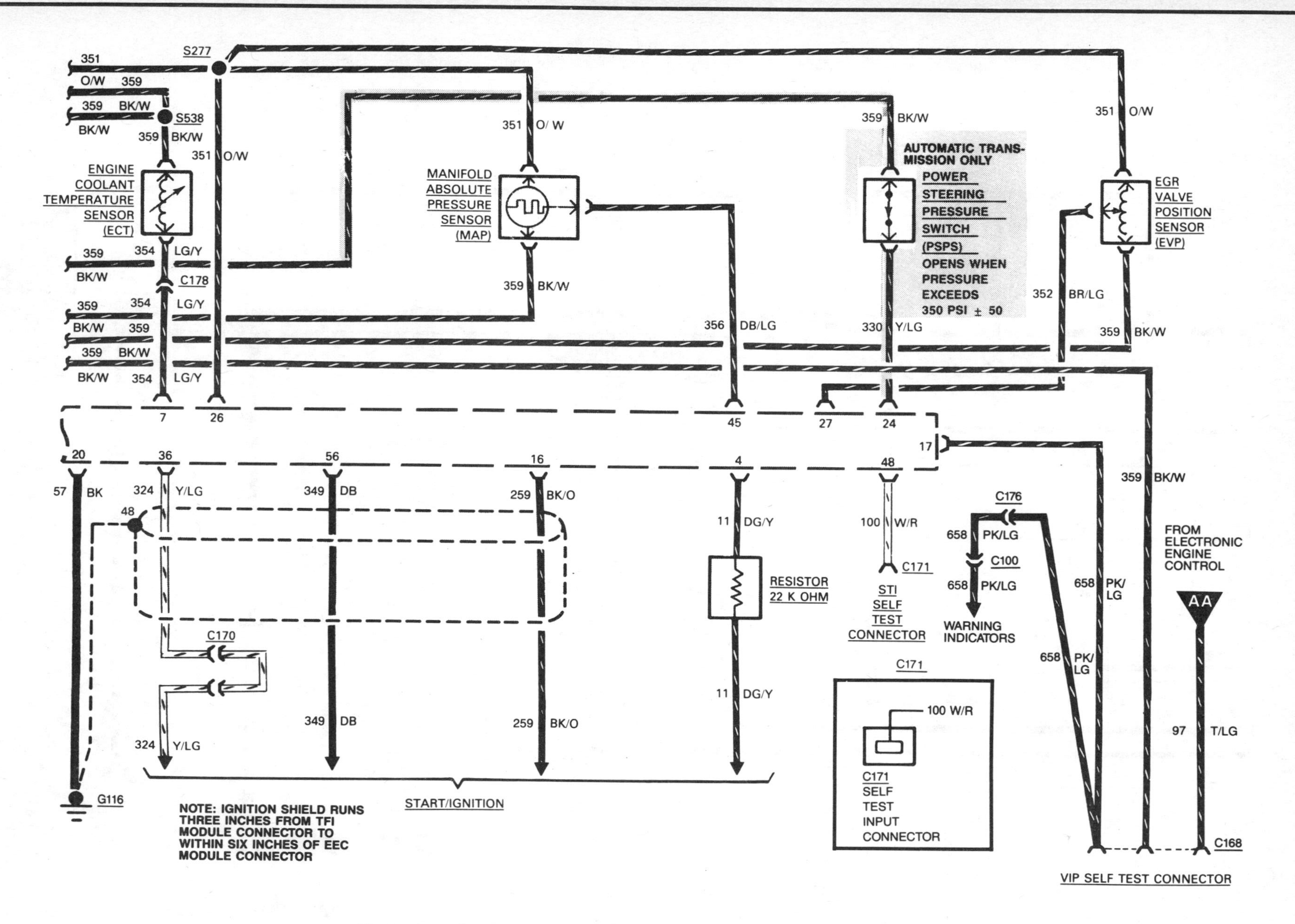

Electronic engine control wiring diagram — 4.9L engine — 1987 on (4 of 4)

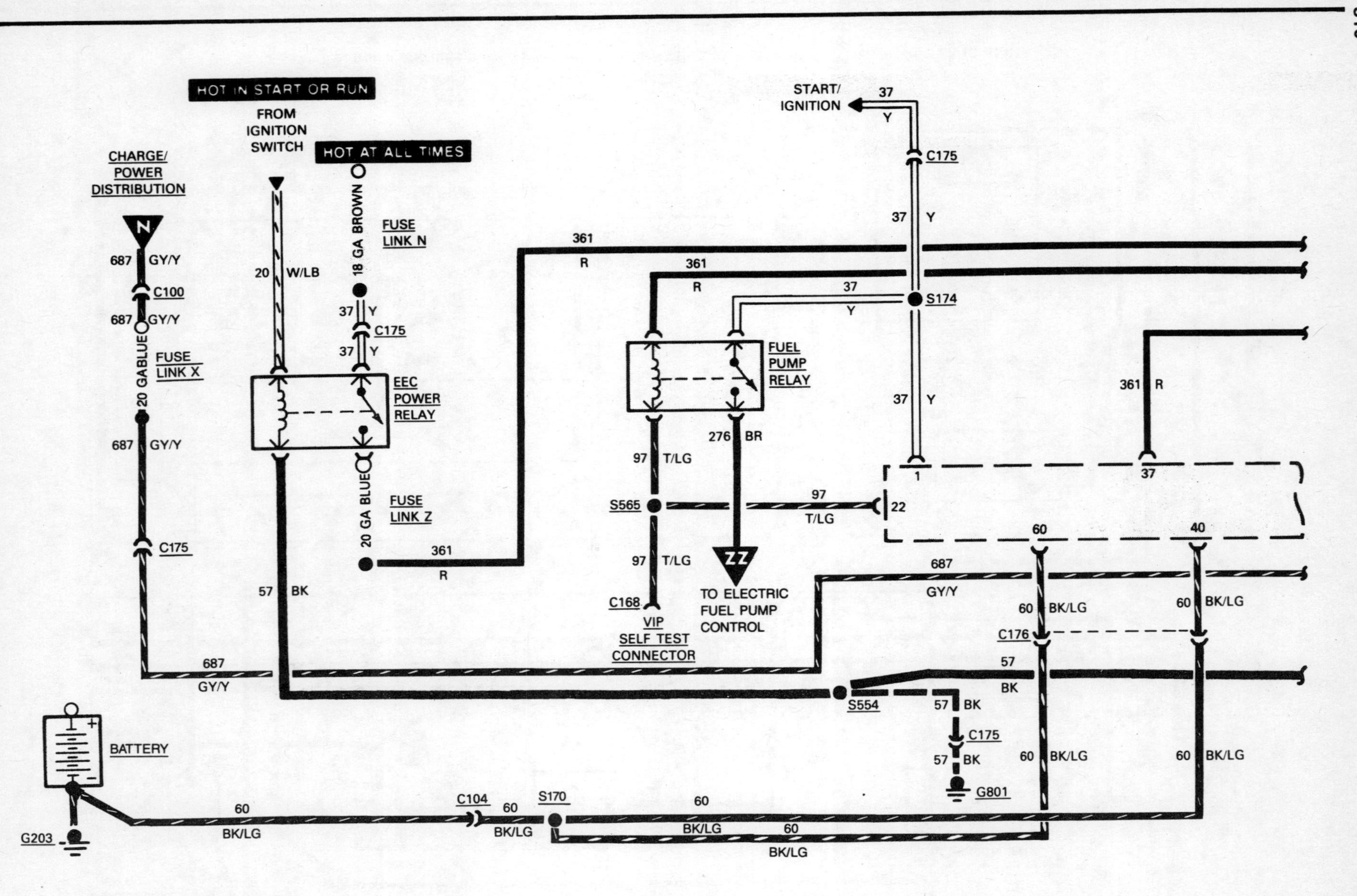

Electronic engine control wiring diagram — 5.0L, 5.8L and 7.5L engines — 1987 on (1 of 4)

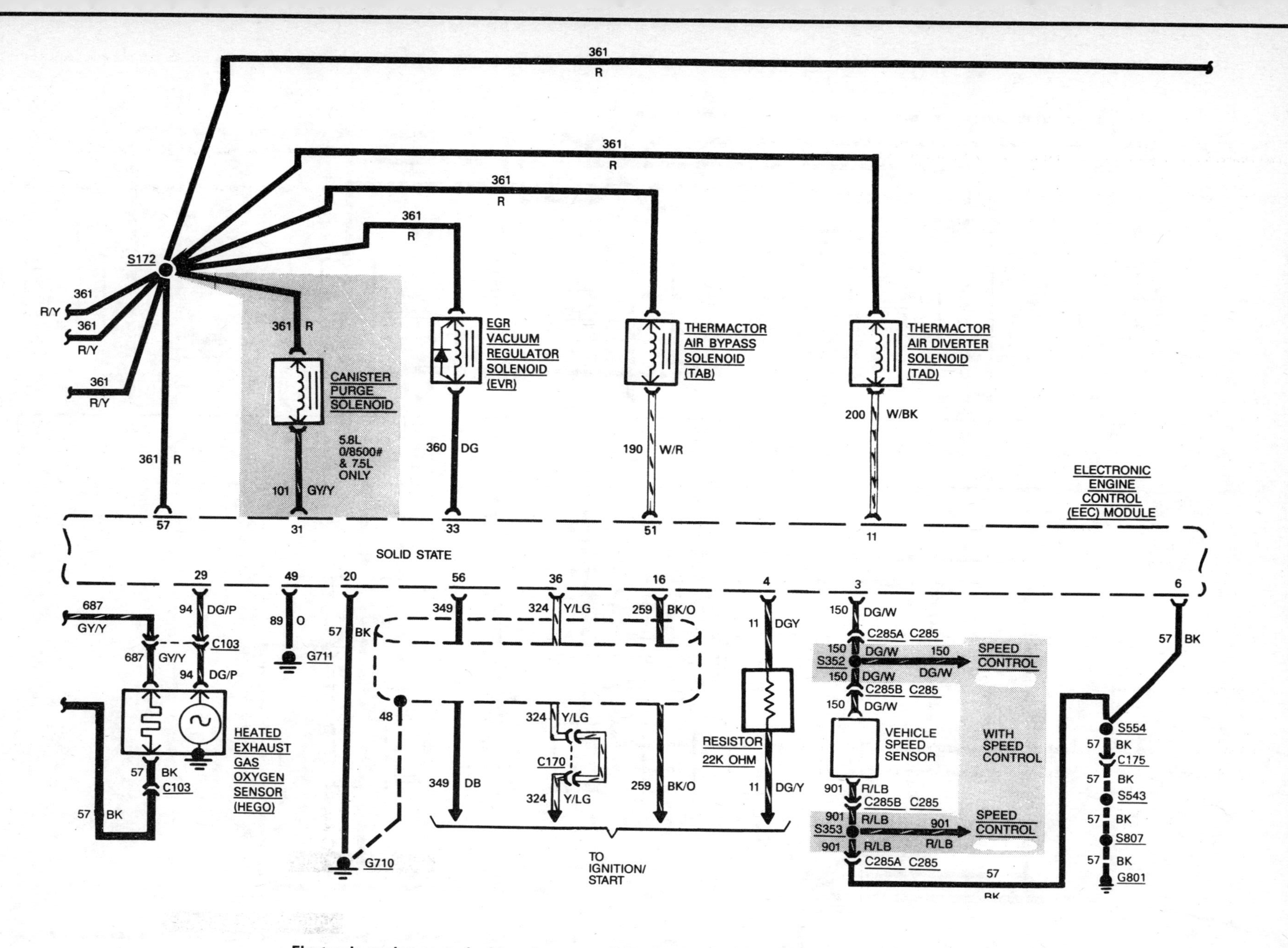

Electronic engine control wiring diagram — 5.0L, 5.8L and 7.5L engines — 1987 on (2 of 4)

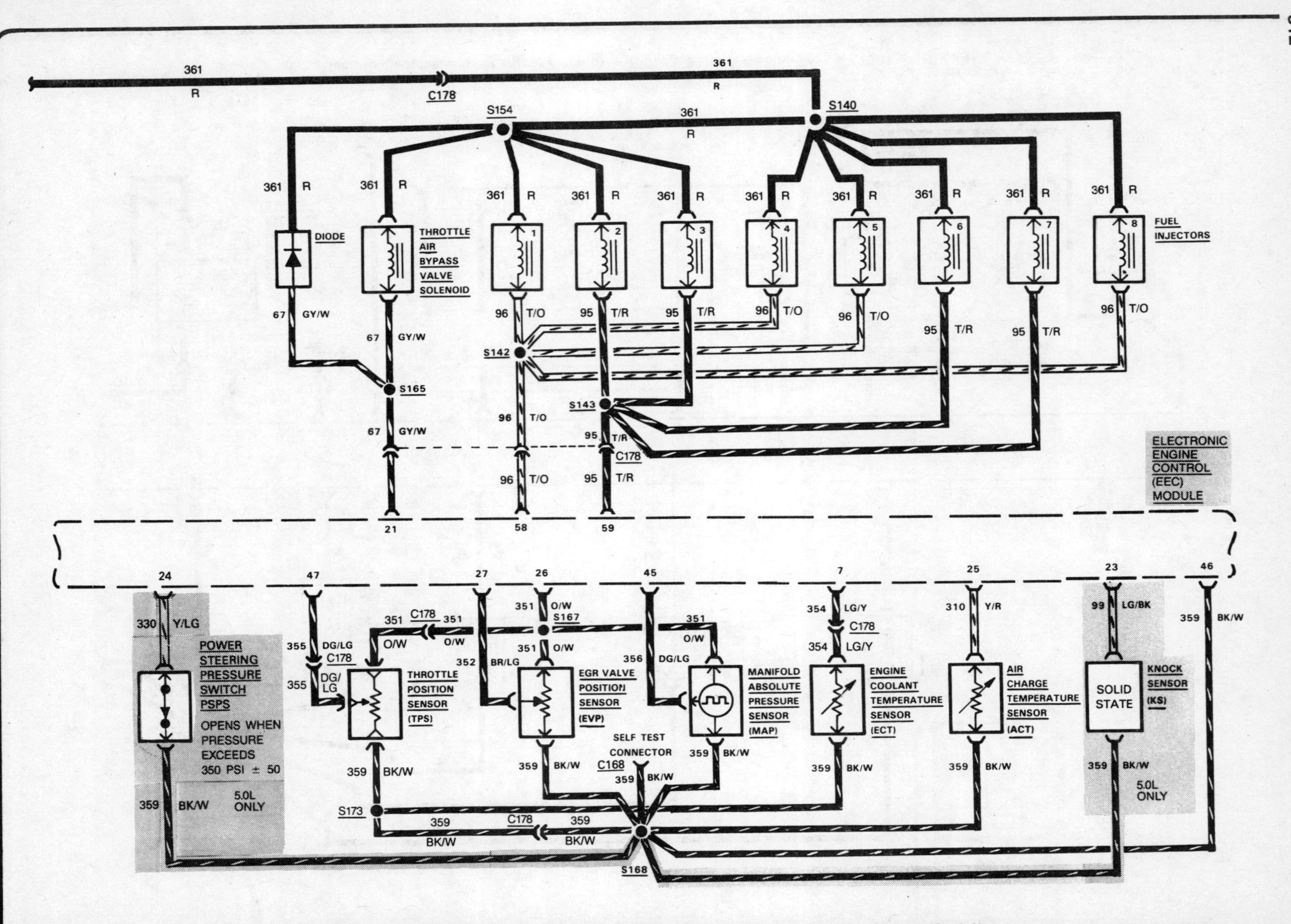

Electronic engine control wiring diagram — 5.0L, 5.8L and 7.5L engines — 1987 on (3 of 4)

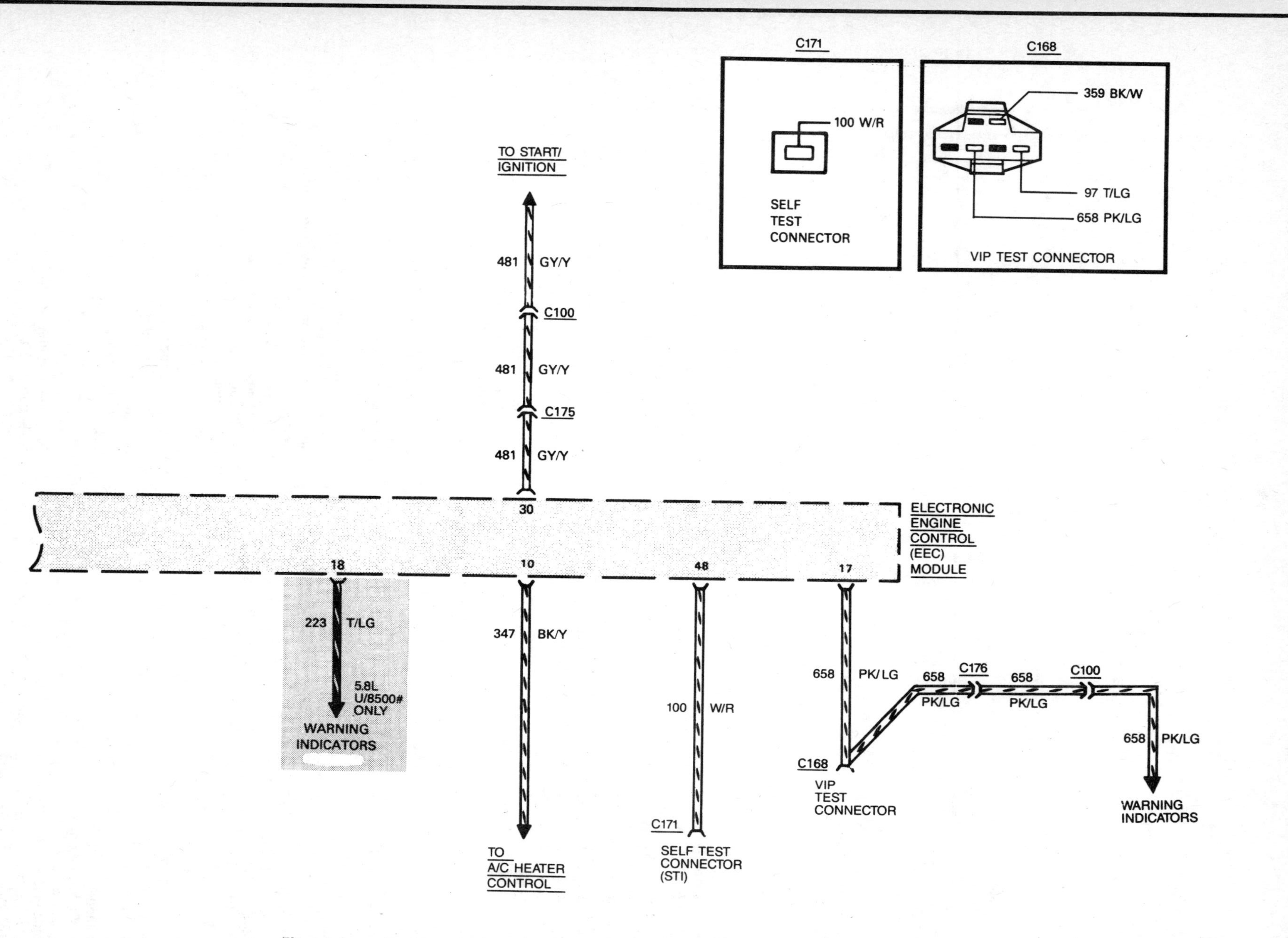

Electronic engine control wiring diagram — 5.0L, 5.8L and 7.5L engines — 1987 on (4 of 4)

Index

D

E

F

H

Haynes Automotive Manuals

NOTE: New manuals are added to this list on a periodic basis. If you do not see a listing for your vehicle, consult your local Haynes dealer for the latest product information.

ACURA

*12020 **Integra** '86 thru '89 & **Legend** '86 thru '90

AMC

Jeep CJ - *see JEEP (50020)*
14020 **Mid-size models,** Concord, Hornet, Gremlin & Spirit '70 thru '83
14025 **(Renault) Alliance & Encore** '83 thru '87

AUDI

15020 **4000** all models '80 thru '87
15025 **5000** all models '77 thru '83
15026 **5000** all models '84 thru '88

AUSTIN-HEALEY

Sprite - *see MG Midget (66015)*

BMW

*18020 **3/5 Series** not including diesel or all-wheel drive models '82 thru '92
*18021 **3 Series** except 325iX models '92 thru '97
18025 **320i** all 4 cyl models '75 thru '83
18035 **528i** & 530i all models '75 thru '80
18050 **1500 thru 2002** except Turbo '59 thru '77

BUICK

Century (front wheel drive) - *see GM (829)*
*19020 **Buick, Oldsmobile & Pontiac Full-size (Front wheel drive)** all models '85 thru '98 **Buick** Electra, LeSabre and Park Avenue; **Oldsmobile** Delta 88 Royale, Ninety Eight and Regency; **Pontiac** Bonneville
19025 **Buick Oldsmobile & Pontiac Full-size (Rear wheel drive)** **Buick** Estate '70 thru '90, Electra'70 thru '84, LeSabre '70 thru '85, Limited '74 thru '79 **Oldsmobile** Custom Cruiser '70 thru '90, Delta 88 '70 thru '85,Ninety-eight '70 thru '84 **Pontiac** Bonneville '70 thru '81, Catalina '70 thru '81, Grandville '70 thru '75, Parisienne '83 thru '86
19030 **Mid-size Regal & Century** all rear-drive models with V6, V8 and Turbo '74 thru '87
Regal - *see GENERAL MOTORS (38010)*
Riviera - *see GENERAL MOTORS (38030)*
Roadmaster - *see CHEVROLET (24046)*
Skyhawk - *see GENERAL MOTORS (38015)*
Skylark '80 thru '85 - *see GM (38020)*
Skylark '86 on - *see GM (38025)*
Somerset - *see GENERAL MOTORS (38025)*

CADILLAC

*21030 **Cadillac Rear Wheel Drive** all gasoline models '70 thru '93
Cimarron - *see GENERAL MOTORS (38015)*
Eldorado - *see GENERAL MOTORS (38030)*
Seville '80 thru '85 - *see GM (38030)*

CHEVROLET

*24010 **Astro & GMC Safari Mini-vans** '85 thru '93
24015 **Camaro V8** all models '70 thru '81
24016 **Camaro** all models '82 thru '92
Cavalier - *see GENERAL MOTORS (38015)*
Celebrity - *see GENERAL MOTORS (38005)*
24017 **Camaro & Firebird** '93 thru '97
24020 **Chevelle, Malibu & El Camino** '69 thru '87
24024 **Chevette & Pontiac T1000** '76 thru '87
Citation - *see GENERAL MOTORS (38020)*
*24032 **Corsica/Beretta** all models '87 thru '96
24040 **Corvette** all V8 models '68 thru '82
*24041 **Corvette** all models '84 thru '96
10305 **Chevrolet Engine Overhaul Manual**
24045 **Full-size Sedans** Caprice, Impala, Biscayne, Bel Air & Wagons '69 thru '90
24046 **Impala SS & Caprice and Buick Roadmaster** '91 thru '96
Lumina - *see GENERAL MOTORS (38010)*
24048 **Lumina & Monte Carlo** '95 thru '98
Lumina APV - *see GM (38035)*
24050 **Luv Pick-up** all 2WD & 4WD '72 thru '82
*24055 **Monte Carlo** all models '70 thru '88
Monte Carlo '95 thru '98 - *see LUMINA (24048)*
24059 **Nova** all V8 models '69 thru '79
*24060 **Nova and Geo Prizm** '85 thru '92
24064 **Pick-ups '67 thru '87** - Chevrolet & GMC, all V8 & in-line 6 cyl, 2WD & 4WD '67 thru '87; Suburbans, Blazers & Jimmys '67 thru '91
*24065 **Pick-ups '88 thru '98** - Chevrolet & GMC, all full-size pick-ups, '88 thru '98; Blazer & Jimmy '92 thru '94; Suburban '92 thru '98; Tahoe & Yukon '98
24070 **S-10 & S-15 Pick-ups** '82 thru '93, **Blazer & Jimmy** '83 thru '94,
*24071 **S-10 & S-15 Pick-ups** '94 thru '96 **Blazer & Jimmy** '95 thru '96
*24075 **Sprint & Geo Metro** '85 thru '94
*24080 **Vans - Chevrolet & GMC,** V8 & in-line 6 cylinder models '68 thru '96

CHRYSLER

25015 **Chrysler Cirrus, Dodge Stratus, Plymouth Breeze** '95 thru '98
25025 **Chrysler Concorde, New Yorker & LHS, Dodge** Intrepid, **Eagle** Vision, '93 thru '97
10310 **Chrysler Engine Overhaul Manual**
*25020 **Full-size Front-Wheel Drive** '88 thru '93
K-Cars - *see DODGE Aries (30008)*
Laser - *see DODGE Daytona (30030)*
*25030 **Chrysler & Plymouth Mid-size** front wheel drive '82 thru '95
Rear-wheel Drive - *see Dodge (30050)*

DATSUN

28005 **200SX** all models '80 thru '83
28007 **B-210** all models '73 thru '78
28009 **210** all models '79 thru '82
28012 **240Z, 260Z & 280Z** Coupe '70 thru '78
28014 **280ZX** Coupe & 2+2 '79 thru '83
300ZX - *see NISSAN (72010)*
28016 **310** all models '78 thru '82
28018 **510 & PL521 Pick-up** '68 thru '73
28020 **510** all models '78 thru '81
28022 **620 Series Pick-up** all models '73 thru '79
720 Series Pick-up - *see NISSAN (72030)*
28025 **810/Maxima** all gasoline models, '77 thru '84

DODGE

400 & 600 - *see CHRYSLER (25030)*
*30008 **Aries & Plymouth Reliant** '81 thru '89
30010 **Caravan & Plymouth Voyager Mini-Vans** all models '84 thru '95
*30011 **Caravan & Plymouth Voyager Mini-Vans** all models '96 thru '98
30012 **Challenger/Plymouth Saporro** '78 thru '83
30016 **Colt & Plymouth Champ (front wheel drive)** all models '78 thru '87
*30020 **Dakota Pick-ups** all models '87 thru '96
30025 **Dart, Demon, Plymouth Barracuda, Duster & Valiant** 6 cyl models '67 thru '76
*30030 **Daytona & Chrysler Laser** '84 thru '89
Intrepid - *see CHRYSLER (25025)*
*30034 **Neon** all models '95 thru '97
*30035 **Omni & Plymouth Horizon** '78 thru '90
*30040 **Pick-ups** all full-size models '74 thru '93
*30041 **Pick-ups** all full-size models '94 thru '96
*30045 **Ram 50/D50 Pick-ups & Raider and Plymouth Arrow Pick-ups** '79 thru '93
30050 **Dodge/Plymouth/Chrysler** rear wheel drive '71 thru '89
*30055 **Shadow & Plymouth Sundance** '87 thru '94
*30060 **Spirit & Plymouth Acclaim** '89 thru '95
*30065 **Vans - Dodge & Plymouth** '71 thru '96

EAGLE

Talon - *see Mitsubishi Eclipse (68030)*
Vision - *see CHRYSLER (25025)*

FIAT

34010 **124 Sport Coupe & Spider** '68 thru '78
34025 **X1/9** all models '74 thru '80

FORD

10355 **Ford Automatic Transmission Overhaul**
*36004 **Aerostar Mini-vans** all models '86 thru '96
*36006 **Contour & Mercury Mystique** '95 thru '98
36008 **Courier Pick-up** all models '72 thru '82
36012 **Crown Victoria & Mercury Grand Marquis** '88 thru '96
10320 **Ford Engine Overhaul Manual**
36016 **Escort/Mercury Lynx** all models '81 thru '90
*36020 **Escort/Mercury Tracer** '91 thru '96
*36024 **Explorer & Mazda Navajo** '91 thru '95
36028 **Fairmont & Mercury Zephyr** '78 thru '83
36030 **Festiva & Aspire** '88 thru '97
36032 **Fiesta** all models '77 thru '80
36036 **Ford & Mercury Full-size,** Ford LTD & Mercury Marquis ('75 thru '82); Ford Custom 500,Country Squire, Crown Victoria & Mercury Colony Park ('75 thru '87); Ford LTD Crown Victoria & Mercury Gran Marquis ('83 thru '87)
36040 **Granada & Mercury Monarch** '75 thru '80
36044 **Ford & Mercury Mid-size,** Ford Thunderbird & Mercury Cougar ('75 thru '82); Ford LTD & Mercury Marquis ('83 thru '86); Ford Torino,Gran Torino, Elite, Ranchero pick-up, LTD II, Mercury Montego, Comet, XR-7 & Lincoln Versailles ('75 thru '86)
36048 **Mustang V8** all models '64-1/2 thru '73
36049 **Mustang II** 4 cyl, V6 & V8 models '74 thru '78
36050 **Mustang & Mercury Capri** all models Mustang, '79 thru '93; Capri, '79 thru '86
*36051 **Mustang** all models '94 thru '97
36054 **Pick-ups & Bronco** '73 thru '79
36058 **Pick-ups & Bronco** '80 thru '96
36059 **Pick-ups, Expedition & Mercury Navigator** '97 thru '98
36062 **Pinto & Mercury Bobcat** '75 thru '80
36066 **Probe** all models '89 thru '92
36070 **Ranger/Bronco II** gasoline models '83 thru '92
*36071 **Ranger** '93 thru '97 & **Mazda Pick-ups** '94 thru '97
36074 **Taurus & Mercury Sable** '86 thru '95
*36075 **Taurus & Mercury Sable** '96 thru '98
*36078 **Tempo & Mercury Topaz** '84 thru '94
36082 **Thunderbird/Mercury Cougar** '83 thru '88
*36086 **Thunderbird/Mercury Cougar** '89 and '97
36090 **Vans** all V8 Econoline models '69 thru '91
*36094 **Vans** full size '92-'95
*36097 **Windstar Mini-van** '95-'98

GENERAL MOTORS

*10360 **GM Automatic Transmission Overhaul**
*38005 **Buick Century, Chevrolet Celebrity, Oldsmobile Cutlass Ciera & Pontiac 6000** all models '82 thru '96
*38010 **Buick Regal, Chevrolet Lumina, Oldsmobile Cutlass Supreme & Pontiac Grand Prix** front-wheel drive models '88 thru '95
*38015 **Buick Skyhawk, Cadillac Cimarron, Chevrolet Cavalier, Oldsmobile Firenza & Pontiac J-2000 & Sunbird** '82 thru '94
*38016 **Chevrolet Cavalier & Pontiac Sunfire** '95 thru '98
38020 **Buick Skylark, Chevrolet Citation, Olds Omega, Pontiac Phoenix** '80 thru '85
38025 **Buick Skylark & Somerset, Oldsmobile Achieva & Calais and Pontiac Grand Am** all models '85 thru '95
38030 **Cadillac Eldorado** '71 thru '85, **Seville** '80 thru '85, **Oldsmobile Toronado** '71 thru '85 **& Buick Riviera** '79 thru '85
*38035 **Chevrolet Lumina APV, Olds Silhouette & Pontiac Trans Sport** all models '90 thru '95
General Motors Full-size Rear-wheel Drive - *see BUICK (19025)*

(Continued on other side)

** Listings shown with an asterisk (*) indicate model coverage as of this printing. These titles will be periodically updated to include later model years - consult your Haynes dealer for more information.*

Haynes North America, Inc., 861 Lawrence Drive, Newbury Park, CA 91320-1514 • (805) 498-6703

Haynes Automotive Manuals (continued)

NOTE: New manuals are added to this list on a periodic basis. If you do not see a listing for your vehicle, consult your local Haynes dealer for the latest product information.

GEO

Metro - *see CHEVROLET Sprint (24075)*
Prizm - *'85 thru '92 see CHEVY (24060), '93 thru '96 see TOYOTA Corolla (92036)*
*40030 **Storm** all models '90 thru '93
Tracker - *see SUZUKI Samurai (90010)*

GMC

Safari - *see CHEVROLET ASTRO (24010)*
Vans & Pick-ups - *see CHEVROLET*

HONDA

42010 **Accord CVCC** all models '76 thru '83
42011 **Accord** all models '84 thru '89
42012 **Accord** all models '90 thru '93
42013 **Accord** all models '94 thru '95
42020 **Civic 1200** all models '73 thru '79
42021 **Civic 1300 & 1500 CVCC** '80 thru '83
42022 **Civic 1500 CVCC** all models '75 thru '79
42023 **Civic** all models '84 thru '91
*42024 **Civic & del Sol** '92 thru '95
*42040 **Prelude CVCC** all models '79 thru '89

HYUNDAI

*43015 **Excel** all models '86 thru '94

ISUZU

Hombre - *see CHEVROLET S-10 (24071)*
*47017 **Rodeo** '91 thru '97; **Amigo** '89 thru '94; **Honda Passport** '95 thru '97
*47020 **Trooper & Pick-up**, all gasoline models Pick-up, '81 thru '93; Trooper, '84 thru '91

JAGUAR

*49010 **XJ6** all 6 cyl models '68 thru '86
*49011 **XJ6** all models '88 thru '94
*49015 **XJ12 & XJS** all 12 cyl models '72 thru '85

JEEP

*50010 **Cherokee, Comanche & Wagoneer Limited** all models '84 thru '96
50020 **CJ** all models '49 thru '86
*50025 **Grand Cherokee** all models '93 thru '98
50029 **Grand Wagoneer & Pick-up** '72 thru '91 Grand Wagoneer '84 thru '91, Cherokee & Wagoneer '72 thru '83, Pick-up '72 thru '88
*50030 **Wrangler** all models '87 thru '95

LINCOLN

Navigator - *see FORD Pick-up (36059)*
59010 **Rear Wheel Drive** all models '70 thru '96

MAZDA

61010 **GLC Hatchback (rear wheel drive)** '77 thru '83
61011 **GLC (front wheel drive)** '81 thru '85
*61015 **323 & Protogé** '90 thru '97
*61016 **MX-5 Miata** '90 thru '97
*61020 **MPV** all models '89 thru '94
Navajo - *see Ford Explorer (36024)*
61030 **Pick-ups** '72 thru '93
Pick-ups '94 thru '96 - *see Ford Ranger (36071)*
61035 **RX-7** all models '79 thru '85
*61036 **RX-7** all models '86 thru '91
61040 **626** (rear wheel drive) all models '79 thru '82
*61041 **626/MX-6 (front wheel drive)** '83 thru '91

MERCEDES-BENZ

63012 **123 Series Diesel** '76 thru '85
*63015 **190 Series** four-cyl gas models, '84 thru '88
63020 **230/250/280** 6 cyl sohc models '68 thru '72
63025 **280 123 Series** gasoline models '77 thru '81
63030 **350 & 450** all models '71 thru '80

MERCURY

See FORD Listing.

MG

66010 **MGB** Roadster & GT Coupe '62 thru '80
66015 **MG Midget, Austin Healey Sprite** '58 thru '80

MITSUBISHI

*68020 **Cordia, Tredia, Galant, Precis & Mirage** '83 thru '93
*68030 **Eclipse, Eagle Talon & Ply. Laser** '90 thru '94
*68040 **Pick-up** '83 thru '96 **& Montero** '83 thru '93

NISSAN

72010 **300ZX** all models including Turbo '84 thru '89
*72015 **Altima** all models '93 thru '97
*72020 **Maxima** all models '85 thru '91
*72030 **Pick-ups** '80 thru '96 **Pathfinder** '87 thru '95
72040 **Pulsar** all models '83 thru '86
*72050 **Sentra** all models '82 thru '94
*72051 **Sentra & 200SX** all models '95 thru '98
*72060 **Stanza** all models '82 thru '90

OLDSMOBILE

*73015 **Cutlass** V6 & V8 gas models '74 thru '88
For other OLDSMOBILE titles, see BUICK, CHEVROLET or GENERAL MOTORS listing.

PLYMOUTH

For PLYMOUTH titles, see DODGE listing.

PONTIAC

79008 **Fiero** all models '84 thru '88
79018 **Firebird** V8 models except Turbo '70 thru '81
79019 **Firebird** all models '82 thru '92
For other PONTIAC titles, see BUICK, CHEVROLET or GENERAL MOTORS listing.

PORSCHE

*80020 **911** except Turbo & Carrera 4 '65 thru '89
80025 **914** all 4 cyl models '69 thru '76
80030 **924** all models including Turbo '76 thru '82
*80035 **944** all models including Turbo '83 thru '89

RENAULT

Alliance & Encore - *see AMC (14020)*

SAAB

*84010 **900** all models including Turbo '79 thru '88

SATURN

87010 **Saturn** all models '91 thru '96

SUBARU

89002 **1100, 1300, 1400 & 1600** '71 thru '79
*89003 **1600 & 1800** 2WD & 4WD '80 thru '94

SUZUKI

*90010 **Samurai/Sidekick & Geo Tracker** '86 thru '96

TOYOTA

92005 **Camry** all models '83 thru '91
92006 **Camry** all models '92 thru '96
92015 **Celica Rear Wheel Drive** '71 thru '85
*92020 **Celica Front Wheel Drive** '86 thru '93
92025 **Celica Supra** all models '79 thru '92
92030 **Corolla** all models '75 thru '79
92032 **Corolla** all rear wheel drive models '80 thru '87
92035 **Corolla** all front wheel drive models '84 thru '92
*92036 **Corolla & Geo Prizm** '93 thru '97
92040 **Corolla Tercel** all models '80 thru '82
92045 **Corona** all models '74 thru '82
92050 **Cressida** all models '78 thru '82
92055 **Land Cruiser** FJ40, 43, 45, 55 '68 thru '82
92056 **Land Cruiser** FJ60, 62, 80, FZJ80 '80 thru '96
*92065 **MR2** all models '85 thru '87
92070 **Pick-up** all models '69 thru '78
*92075 **Pick-up** all models '79 thru '95
*92076 **Tacoma** '95 thru '98, **4Runner** '96 thru '98, **& T100** '93 thru '98
*92080 **Previa** all models '91 thru '95
92085 **Tercel** all models '87 thru '94

TRIUMPH

94007 **Spitfire** all models '62 thru '81
94010 **TR7** all models '75 thru '81

VW

96008 **Beetle & Karmann Ghia** '54 thru '79
96012 **Dasher** all gasoline models '74 thru '81
*96016 **Rabbit, Jetta, Scirocco, & Pick-up** gas models '74 thru '91 & Convertible '80 thru '92
96017 **Golf & Jetta** all models '93 thru '97
96020 **Rabbit, Jetta & Pick-up** diesel '77 thru '84
96030 **Transporter 1600** all models '68 thru '79
96035 **Transporter 1700, 1800 & 2000** '72 thru '79
96040 **Type 3 1500 & 1600** all models '63 thru '73
96045 **Vanagon** all air-cooled models '80 thru '83

VOLVO

97010 **120, 130 Series & 1800 Sports** '61 thru '73
97015 **140 Series** all models '66 thru '74
*97020 **240 Series** all models '76 thru '93
97025 **260 Series** all models '75 thru '82
*97040 **740 & 760 Series** all models '82 thru '88

TECHBOOK MANUALS

10205 **Automotive Computer Codes**
10210 **Automotive Emissions Control Manual**
10215 **Fuel Injection Manual, 1978 thru 1985**
10220 **Fuel Injection Manual, 1986 thru 1996**
10225 **Holley Carburetor Manual**
10230 **Rochester Carburetor Manual**
10240 **Weber/Zenith/Stromberg/SU Carburetors**
10305 **Chevrolet Engine Overhaul Manual**
10310 **Chrysler Engine Overhaul Manual**
10320 **Ford Engine Overhaul Manual**
10330 **GM and Ford Diesel Engine Repair Manual**
10340 **Small Engine Repair Manual**
10345 **Suspension, Steering & Driveline Manual**
10355 **Ford Automatic Transmission Overhaul**
10360 **GM Automatic Transmission Overhaul**
10405 **Automotive Body Repair & Painting**
10410 **Automotive Brake Manual**
10415 **Automotive Detaiing Manual**
10420 **Automotive Eelectrical Manual**
10425 **Automotive Heating & Air Conditioning**
10430 **Automotive Reference Manual & Dictionary**
10435 **Automotive Tools Manual**
10440 **Used Car Buying Guide**
10445 **Welding Manual**
10450 **ATV Basics**

SPANISH MANUALS

98903 **Reparación de Carrocería & Pintura**
98905 **Códigos Automotrices de la Computadora**
98910 **Frenos Automotriz**
98915 **Inyección de Combustible 1986 al 1994**
99040 **Chevrolet & GMC Camionetas** '67 al '87 Incluye Suburban, Blazer & Jimmy '67 al '91
99041 **Chevrolet & GMC Camionetas** '88 al '95 Incluye Suburban '92 al '95, Blazer & Jimmy '92 al '94, Tahoe y Yukon '95
99042 **Chevrolet & GMC Camionetas Cerradas** '68 al '95
99055 **Dodge Caravan & Plymouth Voyager** '84 al '95
99075 **Ford Camionetas y Bronco** '80 al '94
99077 **Ford Camionetas Cerradas** '69 al '91
99083 **Ford Modelos de Tamaño Grande** '75 al '87
99088 **Ford Modelos de Tamaño Mediano** '75 al '86
99091 **Ford Taurus & Mercury Sable** '86 al '95
99095 **GM Modelos de Tamaño Grande** '70 al '90
99100 **GM Modelos de Tamaño Mediano** '70 al '88
99110 **Nissan Camionetas** '80 al '96, **Pathfinder** '87 al '95
99118 **Nissan Sentra** '82 al '94
99125 **Toyota Camionetas y 4Runner** '79 al '95

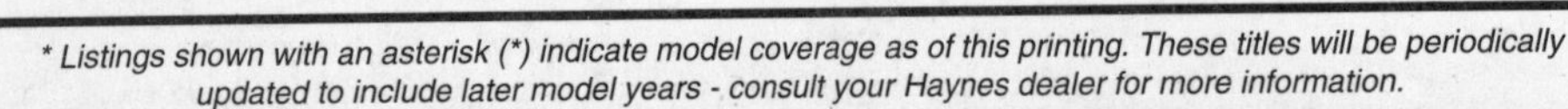
** Listings shown with an asterisk (*) indicate model coverage as of this printing. These titles will be periodically updated to include later model years - consult your Haynes dealer for more information.*

Over 100 Haynes motorcycle manuals also available

5-98

Haynes North America, Inc., 861 Lawrence Drive, Newbury Park, CA 91320-1514 • (805) 498-6703